Pesticide residues in food — 2007

Toxicological evaluations

Sponsored jointly by FAO and WHO
With the support of the International Programme on Chemical Safety (IPCS)

Joint Meeting of the FAO Panel of Experts on Pesticide Residues in Food and the Environment and the WHO Core Assessment Group

Geneva, Switzerland, 18–27 September 2007

The summaries and evaluations contained in this book are, in most cases, based on unpublished proprietary data submitted for the purpose of the JMPR assessment. A registration authority should not grant a registration on the basis of an evaluation unless it has first received authorization for such use from the owner who submitted the data for JMPR review or has received the data on which the summaries are based, either from the owner of the data or from a second party that has obtained permission from the owner of the data for this purpose.

WHO Library Cataloguing-in-Publication Data

Pesticide residues in food : 2007, toxicological evaluations, sponsored jointly by FAO and WHO, with the support of the International Programme on Chemical Safety / Joint Meeting of the FAO Panel of Experts on Pesticide Residues in Food and the Environment and WHO Core Assessment Group, Geneva, Switzerland, 18–27 September 2007.

(WHO pesticide residues series)
1.Pesticide residues - toxicity. 2.No-observed-adverse-effect level. 3.Food contamination. I.FAO Panel of Experts on Pesticide Residues in Food and Environment. II.WHO Core Assessment Group on Pesticide Residues. III. Pesticide residues in food 2007 : evaluations. Part 2, Toxicological.

ISBN 978 92 4 166523 0 (NLM classification: WA 240)

This report contains the collective views of two international groups of experts and does not necessarily represent the decisions nor the stated policy of the Food and Agriculture Organization of the United Nations or the World Health Organization.

The preparatory work for the toxicological evaluations of pesticide residues carried out by the WHO Expert Group on Pesticide Residues for consideration by the FAO/WHO Joint Meeting on Pesticide Residues in Food and the Environment is actively supported by the International Programme on Chemical Safety within the framework of the Inter-Organization Programme for the Sound Management of Chemicals.

The **International Programme on Chemical Safety** (IPCS), established in 1980, is a joint venture of the United Nations Environment Programme (UNEP), the International Labour Organization (ILO), and the World Health Organization (WHO). The overall objectives of the IPCS are to establish the scientific basis for assessing the risk to human health and the environment to exposure from chemicals, through international peer-review processes as a prerequisite for the promotion of chemical safety, and to provide technical assistance in strengthening national capacities for the sound management of chemicals.

The Inter-Organization Programme for the Sound Management of Chemicals (IOMC) was established in 1995 by UNEP, the Food and Agriculture Organization of the United Nations, WHO, the United Nations Industrial Development Organization, and the Organisation for Economic Co-operation and Development (Participating Organizations), following recommendations made by the 1992 United Nations Conference on the Environment and Development to strengthen cooperation and increase coordination in the field of chemical safety. The purpose of the IOMC is to promote coordination of the policies and activities pursued by the Participating Organizations, jointly or separately, to achieve the sound management of chemicals in relation to human health and the environment.

Typeset in India

Printed in Spain

TABLE OF CONTENTS

* First full evaluation

** Evaluated within the periodic review programme of the Codex Committee on Pesticide Residues

2007 Joint Meeting of the FAO Panel of Experts on Pesticide Residues in Food and the Environment and the WHO Core Assessment Group

Geneva, 18–27 September 2007

PARTICIPANTS

FAO Panel of Experts on Pesticide Residues in Food and the Environment

Dr Ursula Banasiak, Federal Institute for Risk Assessment, Thielallee 88-92, D-14195 Berlin, Germany

Professor Eloisa Dutra Caldas, University of Brasilia, College of Health Sciences, Pharmaceutical Sciences Department, Campus Universitário Darci Ribeiro, 70919-970 Brasília/DF, Brazil

Mr Stephen Funk, Health Effects Division (7509P), United States Environmental Protection Agency, 1200 Pennsylvania Avenue NW, Washington DC 20460, USA *(FAO Chairman)*

Mr Denis J. Hamilton, Principal Scientific Officer, Biosecurity, Department of Primary Industries and Fisheries, PO Box 46, Brisbane, QLD 4001, Australia

Mr David Lunn, Programme Manager (Residues–Plants), Dairy and Plant Products Group, New Zealand Food Safety Authority, PO Box 2835, Wellington, New Zealand

Dr Dugald MacLachlan, Australian Quarantine and Inspection Service, Australian Government Department of Agriculture, Fisheries and Forestry, GPO Box 858, Canberra, ACT 2601, Australia

Ms Bernadette C. Ossendorp, Centre for Substances and Integrated Risk Assessment, National Institute of Public Health and the Environment (RIVM), Antonie van Leeuwenhoeklaan 9, PO Box 1, 3720 BA Bilthoven, Netherlands

Dr Yukiko Yamada, Director, Food Safety and Consumer Policy Division, Food Safety and Consumer Affairs Bureau, Ministry of Agriculture, Forestry and Fisheries, 1-2-1 Kasumigaseki, Tokyo 100-8950, Japan

WHO Members

Professor Alan R. Boobis, Experimental Medicine & Toxicology, Division of Medicine, Faculty of Medicine, Imperial College London, Hammersmith Campus, Ducane Road, London W12 0NN, England *(WHO Chairman)*

Dr Les Davies, Chemical Review, Australian Pesticides & Veterinary Medicines Authority, PO Box E240, Kingston ACT 2604, Australia

Dr Vicki L. Dellarco, Office of Pesticide Programs (7509P), United States Environmental Protection Agency, Health Effects Division; 1200 Pennsylvania Avenue NW, Washington, DC 20460, USA *(WHO Rapporteur)*

Dr Helen Hakansson, Institute of Environmental Medicine, Karolinska Institutet, Unit of Environmental Health Risk Assessment, Box 210, Nobels väg 13, S-171 77 Stockholm, Sweden

Dr Angelo Moretto, Department of Environmental and Occupational Health, University of Milan, International Centre for Pesticides and Health Risk Prevention (ICPS), Luigi Sacco Hospital, Via Grassi 74, 20157 Milan, Italy

Professor David Ray, MRC Applied Neuroscience Group, Biomedical Sciences, University of Nottingham, Queens Medical Centre, Nottingham NG7 2UH, England

Dr Roland Solecki, Safety of Substances and Preparations, Coordination and Overall Assessment, Federal Institute for Risk Assessment, Thielallee 88-92, D-14195 Berlin, Germany

Dr Maria Tasheva, National Center of Public Health Protection (NCPHP), 15 Iv. Ev. Geshov boulevard, 1431 Sofia, Bulgaria

Secretariat

Ms Catherine Adcock, Fungicide/Herbicide Toxicological Evaluation Section, Health Evaluation Directorate, Pest Managment Regulatory Agency, 2720 Riverside Drive, AL 6605E Ottawa, Ontario K1A 0K9, Canada *(WHO Temporary Adviser)*

Dr habil. Árpád Ambrus, Hungarian Food Safety Office, Gyali ut 2-6, 1097 Budapest, Hungary *(FAO Temporary Adviser)*

Mr Kevin Bodnaruk, 26/12 Phillip Mall, West Pymble, NSW 2073, Australia *(FAO Temporary Adviser)*

Professor Zongmao Chen, Chairman of Codex Committee on Pesticide Residues, Academician, Chinese Academy of Engineering, Chinese Academy of Agricultural Sciences, No. 1, Yunqi Road, Hangzhou/Zhejiang 310008, China *(WHO Temporary Adviser)*

Dr Myoengsin Choi, International Programme on Chemical Safety, World Health Organization, 1211 Geneva 27, Switzerland *(WHO Staff Member)*

Dr Ian C Dewhurst, Pesticides Safety Directorate, Mallard House, Kings Pool, 3 Peasholme Green, York YO1 7PX, England *(WHO Temporary Adviser)*

Dr Ronald D. Eichner, 13 Cruikshank Street, Wanniassa ACT 2903, Australia *(FAO Temporary Adviser)*

Dr Yibing He, Pesticide Residue Division, Institute for the Control of Agrochemicals, Ministry of Agriculture, Building 22, Maizidian Street, Chaoyang District, Beijing 100026, China *(FAO Temporary Adviser)*

Dr D. Kanungo, Additional DG, Directorate General of Health Services, Ministry of Health and Family Welfare, West, Block No. 1, RK Puram, New Delhi, India *(WHO Temporary Adviser)*

Dr Jeronimas Maskeliunas, Food Standards Officer, Joint FAO/WHO Food Standards Programme, Nutrition and Consumer Protection Division, Food and Agriculture Organization of the United Nations, Viale delle terme di Caracalla, 00153 Rome, Italy *(FAO Staff Member)*

Dr Heidi Mattock, 21bis rue du Mont Ouest, 38230 Tignieu-Jameyzieu, France *(WHO Editor)*

Dr Douglas B. McGregor, Toxicity Evaluation Consultants, 38 Shore Road, Aberdour KY3 0TU, Scotland *(WHO Temporary Adviser)*

Dr Utz Mueller, Principal Toxicologist, Risk Assessment – Chemical Safety, Food Standards Australia New Zealand, PO Box 7186, Canberra, BC ACT 2610, Australia *(WHO Temporary Adviser)*

Dr Rudolf Pfeil, Pesticides and Biocides Division, Federal Institute for Risk Assessment, Thielallee 88-92, D-14195 Berlin, Germany *(WHO Temporary Adviser)*

Dr Prakashchandra V. Shah, United States Environmental Protection Agency, Mail Stop: 7509P, 1200 Pennsylvania Avenue NW, Washington DC 20460, USA *(WHO Temporary Adviser)*

Mr Christian Sieke, Federal Institute for Risk Assessment, Thielallee 88-92, D-14195 Berlin, Germany *(FAO Temporary Adviser)*

Dr Atsuya Takagi, Division of Cellular and Molecular Toxicology, Biological Safety Research Center, National Institute of Health Sciences, 1-18-1 Kamiyoga, Setagaya-ku, Tokyo 158-8501, Japan *(WHO Temporary Adviser)*

Dr Angelika Tritscher, WHO Joint Secretary, International Programme on Chemical Safety, World Health Organization, 1211 Geneva 27, Switzerland *(WHO Joint Secretary)*

Dr Qiang Wang; Institute of Quality and Standard for Agricultural Products, Zhenjiang Academy of Agricultural Sciences, 198 Shiqiao Road, Hangzhou 310021, China *(FAO Temporary Adviser)*

Dr Gerrit Wolterink, Centre for Substances & Integrated Risk Assessment, National Institute of Public Health and the Environment (RIVM), Antonie van Leeuwenhoeklaan 9, PO Box 1, 3720 BA Bilthoven, Netherlands *(WHO Temporary Adviser)*

Dr Yong Zhen Yang, FAO Joint Secretary, FAO Plant Protection Service (AGPP), Viale delle Terme di Caracalla, 00153 Rome, Italy *(FAO Joint Secretary)*

Dr Jürg Zarn, Swiss Federal Office of Public Health, Food Toxicology Section, Stauffacherstrasse 101, CH-8004 Zurich, Switzerland *(WHO Temporary Adviser)*

Abbreviations used

3-MC	3-methylcholanthrene
ACTH	adenocorticotropic hormone
ADI	acceptable daily intake
ai	active ingredient
ALT	alanine aminotransferase
ANT	antarelix
AR	androgen receptor
ARfD	acute reference dose
AST	aspartate aminotransferase
AUC	area under the curve of concentration–time
$BMDL_{10}$	benchmark-dose lower 95% confidence level
BROM	bromocriptine
bw	body weight
CAS	Chemical Abstracts Service
CCFAC	Codex Committee on Food Additives and Contaminants
CCN	Codex classification number (for compounds or commodities)
CCPR	Codex Committee on Pesticide Residues
CI	confidence interval
CSAF	chemical-specific assessment factor
C_{max}	maximum concentration
DβH	dopamine-β-hydroxylase
DA	dopamine
DACT	diaminochlorotriazine
DEA	deethyl-atrazine
DHT	dihydroxytestosterone
DIA	deisopropyl-atrazine
DMCPA	1,2-dimethylcyclopropane-dicarboxylic acid
DMSO	dimethyl sulfoxide
DNCB	1-chloro-2,4-dinitrobenzene
E2	estradiol
EC_{50}	the concentration of agonist that elicits a response that is 50% of the possible maximum
ECG	electrocardiogram
ER	estrogen receptor
EROD	7-ethoxyresorufin *O*-deethylase
F_0	parental generation
F_1	first filial generation
F_2	second filial generation

FAO	Food and Agricultural Organization of the United Nations
FIFRA	Federal Insecticide, Fungicide, and Rodenticide Act
FOB	functional observational battery
GAP	good agricultural practice
GC	gas chromatography
GGT	gamma-glutamyltransferase
GEMS/Food	Global Environment Monitoring System–Food Contamination Monitoring and Assessment Programme
GnRH	gonadotropin-releasing hormone
HA	hypothalamic amenorrhoea
hCG	human chorionic gonadotrophin hormone
HPLC	high-performance liquid chromatography
HPLC-UV	high-performance liquid chromatography with ultraviolet light detector
HR	highest residue in the edible portion of a commodity found in trials used to estimate a maximum residue level in the commodity
HR-P	highest residue in a processed commodity calculated by multiplying the HR of the raw commodity by the corresponding processing factor
IC_{50}	concentration required to inhibit activity by 50%
IEDI	international estimated daily intake
IESTI	international estimate of short-term dietary intake
ISO	International Organization for Standardization
IUPAC	International Union of Pure and Applied Chemistry
JECFA	Joint FAO/WHO Expert Committee on Food Additives
JMPR	Joint Meeting on Pesticide Residues
JMPS	Joint FAO/WHO Meeting on Pesticide Specifications
LC	liquid chromatography
LC50	median lethal concentration
LD50	median lethal dose
LE	Long-Evans
LH	luteinizing hormone
LOAEC	lowest-observed-adverse-effect concentration
LOAEL	lowest-observed-adverse-effect level
LOD	limit of detection
LOQ	limit of quantification
MAFF	Ministry of Agriculture, Fisheries and Food
MCH	mean corpuscular haemoglobin
MCHC	mean corpuscular haemoglobin concentration
MCV	mean corpuscular volume
MEQ	methylethoxyquin
MIC	minimum inhibitory concentration
MMAD	mass median aerodynamic diameter

MRL	maximum residue limit
MS	mass spectrometry
MS/MS	tandem mass spectrometry
MTD	maximum tolerated dose
NMR	nuclear magnetic resonance
NOAEC	no-observed-adverse-effect concentration
NOAEL	no-observed-adverse-effect level
NOEL	no-observed-effect level
OECD	Organization for Economic Co-operation and Development
OR	odds ratio
PCPS	polycystic ovarian syndrome
PHI	pre-harvest interval
PPARα	peroxisome proliferator-induced receptor alpha
ppm	parts per million
PROD	7-pentoxyresorufin *O*-depentylase
QA	quality assurance
STMR	supervised trials median residue
STMR-P	supervised trials median residue in a processed commodity calculated by multiplying the STMR of the raw commodity by the corresponding processing factor
T3	triiodothyronine
T4	thyroxin
TBPS	*tert*-butylbicyclophosphorothionate
TH	tyrosine hydroxylase
TIPA	triisopropylammonium
TLC	thin-layer chromatography
THPI	1,2,3,6-tetrahydrophthalimide
TRR	total radiolabelled residue
TSH	thyroid-stimulating hormone
TMDI	theoretical maximum daily intake
WHO	World Health Organization
w/v	weight for volume
w/w	weight for weight

Introduction

The toxicological monographs and monograph addenda contained in this volume were prepared by a WHO Core Assessment Group that met with the FAO Panel of Experts on Pesticide Residues in Food and the Environment in a Joint Meeting on Pesticide Residues (JMPR) in Geneva, Switzerland, on 18–27 September 2007.

Five of the substances evaluated by the WHO Core Assessment Group (aminopyralid, difenoconazole, dimethomorph, pyrimethanil and zoxamide) were evaluated for the first time. Six compounds (azinphos-methyl, lambda-cyhalothrin, flusilazole, procymidone, profenofos) were re-evaluated within the periodic review programme of the Codex Committee on Pesticide Residues (CCPR). Reports and other documents resulting from previous Joint Meetings on Pesticide Residues are listed in Annex 1.

The report of the Joint Meeting has been published by the FAO as *FAO Plant Production and Protection Paper 191*. That report contains comments on the compounds considered, acceptable daily intakes established by the WHO Core Assessment Group, and maximum residue limits established by the FAO Panel of Experts. Monographs on residues prepared by the FAO Panel of Experts are published as a companion volume, as *Evaluations 2007, Part I, Residues*, in the FAO Plant Production and Protection Paper series.

The toxicological monographs and addenda contained in this volume are based on working papers that were prepared by temporary advisers before the 2007 Joint Meeting. A special acknowledgement is made to those advisers and to the Members of the Joint Meeting who reviewed early drafts of these working papers.

The preparation and editing of this volume was made possible by the technical and financial contributions of the lead institutions of the International Programme on Chemical Safety (IPCS), which supports the activities of the JMPR. The designations employed and the presentation of the material in this publication do not imply the expression of any opinion whatsoever on the part of the Central Unit of the IPCS concerning the legal status of any country, territory, city or area or of its authorities, nor concerning the delimitation of its frontiers or boundaries. The mention of specific companies or of certain manufacturers' products does not imply that they are endorsed or recommended by the IPCS in preference to others of a similar nature that are not mentioned.

Any comments or new information on the biological properties or toxicity of the compounds included in this volume should be addressed to: Joint WHO Secretary of the Joint FAO/WHO Meeting on Pesticide Residues, International Programme on Chemical Safety, World Health Organization, 20 Avenue Appia, 1211 Geneva, Switzerland.

Introduction

TOXICOLOGICAL MONOGRAPHS
AND MONOGRAPH ADDENDA

AMINOPYRALID

First draft prepared by
Jürg Zarn[1] & Alan Boobis[2]

[1]*Food Toxicology Section, Swiss Federal Office of Public Health, Zurich, Switzerland; and*
[2]*Experimental Medicine & Toxicology, Division of Investigative Science, Faculty of Medicine, Imperial College London, London, England*

Explanation

Aminopyralid is the International Organization for Standardization (ISO) approved name for 4-amino-3,6-dichloropyridine-2-carboxylic acid (Chemical Abstracts Service, CAS No. 150114-71-9). It is a post-emergent auxin-type herbicide for the control of a wide variety of broadleaf weed species. Most of the toxicological studies evaluated were performed with the aminopyralid acid. However, some studies were performed with the commercially available aqueous solution of aminopyralid triisopropylammonium (TIPA) salt, called GF-871, since products are also marketed in this form. Throughout the current evaluation, aminopyralid acid is termed aminopyralid and its TIPA salt is termed aminopyralid TIPA. GF-871 is a 41.3–41.9% aqueous solution of aminopyralid TIPA corresponding to approximately 21.7% aminopyralid. If not otherwise stated, doses are given as aminopyralid equivalents.

Aminopyralid has not been evaluated previously by the Meeting. The evaluation of aminopyralid was scheduled for the 2006 JMPR but, owing to incomplete submission of data, was postponed to the present Meeting. Aminopyralid was reviewed at the request of the Codex Committee on Pesticide Residues (CCPR). All pivotal studies with aminopyralid and aminopyralid TIPA were certified as complying with good laboratory practice (GLP).

Figure 1. Chemical structure of aminopyralid

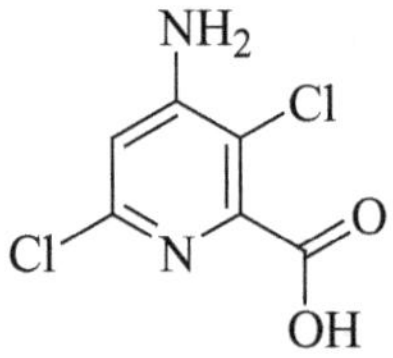

NH_2 Cl Cl N O OH

Figure 2. Chemical structure of TIPA

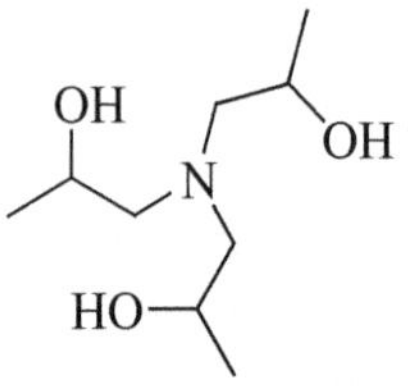

OH OH N HO

TIPA, triisopropylammonium

Evaluation for acceptable daily intake

1. Biochemical aspects

1.1 Absorption, distribution, metabolism and excretion

Rat

The absorption, distribution, metabolism and excretion of aminopyralid were studied in groups of four male Fischer 344 rats given single and multiple oral daily doses of [^{14}C]aminopyralid (purity, 99.5%; radiochemical purity, 98.6%; specific activity, 1.014 GBq/mmol; batch No. F380-135a) by gavage. Animals received either a single higher dose of 1000 mg/kg bw or a lower dose of 50 mg/kg bw. An additional group of four male rats was given non-radiolabelled test material as a daily dose at at 50 mg/kg bw per day (days 1 to 14) and radiolabelled test material as a single dose (day 15) at 50 mg/kg bw per day. Aminopyralid was administered in 0.5% methylcellulose in distilled water. The radiolabelled preparations also contained 5.3% ethanol (v/v) (equivalent to approximately 300 mg/kg bw for in groups receiving higher and lower doses). The study complied with GLP.

Throughout the study, the animals looked healthy and showed no changes in appearance or behaviour. During the 7-day collection period, the percentages of radioactivity excreted in the groups receiving a single lower dose or a single higher dose were comparable, most of the excretion being completed by 24 h. Renal and faecal excretion were slightly greater than 40%, with a tendency to higher renal excretion in the group receiving a single lower dose. In the group receiving multiple lower doses, excretion via the urine amounted to approximately 60% and via faeces was nearly 30%. More details are presented in Table 1. Biliary excretion of aminopyralid was assumed to be very low, on the basis of studies with the closely-related substances triclopyr, picloram, fluroxypyr and 2,4-D. Therefore, the company did not perform experiments to analyse the biliary excretion of aminopyralid and the percentage of absorbed material was calculated as the sum of the renally excreted and residual substance in the tissues and the carcass. On this basis, the percentage of absorbed material was estimated to be 42%, 50% and 59% for the groups receiving the single higher dose, the single lower dose and multiple lower doses, respectively.

The urinary elimination was biphasic, with elimination half-lives $t_{1/2\alpha}$ and $t_{1/2\beta}$ of 3–4 h and 10–12 h, respectively (see Table 2). For both half-lives, there were no significant differences between the three dosing groups, 74.0–93.1% of the administered dose being excreted within the first 24 h after dosing.

As can be seen from Table 3, after a depletion period of 168 h, virtually all of the examined tissues had radioactivity levels of less than 0.01% of the administered dose. Only in the skin of the rats in the group receiving a single higher dose were notable levels detected, but with high interindividual variability (1–66 µg eq/g).

Pooled samples of urine from 0–6 h and 6–12 h, together contained 78–90% of the total radioactivity recovered in the urine and in all dose groups unchanged aminopyralid comprised more than 96% of the radiolabel present. The remainder of the radiolabel in the urine, approximately 4%,

Table 1. Recovery of radioactivity over 7 days in rats given radiolabelled aminopyralid by gavage

Sample	Recovery (% of administered dose)		
	1000 mg/kg bw (single dose)	50 mg/kg bw (single dose)	50 mg/kg bw per day (rpeat dose)
Urine	41.3 ± 8.6	49.8 ± 4.8	58.65 ± 5.5
Faeces	43.4 ± 7.7	43.1 ± 5.8	33.0 ± 4.6
Volatiles	0.01 ± 0.00	NA	NA
Cage wash	9.7 ± 3.1	3.0 ± 0.9	3.4 ± 0.7
Tissues/carcass	0.73 ± 0.72	0.07 ± 0.04	0.09 ± 0.02
Total recovery	95.2 ± 10.5	96.0 ± 1.7	95.0 ± 1.8
Absorbed material	42	50	59

From Liu (2004)
NA, not applicable, no collection performed.

Table 2. Average urinary elimination half-lives in rats given radiolabelled aminopyralid by gavage

Urinary elimination half-life (h)	Dose		
	1000 mg/kg bw (single dose)	50 mg/kg bw (single dose)	50 mg/kg bw per day (repeat dose)
t1/2α	3.8 ± 0.5	2.9 ± 1.2	3.3 ± 0.7
t1/2β	10.9 ± 2.7	10.2 ± 3.3	12.3 ± 3.7

From Liu (2004)

Table 3. Tissue distribution of radiolabel at 168 h after dosing in rats given radiolabelled aminopyralid by gavage

Tissue	Dose					
	1000 mg/kg bw (single dose)		50 mg/kg bw (single dose)		50 mg/kg bw per day (repeat dose)	
	% of dose	µg Eq/g	% of dose	µg Eq/g	% of dose	µg Eq/g
Whole blood	NA	0.608	NA	0.026	NA	0.029
Kidneys	< 0.01	0.541	< 0.01	0.020	< 0.01	0.021
Liver	< 0.01	0.607	< 0.01	0.024	< 0.01	0.027
Perirenal fat	NA	0.072	NA	0.004[a]	NA	0.004[a]
Gastrointestinal tract and contents	0.03	2.56	< 0.01	0.019	< 0.01	0.025
Skin	0.45	27.0	0.03	0.074	0.05	0.148
Spleen	< 0.01	0.609	< 0.01	0.017	< 0.01	0.015[a]
Remaining carcass	0.26	4.45	0.04	0.030	0.05	0.032

From Liu (2004)
NA, not applicable.
[a]Average of 50% of the values for the limit of detection within the group.

comprised three unknown components, which were also found in the dosing formulation and hence were not further characterized. In pooled samples of faeces from all dose groups at 0–24 h and a pooled sample from the group receiving the single higher dose at 24–48 h, only unchanged parent compound was identified. These results suggest that aminopyralid is not metabolized (Liu, 2004).

To compare the patterns of absorption, distribution, metabolism and excretion of aminopyralid and aminopyralid TIPA, two groups of four male Fischer 344 rats received single equimolar doses of either [^{14}C]aminopyralid (batch No. DE3-E1004-77, INV1893) at 50 mg/kg bw or [^{14}C]aminopyralid TIPA at 96 mg/kg bw by oral gavage. [^{14}C]Aminopyralid TIPA was prepared using [^{14}C]aminopyralid (batch No. DE3-E1004-77, INV1893) by adding appropriate amounts of TIPA.

The test material was 94.5% chemically and 98.25% radiochemically pure active ingredient. Both test materials had a specific activity of 1.058 GBq/mmol and were administered in 0.5% methylcellulose in distilled water. Dosing preparations containing aminopyralid (acid form) appeared to be much more like a suspension than did the TIPA salt, which was readily soluble. Aminopyralid was re-suspended before gavage by shaking by hand. Rats were fitted with indwelling jugular vein cannulae and blood samples (14) were taken at intervals until 120 h after dosing. The study complied with GLP.

Throughout the study, the animals looked healthy and showed no changes in appearance or behaviour. Calculated as the sum of renal excretion and remaining radioactivity in tissues and carcass, 46% and 43% of aminopyralid in the acid form and as the TIPA salt were absorbed, respectively. Faecal excretion amounted to about 50% for both compounds. Except for one male with 0.02% of the administered dose in the skin, no radioactivity at above the limit of quantitation of ≤ 0.01% of administered dose (LOQ) was recovered in the tissues of rats treated with aminopyralid. In rats treated with aminopyralid TIPA, traces of radioactivity at slightly greater than the LOQ were identified in the kidney and the spleen. Plasma peak concentrations were achieved at 0.25 h after dosing resulting in 26 and 16 µg acid equivalents/g plasma in rats treated with aminopyralid or aminopyralid TIPA, respectively. For both compounds, urinary excretion was nearly completed within the first 24 h after dosing and the biphasic elimination parameters were comparable.

Table 4. Recovery of radioactivity over 5 days in rats given radiolabelled aminopyralid or aminopyralid TIPA by gavage

Sample	Recovery (% of administered dose)	
	Aminopyralid	Aminopyralid TIPA
Urine	46.3 ± 8.0	42.5 ± 8.8
Faeces	50.6 ± 4.6	54.3 ± 5.6
Cage wash	0.2 ± 0.1	0.2 ± 0.1
Tissues/carcass	0.01 ± 0.01	0.0 ± 0.0
Total recovery	97.1 ± 3.5	97.0 ± 3.1
Absorbed material	46	43

From Domoradzki et al. (2004)
TIPA, triisopropylammonium

Table 5. Average plasma and urinary elimination half-lives in rats given radiolabelled aminopyralid or aminopyralid TIPA by gavage

Half-life (h)	Aminopyralid	Aminopyralid TIPA
Plasma		
$t_{1/2\alpha}$	0.34 ± 0.08	0.51 ± 0.21
$t_{1/2\beta}$	8.8 ± 2.3	13.0 ± 7.5
Urine		
$t_{1/2\alpha}$	2.8 ± 1.1	2.5 ± 0.8
$t_{1/2\beta}$	7.8 ± 2.2	10.7 ± 1.7

From Domoradzki et al. (2004)
TIPA, triisopropylammonium

In pooled samples of urine and faeces of animals treated with aminopyralid, only one labelled peak was found, corresponding to unchanged aminopyralid. On analysis of pooled samples of urine and faeces from rats in the group treated with aminopyralid TIPA, the result was essentially the same but a minor labelled peak was observed, but not further characterized because it accounted for only 0.34% of the administered dose (Domoradzki et al., 2004).

Rabbit

The absorption, distribution, metabolism and excretion of aminopyralid were studied in three groups of three female New Zealand White (NZW) rabbits given single and multiple oral daily doses by gavage. Three non-pregnant animals received a single dose of radiolabelled test material at 371 mg/kg bw, three pregnant animals received a single dose of radiolabelled aminopyralid at 362 mg/kg bw on day 7 of gestation and another three pregnant animals received repeated daily doses of nonlabelled aminopyralid at 279 mg/kg bw from day 7 of gestation until day 21 of gestation, followed by a single dose of radiolabelled aminopyralid at 279 mg/kg bw on day 22 of gestation. The variation between doses administered was the consequence of inadvertently using different dose preparations, but this was judged not to affect the interpretation of the results. For blood sampling, vascular access ports were placed into the jugular vein 2 weeks before mating. All animals were killed 72 h after dosing. Additionally, plasma from rabbits that were not pregnant, from pregnant animals on day 7 of gestation and on day 22 of gestation and for comparison purposes from non-pregnant female Fischer 344 rats was used for studies of plasma protein binding with radiolabelled aminopyralid. The study complied with GLP.

The test material was 94.5% chemically and 98.25% radiochemically pure active ingredient and had a specific activity of 1.058 GBq/mmol (batch No. of labelled compound, TSN 102298, DE3-E1004-77, INV1893).

Renal excretion was the major route of elimination in all three groups. In the group of pregnant rabbits that were pre-treated with nonlabelled aminopyralid, faecal excretion was substantially less, while renal excretion was minimally increased when compared with that in the other groups. The only radioactive peak in pooled samples of urine and faeces was identified as unchanged aminopyralid.

Other than the gastrointestinal tract, which showed some radioactivity at termination, the spleen in the group of pregnant rabbits that were pre-treated with nonlabelled aminopyralid was the only organ showing some radioactivity.

In all dosed groups, radioactivity was detected in the plasma within 10 min after dosing and T_{max} was achieved within 1 h. The plasma elimination half-life was lowest in the group of pretreated pregnant rabbits (4.1 h) and highest in the group of non-pregnant rabbits (7.2 h). There was increased early renal excretion, shorter T_{max} and higher plasma area under the curve of concentration–time (AUC) in pregnant rabbits given repeated doses, indicating more rapid and extensive absorption in these animals when compared with non-pregnant rabbit or pregnant rabbits given a single dose on day 7 of gestation.

Table 6. Recovery of radioactivity in rabbits given radiolabelled aminopyralid by gavage

Sample	Recovery (% of radiolabelled dose)		
	Not pregnant	Pregnant	
		Treated on day 7 of gestation	Treated on days 7–22 of gestation
Expired air	NC	NC	NC
Tissues	0.15 ± 0.05	0.11 ± 0.10	0.11 ± 0.04
Carcass	NA	NA	NA
Cage wash	1.19 ± 1.35	0.36 ± 0.42	0.52 ± 0.45
Urine:			
0–12 h	39.0 ± 11.5	47.5 ± 11.8	61.5 ± 5.7
12–24 h	33.5 ± 7.5	29.5 ± 3.9	21.7 ± 3.3
24–48 h	3.86 ± 0.71	5.33 ± 3.96	1.88 ± 0.26
24–72 h	0.38 ± 0.28	0.23 ± 0.03	0.67 ± 0.23
0–72 h	76.6 ± 7.5	82.5 ± 7.6	85.7 ± 2.5
Faeces:			
0–24 h	19.6 ± 5.5	15.3 ± 6.1	7.1 ± 4.0
24–48 h	0.53 ± 0.42	0.64 ± 0.83	0.33 ± 0.30
48–72 h	0.06 ± 0.01	0.09 ± 0.03	0.08 ± 0.02
0–72 h	20.3 ± 5.8	16.1 ± 6.8	7.54 ± 4.26
Total	98.2 ± 2.9	99.1 ± 2.9	93.9 ± 4.4
Urinary half-life (h)	6.47 ± 0.98	5.96 ± 0.41	6.77 ± 0.40

From Hansen et al. (2005)
NA, not analysed; NC, Not collected

Table 7. Pharmacokinetic parameters in rabbits given radiolabelled aminopyralid by gavage

Plasma parameter	Not pregnant	Pregnant, , treated on day 7 of gestation	Pregnant, treated on days 7–22 of gestation
C_{max} (µg/g plasma)	50 ± 16	50 ± 15	71 ± 5
T_{max} (h)	1.3 ± 0.6	1.6 ± 0.1	0.5 ± 0.4
$t_{½}$ (h)	7.2 ± 2.6	4.4 ± 1.6	4.1 ± 0.7
K_{el} (h^{-1})	0.106 ± 0.045	0.180 ± 0.085	0.171 ± 0.033
AUC (µg•h/ml)	334 ± 13	265 ± 39	441 ± 55

From Hansen et al. (2005)
AUC, area under the curve of concentration–time.

Table 8. Plasma protein binding of radiolabelled aminopyralid in rabbits treated by gavage

Concentration of radiolabelled aminopyralid	Plasma protein binding (%)			
	Rat	Not pregnant	Pregnant, treated on day 7 of gestation	Pregnant, treated on days 7–22 of gestation
10 µg/ml	72.3	68.2	65.7	58.1
33 µg/ml	70.3	67.5	62.7	53.4
154 µg/ml	55.5	53.4	47.3	43.1

From Hansen et al. (2005)

Plasma protein binding of aminopyralid was moderate, about 40–70%, and was greater in rats than in rabbits. In rabbits, it was slightly lower in pregnant animals on day 22 of gestation after pretreatment.

In summary, the pharmacokinetic and plasma-protein-binding studies indicated a somewhat higher bioavailability in late-stage pregnant rabbits than in non-pregnant or early-stage pregnant rabbits, although the difference in terms of unbound compound was not more than twofold (Hansen et al., 2005).

1.2 Effects on enzymes and other biochemical parameters

No information was available.

2. Toxicological studies

2.1 Acute toxicity

(a) Oral administration

Groups of five overnight-fasted male and female Fischer 344 rats received aminopyralid (purity, 94.5%; batch No. TSN102319/F0031-143) at a dose of 5000 mg/kg bw by gavage as two doses at 2500 mg/kg administered with an interval of 1 h, as a 50% mixture in 0.5% aqueous methylcellulose. The rats were observed for mortality, clinical signs and body-weight development during the 14 days after dosing and were then necropsied. The study complied with GLP.

One male rat died on test day 3, its appearance before death being consistent with a moribund condition. Clinical observations in the surviving rats revealed a high incidence of various combinations of perineal, perioral and perinasal soiling, watery faeces and a lower incidence of periocular soiling, decreases in muscle tone, resistance to removal, extensor-thrust, and reactivity to handling were observed. All surviving animals recovered completely by the time of study termination. Four rats had transient body-weight loss during the first week of the study, but all rats gained weight over the duration of the study. On necropsy, the male that died had treatment-related gross findings consisting of haemolysis, gas in the gastrointestinal tract and perineal soiling. Surviving animals had no treatment-related gross pathological changes.

The acute oral median lethal dose (LD_{50}) for aminopyralid was > 5000 mg/kg bw (Brooks & Yano, 2001b).

Groups of five overnight fasted male and female Fischer 344 rats received a single dose of undiluted GF-871 (41.9% aminopyralid TIPA; batch No. TSN103622/E-1175-52) at a dose of 5000 mg/kg bw (equivalent to a dose of aminopyralid of 1085 mg/kg bw) by gavage. The rats were observed for mortality, clinical signs and body-weight development during the next 14 days and were then necropsied. The study complied with GLP.

All animals survived the 14-day observation period and gained weight. Clinical observations consisted of bilateral cloudy eyes in all animals on day 1, lacrimation, watery or soft faeces and soiling of the periocular and/or perineal regions. All clinical signs resolved by test day 4. No treatment-related pathological changes were observed.

The acute oral LD_{50} for aminopyralid TIPA was > 5000 mg/kg bw, equivalent to 1085 mg/kg bw expressed as aminopyralid (Wilson et al., 2002b).

(b) Dermal application

Five male and five female Fischer 344 rats were treated by dermal patch application for 24 h with aminopyralid (purity, 94.5%; batch No. TSN102319/F0031-143) at a dose of 5000 mg/kg bw, moistened with 0.5 ml of 0.5% aqueous methylcellulose. The rats were observed for mortality, clinical signs and body-weight development during the next 14 days and were then necropsied. The study complied with GLP.

All animals survived the 14-day observation period. Clinical observations consisted of perioral soiling in one male and two females on test day 1 or 2 and periocular soiling in one male on test day 1. All animals lost weight by test day 2, but gained weight over the remainder of the study. There were no gross pathological observations.

The acute dermal LD_{50} of aminopyralid was > 5000 mg/kg bw (Brooks & Yano, 2001a).

Five male and five female Fischer 344 rats were treated by dermal patch application for 24 h with GF-871 (41.9% aminopyralid TIPA; batch No. TSN103622/E-1175-52) at a dose of 5000 mg/kg bw (equivalent to aminopyralid at a dose of 1085 mg/kg bw). The rats were observed for mortality, clinical signs and body-weight development during the next 14 days and were then necropsied. The study complied with GLP.

All animals survived the 14-day observation period. After a body-weight loss on test day 2 in all animals, all animals gained weight over the remainder of the study. Clinical observations consisted of perineal soiling in one male and reddening of the skin on the test site in two males. There were no gross pathological observations.

The acute dermal LD_{50} of aminopyralid TIPA was > 5000 mg/kg, equivalent to 1085 mg/kg bw expressed as aminopyralid (Wilson et al., 2002a).

(c) *Exposure by inhalation*

Five male and five female Fischer 344 rats were exposed by nose-only inhalation for 4 h to a dust aerosol containing aminopyralid (purity, 94.5%; batch No. TSN102319/F0031-143) at 5.5 mg/l and with a mass median aerodynamic diameter (MMAD) of 2.5 µm (geometric standard deviation, ± 2.45). The rats were observed for mortality, clinical signs and body-weight development during the 14 days after exposure and were then necropsied. The study complied with GLP.

Immediately after exposure, gasping was observed in one male and bilateral ptosis in four males and five females. There were no other toxicologically significant clinical observations immediately after exposure. During the 14-day observation period, toxicologically significant clinical observations were limited to one female during the first week. This animal was noted with dried red material around the nose, ptosis or complete closure of the eyes and/or yellow material on the urogenital area. All animals were considered normal by day 7. Initially, a minimal body-weight decrease was observed for both sexes (9 out of 10 animals), but this effect was no longer seen on day 3 and in all animals no relevant findings were recorded on terminal necropsy.

The acute median lethal concentration (LC_{50}) after exposure to aminopyralid by inhalation was > 5.5 mg/l (Kiplinger, 2001).

Five male and five female Fischer 344 rats were exposed by nose-only inhalation for 4 h to a dust aerosol containing GF-871 (41.9% aminopyralid TIPA, equivalent to aminopyralid at 1.26 mg/l, E-1175-52) at 5.79 mg/l and with a MMAD of 2.9 µm (geometric standard deviation, ± 1.67). The rats were observed for mortality, clinical signs and body-weight development during the 14 days after exposure and were then necropsied. The study complied with GLP.

All animals survived the treatment and post-exposure period and the only treatment-related signs were extensive soiling and body-weight loss on day 2. Thereafter, all animals gained weight and no treatment-related observations were made upon necropsy.

The acute LC_{50} on inhalation exposure to GF-871 was > 5.79 mg/l, equivalent to aminopyralid at 1.26 mg/l (Landry & Krieger, 2002).

(d) *Dermal and ocular irritation*

One male and two female NZW rabbits were exposed dermally on the intact skin for 4 h to 500 mg of aminopyralid (purity, 94.5%; batch No. F0031-143) moistened with 0.3 ml of 0.5% aqueous methylcellulose. The study complied with GLP.

Over the 3-day study period, no signs of dermal irritation were observed and the animals gained weight normally (Brooks, 2001a).

One male and two female NZW rabbits were exposed dermally on the intact skin for 4 h to 0.5 ml of undiluted GF-871 (41.9% aminopyralid TIPA, equivalent to approximately 108.5 mg of aminopyralid, batch No. TSN103622/ E-1175-52). The study complied with GLP.

The female rabbits showed very slight erythema within the first 48 h and 72 h after unwrapping, respectively. These signs resolved within 7 days. All animals gained weight normally (Brooks & Radtke, 2002a).

Two male and one female NZW rabbits were treated with 100 mg of undiluted aminopyralid (purity, 94.5%; batch No. F0031-143) in powder form in the conjunctival sac of the right eye. Animals were observed for 35 days for ocular and body-weight changes. The study complied with GLP.

One hour after dosing, all rabbits had slight conjunctival redness, moderate chemosis, and moderate discharge. Twenty-four hours after dosing, all rabbits had moderate to marked conjunctival redness, moderate chemosis, moderate to marked discharge, and slight corneal opacity. The two male rabbits also had slight irritation of the iris 24 h after dosing. Forty-eight hours after dosing, all rabbits had moderate to marked conjunctival redness, slight to moderate chemosis, moderate to marked discharge, slight corneal opacity, and slight irritation of the iris. The ocular irritation slowly diminished over time. Corneal vascularization was noted in all three animals starting on day 15 and continuing for one rabbit until day 36. Two rabbits had ocular irritation that continued to study termination. All rabbits gained body weight normally throughout the study. The Meeting concluded that aminopyralid is irritating to the eyes (Brooks, 2001b).

One male and two female NZW rabbits were treated with 0.1 ml of undiluted GF-871 (41.9% aminopyralid TIPA, equivalent to approximately 21.7 mg of aminopyralid; batch No. TSN013622/E-1175-52) in the conjunctival sac of the right eye. Animals were observed for 3 days for ocular and body-weight changes. The study complied with GLP.

One hour after dosing, the male and one of the two females showed slight conjunctival redness, which resolved by test day 2. There were no other signs of ocular irritation. The Meeting concuded that GF-871 causes slight initial irritation to the eyes (Brooks & Radtke, 2002b).

(e) Dermal sensitization

Ten male and ten female Hartley-derived guinea-pigs were used to evaluate the dermal sensitization potential of aminopyralid according to the Magnusson & Kligman method. The study complied with GLP. Aminopyralid (batch No. TSN102319/F0031-143) was administered intradermally at a dose of 5% w/v, followed 1 week later by a topical dose at 100% w/v. Two weeks later, the animals were challenged with 100% aminopyralid, applied topically. Aminopyralid was not a contact sensitizer, since there was no dermal reaction in any of the test animals, but clear reactions with the positive-control compounds 1-chloro-2,4-dinitrobenzene (DNCB) and α-hexylcinnamaldehyde (HCA) (Wilson, 2001).

Ten male and ten female Hartley-derived guinea-pigs were used to evaluate the dermal sensitization potential of GF-871 (41.9% aminopyralid TIPA; batch No. E-1175-52) according to the Magnusson and Kligman method. The study complied with GLP. GF-871 was administered intradermally at a dose of 1% w/v, followed 1 week later by a topical dose of 100% w/v. Two weeks later, the animals were challenged with GF-871 at 100%, applied topically.

GF-871 was not a contact sensitizer, since there was no dermal reaction in any of the test animals, but clear reactions with the positive-control compounds DNCB and HCA (Brooks & Wilson, 2002).

2.2 *Short-term studies of toxicity*

Mouse

In a 4-week feeding study, groups of five male and five female CD-1 mice were fed diets containing aminopyralid (purity, 95.4%; batch No. F-0031-125, TSN102095) to supply doses of 0, 10, 100, 500 or 1000 mg/kg bw per day. Animals were examined twice per day for general health, moribundity and mortality, and weekly detailed clinical examinations were performed. Feed consumption and body-weight development were recorded regularly. At the end of the study, haematological investigations and clinical chemistry including coagulation time and urine analysis parameters were analysed, organ weights recorded and gross pathology and histopathology performed. The study complied with GLP.

Male and female mice at the highest dose had statistically significantly lower white blood cell counts compared with the concurrent controls, but values were within the range for historical controls in females and slightly less than the lower bound in males. There were no histological changes in the bone marrow and no evidence of leukocyte migration from the blood. Two males showed hepatocyte hypertrophy and decreased hepatocellular glycogen. There was no change in liver weight. The Meeting considered that these effects were not of toxicological significance. The no-observed-adverse-effect level (NOAEL) was 1000 mg/kg bw per day, the highest dose tested (Yano & Dryzga, 2000).

Table 9. Acute toxicity of aminopyralid

Test article	Species	Strain	Sex	Route	LD_{50} (mg/kg bw)	LC_{50} (mg/l)	Purity (%)	Reference
Aminopyralid	Rat	F344	M & F	Dermal	> 5000	—	94.5	Brooks & Yano (2001a)
Aminopyralid TIPA	Rat	F344	M & F	Dermal	> 5000, equivalent to > 1085 aminopyralid	—	Not reported	Wilson et al. (2002a)
Aminopyralid	Rat	F344	M & F	Oral	> 5000	—	94.5	Brooks & Yano (2001b)
Aminopyralid TIPA	Rat	F344	M & F	Oral	> 5000, equivalent to > 1085 aminopyralid	—	21.8% aminopyralid equivalents	Wilson et al. (2002b)
Aminopyralid	Rat	F344	M & F	Inhalation	—	> 5.5	94.5	Kiplinger (2001)
Aminopyralid TIPA	Rat	F344	M & F	Inhalation	—	> 5.79, equivalent to > 1.26 aminopyralid	21.8% aminopyralid equivalents	Landry & Krieger (2002)
Aminopyralid	Rabbit	NZW	M & F	Dermal irritation	No irritation	—	94.5	Brooks, (2001a)
Aminopyralid TIPA	Rabbit	NZW	M & F	Dermal irritation	Slight irritation	—	21.8% aminopyralid equivalents	Brooks & Radtke (2002a)
Aminopyralid	Rabbit	NZW	M & F	Ocular irritation	Irritant	—	94.5	Brooks, (2001b)
Aminopyralid TIPA	Rabbit	NZW	M & F	Ocular irritation	Slightly irritant	—	21.8% aminopyralid equivalents	Brooks & Radtke (2002b)
Aminopyralid	Guinea-pig	Hartley-derived albinos	M & F	Dermal sensitization	No sensitization	—	94.5	Wilson, (2001)
Aminopyralid TIPA	Guinea-pig	Hartley-derived albinos	M & F	Dermal sensitization	No sensitization	—	21.8% aminopyralid equivalents	Brooks & Wilson (2002)

F344, Fischer 344; F, female; NZW, M, male; New Zealand White.

In a 13-week feeding study, groups of 10 male and 10 female CD-1 mice were fed diets containing aminopyralid (purity, 94.5%; batch No. F0031-143, TSN102319) to supply doses of 0, 10, 100, 500 or 1000 mg/kg bw per day. Daily cage-side observations, weekly detailed clinical observations, ophthalmological examinations, body weights, feed consumption, haematology and clinical chemistry were evaluated, a gross necropsy was conducted with extensive histopathological examination of tissues and organ weights were recorded. The study complied with GLP.

The only statistically significant finding was a slight increase in plasma sodium concentrations in groups of males at 100, 500 or 1000 mg/kg bw per day (79–181 mmol/l vs 175 mmol/l in males in the control group). The plasma sodium concentrations of the controls in this study were greater than the range for historical controls. The Meeting considered that the changes in plasma sodium concentration were not of toxicological significance, since the effect was minimal and in its severity was not dose-dependant. Sodium was also slightly but significantly increased in females of the group at 500 mg/kg bw per day only. No other treatment-related changes were identified in any of the parameters evaluated. There were no changes in leukocyte number or in hepatic histology.

The NOAEL was 1000 mg/kg bw per day, the highest dose tested (Stebbins et al., 2001).

Rat

In a 4-week feeding study, groups of five male and five female Fischer 344 rats were fed diets containing aminopyralid (purity, 95.4%; batch No. F-0031-125, TSN 102095) to supply doses of 0, 10, 100, 500 or 1000 mg/kg bw per day. Animals were examined twice per day for general health, moribundity and mortality, and weekly detailed clinical examinations were performed. Feed consumption and body-weight development were recorded regularly. At the end of the study, haematological and clinical chemistry (including coagulation time) and urine analysis parameters were analysed, organ weights recorded and gross pathology and histopathology performed. The study complied with GLP.

The increases in prothrombin time (15.3 s in the males at the highest dose vs 13.1 s in males in the control group) and of plasma urea nitrogen (16 mg/dl in females at the highest dose vs 14 mg/dl in the females in the control group) in one sex at the highest dose were statistically significant, but fell well within ranges for historical controls. Three males and two females at 500 mg/kg bw and all rats at the highest dose showed enlarged caeca without any histological findings. Caecal enlargement is often an adaptive physiological response. Although treatment-related effects were identified at the highest dose and to a minor degree also at 500 mg/kg bw per day, the Meeting considered these to be not adverse.

The NOAEL was 1000 mg/kg bw per day, the highest dose tested (Stebbins & Day, 2000).

In a 4-week study of dermal toxicity, groups of 10 male and 10 female Fischer 344 rats received aminopyralid (purity, 94.5%; batch No. F-0031-143, TSN102319) at a dose of 0, 100, 500, or 1000 mg/kg bw per day for 6 h/day on 28 consecutive days. The test substance was applied as a suspension in 0.5% aqueous methylcellulose to a shaved area of at least 10% of the total body surface by semi-occlusion with a gauze dressing, non-absorbent cotton. Approximately 6 h after application, the exposure site was wiped with a water-dampened towel to remove any residual test material. Daily cage-side observations, weekly dermal and detailed clinical observations and ophthalmological examinations were performed, body weights and feed consumption recorded, haematology, clinical chemistry and urine analysis parameters analysed, and a gross necropsy with extensive histopathological examination of tissues and organ weights was conducted. The study complied with GLP.

Males at the lowest and the intermediate doses had statistically significant weight increases of the full but not of the empty caeca. The effect was not dose-dependent and there was no significant change in the group at the highest dose or in any of the groups of females. The effect was therefore judged by the Meeting not to be compound-related. In males at 500 and 1000 mg/kg bw per day, very slight to slight hyperplasia of the epidermis at the test sites was observed. There were no signs of any systemic effects attributable to dermal exposure.

The NOAEL for systemic toxicity was 1000 mg/kg bw per day, the highest dose tested (Stebbins et al., 2002).

In a 13-week feeding study, groups of 10 male and 10 female Fischer 344 rats were fed diets containing aminopyralid (purity, 94.5%; batch No. F0031-143, TSN102319) to supply doses of 0, 10, 100, 500 or 1000 mg/kg bw per day. Additionally, recovery groups of 10 male and 10 female rats were fed diets containing aminopyralid at 0 or 1000 mg/kg bw per day for 13 weeks and were given control feed for an additional 4 weeks to evaluate the reversibility of any effects induced during the 13 weeks of treatment with aminopyralid. Daily cage-side observations, weekly detailed clinical and ophthalmologic examinations were performed, body weights and feed consumption recorded, haematology, clinical chemistry and urine analysis parameters analysed, and a gross necropsy with extensive histopathological examination of tissues and organ weights was conducted. The study complied with GLP.

The data from urine analysis revealed some treatment-related changes. These included decreases in pH at 500 mg/kg bw per day and above, most likely to be a consequence of the urinary excretion of the test substance, which is an acid, and decreases of protein and ketone concentrations in the urine of males and females at the highest dose. A complete reversal of these effects was observed in the recovery group. There were no histopathological correlates of these changes. In male and female rats, absolute and relative weights of full and empty caeca were statistically significantly increased in the groups at 500 and the 1000 mg/kg bw per day; for the empty caeca of females this was only significant at the highest dose. These weight increases essentially returned to normal within the 4-week recovery period. The only treatment-related histopathological alteration was a very slight and diffuse hyperplasia of the mucosal epithelium of the caecum and the ileum in all males of the group at the highest dose. This effect was reversible in the recovery group. These changes are fully consistent with a physiological adaptation of the caecum. In males, slight but statistically significant decreases in thymus weights were identified in the groups at 10, 500 and the 1000 mg/kg bw per day, but a clear dose–response relationship was not evident. These changes were within the range for historical controls. There were no significant changes in prothrombin time or plasma urea nitrogen in any group.

The NOAEL was 1000 mg/kg bw per day, the highest dose tested (Dryzga & Stebbins, 2001).

In a 13-week feeding study, groups of 10 male and 10 female Fischer 344 rats were fed diets containing GF-871 (41.3% aminopyralid TIPA; batch No. 173-162-1A, TSN104110) to supply doses of 0, 465, 1211 or 2421 mg/kg bw per day, or 0, 101, 263 or 525 mg/kg bw per day expressed as aminopyralid. Daily cage-side observations, weekly detailed clinical and ophthalmological examinations were performed, body weights and feed consumption recorded, haematology, clinical chemistry and urine analysis parameters analysed, and a gross necropsy with extensive histopathological examination of tissues and organ weights was conducted. The study complied with GLP.

Treatment-related changes were confined to urine parameters and caecal weights. In the group at the highest dose, slightly higher urine volumes in both sexes and slightly reduced urine specific gravity in females were recorded. Absolute and relative weights of the full caeca were statistically significantly increased in both sexes at the intermediate and the highest dose, while the weights of the empty caeca were only increased in females at the highest dose. These changes were not considered to be toxicologically significant.

The NOAEL was 2421 mg/kg bw per day, or 525 mg/kg bw per day expressed as aminopyralid, the highest dose tested (Stebbins & Dryzga, 2004).

Dog

In a 4-week feeding study, groups of two male and two female beagle dogs were fed diets containing aminopyralid (purity, 95.4%; batch No. F0031-125, TSN102095) at a concentration of (0, 0.15, 0.45, or 1.5%, equal to 0, 62, 193 and 543 mg/kg bw per day. Animals were examined twice per day for general health, moribundity and mortality, and weekly detailed clinical examinations were performed. Feed consumption and body-weight development were recorded regularly. Before treatment and in the final week of treatment, detailed ophthalmic examinations were performed. At the beginning and the end of the study, haematological and clinical chemistry (including coagulation time) and urine analysis parameters were analysed, while after termination, organ weights were recorded and gross pathology and histopathology performed. The study complied with GLP.

Feed intake showed inconsistent reduction in males at the highest dose, but there was no obvious treatment-related effect. Slight changes in some erythrocyte parameters apparent in females on a groupwise comparison were not evident when the comparison was made with pretreatment values. There were no treatment-related effects identified in this study.

The NOAEL was 1.5% aminopyralid in the feed, equal to 543 mg/kg bw per day, the highest dose tested (Stebbins & Baker, 2000).

In a 13-week feeding study, groups of four male and four female beagle dogs were fed diets containing aminopyralid (purity, 94.5%; batch No. F-0031-143, TSN102319) at a concentration of 0, 0.15, 0.75, or 3.0%, equal to 0, 52.7, 232, 929 mg/kg bw per day. Animals were examined twice per day for general health, moribundity and mortality, and weekly detailed clinical examinations were performed. Feed consumption and body-weight development were recorded regularly. Before treatment and in the final week of treatment, detailed ophthalmic examinations were performed. At the beginning, in the middle and at the end of the study, haematological and clinical chemistry (including coagulation time) and urine analysis parameters were analysed, at the end of the study, organ weights were recorded and gross pathology and histopathology performed. The study complied with GLP.

A statistically significant slight increase in absolute and relative liver weights in males and females in the group at the highest dose was observed. The toxicological significance of this effect is questionable since the effect was marginal, within or near the range for historical controls and a histopathological correlate was not identified. All males and females at the highest dose showed slight and diffuse hyperplasia and hypertrophy of the mucosal epithelium of the stomach. The mucosal hyperplasia was characterized by increased numbers of mucous cells and chief cells in the fundus of the stomach. Mucous cell hyperplasia was also noted in the pylorus of the stomach. Hypertrophy of mucous cells, characterized by increased cytoplasmic volume, was most prominent in mucous cells of the pylorus. There was no accompanying degeneration, necrosis or inflammation of the mucosa of the stomach.

The NOAEL was 0.75% aminopyralid in the feed, equal to 232 mg/kg bw per day, on the basis of histological changes in the stomach at the next higher dose (Stebbins & Baker, 2002).

In a 1-year feeding study, groups of four male and four female beagle dogs were fed diets containing aminopyralid (purity, 94.5%; batch No. F-0031-143, TSN102319) at a concentration of 0, 0.03, 0.3, or 3.0%, equal to 0, 9.2 (females), 93.2 (females), 967 (males) mg/kg bw per day, the lowest dose relative to body weight was attained in males of the group at 3.0%, in the other two treated groups this was in the females). Animals were examined twice per day for general health, moribundity and mortality, and weekly detailed clinical examinations were performed. Feed consumption and body-weight development were recorded regularly. Before treatment and before termination, detailed ophthalmic examinations were performed. At the beginning, at weeks 14 (both sexes) and 26 (males)/27 (females) and at termination, haematological and clinical chemistry (including coagulation time) and urine analysis parameters were analysed, and at the end of the study, organ weights were recorded and gross pathology and histopathology performed. The study complied with GLP.

All observed effects were restricted to the groups at the highest dose. At the end of the study, females had a 9% lower body weight. Both sexes showed statistically significant increased relative liver weights (males, 21.6%; and females, 10.6%) and in two males and two females centrilobular to midzonal hepatocyte hypertrophy was found. In two females the stomach mucosa was diffusely thickened and in all animals at the highest dose mucosal hyperplasia and hypertrophy, very slight or slight chronic inflammation and slight lymphoid hyperplasia were observed.

The NOAEL was 0.3%, equal to 93.2 mg/kg bw per day, on the basis of histopathological changes in the gastric mucosa at the highest dose of 3.0% (Stebbins & Day, 2003b).

2.3 Long-term studies of toxicity and carcinogenicity

Mouse

In an 18-month study of toxicity and carcinogenicity, groups of 50 male and 50 female CD-1 mice were fed diets containing aminopyralid (purity, 94.5%; batch No. F0031-143, TSN102319) adjusted to provide doses of 0, 50, 250, or 1000 mg/kg bw per day. Monthly in the first year, then at month 17 and at study termination, animals were examined for clinical signs and from month 6 to

18, monthly for palpable masses. Body weight and feed consumption were recorded monthly, clinical chemistry and haematology parameters at month 12 and at final termination. At study termination, ophthalmological examinations and histopathology on tissues were performed, and organ weights were recorded. The study complied with GLP.

No treatment-related effects were observed in males at any dose. All females receiving aminopyralid showed increased cumulative mortality (Table 11), which was most pronounced in the group at the highest dose (42%) and statistically significant in this group and and the group at the lowest dose (34%). The increased mortality in the group at the lowest dose was not considered to be compound-related as there was no increase in any toxicological correlate, and there was no relationship with dose when considered with values for the group at the intermediate dose (30%). The mean cumulative mortality in females of six groups of historical controls in studies performed by the company was 23% (range, 18–32%). The most common cause of pre-term death in the group at the highest dose was nephropathy. There were 11 female mice with early death or moribund status with nephropathy at 1000 mg/kg bw per day, compared with 4, 2, and 4 mice at 0, 50, or 250 mg/kg bw per day, respectively. However, the overall incidence of nephropathy in females at the highest dose was not increased, nor was its severity. Nevertheless, it is possible that the increased mortality in females at the highest dose was due to exacerbation of the effects of age-related nephropathy in these animals. An increased incidence of bilateral pale kidneys, reduced amount of body fat, haemolysed blood in the gastronintestinal tract, atelectasis of the lung and perineal soiling were found (Table 10). These effects most likely reflect the moribund status of the animals. There were no changes in incidences of tumours, which would be indicative of a carcinogenic potential for aminopyralid.

The NOAEL was 250 mg/kg bw per day on the basis of the increased mortality in females at 1000 mg/kg bw per day (Stebbins & Day, 2003a).

Table 10. Number of females for which pathological observations were made among mice fed diets containing aminopyralid for 18 months

Observation	Dose (mg/kg bw per day)			
	0	50	250	1000
Bilateral pale kidneys	5	6	7	13
Decreased amount of fat	4	5	8	9
Haemolysed blood in gastrointestinal tract	1	5	1	10
Atelectasis of the lung	0	0	2	4
Perineal soiling	2	5	3	8

From Stebbins & Day (2003a)

Table 11. Cumulative mortality in females at study termination among mice fed diets containing aminopyralid for 18 months

Dose (mg/kg bw per day)	Cumulative mortality (%)
0	16
50	34
250	30
1000	42
Mean for historical controls (range)	23 (18–32)

From Stebbins & Day (2003a)

Rat

In a 24-month study of toxicity and carcinogenicity, groups of 65 male and 65 female Fischer 344 rats were fed diets containing aminopyralid (purity, 94.5%; batch No. F0031-143, TSN102319) adjusted to provide aminopyralid at a dose of 0, 5, 50, 500, or 1000 mg/kg bw per day. From each group, five males and five females were pre-selected for evaluation of long-term toxicity, five males and five females for long-term neurotoxicity and five males and five females for evaluation of long-term toxicity and neurotoxicity after 12 months. The remaining 50 animals were treated for another 12 months. The following parameters were measured in the long-term toxicity and/or carcinogenicity groups: daily cage-side observations, monthly detailed clinical observations for the first 12 months then at 15, 18, 21, and 24 months, monthly palpable tumour observations on all rats in the study of carcinogenicity for months 12–24, ophthalmic examinations, body weights, feed consumption, organ weights and gross necropsy with extensive histopathological examination of tissues. Haematology, clinical chemistry, and urine analysis data were collected from the groups after 3, 6, 12, 18, and 24 months of dosing. The study complied with GLP.

The long-term neurotoxicity part of this study is described in section 2.6.

No clinical signs of toxicity were observed and mortality at the end of the study was comparable in all groups. Nevertheless, there seemed to be an effect of treatment on the time to death in males at the highest dose, but it did not attain statistical significance. Body weights in males were statistically significantly reduced at 500 and the 1000 mg/kg bw per day and in the females at 1000 mg/kg bw per day (Table 12). In both sexes, the body-weight reduction developed in the first weeks and remained throughout the study more or less at the same level. Minimally increased feed intake and consequently reduced feed efficiency was observed in males at 1000 mg/kg bw per day. Slightly reduced but not statistically significant body-weight decrements were also observed in males at 5 and 50 mg/kg bw per day, but were not considered to be treatment-related. The males and females at 500 and 1000 mg/kg bw per day at the 12-month interim necropsy and at the end of the study had statistically significantly increased absolute and relative weights of the full and the empty caeca (Table 13). Histopathologically, the incidence of very slight hyperplasia of the caecal mucosa was increased in males and females of the group at the highest dose, being statistically significant only in males. In all groups, no treatment-related effects on haematological parameters were observed and in clinical chemistry, only females at the highest dose showed slightly increased aspartate aminotransferase (AST) activities in the first year of treatment. Urine analysis provided a consistent pattern of treatment-related changes for males and females at 500 and 1000 mg/kg bw per day. The effects included increased urine volumes, decreased specific gravity and protein and ketone content and reduced pH values. As there was no renal histopathological correlate, these changes were most likely to be adaptive effects to the high concentrations of aminopyralid in the diet. The caeca of rats at 500 or 1000 mg/kg bw per day were enlarged owing to the presence of increased amounts of semi-solid contents of similar consistency to that normally found in rats. As only the volume but not the consistency and appearance of the faeces was changed, it is postulated that there was increased colonic water resorption with compensatory renal excretion of the additional water, which led to the somewhat increased urine volume and decreased specific gravity. Decreased urine pH was attributed to renal excretion of aminopyralid, which is an acid.

There were no treatment-related increases in tumour incidences. In females at the highest dose, a decrease in the incidence of pituitary adenomas was observed.

The NOAEL was 500 mg/kg bw per day on the basis of slight but statistically significant body-weight decreases in males at 1000 mg/kg bw per day. Aminopyralid was not carcinogenic to rats under the conditions of this study (Johnson & Dryzga, 2003).

Table 12. Body weights of rats fed diets containing aminopyralid in a 24-month feeding study

Test day	Body weight (g)				
	Dose (mg/kg bw per day)				
	0	5	50	500	1000
Males					
1	150.5	147.8	147.9	149.5	151.3
29	255.0	252.1	251.9	253.9	252.5
91	334.3	330.1	328.1	328.9	325.9
120	353.1	346.4	342.3*	344.8*	339.7*
176	385.1	379.5	375.8*	374.9*	370.1*
260	418.8	413.1	410.1	407.3*	401.1*
364	444.1	436.6	433.9	430.1*	421.4*
540	466.4	454.1	457.4	444.3*	436.2*
729	431.4	420.3	421.8	409.1	412.9
Females					
1	113.7	113.6	113.7	113.1	112.8
29	154.3	156.2	155.0	153.9	151.4
91	182.5	182.8	183.3	182.4	178.0*
120	188.5	188.4	187.7	187.2	183.7*
176	200.5	201.0	200.6	199.1	196.3*
260	214.0	216.4	217.1	214.4	210.8
364	229.1	233.1	231.7	231.1	224.9
540	275.4	278.2	279.7	273.9	264.2
729	294.4	294.0	305.1	289.4	283.5

From Johnson & Dryzga (2003)
* Statistically different from mean for control group according to Dunnett test, alpha = 0.05.

Table 13. Weight of caecum of rats fed diets containing aminopyralid in a 24-month feeding study

Caecal weights	Dose (mg/kg bw per day)				
	0	5	50	500	1000
Males					
Full caecum (g)	5.21	4.98	5.52	9.05§	15.5§
Relative weight (g/100g)	1.27	1.25	1.39	2.33§	4.00§
Empty caecum (g)	2.40	2.38	2.34	3.12§	3.57§
Relative weight (g/100g)	0.584	0.592	0.586	0.807§	0.925§
Females					
Full caecum (g)	4.00	4.17	4.21	5.72§	10.4§
Relative weight (g/100g)	1.44	1.52	1.47	2.12§	3.94§
Empty caecum (g)	1.86	1.89	1.96	2.14*	2.59*
Relative weight (g/100g)	0.671	0.687	0.685	0.795§	0.980

From Johnson & Dryzga (2003)
* Statistically different from mean for control group according to Dunnett test, alpha = 0.05.
** Statistically different from mean for control group according to Wilcoxon test, alpha = 0.05.

2.4 Genotoxicity

Aminopyralid (batch No. F0031-143, TSN102319) and aminopyralid TIPA (batch No. 173-162-1A, TSN104110) were tested for genotoxicity in assays for reverse mutation in *S. typhimurium* and *E. coli*, an assay for HGPRT forward mutation in Chinese hamster ovary cells (CHO), an assay for chromosomal aberration in rat lymphocytes in vitro and an assay for micronucleus formation in mouse bone marrow in vivo. All four studies complied with GLP.

The results of all assays except that for chromosomal aberration in rat lymphocytes were clearly negative (Table 14). Chromosomal aberrations were only found after 24 h of treatment (but not after 4 h) in the absence of metabolic activation (S9) at concentrations that clearly reduced mitotic indices, indicative of a cytotoxic potential for aminopyralid (Table 15).

Table 14. Results of studies of genotoxicity with aminopyralid and aminopyralid TIPA

Test article	End-point	Test object	Concentration	Purity (%)	Result	Reference
In vitro						
Aminopyralid	Reverse mutation	*Salmonella typhimurium* TA98, TA100, TA1535, TA1537 and *E. coli* WP2*uvr*A	100–5000 µg/plate ± S9, in DMSO	94.5	Negative	Mecchi (2001)
Aminopyralid TIPA	Reverse mutation	*S. typhimurium* TA98, TA100, TA1535, TA1537 and *E. coli* WP2*uvr*A	33.3–5000 µg/plate (7.3–1090 µg/plate aminopyralid equivalents) in DMSO ± S9, in DMSO	21.8% aminopyralid equivalents	Negative	Charles & Mecchi (2004)
Aminopyralid	Forward mutation	*Hgprt* locus in CHO cells	31.25–2070 µg/ml ± S9, in DMSO	94.5	Negative	Linscombe et al. (2001)
Aminopyralid TIPA	Forward mutation	*Hgprt* locus in CHO cells	250–4000 µg/ml (54.5–872 µg/ml aminopyralid equivalents) ± S9, in DMSO	21.8% aminopyralid equivalents	Negative	Schisler et al. (2004)
Aminopyralid	Chromosomal aberrations	Rat lymphocytes	32.3–2070 µg/ml ± S9, in DMSO	94.5	Negative +S9 Positive −S9 (see Table 15)	Linscombe et al. (2002)
Aminopyralid TIPA	Chromosomal aberrations	Rat lymphocytes	1000–4000 µg/ml (218–872 µg/ml aminopyralid equivalents) ± S9, in DMSO	21.8% aminopyralid equivalents	Negative	Linscombe et al. (2004)
In vivo						
Aminopyralid	Micronucleus formation	Male CD-1 mice	2 × 500–2000 mg/kg bw on two consecutive days	94.5	Negative	Spencer & Gorski (2002)
Aminopyralid TIPA	Micronucleus formation	Male CD-1 mice	2 × 500–2000 (109–436 aminopyralid equivalents) mg/kg bw on two consecutive days	21.8% aminopyralid equivalents	Negative	Spencer et al. (2004)

CHO, Chinese hamster ovary cells; S9, 9000 × *g* supernatant from rat livers; TIPA, triisopropylammonium

Table 15. Results of assays for chromosomal aberration with aminopyralid in rat lymphocytes

Concentration (μg/ml)	−S9		+S9	
	Effect (%)[a]	RMI (%)	Effect (%)[a]	RMI (%)
4-h treatment				
0	1.0	100	0.5	100
64.7	—	114.5	0.5	82.1
517.5	1.5	113.3	—	69.0
1035	1.0	92.8	0.5	72.6
2070	1.5	63.9	1.0	63.1
24-h treatment				
0	0.5	100	0.5	100
62.5	—	—	2.0	95.0
125	1.0	97.7	—	81.8
750	1.0	60.2	—	—
1000	4.5 [b]	50.4	1.0	81.0
1400	5.0 [b]	33.1	—	—
1700	7.5 [b]	27.8	—	—
2070	—	18.0	2.5	84.3
24-h treatment				
0	2.5	100	—	—
1000	4.5	47.8	—	—
1400	7.0 [b]	37.7	—	—
1700	7.5 [b]	29.0	—	—

From Linscombe et al. (2002)
RMI, relative mitotic index; S9, 9000 × *g* supernatant from rat livers.
[a] Cells with aberrations excluding gaps.
[b] Significantly different (alpha = 0.05) from negative controls.

2.5 *Reproductive toxicity*

(a) Multigeneration studies

Rat

Groups of 30 male and 30 female CD rats were fed diets containing aminopyralid (purity, 94.5%; batch No. F-0031-143, TSN102319) and supplying a dose of 0, 50, 250 or 1000 mg/kg bw per day for approximately 10 weeks before breeding, and continuing during breeding (2 weeks), gestation (3 weeks) and lactation (3 weeks) for each of two generations to evaluate the potential for reproductive toxicity and effects on neonatal growth and survival. A comprehensive evaluation of male and female reproductive systems was conducted, including an evaluation of estrous cyclicity, gonadal function, mating performance, conception, gestation, parturition, lactation and weaning, as well as survival, growth and development of the offspring. In-life observations, body weights, feed consumption and litter data were recorded. In addition, a gross necropsy of the P_1 and P_2 adults was conducted with extensive histopathological examination of reproductive organs and other tissues. The study complied with GLP.

In P_1 and P_2 male and female rats at 250 and the 1000 mg/kg bw per day, the absolute and relative weights of full end empty caeca were increased; this effect did not attain statistical significance in females at 250 mg/kg bw per day. There were no histopathological correlates of these effects. This change was considered to be a physiological adaptation and not of toxicological significance. There were no other treatment-related findings in either the parameters for general health or reproductive performance.

The NOAELs for general toxicity and for reproductive toxicity were both 1000 mg/kg bw per day, the highest dose tested (Marty et al., 2003).

(b) Developmental toxicity

Rat

Groups of 25 time-mated female CD rats received aminopyralid (purity, 94.5%; batch No. F-0031-143, TSN102319) at a dose of 0, 100, 300, or 1000 mg/kg bw per day as a suspension in 0.5% aqueous Methocel (A4M® methylcellulose) in a volume of 4 ml/kg, by oral gavage, starting on day 6 of gestation and continiuning until day 20. Clinical examinations were conducted throughout the study, and body weights and feed consumption were recorded regularly. On day 21, dams were terminated and the fetuses delivered by caesarian section. The weights of gravid uteri, the numbers of corpora lutea, implantation sites, viable fetuses and resorptions were recorded. Fetuses were sexed and body weights measured and they were subjected to external, visceral, craniofacial and skeletal examinations. The study complied with GLP.

In all the parameters examined in this study, no treatment-related findings were found. Under the conditions of this study, aminopyralid did not show evidence of any maternal or embryo/fetal toxicity or teratogenicity.

The NOAEL was 1000 mg/kg bw per day, the highest dose tested (Carney & Tornesi, 2001).

Groups of 25 time-mated female CD rats received GF-871 (41.3% aminopyralid TIPA) at a dose of 0, 484, 1211 or 2421 mg/kg bw per day in a volume of 4 ml/kg, by oral gavage, starting on day 6 of gestation and continuing until day 19. This dosing regime provided aminopyralid at a dose of 0, 105, 263 or 525 mg/kg bw per day. Clinical examinations were conducted throughout the study, and body weights and feed consumption were recorded regularly. On day 21, dams were terminated and the fetuses delivered by caesarian section. The weights of gravid uteri, the numbers of corpora lutea, implantation sites, viable fetuses and resorptions were recorded. Fetuses were sexed and body weights measured and they were subjected to external, visceral, craniofacial and skeletal examinations. The study complied with GLP.

No treatment-related effects on dams or fetuses were observed even at the highest dose tested.

The NOAEL for GF-871 was 2421 mg/kg bw per day, or 525 mg/kg bw per day as aminopyralid equivalents, the highest dose tested (Zablotny & Thorsrud, 2004).

Rabbit

Groups of 26 time-mated female NZW rabbits received aminopyralid (purity, 94.5%; batch No. F-0031-143, TSN102319) at a dose of 0, 25, 100, or 250 mg/kg bw per day by oral gavage suspensions in 0.5% aqueous Methocel (A4M® methylcellulose) in a volume of 4 ml/kg, starting on day 7 of gestation and continuing until day 27 (phase I). Since there was no evidence of maternal toxicity at the highest dose, in a second part of this study with the same design, mated rabbits were given aminopyralid in suspension at a dose of 0, 500, or 750 mg/kg bw per day (phase II). On day 28, dams were terminated and the fetuses were delivered by caesarian section. The weights of gravid uteri, the numbers of corpora lutea, implantation sites, viable fetuses and resorptions were recorded. Fetuses were sexed and body weights measured and they were subjected to external, visceral, craniofacial and skeletal examinations. The study complied with GLP.

Mean body-weight gain was significantly reduced at 500 and 750 mg/kg bw per day (−11.8 g and −70.0 g vs 25 g in the rabbits in the control group) on days 7–10 of gestation. Thereafter, the reduced body-weight gain remained significant for rabbits at the highest dose only. Intermittently in both groups, feed consumption was reduced. In the group at the highest dose, higher incidences of decreased amounts of faeces and of uncoordinated gait shortly after dosing were observed. In one rabbit, the observation was made on the first day of treatment (Table 16).Two animals showed the same effect on the second day of treatment. Uncoordinated gait was also observed in rabbits at 500 mg/kg bw per day, but it was much less pronounced and was first observed on the second day of treatment. In both groups, these effects were seen in 23 out of 26 animals and lasted for only 2 h at a time and were completely resolved thereafter. Also, the degree of severity of uncoordinated gait did not increase with time. Two rabbits at the highest dose were killed after being found to be in a moribund condition on day 17 of gestation; these rabbits showed uncoordinated gait, decreased activity, perineal urine soiling and decreased amounts of faeces. At gross necropsy, examination of these animals showed that both rabbits were pregnant and had pale kidneys, watery and dark caecal contents, and erosions/ulcers in the glandular mucosa of the stomach. The moribund condition of these rabbits was considered to be treatment-related. Another dam at 750 mg/kg bw per day and one dam at 500 mg/kg bw per day died during the treatment period, with signs indicative of complications arising from incorrect gavage dosing. At termination, one dam at 500 mg/kg bw per day had ulcers/erosions in the glandular mucosa of the stomach. Owing to the severity of the clinical signs, all rabbits at the highest dose were terminated before day 20 of gestation and were not available for the evaluation of reproductive and fetal parameters.

Weights of gravid uteri, the numbers of corpora lutea, implantation rates, numbers of viable fetuses and resorptions were not impaired by treatment at any dose. Additionally, no treatment-related changes in the incidences of malformations, numbers of malformed fetuses or numbers of affected litters were observed.

The NOAEL for maternal toxicity was 250 mg/kg bw per day and the NOAEL for developmental toxicity was 500 mg/kg bw per day, the highest dose evaluated (Marty et al., 2002).

Groups of 26 time-mated female NZW rabbits received GF-871 (41.3% aminopyralid TIPA; batch No. 173-162-1A, TSN104110) at a dose of 0, 484, 1211 or 2421 mg/kg bw per day in a volume of 4 ml/kg bw by oral gavage on days 7 to 27 of gestation. This regimen supplied aminopyralid at doses of 0, 105, 263 or 525 mg/kg bw per day.

On day 28 of gestation, dams were terminated and the fetuses delivered by caesarian section. The weights of gravid uteri, the numbers of corpora lutea, implantation sites, viable fetuses and resorptions were recorded. Fetuses were sexed and body weights measured and they were subjected to external, visceral, craniofacial and skeletal examinations. The study complied with GLP.

Three rabbits at the highest dose and one at the intermediate dose were killed between days 14 and 18 owing to severe body-weight loss and poor general condition. In all treated groups, a dose-related reduction in feed consumption was observed, resulting in a statistically significant reduction in body-weight gain only in rabbits at the highest dose. Of rabbits at the highest dose, 19 out of 26 showed uncoordinated gait on several days, usually starting and resolving within 1–2 h after dosing. Six rabbits in this group also had repetitive chewing behaviour. Two animals in the group at the intermediate dose and one at the lowest dose also showed one incident of uncoordinated gait. Uncoordinated gait was not apparent until after dosing for a number of days (6 days) (Table 16).

No organ-weight or pathological changes were observed in dams and reproductive parameters were not impaired in any treated group. In the group at the highest dose, fetal body weights were decreased for females (10.5%), males (7.4%), and both sexes combined (9.8%). The fetal weights were also reduced when compared with those in historical control groups. The incidence, the nature and the lack of a consistent pattern of fetal malformations, found in all groups, do not imply any relationship with treatment.

Therefore, the NOAEL for maternal toxicity with GF-871 was less than 484 mg/kg bw per day, or 105 mg/kg bw per day expressed as aminopyralid equivalents, on the basis of uncoordinated gait in one rabbit in the group at the lowest dose. The NOAEL for developmental toxicity with GF-871 was 1211 mg/kg bw per day, or 263 mg/kg bw per day expressed as aminopyralid equivalents, on the basis of fetal body-weight reduction in the group at the highest dose (Carney & Tornesi, 2004a).

In a supplemental study, groups of 26 time-mated female NZW rabbits were given GF-871 (41.3% aminopyralid TIPA; batch No. 173-162-1A, TSN104110) at a dose of 0, 121 or 363 mg/kg bw per day in a volume of 4 ml/kg bw by oral gavage, from day 7 to day 27 of gestation. This study was designed to identify a NOAEL for maternal toxicity, which was 0, 26 or 78 mg/kg bw per day expressed as aminopyralid.

On day 28 of gestation, dams were terminated and the fetuses were delivered by caesarian section. In-life, body-weight changes, feed consumption and clinical observations were recorded. On necropsy, gross pathology in dams was performed and the pregnancy status was examined. The study complied with GLP.

The only findings were restricted to one rabbit at the lowest dose and three rabbits at the highest dose, which showed uncoordination at one time-point. In the rabbit at the lowest dose, the effect occurred at day 10 of gestation, equivalent to treatment day 4, but this was not compound-related, as the rabbit was suffering from a fatal dosing injury. In the rabbits at the highest dose, the effect occurred on days 14, 25 and 26, equivalent to treatment days 8, 19 and 20, respectively (Table 16). Concomitantly, two of the affected rabbits at the highest dose also showed repetitive chewing behaviour. No other investigated parameters in this or the group at 121 mg/kg bw showed any treatment-related changes.

The NOAEL for maternal toxicity with GF-871 was 121 mg/kg bw per day, or26 mg/kg bw per day expressed as aminopyralid equivalents (Carney & Tornesi, 2004b).

2.6 *Special studies: neurotoxicity*

Rat

In a study of acute neurotoxicity, 10 male and 10 female Fischer 344 rats were given aminopyralid (purity, 94.5%; batch No. TSN102319/F0031-143) as a single dose at 0, 500, 1000 or 2000 mg/kg bw as a suspension in aqueous 0.5% Methocel® (methylcellulose) by oral gavage in a volume of 10 ml/kg bw.

Daily cage-side observations were performed on all rats and clinical examinations on test days 2, 3 and 4. Before dosing and on days 8 and 15 after dosing, body weights were recorded and the rats were tested for motor activity and subjected to a functional observational battery (FOB). At study termination after 2 weeks, five males and females per dose were killed for evaluation of pathological and histopathological changes in central and peripheral nervous tissues. The remaining five males and females per dose were also killed and a standard set of tissues was preserved. The study complied with GLP.

Clinical signs were restricted to increased faecal soiling in males at the highest dose and urine soiling in females at the highest dose. These findings resolved within the study period and whether this soiling was indicative of a neurotoxic potential at the highest dose is not clear.

There were no other signs of neurotoxicity or systemic toxicity.

The NOAEL for acute neurotoxicity was 2000 mg/kg bw, the highest dose tested, and the NOAEL for general toxicity was 1000 mg/kg bw (Marable et al., 2001).

In a 12-month dietary study of neurotoxicity, which was part of the 24-month study of toxicity and carcinogenicity, groups of 10 male and 10 female Fischer 344 rats were fed diets containing aminopyralid (purity, 94.5%; batch No. F0031-143, TSN102319) at concentrations adjusted to provide doses of 0, 5, 50, 500, or 1000 mg/kg bw per day. The study design and parameters investigated were the same as in the main 24-month study. Five males and females per group were pre-selected for neurotoxicity evaluation and five males and females per group to evaluate both toxicity and neurotoxicity after 12 months. Before dosing and at 1, 3, 6, 9 and 12 months, a FOB and an evaluation of motor activity were performed for these rats. At termination, five males and females per group (and all rats in the control group and the group at the highest dose) were subjected to neuropathological evaluation. The study complied with GLP.

In males at the highest dose, a treatment-related increase in defaecation level was observed. This effect might be related to the caecal enlargement observed in the main study at 500 and 1000 mg/kg bw per day in both sexes and which was considered to represent physiological adaptation. There were no other treatment-related changes indicating a neurotoxic potential or systemic toxicity.

The NOAEL for neurotoxicity and for general toxicity was 1000 mg/kg bw per day, the highest dose tested (Maurissen et al., 2003).

Rabbit

In a study in three non-pregnant female NZW rabbits, aminopyralid (purity not stated) was administered at a dose of 500 mg/kg bw by oral gavage in 0.5% Methocel® for 22 days (phase I) and then at 1000 mg/kg bw for another 20 days (phase II). No control group was included in this study as it was intended to look for predefined uncoordination that would not occur in rabbits in the control group. Behavioural observations (open field and sensorimotor coordination) were performed daily throughout the study (rabbits were observed in open field for 15 min after dosing and then subsequently in their home cages at unscheduled times throughout the day), weekly body-weight changes were recorded and daily clinical cage-side observations were made. In each of the two phases of the study, clinical chemistry parameters were investigated and at the end of the study all rabbits were necropsied, with special focus on the gastrointestinal tract and the brain.

During phase I, neither behavioural nor obvious clinical chemistry changes were observed. In phase II, one rabbit showed clear signs of uncoordinated gait on several days starting typically in the first 2 h after dosing. The first observation was made on day 2 of treatment. A second rabbit had similar signs on day 19, while the third rabbit was free of behavioural signs for the whole study. The results of clinical chemistry were not notable. On necropsy, all rabbits showed adherent mucus in the linings of the stomach and the animal with the repeated behavioural signs was free of further histological changes. The other two rabbits had ascites in the abdominal cavity and one showed decreased adipose tissue (Marable & Day, 2004).

Two groups of 10 time-mated NZW rabbits were given GF-871 (41.3% aminopyralid TIPA; batch No. 173-162-IA/TSN104110) at a dose of 0 or 800 mg/kg bw (173.6 mg/kg bw expressed as aminopyralid equivalents) by gavage on day 7 of gestation until day 27 of gestation to evaluate the potential for neuropathological changes. Daily body weights, feed consumption and clinical observations (approximately 1 h after dosing) were recorded (including observations of incoordination) and on day 28 of gestation, an extensive investigation on neurological tissues was performed: brain—cerebrum, thalamus/hypothalamus, midbrain, pons, cerebellum, medulla; pituitary gland, trigeminal ganglia with nerve; spinal cord—cervical swelling (C3–C6), lumbar swelling (L1–L4); dorsal root ganglia (cervical and lumbar); dorsal and ventral roots (cervical and lumbar); peripheral nerves—proximal sciatic, proximal and distal tibial, sural; eyes with optic nerve, skeletal muscle (anterior tibial and gastrocnemius). The study complied with GLP.

One rabbit in the control group was killed after having aborted and one rabbit in the treatment group died accidentally during clinical examination. After an initial body-weight loss, all animals gained weight and at the end no statistically significant difference from the control group was evident.

Three rabbits in the treatment group showed slight to moderate/severe signs of uncoordinated gait at isolated, infrequent time-points, one of these rabbits died accidentally (see above). The first day of this observation was recorded on day 7 (slight), 9 (slight) or 14 (moderate/severe) (this was the animal that died) of gestation in these three rabbits, equivalent to treatment days 1, 3 and 9, respectively (Table 16). Gross pathology and histopathology revealed no treatment-related changes in the central or peripheral nervous system. However, one treated rabbit showed acute degeneration of several muscle fibers. This finding was not accompanied by inflammation or kidney changes (indicative of chronic lesion and high release of myoglobin, respectively).

In summary, no neuropathological potential of GF-871 was evident from this study (Yano & Zablotny, 2005).

Table 16. Day on which first observation was recorded of uncoordination in pregnant rabbits treated with aminopyralid or aminopyralid TIPA[a]

Aminopyralid								Aminopyralid TIPA (GF-871)							
Marty et al. (2002)				Carney & Tornesi (2004b)				Carney & Tornesi (2004a)				Yano & Zablotny (2005)			
Dose as aminopyralid equivalents (mg/kg bw)															
500		750		26		78		105		263		525		173.6	
GD	TD	GD	TD	GD	TD	GD	TD	GD	TD	GD	TD	GD	TD	GD	TD
—	—	—	—	—	—	—	—	—	—	—	—	—	—	—	—
—	—	—	—	—	—	—	—	—	—	—	—	—	—	—	—
—	—	—	—	—	—	—	—	—	—	—	—	—	—	—	—
8	2	7	1	—	—	—	—	—	—	—	—	—	—	—	—
9	3	8	2	—	—	—	—	—	—	—	—	—	—	—	—
10	4	8	2	—	—	—	—	—	—	—	—	—	—	—	—
10	4	9	3	—	—	—	—	—	—	—	—	—	—	7	1
11	5	9	3	—	—	—	—	—	—	—	—	12	6	9	3
11	5	9	3	—	—	—	—	—	—	—	—	13	7	14	9
11	5	9	3	—	—	—	—	—	—	—	—	15	9		
12	6	10	4	—	—	—	—	—	—	—	—	18	12		
12	6	10	4	—	—	—	—	—	—	—	—	18	12		
13	7	10	4	—	—	—	—	—	—	—	—	19	13		
14	8	11	5	—	—	—	—	—	—	—	—	19	13		
15	9	13	7	—	—	—	—	—	—	—	—	20	14		
17	11	13	7	—	—	—	—	—	—	—	—	20	14		
17	11	14	8	—	—	—	—	—	—	—	—	20	14		
19	13	14	8	—	—	—	—	—	—	—	—	22	16		
20	14	14	8	—	—	—	—	—	—	—	—	23	17		
21	15	14	8	—	—	—	—	—	—	—	—	24	18		
22	16	15	9	—	—	—	—	—	—	—	—	24	18		
24	18	15	9	—	—	—	—	—	—	—	—	25	19		
25	19	15	9	—	—	—	—	—	—	—	—	25	19		
26	20	16	10	—	—	14	8	—	—	—	—	25	19		
27	21	18	12	—	—	25	19	—	—	16	10	26	20		
27	21	19	13	10[c]	4[c]	26	20	16	10	18	12	26	20		

GD, day of gestation; TD, treatment day; TIPA, triisopropylammonium.

[a] For each animal in the respective studies, the first day (day of gestation and treatment day) on which uncoordination was observed is given (day 7 of gestation is equivalent to day 1 of treatment)

[b] 26 animals were used in (Marty et al., 2002; Carney & Tornesi, 2004a; Carney & Tornesi, 2004b) and 10 animals in (Yano & Zablotny, 2005)

[c] Not treatment-related, fatal dosing injury.

The same five time-mated NZW rabbits received GF-871 via gavage according to the following dosing regimen: 500 mg/kg bw per day (equivalent to 108.5 mg/kg bw per day expressed as aminopyralid equivalents) from day 7 to day 15 of gestation; 1000 mg/kg bw per day (equivalent to 217.0 mg/kg bw per day expressed as aminopyralid equivalents) from day 16 to day 23 of gestation; or two doses at 1000 mg/kg bw given at an interval of 1 h on day 23 of gestation, in order to make a detailed evaluation of uncoordination in pregnant rabbits. Two additional rabbits served as controls. Rabbits were monitored daily for clinical signs of toxicity, body weight and feed consumption. In addition, expanded clinical observations were performed after dosing on day 23 of gestationby an independent neurotoxicology consultant (Dr J. Ross, Ross Toxicology Services, LLC).

For this study, no detailed results were available but only the following summary.

No clinical signs were noted in the treated rabbits during the first phase of dosing at 500 mg/kg bw per day. However, after increasing the dose to 1000 mg/kg bw per day on day 16 of gestation, uncoordination was noted at least once in all five rabbits, with these signs first appearing as early as day 18 and as late as day 23 of gestation. At the expanded clinical observations on day 23 of gestation, five out of five dosed rabbits exhibited abnormal behaviour, characterized by uncoordination. Hindlimbs and forelimbs were affected. Affected animals were easily displaced in the lateral direction by pushing with the hands. There were no consistent changes in muscle strength as evaluated by observing external ear position, head position and extensor thrust reflexes. There were no autonomic or somatic signs of convulsive activity, such as urination, defaecation, or paroxysmal movements, no signs of myoclonus or tremors. There were no apparent changes in level of consciousness as evaluated by observations of the rabbits' attention to external stimuli in the environment and no apparent change in vestibular function as evaluated by observing head position and spontaneous nystagmus. Optokinetic nystagmus was normal. The most severely affected rabbits had a slowed righting reaction, but this appeared to be secondary to uncoordination, and not due to either weakness or depression of the central nervous system. The primary neurological change was uncoordination of all four limbs (Carney, 2004).

3. Observations in humans

No information was available.

Comments

Biochemical aspects

In a study of the absorption, distribution, metabolism and excretion of radiolabelled aminopyralid administered by oral gavage, male rats were given single doses at 50 or 1000 mg/kg bw, or a single dose at 50 mg/kg bw after 14 days pretreatment with unlabelled aminopyralid at a dose of 50 mg/kg bw per day. Male rats also received radiolabelled aminopyralid TIPA salt as a single oral gavage dose at 96 mg/kg bw (equal to aminopyralid at 50 mg/kg bw). The pharmacokinetic behaviour of aminopyralid and its TIPA salt was very similar. Of the administered dose, 42–59% was absorbed and rapidly excreted in the urine, most within the first 24 h. Excretion was biphasic, with half-lives of 3.0–3.8 h and 10.2–12.3 h. Biliary excretion was assumed to be negligible. After a depletion period of 168 h, virtually all tissue samples showed concentrations of radiolabel of less than 0.01% of the administered dose. Aminopyralid was not metabolized in rats; more than 95% of the radiolabel was accounted for. On the basis of lack of coordination in gait after exposure to aminopyralid and aminopyralid TIPA in rabbits but not in rats, an extensive study of the absorption, distribution, metabolism and excretion of aminopyralid in rabbits was performed. A group of non-pregnant rabbits received radiolabelled aminopyralid as a single dose at 280–370 mg/kg bw. A group of pregnant rabbits ("late-stage") received nonlabelled aminopyralid on days 7–21 of gestation and then radiolabelled aminopyralid as a single dose at 280–370 mg/kg bw on day 22 of gestation. An additional group of pregnant rabbits ("early-stage") received radiolabelled aminopyralid as a single dose at 280–370 mg/kg bw on day 7 of gestation, without pretreatment with unlabelled aminopyralid. Using plasma from animals in these three groups and from a group of non-pregnant rats given radiolabelled aminopyralid as a single dose at 280–370 mg/kg bw, a study of plasma-protein binding was performed. In late-stage pregnant rabbits pretreated with repeated doses of unlabelled aminopyralid, the absorption of aminopyralid (on the basis of lower Tmax, higher AUC and increased renal excretion) was somewhat more rapid and greater than in non-pregnant and early-stage pregnant rabbits. Plasma-protein binding was lower (43–58%) in late-stage pregnant rabbits pretreated with repeated doses of unlabelled aminopyralid than in non-pregnant and early-stage pregnant rabbits (47–68%). The difference in bioavailability (expressed as unbound compound) of aminopyralid was at most twofold. However, interpretation of these results remains ambiguous because of the different dosing regimens used (single dose without pretreatment in non-pregnant and early-stage pregnant rabbits and single dose after pretreatment in late-stage pregnant rabbits). On the basis of renal excretion of radiolabel, absorption of aminopyralid in rabbits was close to 80% or greater, being 20–40% higher than in rats.

Toxicological data

Aminopyralid and aminopyralid TIPA have low acute toxicity in rats when administered orally, dermally or by inhalation. The oral and the dermal LD_{50}s are both > 5000 mg/kg bw, and by inhalation, the LC_{50} is > 5.5 mg/l, the highest dose tested. Aminopyralid is clearly irritating to the eye, while aminopyralid TIPA is only slightly irritating. Aminopyralid is not a dermal irritant, while aminopyralid TIPA is a slightly irritant. In guinea-pigs, aminopyralid and aminopyralid TIPA produced no signs of skin-sensitizing potential, as tested by the Magnusson & Kligman method.

In short-term feeding studies in mice, rats and dogs and in a study of dermal exposure studying rats, animals received aminopyralid at doses of up to 1000 mg/kg bw per day. Body weight was reduced only in female dogs receiving aminopyralid at 967 mg/kg bw per day, the highest dose tested in a 1-year study. Males and females at this dose also showed a slight increase in relative liver weights accompanied by hepatocyte hypertrophy in two out of four animals per sex. In male and female rats at doses of 500 mg/kg bw per day and greater, reversibly increased absolute and relative weights of full and empty caeca were observed and slight mucosal hyperplasia of the caecum and the ileum was found in males at the highest dose. These changes were considered to be a consequence of physiological adaptation. Mucosal hyperplasia of the stomach was observed in all dogs at 967 mg/kg bw per day. Treatment-related clinical chemistry changes were restricted to the urine of rats at 500 mg/kg bw per day and greater, where decreased pH values and decreased concentrations of protein and ketone were found. Changes in the pH of the urine were most likely due to urinary excretion of the unchanged, acid parent compound. Generally, aminopyralid was well tolerated by mice, rats and dogs in short-term studies. The NOAEL for aminopyralid in mice was 1000 mg/kg bw per day (the highest dose tested), 1000 mg/kg bw per day in rats (the highest dose tested) and 93.2 mg/kg bw per day in dogs, on the basis of histopathological changes in the gastric mucosa at the highest dose tested. In a 13-week feeding study in rats with aminopyralid TIPA, the same effects on caecal weights and urine chemistry were observed as with aminopyralid. On the basis of the lack of other effects, the NOAEL for aminopyralid TIPA in rats was 2421 mg/kg bw per day as GF-871, equal to aminopyralid at 525 mg/kg bw per day, the highest dose tested.

In an 18-month study in mice and a 24-month feeding study in rats, diets adjusted to provide aminopyralid at maximal doses of 1000 mg/kg bw per day did not induce any increases in the incidence of neoplastic findings.

Mortality was increased in all groups of female mice receiving aminopyralid, but appeared to be compound-related only in animals at the highest dose. Animals that died showed an increased incidence of age-related nephropathy. Although the overall incidence of nephropathy was not increased in this or any treated group, exacerbation of the effects of age-related nephropathy in these animals may have been responsible for the increase in mortality. Other treatment-related signs of toxicity in this group e.g. pale kidneys, reduced body fat, haemolysed blood in the gatrointestinal tract, atelectasis of the lung and perineal soiling, were most likely a reflection of the moribund state of the animals. The NOAEL was 250 mg/kg bw per day on the basis of increased mortality in females at 1000 mg/kg bw per day. Although the final cumulative mortality in rats was comparable at all doses, the onset of mortality in males at 1000 mg/kg bw per day appeared earlier. As statistical significance and clear treatment-related causes for death were lacking, the Meeting did not consider this finding as being related to treatment.

Slightly reduced body-weight gain was observed in male rats at 500 mg/kg bw per day and above. Increased absolute and relative weights of full and empty caeca were observed in males and females at 500 mg/kg bw per day and above, accompanied by very slight mucosal hyperplasia of the caeca, which was statistically significant only in males at the highest dose. This finding is common to many substances and, due to the lack of further histological changes, was not considered to be an adverse finding. In males and females at 500 and 1000 mg/kg bw per day, increased urine volumes, decreased specific gravity and protein and ketone contents and reduced pH values were observed without any renal histopathological correlate. These changes were not considered to be toxicologically significant, the change in pH most likely reflecting urinary excretion of the acidic parent compound. The NOAEL in this feeding study in rats was 500 mg/kg bw per day on the basis of slight but statistically significant body-weight decreases in males at 1000 mg/kg bw per day.

Aminopyralid was not carcinogenic in mice or rats.

Aminopyralid and its TIPA salt were tested for genotoxicity in an adequate range of tests, including an assay for reverse mutation in *S. typhimurium* and *E. coli*, an assay for forward mutation in *Hgprt* with Chinese hamster ovary cells, an assay for chromosomal aberration in rat lymphocytes in vitro and an assay for micronucleus formation in mouse bone marrow in vivo. In the assay for chromosomal aberration in rat lymphocytes in vitro, an increased frequency of chromosomal aberration was seen only after 24 h of treatment at clearly cytotoxic concentrations of aminopyralid and in the absence of metabolic activation. The results of all other assays showed no evidence for genotoxicity.

The Meeting concluded that aminopyralid and aminopyralid TIPA are unlikely to be genotoxic.

In view of the lack of genotoxicity and the absence of carcinogenicity in rats and mice, the Meeting concluded that aminopyralid is unlikely to pose a carcinogenic risk to humans.

In a two-generation study of reproduction in rats given diets containing aminopyralid, only increases in the weight of full and empty caeca in parental animals treated with aminopyralid were observed and no changes in reproductive parameters were observed. The caecal changes were not considered to be toxicologically significant. The NOAEL for general toxicity and reproductive toxicity was 1000 mg/kg bw per day, the highest dose tested. In pregnant rats and rabbits treated with aminopyralid by gavage, even the highest doses tested, 1000 and 500 mg/kg bw per day, respectively, did not induce any treatment-related malformations in the offspring. When rats and rabbits were treated with aminopyralid TIPA, no treatment-related malformations occurred, but fetal body-weight reduction was observed in only in rabbits at high doses (aminopyralid equivalents, 526 mg/kg bw per day).

In rabbits treated with aminopyralid at doses of 500 mg/kg bw per day and greater, body-weight gain was reduced, and uncoordinated gait was evident immediately after dosing, lasting for approximately 2 h. Neither the severity nor the duration of uncoordinated gait increased as the study progressed. Additionally, two rabbits treated with aminopyralid at the highest dose were killed in a moribund condition on day 17 of gestation, being found to have pale kidneys, watery and dark caecal contents and erosions/ulcers in the glandular mucosa of the stomach. Owing to severe clinical signs observed at 750 mg/kg bw per day, all remaining animals in this study were killed at day 20 of gestation and were not available for evaluation of reproductive performance. Therefore, the NOAEL for maternal toxicity was 250 mg/kg bw per day. In pregnant rabbits treated with aminopyralid TIPA, maternal toxicity was evident as uncoordinated gait at a dose of 78 mg/kg bw per day as aminopyralid equivalents and greater. At higher doses, the clinical effects observed were similar to those seen with aminopyralid. The NOAEL for maternal toxicity was 26 mg/kg bw per day as aminopyralid equivalents.

In a study of acute toxicity and a 1-year study of neurotoxicity in rats treated with aminopyralid, no signs of behavioural changes and no histopathological findings suggesting neurotoxic potential were observed. Faecal and urine soiling at 2000 mg/kg bw in the study of acute toxicity were most likely due to general toxicity rather than indicative of specific neurotoxicity. The NOAELs for neurotoxicity in these two studies were 2000 and 1000 mg/kg bw per day as aminopyralid, respectively, the highest doses tested.

To obtain further insight into the possible mode of action leading to uncoordinated gait in rabbits, pregnant and non-pregnant animals were treated with aminopyralid and aminopyralid TIPA. Again, uncoordinated gait was observed but without any histopathological correlate in the central or peripheral nervous system. Additionally, no changes in consciousness, muscle strength or autonomic and somatic control were observed. For the acute dietary risk assessment, the occurrence of uncoordination after one or two doses in pregnant rabbits treated with aminopyralid may be the only relevant end-point. The NOAEL for aminopyralid for this effect was 250 mg/kg bw per day. In one study with aminopyralid TIPA, slight uncoordination was seen in one animal after 1 day of treatment at a dose of 173.6 mg/kg bw per day as aminopyralid equivalents. In another two studies of developmental toxicity in rabbits receiving aminopyralid TIPA, a dose-related increase in the incidence of uncoordination was found. However, at doses at which uncoordination was observed, i.e. 78, 105, 263 and 525 mg/kg bw per day as aminopyralid equivalents, the effect occurred only after at least six exposures. The NOAEL was 26 mg/kg bw per day as aminopyralid equivalents. The Meeting concluded that the weight of evidence indicated that the NOAEL for acute uncoordination was 250 mg/kg bw as aminopyralid. Effects observed in short-term studies in dogs were not considered relevant for establishing an acute reference dose.

The Meeting concluded that the existing data were adequate to characterize the potential hazard to fetuses, infants and children.

Toxicological evaluation

The Meeting established an ADI of 0–0.9 mg/kg bw based on a NOAEL of 93.2 mg/kg bw per day identified on the basis of histological changes in the gastric mucosa at higher doses in a 1-year study in dogs, and a safety factor of 100.

The Meeting concluded that it was not necessary to establish an ARfD for aminopyralid. The only end-point that might be suitable as a basis for establishing an ARfD for aminopyralid was uncoordinated gait in the studies of developmental toxicity in rabbits. Although this effect was observed after repeated exposure at 78 mg/kg bw with a NOAEL of 26 mg/kg bw, it was observed after one or two exposures only at higher doses, with a NOAEL of 250 mg/kg bw. As this effect was dependent on C_{max} and in view of the kinetics of aminopyralid and the dynamics of this response, the Meeting considered it appropriate to adjust the safety factor. Applying a safety factor of 25 to the NOAEL of 250 mg/kg bw would result in a putative ARfD of 10 mg/kg bw which is greater than the JMPR-recommended cut-off value for an ARfD of 5 mg/kg bw.

Levels relevant to risk assessment

Species	Test material	Study	Effect	NOAEL[b]	LOAEL[b]
Mouse	Aminopyralid	Eighty-week study of toxicity and carcinogenicity[a]	Toxicity	250 mg/kg bw per day	1000 mg/kg bw per day
			Carcinogenicity	1000 mg/kg bw per day[c]	—
Rat	Aminopyralid	Two-year study of toxicity and carcinogenicity[a]	Toxicity	500 mg/kg bw per day	1000 mg/kg bw per day
			Carcinogenicity	1000 mg/kg bw per day[c]	—
	Aminopyralid	Two-generation study of reproductive toxicity[a]	Parental toxicity/	1000 mg/kg bw per day[c]	—
			Offspring toxicity	1000 mg/kg bw per day[c]	—
	Aminopyralid	Developmental toxicity[b]	Maternal toxicity	1000 mg/kg bw per day[c]	—
			Embryo/ fetotoxicity	1000 mg/kg bw per day[c]	—
	Aminopyralid TIPA	Developmental toxicity[d]	Maternal toxicity	525 mg/kg bw per day[c]	—
			Embryo/ fetotoxicity	525 mg/kg bw per day[c]	—
	Aminopyralid	Long-term study of neurotoxicity	Neurotoxicity	1000 mg/kg bw per day[c]	—
Rabbit	Aminopyralid	Developmental toxicity[d]	Maternal toxicity	250 mg/kg bw per day	500 mg/kg bw per day
			Embryo/ fetotoxicity	500 mg/kg bw per day[c]	—
	Aminopyralid TIPA	Developmental toxicity[d]	Maternal toxicity	26 mg/kg bw per day	78 mg/kg bw per day
			Embryo/ fetotoxicity	263 mg/kg bw per day	526 mg/kg bw per day
Dog	Aminopyralid	One-year study of toxicity[a]	Toxicity	93.2 mg/kg bw per day	967 mg/kg bw per day

TIPA, triisopropylammonium.
[a] Dietary administration.
[b] Values for aminopyralid TIPA are given as aminopyralid equivalents.
[c] Highest dose tested.
[d] Gavage administration.

Estimate of acceptable daily intake for humans

0–0.9 mg/kg bw

Estimate of acute reference dose

Unnecessary

Information that would be useful for the continued evaluation of the compound

Results from epidemiological, occupational health and other such observational studies of human exposures

Critical end-points for setting guidance values for exposure to aminopyralid

Absorption, distribution, excretion and metabolism in mammals	
Rate and extent of oral absorption	Rapid, 42–59%
Dermal absorption	No data
Distribution	Extensive
Potential for accumulation	No evidence of accumulation
Rate and extent of excretion	Rapid
Metabolism in animals	Minimal. No metabolites identified.
Toxicologically significant compounds in animals, plants and the environment	Aminopyralid
Acute toxicity	
Rat, LD_{50}, oral	> 5000 mg/kg bw
Rat, LD_{50}, dermal	> 5000 mg/kg bw
Rat, LC_{50}, inhalation	> 5.5 mg/l
Rabbit, dermal irritation	Aminopyralid is not an irritant, aminopyralid TIPA is a slight irritant.
Rabbit, ocular irritation	Aminopyralid is an irritant, aminopyralid TIPA is not an irritant.
Skin sensitization (test method used)	Not a sensitizer (Magnusson & Kligman test)
Short-term studies of toxicity	
Target/critical effect	Histopathological changes in stomach
Lowest relevant oral NOAEL	93.2 mg/kg bw per day (1-year study in dogs)
Lowest relevant dermal NOAEL	1000 mg/kg bw per day, the highest dose tested (4-week study in rats)
Lowest relevant inhalation NOAEC	No studies available
Genotoxicity	
	No genotoxic potential
Long-term studies of toxicity and carcinogenicity	
Target/critical effect	Increased mortality in female mice
Lowest relevant NOAEL	250 mg/kg bw per day (18-month study in mice)
Carcinogenicity	Not carcinogenic
Reproductive toxicity	
Reproduction target/critical effect	No reproductive effects in rats
Lowest relevant reproductive NOAEL	1000 mg/kg bw per day, the highest dose tested
Developmental target/critical effect	No developmental effects in rats and rabbits with aminopyralid; fetal body-weight changes with aminopyralid TIPA.
Lowest relevant developmental NOAEL	263 mg/kg bw per day (rabbit)
Neurotoxicity/delayed neurotoxicity	
	No neurotoxic effects in rats with aminopyralid and aminopyralid TIPA; uncoordinated gait in pregnant rabbits, with both aminopyralid and aminopyralid TIPA.
Lowest relevant NOAEL	26 mg/kg bw per day (repeated doses), 250 mg/kg bw (single dose)
Medical data	
	No data available

Summary

	Value	**Study**	**Safety factor**
ADI	0–0.9 mg/kg bw	Dog, 1-year study	100
ARfD	Unnecessary	—	—

TIPA, triisopropylammonium

References

Brooks, K.J. (2001a) Aminopyralid: acute dermal irritation study in New Zealand White rabbits. Unpublished report No. 011117 from Toxicology & Environmental Research and Consulting, The Dow Chemical Company, Midland, MI, USA. Submitted to WHO by Dow AgroSciences LLC, Indianapolis, IN, USA.

Brooks, K.J. (2001b) Aminopyralid: acute eye irritation study in New Zealand White rabbits. Unpublished report No. 011118 from Toxicology & Environmental Research and Consulting, The Dow Chemical Company, Midland, MI, USA. Submitted to WHO by Dow AgroSciences LLC, Indianapolis, IN, USA.

Brooks, K.J. & Radtke, B.J. (2002a) GF-871: acute dermal irritation study in New Zealand White rabbits. Unpublished report No. DECO HET DR-0368-4864-012 from Toxicology & Environmental Research and Consulting, The Dow Chemical Company, Midland, MI, USA. Submitted to WHO by Dow AgroSciences LLC, Indianapolis, IN, USA.

Brooks, K.J. & Radtke, B.J. (2002b) GF-871: acute eye irritation study in New Zealand White rabbits. Unpublished report No. DECO HET DR-0368-4864-013 from Toxicology & Environmental Research and Consulting, The Dow Chemical Company, Midland, MI, USA. Submitted to WHO by Dow AgroSciences LLC, Indianapolis, IN, USA.

Brooks, K.J. & Wilson, C.W. (2002) GF-871: a dermal sensitization study in Hartley albino guinea pigs — maximization design. Unpublished report No. DECO HET DR-0368-4864-008 from Springborn Laboratories, INC, Spencerville, OH, USA. Submitted to WHO by Dow AgroSciences LLC, Indianapolis, IN, USA.

Brooks, K.J. & Yano, B.L. (2001a) Aminopyralid: acute dermal toxicity study in Fischer 344 rats. Unpublished report No. 011116 from Toxicology & Environmental Research and Consulting, The Dow Chemical Company, Midland, MI, USA. Submitted to WHO by Dow AgroSciences LLC, Indianapolis, IN, USA.

Brooks, K.J. & Yano, B.L. (2001b) Aminopyralid: acute oral toxicity study in Fischer 344 rats. Unpublished report No. 011115 from Toxicology & Environmental Research and Consulting, The Dow Chemical Company, Midland, MI, USA. Submitted to WHO by Dow AgroSciences LLC, Indianapolis, IN, USA.

Carney, E.W. (2004) GF-871: behavioral study in time mated in New Zealand White rabbits. Unpublished report No. DECO HET DR-0368-4864-024 from Toxicology & Environmental Research and Consulting, The Dow Chemical Company, Midland, MI, USA. Submitted to WHO by Dow AgroSciences LLC, Indianapolis, IN, USA.

Carney, E.W. & Tornesi, B. (2001) Aminopyralid: oral gavage developmental toxicity study in CD rats. Unpublished report No. 011061 from Toxicology & Environmental Research and Consulting, The Dow Chemical Company, Midland, MI, USA. Submitted to WHO by Dow AgroSciences LLC, Indianapolis, IN, USA.

Carney, E.W. & Tornesi, B. (2004a) GF-871: oral gavage developmental toxicity study in New Zealand White rabbits. Unpublished report No. DECO HET DR-0368-4864-020 from Toxicology & Environmental Research and Consulting, The Dow Chemical Company, Midland, MI, USA. Submitted to WHO by Dow AgroSciences LLC, Indianapolis, IN, USA.

Carney, E.W. & Tornesi, B. (2004b) Supplemental report for GF-871: Oral gavage developmental toxicity study in New Zealand White rabbits. Unpublished report No. DECO HET DR-0368-4864-020-S from Toxicology & Environmental Research and Consulting, The Dow Chemical Company, Midland, MI, USA. Submitted to WHO by Dow AgroSciences LLC, Indianapolis, IN, USA.

Charles, G.D. & Mecchi, M.S. (2004) GF-871: *Salmonella-Escherichia coli*/mammalian-microsome reverse mutation assay preincubation method with a confirmatory assay with GF-871. Unpublished report No. DECO HET DR-0368-4864-022 from Covance Laboratories Inc., Vienna, VA, USA. Submitted to WHO by Dow AgroSciences LLC, Indianapolis, IN, USA.

Domoradzki, J.Y., Rick, D.L. & Clark, A.J. (2004) Aminopyralid, triisopropanolamine salt: dissociation and metabolism in male Fischer 344 rats. Unpublished report No. DECO HET DR-0293-9028-080 from Toxicology & Environmental Research and Consulting, The Dow Chemical Company, Midland, MI, USA. Submitted to WHO by Dow AgroSciences LLC, Indianapolis, IN, USA.

Dryzga, M.D. & Stebbins, K.E. (2001) Aminopyralid: 13-week dietary toxicity with recovery study in Fischer 344 rats. Unpublished report No. 001221R from Toxicology & Environmental Research and Consulting, The Dow Chemical Company, Midland, MI, USA. Submitted to WHO by Dow AgroSciences LLC, Indianapolis, IN, USA.

Hansen, S.C., Mendrala, A.L., Markham, D.A. & Saghir, S. (2005) DE-750: metabolism and pharmacokinetics of ^{14}C-DE-750 in non-pregnant and pregnant New Zealand White rabbits following single and repeated oral administration. Unpublished report No. DECO HET DR-0293-9028-082 from Toxicology & Environmental Research and Consulting, The Dow Chemical Company, Midland, MI, USA. Submitted to WHO by Dow AgroSciences LLC, Indianapolis, IN, USA.

Johnson, K.A. & Dryzga, M.D. (2003) Aminopyralid: two-year chronic toxicity/oncogenicity and chronic neurotoxicity study in Fischer 344 rats. Unpublished report No. 011049 from Toxicology & Environmental Research and Consulting, The Dow Chemical Company, Midland, MI, USA. Submitted to WHO by Dow AgroSciences LLC, Indianapolis, IN, USA.

Kiplinger, G.R. (2001) Aminopyralid: acute inhalation toxicity study in the Fischer 344 rat. Unpublished report No. WIL-406012 from WIL Research Laboratories, Inc., Ashland, OH, USA. Submitted to WHO by Dow AgroSciences LLC (report No. 011097) , Indianapolis, IN, USA.

Landry, T.D. & Krieger, S.M. (2002) GF-871: acute liquid aerosol inhalation toxicity study in Fischer 344 rats. Unpublished report No. DECO HET DR-0368-4864-005 from Toxicology & Environmental Research and Consulting, The Dow Chemical Company, Midland, MI, USA. Submitted to WHO by Dow AgroSciences LLC, Indianapolis, IN, USA.

Linscombe, V.A., Jackson, K.M. & Schisler, M.R. (2004) Evaluation of GF-871 in an in vitro chromosomal aberration assay utilizing rat lymphocytes. Unpublished report No. DECO HET DR-0368-4864-015 from Toxicology & Environmental Research and Consulting, The Dow Chemical Company, Midland, MI, USA. Submitted to WHO by Dow AgroSciences LLC, Indianapolis, IN, USA.

Linscombe, V.A., Jackson, K.M., Schisler, M.R. & Beuthin, D.J. (2002) Evaluation of aminopyralid in an *in vitro* chromosomal aberration assay utilizing rat lymphocytes. Unpublished report No. 011040 from Toxicology & Environmental Research and Consulting, The Dow Chemical Company, Midland, MI, USA. Submitted to WHO by Dow AgroSciences LLC, Indianapolis, IN, USA.

Linscombe, V.A., Schisler, M.R. & Beuthin, D.J. (2001) Evaluation of aminopyralid in the Chinese hamster ovary cell/hypoxanthine-guanine-phosphoribosyl transferase (CHO/HGPRT) forward mutation assay. Unpublished report No. 011037 from Toxicology & Environmental Research and Consulting, The Dow Chemical Company, Midland, MI, USA. Submitted to WHO by Dow AgroSciences LLC, Indianapolis, IN, USA.

Liu, J. (2004) [^{14}C]Aminopyralid: absorption, distribution, metabolism, and excretion in male Fischer 344 rats. Unpublished report No. 47456 from ABC Laboratories, Inc., Columbia, MO, USA. Submitted to WHO by Dow AgroSciences LLC (Report No. 021200), Indianapolis, IN, USA.

Marable, B.R., Andrus, A.K. & Stebbins, K.E. (2001) Aminopyralid: acute neurotoxicity study in Fischer 344 rats. Unpublished report No. 011073R from Toxicology & Environmental Research and Consulting, The Dow Chemical Company, Midland, MI, USA. Submitted to WHO by Dow AgroSciences LLC, Indianapolis, IN, USA.

Marable, B.R. & Day, S.J. (2004) Aminopyralid: oral gavage pilot study in non-pregnant New Zealand White Rabbits. Unpublished report No. DECO HET DR-0293-9028-079 from Toxicology & Environmental Research and Consulting, The Dow Chemical Company, Midland, MI, USA. Submitted to WHO by Dow AgroSciences LLC, Indianapolis, IN, USA.

Marty, M.S., Liberacki, A.B. & Thomas, J. (2002) Aminopyralid: oral gavage developmental toxicity study in New Zealand White rabbits. Unpublished report No. 011047 from Toxicology & Environmental Research and Consulting, The Dow Chemical Company, Midland, MI, USA. Submitted to WHO by Dow AgroSciences LLC, Indianapolis, IN, USA.

Marty, M.S., Zablotny, C.L. & Thomas, J. (2003) Aminopyralid: two-generation dietary reproduction toxicity study in CD rats. Unpublished report No. 011205 from Toxicology & Environmental Research and

Consulting, The Dow Chemical Company, Midland, MI, USA. Submitted to WHO by Dow AgroSciences LLC, Indianapolis, IN, USA.

Maurissen, J.P., Andrus, A.K., Johnson, K.A. & Dryzga, M.D. (2003) Aminopyralid: chronic neurotoxicity study in Fischer 344 rats. Unpublished report No. 011049N from Toxicology & Environmental Research and Consulting, The Dow Chemical Company, Midland, MI, USA. Submitted to WHO by Dow AgroSciences LLC, Indianapolis, IN, USA.

Mecchi, M.S. (2001) *Salmonella - Escherichia coli*/mammalian-microsome reverse mutation assay preincubation method with a confirmatory assay with aminopyralid. Unpublished report No. 22338-0-422OECD from Covance Laboratories Inc., Vienna, VA, USA. Submitted to WHO by Dow AgroSciences LLC (report No. 011058R), Indianapolis, IN, USA.

Schisler, M.R., Linscombe, V.A. & Seidel, S. (2004) Revised report for: evaluation of GF-871 in the Chinese hamster ovary cell/hypoxanthine-guanine-phosphoribosyl transferase (CHO/HGPRT) forward mutation assay. Unpublished report No. DECO HET DR-0368-4864-016 REV from Toxicology & Environmental Research and Consulting, The Dow Chemical Company, Midland, MI, USA. Submitted to WHO by Dow AgroSciences LLC, Indianapolis, IN, USA.

Spencer, P.J. & Gorski, T.A. (2002) Evaluation of aminopyralid in the mouse bone marrow micronucleus test. Unpublished report No. 011125 from Toxicology & Environmental Research and Consulting, The Dow Chemical Company, Midland, MI, USA. Submitted to WHO by Dow AgroSciences LLC, Indianapolis, IN, USA.

Spencer, P.J., Linscombe, V.A. & Grundy, J. (2004) Evaluation of GF-871 in the mouse bone marrow micronucleus test. Unpublished report No. DECO HET DR-0368-4864-017 from Toxicology & Environmental Research and Consulting, The Dow Chemical Company, Midland, MI, USA. Submitted to WHO by Dow AgroSciences LLC, Indianapolis, IN, USA.

Stebbins, K.E. & Baker, P.C. (2000) Aminopyralid: 4-week dietary toxicity study in beagle dogs. Unpublished report No. 001030 from Toxicology & Environmental Research and Consulting, The Dow Chemical Company, Midland, MI, USA. Submitted to WHO by Dow AgroSciences LLC, Indianapolis, IN, USA.

Stebbins, K.E. & Baker, P.C. (2002) Aminopyralid: 13-week dietary toxicity study in beagle dogs. Unpublished report No. 001239R from Toxicology & Environmental Research and Consulting, The Dow Chemical Company, Midland, MI, USA. Submitted to WHO by Dow AgroSciences LLC, Indianapolis, IN, USA.

Stebbins, K.E. & Day, S.J. (2000) Aminopyralid: 4-week repeated dose dietary toxicity study in Fischer 344 rats. Unpublished report No. 001031 from Toxicology & Environmental Research and Consulting, The Dow Chemical Company, Midland, MI, USA. Submitted to WHO by Dow AgroSciences LLC, Indianapolis, IN, USA.

Stebbins, K.E. & Day, S.J. (2003a) Aminopyralid: oncogenicity dietary study in CD-1 mice. Unpublished report No. 011163 from Toxicology & Environmental Research and Consulting, The Dow Chemical Company, Midland, MI, USA. Submitted to WHO by Dow AgroSciences LLC, Indianapolis, IN, USA.

Stebbins, K.E. & Day, S.J. (2003b) Aminopyralid: one-year dietary toxicity study in beagle dogs. Unpublished report No. 021027 from Toxicology & Environmental Research and Consulting, The Dow Chemical Company, Midland, MI, USA. Submitted to WHO by Dow AgroSciences LLC, Indianapolis, IN, USA.

Stebbins, K.E., Day, S.J. & Thomas, J. (2001) Aminopyralid: 13-week dietary toxicity study in CD-1 mice. Unpublished report No. 001240 from Toxicology & Environmental Research and Consulting, The Dow Chemical Company, Midland, MI, USA. Submitted to WHO by Dow AgroSciences LLC, Indianapolis, IN, USA.

Stebbins, K.E. & Dryzga, M.D. (2004) GF-871: 90-day dietary toxicity study in Fischer 344 Rats. Unpublished report No. DECO HET DR-0368-4864-018 from Toxicology & Environmental Research and Consulting, The Dow Chemical Company, Midland, MI, USA. Submitted to WHO by Dow AgroSciences LLC, Indianapolis, IN, USA.

Stebbins, K.E., Thomas, J. & Day, S.J. (2002) Aminopyralid: 28-day dermal toxicity study in Fischer 344 rats. Unpublished report No. 011219 from Toxicology & Environmental Research and Consulting, The Dow Chemical Company, Midland, MI, USA. Submitted to WHO by Dow AgroSciences LLC, Indianapolis, IN, USA.

Wilson, C.W. (2001) Aminopyralid: a dermal sensitization study in Hartley albino guinea pigs. Unpublished report No. 3504.155 from Springborn Laboratories, INC, Spencerville, OH, USA. Submitted to WHO by Dow AgroSciences LLC (Report No. 011034), Indianapolis, IN, USA.

Wilson, D.M., Brooks, K.J. & Radtke, B.J. (2002a) GF-871: acute dermal toxicity study in Fischer 344 rats. Unpublished report No. DECO HET DR-0368-4864-014 from Toxicology & Environmental Research and Consulting, The Dow Chemical Company, Midland, MI, USA. Submitted to WHO by Dow AgroSciences LLC, Indianapolis, IN, USA.

Wilson, D.M., Brooks, K.J. & Radtke, B.J. (2002b) GF-871: acute oral toxicity study in Fischer 344 rats. Unpublished report No. DECO HET DR-0368-4864-011 from Toxicology & Environmental Research and Consulting, The Dow Chemical Company, Midland, MI, USA. Submitted to WHO by Dow AgroSciences LLC, Indianapolis, IN, USA.

Yano, B.L. & Dryzga, M.D. (2000) Aminopyralid: 4-week repeated dose dietary toxicity study in CD-1 mice. Unpublished report No. 001048 from Toxicology & Environmental Research and Consulting, The Dow Chemical Company, Midland, MI, USA. Submitted to WHO by Dow AgroSciences LLC, Indianapolis, IN, USA.

Yano, B.L. & Zablotny, C.L. (2005) GF-871: oral gavage neuropathology study in time-mated New Zealand White rabbits. Unpublished report No. DECO HET DR-0368-4864-023 from Toxicology & Environmental Research and Consulting, The Dow Chemical Company, Midland, MI, USA. Submitted to WHO by Dow AgroSciences LLC, Indianapolis, IN, USA.

Zablotny, C.L. & Thorsrud, B.A. (2004) GF-871: an oral developmental toxicity study in Sprague Dawley rats. Unpublished report No. DECO HET DR-0368-4864-021 from Charles River Laboratories, Springborn Division, INC, Spencerville, OH, USA. Submitted to WHO by Dow AgroSciences LLC, Indianapolis, IN, USA.

ATRAZINE

First draft prepared by
Rudolf Pfeil,[1] Vicki Dellarco[2] and Les Davies[3]

[1]Toxicology of Pesticides, Federal Institute for Risk Assessment, Berlin, Germany;
[2]Office of Pesticide Programs, United States Environmental Protection Agency, Health Effects Division, Washington, USA; and
[3]Chemical Review, Australian Pesticides and Veterinary Medicines Authority, Canberra, Australia

Explanation

Atrazine, 6-chloro-N^2-ethyl-N^4-isopropyl-1,3,5-triazine-2,4-diamine (International Union of Pure and Applied Chemistry, IUPAC) (Chemical Abstracts Service, CAS No. 1912-24-9), is a selective systemic herbicide of the chlorotriazine class, which is used for the control of annual broadleaf and grassy weeds. It acts as a photosynthetic electron transport inhibitor at the photosystem II receptor site. Atrazine and its chloro-*s*-triazine metabolites deethyl-atrazine (DEA), deisopropyl-atrazine (DIA) and diaminochlorotriazine (DACT) have been found in surface and groundwater as a result of the use of atrazine as a pre-emergent or early post-emergent herbicide. Hydroxyatrazine is more commonly detected in groundwater than in surface water. The relative order of concentrations of these substances measured in rural wells in the USA was generally as follows: atrazine ~ DEA ~DACT > DIA > hydroxyatrazine. However, concentrations of DEA that are severalfold those of the parent compound have been reported.

Atrazine was evaluated previously by WHO, a tolerable daily intake (TDI) of 0.0005 mg/kg bw being established in the 1993 *Guidelines for Drinking-water Quality* on the basis of a no-observed-adverse-effect level (NOAEL) of 0.5 mg/kg bw per day in a study of carcinogenicity in rats and using a safety factor of 1000 (100 for inter- and intraspecies variation and 10 to reflect potential carcinogenic risk to humans).

Atrazine has not been previously evaluated by JMPR, and no acceptable daily intake (ADI) had been established. For that reason, the WHO Drinking-water Guidelines Programme recommended that atrazine should be evaluated toxicologically by JMPR.

The database on atrazine was extensive, consisting of a comprehensive set of studies with atrazine and its four key metabolites, which complied with good laboratory practice (GLP), as well as a large number of published studies. The present Meeting did not aim to perform a review of the database de novo, but to summarize the key studies, focusing on the issues of carcinogenicity, endocrine disruption (especially neuroendocrine mode of action) and immunotoxicity. Reference was made to a number of reviews made by national and international agencies and organizations in recent years.

Summary data were provided on the occurrence of atrazine residues in water, abstracted from monitoring data collected in areas where atrazide is widely used. Globally, atrazine is most commonly used as a pre-emergent or early post-emergent herbicide in corn in the USA and Brazil. Much smaller amounts are used in sorghum (Australia and the USA), sugar-cane (Brazil and the USA), oilseed crops (Australia) and tree plantations (Australia). The non-agricultural uses of atrazine (e.g. railways, road embankments, turf, home garden) have largely been removed from atrazine labels in recent years.

Because atrazine is somewhat persistent in the environment and is reasonably mobile in soils, a number of studies have monitored the concentrations of atrazine in groundwater and surface water over the last two decades. Recent monitoring data show declining levels and incidences of detections of atrazine and its chloro-*s*-metabolites (DIA; DEA and DACT) compared with data collected in the early 1990s; this reflects restrictions on the use of atrazine that were introduced in the late 1990s and early 2000s and the introduction of the "Good Farming Practice Programme" (GFPP) in the European Union (EU) and other parts of the world and "Best Management Practices" (BMPs) in the USA. Therefore, older monitoring data generally represent an overestimate of environmental concentrations likely to arise from current use practices. In surface water, the concentrations of the chlorotriazine metabolites of atrazine are generally less than those of atrazine itself, while the concentrations of these metabolites in rural wells are more similar to those of atrazine.

Consideration of data from monitoring carried out in a number of countries indicate that concentrations of atrazine and its chloro-*s*-metabolites in groundwater and surface water rarely exceed the current WHO guideline value of 2 ppb; levels are commonly well below the EU "parametric value" for pesticide residues in drinking-water of 0.1 ppb. Specific monitoring of drinking-water indicates that:

- recent data from the USA show that in no public-water supplies does the concentration of atrazine exceed the United States Environmental Protection Agency (EPA) Drinking Water Levels of Comparison (DWLOCs) for any age group;
- concentrations of atrazine in the United Kingdom (UK) are less than 0.1 ppb; and
- in Canada, concentrations are lower than in "raw" groundwater or surface water.

Hydroxyatrazine is a plant metabolite of atrazine. It can also be formed in acidic and humitic soils. It is reported to bind to soil to a greater extent than parent atrazine or the chlorinated metabolites. Hydroxyatrazine has been less frequently measured in monitoring programmes than parent atrazine or the chlorinated metabolites; however, in studies in which hydroxyatrazine has been measured, it was generally detected less frequently than were the chlorinated metabolites, and commonly at lower concentrations. The available data suggest that hydroxy compounds are unlikely to significantly contaminate surface water.

Evaluation for acceptable daily intake

Unless otherwise stated, the studies evaluated in this monograph were performed by laboratories that were certified for GLP and that complied with the relevant Organisation for Economic Co-operation and Development (OECD) test guidelines or similar guidelines of the EU or United States Environmental Protection Agency. As these guidelines specify the clinical pathology tests normally performed and the tissues normally examined, only significant exceptions to these guidelines are reported here, to avoid repetitive listing of study parameters.

1. Biochemical aspects

1.1 Absorption, distribution and excretion

Rats

In a study on absorption, distribution and elimination, male Sprague-Dawley rats received daily oral doses of [^{14}C]atrazine ([U^{14}C]-triazine labelled; radiochemical purity, ≥ 97.5%) for up to 7 days at 0.1 mg per rat (equal to approximately 0.4 mg/kg bw) or 1.0 mg per rat (equal to approximately 4.0 mg/kg bw). Groups of three rats were killed 5, 7, 9, 10, 14 and 18 days after the start of dosing, and blood and tissues were sampled for radioactivity. From the rats killed after 18 days (11 days after last dose), faeces and urine were collected for periods of 24 h; tissues were analysed for radioactivity, and a final material balance was calculated. A material balance of 103.9% and 93.4% was achieved for the doses at 0.4 and 4.0 mg/kg bw, respectively. Recovery of radiolabel was approximately 67–73% in urine, 24–28% in faeces, and 2–3% in the tissues, with the proportions remaining constant between the doses. Approximately 95% of the administered dose was eliminated within 48 h of the final dose; however, 11 days after the final dose appreciable residues were still present in muscle and erythrocytes. Elimination half-lives of 4 days for most tissues, 10 days for brain, and 25–30 days for erythrocytes and muscle were derived. Binding to erythrocytes was postulated, with the decline in this tissue correlating to the half-life for erythrocyte turnover (Ballantine et al., 1985).

In a study on absorption, distribution and elimination, groups of five male and five female Sprague-Dawley rats received [^{14}C]atrazine ([U^{14}C]-triazine labelled; radiochemical purity, $\geq$ 98%) as a single dose at 1.0 or 100 mg/kg bw, or as a single dose at 1.0 mg/kg bw after receiving unlabelled atrazine as daily doses at 1.0 mg/kg bw for 14 days. Faeces and urine were collected for up to 168 h (7 days) after administration of the radiolabel. After 7 days, rats were killed and selected tissues and carcass were harvested. The material balance achieved 103% for rats receiving single doses at 1.0 or 100 mg/kg bw, but only 88% for the rats receiving repeated doses of non-radiolabelled atrazine. Recovery of radiolabel was approximately 74% in the urine and approximately 19% in the faeces, most elimination occurring within 48 h; the residues in the tissues accounted for 4.7–7.0% of the administered dose and were predominantly associated with erythrocytes. In most tissues, a lower proportion of the administered dose remained after a dose of 100 mg/kg bw than 1 mg/kg bw (4.7% versus 7.0%); this was proposed to be evidence of tissue saturation at higher doses. The kinetics of atrazine elimination were determined using urinary radioactivity excretion data, which best fitted a linear two-compartment open model for each of the three groups. The half-lives of the alpha (distribution and elimination) phase, and of the beta (whole-body renal elimination phase) were 6.9 h and 31.1 h, respectively. The half-life of renal elimination from the central compartment was 12.4 h (Orr et al., 1987).

In a study on absorption, distribution and elimination, pairs of female Sprague-Dawley CD rats were given [^{14}C]atrazine ([U^{14}C]-triazine labelled; radiochemical purity, 97.9%; purity of unlabeled atrazine, 98.8%) by gavage daily at a dose of 1, 3, 7, 10, 50 or 100 mg/kg bw for 10 days. Blood samples were collected at 24-h intervals, and tissue samples were collected at necropsy. Urine and faeces were collected over successive 24-h periods. One rat of each pair was killed at 3 h and the second at 72 h after the tenth dose. Proportions excreted in the urine (70–76%) and faeces (13–15%) did not vary with dose. Concentrations of atrazine in plasma rose until day 8 or 9 then achieved plateau values. Concentrations of atrazine in erythrocytes continued to increase throughout the 10-day dosing period; it was estimated that steady-state would be achieved only after 30 days. Half-lives for elimination from plasma or erythrocytes were 38.6 h and 8.1 days, respectively. Tissue concentrations 3 h after the tenth dose were linearly related to plasma concentrations obtained at the same time-point for the dose range of 1–100 mg/kg bw (Khars & Thede, 1987).

In a study on absorption, distribution and elimination, groups of six Sprague-Dawley rats (three males and three females) received [^{14}C]atrazine ([U^{14}C]-triazine labelled; radiochemical purity, > 99.7%; purity of unlabeled atrazine, 99.0%) as a single dose at 1 or 100 mg/kg bw. Blood samples were taken at 17 intervals within 168 h to determine blood-level kinetics. A further two groups of 12 male rats received a single dose at 1 or 100 mg/kg bw after which three rats per group were killed at 2, 48, 168 and 336 h to determine tissue residues. From three of these males that were killed at 168 h, urine and faeces were collected at successive time-points. A further group of four males were bile-duct cannulated, given a single dose of 1 mg/kg bw and then sampled for bile, urine and faeces during 48 h. Another group of four male rats was given a single dose at 100 mg/kg bw, and one rat was killed at each time-point, 24, 72, 168 and 336 h, for whole-body autoradiography. Excreta from this study were also used for the characterization of metabolites (see section 1.2).

Maximum concentrations in the blood were achieved at 2 h at the lower dose and 24 h at the higher dose. Areas under the curve of blood concentration–time (AUC) were not different between the sexes. The half-life in blood, assuming first-order kinetics, was approximately 150 h, independent of sex and dose. Results from bile-cannulated animals showed approximately 88% absorption of the lowest dose, on the basis of the excretion via urine (65%) and bile (7%) over 48 h and the amount remaining in the carcass (16%); elimination via the faeces was about 3%. At the highest dose, elimination via the urine and faeces during 168 h accounted for approximately 66% and 20%, respectively, with approximately 85% excreted within the first 48 h.

Tissue concentrations from rats killed at four different time-points showed that the highest concentrations of atrazine occurred in the liver, kidney and erythrocytes. Elimination from the tissues was judged to be biphasic, and half-lives ranged from 59 to 300 h (assuming first-order kinetics). Elimination from the erythrocytes was monophasic, with a half-life of 320 h. There were no significant differences between the sexes or doses.

Whole-body autoradiography supported the previous toxicokinetic and tissue-residue data, with slow elimination of radioactivity from well-perfused tissues (liver, kidney, lungs, heart and spleen) owing to the presence of erythrocytes (Paul et al., 1993).

In a study on absorption, distribution, metabolism and elimination, two groups of four male Fischer 344 rats received [^{14}C]atrazine ([U^{14}C]-triazine labelled; radiochemical purity, > 98%; purity of unlabeled atrazine, 99.6%) as a single oral dose at 30 mg/kg bw with or without pre-treatment with non-radiolabeled tridiphane (purity, > 99%) as a single oral dose at 60 mg/kg bw. The plasma time-course for the radiolabel exhibited a mono-exponential decrease for both treatment groups, with an absorption and elimination half-life of approximately 3 h and 11 h, respectively. About 93% of the administered radioactivity was recovered 72 h after dosing; with approximately 67% found in the urine and approximately 18% in the faeces, and less than 10% in the carcass, skin and erythrocytes. There were no appreciable differences in the metabolite distribution between treatment groups, and the major urinary metabolite of atrazine was found to be 2-chloro-4,6-diamino-1,3,5-triazine (64–67% of total urinary radioactivity). *S*-(2-Amino-4-methylethylamino-1,3,5-triazin-6-yl)-mercapturic acid (13–14%), and *S*-(2,4-diamino-1,3,5-triazin-6-yl)-mercapturic acid (9%) were tentatively identified based on similar high-performance liquid chromatography (HPLC) retention times. The data indicated that there were no meaningful differences in the absorption, distribution, metabolism, and excretion between rats given only [^{14}C]atrazine and those given both tridiphane and [^{14}C]atrazine (Timchalk et al., 1990).

The slow elimination of [^{14}C]atrazine from rat tissue may in part be related to the extent of blood perfusion of the tissue because a metabolite of *s*-triazines binds covalently to Cys-125 in the β-chain of rodent haemoglobin rather than to the usually more reactive Cys-93, which is present in most mammalian haemoglobins. The metabolite reacted also (but to a lesser extent) with haemoglobin from chicken (Cys-126), but not with haemoglobin from humans, dogs, sheep, cows or pigs, which lack Cys-125 in their haemoglobin (Hamboeck et al., 1981). In Sprague-Dawley rats exposed to atrazine, it was shown using matrix-assisted laser desorption ionization-time-of-flight mass spectrometry (MALDI-TOF MS) that the adduct to Cys-125 of the β-chain of haemoglobin was DACT (G 28273) and not parent atrazine or the mono-dealkylated metabolites G 28279 or G 30033 (Dooley et al., 2006).

Monkeys

In a study on the clearance of atrazine from the blood, four female Rhesus monkeys (aged approximately 20–30 years) received [^{14}C]atrazine ([U^{14}C]-triazine labeled; radiochemical purity, 98.1%; purity of unlabeled atrazine, 96.5%) intravenously as a single dose at 0.26 mg per monkey. Samles of urine, faeces and blood were collected for 168 h after treatment. The principal route of excretion was via the urine with approximately 63% of the administered dose being excreted within

24 h. Clearance of radioactivity from the blood was best described by a two-compartment model with half-lives of 1.5 and 17.7 h, respectively; the half-life for renal clearance was 20.8 h. By the end of the 7-day collection period, about 85% of the administered dose was found in the urine and 12% in the faeces (Hui et al., 1996a).

In a study on absorption, distribution, metabolism and elimination, three groups of four female Rhesus monkeys (aged 6–25 years) received capsules containing [^{14}C]atrazine ([U^{14}C]-triazine labelled; radiochemical purity, 96.8%; purity of unlabeled atrazine, 98.7%) as a single dose at 1, 10 or 100 mg per monkey. Samples of urine, faeces and blood were collected for 168 h after treatment. Maximum blood concentrations were achieved after 2, 8 and 24 h at the lowest, intermediate and highest dose, respectively. On the basis of the comparison of AUCs after oral and intravenous administration (Hui et al., 1996a), the bioavailabilities were estimated at 92%, 75% and 78% for the lowest, intermediate and highest dose, respectively. The terminal half-lives for elimination from plasma were 31.9 h, 21.6 h and 19.7 h for the lowest, intermediate and highest dose, respectively. By the end of the 7-day collection period, excretion of radioactivity in the urine was about 57%, 58% or 53% for the lowest, intermediate and highest dose, respectively, and in faeces about 21%, 25% or 35% for the lowest, intermediate and highest dose, respectively (Hui et al., 1996b).

1.2 Biotransformation

The metabolic pathway of atrazine in the rat is illustrated in Figure 1. The experimental evidence supports a metabolic pathway dominated by oxidative removal of the alkyl side-chains with 2-chloro-4, 6-diamino-*s*-triazine (G 28273) being the major metabolite. The 2 carbon-chlorine bond is stable to enzymatic hydrolysis, but is subject to conjugation via the action of glutathione-*S*-transferase. Action on sulfur-containing metabolites gives 2-sulfhydryl-*s*-triazines which in turn are subject to methylation followed by oxidation to the corresponding *S*-oxides. Oxidation of primary positions of the alkyl side-chains to carboxyl functions is a minor alternative metabolic route.

The different species including humans appear to share this common pathway (Figure 2); however, differences in the kinetics of each step may be inferred. No difference in metabolism was found in those cases where the sexes were compared. Although in-vitro data suggest that metabolism to the bi-dealkylated metabolite by cultured hepatocytes from women is minimal, this data were not confirmed by analysis of urine in vivo in two studies in men in which the bi-dealkylated form was a major component, as in the rat. The possibility of significant post-hepatic modification in humans cannot therefore be excluded.

Rats

From a study in Sprague-Dawley rats described previously (section 1.1; Paul et al., 1993), the excreta from male rats treated with atrazine as a single dose at 100 mg/kg bw, and from bile-duct cannulated male rats treated with a single dose at 1 mg/kg bw were used for the separation and identification of the metabolites. Urine and faeces were collected for up to 168 h, and bile for up to 48 h. Metabolites were analysed by two-dimensional thin-layer chromatography (TLC). The pattern of metabolites was complex, with 26, 12 and 9 different metabolites identified in the urine, faeces and bile, respectively. The major metabolite DACT (G 28273) accounted for approximately 26% of the administered dose in the urine, 1.6% in bile and 1.3% in faeces. Approximately 1% of the administered dose was the mercapturic derivative of the DACT (CGA 10582) in the urine. Lesser amounts of the mono-dealkylated metabolites G 28279 and G 30033 were found in the urine, bile or faeces. A trace amount of parent atrazine was found in the faeces (Paul et al., 1993).

In a study on comparative metabolism of two *S*-triazines, [^{14}C]atrazine ([U^{14}C]-triazine labeled; radiochemical purity, 96.4–97.0%; purity of unlabeled atrazine, 97.5–98.9%) was given to female Sprague-Dawley rats as a single oral dose at 1 or 76 mg/kg bw. Urine and faeces were collected at 24-h intervals for 3 days. At 72 h after treatment, the rats were killed and brains were analysed for residual radioactivity. Within 72 h, approximately 69% of the radioactivity was eliminated in the urine and approximately 16% in the faeces. Residues of atrazine equivalents in brain were 0.167 and 3.85 ppm for the lower and higher dose, respectively. No parent compound or its dealkylated metabolites were found in brain. Sample of urine analysed by cation exchange chromatography, HPLC and two-dimensional TLC showed that DACT (G 28273) was the major metabolite of atrazine, accounting for approximately 24–29% of the administered dose in urine for 0–24 h, while another chloro-*s*-triazine metabolite (metabolite UD, consisting of a mixture of chloro sidechain acids) accounted for approximately 6% of the administered dose. The total of the mercapturate of atrazine (CGA-359008) and the mercapturate metabolites (CGA-246059, CGA-63079 and CGA-10582) accounted for approximately 2–3% of the administered dose (Simoneaux, 2001).

Monkeys

In a study described previously (section 1.1, Hui et al., 1996a), female Rhesus monkeys received [^{14}C]atrazine intravenously as a single dose at 0.26 mg per monkey, and samples of urine, faeces and blood were collected during 168 h after treatment. Atrazine metabolites identified in the urine by TLC included 10.7% DEA (G 30033), 13.7% DIA (G 28279), 4.4% atrazine mercapturate and 5.6% di-dealkylated atrazine mercapturate. Plasma contained trace amounts of atrazine in addition to G 30033, G 28279 and G 28273. The mercapturic acid of atrazine, as determined by enzyme immunoassay (EIA), was detected primarily in urine samples collected during the first 24 h after dosing. All four potential mercapturic acid conjugates of chlorotriazine moieties were detected in the urine using the LC/MS/MS method (liquid chromatography/mass spectroscopy). The total mercapturic acid residues in 0–24 h composite urine samples ranged from 15 to 25 ppb and the individual mercapturic acids ranged from 1 to 14 ppb. Total accountability of the mercapturates ranged from 1.6% to 3.0% of the total radioactive residues (TRR). The amount of atrazine mercapturate (CGA 359008) present in the composite sample as determined by EIA was an average of 16 times greater than the amount determined by LC/MS/MS. Based upon preliminary characterization data, it was postulated that one or more unknown chlorotriazine compounds altered in one of the side-chain alkyl groups could react with the EIA reagent to explain the higher results obtained by this method compared with the more specific LC/MS/MS method (Hui et al., 1996a).

In a study described previously (section 1.1, Hui et al., 1996b), female Rhesus monkeys received [^{14}C]atrazine orally as a single dose at 1, 10 and 100 mg per monkey, and samples of urine, faeces and blood were collected during 168 h after treatment. Samples of urine were analysed by TLC and gas chromatography/mass selective detection (GC/MSD) for atrazine and its dealkylated metabolites DEA (G 30033), DIA (G 28279), and DACT (G 28273). Both analytical methods gave similar amounts of chlorotriazine residues. An average of approximately 11% of the administered dose in all dose groups was found as chlorotriazine residues when measured in urine samples for 0–24 h, the major metabolites in urine were G 30033 and G 28273. Atrazine mercapturate (CGA 359008) accounted for 1% or less of the total radioactive residues in urine (Hui et al., 1996b).

Humans

In a study on kinetics and metabolism, six adult male volunteers received atrazine (purity not reported) as a single oral dose of 0.1 mg/kg bw. Urine was collected over 168 h. Blood samples were obtained from one man at 0, 2, 3, 4, 5, 6, 8, 24, 32, 72 and 168 h after dosing. Samples of urine and blood were analysed for atrazine and its dealkylated metabolites DEA (G 30033), DIA (G 28279) and DACT (G 28273) using a gas chromatography method with stated detection limit of 0.005 ppm. Atrazine and G 28279 were detected in whole blood but were below the limit of quantification. The metabolite G 30033 appeared rapidly in the plasma reaching a peak after 2 h to decline rapidly thereafter with a half-life of 2.8 h. The disappearance of G 30033 corresponded with an increased plasma concentration of the metabolite G 28273 which reached a peak at 5 h after dosing and was eliminated with a half-life of 17.8 h. This suggests a step-wise dealkylation of atrazine to DEA and then to DACT. The elimination of both metabolites from the blood was consistent with one-compartment first-order kinetics at least up to 32 h after dosing. Analysis of urine identified only approximately 14.5% of the administered dose, with consistent results for all six men. The chloro-metabolites G 30033, G 28279 and G 28273 accounted for 5.4%, 1.4% and 7.7% of the administered dose, respectively. For metabolite G 28273, urinary kinetics suggested a single-compartment first-order model with a half-life of 11.5 h. For the metabolites G 30033 and G 28279, a two-compartment first-order model suggested a fast and slow phase of 2.3 h or 8.4 h and 2.4 h or 36.2 h, respectively (Davidson, 1988).

In a biological monitoring study on a group of six manufacturing workers exposed to atrazine (10–700 μmol per workshift), total urinary excretion of atrazine plus three metabolites accounted for 1–2% of the external dose, with 50% of the amount excreted in the first 8 h following the workshift. About 80% of the excreted metabolites accounted for DACT (G 28273), 10% for DIA (G 28279), 8% for DEA (G 30033), and only 1–2% was atrazine (Catenacci et al., 1993).

In a study on comparative metabolism of atrazine in vitro, cultures of hepatocytes from CD1 mice, Sprague-Dawley rats, Fischer 344 rats, guinea-pigs, goats and Rhesus monkeys, and two women were used. The isolated cells were assayed for viability and cultured with [^{14}C]atrazine (purity, 97%) at a concentration of 1, 5, 10, 25 or 100 ppm for up to 24 h. Analysis of the metabolites was by thin-layer and cation-exchange chromatography. The different species appeared to share common pathways for the metabolism of atrazine, dealkylation of the ethyl- and isopropyl moities but displayed significant differences in the kinetics of phase I and phase II metabolism of atrazine. Human and CD1 mice hepatocytes showed a marked preference for rapid *N*-deethylation, while in rats and Rhesus monkeys *N*-deisopropylation predominanted over *N*-deethylation. Sprague-Dawley rat, Fischer rat, CD1 mouse and Rhesus monkey hepatocytes rapidly metabolized atrazine to the di-dealkylated metabolite DACT (G 28273) and then further metabolized this intermediate to different conjugates. For guinea-pig, goat and human hepatocytes, the primary products were the mono-dealkylated metabolites DEA (G 30033) and DIA (G 28279) with only traces of the bis-dealkylated metabolite DACT. The bis-dealkylated metabolite DACT accounted for 50–70% of total radiolabel in 24-h incubations with hepatocytes from CD1 mice, rats (both strains) and Rhesus monkeys, but only 3–10% in incubations with guinea-pig, goat and human hepatocyte cultures. The percentages of the different metabolic products of atrazine produced by hepatocytes from different species are presented in Table 1 (Thede, 1988).

2. Toxicological studies

2.1 Acute toxicity

(a) Lethal doses

In a study of acute oral toxicity, groups of five male and five female Tif:RAI rats received atrazine as a single dose at 600–6000 mg/kg bw suspended in 2% carboxymethyl cellulose Clinical symptoms occurred within 2 h after treatment in all groups and included sedation, dyspnoea, exophthalmus, curved body position and ruffled fur. The surviving animals recovered between days 7 and 8 after treatment. The median lethal dose (LD_{50}) was 1869 mg/kg bw (Sachsse & Bathe, 1975b).

Table 1. Metabolism of atrazine by hepatocytes from different species in vitro

Metabolite	Species						
	Sprague-Dawley rat	F-344 rat	Mouse	Rhesus monkey	Goat	Guinea-pig	Human
Atrazine (%)	ND	ND	ND	2	60	4	19
DIA (%)	14	12	3	25	18	30	18
DEA (%)	5	6	6	10	20	32	50
DACT (%)	50	54	60	50	9%	10	< 5
Ratio of DEA to DIA	1 : 3	1 : 2	2 : 1	1 : 2.5	1 : 1	1 : 1	3 : 1
Conjugated metabolites (%)	34	29	30	9	< 1	25	11

From Thede (1988)
DACT, diaminochlorotriazine; DEA, deethyl-atrazine; DIA, deisopropyl-atrazine; ND, not detected.

In a study of acute oral toxicity conducted in compliance with GLP and US EPA test guidelines, groups of five male and five female HSD:(SD) rats received atrazine as a single dose at 4000–5500 (males) or 2000–5500 (females) mg/kg bw suspended in 2% carboxymethyl cellulose. Clinical symptoms included hypoactivity, ataxia, emaciation, lacrimation, nasal discharge, piloerection, polyuria and salivation. The surviving rats recovered within 3–6 days. No organ-specific lesions were discernible on macroscopic examination of the animals. Discoloration of the gastrointestinal tract in the rat may be related to local irritation. The male, female and combined oral LD_{50} and the 95% confidence limits (CL) were 3520 mg/kg bw (2290–5400), 3000 mg/kg bw (2090–4300) and 3090 mg/kg bw (2170–4420), respectively (Kuhn, 1991a).

In a study of acute dermal toxicity, groups of five male and five female Tif:RAIf rats received atrazine as a single application at a dose of 2150 or 3170 mg/kg bw suspended in 2% carboxymethyl cellulose on the shaved back skin. No mortality occurred during the 14-day observation period. There were no treatment-related clinical signs or gross changes in the organs. The dermal LD_{50} was > 3100 mg/kg bw in males and females (Sachsse & Bathe, 1976).

In a study of acute dermal toxicity conducted in compliance with GLP and OECD test guidelines, one group of five male and five female Tif:RAIf rats received atrazine as a single application at a dose of 2000 mg/kg bw moistened with 0.5% carboxymethyl cellulose on the shaved skin for 24 h. No mortality occurred. Clinical symptoms including piloerection and hunched body position were seen on the first 5 days of application. No organ-specific lesions were discernible on macroscopic examination of the animals. The dermal LD_{50} was > 2000 mg/kg bw in males and females (Hartmann, 1993).

In a study of acute toxicity after exposure by inhalation that was conducted in compliance with GLP and OECD test guidelines, one group of five male and five female Tif:RAIf rats was exposed to atrazine at an average air concentration of 5148 mg/m^3 for 4 h. No mortality occurred. Rats showed piloerection, hunched posture, dyspnea, and reduced locomotor activity. All animals recovered within 5 days. No organ abnormalities were recorded. The median lethal concentration (LC_{50}) was > 5148 mg/m^3 (5.15 mg/l) (Hartmann, 1989).

In a study of acute toxicity after exposure by inhalation that was conducted in compliance with GLP and US EPA test guidelines, one group of five male and five female Sprague-Dawley rats was

Figure 1. Proposed metabolic pathway for atrazine in the rat

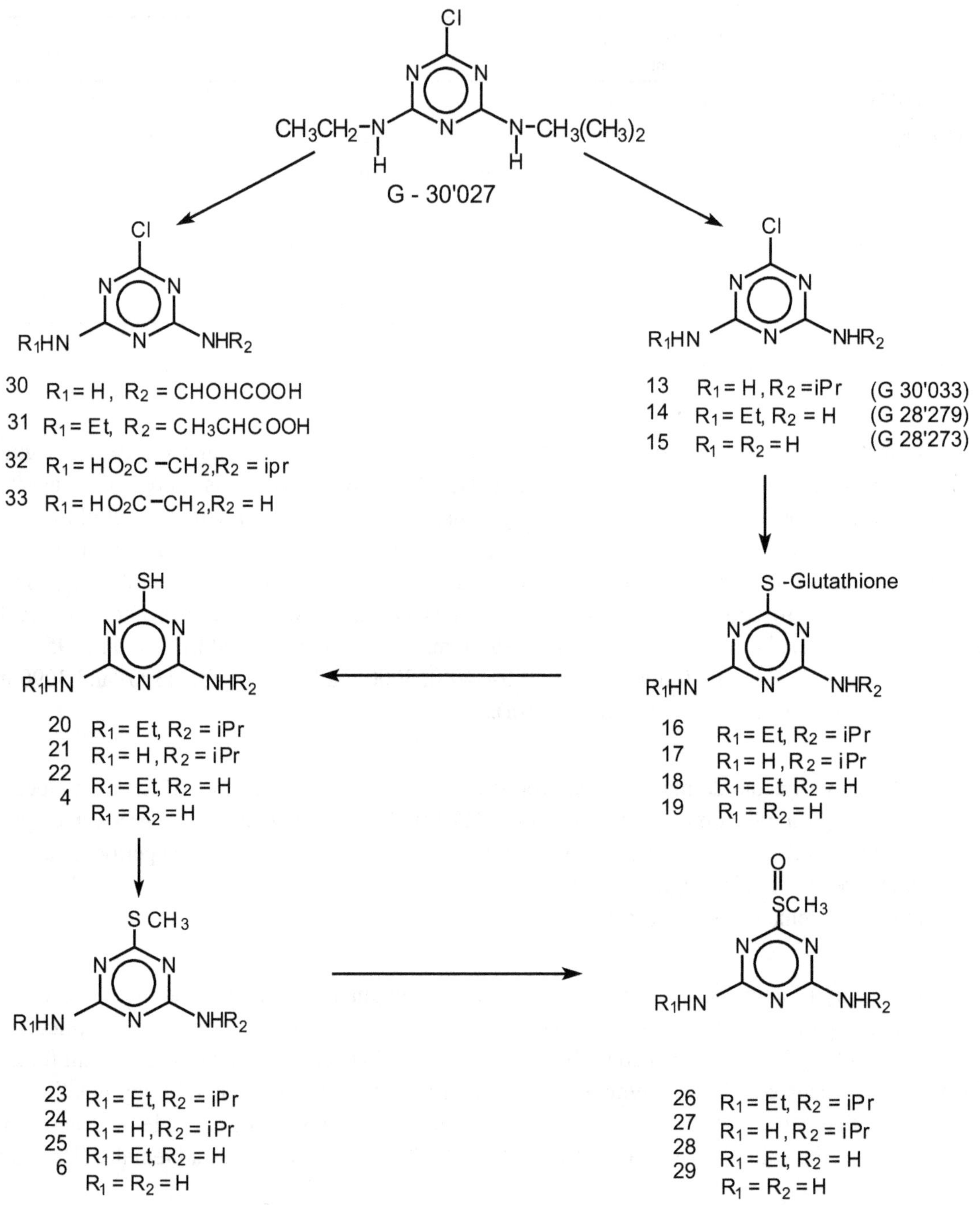

Cl
N N
CH_3CH_2-N N N-$CH_3(CH_3)_2$
H H
G - 30'027

Cl
N N
R_1HN N NHR_2

30 R_1 = H, R_2 = CHOHCOOH
31 R_1 = Et, R_2 = $CH_3CHCOOH$
32 R_1 = HO_2C-CH_2, R_2 = ipr
33 R_1 = HO_2C-CH_2, R_2 = H

Cl
N N
R_1HN N NHR_2

13 R_1 = H, R_2 = iPr (G 30'033)
14 R_1 = Et, R_2 = H (G 28'279)
15 R_1 = R_2 = H (G 28'273)

S -Glutathione
N N
R_1HN N NHR_2

16 R_1 = Et, R_2 = iPr
17 R_1 = H, R_2 = iPr
18 R_1 = Et, R_2 = H
19 R_1 = R_2 = H

SH
N N
R_1HN N NHR_2

20 R_1 = Et, R_2 = iPr
21 R_1 = H, R_2 = iPr
22 R_1 = Et, R_2 = H
4 R_1 = R_2 = H

S CH_3
N N
R_1HN N NHR_2

23 R_1 = Et, R_2 = iPr
24 R_1 = H, R_2 = iPr
25 R_1 = Et, R_2 = H
6 R_1 = R_2 = H

O
SCH_3
N N
R_1HN N NHR_2

26 R_1 = Et, R_2 = iPr
27 R_1 = H, R_2 = iPr
28 R_1 = Et, R_2 = H
29 R_1 = R_2 = H

exposed to atrazine at an average air concentration of 5820 mg/m^3 for 4 h. No mortality occurred. Rats showed reduced activity, lacrimation, nasal discharge, piloerection, polyuria, ptosis and salivation. No organ abnormalities were recorded. The LC_{50} was > 5820 mg/m^3 (5.82 mg/l) (Holbert, 1991).

In a study of acute toxicity after intraperitoneal administration, groups of five male and five female Tif:RAI rats received atrazine at doses of 147–1000 mg/kg bw as a suspension in a 2% solution of carboxymethyl cellulose by intraperitoneal injection. Within 2 h after treatment all rats showed sedation, dyspnea, exophthalmus, curved body position and ruffled fur, and at doses of 600 mg/kg bw

Figure 2. Proposed metabolic pathway for atrazine in primates and humans

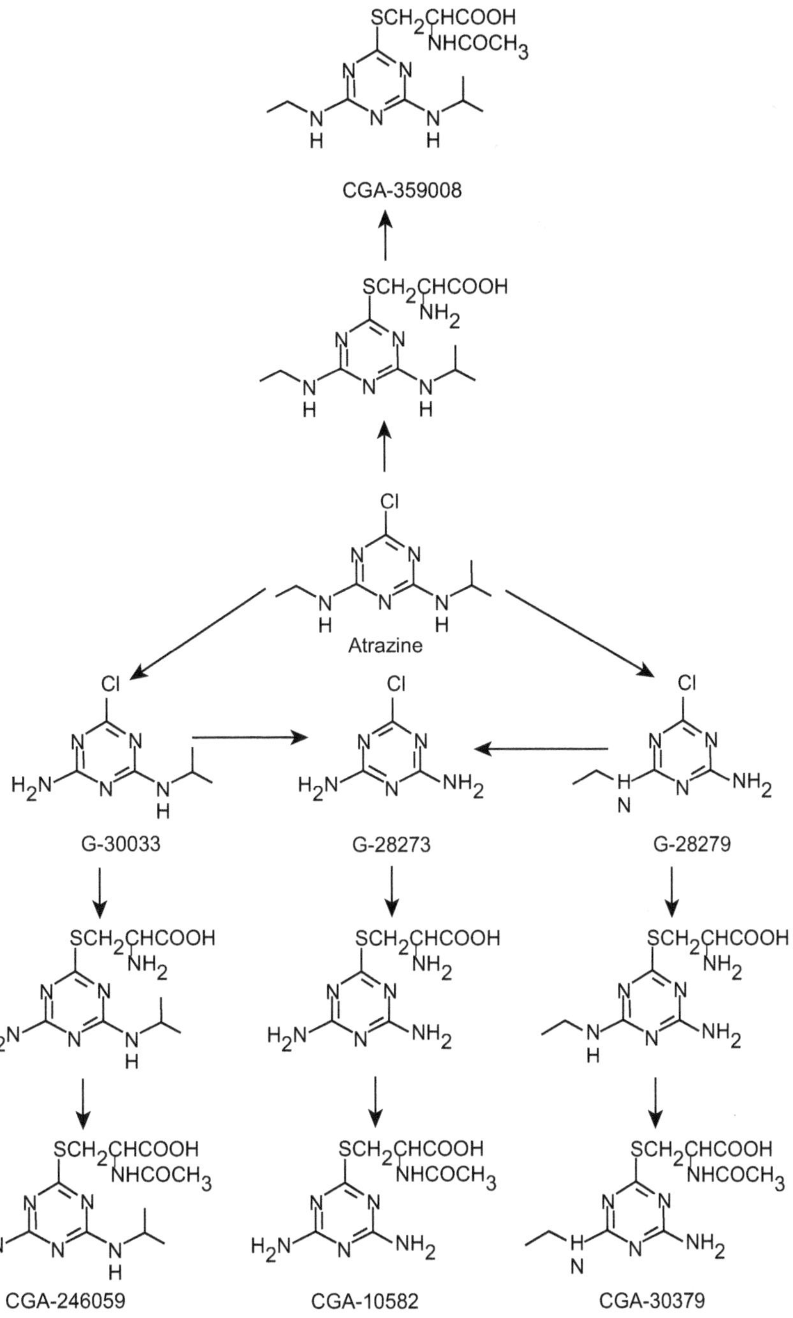
SCH2CHCOOH
NHCOCH3
CGA-359008
SCH2CHCOOH
NH2
Cl
Atrazine
G-30033
G-28273
G-28279
CGA-246059
CGA-10582
CGA-30379

and greater, also showed chromodacryorrhoea and salivation. No treatment-related gross changes in organs were seen at necropsy. The LD_{50} was 235 mg/kg bw (Sachsse and Bathe, 1975a).

(b) Dermal irritation

In a study on skin irritation potential, atrazine powder (0.5 g) was applied to intact and abraded skin of three male and three female Himalayan rabbits. The treated area was occluded for 24 h. A very slight erythema was observed in three rabbits (abraded) at 24 h after application and in one rabbit at 72 h. The study differed from present OECD test guidelines in that the test material was not moistened, the exposure period was for 244 h not 4 h; that abraded skin sites were used in addition

Figure 3. Primary metabolic pathway for hydroxyatrazine in the rat

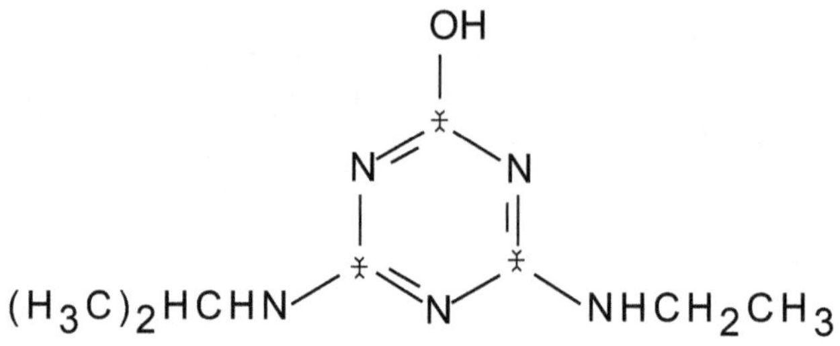

(^{14}C)-hydroxyatrazine (G-34048)

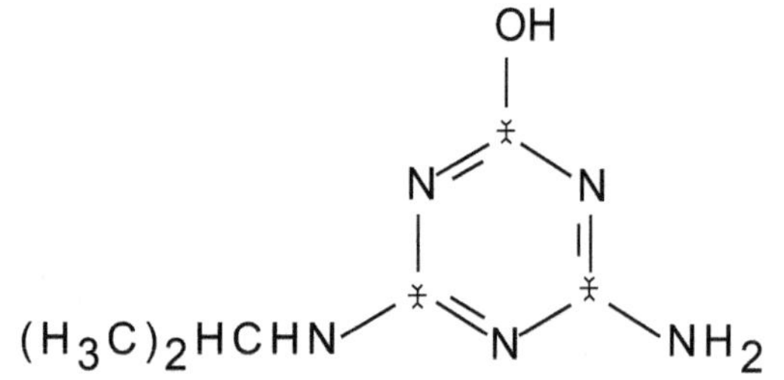

[^{14}C]-GS-17794

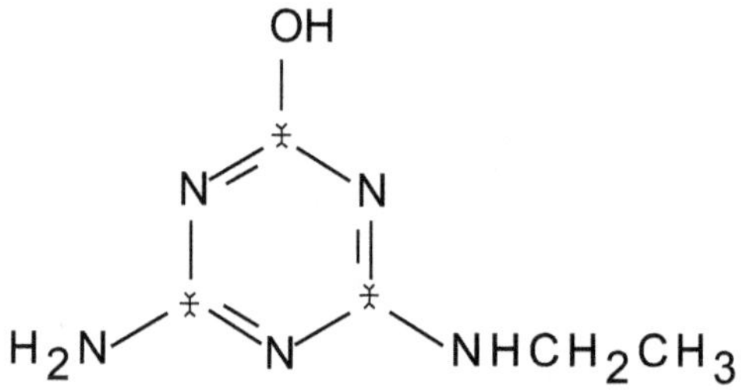

[^{14}C]-GS-17792

Table 2. Results of studies of acute toxicity with atrazine

Species	Strain	Sex	Route	Purity (%)	LD_{50} (mg/kg)	LC_{50} (mg/l air)	Reference
Rat	Tif:RAI	Males & females	Oral (in 2% CMC)	NR	1869	—	Sachsse & Bathe (1975b)
Rat	HSD:(SD)	Males & females	Oral (40% w/v in water)	97.7	3090	—	Kuhn (1991a)
Mouse	Tif:MAG	Males & females	Oral (in 2% CMC)	NR	3992	—	Sachsse & Bathe (1975c)
Mouse	HSD:(ICR)	Males & females	Oral (in 0.5% CMC)	97.7	> 1332	—	Kuhn (1988)
Rat	Tif:RAIf	Males & females	Dermal (in 2% CMC)	NR	> 3100	—	Sachsse & Bathe (1976)
Rat	Tif:RAIf	Males & females	Dermal (in 0.5% CMC)	97.1	> 2000	—	Hartmann (1993)
Rat	Tif:RAIf	Males & females	Inhalation	96.7	—	> 5.15	Hartmann (1989)
Rat	HSD:(SD)	Males & females	Inhalation	97.4	—	> 5.82	Holbert (1991)
Rat	Tif:RAI	Males & females	Intraperitoneal (in 2% CMC)	NR	235	—	Sachsse & Bathe (1975a)

CMC, carboxymethyl cellulose; NR, not reported.

to intact skin; and that no reading was made at 48 h after application. However, the study is adequate for the purpose intended. Very slight erythema was recorded on abraded skin of three animals and, therefore, atrazine is classified as a mild skin irritant under EPA guidelines but is not an irritant under OECD or EC criteria (Ullmann, 1976b).

(c) Ocular irritation

In a study on eye irritation potential, atrazine powder (0.1 g) was applied into the left eyes of three male and three female Himalayan rabbits, and after 30 s was rinsed out from three of these eyes. The eyes were closely examined for reaction at intervals after application. The study deviates from present OECD test guideline in that no recording was performed at 1 h. In view of the results, this does not detract from the acceptability of the study. Atrazine caused no reaction to the eye at 24, 48 or 72 h, and therefore is non-irritant to the eye (Ullmann, 1976a).

(d) Dermal sensitization

In a study on skin sensitization potential that was conducted in compliance with GLP and OECD test guidelines (optimization test), 10 male and 10 female Pirbright White (Tif:DHP) guinea-pigs received a total of 10 intradermal injections (each of 0.1 ml of 0.1% atrazine in 20% ethanol and 80% physiological saline) during the induction period, while the control group (10 males and 10 females) received the vehicle only. Two weeks later, a challenge dose of 0.1 ml of 0.1% atrazine in a mixture of 20% ethanol and 80% physiological saline was injected intradermally. Ten days later, a sub-irritant level dose of 30% atrazine in vaseline was applied on the skin under occlusive patches for 24 h. Twenty-four and 48 h after removal of the patches, 10 out of 19 animals showed skin reactions indicating sensitization potential in the test system (Maurer, 1983).

In a study on skin sensitization potential that was conducted in compliance with GLP and OECD test guidelines (Magnusson & Kligman test), 10 male and 10 female Pirbright White (Tif:DHP) guinea-pigs received three pairs of intradermal injections (adjuvant/saline 1 : 1, atrazine in oleum arachidis 1% mixture, and atrazine in adjuvant/saline mixture 1%). One week later, 0.4 g atrazine incorporated in vaseline (30%) was applied epidermally to the neck of the animals for 48 h. After a 2-week rest period (week 3 and 4) the guinea-pigs were challenged by application of 0.2 g of atrazine at a concentration of 30% incorporated in vaseline for 24 h. Twenty-four and 48 h after removal of the patches, 65% and 70% of the guinea-pigs showed skin reactions indicating sensitization potential in the test system (Schoch, 1985).

In a study on skin sensitization potential, repeated-insult patch tests were conducted on 50 (otherwise undescribed) humans given 0.5 ml of a 0.5% suspension of atrazine 80W formulation in water; the composition of the formulation was not given. Subjects received 15 consecutive 24-h, alternate-day treatments followed by a 14-day rest period before a challenge dose. None of the subjects reacted to any application or to the challenge. The tested spray dilution of atrazine did not appear to be sensitizing to humans (Shelanski & Gittes, 1965).

2.2 Short-term studies of toxicity

Rats

In a study of oral toxicity, groups of 10 male and 10 female Tif:RAIf (Sprague-Dawley derived) rats were fed diets containing atrazine (purity, 97.1%) at a concentration of 10, 50 or 500 ppm, equal to 0, 0.6, 3.3 and 34.0 mg/kg bw per day in males and 0, 0.66, 3.35 and 35.3 mg/kg bw per day in females, for 13 weeks. An additional 10 males and 10 females were allocated to the groups at the

Table 3. Dermal irritation and sensitization potential of atrazine

Species	Strain	Sex	End-point (method)	Purity (%)	Result	Reference
Rabbit	Himalayan	Males & females	Skin irritation	NR	Not irritating	Ullmann (1976b)
Rabbit	Himalayan	Males & females	Eye irritation	NR	Not irritating	Ullmann (1976a)
Guinea-pig	Pirbright White (Tif:DHP)	Males & females	Skin sensitization (optimization test)	98.2	Sensitizing	Maurer (1983)
Guinea-pig	Pirbright White (Tif:DHP)	Males & females	Skin sensitization (Magnussen & Kligman test)	98	Sensitizing	Schoch (1985)
Human	—	NR	Repeated-insult patch test	Atrazine 80 W formulation	Not sensitizing	Shelanski & Gittes (1965)

NR, not reported.

highest dose and the control group to evaluate an additional 4-week recovery period. The study was conducted in compliance with GLP and US EPA and OECD test guidelines.

There were no major treatment-related clinical signs and no deaths. A statistically significant decrease in body-weight gain, compared with controls, was found in males (−15%) and in females (−9%) of the group at 500 ppm. A slight, non-statistically significant decrease (−8%) in body-weight gain was also reported in males at 50 ppm. Food consumption was decreased in males (−10%) and females (−5%) of the group at 500 ppm and for males at 50 ppm (−5%). No treatment-related changes were found for any haematological or clinical chemistry parameters at any dose. The mean absolute liver and kidney weights were lower in the males at 500 ppm and liver weight relative to body weight was higher in females at 500 ppm than in the controls. Histologically, deposition of haemosiderin pigment in the spleen was found at an increased incidence and severity at 500 ppm. In females, this finding was still present at a higher incidence after the recovery period.

The NOAEL was 50 ppm, equal to 3.3 mg/kg bw per day, on the basis of decreased body weight and increased splenic haemosiderosis at 500 ppm (Bachmann, 1994).

Dogs

In a study of oral toxicity in beagle dogs, groups of four males and four females (lowest and intermediate dose) or six males and six females (control and highest dose) were fed diets containing atrazine (purity, 97%) at a concentration of 0, 15, 150 or 1000 ppm, equal to 0, 0.5, 5 and 33.7 mg/kg bw per day, for 52 weeks. Fixed portions of 400 g of diet were offered during a 3-h period each day. Two males and two females in the control group and in the group at highest dose were designated to be placed in a recovery group. Investigations included clinical pathology, ophthalmoscopic, auditory and electrocardiographic examinations at 3-monthly intervals throughout the study. The study was conducted in compliance with GLP and US EPA and OECD test guidelines.

Two dogs at the highest dose (one male on day 250, one female on day 113) and one male at the intermediate dose (on day 75) were killed in a moribund condition during the course of the study. The two dogs found in a moribund condition at the highest dose may have been affected by treatment-related cardiac dysfunction since compound-related myocardial lesions were observed in these animals. Treatment-related symptoms were restricted to dogs at the highest dose. Clinical symptoms at 1000 ppm included cachexia, ascites and laboured and shallow breathing. The ascites was considered related to treatment-induced myocardial degeneration, as was the laboured breathing. Marked reduction in body-weight gain (males, 61%; and females, 55% of control values) was recorded in dogs at the highest dose, and food consumption was consistently lower in the group at the highest dose when compared with controls.

Electrocardiographic changes occurred in dogs at the highest dose and included slight to moderate increases in heart rate (primarily in males), moderate decreases in height of the P-wave, PR and QT values. Atrial premature complexes and atrial fibrillation were found in one female.

Haematology data indicated a slight but significant decrease in erythrocyte parameters in males and an increase in the number of platelets in males and females at 1000 ppm. Treatment-related changes in clinical chemistry were restricted to dogs at 1000 ppm and included a slight decrease in total plasma protein and albumin, considered to be secondary to the reduction in food consumption.

Necropsy revealed heart lesions in the group at the highest dose, these consisting of moderate to severe dilatation of right and/or left atria, and in some dogs, a fluid-filled pericardium and enlarged heart. In females at the highest dose, there was a treatment-related decrease in absolute heart weight. Histopathological findings were restricted to the heart in the group at the highest dose and included myolysis and atrophy of myocardial fibres, and oedema of the heart. These findings correlated with the clinical symptoms, the electrocardiogram results and with the gross pathology observations.

The NOAEL was 150 ppm, equal to 5 mg/kg bw per day, on the basis of decreased body-weight gain and cardiac effects at 1000 ppm (O'Connor et al., 1987).

Rabbits

In a short-term study of dermal toxicity, groups of five male and five female New Zealand White rabbits were given atrazine (purity, 97.6%) at a dose of 0, 10, 100 or 1000 mg/kg bw per day under semi-occluded conditions to the skin for 6 h per day for at least 25 consecutive days. The study was conducted in compliance with GLP and US EPA test guidelines.

At 1000 mg/kg bw per day, systemic toxicity (faecal changes, body-weight loss, decrease in food consumption) was recorded in males and females during treatment, while females at 100 mg/kg bw per day showed a transient decrease in body-weight gain during the first week of treatment. Also at 1000 mg/kg bw per day, changes in haematology and clinical chemistry parameters (slight depression of erythrocyte count and haemoglobin concentration with a minimal increase in the percentage of reticulocytes, decrease in total serum albumin and chloride values) and slight changes in absolute and relative organ weights were observed. Local effects in females at 1000 mg/kg bw per day included slight dermal irritation, an increased incidence of minimal to moderate acanthosis and focal subacute lymphocytic inflammation of the skin.

The NOAEL for systemic toxicity was 100 mg/kg bw per day on the basis of decreased body-weight gain and food intake, a slight reduction in erythrocyte parameters and increased spleen weight at 1000 mg/kg bw per day (Huber et al., 1989).

2.3 Long-term studies of toxicity and carcinogenicity

Mice

In an early study of carcinogenicity, atrazine (from commercial sources, purity not given) was administered to groups of 18 male and 18 female mice of each of the C57BL/6 × C3H/Anf and C57BL/6 × AKR strains (a total of 36 males and 36 females). At the beginning of the study, mice aged 7 days (pre-weanling) were given atrazine at a dose of 21.5 mg/kg bw per day by gavage in 0.5% gelatin, and were switched at weaning (age 4 weeks) to diet containing atrazine at 82 ppm (cited to be approximately equivalent to the dose given by gavage, but estimated in review to be approximately half this dose). Doses were not adjusted for body-weight gain, either when given by gavage or by dietary administration. The dose was selected as the highest not causing mortality in a pilot screening

Table 4. Results of studies of carcinogenicity with atrazine

Species	Strain	Sex	Dietary concentration (ppm)	Carcinogenic effect	Reference
Mouse	C57BL/6 x C3H/Anf and C57BL/6 x AKR	Males & females	21.5 mg/kg bw per day (days 7–28) thereafter 82 ppm	Negative	Innes et al. (1969)
Mouse	CD1	Males & females	0, 10, 300, 1000	Negative	Sumner (1981)
Mouse	CD1	Males & females	0, 10, 300, 1500, 3000	Negative	Hazelette & Green (1987)
Rat	Sprague-Dawley	Males & females	0, 10, 100, 1000	Males: negative Females: positive (increased incidence) at 10 and 1000 ppm	Spindler & Sumner (1981)
Rat	Sprague-Dawley	Males & females	0, 10, 70 500, 1000	Males: negative Female: positive (increased incidence) at 70–1000 ppm	Mayhew (1986)
Rat	Sprague-Dawley	Males & females	0, 10, 50, 500	Negative	Rudzki et al. (1991)
Rat	Sprague-Dawley	Females	0, 70, 400	Positive: (earlier onset) at 400 ppm No effect on incidence	Thakur (1991a)
Rat	Sprague-Dawley	Females	0, 70, 400	Positive: (earlier onset) at 400 ppm No effect on incidence	Thakur (1992a)
Rat	Sprague-Dawley	Females	0, 15, 30, 50, 70, 400	Positive: (increased incidence and earlier onset) at 400 ppm	Pettersen & Turnier (1995)
Rat	Sprague-Dawley	Females (intact)	0, 25, 50, 70, 400	Positive: (increased incidence) at 50 to 400 ppm, (earlier onset) at 400 ppm	Morseth (1998a)
Rat	Sprague-Dawley	Females (ovex)	0, 25, 50, 70, 400	Negative	Morseth (1998a)
Rat	Fischer 344	Males & females	375, 750	Inconclusive	Pintér et al. (1990)
Rat	Fischer 344	Females	0, 10, 70, 200, 400	Negative	Thakur (1991b)
Rat	Fischer 344	Males & females	0, 10, 70, 200, 400	Negative	Thakur (1992b)

procedure. A limited necropsy (cranial cavity not opened) and limited histopathology (including all macroscopic lesions) were conducted on all animals. This study was part of a screening procedure involving 120 chemicals and nearly 20 000 mice, and logistical considerations restricted the procedures conducted (including reporting). Inclusion of carcinogens acting as positive controls proved the responsiveness of the test procedure. Atrazine did not cause a significant increase in the incidence of tumours (Innes et al., 1969).

In a study of carcinogenicity, groups of 60 male and 60 female CD1 mice were given diets containing atrazine (purity, 96.4%) at a concentration of 0, 10, 300 or 1000 ppm for 21 months (males) or 22 months (females).The study data were audited by the sponsor and declared to be valid and useful for the purposes of evaluating carcinogenicity. However, various data were missing (e.g. body-weight gain for the first 2 months of the study, food consumption throughout the study) and some mice (evaluated as roughly equivalent numbers in each group) received treatment with 1% rotenone for a mite infestation. During the early months of the study, diets were mixed sufficiently infrequently that substantial degradation of atrazine (20–30%) may have occurred before feeding. In some tissues, substantial numbers of samples were lost to autolysis. There were other practices that did not meet modern guideline requirements. Histopathology was performed by an independent laboratory. Treatment with atrazine caused a significant reduction in the body weights of mice at 1000 ppm. Survival among males was not affected by treatment although survival was significantly lower in females at 1000 ppm. The cause of death of these female decedents does not appear to have been documented. However, eventual survival and duration of this study was adequate for the evaluation of carcinogenicity. Histopathology revealed no evidence of a treatment-related change in inflammatory, degenerative, proliferative or neoplastic lesions in treated mice. Atrazine was not carcinogenic in mice, in this study suitable to be used as supportive data.

The NOAEL was 300 ppm, equivalent to approximately 30 mg/kg bw per day, on the basis of reduced survival and reduced body weight at 1000 ppm (Sumner, 1981).

In a study of carcinogenicity performed according to US EPA guidelines and in compliance with GLP, groups of 60 male and 60 female CD1 mice were given diets containing atrazine technical (purity, 97.6%) at a concentration of 0, 10, 300, 1500, or 3000 ppm for 91 weeks. Mean daily intakes were equal to 0, 1.2, 38.4, 194.0 and 385.7 mg/kg bw per day in males and 0, 1.6, 47.9, 246.9 and 482.7 mg/kg bw per day in females. Blood samples for haematology were collected from all animals at necropsy, and blood smears prepared from all survivors at weeks 52 and 78. Survival was not affected by treatment in males (59%, 55%, 60%, 65% and 60% at 0, 10, 300, 1500 and 3000 ppm) while decreased survival was reported for the females at 3000 ppm (43%, 39%, 43%, 45% and 25% at 0, 10, 300, 1500 and 3000 ppm). No treatment-related clinical signs occurred during the study. Lower mean body weights and body-weight gains were recorded for males and females at 300, 1500 and 3000 ppm, which correlated with lower food consumption at 1500 and 3000 ppm. Haematology investigations revealed a slight reduction in mean haemoglobin concentration, erythrocyte volume fraction and in erythrocyte count in males at 1500 and in males and females at 3000 ppm. Analysis of organ weight data did not reveal variations with toxicological significance. A lower brain weight in males and females at 3000 ppm was not accompanied by any histopathological correlates. Histopathology revealed an increased incidence of cardiac thrombi primarily in the atria of decedents in males at 3000 ppm and in females at 1500 and 3000 ppm. No cardiac changes were apparent to account for this change. No evidence of compound-related neoplastic lesions was found in this study.

The NOAEL was 10 ppm, equal to 1.2 and 1.6 mg/kg bw per day in males and females, respectively, on the basis of lower body weight/body-weight gain at 300 ppm and greater (Hazelette & Green, 1987).

Rats

In a study of carcinogenicity conducted before GLP, groups of 60 male and 60 female Sprague-Dawley rats were fed diets containing atrazine technical (purity, 98.0%) at a concentration of 0, 10, 100 or 1000 ppm, equal to 0, 0.35, 4 and 40 mg/kg bw per day in males and 0, 0.45, 5.5 and 60 mg/kg bw per day in females, for up to 2 years. Accuracy of diet preparation was confirmed by analysis on only three occasions during the study. Clinical pathology was not investigated in rats receiving the intermediate or lowest dose.

Body-weight gain was reduced significantly at 1000 ppm throughout the study reaching 22% and 27% lower than that of the controls for males and females, respectively. Weight gain for the rats at 100 ppm was reduced to a lesser extent (about 10% lower than that of controls). A decrease in erythroid parameters was reported at the highest dose, and no measurements were performed at the lower doses. Clinical chemistry data showed slight variations in the serum transaminases and cholesterol of females at the highest dose at different occasions. Gross pathology investigations revealed a high incidence of enlarged pituitaries and of dermal/subdermal masses in females from all groups including controls. Histopathology of mammary tissues showed a statistically significant, but not dose-related increase in the incidence of benign fibroadenomas in females at the lowest (10 ppm) and highest (1000 ppm) dose, but not at the intermediate dose of 100 ppm (incidences were: 11 out of 54, 20 out of 52, 14 out of 54, and 22 out of 49 at 0, 10, 100 and 1000 ppm, respectively). Atrazine gave positive results for carcinogenicity in this study, which is suitable to be used for supportive data only. Because of data deficiencies (parameters not investigated in groups at the intermediate dose), a NOAEL was not identified in this study (Spindler & Sumner, 1981).

In a combined long-term study of toxicity and carcinogenicity conducted in compliance with GLP and EPA guidelines, groups of 70 male and 70 female Sprague-Dawley rats were fed diets containing atrazine (purity, 95.8%) at a concentration of 0, 10, 70, 500 or 1000 ppm, equal to 0, 0.4, 2.6, 19.9 and 41.7 mg/kg bw per day in males and 0, 0.5, 3.5, 29.5 and 64.7 mg/kg bw per day in females, for 2 years. Dietary exposure was verified by analysis. The control group and the group at the highest dose contained an additional 20 males and 20 females for interim kill, 10 males and 10 females after 12 months and 10 males and 10 females after a further 1-month recovery period. For the long-term evaluation of toxicity, 20 males and 20 females per group were used, and the study design contained appropriate end-points to satisfy guideline requirements.

The males at the highest dose (1000 ppm) showed an increased survival (67% vs 44% in controls), while a significantly decreased survival was recorded in the females at 1000 ppm when compared with the controls (26% in the treated females vs 50% in the controls) at the study termination.

Increased irritability was seen in the males, and pallor in females, at 500 or 1000 ppm. An increase in numbers of palpable masses was apparent in the females at 70, 500, and 1000 ppm. Body-weight gains of rats at 500 ppm (weight gain after 12 months, −10% in males; −23% in females) and 1000 ppm (weight gain after 12 months, −22% in males; −34% in females) were significantly decreased throughout the study. On the basis of these weight-gain retardations, doses of 500 and 1000 ppm were considered to be excessive for carcinogenicity testing. Food consumption was slightly decreased. Haematology investigations revealed lower values for erythrocyte parameters in females at 1000 ppm. Sporadic changes in some clinical chemistry parameters, namely lower glucose and transient decrease in triglyceride concentration, were recorded in rats at 1000 ppm. The number of females with mammary gland tumours was increased in groups at 70, 500 and 1000 ppm, confirmed by histology peer review (Table 5). Time-to-tumour analyses are not reported.

Also seen was an increase in mammary acinar hyperplasia in males at the highest dose. Other non-neoplastic lesions increased with treatment were confined to the highest dose and included renal pelvic calculi, and prostate epithelial hyperplasia, in males; and urothelial hyperplasia in kidney and urinary bladder, degeneration of the rectus femoris muscle, and hepatic centrilobular necrosis in females. An increase in testicular interstitial cell tumours in males at 1000 ppm was not considered to be treatment-related, since the increased incidence was within the range of spontaneous occurrence and could also be attributed in part to the increased survival of the rats at 1000 ppm.

The 12- and 13-month interim kills were not considered for identification of the NOAEL, since these comprised 20 rats in the control group at low risk of tumours and would dilute the number of tumours in the controls relative to the group at the lowest dose, which had no scheduled interim kill. It is noted that at 10 ppm the incidence of mammary tumours was also higher than that in controls (63% vs 53%, excluding the interim kill), although not achieving statistical significance. This value was outside the cited range for historical controls of 40–51%. However, the range of values for historical controls (on the basis of four studies only) may be considered to be unusually narrow, a 40–60% range being considered to be more representative.

Table 5. Incidence of mammary tumours in female rats at terminal kill

Mammary tumours	Dietary concentration (ppm)				
	0	10	70	500	1000
12- & 13-month kill, including early decedents					
No. of animals examined	22	5	1	5	25
Adenocarcinoma	0	1	1	0	8 (32%)
All mammary tumours	0	1	1	1	9 (36%)
Terminal kill (including deaths occuring between 12 and 24 months)					
No. of animals examined	66	64	68	65	64
Adenocarcinoma	15 (23%)	15 (23%)	26* (38%)	27* (42%)	35** (55%)
Carcinosarcoma	0	0	0	0	2 (3.1%)
Fibroadenoma	29 (44%)	29 (45%)	35 (51%)	38 (58%)	42** (66%)
Adenoma	1 (1.5%)	0	1 (1.5%)	1 (1.5%)	2 (3.1%)
Malignant + benign tumours	35 (53%)	39 (63%)	47 (69%)	47* (72%)	56** (89%)
All rats combined (original report)					
No. of animals examined	88	69	69	70	89
Benign tumours	29 (33%)	29 (42%)	36 (52%)	39 (56%)	46** (52%)
Malignant tumours	15 (17%)	16 (23%)	27* (39%)	27** (39%)	45** (51%)
Malignant + benign tumours	35 (40%)	40 (58%)	48 (70%)	48** (69%)	65** (73%)
All rats combined (peer review)[a]					
No. of animals examined	88	69	69	70	89
Benign tumours	31 (35%)	35 (51%)	38 (55%)	62 (89%)	58 (65%)
Malignant tumours	15 (17%)	16 (23%)	32 (46%)	35 (50%)	56 (63%)

From Mayhew (1986)
* $p < 0.05$; ** $p < 0.01$.
[a] No statistical analysis reported.

The NOAEL was 70 ppm, equal to 2.6 mg/kg bw per day, in males, on the basis of decreased body-weight gain at 500 ppm and greater, and 10 ppm, equal to 0.5 mg/kg bw per day, in females, on the basis of an increased incidence of mammary tumours in females at doses of 70 ppm and greater (Mayhew, 1986).

In a long-term study of toxicity conducted in compliance with GLP and EPA guidelines, groups of 50 male and 50 female Sprague-Dawley rats were fed diets containing atrazine (purity, 97.6%) at a concentration of 0, 10, 50 and 500 ppm, equal to 0, 0.5, 2.3, and 23.6 mg/kg bw per day in males and 0.7, 3.5 and 37.6 mg/kg bw per day in females, for 52 weeks (males) or 104 weeks (females) commencing in utero. The rats were male and female offspring culled from the $F1_a$ generation of a previously conducted two-generation study of reproductive toxicity. Of the 50 males and 50 females per group, 10 males and 10 females were killed after 8 weeks and a further 10 females at 35 weeks. Of the remaining 30 females in the group at the highest dose, 10 were placed on withdrawal after 65 weeks, and 10 females in the control group were placed on diet containing atrazine at the highest dose, 500 ppm. Dietary exposure was verified by regular analysis of diets. Examinations conducted included ophthalmoscopy, blood and urine analyses, and estrous smears. Sections of pituitary were stained immunochemically for prolactin, follicle-stimulating hormone (FSH) and luteinizing hormone (LH).

No treatment-related effect on survival occurred. Body-weight gain and food consumption were reduced at 500 ppm. Reductions in values for erythrocyte parameters late in the study and a persistent increase in serum cholesterol concentration occurred in females at 500 ppm. Histopathology investigations revealed an increased incidence of pituitary gland adenomas in females at 500 ppm (85% vs 45% in the controls) at termination (the control value appeared to be low in view of data for historical controls, presented in the report, of 80–89%). The incidence of pituitary adenoma was 89% in the controls placed on diet containing atrazine at 500 ppm at week 65, and 75% in the females at 500 ppm placed on control diet at week 65. All these incidences are reported among small numbers of animals and the significance is therefore obscure. Controls and females at the highest dose killed after 104 weeks had similar incidences of mammary tumours. A higher incidence of mammary tumours occurred in females from the control group that were placed on diet containing atrazine at 500 ppm at week 65 than occurred in rats exposed in utero or in the females at the highest dose placed on control diet at week 65. However, only 10 females per group were used in this part of the study. There was no obvious effect of treatment on the proportions of prolactin-, LH- or FSH-reactive cells in the pituitary.

Atrazine did not appear to be carcinogenic in female Sprague-Dawley rats in this study of exposure in utero that was of supportive value (because of low numbers of rats at termination). The NOAEL was 50 ppm, equal to 2.3 mg/kg bw per day in males and 3.5 mg/kg bw per day in females, on the basis of body weight and haematological effects at 500 ppm (Rudzki et al., 1991).

In a supplementary study of carcinogenicity, designed to investigate the effect of atrazine on the estrous cycle, selected hormone levels and the development of mammary, ovary, uterine and pituitary tumours, groups of 70 female Sprague-Dawley rats [Crl:CD BR] were fed diets containing atrazine (purity, 97.0%) at a concentration of 0, 70 and 400 ppm, equivalent to 0, 3.5 and 20 mg/kg bw per day, for 104 weeks. Interim terminations of 10 females per group were conducted at 1, 3, 9, 12, 15 and 18 months of the study, and all remaining survivors killed after 2 years. The terminations were conducted at proestrus as determined by vaginal smear, unless this phase was not reached within 7 days of the scheduled termination time-point. In order to avoid stress-related hormone changes, rats were killed by decapitation without anaesthesia. Samples of blood were collected post mortem for hormone analysis. The pituitary, mammary glands, uterus and ovaries from all rats were examined microscopically. The study was conducted in compliance with GLP and OECD test guidelines.

Statistical analysis indicated a significant trend toward early death at 400 ppm, since five deaths in the group at this dose occurred before the earliest death in the control group. At termination, however, there were no treatment-related differences in mortality. A decrease in body-weight gain associated with decreased food consumption was observed at 400 ppm. At necropsy, the incidences of fluid-filled uteri at 70 and 400 ppm and of mammary galactocoeles at 400 ppm were increased, particularly during the first 12 months of the study (Table 6).

At 400 ppm, the onset-time for initial palpation of masses in the mammary region (confirmed histologically as mammary fibroadenomas and/or carcinomas) was decreased when compared with the control group. Because of early onset-time, there was a significant positive trend in the incidence of mammary fibroadenomas or carcinomas analysed either separately or combined, and a significant increase using pairwise comparison for onset rate of fibroadenomas and carcinomas combined in the group at 400 ppm. The incidence of pituitary adenomas was increased at 400 ppm after 12 months of treatment, but not at the end of the study, indicating that the onset-time was shortened. No correlation between rats with mammary tumours and those with pituitary tumours was apparent.

Table 6. Selected findings from a study in rats fed diets containing atrazine for 104 weeks

Finding	Dietary concentration (ppm)		
	0	70	400
Unscheduled deaths, weeks 1–55 (No. of rats)	0	0	5
Unscheduled deaths, weeks 1–104 (No. of rats)	5	6	8
Palpable mammary masses,[a] weeks 1–54 (No. of rats)	6	3	14
Palpable mammary masses,[a] weeks 55–104 (No. of rats)	10	13	8
Fluid-filled uterus, weeks 1–54 (No. of rats/No. examined histopathologically)	4/40	10/40	14/40
Fluid-filled uterus, weeks 1–104 (No. of rats/No. examined histopathologically)	5/70	15/70	15/70
Mammary galactocoele, weeks 1–54 (No. of rats/No. examined histopathologically)	6/39	9/38	20/41
Mammary galactocoele, weeks 1–104 (No. of rats/No. examined histopathologically)	29/69	30/67	41/69
Mammary fibroadenoma, weeks 1–54 (No. of rats/No. examined histopathologically)	1/39	0/38	4/41
Mammary carcinoma, weeks 1–54 (No. of rats/No. examined histopathologically)	0/39	1/38	6/41
Mammary fibroadenoma, weeks 1–104 (No. of animals)	8/69	12/67	13/69
Mammary carcinoma, weeks 1–104 (No. of rats/No. examined histopathologically)	9/69	4/67	11/69
Pituitary adenoma, weeks 1–54 (No. of rats/No. examined histopathologically)	2/39	2/40	8/43
Pituitary adenoma, weeks 1–104 (No. of rats/No. examined histopathologically)	22/68	16/69	20/69

From Thakur (1991a)
[a] Histologically confirmed as tumours.

The NOAEL was 70 ppm, equivalent to 3.5 mg/kg bw per day, on the basis of increased mortality, decreased body-weight gain and an earlier onset of mammary and pituitary tumours at 400 ppm (Thakur, 1991a).

From the rats in the supplementary study of carcinogenicity described above (Thakur, 1991a), data on estrous cycles, vaginal cytology and selected serum hormone levels were also collected from rats killed at 1, 3, 9, 12, 15, 18 and 24 months. Results were evaluated as a function of treatment and as a function of treatment over time.

Normal age-related changes noted in rats in the control group included an increase in the number of days when high density cornified cells were evident in vaginal smears during 12 months, with a decrease in this index between 12 and 18 months. Beginning at approximately age 1 year, episodes of persistent vaginal estrus occurred, resulting in an increase in percentage of total days in estrus at the expense of days in diestrus. Serum estradiol (E2) concentrations increased between 1 and 9 months of the study, followed by decreased levels between months 9 and 18.

Atrazine-treated rats had a dose-related increase in percentage days in estrus at each interval between 9 and 18 months. Compared with rats in the control group, the cornified cell index in treated rats showed a similar increase to that in controls until 12 months, but a slower decline between 12 and 18 months. Episodes of persistent vaginal estrus were observed earlier in treated groups. Although serum concentrations of E2 increased in the control group and in the groups treated with atrazine, the magnitude of the increase was greater than that in the control group between 1 and 9 months in the group at 400 ppm. Overall, treatment with atrazine accelerated the development of typical age-related changes in female Sprague-Dawley rats. As a result, rats treated with atrazine were exposed to persistent endogenous estrogens for longer periods of time (Eldridge et al., 1993a).

From the rats in the supplementary study of carcinogenicity described above (Thakur, 1991a), the ovaries, uterus, vagina, mammary gland and pituitary were re-evaluated microscopically for specific indications of reproductive senescence, which might relate to the onset-time of hormonally-mediated mammary tumours.

The histomorphological evaluation of the reproductive organs and mammary and pituitary glands revealed no evidence of any exogenous estrogenic effect associated with atrazine. This observation was on the basis of the absence of histomorphological changes which occur with administration of exogenous estrogen. Vaginal epithelial morphology in treated and control groups was similar. There was no increased mitotic activity in the basal layer, no significant increase in cornification (an estrogenic response), no suppression of mucification, and no increase in exaggerated rete pegs in treated animals when compared with that in controls. The lack of exogenous estrogenicity was also evident in senescent rats that had progressed into the phase of extended diestrus (pseudopregnancy) when increased endogenous progesterone activity is expressed. E2 directly antagonizes progesterone and if exogenous estrogenicity associated with atrazine had been present, extended diestrus would have been suppressed or inhibited.

Ovarian histomorphology showed that anovulatory cycles (ovaries without corpora lutea) were slightly increased at 400 ppm at 3 months and were present in all rats at 9 months. A similar trend in anovulatory cycles was present in rats in the control group and at 70 ppm but was less frequent. Thus, the period between 3 and 9 months was critical in the development of irregular estrous cycles and development of the first stages of reproductive senescence. Morphological alterations in ovarian follicular and corpora lutea development and interstitial clear cells of rats at 400 ppm indicated a treatment-related interference of the LH surge and prolonged exposure of LH-responsive cells in the interstitial gland to endogenous LH. These changes, exacerbated in rats at 400 ppm between 9 and 12 months, suggested that atrazine had modulated LH secretion.

Changes in the mammary glands, characterized by increased acinar/lobular development, secretory activity with duct ectasia, and galactocoele formation, occurred as early as 3 months and were more frequent and severe at 400 ppm than at 70 ppm or in the control group for 1 year. After 1 year, the mammary changes were balanced in all groups. The mammary-gland changes in rats treated with atrazine were similar in type and degree to those observed during early and later phases of reproductive senescence in rats in the control group. They have been shown to occur by imbalances of endogenous estrogen, progesterone and prolactin.

The Meeting concluded that atrazine induced the earlier appearance of reproductive senescence in Sprage-Dawley rats by gradually interfering with ovulation and causing estrous cycles characterized by extended periods of proestrus/estrus. Although changes characteristic of reproductive senescence occurred earlier in rats at 400 ppm, the stages of senescence tended to equalize with those of the rats at 70 ppm and in the control group after 1 year of treatment (McConnell, 1995).

In a supplementary study of carcinogenicity designed to evaluate the oncogenic potential of atrazine in the ovaries, pituitary, uterus and mammary gland, groups of 60 female Sprague-Dawley rats were fed diets containing atrazine (purity, 97%) at a concentration of 0, 70 and 400 ppm, equivalent to 0, 3.5 and 20 mg/kg bw per day, for 104 weeks. At termination, all surviving rats were killed and uterus, ovary, pituitary, and mammary glands from all rats were examined microscopically for oncogenic effects. The study was conducted in compliance with GLP guidelines.

No treatment-related increases in clinical signs were noted in the study. A slight reduction in survival was found in the group at the highest dose. Body-weight gains were statistically significantly reduced relative to controls (12–13%) at 400 ppm during weeks 0–76. Non-neoplastic lesion findings were comparable in controls and treatment groups. At 400 ppm, the onset-time for palpable masses in the mammary region (confirmed histologically as mammary fibroadenomas and/or carcinomas) was decreased when compared with the control group, although no statistically significant increase in mammary tumours was observed at the end of the study. The incidences of pituitary tumours did not indicate any larger numbers or earlier onset in the treated groups compared with the controls (Table 7).

The NOAEL was 70 ppm, equivalent to 3.5 mg/kg bw per day, on the basis of increased mortality, decreased body-weight gain and an earlier onset of mammary tumours at 400 ppm (Thakur, 1992a).

In a supplementary long-term study of toxicity conducted to determine the effects of atrazine on the mammary and pituitary glands, the estrous cycle, and plasma levels of E2, LH, progesterone and prolactin, groups of 55 female CD (Sprague-Dawley derived) rats were fed diets containing atrazine (purity, 97.1%) at a concentration of 0, 15, 30, 50, 70 and 400 ppm, equal to 0, 0.8, 1.7, 2.8, 4.1 and 23.9 mg/kg bw per day, for up to 12 months. The study was conducted in compliance with GLP and EPA guidelines. For estrous cycle determinations and plasma hormone concentration analysis, interim kills of 10 rats per group were conducted after 3, 6 and 9 months of treatment. Fifteen rats per group were sampled continuously throughout the study, i.e. at 3, 6, 9 and 12 months and subsequently killed for pathology evaluation at 12 months. Survivors of the remaining 10 females per group were killed after 12 months, following estrous cycle determinations and blood sample collection.

Table 7. Summary of selected findings of a study of carcinogenicity in rats fed diets containing atrazine for 104 weeks

Finding	Dietary concentration (ppm)		
	0	70	400
Mortality (No. of rats)	29	35	38*
Palpable mammary masses,[a] weeks 1–52 (No. of rats)	2	3	9*
Palpable mammary masses,[a] weeks 53–104 (No. of rats)	44	31	40
Mammary tumours (No. of rats/No. examined histopathologically)	46/60	34/59	49/60
Pituitary tumours (No. of rats/No. examined histopathologically)	44/58	46/58	47/60
Pituitary and mammary tumours (No. of animals)	34	26	39

From Thakur (1992a)
[a] Histologically confirmed as tumours.
* $p < 0.05$.

Table 8. Incidence of mammary gland tumours in groups of 55 rats fed diets containing atrazine for 12 months

Mammary gland tumour	Dietary concentration (ppm)					
	0	15	30	50	70	400
Adenocarcinoma	1	2	0	1	1	6
Adenoma	0	0	1	0	1	1
Fibroadenoma	2	2	2	1	4	4
Mammary tumours, combined	3	4	3	2	6	10*

From Pettersen & Turnier (1995)
* $p < 0.05$.

There were no effects on clinical signs, survival or organ weights. A slight retardation of body-weight gain (82–89% of controls) and feed consumption (88–95% of controls) were seen at the highest dose. The results of plasma hormone analyses were not reported. There were no statistically significant differences in the incidence of mammary gland adenomas, fibroadenomas or adenocarcinomas at up to 400 ppm. When the incidences of mammary adenomas, fibroadenomas and adenocarcinomas were combined for analysis, a statistically significant increase in the incidence of rats with mammary tumors was observed at 400 ppm (Table 8). There were no significant trends or group differences for onset-time of mammary gland adenomas, fibroadenomas or adenocarcinomas. However, a significant positive trend for onset-time existed when mammary gland adenomas and fibroadenomas were combined or when adenomas, fibroadenomas and adenocarcinomas were combined. No trend was evident when the group at the highest dose (400 ppm) was excluded from the analysis.

The NOAEL for incidence of mammary tumours and onset-time was 70 ppm, equal to 4.1 mg/kg bw per day (Pettersen & Turnier, 1995).

In a supplementary study of carcinogenicity designed to determine the incidence and onset of mammary tumours, groups of 80 ovariectomized or intact female Sprague-Dawley Crl:CD BR rats were fed diets containing atrazine (purity, 97.1%) at a concentration of 0, 25, 50, 70 or 400 ppm, equal to 0, 1.2, 2.5, 3.5 and 20.9 mg/kg bw per day in ovariectomized females and 0, 1.5, 3.1, 4.2 and 24.4 mg/kg bw per day in intact females, respectively. After 52 weeks of treatment, 20 rats in each group were killed and necropsied, and the surviving rats were killed after 104 weeks of treatment. Daily examinations for mortality, morbidity, and indications of toxic effects were performed. There were no clinical pathology investigations. Palpation for tissue masses in rats was performed before the initiation of dosing and weekly thereafter. Necropsies and complete histological examinations were performed on all rats. The study was conducted in compliance with GLP guidelines.

Survival and body-weight gain of ovariectomized rats was markedly better than that of intact rats, and there was no dose-related effect on survival among ovariectomized females. However, there was significantly impaired survival among intact females at the highest dose. Body-weight gain at the highest dose was impaired in intact and in ovariectomized rats, demonstrating a maximum tolerated dose (MTD) to be achieved.

In the ovariectomized rats, there were no treatment-related increases in mammary-gland proliferative changes and mammary tumours were not present in any of the rats. The lack of mammary tumours in ovariectomized rats provides evidence that the mode of action of atrazine is neither a direct genotoxic nor an estrogenic effect on the mammary gland. Rather, an indirect hormonally-mediated effect involving the ovary is implied.

In sexually intact rats, the incidence of palpable masses and of histological mammary neoplasia was higher than in ovariectomized females, and a statistically significant increased incidence of mammary neoplasia was seen at doses of 50 ppm and greater compared with the concurrent control group. When compared with the data for historical controls from the same laboratory (on the basis of 14 studies), only in the group at 400 ppm did incidences exceed the pooled incidence values for either fibroadenoma or carcinoma. An earlier onset and increased incidence of mammary carcinoma were observed only in the intact rats at 400 ppm (Table 9). The study authors therefore concluded that increases in mammary tumours at 50 and 70 ppm did not represent part of a carcinogenic dose–response trend.

However, the comparison with data for historical controls considers carcinoma and fibroadenoma in isolation but not as a combined analysis (i.e. all rats bearing mammary tumours). Also, the spacing of doses between the two intermediate doses (50 and 70 ppm) is small in terms of normal design for studies of carcinogenicity, and is possibly too small to prudently allow for dose–response differentiation. Were the groups at 50 and 70 ppm to be combined, this would result in a statistically significantly increased incidence of tumours, which would be part of a clear trend.

Table 9. Selected findings of a study of carcinogenicity in rats fed diets containing atrazine

Finding	Dietary concentration (ppm)				
	0	25	50	70	400
Intact females					
Body weight (g), week 52	412	428	414	419	390*
Survival, week 104	26/60	19/60	17/59	19/60	13*/60
No. of rats with palpable masses	29/80	34/80	43/80	39/80	46/80
Palpable masses; week of first observation	29	28	38	32	14
Mammary tumours, weeks 1–53:					
Fibroadenomas	0/22	2/22	2/23	2/23	1/25
Carcinomas	2/22	2/22	0/23	2/23	6/25
Total No. with mammary neoplasia	2/22	3/22	2/23	4/23	6/25
Mammary tumours, weeks 1–104:					
Fibroadenomas	16/80 (20%)	25/80 (31%)	33/78** (42%)	29/80* (36%)	25/80* (31%)
Adenomas	0/80	0/80	1/78 (1.3%)	0/80	0/80
Carcinomas	12/80 (15%)	18/80 (23%)	20/78 (26%)	14/80 (18%)	27/80** (34%)
Adenomas/carcinomas	12/80 (15%)	18/80 (23%)	21/78* (27%)	14/80 (18%)	27/80** (34%)
Total No. with mammary neoplasia	24/80 (30%)	34/80 (43%)	44/78** (56%)	38/80* (48%)	43/80** (54%)
Ovariectomized females					
Body weight (g), week 52	523	528	524	526	479*
Survival, week 104	44/60	42/60	46/59	45/60	44/60
No. of rats with palpable masses	3/80	2/80	2/80	2/80	6/80
Palpable masses; week of first observation	35	87	75	91	56
Total No. with mammary neoplasia	0/64	0/66	0/70	0/71	0/72

From Morseth (1996d, 1998)
* $p < 0.05$; ** $p < 0.01$.
Data for historical controls from 14 studies: adenomas, 1.3% (0–5.6%); carcinomas, 21.5% (6.0–32.8%); fibroadenomas, 47.8% (35.3–65.0%).

The NOAEL for mammary carcinogenicity was 25 ppm, equal to 1.5 mg/kg bw per day, on the basis of a statistically significant increase in mammary neoplasia in intact female rats at 50 ppm, equal to 3.1 mg/kg bw per day, and greater compared with the concurrent control group. The carcinogenic effect of atrazine was abolished in ovariectomized rats (Morseth, 1996d, 1998).

In a study of carcinogenicity reported in a published paper, groups of 53–56 male and 50–55 female Fischer 344 rats received diets containing atrazine (purity, 98.9%) for the lifespan of most of the rats (126 weeks in males, 123 weeks in females). Dietary concentrations of 500 and 1000 ppm were used at commencement and were reduced to 350 and 750 ppm after 8 weeks because of excessive retardation in body-weight gain. Histological examination was conducted on a full spectrum of tissues, except that in decedents, results were only considered if a satisfactory evaluation was possible. There was no clinical pathology, diet analysis or evidence of GLP quality-assurance auditing, and the level of detail presented in the paper would be inadequate, in isolation, for regulatory submission.

Body-weight gain was retarded by treatment in both groups of treated males and females. This however appeared to be dose-related only in females, and the growth curves of females were otherwise abnormal in that suppression of normal weight gain (unremarked and cause not obvious) would appear to have occurred between weeks 10 and 60. Water consumption was increased before the reduction in dose but was normal subsequently. The rate of mortality in both groups of treated males was less than in controls, hence more treated animals survived for longer than did controls. Because of the lifetime design of the study, a large number of the rats died during the study, with potential tissue loss to autolysis tending to be more in males in the control group due to the altered survival rate. Statistically significantly increased incidences were reported for benign mammary tumours (fibroma, fibroadenoma and adenoma) in males at the highest dose (1 out of 48, 1 out of 51 and 9 out of 53 at 0, 375 and 750 ppm, respectively), for uterine adenocarcinomas in females at the highest dose (6 out of 45, 8 out of 52 and 13 out of 45 at 0, 375 and 750 ppm, respectively), and for combined leukaemias and lymphomas in females at the highest dose (12 out of 44, 16 out of 52 and 22 out of 51 at 0, 375 and 750 ppm, respectively). However, clear interpretation of these results was compromised by the increased survival at the highest dose (also confirmed for females by additional analysis).

Although this study is inadequate to permit conclusions on the carcinogenicity of atrazine due to significant flaws in design and reporting, the study authors stated that the results are suggestive evidence for tumorigenic activity of atrazine in Fischer 344 rats (Pintér et al., 1990).

To conclude, the increased incidence of combined leukaemias and lymphomas is not indicative of a carcinogenic effect as the combining of these tumour types is inappropriate and neither tumour type displayed a significantly increased incidence when evaluated alone. Also, an additionally conducted survival-adjusted analysis of tumour prevalence did not indicate any treatment-related statistically significant differences in the incidence of benign, malignant nor combined mammary tumours for either males or females (Liu & Thakur, 1999).

In a supplementary study of carcinogenicity designed to investigate the effect of atrazine on the estrous cycle, selected hormone concentrations and the development of mammary, ovary, uterine and pituitary tumours, groups of 70 female Fischer 344 rats, CDF(F344)/CrlBR, were fed diets containing atrazine (purity, 97.0%) at a concentration of 0, 10, 70, 200 and 400 ppm, equivalent to 0, 0.5, 3.5, 10 and 20 mg/kg bw per day, for 104 weeks. Interim kills of 10 females per group were conducted at 1, 3, 9, 12, 15 and 18 months of the study, and all remaining survivors killed after 2 years. The terminations were conducted at proestrus as determined by vaginal smear, unless this phase was not reached within 7 days of the scheduled termination time-point. In order to avoid stress-related hormone changes, the rats were killed by decapitation without anaesthesia. Samples of blood were collected post mortem for hormone analysis. The pituitary, mammary glands, uterus and ovaries from all animals were examined microscopically. The study was conducted in compliance with GLP and OECD test guidelines.

There were no treatment-related effects on survival or on the incidence of clinical observations. A dose-related and statistically significant decrease in mean body-weight gain was observed at 200 ppm (3–10%) and 400 ppm (10–15%) at several time-points. There was no increase in mammary tumours or any other type of tumour at any dose. The NOAEL was 70 ppm, equivalent to 3.5 mg/kg bw per day, on the basis of decreased body-weight gain at 200 ppm and greater (Thakur, 1991b).

From the study described above (Thakur, 1991b), data on estrous cycles, vaginal cytology and selected serum hormone concentrations were also collected from rats killed at 1, 3, 9, 12, 15, 18 and 24 months. Results were evaluated as a function of treatment and as a function of treatment over time.

There were normal age-related changes in rats in the control group and in rats treated with atrazine, which included a shift in estrous cycle patterns toward an enhancement of percent of total days in proestrus at the expense of days in estrus. In rats in the control group as well as treated rats, E2 concentrations increased during the first 9 months of the study and decreased thereafter. By contrast, progesterone concentrations rose in a much more prolonged pattern, with the highest mean for each group always ocurring at 15 or 18 months. There were no consistent treatment-related changes in serum hormone concentrations compared with controls for any of the hormones tested, nor were there differences from control age-related patterns in animals fed up to 400 ppm atrazine. The normally occurring senescence of reproductive cycling parameters was not affected by treatment (Eldridge et al., 1993b).

From rats in the study described above (Thakur, 1991b), the ovaries, uterus, vagina, mammary gland and pituitary were re-evaluated microscopically for specific indications of reproductive senescence, which might relate to the onset-time of hormonally-mediated mammary tumours.

Ovarian histomorphology showed that the great majority of rats in all groups, the control group and dosed groups, maintained corpora lutea throughout most of the study. Only at the final, 24-month, time-point were there dramatic decreases in corpora lutea numbers. At this time-point, the rats treated with atrazine did not show decreases in corpora lutea numbers that were any more severe than those observed in rats in the control group. The reduction in numbers of corpora lutea at this late time-point appeared to be a consequence of a natural progression of the rats from persistent diestrus into acyclicity. All rats in all groups receiving atrazine maintained moderate numbers of secondary, antral and atretic follicles throughout the study, including at the 24-month time-point. The data on ovarian histomorphology indicated that treatment with atrazine did not alter the number of rats in repetitive pseudopregnancy/persistent diestrus. Estrous cycling in F344 females treated with atrazine was not altered.

Mammary gland histomorphology showed that rats in all groups receiving atrazine displayed histomorphological alterations in the mammary gland that would be expected in normally ageing F344 female rats, i.e. there was some evidence of lobular/acinar development with secretory activity and occasional galactocoeles in all groups receiving atrazine at 15, 18 and 24 months.

The histomorphological alterations in the mammary gland that were seen are those that would be expected in normally ageing F344 female rats. Exposure to atrazine did not increase the severity of any histomorphological findings in the mammary gland or decrease their time of onset (McConnell, 1995).

In a study of carcinogenicity , which complied with GLP and EPA guidelines, groups of 60 male and 60 female Fischer 344 rats, Crl:CDF (F344), were fed diets containing atrazine (purity, 97.0%) at a concentration of 0, 10, 70, 200 and 400 ppm, equivalent to 0, 0.5, 3.5, 10 and 20 mg/kg bw per day, for 104 weeks.

Administration of atrazine did not affect survival nor were any clinical signs apparent. Body weight and body-weight gain were significantly reduced in males and females at 200 and 400 ppm throughout most of the study. Food consumption in the group of males at the highest dose was significantly reduced throughout the study, and significantly reduced in females at the highest dose for the first 13 weeks of exposure. Histopathologically, the incidence of pituitary tumours in the group at the highest dose was slightly lower in both sexes than that of the controls. The incidence of mammary tumours among females increased slightly with dose (7, 8, 12, 17 and 13% at 0, 10, 70, 200 and 400 ppm), but these values were within the range for historical controls and did not attain statistical significance. There were no other histological changes considered to be related to treatment. Atrazine was not carcinogenic in Fischer 344 rats in an adequately-conducted study. The NOAEL was 70 ppm, equivalent to 3.5 mg/kg bw per day, on the basis of decreased body-weight gain at 200 ppm and greater (Thakur, 1992b).

2.4 *Genotoxicity*

Atrazine was tested for possible mutagenic potential in nearly 100 tests conducted either by the sponsor or independently in external laboratories. These studies covered different end-points in both eukaryotes and prokaryotes in vivo and/or in vitro. A summary of selected studies conducted in vitro is presented in Table 10, and selected studies conducted in vivo in Table 11.

Brusick (1994) had reviewed a large number of reports and publications (non-plant studies) from the years 1977 to 1992 on the genetic toxicity of atrazine, the results of which were positive in six cases, using two approaches. One was the "expert judgement" in which conflicting results were resolved by thoroughly reviewing each study and critically assessing the detailed data. The second approach used a computer-assisted "weight-of-the-evidence" method. The conclusions reached about the genotoxicity of atrazine were "equivocal" using the first method and "negative" using the second. The weight-of-the-evidence (computer) model, which was originally developed by the International Commission for Protection against Environmental Mutagens and Carcinogens (ICPEMC), was considered by Brusick to be more relevant for assessing atrazine using such a large database. The author concluded also that the positive responses reported for atrazine in plant systems cannot be assumed to be relevant to effects in mammals, especially when differences in metabolite profiles between plants and mammals are known to exist.

In a later review of the genetic toxicity of atrazine by Brusick (2000), data from 21 additional publications from the years 1993 to 2000 were evaluated and summarized. These studies included both standard and non-standard assays. Of the 17 studies using conventional assays for genetic toxicology, 12 were reported as giving negative results while five gave positive results. On the basis of a critical assessment of these four in-vitro assays and one in-vivo assay reported in four papers (Guigas et al., 1993; Della Croce et al., 1996; Gebel et al., 1997; Lioi et al., 1998), it was evident that all had serious deficiencies and the positive responses were attributed to inappropriate study design or technique. Taking these limitations into consideration, the weight of evidence indicates that atrazine is not genotoxic.

Additional studies using novel, but not-yet-validated techniques including single-cell gel electrophoresis (Comet assay), flow cytometry, and differential gene expression, have reported positive findings. These techniques must be carefully evaluated, validated and confirmed by independent investigators before they can be used in decision-making regarding the genotoxic potential of any chemical including atrazine (Brusick, 2000).

2.5 *Reproductive toxicity*

(a) *Multigeneration studies*

In a two-generation study of reproductive toxicity , which complied with GLP and US EPA test guidelines, groups of 30 male and 30 female Sprague-Dawley rats were fed diets containing atrazine technical (purity, 97.6%) at a concentration of 0, 10, 50 and 500 ppm, for 10 weeks before mating and further during mating, gestation and rearing of offspring. Thirty male and 30 female F_1 pups from each group were selected and after 12 weeks exposure to the treated diets were mated to derive the F_2 generation. F_2 pups were killed 21 days after birth. Histology of reproductive organs and tissues was conducted from 30 rats of each sex from the control groups and groups at the highest dose of the F_0 and F_1, and five of each sex of the control group and group at the highest dose for the F_2 generations. The mean daily intakes at 0, 10, 50 and 500 ppm in the period before mating were 0, 0.7, 3.6 and 36.1 mg/kg bw per day for F_0 males and 0, 0.8, 4.0 and 40.7 mg/kg bw per day for F_0 females,

Table 10. Selected studies of genotoxicity with atrazine in vitro

End-point	Test system	Concentration or dose	Purity (%)	Results	References
Gene mutation	*S. typhimurium* TA98, TA100, TA1535, TA1537, TA1538	50–10 000 µg/plate	98.8	Negative	Sutou et al. (1979)
Gene mutation	*S. typhimurium* TA98, TA100, TA1535, TA1537	10–810 µg/plate	NR	Negative	Arni (1978)
Gene mutation	*S. typhimurium* TA98, TA100, TA1535, TA1537	20–5 000 µg/plate	98.2	Negative	Deparade (1986)
Gene mutation	*S. typhimurium* TA97, TA98, TA100	2–2 157 µg/plate	> 99	Negative	Butler & Hoagland (1989)
Gene mutation	*S. typhimurium* TA98, TA100	0.86 mg/ml	NR	Negative	De Veer et al. (1994)
Gene mutation	*S. typhimurium* TA98, TA100	10–1 000 µg/plate	42.7[a]	Negative	Della Croce et al. (1996)
Gene mutation	*S. typhimurium* TA98, TA100, TA102, TA1535, TA1537	1–1 000 µg/plate	98.9	Negative	Ruiz & Marzin (1997)
Gene mutation	*E. coli* B/r WP2 Tyr⁻ Hcr⁻	50–5 000 µg/plate	98.8	Negative	Sutou et al. (1979)
Gene mutation	Chinese hamster V79 cells; HGPRT	1.25–10 mmol/l	NR	Negative	Adler (1980)
Gene mutation	CHO cells; HGPRT	0.023–0.23 µg/ml	NR	Positive	Guigas et al. (1993)
Chromosomal aberration	CHO cells	1.25–10 mmol/l	NR	Negative	Adler (1980)
Chromosomal aberration	Human lymphocytes	0.01–1 µg/ml	NR	Positive	Meisner et al. (1992)
Chromosomal aberration	Human lymphocytes	5–51 µmol/l	≥ 98	Positive	Lioi et al. (1998)
Chromosomal aberration	Human lymphocytes	5–100 µg/ml	98.4	Negative	Ribas et al. (1998)
Chromosomal aberration	Human lymphocytes	0.5–50 µg/ml	97.7	Negative	Kligerman et al. (2000a)
Micronucleus formation	Human lymphocytes	1–100 µg/ml	98.4	Negative	Surrallés et al. (1995)
Micronucleus formation	Human lymphocytes	5–200 µg/ml	98.4	Negative	Ribas et al. (1998)
DNA damage and repair (Rec assay)	*B. subtilis* H17/M45	100–10 000 µg/well	98.8	Negative	Sutou et al. (1979)
SOS chromotest	*E. coli* PQ37	NR	98.9	Negative	Ruiz & Marzin. (1997)
UDS	Rat primary hepatocytes	1.2–150 µg/ml	98.2	Negative	Puri (1984a)
UDS	Human fibroblasts CRL 1121	1.2–150 µg/ml	98.2	Negative	Puri (1984b)

End-point	Test system	Concentration or dose	Purity (%)	Results	References
UDS	Rat primary hepatocytes	15.5–1 670 µg/ml	97.1	Negative	Hertner (1992)
SCE	CHO cells	1.25–10 mmol/l	NR	Negative	Adler (1980)
SCE	CHO cells; HGPRT	0.023–0.23 µg/ml	NR	Negative	Guigas et al. (1993)
SCE	Chinese hamster V79 cells	0.86 mg/ml	NR	Negative	De Veer et al. (1994)
SCE	Human lymphocytes	5–50 µmol/l	98.7	Negative	Dunkelberg et al. (1994)
SCE	Human lymphocytes	5–51 µmol/l	≥ 98	Positive	Lioi et al. (1998)
SCE	Human lymphocytes	5–200 µg/ml	98.4	Negative	Ribas et al. (1998)
SCE	Human lymphocytes	0.5–50 µg/ml	97.7	Negative	Kligerman et al. (2000a)
Comet assay	Human lymphocytes	50–200 µg/ml −S9 50–200 µg/ml +S9	98.4	Positive Negative	Ribas et al. (1995)
Micronucleus formation[b]	Human lymphocytes	1–100 µg/ml	98.4	Negative	Surrallés et al. (1995)
Mitotic gene conversion	*S. cerevisiae* D7	150–350 mmol/l (stationary phase) 1–10 mmol/l (log phase)	42.7[a]	Negative Positive	Della Croce et al. (1996)

NR, not reported; SCE, sister-chromatid exchange; UDS, unscheduled DNA synthesis.
[a] Formulation tested.
[b] Used as biomarker of excision repair.

Table 11. Selected studies of genotoxicity with atrazine in vivo

End-point	Test object	Concentration or dose	Purity (%)	Results	References
Somatic cells					
Chromosomal aberration	Chinese hamster bone marrow	500 mg/kg bw; oral, single dose	NR	Negative	Basler & Röhrborn (1978)
Micronucleus formation	Mouse bone marrow	2000 mg/kg bw; oral, single dose	NR	Negative	Ehling (1979)
Chromosomal aberration	Mouse bone marrow	1500, 2000 mg/kg bw; oral, single dose	NR	Positive	Adler (1980)
Nucleus anomaly test	Chinese hamster bone marrow	282, 564, 1128 mg/kg bw per day; oral, on two consecutive days	NR	Negative	Hool (1981a)
Chromosomal aberration	Mouse bone marrow	6 mg/kg, intraperitoneal, single dose 1 ppm (drinking-water) for 7 weeks	NR	Negative	Chollet et al. (1982)
Chromosomal aberration	Mouse bone marrow	562.5, 1125, 2250 mg/kg bw; oral, single dose	98.2	Negative	Ceresa (1988b)
Chromosomal aberration	Mouse bone marrow	20 ppm (drinking-water) over 30 or 90 days	NR	Negative	Meisner et al. (1992)

End-point	Test object	Concentration or dose	Purity (%)	Results	References
Chromosomal aberration	Mouse bone marrow	Males: 900–1750 mg/kg bw; Females: 1400, 1750 mg/kg bw; oral, single dose	98.7	Negative in males Positive in females	Gebel et al. (1997)
Chromosomal aberration	Mouse bone marrow	125, 250, 500 mg/kg bw; intraperitoneal, two injections 24 h apart	97.7	Negative	Kligerman et al. (2000b)
DNA damage	Rat stomach, liver and kidney	875 mg/kg bw; oral, single dose 350 mg/kg bw per day; oral, 5 or 15 days	NR	Positive	Pino et al. (1988)
DNA damage	Rat lung	875 mg/kg bw; oral, single dose 350 mg/kg bw per day; oral, 5 or 15 days	NR	Negative	Pino et al. (1988)
DNA damage	Mouse blood leukocytes	125, 250, 500 mg/kg bw; intraperitoneal	97.7	Equivocal	Tennant et al. (2001)
Germ cells					
Chromosomal aberration	Mouse spermatogonia	444, 1332 mg/kg bw per day; oral, on five consecutive days	NR	Negative	Hool (1981b)
Chromosomal aberration	Mouse spermatogonia	6 mg/kg bw, intraperitoneal, single dose 1 ppm (drinking-water) for 7 weeks	NR	Negative	Chollet et al. (1982)
Chromosomal aberrations	Mouse spermatocytes	444, 1332 mg/kg bw; oral, five doses over 10 days	NR	Negative	Hool (1981c)
Dominant lethal mutation	Mouse spermiogenesis	1500, 2000 mg/kg bw; oral, single dose	NR	Positive	Adler (1980)
Dominant lethal mutation	Mouse spermiogenesis	444, 1332 mg/kg bw; oral, single dose	98.9	Negative	Hool (1981d)
Dominant lethal mutation	Mouse spermiogenesis	6 mg/kg, intraperitoneal, single dose	NR	Negative	Chollet et al. (1982)
Dominant lethal mutation	Mouse spermiogenesis	500, 1000, 2000, 2400 mg/kg bw; oral, single dose	97.1	Negative	Hertner (1993)
Sperm head morphology	Mouse sperm	600 mg/kg bw per day; intraperitoneal, on five consecutive days	97.2	Negative	Osterloh et al. (1983)

NR, not reported.

respectively, and 0, 0.7, 3.5 and 35.0 mg/kg bw per day for F_1 males and 0, 0.8, 3.8 and 37.5 mg/kg bw per day for F_1 females, respectively.

No treatment-related mortality or signs of toxicity were observed during the study. Parental body weights, body-weight gain, and food consumption were statistically significantly reduced at 500 ppm (the highest dose tested) in both sexes and both generations throughout the study. Compared with controls, body weights for F_0 males and females at the highest dose at 70 days into the study were decreased by 12% and 15%, respectively, while body weight of the F_1 generation for the same period was decreased by 15% and 13% for males and females, respectively.

Litter parameters, including litter size, sex ratio, and postnatal survival indices, of the F_1 and F_2 generations were not affected by treatment with atrazine at any dose. There were no clinical symptoms, necropsy findings or malformations that were considered to be related to treatment. Slightly lower body weights (8–10%) were noted in both generations of male pups at 500 ppm on postnatal day 21.

No treatment-related pathological or micropathological findings were noted in any of the reproductive organs of F_0 or F_1 generations. Relative testes weights were increased in F_1 and F_2 males at 500 ppm as a result of decreased terminal body weights.

The NOAEL for parental toxicity was 50 ppm, equal to 3.6 and 4.0 mg/kg bw per day in males and females, respectively, on the basis of decreased body-weight gains and food consumption at 500 ppm. The NOAEL for reproductive toxicity was 50 ppm on the basis of decreased body weights of male pups at 500 ppm on postnatal day 21 (Mainiero et al., 1987).

(b) Studies of developmental toxicity

Rats

In a study of prenatal developmental toxicity, which complied with GLP and US EPA test guidelines, groups of 27 pregnant female Crl:COBS CD(SD)BR rats were given atrazine (purity, 96.7%; suspended in 3% corn starch with 0.5% Tween 80) at a dose of 0, 10, 70 and 700 mg/kg bw per day by oral gavage on days 6 to 15 of gestation.

A total of 21 dams treated at 700 mg/kg bw per day died between days 13 and 20. All other dams survived the study. Treatment-related clinical signs were confined to the group at 700 mg/kg bw per day and included salivation, oral/nasal discharge, ptosis, swollen abdomen and blood on the vulva. Necropsy findings included enlargement of the stomach. A marked reduction in food intake was recorded for the group at 700 mg/kg bw per day and was associated with a marked loss of maternal body weight. In the group at 70 mg/kg bw per day, significant reductions of food consumption and body-weight gain were reported at days 6–7 or 6–10 of gestation, respectively.

Fetal toxicity, attributed to maternal toxicity, was observed in the groups at 70 and 700 mg/kg bw per day and manifested as a marked reduction of fetal weight at 700 mg/kg bw per day (no skeletal examination was performed in this group). A marginal increase in numbers of runted pups at all doses was within the range of data for historical controls. Skeletal variations in the group at 70 mg/kg bw per day included incomplete ossification of skull, teeth, metacarpals and hindpaw distal phalanges, and bipartite metacarpals. These findings can be considered to be developmental delays attributable to maternal toxicity, although it should be noted that maternal toxicity at 70 mg/kg bw per day was minimal and fetal weights were not different to those of the controls (Table 12).

There was no evidence of teratogenicity.

The NOAEL for maternal toxicity was 10 mg/kg bw per day on the basis of decreased body-weight gain and food intake at 70 mg/kg bw per day and greater. The NOAEL for developmental toxicity was 10 mg/kg bw per day on the basis of incomplete ossification at several sites at 70 mg/kg bw per day and greater (Infurna, 1984).

In a study of prenatal developmental toxicity, which complied with GLP and US EPA test guidelines, groups of 26 pregnant female Crl:COBS CD(SD)BR rats were given atrazine (purity, 97.6%; suspended in 3% corn starch with 0.5% Tween 80) at a dose of 0, 5, 25 and 100 mg/kg bw per day by oral gavage on days 6 to 15 of gestation.

One female at the highest dose died without obvious cause on day 20 of gestation. At the highest dose, an increased incidence of salivation was seen. Also at the highest dose, food consumption

Table 12. Relevant findings in a study of developmental toxicity in rats given atrazine by gavage

Finding	Dose (mg/kg bw per day)			
	0	10	70	700
Salivation (No. of rats with finding/No. of rats in group)	0/27	0/27	1/27	13/27**
Ptosis (No. of rats with finding/No. of rats in group)	0/27	0/27	1/27	11/27**
Abdominal area swollen (No. of rats with finding/ No. of rats in group)	0/27	0/27	1/27	8/27**
Blood on vulva (No. of rats with finding/No. of rats in group)	0/27	0/27	1/27	11/27**
Death (No. of rats with finding/No. of rats in group)	0/27	0/27	1/27	21/27**
Food consumption (g):				
Day 6	23	23	17**	13**
Day 7	23	22	20**	13**
Body-weight gain (g):				
Days 6–10	13	16*	6**	−1**
Days 0–20	61	59	53	−25**
No. of pregnant rats	24	23	25	26
No. of litters examined	23	23	25	5
Mean No. of resorptions	0.8	0.9	0.9	1.3
Postimplantation loss (%)	9.7	5.9	6.1	20.9
Mean No.of live fetuses	12.7	13.7	14.0	13.4
Mean fetal weights (g), males/females	3.4/3.3	3.6/3.3	3.4/3.2	1.9*/1.8*
Skeletal examinations (fetuses/litters)	203/23	217/23	244/25	—[a]
Skull not completely ossified (fetuses/litters)	23/9	7/4	47**/16**	—
Hyoid not ossified (fetuses/litters)	20/10	9/5	44**/12*	—
Teeth not ossified (fetuses/litters)	2/1	4/4	34**/10**	—
Forepaw:				
Metacarpals not ossified (fetuses/litters)	117/19	106/20	184*/23*	—
Metacarpals bipartite (fetuses/litters)	0	0	3*/3*	—
Hindpaw: distal phalanx not ossified (fetuses/litters)	0	0	6*/3*	—

From Infurna (1984)

[a] No skeletal examination was performed due to extremely reduced fetal weights and subsequent reduced ossification.

* $p < 0.05$; ** $p < 0.01$.

was significantly reduced on days 6–8 and 8–12 of gestation, while body weight was significantly reduced on days 8 to 16 of gestation. Corrected body-weight gain for the entire gestation period was reduced by 20%.

There were no treatment-related effects on any reproductive parameter examined, on fetal sex ratio, mean fetal weights or the incidences of gross, visceral and skeletal malformations. A significantly increased incidence of incomplete ossification of various skull bones (hyoid, interparietal, occipital and parietal bones) was observed at the highest dose. A significantly increased incidence of incomplete ossification of the interparietals was also observed in the groups at the intermediate and lowest dose, but there was no dose–response relationship when evaluated by litters (Table 13).

There was no evidence of teratogenicity.

The NOAEL for maternal toxicity was 25 mg/kg bw per day on the basis of decreased body-weight gain and food intake at 100 mg/kg bw per day. The NOAEL for developmental toxicity was

Table 13. Relevant findings of a study of developmental toxicity in rats given atrazine by gavage

Finding	Dose (mg/kg bw per day)			
	0	5	25	100
Salivation (No. of rats with finding/No of rats in group)	0/26	0/26	0/26	18/26
Food consumption (g):				
Days 6–8	24.40	23.52	23.02	16.98*
Days 8–12	25.13	24.18	23.73	21.89*
Body-weight gain (g):				
Days 6–8	7.54	6.92	6.48	−3.91*
Days 6–16	57.23	55.88	56.36	46.95*
No. of pregnant rats	26	25	25	22
No. of litters examined	26	25	24	21
Mean No. of resorptions	0.58	0.80	0.50	0.67
Postimplantation loss (%)	4.09	5.52	3.40	4.47
Mean No. of live fetuses	13.42	13.80	14.46	15.43*
Mean fetal weights (g), males/females	3.52/3.31	3.63/3.41	3.56/3.42	3.49/3.32
Hyoid not completely ossified (fetuses/litters)	20/10	26/9	27/13	36*/15*
Interparietal not completely ossified (fetuses/litters)	16/10	42*/15*	43*/14*	73*/20*
Occipitals not completely ossified (fetuses/litters)	14/10	26/13	22/10	35*/16*
Parietals not completely ossified (fetuses/litters)	4/3	9/6	7/3	14*/9*

From Giknis (1989)
* $p < 0.05$.

25 mg/kg bw per day on the basis of incomplete ossification of skull bones at 100 mg/kg bw per day (Giknis, 1989).

Rabbits

In a study of prenatal developmental toxicity, which complied with GLP and US EPA test guidelines, groups of 19 inseminated female New Zealand White rabbits were given atrazine (purity, 96.3%; suspended in 3% corn starch with 0.5% Tween 80) at a dose of 0, 1, 5 or 75 mg/kg bw per day by oral gavage on days 7 to 19 of gestation.

Three females at the lowest dose died during the study; one, possibly two, being the result of misdosing; one doe at 5 mg/kg bw per day and two does at 75 mg/kg bw per day were killed because they were aborting. At 75 mg/kg bw per day, all dams exhibited stool changes (little, none, and/or soft), and showed marked reductions in food consumption and body-weight loss, with rebound after cessation of dosing. The slight reductions in food intake and body-weight gain at 1 and 5 mg/kg bw per day were not considered to be adverse.

Embryotoxicity was evident at 75 mg/kg bw per day and included increased numbers of resorptions and postimplantation losses, and thus decreased numbers of viable fetuses (Table 14). Fetotoxicity was also observed at 75 mg/kg bw per day and included lower fetal weights and delayed ossification of forelimb metacarpals and middle phalanges, and hindlimb patellae, tali and middle phalanges. Treatment-related malformations were not observed.

There was no evidence of teratogenicity.

The NOAEL for maternal toxicity was 5 mg/kg bw per day on the basis of clinical signs, abortion, and decreased food intake and body-weight gain at 75 mg/kg bw per day. The NOAEL

Table 14. Relevant findings in a study of developmental toxicity in rabbits given atrazine by gavage

Finding	Dose (mg/kg bw per day)			
	0	1	5	75
Little, none and/or soft stool (No. of rabbits with finding/ No. of rabbits in group)	9/19	4/19	10/19	19/19**
Blood on vulva (No. of rabbits with finding/No. of rabbits in group)	0/19	1/19	0/19	4/19*
Abortion (No. of rabbits with finding/No. of rabbits in group)	0/19	(1)/19[a]	1/19	2/19
Death (No. of rabbits with finding/No. of rabbits in group)	0/19	3/19	0/19	0/19
Food consumption (g):				
Day 7	183	164	185	76**
Day 13	182	155*	163	1**
Day 17	182	157	134*	6**
Day 19	164	136	129*	14**
Body-weight change (g):	113	76	63	−522**
Days 7–14				
Days 14–19	120	105	45*	−204**
Days 0–29[b]	−73	−113	−143	−393**
No. of pregnant rabbits	16	17	16	18
No. of litters examined	16	14	15	15
Mean No. of resorptions	1.3	1.4	1.4	4.8**
Postimplantation loss (%)	12.0	11.4	13.0	42.6**
Mean No. live fetuses	8.8	8.9	9.1	5.9*
Mean fetal weights (g), males/females	46.1/44.0	44.0/43.3	43.2/43.1	35.7*/35.8*
Forepaw:				
Metacarpal not ossified (fetuses/litters)	1/1	0	1/1	7*/5*
Proximal phalanx not ossified (fetuses/litters)	0	0	1/1	14**/7**
Hindpaw:				
Patella not ossified (fetuses/litters)	5/5	9/4	12/5	35**/10**
Talus not ossified (fetuses/litters)	0	0	0	10**/5**
Middle phalanx not ossified (fetuses/litters)	0	0	2/1	4**/4**

From Arthur (1984)

[a] One dam found dead was possibly aborting.

[b] Terminal body weight minus uterus, placentas and fetuses.

* $p < 0.05$; ** $p < 0.01$.

for developmental toxicity was 5 mg/kg bw per day on the basis of increased resorptions, reduced litter size and incomplete ossification of appendicular skeletal elements at 75 mg/kg bw per day (Arthur, 1984).

2.6 *Special studies*

(a) Studies on metabolites

A number of metabolites of atrazine, such as the chlorometabolites (DEA, DIA, DACT) and hydroxyatrazine, can be found in drinking-water and in the diet.

Dealkylation reactions at the 4 and 6 positions resulting in the formation of either of the mono-dealkylated metabolites (desethyl-atrazine, desisopropyl-atrazine), which in turn can be further dealkylated to DACT, may occur in animals, plants and bacteria. Hydroxyatrazine is the major metabolite in plants, but is only a minor metabolite in animals in which the varying forms of the delkylated chlorometabolites dominate instead. The metabolism of atrazine to hydroxyatrazine in plants is a detoxification reaction as the phytotoxicity of hydroxyatrazine is greatly reduced compared with the parent compound.

A limited toxicology database is available for these four metabolites and the studies are summarized below.

(i) Deethyl-atrazine, DEA (G 30033)

DEA was of moderate acute oral toxicity in rats (LD_{50}, 1110 mg/kg bw). In short-term studies in rats given DEA at dietary concentrations of up to 1500 ppm, effects included reduced body-weight gain and food consumption, decreased erythrocyte parameters and lymphoid atrophy of the thymus. The overall NOAEL was 50 ppm, equal to 3.2 mg/kg bw per day. In a 13-week study in dogs given DEA at dietary concentrations of up to 1000 ppm, effects included reduced body weight and food consumption, decreased erythrocyte parameters and renal tubular epithelial hyperplasia/basophilia. The NOAEL was 100 ppm, equal to 3.7 mg/kg bw per day. DEA was not genotoxic in a battery of tests including assays for point mutation and DNA repair in vitro and testing for clastogenicity in vivo. In a study of prenatal developmental toxicity in rats, the NOAEL for maternal toxicity was 5 mg/kg bw per day on the basis of decreased body-weight gain and food intake at 25 mg/kg bw per day and greater. The NOAEL for developmental toxicity was 25 mg/kg bw per day on the basis of increased incidences of fused sternebrae and incomplete ossification of the proximal phalanx of posterior digit 5 at 100 mg/kg bw per day. There was no evidence of teratogenicity. In a special study on the effects of DEA on pubertal development in male rats, atrazine equimolar doses of ≥ 25 mg/kg bw per day delayed preputial separation, with a NOAEL of 12.5 mg/kg bw per day.

In a study of acute oral toxicity, which complied with GLP and the US EPA test guidelines, groups of HSD:(SD) rats received DEA (purity, 95.7%; suspended in 2% carboxymethyl-cellulose) as a single dose at 250, 500, 2000, 3500 or 5050 mg/kg bw by gavage. Each group consisted of five males and five females with the exception of the group at the lowest dose, which consisted of five females. The rats were observed for clinical signs and mortality for 14 days. The LD_{50}s for males, females and both sexes combined were 1890, 669 and 1110 mg/kg bw, respectively (Kuhn, 1991c).

In a short-term study of oral toxicity, which complied with GLP and the test guidelines of OECD and US EPA, groups of 10 male and 10 female KFM-WIST rats were fed diets containing DEA (purity, 99.3%) at a concentration of 0, 50, 500 or 1500 ppm, equal to 0, 4.5, 49.3 and 145.9 mg/kg bw per day in males and 0, 4.6, 49.0 and 132.8 mg/kg bw per day in females, for 4 weeks.

Clinical signs of excitement were observed in the rats at the highest dose during the fourth week of treatment. There was a decrease in food consumption that was apparent at 1500 ppm with an average reduction over the 4-week treatment period of 24% in males and 19% in females. A dose-related reduction of body weight with statistically significant differences from the second week of treatment was observed for the males and females at the highest dose and for the males at the intermediate dose. At the end of the 4-week treatment, body weights were significantly decreased in both sexes at 500 ppm (17% in males, 9% in females) and at 1500 ppm (30% in males, 19% in females) when compared with rats in the control group. Slight changes in haematological and clinical biochemistry parameters were observed in the groups at the intermediate and highest dose. Thymus : body weight ratios were decreased in females at the lowest, intermediate and highest dose (15%, 19% and 26%, respectively) and in males at the highest dose (33%). A significantly increased incidence of minimal to slight lymphoid atrophy of the thymus was observed in both sexes at 1500 ppm (the incidences at 0, 50, 500 and 1500 ppm were 1, 2, 2, and 9 in males and 2, 3, 5 and 10 in females, respectively).

The NOAEL was 50 ppm, equal to 4.5 and 4.6 mg/kg bw per day in males and females, respectively, on the basis of reduced body weights and lymphoid atrophy of the thymus at 500 ppm and greater (Duchosal et al., 1990b).

In a short-term study of oral toxicity, which complied with GLP and the test guidelines of OECD and US EPA, groups of 10 male and 10 female Tif:RAIf rats were fed diets containing DEA (purity, 95.7%) at a concentration of 0, 10, 50 or 500 ppm, equal to 0, 0.68, 3.20 and 35.2 mg/kg bw per day in males and 0, 0.73, 3.35 and 38.7 mg/kg bw per day in females, for 90 days.

No mortalities or clinical symptoms were observed and ophthalmological examination did not reveal treatment-related findings. Body-weight gains of males and females were reduced by 21% and 19% at 500 ppm. Food consumption was reduced by 10% and 7% in males and females of the same group, respectively. Haematology data revealed a slightly lower mean cell volume (MCV) and erythrocyte volume fraction and increased mean cell haemoglobin concentration (MCHC) values in females at 500 ppm. Alkaline phosphatase activity was increased in females at 500 ppm. A 12% increase in relative liver weight in females at the highest dose was without any histopathological correlates.

The NOAEL was 50 ppm, equal to 3.2 and 3.35 mg/kg bw per day in males and females, respectively, on the basis of reduced body weight and food efficiency at 500 ppm (Gerspach, 1991).

In a short-term study of oral toxicity, which complied with GLP and US EPA test guidelines, groups of four male and four female beagle dogs were fed diets containing DEA (purity, 95.7%) at a concentration of 0, 15, 100 or 1000 ppm, equal to 0, 0.56, 3.71 and 28.85 mg/kg bw per day in males and 0, 0.51, 3.88 and 32.18 mg/kg bw per day in females, for 13 weeks.

Treatment did not induce mortality or clinical signs of toxicity. At 1000 ppm, a slight body-weight loss was observed in males at days 7 to 42 and a decreased body-weight gain in females lasted the entire study period. Also, food consumption was 73% and 68% that of controls in males and females. Haematology revealed a slight reduction in erythrocyte parameters in dogs at 1000 ppm. No significant changes were recorded in clinical chemistry or urine analysis data.

Decreased uterus and thymus weights were observed in females receiving the highest dose and heart weight was decreased in males at the highest dose. Histopathology investigations confirmed these changes; females at the highest dose displayed thymic atrophy and anovulatory uterine atrophy that were considered to be secondary to reductions in food intake and body weights. Haemorrhagic inflammation with angiomatous hyperplasia in the right atrial wall of the heart was noted in one male at 1000 ppm (without any associated electrocardiographic changes), while the electrocardiographic findings (paroxysmal atrial fibrillation) in one female at 1000 ppm was not associated with histopathological changes in the heart. Mild renal tubular epithelial hyperplasia/basophilia was observed in three males and two females at 1000 ppm.

The NOAEL was 100 ppm, equal to 3.71 and 3.88 mg/kg bw per day in males and females, respectively, on the basis of reduced body weight and renal tubular hyperplasia at 1000 ppm (Rudzki et al., 1992).

In an assay for reverse mutation in bacteria, DEA (purity, 99.3%) did not induce gene mutations at the histidine locus of *S. typhimurium* and at the tryptophan locus in *E. coli* when tested at concentrations of up to 5000 μg/plate (Deparade, 1989).

In a test for DNA repair in vitro, DEA (purity, 99.3%) did not induce unscheduled DNA synthesis in primary rat hepatocytes exposed at concentrations of up to 1000 µg/ml (Geleick, 1991a).

In an assay for micronucleus formation in mouse bone marrow, DEA (purity, 99.3%) gave negative results for induction of micronuclei in the polychromatic erythroctes (PCE) of Tif:MAGF mice treated once orally at doses ranging from 120 to 480 mg/kg bw (Ogorek, 1991b).

In a study of prenatal developmental toxicity, which complied with GLP and the test guidelines of OECD and US EPA, groups of 24 mated female Tif:RAIf rats were given DEA (purity, 95.7%; suspended in 3% corn starch) at a dose of 0, 5, 25 or 100 mg/kg bw per day by oral gavage on days 6 to 15 of gestation.

All dams survived until the end of the experiment and no treatment-related clinical observations were made. Maternal toxicity was observed at 25 and 100 mg/kg bw per day in a dose-related manner as evidenced by effects on body weight/weight gain and food consumption. Body weights were significantly decreased at 100 mg/kg bw per day on days 7–20 of gestation. Body-weight gains were significantly decreased on days 6–11 of gestation at 25 and 100 mg/kg bw per day (83% and 41% of value for controls, respectively) and on days 11–16 of gestation at 100 mg/kg bw per day (87% of value for controls). Corrected body-weight gains were non-significantly decreased at 25 and 100 mg/kg bw per day (73% and 80% of value for controls, respectively). Food consumption was significantly decreased on days 6–11 of gestation at 25 and 100 mg/kg bw per day (91% and 70% of value for controls, respectively).

There were no treatment-related effects on any reproductive parameter examined, on fetal sex ratio, mean fetal weights or the incidences of gross, visceral and skeletal malformations. At 100 mg/kg bw per day, fetal and litter incidences of fused sternebrae 1 and 2 were significantly increased, as was the fetal incidence of poor ossification of the proximal phalanx of posterior digit 5 (Table 15).

There was no evidence of teratogenicity.

The NOAEL for maternal toxicity was 5 mg/kg bw per day on the basis of decreased body-weight gain and food intake at 25 mg/kg bw per day. The NOAEL for developmental toxicity was 25 mg/kg bw per day on the basis of increased incidences of fused sternebrae and incomplete ossification of the proximal phalanx of posterior digit 5 at 100 mg/kg bw per day (Marty, 1992b).

(ii) Deisopropyl-atrazine, DIA (G 28279)

DIA was of moderate acute oral toxicity in rats (LD_{50}, 1240 mg/kg bw). In short-term studies in rats given DIA at dietary concentrations of up to 2000 ppm, effects included reduced body-weight gain and food consumption, extramedullary haematopoiesis in liver and spleen, and histomorphological changes in adrenals, thyroid and pituitary. The overall NOAEL was 50 ppm, equal to 3.2 mg/kg bw per day. In a 14-week study in dogs given DIA at dietary concentrations of up to 1000 ppm, effects consisted of reduced body weight and food consumption. The NOAEL was 100 ppm, equal to 3.8 mg/kg bw per day. DIA was not genotoxic in a battery of tests including assays for point mutation and DNA repair in vitro and testing for clastogenicity in vivo. In a study of prenatal developmental toxicity in rats, the NOAEL for maternal toxicity was 5 mg/kg bw per day on the basis of decreased body-weight gain and food intake at 25 mg/kg bw per day and greater. The NOAEL for developmental toxicity was 5 mg/kg bw per day on the basis of increased incidences of fused sternebrae at 25 mg/kg bw per day and greater. There was no evidence of teratogenicity. In a special study on the effects of DIA on pubertal development in male rats, atrazine equimolar doses of ≥ 25 mg/kg bw per day delayed the preputial separation, with a NOAEL of 12.5 mg/kg bw per day.

In a study of acute oral toxicity, which complied with GLP and the US EPA test guidelines, groups of HSD:(SD) rats received DIA (purity, not reported; suspended in 2% carboxymethyl-cellulose) as a

Table 15. Relevant findings in a study of prenatal developmental toxicity in rats given DEA by gavage

Finding	Dose (mg/kg bw per day)			
	0	5	25	100
Food consumption (g/day):				
Days 6–11	23	22	21*	16**
Days 11–16	26	25	26	24
Body-weight gain (g):				
Days 6–11	26.0	24.3	21.5*	10.7**
Days 11–16	39.7	39.8	37.3	34.5*
Net body-weight change (g) from day 6	37.1	31.9	27.2	29.5
No. of animals pregnant	23	23	22	24
Postimplantatation loss (mean)	0.6	0.8	0.9	1.1
Postimplantation loss (%)	4.0	5.1	6.3	8.8
Mean No. of live fetuses	14.2	14.5	14.2	12.6
Mean fetal weights (g), males/females	5.7/5.3	5.6/5.3	5.8/5.4	5.6/5.3
Skeletal examination (No. of fetuses/litters)	168/23	174/23	164/22	158/24
Fused sternebrae 1 and 2[a]	2/2	1/1	2/2	32**/14**
Total skeletal anomalies[a]	5/5	5/5	5/5	41/18**
Poster. digit 5, proximal phalanx; absent ossification[a]	16/11	23/12	8/6	28*/16
Total skeletal variations[a]	168/23	173/23	163/22	158/24

From Marty (1992b)
DEA, deethyl-atrazine.
[a] No. of fetuses/litters.
* $p < 0.05$; ** $p < 0.01$.

single dose at 250, 500, 2000, 3500 or 5050 mg/kg bw by gavage. Each group consisted of five males and five female with the exception of the group at the lowest dose that consisted of five females. The rats were observed for clinical signs and mortality for 14 days. The LD_{50}s for males, females and both sexes combined were 2290, 810 and 1240 mg/kg, respectively (Kuhn, 1991d).

In a short-term study of oral toxicity, which complied with GLP and the test guidelines of OECD and US EPA, groups of 10 male and 10 female KFM-WIST rats were fed diets containing DIA (purity, 97.4%) at a concentration of 0, 50, 500 or 2000 ppm, equal to 0, 4.6, 46.5 and 161.2 mg/kg bw per day in males and 0, 4.6, 49.7 and 164.7 mg/kg bw per day in females, for 4 weeks.

There was no mortality. Clinical signs of restlessness were observed in the rats at the highest dose during the fourth week of treatment. At 2000 ppm, food consumption was reduced by approximately 60% during the first week of treatment and by about 40% over the 4-week treatment period. Body weights were significantly reduced at 500 ppm from the third week and from the second week at 2000 ppm; after 4 weeks the body weights at 2000 ppm were reduced by 34% (males) and 28% (females) when compared with rats in the control group. Slight changes in haematological and clinical biochemistry parameters were observed in the groups at the intermediate and highest dose, with indication of a compensated haemolytic anaemia in the group at the highest dose. Absolute and relative thymus weights were reduced at 500 ppm in males and at 2000 ppm in both sexes. No treatment-related histopathological findings were noted in the groups at the lowest and intermediate dose. At 2000 ppm, mineralized deposits were noted in the renal pelvis of four males and two females, and a minimal erosion of the gastric mucosa was noted in one male.

The NOAEL was 50 ppm, equal to 4.6 mg/kg bw per day in males and females, on the basis of reduced body weights at 500 ppm and greater (Duchosal et al., 1990a).

In a short-term study of oral toxicity, which complied with GLP and the test guidelines of OECD and US EPA, groups of 10 male and 10 female Tif:RAIf rats were fed diets containing DIA (purity, 96.7%) at a concentration of 0, 10, 50 or 500 ppm, equal to 0, 0.60, 3.20 and 34.9 mg/kg bw per day in males and 0, 0.64, 3.34 and 37.5 mg/kg bw per day in females, for 13 weeks.

No mortalities or treatment-related clinical symptoms were observed during the study. At 500 ppm, body-weight gains of males and females were reduced by 16% and 21%, respectively, while food consumption of females was slightly reduced by 7% during weeks 1–7. There were no treatment-related changes in haematological or clinical chemistry parameters. A 13% increase in relative liver weights in females at 500 ppm was accompanied by extramedullary haematopoiesis in the liver and spleen. Histopathological findings in males at the highest dose included fatty changes in the adrenal cortex, hypertrophy of thyroid follicular epithelium and hypertrophy of thyroid-stimulating hormone (TSH)-producing cells in the pituitary gland.

The NOAEL was 50 ppm, equal to 3.2 and 3.34 mg/kg bw per day in males and females, respectively, on the basis of reduced body weight and histopathological changes in the liver, spleen, adrenals, thyroid and pituitary at 500 ppm (Schneider, 1992).

In a short-term study of oral toxicity, which complied with GLP and US EPA test guidelines, groups of four male and four female beagle dogs were fed diets containing DIA (purity, 96.7%) at a concentration of 0, 15, 100, 500 or 1000 ppm, equal to 0, 0.6, 3.8, 18.9 and 33.4 mg/kg bw per day in males and 0, 0.6, 3.8, 18.0 and 33.3 mg/kg bw per day in females, for 14 weeks.

No mortalities or treatment-related clinical symptoms were observed during the study. Males at 1000 pm and females at 500 and 1000 ppm exhibited significant decreases in food consumption and decreased body-weight gain relative to controls and to respective baseline values. Mean food efficiency was negative in males at 1000 ppm and in females at 500 and 1000 ppm. Haematology and clinical chemistry parameters were unaffected by treatment. Gross and histopathological examinations revealed no treatment-related findings. Absolute and relative (to brain) heart weights were significantly decreased in males at 500 and 1000 ppm, but there were no histopathological or functional (electrocardiography) correlates indicating myocardial toxicity.

The NOAEL was 100 ppm, equal to 3.8 mg/kg bw per day in males and females, on the basis of decreased body weight and heart weight at 500 ppm and greater (Thompson et al., 1992).

In an assay for reverse mutation in bacteria, DIA (purity, 97.4%) did not induce gene mutations at the histidine locus of *S. typhimurium* and at the tryptophan locus in *E. coli* when tested at concentrations of up to 5000 µg/plate (Deparade, 1990).

In a test for DNA repair in vitro, DIA (purity, 97.4%) did not induce unscheduled DNA synthesis in primary rat hepatocytes exposed at concentrations of up to 800 µg/ml (Geleick, 1991b).

In an assay for micronucleus formation in mouse bone marrow, DIA (purity, 97.4%) gave negative results for micronucleus formation in the PCE of Tif:MAGF mice treated once orally at doses ranging from 120 to 480 mg/kg bw (Ogorek, 1991a).

In a study of prenatal developmental toxicity, which complied with GLP and the test guidelines of OECD and US EPA, groups of 24 mated female Tif:RAIf rats were given DIA (purity, 97.4%; suspended in 3% corn starch) at a dose of 0, 5, 25 or 100 mg/kg bw per day by oral gavage on days 6 to 15 of gestation.

All dams survived until the end of the experiment and no treatment-related clinical symptoms were observed. Body-weight gains were decreased by 27% at 25 mg/kg bw per day and by 70% at 100 mg/kg bw per day during days 6–11 of gestation, while food consumption was decreased by 9% at 25 mg/kg bw per day and by 20% at 100 mg/kg bw per day during days 6–11 gestation.

There were no treatment-related effects on any reproductive parameter examined, on fetal sex ratio, mean fetal weights or the incidences of gross, visceral and skeletal malformations. At 50 and 100 mg/kg bw per day, fetal and litter incidences of fused sternebrae 1 and 2 were significantly increased (Table 16). Significantly increased incidences of incomplete or absent ossification were noted at 100 mg/kg bw per day, which included poor ossification of sternebra 2, absent ossification of proximal phalanx of posterior digit 2, 3, 4 and 5, and absent ossification of metatarsal 1 (Table 16)

There was no evidence of teratogenicity.

The NOAEL for maternal toxicity was 5 mg/kg bw per day, on the basis of decreased body-weight gain and food intake at 25 mg/kg bw per day and greater. The NOAEL for developmental toxicity was 5 mg/kg bw per day, on the basis of increased incidences of fused sternebrae 1 and 2 at 25 mg/kg bw per day and greater (Marty, 1992a).

Table 16. Relevant findings in a study of prenatal developmental toxicity in rats given DIA by gavage

Finding	Dose (mg/kg bw per day)			
	0	5	25	100
Food consumption (g/day):				
Days 6–11	23	22	21**	16**
Days 11–16	26	25	25	23**
Body-weight gain (g), days 6–11	26.1	24.6	19.0	7.7**
Net weight change (g) from day 6	39.9	32.1	32.0	22.1**
No. of animals pregnant	23	21	24	24
Postimplantatation loss, mean	0.8	0.6	1.2	1.0
Postimplantation loss (%)	5.8	3.8	9.8	8.2
Mean No. of live fetuses	14.0	14.3	13.3	14.0
Mean fetal weights (g), males/females	5.6/5.3	5.6/5.3	5.7/5.3	5.5/5.2
Skeletal examination (No. of fetuses/litters)	160/22	155/21	165/22	172/23
Fused sternebrae 1 and 2[a]	0	0	9**/6*	29**/16**
Asymmetrically shaped sternebra 6[a]	0	0	1/1	4/3
Total skeletal anomalies[a]	2/2	4/4	14/11**	38/18**
Metatarsal 1; absent ossification[a]	14/9	6/6	22/9	34**/14
Posterior digit 2, proximal phalanx; absent ossification[a]	37/13	27/15	51/15	66**/18
Posterior digit 3, proximal phalanx; absent ossification[a]	23/11	12/9	29/14	42*/14
Posterior digit 4, proximal phalanx; absent ossification[a]	24/13	10*/8	28/13	43*/15
Posterior digit 5, proximal phalanx; absent ossification[a]	69/15	58/17	88/20	113**/22
Total skeletal variations[a]	160/22	155/21	164/22	172/23

From Marty (1992a)
DIA, deisopropyl-atrazine.
[a] No. of fetuses/litters.
* $p < 0.05$; ** $p < 0.01$.

(iii) Diaminochlorotriazine, DACT (G 28273)

DACT was of low acute oral toxicity in rats (LD_{50}, 2310 to 5460 mg/kg bw). In a short-term study in rats given DACT at dietary concentrations of up to 500 ppm, effects included reduced body-weight gain and food consumption and an increased number of females with shortened or prolonged estrous cycles and with persistent estrus or diestrus. The NOAEL was 10 ppm, equal to 0.7 mg/kg bw per day. In a 13/52-week study in dogs given DACT at dietary concentrations of up to 750–1500 ppm, effects included clinical signs and mortality, cardiac damage and failure, liver toxicity and decreased erythrocyte parameters. The NOAEL was 100 ppm, equal to 3.5 mg/kg bw per day. DACT was not genotoxic in a battery of tests, including in-vitro assays for point mutation and DNA repair and tests for clastogenicity in vivo. In a study of prenatal developmental toxicity in rats, the NOAEL for maternal toxicity was 2.5 mg/kg bw per day on the basis of decreased body-weight gain at 25 mg/kg bw per day and greater. The NOAEL for developmental toxicity was 2.5 mg/kg bw per day on the basis of increased incidences of incompletely ossified interparietals or parietals and unossified hyoids at 25 mg/kg bw per day and greater. There was no evidence of teratogenicity. In special studies on the effects of DACT on pubertal development in male and female rats, atrazine equimolar doses of $\geq$ 12.5 or $\geq$ 50 mg/kg bw per day delayed preputial separation or vaginal opening, respectively, with NOAELs of 6.25 or 25 mg/kg bw per day, respectively.

In a study of acute oral toxicity, which complied with GLP and the US EPA test guidelines, Blu:(SD) rats received DACT (purity, not reported; 16.65% w/v suspension in corn oil) as a single dose at 2463, 3547 or 4256 mg/kg bw by gavage. The group at the lowest dose contained five females and the group at the highest dose contained five males; 10 male and 10 females were used in the group at the intermediate dose. The rats were observed for clinical signs and mortality for 14 days. The oral LD_{50}s for males, females and both sexes combined were 3690, 2360 and 2310 mg/kg bw, respectively (Mehta, 1980a).

In a study of acute oral toxicity, which complied with GLP and the US EPA test guidelines, Blu:(SD) rats received DACT (purity, not reported; 33.3% w/v suspension in corn oil) as a single dose at 2491, 3547, 5050, or 7189 mg/kg bw by gavage. The group at the lowest dose contained five males, and five males and five females were used in all other groups. The rats were observed for clinical signs and mortality for 14 days. The oral LD_{50}s for males, females and both sexes combined LD_{50} were 11 300, 5230 and 5460 mg/kg bw, respectively (Mehta, 1980b).

In a study of acute oral toxicity, which complied with GLP and the US EPA test guidelines, HSD:(SD) rats received DACT (purity, not reported; 40% w/v suspension in deionized water) as a single dose at 4000, 5050, or 5500 mg/kg bw. The groups at the lowest dose and the highest dose contained five females, and the group at the intermediate dose contained five males and five females. The rats were observed for clinical signs and mortality for 14 days. The LD_{50} was > 5050 mg/kg bw for males and > 5500 mg/kg bw for females (Kuhn, 1991b).

In a short-term study of oral toxicity , which complied with GLP and the test guidelines of OECD and US EPA, groups of 15 male and 15 female Crl:CD(SD)BR rats were fed diets containing DACT (purity, 98.2%) at a concentration of 0, 10, 100, 250 or 500ppm, equal to 0, 0.7, 6.7, 16.7 and 34.1 mg/kg bw per day in males and 0, 0.7, 7.6, 19.7 and 40.2 mg/kg bw per day in females, for 90 days. In addition to the standard examinations, the estrous cycles and hormone concentrations of female rats were evaluated.

No mortalities or treatment-related clinical symptoms were observed during the study. There was a statistically significant decrease in body-weight gain in males at 500 ppm and in females at

250 and 500 ppm. Body-weight gain at week 12 was decreased by 19% in males at 500 ppm and by 15% and 17% in females at 250 and 500 ppm, respectively. Food consumption was generally comparable between all groups. There were no biologically significant effects on haematology, clinical chemistry and urine analysis at any dose. Organ-weight changes were unremarkable and tended to be associated with body-weight gain reductions. Macro- and micropathological examinations revealed no treatment-related findings.

At dietary concentrations of 100 ppm or greater, the proportion of rats exhibiting normal, 4- to 5-day estrous cycles tended to be reduced, while the incidence of persistent estrus tended to be increased, with the consequence of a reduction of the mean number of estrous cycles during the observation period. These effects, which were more pronounced on days 70–85 than during days 42–56, were statistically significant only at dietary concentrations of 250 ppm or greater after 70 days of treatment (Table 17). A more exact determination of effects at lower doses was precluded by the greater variability between individual rats in data on estrous cycle in the treated groups. There were no treatment-related effects on plasma concentrations of E2, progesterone, prolactin and corticosterone.

The NOAEL was 100 ppm, equal to 7.6 mg/kg bw per day in females, on the basis of decreased body-weight gain and significant effects on the estrous cycle at 250 ppm and greater (Pettersen et al., 1991; Terranova, 1991).

In a combined short-term and long-term study of oral toxicity, which complied with GLP and US EPA test guidelines, groups of 8–10 male and 8–10 female beagle dogs were fed diets containing DACT (purity, 98.7%) at a concentration of 0, 5, 100 or 1500 ppm for 13 or 52 weeks. Because of severe toxicity at the highest dose, which was evident after 6 weeks of treatment, the diet of the group of dogs at the highest dose was chanegd to one containing DACT at 750 ppm. Females tolerated this dose and received diet at 750 ppm until termination at 13 or 52 weeks. Since males continued to exhibit signs of toxicity at 750 ppm, they were fed untreated diet for weeks 9–13. At 13 weeks, two to six dogs per group were killed for interim study. Two females at the highest dose were then placed

Table 17. Data on the estrous cycle in rats given DACT in a short-term study of oral toxicity

Parameter	Dietary concentration (ppm)				
	0	10	100	250	500
Days 42–56					
Four- to five-day cycles[a]	14/15	12/15	9/15	8/15	8/15
Persistent estrus[a]	0/15	1/15	3/15	5/15	4/15
Persistent diestrus[a]	2/15	2/15	3/15	4/15	1/15
Mean No. of cycles (days)	2.60 ± 0.51	2.40 ± 0.74	2.07 ± 0.88	1.87 ± 0.83	2.00 ± 1.00
Days 70–85					
Four- to five-day cycles[a]	15/15	13/15	14/15	7/15*	6/15*
Persistent estrus[a]	0/15	1/15	2/15	8/15*	4/15
Persistent diestrus[a]	0/15	3/15	1/15	3/15	5/15
Mean No. of cycles (days)	2.80 ± 0.41	2.33 ± 0.90	2.47 ± 0.83	1.80 ± 1.01**	1.67 ± 1.11**

From Terranova (1991)
DACT, diaminochlorotriazine.
[a] Total may not equal 15 since a rat might display a regular 4- to 5-day cycle followed by persistent estrus or persistent diestrus.
* $p < 0.05$; ** $p < 0.01$.

on control diet for a 39-week recovery period, while four males at the highest dose were placed again on a diet containing DACT at 750 ppm until termination at 52 weeks. The mean daily intakes at 0, 5, 100 or 1500/750 ppm were equal to doses of 0, 0.2, 3.5 and 23.8 mg/kg bw per day for males and 0, 0.2, 3.3 and 29.9 mg/kg bw per day for females, respectively.

Five males and two females in the group at the highest dose were kiled in a moribund condition, mainly during the first 9 weeks. Clinical observation revealed tremors in all dogs at the highest dose during weeks 5–15. Additionally, inactivity, paleness, abdominal distension and emaciation were noted. Dogs that were killed moribund displayed inappetence, hypothermia, laboured breathing, hunched posture, and abnormal gait.

In the group at the highest dose, treatment-related physical examination findings consistent with impaired heart function occurred from week 6 onwards and included irregular (rapid) heart rate, pericardial thrill, pulse deficit, abdominal ascites and emaciation. Electrocardiographic abnormalities (atrial fibrillation) ocurred in three males and two females at week 5, the first electrocardiographic evaluation during the treatment period, and were diagnosed in a total of six males and four females in the group at the highest dose. Ophthalmological examinations revealed no treatment-related findings.

Dogs in the group at the highest dose lost weight during the first 6 weeks of treatment. After the concentration of DACT in the diet was reduced to 750 ppm, the dogs maintained their body weight. Body-weight gain of females in the group improved after cessation of test article treatment and was greater than the body-weight gain of females in the control group for the same period. Food consumption was decreased in males at the highest dose during the first 6 months of the study, except during the period when the dogs were fed the control diet. In females at the highest dose, the reduction of food consumption was not as pronounced as in males and was comparable to that of controls from week 18 onwards.

Moderate anaemia accompanied by an increased number of reticulocytes indicating an increased erythropoiesis was noted in dogs receiving the highest dose. The decrease of erythrocyte counts, erythrocyte volume fraction and haemoglobin concentration was significant in females at weeks 13 and 26. By week 52 the number of reticulocytes was within the normal range and the erythrocyte parameters had returned to nearly normal levels. Decreased calcium and increased lactate dehydrogenase and albumin concentrations were observed in the group at the highest dose. There were no effects of treatment on urinary parameters.

Mean absolute and relative liver, spleen and kidney weights were increased in dogs at the highest dose when compared with those of the controls at 13 and 52 weeks. At necropsy, heart and liver lesions were observed in dogs at the highest dose. Accumulations of fluids in the abdominal or thoracic cavity or the pericardium were secondary to the impaired heart function. Histopathology revealed chronic myocarditis in several dogs at the highest dose. The right atrium was most often affected, although the ventricles and papillary muscles were also affected in several animals. Males were more severely affected than females. Liver lesions included centrilobular fibrosis/atrophy, bile stasis, necrosis, haemorrhage, haemosiderosis and inflammation. Hyperplasia of the bone marrow, thymus atrophy and hypospermatogenesis of the testes were recorded at the interim kill, primarily in male dogs receiving the highest dose.

The effects observed in females at the highest dose were reversible 3 months after the cessation of treatment. The two females in the recovery group did not exhibit any clinical, electrocardiographic, gross or microscopic evidence of cardiac abnormalities. Likewise, no effects on haematological or biochemistry parameters were recorded at termination.

The NOAEL was 100 ppm, equal to 3.5 and 3.3 mg/kg bw per day in males and females, respectively, on the basis of clinical, electrocardiographic, gross and microscopic evidence of impaired heart function at 1500/750 ppm (Thompson et al., 1990).

In an assay for reverse mutation in bacteria, DACT (purity, 97%) did not induce gene mutations at the histidine locus of *S. typhimurium* when tested at concentrations of up to 5000 μg/plate (Deparade, 1987).

In two tests for DNA repair in vitro, DACT (purity, 97%) did not induce unscheduled DNA synthesis in primary rat hepatocytes exposed at concentrations of up to 400 μg/ml (Hertner, 1988b) and in human fibroblasts exposed at concentrations of up to 600 μg/ml (Meyer, 1987).

In an assay for micronucleus formation in mouse bone marrow , DACT (purity, 97%) gave negative results for micronucleus formation in the PCE of Tif:MAGF mice treated once orally at doses ranging from 1250 to 5000 mg/kg bw (Strasser, 1988).

In a study of prenatal developmental toxicity, which complied with GLP and the US EPA test guidelines, groups of 26 mated female Crl:COBS CD(SD)BR rats were given DACT (purity, 98.7%; suspended in 3% corn starch) at a dose of 0, 2.5, 25, 75 or 150 mg/kg bw per day by oral gavage on days 6 to 15 of gestation.

No mortalities or treatment-related clinical observations were noted. Food consumption was significantly decreased throughout the treatment period at 150 mg/kg bw per day and during days 6–8 of gestation at 75 mg/kg bw per day, while a non-significant decrease was noted at 25 mg/kg bw per day during the first 3 days of treatment. Body-weight loss was observed during the first 3 days of treatment (days 6–8 of gestation) in females at 75 and 150 mg/kg bw per day, and body weight was significantly decreased in the dams at the highest dose during the rest of the treatment period. Total (−33%) and corrected (−43%) body-weight gain (days 0–20 of gestation) was also significantly decreased at the same dose. A tendency towards a reduced body-weight gain was observed at 75 mg/kg bw per day for days 6–16 of gestation (–28%) and for days 6–8 of gestation (−32%) at 25 mg/kg bw per day. These decreases were of transient nature; total and corrected body-weight gains in the group at 25 mg/kg bw per day were comparable to those of rats in the control group.

The number of corpora lutea and implantation sites was comparable between all groups. The number of resorptions was significantly increased in the group at the highest dose. Fetal body weights were significantly decreased in the groups at 75 and 150 mg/kg bw per day.

No treatment-related malformations were observed, although incidental malformations included an umbilical hernia observed in one fetus at the lowest dose and one fetus at the highest dose; one fetus at the lowest dose had a protruding tongue, one fetus at 75 mg/kg bw per day had a filamentous tail, and one fetus at the highest dose was acaudate. Visceral malformations were restricted to a single fetus at 25 mg/kg bw per day. A significantly increased incidence of visceral variations was observed in fetuses at the highest dose (pitted kidneys and absent renal papillae). No skeletal malformations were observed at any dose. There was a dose-related increase of incomplete ossification of several bones at 75 and 150 mg/kg bw per day and in three skull bones (interparietal, parietals; hyoid [unossified]) at 25 mg/kg bw per day and greater (Table 18).

There was no evidence of teratogenicity.

The NOAEL for maternal toxicity was 2.5 mg/kg bw per day on the basis of decreased body-weight gain during the initial 3 days of treatment at 25 mg/kg bw per day. The NOAEL for developmental toxicity was 2.5 mg/kg bw per day on the basis of increased incidences of incompletely ossified interparietals or parietals and unossified hyoids at 25 mg/kg bw per day and greater (Hummel et al., 1989).

(iv) Hydroxyatrazine (G 34048)

Hydroxyatrazine was of low acute oral toxicity in rats (LD_{50}, > 5050 mg/kg bw). In short-term studies in rats given hydroxyatrazine at dietary concentrations of up to 750 ppm, effects included

Table 18. Relevant findings in a study of prenatal developmental toxicity in rats given DACT by gavage

Finding	Dose (mg/kg bw per day)				
	0	2.5	25	75	150
Food consumption (g/day):					
Days 6–8	22.95	22.80	20.62	18.48*	11.98*
Days 8–12	22.54	24.73	23.70	21.85	15.63*
Body-weight gain (g):					
Days 6–8	11.68	11.39	7.92	−1.64*	−11.26*
Days 8–12	21.18	23.00	22.20	17.12	5.70*
No. of pregnant rats	22	23	25	25	23
Mean No. of resorptions	0.77	0.48	1.04	0.84	2.61*
Postimplantatation loss (mean)	0.77	0.48	1.04	0.84	2.70*
Postimplantation loss (%)	5.56	3.59	7.15	6.16	18.86*
Mean No. of live fetuses	13.18	12.61	13.20	13.60	11.26
Mean fetal weights (g), males/females	3.45/3.29	3.45/3.32	3.43/3.29	3.14*/3.03*	2.79*/2.68*
Visceral examination (No. of fetuses/litters)	141/22	140/23	160/25	166/25	126/23
Renal papilla absent (No. of fetuses/litters)	5/3	7/6	8/5	12/8	28*/11*
Kidneys pitted (No. of fetuses/litters)	0	0	0	0	6*/3*
Skeletal examination (No. of fetuses/litters)	149/22	150/23	170/25	174/25	133/23
Hyoid not ossified (No. of fetuses/litters)	7/5	7/5	26*/10*	49*/16*	53*/15*
Interparietal not completely ossified (No. of fetuses/litters)	27/10	28/11	60*/20*	92*/23*	86*/21*
Parietals not completely ossified (No. of fetuses/litters)	5/3	14/5	18*/10*	20*/10*	20*/8*

From Hummel et al. (1989)
DACT, diaminochlorotriazine.
* $p < 0.05$; ** $p < 0.01$;

reduced body-weight gain, increased water consumption, changes in clinical chemistry and urine analysis parameters as well as macroscopic and microscopic lesions in the kidney, due to the low solubility of hydroxyatrazine in water resulting in crystal formation and consequent inflammatory response. The overall NOAEL was 100 ppm, equal to 6.3 mg/kg bw per day. In a 13-week study in dogs given hydroxyatrazine at dietary concentrations of up to 6000 ppm, effects included reduced body-weight gain and food consumption, changes in clinical chemistry and urine analysis parameters and macroscopic and microscopic lesions in the kidney. The NOAEL was 150 ppm, equal to 5.8 mg/kg bw per day. In a 2-year study of toxicity and carcinogenicity in rats given hydroxyatrazine at dietary concentrations of up to 400 ppm, effects included clinical signs and increased mortality, reduced body-weight gain and food consumption, increased water consumption, changes in haematological, clinical chemistry and urine analysis parameters and macroscopic and microscopic lesions in the kidney. The NOAEL was 25 ppm, equal to 1 mg/kg bw per day. There was no evidence of carcinogenicity. Hydroxyatrazine was not genotoxic in a battery of tests including assays for point mutation and DNA repair in vitro and tests for clastogenicity in vivo. In a study of prenatal developmental toxicity in rats, reduced food consumption and body-weight gain in dams and increased incidences of incompletely ossified hyoid and interparietal bones and not ossified forepaw metacarpals and proximal phalanges in fetuses were seen at 125 mg/kg bw per day, and the NOAEL was 25 mg/kg bw per day for both maternal and developmental toxicity. In a special study on the effects of hydroxyatrazine on pubertal development in female rats, atrazine equimolar doses of up to 200 mg/kg bw per day did not significantly delay vaginal opening.

In a study of acute oral toxicity, which complied with GLP and the US EPA test guidelines, a group of five male and five female HSD:(SD) rats received hydroxyatrazine (purity, 97.1%; suspended in 2% carboxymethyl-cellulose) as a single dose at 5050 mg/kg bw by gavage. The rats were observed for clinical signs and mortality for 14 days. The oral LD_{50} was > 5050 mg/kg bw (Kuhn, 1991e).

In a short-term study of oral toxicity, which complied with GLP and US EPA test guidelines, groups of 10 male and 10 female Sprague-Dawley rats were fed diets containing hydroxyatrazine (purity, not reported) at a concentration of 0, 10, 250, 500 or 750 ppm, equal to 0, 0.92, 22.5, 44.4 and 67.3 mg/kg bw per day in males and 0, 1.0, 25.2, 47.3 and 67.4 mg/kg bw per day in females, for at least 29 days.

No mortalities were observed during the study. An increased incidence of diarrhoea was observed in males at 250 ppm and greater. Body weight and body-weight gain were decreased in males at 500 ppm and greater and in females at 750 ppm, and food consumption was reduced in males at 250 ppm and greater and in females at 500 ppm and greater. Some changes in haematological and clinical chemistry parameters were observed in both sexes at 500 ppm and greater. At necropsy, an increased incidence of mottled and rough kidneys was seen in males at 500 ppm and greater and in females at 750 ppm.

The NOAEL was 10 ppm, equal to 0.9 and 1.0 mg/kg bw per day in males and females, respectively, on the basis of increased incidence of diarrhoea and reduced food consumption at 250 ppm and greater (Hazelette & Arthur, 1989a).

In a short-term study of oral toxicity, which complied with GLP and US EPA test guidelines, groups of 15 male and 15 female Sprague-Dawley (Crl:CD Br) rats were fed diets containing hydroxyatrazine (purity, 97.1%) at a concentration of 0, 10, 100, 300, or 600 ppm, equal to 0, 0.64, 6.3, 18.89 and 37.47 mg/kg bw per day in males and 0, 0.75, 7.35, 22.73 and 45.64 mg/kg bw per day in females, for 13 weeks.

No mortality occurred throughout the study and no treatment-related clinical symptoms were observed. Body-weight gain was decreased by about 12.5% in both sexes at 600 ppm, while food consumption was slightly decreased in males at 600 ppm. An increase in water consumption was observed in both sexes at 600 ppm. Slight decreases in erythrocyte parameters (erythrocyte count, haemoglobin concentration and erythrocyte volume fraction) and increases in serum urea nitrogen, creatinine, sodium and chloride concentrations were seen at 600 ppm, while increased chloride concentrations were also observed in females at 300 ppm. Mean urine volume was increased in males at 300 ppm and in both sexes at 600 ppm, and mean specific gravity in urine was decreased in females at 600 ppm.

At necropsy, mean absolute and relative kidney weights were increased in both sexes at 600 ppm (absolute weights, +33.7 and +33.2%; relative organ-to-body weights, +44.3 and +45.1% in males and females, respectively). Rough or pitted kidneys, with or without pale and tan discoloration, were noted in all males and in 14 out of 15 females at 600 ppm. Similar findings were noted in four males and two females at 300 ppm. Treatment-related histopathological findings were restricted to the kidneys at 300 ppm and greater. All rats at the highest dose had marked tubular dilatation and basophilia, extensive chronic hyperplastic inflammation in the interstices, and cellular casts. Anisotropic crystals, later identified as hydroxyatrazine, were noted in the papillary tubules of 11 males and 13 females of the group at the highest dose. Less severe lesions (minimal tubular dilatation and tubular basophilia, minimal subacute interstitial inflammation) were observed in 7 males and 11 females at 300 ppm.

The NOAEL was 100 ppm, equal to 6.3 and 7.35 mg/kg bw per day in males and females, respectively, on the basis of kidney toxicity (tubular and interstitial chronic nephropathy) at 300 ppm and greater (Rudzki et al., 1989).

In a short-term study of oral toxicity, which complied with GLP and US EPA test guidelines, groups of four male and four female beagle dogs were fed diets containing hydroxyatrazine (purity, 97.1%) at a concentration of 0, 15, 150, 1500, or 6000 ppm, equal to 0, 0.6, 5.8, 59.6, and 247.7 mg/kg bw per day in males and 0, 0.6, 6.2, 63.9, and 222.1 mg/kg bw per day in females, for 13 weeks.

There were no mortalities. A treatment-related reduction in body-weight gain (−29% in males, -31-37% in females), accompanied by some initial reduction in food consumption was observed at 1500 ppm and greater. Treatment-related clinical chemistry changes were restricted to females at 6000 ppm that displayed increased blood urea nitrogen (BUN) and/or creatinine concentrations. Urine analysis performed at day 85 revealed an increased urine volume (about twofold) and a decreased specific gravity at 1500 ppm and greater.

There were no treatment-related effects on organ weights, although at necropsy, pitted or rough kidneys were observed in three out of four males and one out of four females at 1500 ppm and four out of four males and two out of four females at 6000 ppm. Treatment-related histopathological lesions were restricted to the kidneys at 1500 ppm and greater. Minimal to marked multifocal, chronic nephropathy characterized by tubular dilation, atrophy and basophilia was observed, often in the presence of a prominent chronic interstitial fibrosis and lymphocytic infiltration. The kidney lesions appeared predominantly in the cortex, while medullary areas were occasionally involved. Intratubular crystalline casts were observed in the renal papilla in all males at 1500 ppm and greater and in three out of four females in each group at 1500 and 6000 ppm.

The NOAEL was 150 ppm, equal to 5.8 and 6.2 mg/kg bw per day in males and females, respectively, on the basis of decreased body-weight gain and kidney toxicity (tubular and interstitial chronic nephropathy) at 1500 ppm and greater (Chau et al., 1990).

In a combined long-term study of toxicity and carcinogenicity, which complied with GLP and the test guidelines of OECD and US EPA, groups of 80 male and 80 female (control and highest dose) or 70 male and 70 female (all other groups) Crl:CD(SD)BR rats were fed diets containing hydroxyatrazine (purity, 97.1%) at a concentration of 0, 10, 25, 200, or 400 ppm, equal to 0, 0.388, 0.962, 7.75 and 17.4 mg/kg bw per day in males and 0, 0.475, 1.17, 9.53 and 22.3 mg/kg bw per day in females, for 24 months. Ten males and 10 females (intermediate dose) or 20 males and 20 females (control and highest dose) were scheduled for interim kill at 12 months.

Mortality attributable to severe renal failure was markedly increased at 400 ppm (survival rate by week 52, 96%, 94%, 94%, 94% and 75% in males; 98%, 97%, 97%, 97%, and 76% in females at 0, 10, 25, 200 and 400 ppm, respectively) and, thus, rats at the highest dose were killed after 18 months. Treatment-related clinical signs were limited to the group at 400 ppm and included emaciation, polyuria, general pallor, piloerection and tremors. Body weight and body-weight gain were significantly decreased at 400 ppm throughout the study. Food consumption was decreased at 200 ppm and greater, while food efficiency was decreased only at 400 ppm. Water consumption was increased at 200 ppm and greater, but only for the first year.

Treatment-related changes in haematology parameters (decreases in erythrocyte count, haemoglobin concentration, erythrocyte volume fraction), clinical chemistry parameters (increased calcium, phosphorus, gamma-glutamyl transferase, BUN and creatinine concentrations; decreased glucose, total protein and albumin plasma concentrations) and urine analysis parameters (decreased pH, specific gravity and colour intensity; increased urine volume; presence of crystalline hydroxyatrazine sediments) were observed at 400 ppm.

Necropsy of rats at interim and decedent rats at 400 ppm revealed discoloration and enlargement of renal lymph nodes and calculi, cysts, dilated pelvis and rough pitted surface of the kidney. Additionally, in males calculi and thickened walls were recorded in the urinary bladder. Absolute and relative kidney weights were increased in males (+19% and +50%) and females (+10% and +51%) at 400 ppm at 12 months only.

Treatment-related histopathological changes in the kidneys were noted at 200 ppm and greater at all killing intervals and included the deposition of crystalline material within collecting ducts, renal pelvises and occasional distal tubules (summarized in Table 19 as "dilatation with crystal deposits"). Tubules and collecting ducts that contained the crystalline material were dilated and either devoid of epithelium or lined by hyperplastic tubular epithelium. The tubular changes were often accompanied by acute intratubular inflammatory infiltration and by thickening and fibrosis of the papillary interstitium. In kidneys that contained pelvic aggregates of the crystalline material, multifocal transitional cell erosion and/or ulceration of the renal transitional epithelium was noted.

In the renal papillae, swelling and increased eosinophilia of interstitial cells, which was accompanied by the interstitial accumulation of a hyaline basophilic material (acidic sulfated mucosubstances that make up the ground substance of the interstitial matrix) was significantly increased in incidence/severity in males at 200 ppm and greater and in females at at 25 ppm and greater. However, the toxicological significance of the minimal to moderate accumulation of this matrix in females at 25 ppm, in the absence of any other signs of renal damage or impaired renal function, was highly questionable.

In males at 400 ppm and in females at 200 ppm and greater, papillary lesions were accompanied by cortical changes consistent with chronic progressive nephropathy. These consisted of thickening of tubular and glomerular basement membranes, tubular dilatation with accumulation of proteinaceous material, chronic interstitial nephritis, hyaline droplet accumulation within proximal tubular cells, pronounced glomerulosclerosis and infrequent tubular epithelial basophilia or hyperplasia.

In both sexes at 400 ppm, nephropathy was often accompanied by mineralization of renal (tubular epithelia and basement membranes) as well as extrarenal (e.g. aorta, heart, and lungs) tissues (metastatic mineralization).

There was no increase in the incidence of any tumour or decrease in any tumour-onset time that was attributable to treatment.

Table 19. Relevant histopathological findings in the kidney of rats fed diets containing hydroxyatrazine in a long-term study of toxicity

Finding	Dietary concentration (ppm)									
	Males					Females				
	0	10	25	200	400	0	10	25	200	400
Kidney, No. examined	79	69	70	70	80	79	70	68	69	80
Dilatation with crystal deposits	0	0	0	5**	79**	0	0	0	15**	78**
Nephropathy, progressive	75	67	64	65	80**	36	32	34	50**	79**
Papilla, accumulation, interstitial, matrix	4	3	2	32**		17	10	26*	26	0**
Papilla, fibrosis, interstitial	1	2	1	11**	80**	0	0	0	20**	79**
Pelvis, dilatation with crystal deposits	0	0	0	5	60**	0	0	0	9*	40**

From Chow & Hart (1995)
* $p < 0.05$; ** $p < 0.01$.

Hydroxyatrazine was not carcinogenic in rats. The NOAEL was 25 ppm, equal to 0.96 and 1.2 mg/kg bw per day in males and females, respectively, on the basis of kidney toxicity at 200 ppm and greater (Chow & Hart, 1995).

In an assay for reverse mutation in bacteria, hydroxyatrazine (purity, 99%) did not induce gene mutations at the histidine locus of *S. typhimurium* when tested at concentrations of up to 5000 μg/plate (Deparade, 1988).

In two tests for DNA repair in vitro, hydroxyatrazine (purity, 96–99%) did not induce unscheduled DNA synthesis in primary rat hepatocytes exposed at concentrations of up to 1500 μg/ml (Hertner, 1988a) and in human fibroblasts exposed at concentrations of up to 1500 μg/ml (Meyer, 1988).

In an assay for micronucleus formation in mouse bone marrow, hydroxyatrazine (purity, 99%) did not induce micronucleus formation in the PCE of Tif:MAGF mice treated once orally at doses ranging from 1250 to 5000 mg/kg bw (Ceresa, 1988a).

In a study of prenatal developmental toxicity, which complied with GLP and US EPA test guidelines, groups of 26 pregnant female Crl:COBS CD(SD)BR rats were given hydroxyatrazine (purity, 96.7%; suspended in 3% corn starch with 0.5% Tween 80) at a dose of 0, 5, 25 or 125 mg/kg bw per day by oral gavage on days 6 to 15 of gestation.

No mortalities or treatment-related clinical observations were noted. Food consumption was significantly decreased at 125 mg/kg bw per day during days 8–12 of gestation (−12.1%) and during the entire dosing period (−8.4%). Body-weight gain at 125 mg/kg bw per day was decreased by about 24% during days 8–12 of gestation, while corrected body-weight gain was decreased by about 11%.

There were no treatment-related effects on any reproductive parameter examined. Fetal body weights were slightly, but significantly lower at 125 mg/kg bw per day (males, −3.9%; females, −5.0%). There were no treatment-related external, visceral or skeletal malformations. At 125 mg/kg bw per day, there were significantly increased fetal and litter incidences of not completely ossified hyoids, not completely ossified interparietals, and not ossified forepaw metacarpals and proximal phalanges (Table 20).

There was no evidence of teratogenicity.

The NOAEL for maternal toxicity was 25 mg/kg bw per day on the basis of decreased food consumption and body-weight gain at 125 mg/kg bw per day. The NOAEL for developmental toxicity was 25 mg/kg bw per day on the basis of increased incidences of incompletely ossified hyoid and interparietal bones and not ossified forepaw metacarpals and proximal phalanges at 125 mg/kg bw per day (Lindsay et al., 1989).

(b) Studies on site and mechanism of action in the central nervous system

In a in-vitro study on the mechanism by which chlorotriazines interfere with hypothalamic control of the gonadotrophin releasing hormone (GnRH) release and luteinizing hormone (LH) surge, the ability of atrazine and its metabolites DIA, DEA and DACT to interact with gamma-aminobutyric acid A-type ($GABA_A$) receptors in rat cortical membranes was examined by measuring their effects on binding of the following prototypical ligands to their recognition sites on $GABA_A$ receptors: [^{3}H]muscimol, which binds to the GABA binding site; [^{3}H]Ro15-4513, which binds to the benzodiazepine site on the $GABA_A$ receptor; and [^{35}S]*tert*-butylbicyclophosphorothionate (TBPS), which binds to the picrotoxin binding site in the chloride channel of $GABA_A$ receptors.

Atrazine significantly inhibited the binding of [^{3}H]Ro15-4513 at concentrations of 30 μmol/l and greater, and the half-maximal inhibitory concentration (IC_{50}) was calculated to be 305 μmol/l. The chlorotriazine metabolites, however, were without significant effect on Ro15-4513-binding when

Table 20. Relevant findings in a study of prenatal developmental toxicity in rats given hydroxyatrazine by gavage

Finding	Dose (mg/kg bw per day)			
	0	5	25	100
Food consumption (g/day), days 8–12	23.4	23.7	22.6	20.6*
Total food consumption (g), days 6–16	229.4	236.7	223.3	210.1*
Body-weight gain (g), days 8–12	23.5	20.9	21.1	18.0
Net body-weight change (g), days 0–20	77.2	81.9	71.1	68.6
No. of rats pregnant	25	23	23	22
Postimplantatation loss (mean)	0.5	1.1	1.0	0.7
Postimplantation loss (%)	5.6	7.4	11.6	4.9
Mean No. of live fetuses	13.40	13.30	12.30	14.14
Mean fetal weights (g), males/females	3.61/3.43	3.71/3.51	3.55/3.35	3.47*/3.26*
Skeletal examination (No. of fetuses/litters)	177/25	159/23	147/22	160/22
Hyoid not completely ossified (No. of fetuses/litters)	11/5	26/9	14/8	27*/12*
Interparietal not completely ossified (No. of fetuses/litters)	35/14	57/19	42/14	70*/20*
Metacarpal not ossified (No. of fetuses/litters)	67/21	58/18	61/19	96*/21*
Proximal phalange not ossified (No. of fetuses/litters)	169/24	153/23	139/22	159*/22*

From Lindsay et al. (1989)
* $p < 0.05$; ** $p < 0.01$.\

compared with controls. All substances were without effect on [^{3}H]muscimol or [^{35}S]TBPS binding. The results suggested that atrazine modulates benzodiazepine, but not the muscimol (GABA receptor site) or TBPS (chloride channel), binding sites on $GABA_A$ receptors (Shafer et al., 1999).

In a study on the effect of chlorotriazines on catecholamine metabolism in vitro using pheochromocytoma (PC12) cells, intracellular norepinephrine (NE) and dopamine (DA) concentrations and spontaneous NE release were measured after treatment with atrazine (0, 12.5, 25, 50, 100 and 200 µmol/l) for 6, 12, 18, 24 and 48 h.

Intracellular DA concentration was significantly decreased at 12.5 µmol/l and greater, while intracellular NE concentration was significantly decreased at 100 µmol/l and greater. Similarly, there was a dose-dependent inhibition of NE release at 50 µmol/l and greater. The $GABA_A$-receptor agonist, muscimol (at 0, 0.01, 0.1, and 1.0 µmol/l) had no effect on either the release or on intracellular catecholamine concentrations from 6 h until 24 h of treatment. Cell viability was somewhat lower at 100–200 µmol/l, but the reduction in viability was significant only at 200 µmol/l at 24 h. The data indicated that atrazine inhibits the cellular synthesis of DA mediated by the tyrosine hydroxylase, and NE mediated by dopamine beta-hydroxylase, resulting in an inhibition of NE release. Thus, atrazine presumably acts at the enzymatic steps or sites of DA biosynthesis to modulate monoaminergic activity in PC12 cells (Das et al., 2000).

In a subsequent study on the effect of chlorotriazine metabolites on catecholamine metabolism in vitro using PC12 cells, intracellular NE and DA concentrations and spontaneous NE release were measured after treatment with hydroxyatrazine (0–400 µmol/l), DEA (0–200 µmol/l), DIA (0–200 µmol/l) and DACT (0–160 µmol/l) for 3–24 h.

Hydroxyatrazine significantly decreased intracellular DA and NE concentrations in a dose- and time-dependent manner, and caused also a significant inhibition of NE release from cells. In contrast,

DEA and DIA significantly increased intracellular DA concentrations at 50 μmol/l and greater from 12 to 24 h. Intracellular NE was significantly reduced at 50 μmol/l and greater of DEA at 24 h, while DIA had no effect. NE release was decreased at 100 μmol/l and greater for both DEA and DIA. DACT significantly increased intracellular DA and NE concentrations at 160 μmol/l, while NE release was increased at 40 μmol/l and greater. Cell viability was reduced by 10–12% in the presence of hydroxyatrazine at 200–400 μmol/l, while for the other metabolites, viability was reduced by only 2–5% at the highest concentrations. The data suggested that the catecholamine neurons may be a target for the chlorotriazines and/or their metabolites, that the metabolites produce a unique pattern of catecholamine response, and that all of the changes were seen within the same range of doses (Das et al., 2001).

In a subsequent study on mechanisms responsible for chlorotriazine-induced alterations in catecholamines in PC12 cells in vitro, the effect of atrazine (100 μmol/l for 1–36 h) on the protein expression of the enzymes responsible for synthesis of DA—tyrosine hydroxylase (TH)—and NA—dopamine-β-hydroxylase (DβH)—was determined.

Atrazine decreased intracellular DA and NE concentration and NE release, and the protein expression of TH and DβH. Siumultaneous exposure to the essential cofactors for TH (iron and tetrahydrobiopterine) was ineffective in altering cellular DA. Agents known to enhance TH and DβH transcription, phosphorylation or activity (e.g. 8-bromo cAMP, forskolin or dexamethasone) reversed the inhibitory effects of atrazine on NE. The data indicated that atrazine affects DA and NE synthesis, probably via an alteration of the enzymes TH and DβH (Das et al., 2003).

In a study on the effects of atrazine on the brain monoamine systems, groups of male Long-Evans (LE) rats received diets containing atrazine (purity, 98%) at a dose of 0, 5 or 10 mg/kg bw per day for 6 months, and locomotor activity (at 2, 3 and 6 months and 2 weeks after cessation of exposure), monoamine levels (in hypothalamus, prefrontal cortex, striatum, nucleus accumbens) and the numbers of TH-positive and TH-negative dopaminergic neurons (in substatia nigra pars compacta, ventral tegmental area) were determined. At 10 mg/kg bw per day, rats exhibited an enhanced locomotor activity response to a *d*-amphetamine challenge (evaluated at 2 months only), while increased locomotor activity was present at 3 and 6 months and at termination (2 weeks after cessation of exposure). Also at termination, the levels of various monoamines were decreased (e.g. DA, −20% in striatum; serotonin, −15% in hypothalamus; NE, −20% in prefrontal cortex), and the numbers of TH-positive and TH-negative dopaminergic neurons were reduced in both dopaminergic tracts (9–13%) at 10 mg/kg bw per day. Acute exposures of male rats to atrazine given intraperitoneally at doses of 100 and 200 mg/kg bw reduced basal and potassium-evoked striatal release of DA (Rodriguez et al., 2005).

In studies conducted in vivo to further examine effects of atrazine on the concentration of DA and NE in three different hypothalamic nuclei, administration of atrazine to intact female rats at doses of 25 and 75 mg/kg bw per day by gavage during one complete 4-day estrous cycle resulted in an increased DA concentration in the median eminence region (presumably the tuberoinfundibular dopaminergic neurons) and the medial preoptic area, when compared with controls, while NE concentration was not different in any of the three nuclei. Treatment at doses of 25 and 75 mg/kg bw per day also resulted in an increased concentration of GnRH in the median eminence when compared with controls (Cooper et al., 2007).

(c) Studies on effects on estrous cycle or LH surge

Rats

In a study on the effects of atrazine on ovarian function in the rat, female LE hooded and Sprague-Dawley rats aged approximately 15 weeks were given atrazine (purity, 97.1%) at a dose of

0, 75, 150 or 300 mg/kg bw per day by gavage for 21 days. Only rats that displayed regular 4-day estrous cycles for 2 weeks were used in the study. Blood for serum hormone concentrations (E2, progesterone) was taken from rats that displayed a pattern of vaginal diestrus for 10 days. After the 21-day treatment period, all females were ovariectomized (the day of ovariectomy was selected by using the vaginal smear pattern) and the ovaries were examined microscopically.

In both strains, dosing at 75 mg/kg bw per day disrupted the 4-day ovarian cycle; however, no distinct alteration (i.e. irregular cycles but not persistent estrus or diestrus) was apparent at this dose. At 150 mg/kg bw per day, atrazine induced repetitive pseudopregnancies in females of both strains. The highest dose tested (300 mg/kg bw per day) also induced repetitive pseudopregnancies in the Sprague-Dawley females, while the ovaries of the LE hooded females appeared regressed and the smear cytology was indicative of the anestrous condition.

A NOAEL was not identified; however, the doses employed were in excess of those used in long-term feeding studies in which an early onset of mammary gland tumours was noted. These data demonstrate that atrazine can disrupt ovarian function and bring about major changes in the endocrine profile of the female rat (Cooper et al., 1996).

In a pilot study designed to determine the validity of a proposed protocol for testing the effect of exposure to atrazine on the proestrous afternoon LH surge, E2 as estradiol benzoate was given to 70 ovariectomized female Sprague-Dawley rats (Crl:CD BR) through a subcutaneously surgically implanted capsule. Serum concentrations of LH, E2, and prolactin were measured 3 days later at 2-h intervals. The results showed that nearly all of the rats had E2 concentrations within the desired range of 75–150 pg/ml in the hour preceeding and during the rise of LH. There was a peak in LH secretion at 16:00 biological time, while prolaction showed a peak at 14:00 biological time and a return to baseline values by 24:00 biological time. The results were consistent with numerous published reports of LH and prolactin surges in intact cycling female rats, thus demonstrating the validity of the experimental method (Morseth, 1996a).

In a study of method validation designed to determine the optimum experimental methods to be used for testing the integrity of the LH surge in atrazine-treated female rats, E2 (as estradiol benzoate) was given to 20 ovariectomized female Sprague-Dawley (Crl:CD BR) rats through a subcutaneously surgically implanted capsule. Ten rats were used as a control group for the vehicle (0.5% aqueous carboxymethyl-cellulose), and ten rats were given atrazine (purity, 97.1%) by oral gavage for 3 days, beginning the day after surgery. The rats were subsequently bled at designated intervals (11:00, 13:00, 15:00, 18:00 and 22:00 biological time), and serum LH and prolactin concentrations were measured. Results indicated that the LH surge was attenuated in rats treated with atrazine at 15:00, the time range in which the peak LH surge is expected to occur in young intact rats. Prolactin concentrations rose over the course of the day, but failed to return to the expected baseline level late in the day in the control group and in the rats treated with atrazine, which was considered to be a response to the stress of repeated jugular bleeding (Morseth, 1996b).

In a study designed to evaluate the effects of atrazine on the pre-ovulatory LH surge and on the estrous cycle, groups of 90 female Sprague-Dawley (Crl:CD BR) rats received diets containing atrazine (purity, 97.1%) at a concentration of 0, 25, 50 or 400 ppm (equal to 0, 1.8, 3.65 and 29.44 mg/kg bw per day) for 26 weeks. The study was conducted in compliance with GLP guidelines. Vaginal smears were evaluated for 2-week periods each 4 weeks. Ten days before beng killed, the rats were ovariectomized, and 3 days before being killed a capsule releasing E2 was implanted subcutaneously. The rats were killed by decapitation in groups of 10 or 15 at designated intervals (11:00, 14:00, 16:00, 18:00, 20:00 and 23:00 biological time), and blood samples were analysed for LH, prolactin and E2. Serial blood samples were drawn without anaesthesia at the same time-point from an additional group of 10 rats and analysed for LH. Rats were necropsied and histology (limited to mammary tissue, uterus, vagina and ovaries by the study plan) was not reported. Rats showing abnormal E2 data were excluded from the results.

There were no compound-related effects in mortality or clinical signs. Body weight, body-weight gain and food consumption were significantly decreased in rats at the highest dose tested compared with controls.

The percentage of days in estrus was significantly increased during the 21–22 and 25–26-week periods at 400 ppm. The percentage of days in estrus was also increased during the 21–22 and 25–26-week periods at 50 ppm, but the increase was only significant for the 21–22-week period (Table 21).

All the rats evaluated had E2 concentrations that were within the acceptable range to prime the LH surge in the hour preceding and during the rise of LH. The baseline (11:00) LH values were similar among groups. In non-repeat bled rats, LH surges comparable to those in controls were observed at 25 and 50 ppm, while rats at 400 ppm failed to have an LH surge, with LH values never rising above baseline. In repeat-bled rats, the LH surge was also severely attenuated at 400 ppm and less so at 50 ppm (maximum increase over baseline was 157% compared with maximum increase over baseline of of 273% in controls). Prolactin concentrations increased over the course of the sampling period in all groups, with peak values noted at 14:00 and 16:00, and began to return to their baseline values by 20:00 biological time. Treatment with atrazine had no effect on prolactin concentrations.

Most of the rats at termination had distended uteri with fluid-filled lumen, a condition commonly seen in ovarietomized, E2-stimulated rats. Selected tissues were saved for histopathology, but the absence of the histopathology report was not considered to be of significance when interpreting the results.

Table 21. Selected findings in a study designed to evaluate the effects of atrazine on the pre-ovulatory LH surge and on the estrous cycle in rats

Finding	Dietary concentration (ppm)			
	0	25	50	400
Body-weight gain (g), weeks 1–26	133	138	131	114*
Food consumption (g), weeks 1–25	3438	3462	3455	3309*
Percentage of days in diestrus/estrus				
Weeks 1–2	58/22	57/22	56/22	61/21
Weeks 13–14	49/31	53/28	49/31	44/40*
Weeks 17–18	47/34	49/33	47/36	41/45*
Weeks 21–22	51/32	43/41	39**/45*	37**/51**
Weeks 25–26	40/47	42/48	34/54	29*/63*
LH (pg/ml), non-repeat bled rats:				
11:00	1900	1816	1581	1863
18:00	3456	3235	3175	1358*
20:00	2327	2249	1899	1308*
LH (pg/ml), repeat-bled rats:	909	1075	972	1005
11:00				
18:00	3336	3631	2500	858*
20:00	3388	2510	2409	1042*

From Morseth (1996c)
LH, luteinizing hormone.
* $p < 0.05$; ** $p < 0.01$.

The NOAEL was 25 ppm, equal to 1.8 mg/kg bw per day, on the basis of estrous cycle alterations and attenuation of the LH surge at 50 ppm (Morseth, 1996c).

In a study on the effects of atrazine on the hypothalamic control of pituitary-ovarian function in the rat, the estrogen-induced surges of LH and prolactin were examined in ovariectomized Sprague-Dawley and LE rats treated with atrazine (purity, 97.1%) by gavage for 1, 3 or 21 days. Rats treated for 1 day were ovariectomized and received a subcutaneous estrogen implant on day 0, and then atrazine at a dose of 0, 50, 100, 200 or 300 mg/kg bw on day 3, and were killed at 0, 1, 3 or 6 h after treatment. Rats treated for 3 days were ovariectomized and received a subcutaneous estrogen implant on day 0, and then atrazine at a dose of 0, 50, 100, 200 or 300 mg/kg bw per day on days 1–3, and were killed at 0, 1, 3 or 6 h after treatment on day 3. Rats treated for 21 days were ovariectomized on day 0, and then received atrazine at a dose of 0, 75, 150 or 300 mg/kg bw per day on days 1–21, followed by a subcutaneous estrogen implant on day 21 and were killed on day 24 at the time of expected peak LH concentration.

Atrazine at the highest single dose (300 mg/kg bw) significantly suppressed the LH and prolactin surge in ovariectomized LE rats, but not in Sprague-Dawley rats. Treatment with atrazine on three consecutive days suppressed the estrogen-induced LH and prolactin surges in ovariectomized LE females in a dose-dependent manner at 50 mg/kg bw and greater, but this same treatment was without effect on serum LH and prolactin in Sprague-Dawley females. Also in LE females, exposure to atrazine at a dose of 50 mg/kg bw and greater inhibited the decrease in pituitary prolactin concentration that was observed in rats in the control group. After 21 days of treatment with atrazine, there was a significant dose-dependent suppression of serum LH and prolactin in both strains at all doses, while the concentration of prolactin in the pituitary was significantly increased in both strains at all doses.

In a further experiment conducted to determine the effect of a single dose of atrazine on ovulation and ovarian cycling, intact female LE rats displaying regular 4-day estrous cycles received atrazine at a dose of 0, 75, 150 or 300 mg/kg bw on the day of vaginal proestrus, and vaginal smears were examined for 3 weeks in half of the rats in each group, while in the remaining rats, oocytes were collected and quantified on the day of vaginal estrus. Atrazine administered at the highest dose of 300 mg/kg bw induced a pseudopregnancy in seven of nine females, but was without effect on ovulation.

Three further experiments were performed to determine whether the brain, pituitary or both organs were target sites for atrazine. These included examination of the ability of: (a) the pituitary lactotrophs to secrete prolactin, using hypophysectomized females bearing pituitary autotransplants (ectopic pituitaries) in female LE rats receiving atrazine as a single dose at 300 mg/kg bw; (b) the synthetic GnRH to induce LH secretion in ovariectomized LE rats treated with atrazine at a dose of 300 mg/kg bw for 3 days; and (c) atrazine (administered in vivo or in vitro) to suppress LH and prolactin secretion from pituitaries, using a flow-through perfusion procedure. The results indicated that: (a) the secretion of prolactin by the pituitary was not altered by atrazine if the gland was removed from the influence of central nervous sytem factors; (b) the effect of atrazine on LH secretion was not a result of a direct impairment of LH release from the pituitary; and (c) direct exposure of the pituitary had no effect on the release of LH and prolactin.

In summary, the experiments demonstrated a clear effect of atrazine on the estrogen-induced LH and prolactin surge in female LE and Sprague-Dawley rats. LE rats appeared to be more sensitive to the hormone-suppressive effects of atrazine than Sprague-Dawley rats. The results support the hypothesis that the effects of atrazine on LH and prolactin secretion are mediated via a hypothalamic site of action (Cooper et al., 2000).

In a study designed to compare the effects of atrazine and its metabolite DACT on the the pre-ovulatory LH surge, groups of 20 female Sprague-Dawley [Crl:CD BR] rats were given atrazine (purity, 97.1%) or DACT (purity, 96.8%) at a dose of 2,5, 5, 40 or 200 mg/kg bw per day by oral gavage for 4 weeks; a group of 40 females served as a control group for the vehicle (0.5% carboxymethyl cellulose). The study was conducted in compliance with GLP guidelines. On day 22, all rats were ovariectomized, and on day 28, a capsule releasing E2 was implanted subcutaneously. On day 31, all rats were dosed at about 06:30 (10:30 biological time), and serial blood samples were collected from each rat at designated time-points (13:00, 16:00, 18:00, 20:00, 22:00 and 24:00 biological time) for analysis of LH. The maximum LH amplitude (LH_{max}), the time to maximum concentration of LH (T_{max}), and the area under the LH curve (AUC) were evaluated. After the final blood sample, the rats were killed and mammary tissues, uterus, vagina and pituitary were examined and preserved. Administration of both atrazine and DACT was associated with body-weight losses during the first week of treatment at doses of 40 mg/kg bw per day and greater, while body-weight gain during the study was decreased at doses of 40 mg/kg bw per day or greater for DACT and 200 mg/kg bw per day for atrazine, respectively. Both LH_{max} and AUC were significantly decreased in rats given DACT at 200 mg/kg bw per day, but not in rats given atrazine. There was no effect of treatment on T_{max} for either substance (Minnema, 2001).

In a subsequent study designed to compare the effects of atrazine and its metabolite DACT on the the pre-ovulatory LH surge, groups of 16 (dose groups) or 32 (control group) female Sprague-Dawley [Crl:CD BR] rats received diets containing atrazine (purity, 97.1%) at a concentration of 0, 25, 50, 70 or 400 ppm (equal to 0, 1.8, 3.4, 4.9 or 28.2–29.1 mg/kg bw per day) or DACT (purity, 96.8%) at atrazine molar equivalent concentrations of 0, 17, 34, 48 or 270 ppm (equal to 0, 1.2, 2.4, 3.4 or 18.8–19.7 mg/kg bw per day), respectively. Sixteen rats in each treatment group and 32 rats in the control group were designated for assessment of the plasma LH surge during weeks 30–31 (after at least 29 weeks of treatment). Fifty rats in the control group and 20 rats in each of the groups at the highest dose were designated for histopathology examination after 52 weeks of treatment. The study was conducted in compliance with GLP guidelines. During weeks 30 and 31, all surviving rats designated for plasma LH-surge assessment were ovariectomized. Six days later, a silastic E2 capsule was implanted subcutaneously; the LH surge was examined 3 days later. For the LH-surge measurement, blood samples were collected from each rat at designated time-points (13:00, 16:00, 18:00, 20:00, 22:00 and 24:00 biological time). At the highest doses of atrazine or DACT, the mean body-weight gains were 72% or 84% of those in the control group after 29 weeks of treatment, respectively, and 65% or 95% of those in the control group after 52 weeks of treatment, respectively. Exposure of rats to DACT at the highest dose resulted in a significant reduction in the estrogen-induced LH surge, while there was no such effect for atrazine.

The NOAEL for effects on LH was 400 ppm (equal to 29.1 mg/kg bw per day) for atrazine and 48 ppm (equal to 3.4 mg/kg bw per day) for DACT (Sielken & Holden, 2002).

On the basis of the analysis of the vaginal smears (percentage days in diestrus, percentage days in diestrous blocks, percentage days in estrus, and percentage days in estrous blocks), treatment with atrazine or DACT did not affect the estrous cycle. Microscopic examination of the female reproductive organs revealed no effects associated with treatment with atrazine or DACT, although a non-significant increase in the percentage of pituitary adenomas was noted at the highest doses of atrazine and DACT when compared with controls (19% and 22% vs 8%, respectively). Statistically significant increases in the incidences of mammary carcinoma and mammary fibroadenoma-carcinoma were noted at the highest dose of DACT (17% and 23%, respectively), when compared with those for the controls (5% and 7%, respectively).

Overall, the NOAEL for atrazine was 70 ppm, equal to 4.9 mg/kg bw per day, on the basis of decreased body-weight gain at 400 ppm. The NOAEL for DACT was 48 ppm, equal to 3.4 mg/kg bw per day, on the basis of decreased body weight, attenuation of the LH surge and increased incidences of mammary carcinoma and mammary fibroadenoma-carcinoma at 270 ppm (Minnema, 2002; Sielken & Holden, 2002).

In a study on the effect of atrazine on LH secretion in the intact proestrus female rat, LE rats displaying regular 4-day estrous cycles reveived atrazine (purity, not reported) at a dose of 0, 6.25, 12.5 or 25 mg/kg bw per day by gavage through a full ovarian cycle, beginning at vaginal estrus. The rats were maintained on a 14-h light : 10-h dark schedule (lights on at 05:00) and dosing was done at 12:00. The last dose was administered on the day of vaginal proestrus and groups were killed at 2-h intervals until lights out (12:00, 14:00, 16:00, 18:00 and 20:00). Atrazine, at all doses administered, resulted in a significant decrease in peak LH concentration when determined at 18:00. However, the effect was of a similar extent at all doses tested, and the study authors believed that the flat dose–response relationship reflected the effect of atrazine on LH within this range of dosing (Cooper et al., 2007).

Monkeys

In a study on the effect of atrazine on pituitary hormone secretion in a non-human primate model, groups of six female ovariectomized Rhesus monkeys received atrazine (purity, 97.2%; in 0.5% carboxymethyl cellulose) at a dose of 0 or 25 mg/kg bw per day by oral gavage for 30 days, and were then kept an additional 60 days for recovery. All monkeys received a subcutaneous injection of estradiol benzoate (330 µg) on day 5 before treatment, on days 5 and 26 of treatment and on day 26 of recovery. Blood samples for hormone analysis (LH, FSH, prolactin, progesterone, E2, cortisol) were collected from each monkey approximately 12 h after administration of E2 (as estradiol benzoate), and then at 6-h intervals (up to 102 h after administration of E2).

No treatment-related clinical signs were observed. Body weight was not affected by treatment except in one monkey treated with atrazine that lost 1 kg (19%) of its initial body weight over the course of the study. This monkey displayed a marked suppression of LH and FSH surges after 5 or 26 days of treatment, which was confounded with the effects of other non-specific stressors. In the remaining monkeys, no treatment-related effects on concentrations of LH, FSH, E2, progesterone or prolactin were observed. However, a non-specific stress response in treated monkeys appeared to be associated with intolerance to high doses of atrazine. This resulted in the inability to acclimatize to experimental conditions as demonstrated by smaller reductions in cortisol in monkeys treated with atrazine when compared with controls. In conclusion, the primate model for assessing effects on LH secretion is limited by intra- and inter-animal variability in the normal response to estrogen-induced LH secretion, which is confounded by apparent stress effects related to the required experimental design (Osterburg & Breckenridge, 2004; Simpkins & Eldridge, 2004).

(d) Studies on reproductive and developmental effects

In a study designed to examine the effects of atrazine on suckling-induced prolactin release in Wistar rats, dams were given atrazine (purity, 98%) at a dose of 0, 6.25, 12.5, 25, or 50 mg/kg bw twice per day by gavage on postnatal days 1–4, or the DA-receptor agonist bromocriptine (BROM, which is known to suppress the release of prolactin) twice per day by subcutaneous injection at a dose of 0.052, 0.104, 0.208, or 0.417 mg/kg bw. Serum prolactin concentrations were measured on postnatal day 3 using a serial sampling technique and in-dwelling cardiac catheters. The study hypothesis was that early lactational exposure to agents that suppress the suckling-induced release of prolactin would lead to a disruption in the development of the tuberoinfundibular neurons in the pups (presumably due to the lack of prolactin derived from the dam's milk), with the consequence of impaired regulation of prolactin secretion, hyperprolactinaemia before puberty and prostatitis in the adult male offspring.

A significant rise in the release of prolactin in serum was noted in all females in the control group within 10 min of the initiation of suckling. At 50 mg/kg bw, atrazine inhibited the suckling-induced release of prolactin in all females, while atrazine at a dose of 25 and 12.5 mg/kg bw inhibited this release in some dams but had no discernible effect in others. Give twice per day at a at a dose of 6.25 mg/kg bw (i.e. 12.5 mg/kg bw per day), atrazine was without effect. BROM, used as a positive control, also inhibited the suckling-induced release of prolactin at doses of 0.104 to 0.417 mg/kg bw, with no effect at 0.052 mg/kg bw. To examine the effect of postnatal exposure to atrazine and BROM on the incidence and severity of inflammation of the lateral prostate of the offspring, adult males were examined at 90 and 120 days. While no effect was noted at age 90 days, by 120 days, the incidence and severity of prostate inflammation were both increased in offspring of dams treated with atrazine at 25 and 50 mg/kg bw. Atrazine at a dose of 12.5 mg/kg bw and BROM at a dose of 0.208, or 0.417 mg/kg bw increased the incidence, but not the severity, of prostatitis. Combined treatment with ovine prolactin and atrazine at 25 or 50 mg/kg bw on postnatal day 1–4 reduced the incidence of inflammation observed at 120 days, indicating that this increase in inflammation, seen after atrazine alone, resulted from the suppression of prolactin in the dam.

To determine whether there was a critical period for these effects, dams were given atrazine at a dose of 25 or 50 mg/kg bw twice per day on postnatal days 6–9 and postnatal days 11–14. Inflammation was increased in offspring from dams treated on postnatal days 6–9, but this increase was not statistically significant. Dosing on postnatal day 11–14 was without effect. These data demonstrated that atrazine suppresses the suckling-induced release of prolactin and that this suppression results in lateral prostate inflammation in the offspring. The critical period for this effect is postnatal days 1–9 (Stoker et al., 1999).

In a study on the effects of atrazine on implantation and embryo viability during early gestation in rats, groups of rats of one of four strains (Holtzman; Sprague-Dawley; LE; Fischer 344, F344) were given atrazine (purity, 97.1%) at a dose of 0, 50, 100 or 200 mg/kg bw per day by gavage during days 1–8 of gestation during either the light or dark period of a 14 : 10 light : dark cycle. All rats were necropsied on day 8 or 9 of pregnancy. At a dose of 200 mg/kg bw per day, atrazine reduced body-weight gain in all except one group (F344 diurnal dosing), and nocturnal dosing resulted in significant effects on body-weight gain at lower doses than did diurnal dosing. F344 rats showed a significant increase in pre-implantation loss after nocturnal dosing at 100 and 200 mg/kg bw per day, with no effect in other strains, while Holtzman rats showed a significant increase in postimplantation loss after diurnal and nocturnal dosing at 100 and 200 mg/kg bw per day, with no effect in other strains. Only in Holtzman rats was there a significant decrease in serum progesterone concentrations at 100 and/or 200 mg/kg bw per day in both intervals of dosing, while serum E2 concentration was significantly increased only in Sprague-Dawley rats at 200 mg/kg bw per day and by diurnal dosing. A significant reduction in serum LH concentration was seen in several groups, but there was no effect in Sprague-Dawley rats. In conclusion, F344 rats were most susceptible to the pre-implantation effects of atrazine, while Holtzman rats appeared to be most sensitive to the postimplantation effects. LE and Sprague-Dawley rats were least sensitive to the effects of atrazine during very early pregnancy (Cummings et al., 2000).

In a study on the effects of atrazine on the early postimplantation phase of pregnancy, F344, Sprague-Dawley and LE rats were given atrazine (purity, 97.1%) at a dose of 0, 25, 50, 100 or 200 mg/kg bw per day by gavage on days 6 to 10 of gestation. The dams were allowed to deliver and litters were examined postnatally.

Significant maternal body-weight losses on days 6–7 of gestation were seen in F344 rats at a dose of 50 mg/kg bw per day and greater, in Sprague-Dawley rats at a dose of 25 mg/kg bw per day and greater and in LE rats at a dose of 100 mg/kg bw per day and greater. The F344 strain was the most sensitive to effects on pregnancy, showing full-litter resorption at a dose of 50 mg/kg bw per day and greater. In surviving F344 litters, prenatal loss was increased at 200 mg/kg bw per day. In Sprague-Dawley and LE rats, full-litter resorption occurred only at 200 mg/kg bw per day (Table 22).

Delayed parturition was seen at a dose of 100 mg/kg bw per day and greater in F344 and Sprague-Dawley rats.

In F344 rats given atrazine at at a dose of 200 mg/kg bw per day on days 11 to 15 of gestation (after the LH-dependent period of pregnancy), no full-litter resorptions were seen. These findings suggest that the induction of full-litter resorption by atrazine is maternally mediated, and consistent with loss of LH support of the corpora lutea (Narotsky et al., 2001).

(e) Studies on female pubertal development

In a study on the effects of atrazine on pubertal development and thyroid function in rats, groups of 15 female Wistar rats were given atrazine (purity, 97.1%) at a dose of 0, 12.5, 25, 50, 100 or 200 mg/kg bw per day by gavage on postnatal days 22 to 41. An additional control group was included that was pair-fed with the group at 200 mg/kg bw per day in order to detect any effects caused by the reduced food consumption observed at this dose.

Body weight on postnatal day 41 was reduced by 11.6% at 200 mg/kg bw per day, but at 50 and 100 mg/kg bw per day was not different from that of the controls. Adrenal, kidney, pituitary, ovary, and uterine weights were also reduced at 200 mg/kg bw per day. The day of vaginal opening was significantly delayed by 3.4, 4.5 or more than 6.8 days at 50, 100, and 200 mg/kg bw per day, respectively. Although body weight in the pair-fed controls was reduced to the same extent as at 200 mg/kg bw per day, vaginal opening was not significantly delayed. Serum triiodothyronine (T3), thyroxine (T4), and TSH concentrations were unaltered by atrazine, which was consistent with the fact that no histological changes were observed in the thyroid.

Estrous cyclicity was monitored in a second group of females from the day of vaginal opening to postnatal day 49. The number of females displaying regular 4- or 5-day estrous cycles during the first 15-day interval after vaginal opening was decreased at 100 and 200 mg/kg bw per day and in the pair-fed controls. Irregular cycles were characterized by extended periods of diestrus. By the end of the second 15-day interval (postnatal days 57–71), no effects on estrous cyclicity were observed.

The data indicated that atrazine can delay the onset of puberty and alter estrous cyclicity in the female Wistar rat. Reduced food consumption and body weight did not account for the delay in vaginal opening, because this effect was not observed in the pair-fed controls. In addition, no effect on estrous cyclicity was observed at 100 mg/kg bw per day where no significant reduction in body weight was observed.

The NOAEL was 25 mg/kg bw per day, on the basis of delayed vaginal opening at 50 mg/kg bw per day and greater (Laws et al., 2000).

Table 22. Summary of ED_{10} estimates and benchmark doses for full-litter resorption in rats given atrazine by gavage

Strain	NOAEL (mg/kg bw per day)	LOAEL (mg/kg bw per day)	Dams with full-litter resorption at LOAEL (%)	ED_{10} (mg/kg bw per day)	Benchmark dose (mg/kg bw per day)
Fischer 344	25	50	36	15.1	11.7
Sprague-Dawley	100	200	67	170.3	97.9
Long-Evans	100	200	67	170.6	103.0

From Narotsky et al. (2001)
ED10, estimated dose that causes full litter resorption in 10% of the dams; LOAEL, lowest-observed-adverse-effect level; NOAEL, no-observed-adverse-effect level.

In a study on the effects of atrazine on the sexual maturation of female rats, groups of 8–10 female Wistar and Sprague-Dawley rats were given atrazine (purity, 98.2%) at a dose of 0, 10, 30 or 100 mg/kg bw per day by gavage from postnatal day 21 until postnatal day 43 (Wistar rats) or postnatal day 46 (Sprague-Dawley rats). A separate group of Wistar rats was given the GnRH agonist, antarelix (ANT), which is known to block the release of LH from the pituitary and delay vaginal opening. Uterine weights were determined at postnatal days 30, 33, 43 (Wistar rats) and 46 (Sprague-Dawley rats), and the time of vaginal opening was assessed.

In rats exposed to ANT, uterine growth and vaginal opening were completely prevented. In Wistar rats exposed to atrazine at 100 mg/kg bw per day, vaginal opening was significantly delayed (by 3 days), while uterine growth was delayed at postnatal days 30 and 33, but this growth inhibition had been overcome by postnatal day 43. In Sprague-Dawley rats exposed to atrazine at 30 and 100 mg/kg bw per day, vaginal opening was significantly delayed (by 2.5 and 3 days, respectively), while uterine weights were unaffected at postnatal day 46.

The NOAEL was 10 mg/kg bw per day in Sprague-Dawley rats and 30 mg/kg bw per day in Wistar rats on the basis of delayed vaginal opening at 30 or 100 mg/kg bw per day, respectively (Ashby et al., 2002).

In a study on the effects of atrazine metabolites on pubertal development and thyroid function in female rats, groups of 15 female Wistar rats were given DACT (purity, 96.8%) at a dose of 0, 16.7, 33.8, 67.5 or 135 mg/kg bw per day or hydroxyatrazine (purity, 97.1%) at a dose of 0, 22.8, 45.7, 91.5 or 183 mg/kg bw per day by gavage on postnatal days 22–41. These doses were equivalent to atrazine equimolar doses of 0, 25, 50, 100 and 200 mg/kg bw per day.

The females were monitored daily for vaginal opening and killed on postnatal day 41. DACT significantly delayed vaginal opening by 3.2, 4.8, and 7.6 days at doses of 33.8, 67.5, and 135 mg/kg bw per day, respectively. The NOAEL for DACT of 16.7 mg/kg bw per day was identical to the equimolar NOAEL of 25 mg/kg bw per day for atrazine. For hydroxyatrazine, no significant delays in pubertal development were observed in two separate studies with doses ranging up to 183 mg/kg bw per day, identical to the equimolar NOAEL for atrazine of 200 mg/kg bw per day. No significant or dose-related effects were observed on serum thyroid hormone concentrations (T3, T4 and TSH) or thyroid histopathology (Laws et al., 2003).

(f) Studies on male pubertal development

In two separate studies, the effects of atrazine on pubertal development in male rats were investigated. In the first study, groups of 9–12 male Sprague-Dawley rats were given atrazine (purity, 96.1%) at a dose of 0, 1, 2.5, 5, 10, 25, 50, 100 or 200 mg/kg bw per day on postnatal days 22–47. Since the first study showed that atrazine significantly suppressed body-weight gain at doses of 100 mg/kg bw per day and greater, a second study was performed. One group received atrazine at a dose of 100 mg/kg bw per day, another group received the vehicle (0.5% carboxymethylcellulose) and was fed the mean daily intake of food consumed by the atrazine-treated group on the previous day, and a further group received the vehicle and was fed ad libitum.

There was no effect on any of the measured variables (body weight, food consumption, day of preputial separation, serum testosterone, serum LH, intratesticular testosterone, ventral prostate weight, and seminal vesicle weight) at doses of up to 50 mg/kg bw per day, while doses of 100 mg/kg bw per day or more significantly reduced body-weight development (−9% and −21% at 100 and 200 mg/kg bw per day, respectively). In the second study in the additional pair-fed control group, body-weight was 10% lower than that of the control group fed ad libitum and was comparable to that of the group at 100 mg/kg bw per day. Lower values for serum and intratesticular testosterone concentrations were seen at doses of 100 mg/kg bw per day and greater; similar changes were also observed in the pair-fed group. Reduced ventral prostate and seminal vesicle weights (both organs being androgen-dependent) paralleled the lower testosterone concentrations. Again, decreased ventral prostate and seminal vesicle weights were noted in pair-fed rats. Additionally, lower LH concentrations were observed in groups treated with atrazine at 100 mg/kg bw per day and greater, as well as in pair-fed rats. A delay in preputial separation by about 3 and 4 days was observed at doses of 100 and 200 mg/kg bw per day, respectively (Trentacoste et al., 2001).

In a study on the effects of atrazine on pubertal development and thyroid function in male rats, male Wistar rats were given atrazine (purity, 97.1%) at a dose of 0, 6.25 (preputial separation study only) 12.5, 25, 50, 100, 150, or 200 mg/kg bw per day by gavage on postnatal days 23–53. An additional control group was included that was pair-fed with the group at 200 mg/kg bw per day in order to detect any effects caused by the reduced food consumption observed at this dose. Preputial separation was monitored from postnatal day 33 onwards. The majority of the males were killed on postnatal day 53. The pituitary, testes, ventral and lateral prostates, epididymides, and seminal vesicles with coagulating gland were removed and weighed. Blood was analysed for TSH, T4, T3, LH, prolactin, E2 and estrone. Additionally, concentrations of LH and prolactin were determined in the anterior pituitaries. A subgroup of males in the control group and males at the highest dose was killed on day 45 as described in the previous study for determination of LH receptors as well as serum and intratesticular testosterone content. Finally, another subgroup of rats was killed on postnatal day 120 to examine the reversibility of the effects.

Body weights at 200 mg/kg bw per day were significantly decreased (−17% at postnatal day 53) and returned to normal at postnatal day 120, while body weights of pair-fed rats were decreased to a comparable extent (−14% at postnatal day 53). A dose-dependent decrease in ventral prostate weights was observed with atrazine at doses of 50 mg/kg bw per day and greater and in the pair-fed group; this effect was still seen at postnatal day 120 at 200 mg/kg bw per day. Seminal vesicle and epididymal weights were decreased on postnatal day 53 at 200 mg/kg bw per day and in the pair-fed group. The latter effects were no longer seen at postnatal day 120. There was no effect of treatment on testes weight.

Preputial separation was significantly delayed at 12.5, 50, 100, 150, and 200 mg/kg bw per day (by 2.3, 1.7, 1.7, 1.7, and 3 days, respectively), while there was no such effect in a separate group at a dose of 6.25 mg/kg per day. Preputial separation was also delayed (by 2 days) in the pair-fed control group, although significantly less than in the group at 200 mg/kg bw per day.

Serum testosterone concentrations were decreased at doses of 25 mg/kg bw per day and greater at postnatal day 53 and at 200 mg/kg bw per day at postnatal day 45; however, the decreases were not statistically significant. Intratesticular testosterone concentration was significantly lower at 200 mg/kg bw per day on postnatal day 45, but not on postnatal day 53. Testosterone concentrations in the pair-fed rats were not significantly different to those in rats in the control group that had access to food ad libitum. There was a dose-dependent increase in serum E2 and estrone concentrations on postnatal day 53; however, this increase was statistically significant only at 200 mg/kg bw per day, while concentrations in the pair-fed group were comparable to those in the control group that had access to food ad libitum.

There was a significant trend for a dose-dependent decrease in serum LH concentrations on postnatal day 53; however, mean LH concentrations were not significantly different from those of the controls. Prolactin was reduced at doses of 150 mg/kg bw per day and greater, but the differences failed to reach statistical significance. The number of LH receptors in the testes was not altered by treatment. Serum LH and prolactin concentrations were comparable in the pair-fed group and the rats in the control group that had access to food ad libitum. Serum T3 concentration was significantly increased at 200 mg/kg bw per day, while serum TSH and T4 concentrations were unaffected by treatment. Serum TSH, T3, T4, LH and prolactin concentrations in the pair-fed group were comparable to those in the control group that had access to food ad libitum. There were no histopathological lesions observed in the thyroid gland of rats at the highest dose.

The results indicated that atrazine delays puberty and the development of the reproductive tract in the male rat. The mode of action appears to be in altering the secretion of steroids, probably due to disruption of control of pituitary function by the central nervous sysytem.

The NOAEL was 6.25 mg/kg bw per day on the basis of delayed preputial separation at 12.5 mg/kg bw per day and greater (Stoker et al., 2000).

In a subsequent study on the effects of atrazine metabolites on pubertal development and thyroid function in male rats, groups of 8–18 (dose groups) or 38 (control group) male Wistar rats were given DEA at a dose of 10.8, 21.7, 43,4, 86.8 or 173.9 mg/kg bw per day, or DIA at a dose of 10.4, 20.8, 40.1, 80.3 or 160.9 mg/kg bw per day, or DACT (purity, 97–98%) at a dose of 4.4, 8.4, 16.9, 33.8, 84.3 or 135.3 mg/kg bw per day by gavage on postnatal days 23–53. These doses were equivalent to atrazine equimolar doses of 0, 6.25 (DACT only), 12.5, 25, 50, 100 and 200 mg/kg bw per day.

Preputial separation was significantly delayed by treatment with DEA (atrazine equimolar dose, 25, 100 and 200 mg/kg bw per day), DIA (atrazine equimolar dose, 25, 100 and 200 mg/kg bw per day) and DACT (atrazine equimolar dose, 12.5–200 mg/kg bw per day). When the rats were killed on postnatal day 53, treatment with DEA (atrazine equimolar dose, 100 and 200 mg/kg bw per day), DIA (atrazine equimolar dose, 50–200 mg/kg bw per day) or DACT (atrazine equimolar dose, 200 mg/kg bw per day) caused a significant reduction in ventral protate weight, while only the highest doses of DEA and DIA resulted in a significant decrease in the lateral prostate weight. Seminal vesicle weight was reduced by DEA (atrazine equimolar dose, 25, 100 and 200 mg/kg bw per day), DIA (atrazine equimolar dose, 100 and 200 mg/kg bw per day) and DACT (atrazine equimolar dose, 100 and 200 mg/kg bw per day). Epididymal weights were reduced in the groups receiving DEA (atrazine equimolar dose, 200 mg/kg bw per day), DIA (atrazine equimolar dose, 200 mg/kg bw per day) and DACT (atrazine equimolar dose, 100 and 200 mg/kg bw per day). Serum testosterone concentration was reduced only in rats receiving DIA at the two higher doses. Serum estrone concentration was increased in the groups receiving DACT at the two higher doses, while serum E2 concentration was not changed relative to controls in any group. No differences were observed in any of the measures of thyroid activity.

The results indicated that the three chlorinated metabolites of atrazine delay puberty in a manner similar to atrazine, by affecting the control of the pituitary/gonadal axis by the central nervous system and subsequent development of the reproductive tract. The NOAELs for DEA, DIA, and DACT (expressed in atrazine equimolar doses) were 12.5, 12.5 and 6.25 mg/kg bw per day, respectively, on the basis of delayed preputial separation (Stoker et al., 2002).

(g) Studies on development of the mammary gland

In a study on the effects on development of the mammary gland in rats exposed in utero to atrazine, groups of 20 timed-pregnant LE rats were given atrazine (purity, 97.1%) at a dose of 0 (vehicle control) or 100 mg/kg bw per day by gavage on days 15–19 of gestation. On postnatal day 1, half of all litters were cross-fostered, creating four treatment groups: control–control, atrazine–control, control–atrazine, and atrazine–atrazine (dam–milk source, respectively). A significant delay in vaginal opening and increase in body weight at vaginal opening was seen only in the litters receiving milk from dams exposed to atrazine. However, mammary glands of female offspring (two per dam) in all groups exposed to atrazine (atrazine–control, control–atrazine, and atrazine–atrazine) displayed significant delays in epithelial development. These changes were detected as early as postnatal day 4 and stunted development was evident until postnatal day 40. Further, at all developmental stages examined, offspring in the atrazine–atrazine group exhibited the least developed glands. These delays in pubertal end-points did not appear to be related to body weight or concentrations of endocrine hormone (Rayner et al., 2004).

In a subsequent study conducted to determine whether fetal development of the mammary gland was sensitive to treatment with atrazine during specific periods of development, timed-pregnant LE rats (n = 8 per group per block) were given atrazine (purity, 97.1%) at a dose of 0 or 100 mg/kg bw per day by gavage for 3- or 7-day intervals during gestation (days 13–15, 15–17, 17–19, or 13–19 of gestation), and their offspring were evaluated for changes. Mammary glands taken from pups exposed prenatally to atrazine displayed significant delays in epithelial development as early as postnatal day 4 compared with controls, and showed continued developmental delays at later time-points that varied by time of exposure. However, the most persistent and severe delays were seen in the groups exposed to atrazine during days 17–19 and 13–19 of gestation, which demonstrated statistically similar levels of growth retardation (Rayner et al., 2005).

In a study on development of the mammary gland in rats exposed in utero to atrazine, groups of seven or eight timed-pregnant LE rats were given atrazine (purity, 97.1%) at a dose of 0 or 100 mg/kg bw per day by gavage on days 15–19 of gestation. Delayed development of the mammary gland was detected in the pups on postnatal day 4 and persisted to postnatal day 66. Immunohistochemistry of mammary sections from postnatal day 41 demonstrated increased levels of staining for the estrogen receptor (ER) in the gland stroma, the epithelium and the stroma surrounding the epithelial layer of the terminal end buds of pups exposed to atrazine. Also, the level of progesterone-receptor staining was higher in the ductal epithelium and fat cell nuclei of rats exposed to atrazine in utero (Moon et al., 2007).

In a study on development of the mammary gland in rats exposed in utero to a mixture of atrazine metabolites containing atrazine, hydroxyatrazine, DACT, DEA and DIA, groups of six timed-pregnant LE rats were given the metabolite mixture at a dose of 0.09, 0.87 or 8.73 mg/kg bw per day by gavage on days 15–19 of gestation, using groups given atrazine at a dose of 0 or 100 mg/kg bw per day as negative and positive controls, respectively. Exposure to the metabolite mixture had no statistically significant effect on body-weight gain in dams during the dosing period, weight loss in pups on postnatal day 4, or pubertal timing, being effects that are seen with atrazine alone. However, as with atrazine, development of the mammary gland was delayed, when evaluated by whole-mount analysis, as early as postnatal day 4 in all treatment groups (Enoch et al., 2007).

Although alterations were seen in the development of the mammary gland after administration of atrazine alone or in combination with its metabolites (DACT, DEA, DIA, hydroxyatrazine), it is uncertain whether this morphological observation leads to an adverse consequence or a functionally relevant toxic effect. The scoring system for the observations of the mammary gland is not a validated and standardized procedure and it was not reported whether the scoring was done blind. Furthermore, the dose–response relationship appears flat or variable in the study in which a mixture of atrazine and its metabolites was administered. Because no data on the individual substances were generated, it is unclear what type of interactions may be occurring between atrazine and its metabolites in the mixture. Further work is needed to clarify and repeat these observations before concluding that exposure to a mixture of atrazine and its metabolites can cause alterations in development of the mammary gland at doses as low as 0.09 mg/kg bw per day, which would lead to an adverse public health consequence.

(h) Studies on estrogenic and anti-estrogenic potential

In tests for estrogenic bioactivity, groups of five or six ovariectomized adult female Sprague-Dawley rats were given atrazine (purity, 97.7%) or its metabolite DACT (purity, 98.2%) at an oral dose of 20, 100 or 300 mg/kg bw per day for three consecutive days. Uterine weight did not increase statistically significantly. At the highest dose of atrazine or or DACT, loss of body weight was seen. When the test substances were administered concomitantly with subcutaneous injections of E2 at a dose of 2 µg per rat, statistically significant decreases in uterine weight were seen with both atrazine and DACT at doses of 100 mg/kg bw per day and greater.

In further tests by the same research group, immature female Sprague-Dawley rats (age 23 days) were given atrazine or DACT at a dose of 1, 10, 50, 100 or 300 mg/kg bw per day by gavage for two consecutive days, and received E2 at a dose of 0.15 µg per rat by subcutaneous injection on the second day. Thymidine incorporation into uterine DNA was significantly reduced at doses of 50 mg/kg bw per day and greater.

Again, in further tests by the same research group, ovariectomized adult female Sprague-Dawley rats were given atrazine or DACT at a dose of 50 or 300 mg/kg bw per day by gavage for two consecutive days, and received E2 at a dose of 1 µg per rat by subcutaneous injection each day. Progesterone receptor-binding capacity in cytosol fractions from uteri was significantly reduced at 300 mg/kg bw per day. Uterine progesterone receptor levels were not stimulated in rats that received atrazine or DACT at doses of up to 300 mg/kg bw per day without E2 injections.

The results indicated that atrazine and DACT possess no intrinsic estrogenic activity, but that they are capable of weak inhibition of estrogen-stimulated responses in the rat uterus (Eldridge et al., 1994; Tennant et al., 1994a).

In a study on the interaction with ER binding, competitive co-incubation of atrazine (purity, 96.9%) or DACT (purity, 98.2%) and radiolabeled estrogen with rat uterine cytosol containing ER failed to demonstrate significant displacement of estrogen binding by atrazine. However, when cytosols were pre-incubated with atrazine, and tracer was added to the chilled incubation medium, there was a significant reduction of [^{3}H]E2 binding. Competition was very weak, with IC_{50} estimates of 20 µmol/l for atrazine and 100 µmol/l for DACT, which were four to five orders of magnitude greater than the approximate IC_{50} of E2.

Ex-vivo results in the same report indicated that the uterine ER-binding capacity was reduced by approximately 30% when ovariectomized adult female Sprague-Dawley rats were given atrazine or DACT at a dose of 300 mg/kg bw per day by gavage for two consecutive days; a dose of 50 mg/kg bw per day had no effect.

The result confirmed that atrazine and DACT may be capable of a very weak interaction with estrogen receptors, but only at extremely high concentrations (Tennant et al., 1994b).

The potential estrogenic activities of atrazine were investigated in vivo using the immature female Sprague-Dawley rat uterus and in vitro using the estrogen-responsive MCF7 human breast cancer cell line and the estrogen-dependent recombinant yeast strain PL3. Rats that were dosed with atrazine only at 50, 150, or 300 mg/kg bw per day for three consecutive days did not exhibit any significant increases in uterine wet weight, although decreases in cytosolic progesterone-receptor binding levels and uterine peroxidase activity were observed. 17β-estradiol (E2)-induced increases in uterine wet weight were not (statistically) significantly affected by co-treatment with atrazine; however, some dose-independent decreases in E2-induced cytosolic progesterone-receptor binding and uterine peroxidase activity were observed. In vitro, atrazine did not affect basal or E2-induced MCF7 cell proliferation or the formation of nuclear progesterone receptor–DNA complexes as determined by gel electrophoretic mobility shift assays. In addition, atrazine did not display agonist activity or antagonize E2-induced luciferase activity in MCF7 cells transiently transfected with a Gal4-human estrogen receptor chimera (Gal4-HEGO) and a Gal4-regulated luciferase reporter gene (17m5-G-Luc). Moreover, the estrogen-dependent PL3 yeast strain was not capable of growth on minimal media supplemented with atrazine in place of E2. Collectively, the results indicated that atrazine did not exhibit estrogenic activity; however, antiestrogenic activity via competition for the estrogen receptor cannot be excluded (Connor et al., 1996, 1998).

In a study to investigate the effects of chlorotriazines on the E2 (as estradiol benzoate)/ progesterone-induced LH surge and to determine whether such changes correlate with impaired estrogen receptor (ER) function, adult ovariectomized female Sprague-Dawley rats were given atrazine (purity, 97.1%) at a dose of 0, 30, 100 or 300 mg/kg bw per day or DACT (purity, 96.8%) at a dose of 0 or 77 mg/kg bw per day (a dose that achieved an AUC equivalent to dosing with atrazine at 300 mg/kg bw per day) by oral gavage for five consecutive days. Atrazine caused a dose-dependent suppression of the E2/progesterone-induced LH surge (dosing at 300 mg/kg bw per day completely blocked the LH surge), while DACT to a lesser degree suppressed total plasma LH and peak LH-surge concentrations in E2/progesterone-primed rats by 60% and 58%, respectively. Treatment with DACT also decreased the release of LH from the pituitary in response to exogenous GnRH by 47% compared with the controls. Total plasma LH secretion was reduced by 37% compared with controls, suggesting that, in addition to potential hypothalamic dysfunction, pituitary function is also altered. To further investigate the mechanism by which hypothalamic function might be altered, potential anti-estrogenicity of atrazine and DACT were assessed by evaluating ER function in treated rats. Using an assay for receptor binding in vitro, atrazine (but not DACT) inhibited binding of [^{3}H]E2 to ER at concentrations between 10^{-4} and 10^{-3} mol/l. In contrast, in female rats given atrazine under dosing conditions that suppressed the LH surge, the levels of unoccupied ER and the estrogen-induced up-regulation of progesterone receptor mRNA were not altered. The results indicated that although atrazine is capable of binding to ER in vitro, the suppression of LH after treatment with high doses of atrazine was not due to alterations of hypothalamic ER function (McMullin et al., 2004).

(i) Studies on aromatase activity or gene expression

Studies In vitro have indicated that atrazine (purity, not reported) in 0.1% dimethyl sulfoxide (DMSO) increased aromatase (CYP19) activity and expression in the human adrenocorticocarcinoma cell line H295R and in the human placental choriocarcinoma cell line JEG3, but not in the human breast-cancer cell line MCF7 or the rat Leydig-cell cancer cell line R2C (Sanderson et al., 2000, 2001; Heneweer et al., 2004). In the majority of these studies, atrazine induced aromatase activity to an apparent maximum of about 2- to 3-fold and increased CYP mRNA levels between 1.5- and 2-fold at concentrations of 30 μmol/l (6.5 ppm). DEA and DIA gave weaker responses than did atrazine, but DACT, which is the major metabolite of atrazine in rats, mice and Rhesus monkeys, had no effect, and neither did hydroxyatrazine, which is the major plant metabolite (Sanderson et al., 2001). In KGN human ovary granulosa-like carcinoma cells, atrazine had no effect on aromatase activity at concentrations of up to the maximum tested, 50 μmol/l (Ohno et al., 2004).

Atrazine significantly inhibited phosphodiesterase activity in vitro in homogenates of bovine and swine tissues (Roberge et al., 2004, 2006). Inhibition of phosphodiesterase activity resulted in elevated concentrations of cAMP, increased CYP19 mRNA levels and increased aromatase activity (Sanderson et al., 2002). However, atrazine increased aromatase expression only in cell and tissue types that use the steroidogenic factor 1-dependent aromatase promoter II, e.g. the human adrenocorticocarcinoma cell line H295R, while KGN human ovary granulosa-like carcinoma cells were not responsive (Fan et al., 2007).

Although it has been reported that atrazine induces aromatase (CYP19) mRNA and aromatase activity in certain human cell lines, their biological significance remains in question because of the complex, tissue-specific manner through which aromatase is regulated. The aromatase enzyme expressed in all tissue types is identical, but the promoters, signalling pathway, and proteins involved in initiation of transcription vary between tissue types (Simpson et al., 2002). Also, the results of studies in rats did not support any atrazine-induced upregulation of aromatase expression in brain, testes, or mammary gland.

In a study on development of the mammary gland in vivo that has been described above, in the offspring of rats exposed to atrazine on days 15–19 of gestation the mammary gland expressed significantly less aromatase than did controls at postnatal day 33, while no significant differences were observed between groups at postnatal day 40 (Rayner et al., 2004).

In studies on the effects of atrazine on steroidogenesis, male Wistar rats aged 60 days were given atrazine (purity, 97.1%) at a dose of 0, 50 or 200 mg/kg bw per day by gavage once per day for 1, 4 or 21 days. The rats were killed at 3, 6 or 24 h after the single dose, or 3 h after the last repeat dose (4 or 21 days). Serum estrone, E2, testosterone, androstenedione, progesterone, corticosterone, and hypothalamic and testicular CYP19 mRNA were measured for each time-point. After one dose of atrazine at 200 mg/kg bw, concentrations of progesterone and corticosterone were increased at 3 h, and corticosterone and estrone were elevated at 24 h; androstenedione and testosterone were increased at 6 h for both doses. After 4 or 21 days of treatment at 200 mg/kg bw, concentrations of corticosterone, estrone and E2 were elevated, but androstenedione and testosterone remained at concentrations similar to those of the controls. To determine the effect of atrazine on adrenal steroidogenesis, castrated males treated with atrazine were examined. Despite reduced androstenedione and testosterone concentrations in castrated 200 mg/kg males, increased serum corticosterone, estrone and E2 mirrored previous results. Elevated progesterone was also observed. No change in aromatase CYP19 mRNA was detected by real-time reverse transcription polymerase chain reaction (RT-PCR) in the testicular or hypothalamic tissues at any time-point. These data suggested that elevated concentrations of estrone and E2 were not strictly caused by increased testicular steroidogenesis, altered aromatase mRNA, or substrate availability, but did not rule out a change in steroid metabolism or elimination. Although no CYP19 mRNA was detected in the adrenals, a stress-induced adrenal response may be partially responsible for the increase in steroids. However, data from an assay with minced testes in vitro showed an increase in testosterone after a 4 h exposure to atrazine. Together, these data demonstrated that atrazine can alter the steroidogenic pathway in male Wistar rats (Modic, 2004; Modic et al., 2004).

(j) Studies on adrenal steroidogenesis

In a study to evaluate adenocorticotropic hormone (ACTH), corticosterone, progesterone and testosterone after a single dose of atrazine, male Wistar rats (age 60 days) were acclimatized to dosing by gavage (with methylcellulose) for 7 days. On day 8, the rats were divided into groups of 10, given a single dose of atrazine at 0, 5, 50, 100, or 200 mg/kg bw by gavag, and killed 5, 15, 30, 60 or 180 min later. A dose-dependent increase in plasma ACTH was observed 15 and 30 min after treatment; maximal concentrations of ACTH were observed at 15 min with increases of 2.5, 4.9 and 9.6-fold in the groups at 50, 100 and 200 mg/kg bw, respectively. Dose-dependent increases in serum corticosterone and progesterone concentrations were observed at 15 and 30 min in the groups at 50, 100, and 200 mg/kg bw with the maximal responses (increases of 9–12-fold) occurring at 30 min after dosing. Corticosterone and progesterone concentrations remained elevated in males at 200 mg/kg bw for 180 and 60 min, respectively. Increased serum testosterone concentration was observed at 30 min in the groups at 100 and 200 mg/kg bw, and at 60 min in the groups at 50–200 mg/kg bw. Thus, the atrazine-induced increase in steroidogenesis is the result of increased ACTH secretion, either through a direct effect on the pituitary or through the release of corticotropin-releasing factor (CRF) in the central nervous system (Laws et al., 2006).

(k) Studies of immunotoxicity

The potential immunotoxicity of atrazine has been evaluated in a variety of mammalian and non-mammalian animal models (Table 23). Considered overall, the reports indicate that modulation of the immune system occurs after exposure to atrazine, albeit at doses greater than those known to disrupt neuroendocrine function and suppress LH and prolactin release.

Table 23. Selected published studies of immunotoxicity with atrazine

Test system	Dose/concentration	Finding	Comment	Reference
Lymphatic response in chicks (age 3 days)	150 ppm in feed for 21 days	Increased thymus and bursa weight that correlated with glycogen content.	All parameters returned to control levels by day 21.	Giurgea & Koszta (1979)
E. coli challenge test in Wistar rats	2 or 150 mg/kg bw per day for 60 days	Stimulated immunological reactions (antibody titre, gamma-globulin concentration, leukocyte count)	Inconsistent dose–response relationship; NOAEL could not be identified..	Giurgea et al. (1981)
Mouse leukocyte immunosuppression and phagocytic impairment	Single oral doses at 27.3, 109.4, 437.5 and 875.0 mg/kg bw	Transient and reversible suppression of humoral-mediated and cell-mediated responses; activated macrophage phagocytic activity.	Absence of a dose–response relationship; authors did not attribute the changes to a direct effect of atrazine on immune system	Fournier et al. (1992)
Haematopoietic system (progenitor cells in bone marrow) in atrazine-treated mice	Single intraperitoneal dose at 58.65 mg/kg bw	CFU-S and CFU-GM in bone marrow and reticulocyte count in blood reduced for 6–8 days after treatment; leukocyte count unaffected by treatment.	Transient response to a single high dose. NOAEL not identified.	Mencoboni et al. (1992)
Long-term exposure of laboratory and wild mice	10 ppb in water for 22–103 days	No effect on body weight, spleen weight or lymphocyte plaque-forming ability when challenged with foreign protein	Atrazine in combination with other chemicals not considered here	Porter et al. (1999)
Effect on cytokine production in vitro in mononuclear cells from humans	0.03, 0.3 and 3 μmol/l (6.5, 65 and 647 ppb) in 1% DMSO	Cytokine (IFN-γ, IL-5, TNF-α) production significantly reduced by up to 50–70% at $\geq$ 0.3 μmol/l	Effect of 1% DMSO on cellular uptake of atrazine unknown	Hooghe et al. (2000)
Cytokine production in vitro in mononuclear cells from humans	3 μmol/l (0.65 ppm) in 1% DMSO	Cytokine (IFN-γ, IL-5, TNF-α) production reduced by up to 40–60%; no effect on IL-8.	Authors concluded that the effect was not mediated via the glucocorticoid receptor.	Devos et al. (2003
Exposure of sheep leukocytes in vitro	10^{-6} – 10^{-1} mol/l (0.2–21 570 ppm) in 1% DMSO; added to leukocyte suspensions in 1% volumes	Decreased lymphocyte activation with PHA at concentrations of 10^{-1} – 10^{-2} mol/l (i.e. concentration in leukocyte suspensions: 22–216 ppm)	Limited biological relevance because concentration was at or above the limit of solubility for atrazine in water (33 ppm)	Pistl et al. (2003)

Immunological effects in $B6C3F_1$ mice	Single intraperitoneal. dose at 100, 200 or 300 mg/kg bw	Decrease in percentage of $CD4^+CD8^+$ cells in thymus and $B220^+$ cells in spleen; decreased splenic NK cell activity; decreased IgG1 and IgG2a response to KLH	Non-specific stress, as indicated by plasma corticosterone AUC, was predictive of the immune response	Pruett et al. (2003)
Study of developmental immunotoxicity in Sprague-Dawley rats	Oral gavage at 35 mg/kg bw per day from day 10 of gestation to postnatal day 23	Decrease in pup survival (postnatal days 2–14) and body weight (males, postnatal day 7). Primary antibody (IgM) response to SE and delayed-type hypersensitivity response to BSA decreased in males only.	Effects in males only, may be related to increased pup mortality or reduced pup body weight; NOAEL not identified.	Rooney et al. (2003)
Effect on human NK cell cytotoxic function (tumour-cell lysis) in vitro	10 µmol/l (2.2 ppm)	Decreased cytotoxic function of NK cells (by 63–83%) and of T/NK cells (by 61–65%) after 24 h and 6-day incubation	An effect attributable to high concentration in vitro	Whalen et al. (2003)
Cytokine production in vitro in mononuclear cells from humans	3 µmol/l (0.65 ppm)	Cytokine (IFN-γ, IL-5) production reduced by 12.3 and 14.8%, respectively; no effect on TNF-α, IL-4, IL-6 and IL-13.	Minimum effect that is unlikely to impact the allergic response; NOAEL not identified.	Devos et al. (2004)
Immunological effects in juvenile male C57BL/6 mice (age 1 month)	0, 5, 25, 125 or 250 mg/kg bw per day for 14 days	Decrease of thymus and spleen weights and organ cellularity at ≥ 25 mg/kg; transient changes in thymic and splenic subpopulations at ≥ 5 or ≥ 25 mg/kg, respectively.	Most effects no longer present at 7 weeks after treatment; the decreases of thymic T-cell populations at ≥ 5 mg/kg bw per day (at day 1 only) are not predictive of impaired function (i.e. not considered to be an adverse effect).	Filipov et al. (2005)
Immunological effects in female $B6C3F_1$ mice (age 4–6 weeks)	0, 25, 250 or 500 mg/kg bw per day for 14 days	Decrease of thymus and spleen weights, total spleen cell numbers and fixed macrophage function; increased number of splenic CD8+ T cells, increased cytotoxic T cell and mixed leukocyte responses, reduced host resistance to tumour challenge	Spleen weight reduced at ≥ 25 mg/kg bw per day Cytotoxic T-cell response increased at ≥ 25 mg/kg bw per day	Karrow et al. (2005)
Immunological effects in female $B6C3F_1$ mice	0, 75, 150, 225 or 300 mg/kg by intraperitoneal injection.	Effects on spleen and thymus of all stressors. Corticosterone comparable to atrazine	See Pruett et al. (2003)	Schwab et al. (2005)
Immunological effects in offspring of Balb/c mice exposed on day 10 of gestation to postnatal day 11	0.7 mg/dam per day, equal to 23–35 mg/kg bw per day (released from a subcutaneous implanted capsule)	Increase of HKSP-specific IgM-secreting B cells; increase in both T cell proliferation and cytolytic activity	Effects observed only in males; no changes in the number of CD8+ T-cell, CD4+ T-cell or B220+ B-cell subpopulations	Rowe et al. (2006)

BSA, bovine serum albumin; CFU-GM, colony-forming unit-granulocyte-macrophage; CFU-S, colony-forming unit-spleen; DMSO, dimethyl sulfoxide; IFN, interferon; IL, interleukin; KLH, keyhole limpet haemocyanin; NK, natural killer; PHA, phytohaemagglutinin; SE, sheep erythrocytes; TNF, tumour necrosis factor;

The immune system of adult mice appears to be relatively insensitive to atrazine. For example, exposure of female B6C3F_1 mice to 250 or 500 mg atrazine/kg bw per day for 14 days did not affect antibody synthesis, natural killer (NK) cell activity or lymphocyte proliferation, but did decrease body and lymphoid organ weights and shifted percentages of lymphocyte subpopulations (National Toxicology Program, 1994). The number of spleen cells producing antibody was increased by 35% in groups of mice exposed to atrazine at the lowest dose (25 mg/kg bw per day), but was similar to control values at higher doses. The authors dismissed this result as biologically irrelevant, because the antibody titre was not significantly decreased at this dose.

However, a recent peer-reviewed version of the original 1994 NTP report by Munson et al. (Karrow et al., 2005) concluded that the data at the lowest dose may represent an actual increase in numbers of antibody-producing cells in the spleen. Whether or not this type of enhancement is beneficial or detrimental to the individual has yet to be determined. With atrazine at the highest dose tested (500 mg/kg bw per day), resistance to an natural killer cell-dependent tumour-cell challenge was decreased, but resistance to bacterial challenge (*Listeria monocytogenes*) was not affected. However, animals at the highest dose experienced a 3% loss of body weight during dosing, vs a 10% body-weight gain in controls, indicating marked overt toxicity at the highest dose, thus calling into question the biological relevance of the observed immune effects at 500 mg/kg bw per day. Furthermore, given the lack of effects on natural killer cell activity, even at the highest dose, it is difficult to conclude that immune suppression was directly responsible for the apparent increased susceptibility to a tumour-cell challenge.

In a separate study, the antibody response of C57Bl/6 female mice was suppressed 7 days after a single exposure to a wide range of atrazine doses (27.3–875 mg/kg bw). However, suppression was not dose–responsive, and no dose-related pattern of suppression or recovery was apparent 2, 3 or 6 weeks after exposure (Fournier et al., 1992). Although the relatively short half-life of atrazine may explain why these effects were not persistent, it does not provide a satisfactory explanation for the lack of a dose–response relationship. Chemicals that are immunotoxic in adult animals generally induce similar effects, but at lower doses or for a prolonged period of time, if exposure occurs during development and maturation of the immune system (Luebke et al., 2006).

The effects of exposure to atrazine during immune system ontogeny were therefore evaluated in rats and mice. In replicate studies, exposure to atrazine at a dose of 35 mg/kg bw per day from day 10 of gestation until postnatal day 23 was found to suppress cellular and antibody-mediated immune function in male, but not female rats (Rooney et al., 2003). However, preliminary data from recent studies of dose–response did not corroborate suppressed function in developmentally exposed rats. The basis for this discrepancy is as yet unknown (Robert W. Luebke, personal communication). Studies of developmental exposure in mice indicated that the percentage of antibody-producing cells in the spleens of immunized male offspring of mice implanted with time-release atrazine pellets between days 10 and 12 of gestation was increased by approximately 33%; female offspring were not affected (Rowe et al., 2006). Daily doses were in the same range as those used by Rooney et al. (2003). The results for studies of developmental exposure results suggest that the immature immune system of males may be more sensitive to the effects of atrazine than that of the adult. Nevertheless, studies corroborating previous findings and those that incorporate a range of doses will be necessary to adequately determine NOAELs and LOAEL for immune function, and the relative sensitivity of the immune and endocrine systems in developing rodents.

3. Observations in humans

A large number of studies have been published on the association between exposure to triazines, including atrazine, and cancer epidemiology. Concerning the studies reported until 1999, three independent reviews (Neuberger, 1996; Sathiakumar & Delzell, 1997; IARC, 1999) have identified ten case–control studies and two published cohort studies of workers exposed to triazines at manufacturing plants.

Neuberger (1996) reported that of the ten case–control studies published, six of which considered atrazine, none indicated any statistically significant association between atrazine and cancer. Two studies indicated marginally significant associations between triazines and cancer (odds ratio, OR, 1.6; 95% confidence interval, CI, 1.0–2.6; and OR, 2.7; 90% CI, 1.0–6.9). The author concluded:

> ...on the basis of the data to date... there is no convincing evidence of a causal association between atrazine and/or triazine(s) and colon cancer, soft tissue sarcoma, Hodgkin's disease, multiple myeloma, or leukaemia... There is a suggestion of a possible association between atrazine and/or triazine(s) with ovarian cancer and non-Hodgkin's lymphoma. However, the ovarian cancer study needs to be replicated and the NHL studies fall short of providing conclusive evidence of risk because the results could be due to chance, bias, or confounding.

Sathiakumar & Delzell (1997) assessed the relation between triazines and non-Hodgkin lymphoma in four independent population-based case–control studies, reporting OR of between 1.2 and 2.5, and concluded that these weak statistical associations may have been produced by chance and/or confounding by other agricultural exposures. Furthermore, a pooled analysis of three case–control studies and the combined analysis of two retrospective follow-up studies did not demonstrate the types of dose–response relationship or induction-time patterns that would be expected if triazines were causal factors. The authors concluded that the available epidemiological studies, singly and collectively, did not provide any consistent, convincing evidence of a causal relationship between exposure to triazine herbicides and cancer in humans.

The International Agency for Research on Cancer (IARC, 1999) summarized their evaluation of data on human carcinogenicity as follows:

> A combined analysis of results of two cohort studies of agricultural chemical production workers in the United States showed decreased mortality from cancers at all sites combined among the subset of workers who had had definite or probable exposures to triazine. Site-specific analyses in this subset of workers yielded no significant findings; a non-significant increase in the number of deaths from non-Hodgkin's lymphoma was seen, but was on the basis of very few observed cases.
>
> A pooled analysis of the results of three population-based case–control studies of men in Kansas, eastern Nebraska and Iowa-Minnesota, United States, in which the risk for non-Hodgkin's lymphoma in relation to exposure to atrazine and other herbicides on farms was evaluated, showed a significant association; however, the association was weaker when adjustment was made for reported use of phenoxyacetic acid herbicides or organophosphate insecticides. In all these studies, the farmers tended to have an increased risk for non-Hodgkin's lymphoma, but the excess could not be attributed to atrazine
>
> Less information was available to evaluate the association between exposure to atrazine and other cancers of the lymphatic and haematopoietic tissues. One study of Hodgkin's disease in Kansas, one study of leukaemia in Iowa-Minnesota and one study of multiple myeloma from Iowa gave no indication of excess risk among persons handling triazine herbicides.
>
> In a population-based study in Italy, definite exposure to triazines was associated with a two to threefold increase of borderline significance in the risk for ovarian cancer. The study was small (65 cases, 126 controls), and potential confounding by exposure to other herbicides was not controlled in the analysis.

Therefore, on the basis of the findings described above the IARC (1999) concluded: "There is inadequate evidence in humans for the carcinogenicity of atrazine."

In a later review of cancer epidemiology after exposure to atrazine for the United States EPA (Blondell & Dellarco, 2003), four studies concerning prostate cancer were evaluated: a nested case–control study among workers in a plant manufacturing triazine in Louisiana (Hessel et al., 2004), the Agricultural Health Study which is a prospective cohort study of 55 332 male pesticide applicators from Iowa and North Carolina (Alavanja et al., 2003), an ecological study conducted in California (Mills, 1998), and a nested case–control study in Californian farm workers exposed to simazine (a triazine derivative similar to atrazine) and several other pesticides (Mills & Yang, 2003).

In the studies in California, a borderline statistically significant correlation was found between use of atrazine and prostate cancer in black males, but not among Hispanic, white, or Asian males (Mills, 1998), and a borderline significant association was found between high use of simazine and prostate cancer (Mills & Yang, 2003). However, both studies suffered from aggregation bias because there was no or only a crude measure of exposure and the results should thus not be considered for reaching conclusions about causation.

In the nested case–control study (Hessel et al., 2004), an elevated incidence of prostate cancer was found in active employees who received intensive screening for prostate specific antigen (PSA), but there was no increase in the incidence of advanced tumours or mortality, and proximity to atrazine-manufacturing plants did not appear to be correlated with risk. Thus, the increase in incidence of prostate cancer was probably attributable to increased detection because of the intensive screening programme for PSA (MacLennan et al., 2002).

The largest and most reliable study (Alavanja et al., 2003) showed no association of atrazine exposure with prostate cancer in cohort analysis of pesticide applicators. The overall conclusion of the reviewers was that studies in manufacturing and farming populations do not support a finding that atrazine is a likely cause of prostate cancer.

In a study of workers in a plant manufacturing triazine in Louisiana (MacLennan et al., 2003), a borderline significant result was found for non-Hodgkin lymphoma on the basis of 4 observed deaths vs 1.1 deaths expected. The study authors noted, however, that "one of the decedents whose death certificate included a diagnosis of non-Hodgkin lymphoma had medical records including a biopsy report that indicated a diagnosis of poorly differentiated nasopharyngeal cancer. This case was not removed from our analysis." This acknowledgment of bias on the basis of a misclassified case means that the borderline statistically significant finding would no longer be significant if the case were excluded. Therefore, the overall conclusion of the reviewers was that this evidence is not sufficient to support a finding that atrazine is a likely cause of non-Hodgkin's lymphoma.

Additional evaluations of the cancer incidences in the Agricultural Health Study did not find any clear associations between exposure to atrazine and any cancer analysed (Rusiecki et al., 2004). However, the authors pointed out that further studies were warranted for tumour types for which there was a suggestion of trend (lung, bladder, non-Hodgkin lymphoma, and multiple myeloma). It should be noted that the neuroendocrine mode of action of atrazine cannot account for the biological plausibility of these tumours.

An evaluation of risk of breast cancer among farmers' wives in the Agricultural Health Study (30 454 participants with no history of breast cancer before cohort enrolment in 1993–1997) did not find any association of increased incidence of breast cancer with the use of atrazine; however, reduced risk of breast cancer among postmenopausal women were linked to their use of atrazine (relative risk, RR, 0.4; 95% CI, 0.1–1.0) (Engel et al., 2005).

The incidence of cancer among pesticide applicators exposed to cyanazine (a triazine derivative similar to atrazine) in the Agricultural Health Study has also been unremarkable (Lynch et al., 2006).

Current evidence was not persuasive as to an association between ovarian cancer and exposure to triazine. In a population-based case–control study of incident cases ($n = 256$) and control subjects selected by random digit-dialled techniques (n = 1122) assessing whether there was an increased risk of ovarian cancer associated with occupational exposure to triazine herbicides, no evidence of a dose–response relationship for triazines and ovarian cancer was found (Young et al., 2005).

In an ecological study using regression analysis to evaluate the incidence of breast cancer in Hispanic females in California (23 513 cases diagnosed with breast cancer during the years 1988–1999) at the county level as a function of use of organochlorine and triazine pesticides, no significant associations were found for atrazine and simazine (Mills & Yang, 2006).

In another large population-based study (using 3275 incident cases of breast cancer in women aged 20–79 years from 1987 to 2000 and living in rural areas of Wisconsin, and 3669 matched controls), there was no increased risk of breast cancer for women exposed to atrazine at concentrations of 1.0–2.9 ppb in drinking-water (OR, 1.1; 95% CI, 0.9–1.4) when compared with women with the lowest exposure to atrazine (< 0.15 ppb). Evaluation of a possible risk for women exposed to atrazine at concentrations at or above the statutory action levels of ≥ 3 ppb (OR, 1.3; 95% CI, 0.3–5.0) was limited by the small numbers in this category (McElroy et al., 2007).

Comments

Biochemical aspects

After oral administration to rats, ^{14}C-labelled atrazine was rapidly and almost completely absorbed, independent of dose and sex. Radioactivity was widely distributed throughout the body. Excretion was more than 93% of the administered dose within 7 days, primarily via the urine (approximately 73%) and to a lesser extent via the faeces (approximately 20%; approximately 7% via bile), with more than 50% being excreted within the first 24 h. The elimination half-life of radiolabel from the whole body was 31.3 h in rats; this prolonged half-life was caused by covalent binding of atrazine to cysteine sulfhydryl groups in the β-chain of rodent haemoglobin. Seven days after administration of a single low dose (1 mg/kg bw), tissue residues represented 6.5–7.5% of the dose, with the highest concentrations in erythrocytes (≤ 0.63 ppm), liver (≤ 0.50 ppm) and kidneys (≤ 0.26 ppm). Atrazine was extensively metabolized; more than 25 metabolites have been identified in rats. The major metabolic pathways were stepwise dealkylation via either DIA or DEA to DACT, the major metabolite. Dechlorination involving conjugation with glutathione was a minor pathway. The biotransformation of atrazine in rats and humans was qualitatively similar.

Toxicological data

Atrazine was of low acute toxicity in rats exposed orally (LD_{50}, 1870–3090 mg/kg bw), dermally (LD_{50}, > 2000 mg/kg bw) or by inhalation (LC_{50}, > 5.8 mg/l). Atrazine was not a skin irritant or an eye irritant in rabbits. Although spray dilutions of atrazine did not appear to be sensitizing in humans, atrazine was a skin sensitizer in tests in guinea-pigs (Magnusson & Kligman, Maurer optimization test).

In short-term studies of toxicity in rats, dogs and rabbits, the consistent toxic effects noted across species included reduced body-weight gain and food intake and a slight decrease in erythrocyte parameters. Also in rats, liver weights and splenic haemosiderin deposition were increased, while in dogs there was marked cardiac toxicity.

In a 90-day study of toxicity in rats, the NOAEL was 50 ppm, equal to 3.3 mg/kg bw per day, on the basis of decreased body-weight gain and increased splenic haemosiderin deposition at 500 ppm.

In a 52-week study of toxicity in dogs, the NOAEL was 150 ppm, equal to 5 mg/kg bw per day, on the basis of decreased body-weight gain and marked cardiac toxicity at 1000 ppm, equal to 33.7 mg/kg bw per day.

In a 25-day study in rabbits treated dermally, the NOAEL for systemic toxicity was 100 mg/kg bw per day on the basis of decreased body-weight gain and food intake, a slight reduction in erythrocyte parameters and increased spleen weight at 1000 mg/kg bw per day.

Atrazine was tested for genotoxicity in a large number of studies covering an adequate range of end-points, including assays for gene mutation in bacteria and eukaryotic cells in vitro, for DNA damage and repair in bacteria and mammalian cells (rat hepatocytes, human fibroblasts) in vitro, and for chromosomal aberration in vitro and in somatic and germ cells in vivo. Mostly negative results were obtained in standard assays. In a few published studies, positive responses were reported. However, a number of reviews by national and international agencies (United States Environmental Protection Agency, European Union, International Agency for Research on Cancer) have concluded that, on the basis of the weight of evidence, atrazine is not genotoxic.

The Meeting agreed that it is unlikely that atrazine is genotoxic.

Long-term studies of toxicity and carcinogenicity were conducted in mice and rats. As in short-term studies, reduced body-weight gain and food intake and a decrease in erythrocyte parameters were noted consistently. Additionally, reduced survival of females and cardiovascular effects (atrial thrombi) in both sexes were observed in mice at high doses.

In three studies of carcinogenicity in mice, no treatment-related carcinogenic effects were observed at dietary concentrations of up to 3000 ppm, equal to about 386 and 483 mg/kg bw per day in males and females, respectively. Overall, the NOAEL was 10 ppm, equal to 1.2 mg/kg bw per day, on the basis of lower body weight/body-weight gain at 300 ppm, equal to 38.4 mg/kg bw per day, and greater.

In two studies of carcinogenicity in F344 rats fed diets containing atazine at concentrations of up to 400 ppm, equivalent to about 20 mg/kg bw per day, there was no effect at any dose on the onset or incidence of tumours. The NOAEL was 70 ppm, equivalent to about 3.5 mg/kg bw per day, on the basis of decreased body weight at concentrations of 200 ppm and greater. In a non-guideline study of carcinogenicity in F344 rats, there was a significant increase in the incidence of benign mammary tumours in males and in uterine adenocarcinomas in females at the highest dose of 750 ppm, equivalent to about 38 mg/kg bw per day; however, interpretation of the result was limited by increased survival at the highest dose, and a survival-adjusted analysis of tumour prevalence did not indicate any significant increase in the incidence of benign, malignant or combined mammary tumours.

In seven studies of carcinogenicity in Sprague-Dawley rats fed diets containing atrazine at concentrations of up to 1000 ppm (equal to about 42 and 65 mg/kg bw per day in males and females, respectively), an increased incidence of mammary tumours (adenomas, carcinomas, fibroadenomas) with or without an earlier onset (relative to controls) was observed in four studies, while in two studies there was an earlier onset of mammary tumours without any increase in their overall lifetime incidence. An earlier onset of pituitary tumours was also observed in one study, with no increase in incidence at term. Overall, the NOAEL for mammary carcinogenicity was 25 ppm, equal to 1.5 mg/kg bw per day, on the basis of a statistically significant increased incidence in mammary tumours at 50 ppm, equal to 3.1 mg/kg bw per day.

In a study of carcinogenicity in ovariectomized Sprague-Dawley rats, neither increases in mammary-gland proliferative changes nor mammary tumours were seen at dietary concentrations of up to 400 ppm (equal to about 21 mg/kg bw per day), suggesting that the carcinogenic mode of action of atrazine in Sprague-Dawley rats is related to ovarian function.

In a mechanistic 6-month study in Sprague-Dawley rats, attenuation of the LH surge and subsequent disruption of the estrous cycle (characterized by an increase in days in estrus) were observed at ≥ 50 ppm (equal to 3.65 mg/kg bw per day), with a NOAEL of 25 ppm (equal to 1.8 mg/kg bw per day). The NOAEL and LOAEL for these effects were comparable to those identified in the studies of carcinogenicity. The effects on the LH surge and disruption of the estrous cycle were further supported by a number of short-term mechanistic studies. Additional experiments suggested that the effects of atrazine on LH and prolactin secretion are mediated via a hypothalamic site of action.

The postulated mode of action for atrazine-induced mammary tumours in female Sprague-Dawley rats involved disruption of the hypothalamic–pituitary–ovary axis. Atrazine modifies catecholamine function and the regulation of GnRH pulsatility in the rat hypothalamus, with the consequence that the pulse of LH released from the pituitary gland is of insufficient amplitude or duration to trigger the ovulation. The failure to ovulate results in persistent secretion of estrogen, which provides a feedback to the pituitary leading to increased secretion of prolactin. As a result, atrazine accelerates the normal reproductive ageing process in female Sprague-Dawley rats whereby reproductive senescence is characterized by persistent exposure to estrogen and prolactin. In contrast, women respond to reduced levels of LH by reductions in estrogen concentrations. Thus, the Meeting considered that the mode of carcinogenic action in certain susceptible rat strains is not relevant for risk assessment in humans.

Investigations of other modes of action did not provide any evidence that atrazine had intrinsic estrogenic activity or that it increased aromatase activity in vivo.

The Meeting concluded that atrazine is not likely to pose a carcinogenic risk to humans.

Although carcinogenicity in humans was not a concern owing to the rat-specific mode of action, alterations in neurotransmitter and neuropeptide function regulating LH and secretion of prolactin may potentially induce adverse effects during critical periods of development (as found in special studies showing pregnancy loss, delayed puberty in males and females, and decreased suckling-induced prolactin release in lactating dams). Unlike the carcinogenic effects, the developmental effects do not appear to be specific to certain strains of rats and the Meeting therefore considered these effects to be relevant for risk assessment in humans.

In special studies of reproductive toxicity, exposure of rats during early pregnancy (i.e. the LH-dependent period) caused increased pre- or postimplantation losses, including full-litter resorptions. Effects were seen at doses of 50 mg/kg bw per day and greater after treatment on days 6–10 of gestation, with a NOAEL of 25 mg/kg bw per day. In contrast, exposure on days 11–15 of gestation (after the LH-dependent period of pregnancy) at a dose of 200 mg/kg bw per day did not induce full-litter resorptions.

Suppression of the suckling-induced release of prolactin in lactating rats was seen with atrazine at doses of 25 mg/kg bw per day and greater, with a NOAEL of 12.5 mg/kg bw per day. Treatment of lactating rats on postnatal days 1–4 affected the development of tuberoinfundibular dopaminergic neurons in the pups (presumably due to the lack of prolactin derived from the dam's milk), with the consequence of impaired regulation of prolactin secretion, hyperprolactinaemia before puberty and prostatitis in the adult male offspring.

A delay in sexual development was observed in female rats after exposure on postnatal days 21–46 at doses of 30 mg/kg bw per day and greater, with a NOAEL of 10 mg/kg bw per day, and in male rats after exposure on postnatal days 23–53 at doses of 12.5 mg/kg bw per day and greater, with a NOAEL of 6.25 mg/kg bw per day.

In a standard two-generation study of reproduction (conducted according to earlier guidelines, which did not include end-points such as estrous cyclicity and sexual development) in rats, there was no effect on fertility at 500 ppm, the highest dose tested. The NOAEL for parental toxicity was 50 ppm, equal to 3.6 mg/kg bw per day, on the basis of decreased body-weight gains and food consumption at 500 ppm, equal to 36.1 mg/kg bw per day. The NOAEL for reproductive toxicity was 50 ppm on the basis of decreased body weights of male pups at postnatal day 21 at 500 ppm.

In two studies of prenatal developmental toxicity in rats given atrazine on days 6–15 of gestation, the NOAELs for maternal toxicity were 10 or 25 mg/kg bw per day on the basis of decreased body-weight gain and food intake at 70 or 100 mg/kg bw per day, respectively. The NOAELs for developmental toxicity were 10 or 25 mg/kg bw per day on the basis of incomplete ossification at several sites at 70 or 100 mg/kg bw per day, respectively. In a study of prenatal developmental toxicity in rabbits given atrazine on days 7–19 of gestation, the NOAEL for maternal toxicity was 5 mg/kg bw per day on the basis of clinical signs, abortion and decreased food intake and body-weight gain at 75 mg/kg bw per day. The NOAEL for developmental toxicity was 5 mg/kg bw per day on the basis of increased resorptions, reduced litter size and incomplete ossification at 75 mg/kg bw per day. In rats and rabbits, the developmental effects were observed only at maternally toxic doses.

The Meeting concluded that atrazine was not teratogenic.

Studies using a variety of test systems in vitro and in vivo indicated that modulation of the immune system occurs after exposure to atrazine. However, effects suggestive of impaired function of the immune system were only observed at doses greater than those shown to affect neuroendocrine function, leading to disruption of the estrous cycle or developmental effects.

A range of epidemiological studies (including cohort studies, case–control studies, and ecological or correlational studies) assessed possible relationships between atrazine or other triazine herbicides and cancer in humans. For some cancer types, such as prostate or ovarian cancer and non-Hodgkin's lymphoma, the increased risks reported in single studies could either be explained by the methodology used, or had not been confirmed in more reliable studies. Thus, the weight of evidence from the epidemiological studies did not support a causal association between exposure to atrazine and the occurrence of cancer in humans.

The Meeting concluded that the existing database on atrazine is adequate to characterize the potential hazards to fetuses, infants and children.

Metabolites of atrazine

The toxicity profiles and mode of action of the chloro-*s*-triazine metabolites were similar to those of atrazine; the potency of these metabolites appeared to be similar to that of the parent compound with regard to their neuroendocrine-disrupting properties.

Like atrazine, the chloro-*s*-triazine metabolites were of moderate or low acute oral toxicity in rats; LD_{50}s were 1110, 1240 and 2310–5460 mg/kg bw for DEA, DIA and DACT, respectively.

Like atrazine, its chloro-*s*-triazine metabolites delayed sexual development of male rats exposed on postnatal days 23–53 at atrazine molar equivalent doses of 25 mg/kg bw per day and greater (DEA, DIA) and 12.5 mg/kg bw per day and greater (DACT), with NOAELs of 12.5 and 6.25 mg/kg bw per day, respectively. Exposure of female rats to DACT on postnatal days 22–41 delayed sexual development at atrazine molar equivalent doses of 50 mg/kg bw per day and greater, and the NOAEL was 25 mg/kg bw per day. Doses at which these effects occurred were similar to those observed for parent atrazine.

In short-term feeding studies in rats, the main effects of the chlorinated metabolites were similar to those of atrazine and included reduced body-weight gain and decreased erythrocyte parameters, and also for DACT-induced disruption of the estrous cycle. The NOAELs were 50 ppm (equal to 3.2 mg/kg bw per day) for DEA and DIA, and 100 ppm (equal to 7.6 mg/kg bw per day) for DACT.

In a 29/52-week study with DACT in Sprague-Dawley rats, effects comparable to those observed with atrazine (attenuation of the LH surge, increased incidences of mammary tumours) were seen at 270 ppm; the NOAEL was 48 ppm, equal to 3.4 mg/kg bw per day. No long-term studies were performed with DEA or DIA.

In short-term feeding studies in dogs, the main effects of the chlorinated metabolites were similar to those of atrazine and included reduced body-weight gain and decreased erythrocyte parameters, while DEA and DACT showed cardiac toxicity. The NOAELs were 100 ppm, equal to 3.7, 3.8 and 3.5 mg/kg bw per day, for DEA, DIA and DACT, respectively.

DEA, DIA and DACT did not show genotoxicity in an adequate range of tests in vitro and in vivo.

In studies of prenatal developmental toxicity in rats, the chlorinated metabolites induced increased incidences of fused sternebrae and/or incomplete ossification at doses of 25 to 100 mg/kg bw per day; the NOAELs for developmental toxicity were 25, 5 and 2.5 mg/kg bw per day for DEA, DIA and DACT, respectively. The effects were seen only at doses that also produced maternal toxicity.

The metabolite hydroxyatrazine does not have the same mode of action or toxicity profile as atrazine and its chlorometabolites. The main effect of hydroxyatrazine was kidney toxicity (owing to its low solubility in water, resulting in crystal formation and a subsequent inflammatory response), and there was no evidence that hydroxyatrazine has neuroendocrine-disrupting properties. Also, the acute oral toxicity of hydroxyatrazine in rats (LD_{50}, > 5050 mg/kg bw) was lower than that of atrazine or its chlorometabolites.

In short-term feeding studies, the main effects of hydroxyatrazine in rats included reduced body-weight gain, increased water consumption, changes in clinical chemistry and urine analysis parameters, and kidney lesions. The overall NOAEL was 100 ppm, equal to 6.3 mg/kg bw per day. In dogs, effects included reduced body-weight gain and food consumption, changes in clinical chemistry and urine analysis parameters, and kidney lesions; the NOAEL was 150 ppm, equal to 5.8 mg/kg bw per day.

In a 2-year study of toxicity and carcinogenicity in rats, the effects of hydroxyatrazine included clinical signs and increased mortality, reduced body-weight gain and food consumption, increased water consumption, changes in haematological, clinical chemistry and urine analysis parameters, and kidney lesions. The NOAEL was 25 ppm, equal to 1.0 mg/kg bw per day. There was no evidence of carcinogenicity.

Hydroxyatrazine did not show genotoxicity in an adequate range of tests in vitro and in vivo.

In a study of prenatal developmental toxicity in rats, the effects of hydroxyatrazine consisted of reduced food consumption and body-weight gain in dams and increased incidences of incomplete and absent ossification in fetuses at 125 mg/kg bw per day; the NOAEL was 25 mg/kg bw per day for maternal and developmental toxicity. Exposure of female rats on postnatal days 22–41 at atrazine molar equivalent doses of up to 200 mg/kg bw per day did not delay sexual development.

Toxicological evaluation

Drinking-water may contain metabolites of atrazine as well as atrazine itself. The chloro-*s*-triazine metabolites DEA, DIA and DACT share the same mode of action as atrazine and have a similar toxicological profile and hence the Meeting decided to establish a group ADI and acute reference dose (ARfD). Hydroxyatrazine, the plant and soil degradate, was not included because its mode of action and toxicological profile are different to those of atrazine and its chloro-*s*-triazine metabolites.

The Meeting established a group ADI of 0–0.02 mg/kg bw on the basis of the NOAEL for atrazine of 1.8 mg/kg bw per day identified on the basis of LH-surge suppression and subsequent disruption of the estrous cycle seen at 3.6 mg/kg bw per day in a 6-month study in rats, and using a safety factor of 100. The Meeting considered that this NOAEL was protective for the consequences of neuroendocrine and other adverse effects caused by prolonged exposure to atrazine and its chloro-*s*-triazine metabolites.

The Meeting established a group ARfD of 0.1 mg/kg bw on the basis of the NOAEL for atrazine of 12.5 mg/kg bw per day identified on the basis of impaired suckling-induced prolactin secretion in dams and subsequent alterations in development of the central nervous system and prolactin regulation in male offspring in a special 4-day study in rats, and using a safety factor of 100. This ARfD was supported by the results of other studies of developmental toxicity with atrazine and its chlorometabolites, from which overall NOAELs/LOAELs of 25/50 mg/kg bw per day in rats and 5/75 mg/kg bw per day in rabbits were identified on the basis of effects that might occur after a single exposure (i.e. postimplantation loss, fused sternebrae). The study in rabbits (in which there was a 15-fold difference between NOAEL and LOAEL) was not selected as the basis for the ARfD, because examination of the studies in rats indicated that the dose selected for the ARfD would be adequately protective for these end-points in rabbits.

For hydroxyatrazine, the Meeting established an ADI of 0–0.04 mg/kg bw on the basis of the NOAEL of 1.0 mg/kg bw per day identified on the basis of kidney toxicity (caused by low solubility in water resulting in crystal formation and a subsequent inflammatory response) at 7.8 mg/kg bw per day in a 24-month study in rats, and using a safety factor of 25. A modified safety factor on the basis of kinetic considerations was deemed appropriate since the critical effect of hydroxyatrazine is dependent on its physicochemical properties and the interspecies variability for such effects is lower than for AUC-dependent effects.

The Meeting concluded that it was not necessary to establish an ARfD for hydroxyatrazine in view of its low acute toxicity, the absence of relevant developmental toxicity that could be a consequence of acute exposure, and the absence of any other toxicological effects that would be likely to be elicited by a single dose.

Levels relevant to risk assessment

(a) Atrazine

Species	Study	Effect	NOAEL	LOAEL
Mouse	Long-term studies of carcinogenicity[a,d]	Toxicity	10 ppm, equal to 1.2 mg/kg bw per day	300 ppm, equal to 38.4 mg/kg bw per day
		Carcinogenicity	3000 ppm, equal to 385.7 mg/kg bw per day[c]	—
Rat	Thirteen-week study of toxicity[a]	Toxicity	50 ppm, equal to 3.3 mg/kg bw per day	500 ppm, equal to 34.0 mg/kg bw per day
	Two-year studies of toxicity and carcinogenicity[a,d] (Sprague-Dawley rats)	Toxicity	70 ppm, equal to 2.6 mg/kg bw per day	500 ppm, equal to 19.9 mg/kg bw per day
		Carcinogenicity	25 ppm, equal to 1.5 mg/kg bw per day	50 ppm, equal to 3.1 mg/kg bw per day[e]
	Two-year studies of toxicity and carcinogenicity[a,d] (Fischer 344 rats)	Toxicity	70 ppm, equal to 3.5 mg/kg bw per day	200 ppm, equal to 10 mg/kg bw per day
		Carcinogenicity	400 ppm, equal to 20 mg/kg bw per day[c]	—
	Multigeneration study of reproductive toxicity[a]	Fertility	500 ppm, equal to 36.1 mg/kg bw per day[c]	—
		Parental toxicity	50 ppm, equal to 3.6 mg/kg bw per day	500 ppm, equal to 36.1 mg/kg bw per day
		Offspring toxicity	50 ppm, equal to 3.6 mg/kg bw per day	500 ppm, equal to 36.1 mg/kg bw per day
	Developmental toxicity[b,d]	Maternal toxicity	10 mg/kg bw per day	70 mg/kg bw per day
		Embryo/fetotoxicity	10 mg/kg bw per day	70 mg/kg bw per day

	Special 6-month study[a]	Endocrine disruption (LH surge)	25 ppm, equal to 1.8 mg/kg bw per day	50 ppm, equal to 3.65 mg/kg bw per day
	Special 4-day study[b]	Endocrine disruption (prolactin release)	12.5 mg/kg bw per day	25 mg/kg bw per day
	Special 5-day study[b]	Postimplantation loss	25 mg/kg bw per day	50 mg/kg bw per day
	Special 25-day study[b]	Female pubertal delay	10 mg/kg bw per day	30 mg/kg bw per day
	Special 30-day study[b]	Male pubertal delay	6.25 mg/kg bw per day	12.5 mg/kg bw per day
Rabbit	Developmental toxicity[b]	Maternal toxicity	5 mg/kg bw per day	75 mg/kg bw per day
		Embryo/fetotoxicity	5 mg/kg bw per day	75 mg/kg bw per day
Dog	One-year study of toxicity[a]	Toxicity	150 ppm, equal to 5 mg/kg bw per day	1000 ppm, equal to 33.7 mg/kg bw per day

[a] Dietary administration.
[b] Gavage administration.
[c] Highest dose tested.
[d] Results of two or more studies combined.
[e] Mammary gland tumours—not relevant to humans.

(b) Deethyl-atrazine (DEA)

Species	Study	Effect	NOAEL	LOAEL
Rat	Thirteen-week study of toxicity[a]	Toxicity	50 ppm, equal to 3.2 mg/kg bw per day	500 ppm, equal to 35.2 mg/kg bw per day
	Developmental toxicity[b]	Maternal toxicity	5 mg/kg bw per day	25 mg/kg bw per day
		Embryo- and feto-toxicity	25 mg/kg bw per day	100 mg/kg bw per day
	Special 30-day study[b]	Male pubertal delay	12.5 mg/kg bw per day[c]	25 mg/kg bw per day[c]
Dog	Thirteen-week study of toxicity[a]	Toxicity	100 ppm, equal to 3.7 mg/kg bw per day	1000 ppm, equal to 28.9 mg/kg bw per day

[a] Dietary administration.
[b] Gavage administration.
[c] Atrazine molar equivalent dose.

(c) Deisopropyl-atrazine (DIA)

Species	Study	Effect	NOAEL	LOAEL
Rat	Thirteen-week study of toxicity[a]	Toxicity	50 ppm, equal to 3.2 mg/kg bw per day	500 ppm, equal to 34.9 mg/kg bw per day
	Developmental toxicity[b]	Maternal toxicity	5 mg/kg bw per day	25 mg/kg bw per day
		Embryo/fetotoxicity	5 mg/kg bw per day	25 mg/kg bw per day
	Special 30-day study[b]	Male pubertal delay	12.5 mg/kg bw per day[c]	25 mg/kg bw per day[c]
Dog	Thirteen-week study of toxicity[a]	Toxicity	100 ppm, equal to 3.8 mg/kg bw per day	500 ppm, equal to 18.0 mg/kg bw per day

[a] Dietary administration.
[b] Gavage administration.
[c] Atrazine molar equivalent dose.

(d) Diaminochlorotriazine (DACT)

Species	Study	Effect	NOAEL	LOAEL
Rat	Thirteen-week study of toxicity[a]	Endocrine disrup-tion (estrous cycle)	100 ppm, equal to 7.6 mg/kg bw per day	250 ppm, equal to 19.7 mg/kg bw per day
	Developmental toxicity[b]	Maternal toxicity	2.5 mg/kg bw per day	25 mg/kg bw per day
		Embryo/fetotoxicity	2.5 mg/kg bw per day	25 mg/kg bw per day
	Special study, 29/52 weeks[a]	Endocrine disruption (LH surge)	48 ppm, equal to 3.4 mg/kg bw per day	270 ppm, equal to 18.8 mg/kg bw per day
	Special 19-day study[b]	Female pubertal delay	25 mg/kg bw per day[c]	50 mg/kg bw per day[c]
	Special 30-day study[b]	Male pubertal delay	6.25 mg/kg bw per day[c]	12.5 mg/kg bw per day[c]
Dog	Thirteen-/52-week study of toxicity[a]	Toxicity	100 ppm, equal to 3.5 mg/kg bw per day	1500/750 ppm, equal to 23.8 mg/kg bw per day

LH, luteinizing hormone.
[a] Dietary administration.
[b] Gavage administration.
[c] Atrazine molar equivalent dose.

(e) Hydroxyatrazine

Species	Study	Effect	NOAEL	LOAEL
Rat	Thirteen-week study of toxicity[a]	Toxicity	100 ppm, equal to 6.3 mg/kg bw per day	300 ppm, equal to 18.9 mg/kg bw per day
	Two-year study of toxic-ity and carcinogenicity[a] (Sprague-Dawley rats)	Toxicity	25 ppm, equal to 1.0 mg/kg bw per day	200 ppm, equal to 7.8 mg/kg bw per day
		Carcinogenicity	400 ppm, equal to 17.4 mg/kg bw per day[c]	—
	Developmental toxicity[b]	Maternal toxicity	25 mg/kg bw per day	125 mg/kg bw per day
		Embryo/fetotoxicity	25 mg/kg bw per day	125 mg/kg bw per day
	Special 19-day study[b]	Female pubertal delay	200 mg/kg bw per day[c,d]	—
Dog	Thirteen-/52-week study of toxicity[a]	Toxicity	150 ppm, equal to 5.8 mg/kg bw per day	1500 ppm, equal to 59.6 mg/kg bw per day

[a] Dietary administration.
[b] Gavage administration.
[c] Highest dose tested.
[d] Atrazine molar equivalent dose.

Estimate of acceptable daily intake for humans

Group ADI for atrazine, deethyl-atrazine (DEA), deisopropyl-atrazine (DIA) and diaminochlorotriazine (DACT)

0–0.02 mg/kg bw

Hydroxyatrazine

0–0.04 mg/kg bw

Estimate of acute reference dose

Group ARfD for atrazine, deethyl-atrazine(DEA), deisopropyl-atrazine (DIA) and diaminochlorotriazine (DACT)

0.1 mg/kg bw

Hydroxyatrazine

Unnecessary

Information that would be useful for the continued evaluation of the compound

Results from epidemiological, occupational health and other such observational studies of human exposures

Critical end-points for setting guidance values for exposure to atrazine

Absorption, distribution, excretion and metabolism in animals	
Rate and extent of oral absorption	Rapid, > 80% in rats
Distribution	Widely distributed
Rate and extent of excretion	> 50% in 24 h and > 93% within 7 days; approximately 73% via urine, approximately 20% via faeces (approximately 7% via bile)
Potential for accumulation	Low; binding to rat haemoglobin, not relevant to humans
Metabolism in mammals	Extensive (> 95%) to at least 25 metabolites; major pathway is *N*-dealkylation
Toxicologically significant compounds in animals, plants and the environment	Parent compound, chloro-*s*-triazine metabolites DEA, DIA, DACT (animals, environment), hydroxyatrazine (plants, environment)
Acute toxicity	
Rat, LD_{50}, oral	1870–3090 mg/kg bw
Rat, LD_{50}, dermal	> 2000 mg/kg bw
Rat, $LC_{50,}$ inhalation	> 5.8 mg/l
Rabbit, skin irritation	Not an irritant
Rabbit, eye irritation	Not an irritant
Guinea-pig, skin sensitization	Sensitizer (Magnusson & Kligman; Maurer optimization test)
Short-term studies of toxicity	
Target/critical effect	Reduced body-weight gain, ovaries (inhibition of ovulation), cardiotoxicity (in dogs only)
Lowest relevant oral NOAEL	3.3 mg/kg bw per day (90-day study in rats)
Lowest relevant dermal NOAEL	100 mg/kg bw per day (25-day study in rabbits)
Lowest relevant inhalation NOAEC	No data
Genotoxicity	
	Unlikely to be genotoxic in vivo
Long-term studies of toxicity and carcinogenicity	
Target/critical effect	Ovaries (inhibition of ovulation) and related endocrine changes
Lowest relevant NOAEL	1.8 mg/kg bw per day (6-month –study of LH-surge in Sprague-Dawley rats)
Carcinogenicity	No relevant carcinogenicity
Reproductive toxicity	
Reproductive target/critical effect	Reduced body-weight gain in pups at parentally toxic doses
Lowest relevant reproductive NOAEL	3.6 mg/kg bw per day

Developmental target/critical effect	Increased resorptions and incomplete ossification at maternally toxic doses; delayed sexual development
Lowest relevant developmental NOAEL	6.25 mg/kg bw per day (rat; male pubertal development) 5 mg/kg bw per day (rabbit)
Neurotoxicity	
	No evidence of neurotoxicity in standard tests for toxicity; however, neuroendocrine mode of action has been established for atrazine and its chloro-*s*-triazine metabolites
Other toxicological studies	
Studies on metabolites	DEA, DIA, DACT have the same neuroendocrine mode of action and similar potency to atrazine Hydoxyatrazine has a different mode of action and toxicity profile to atrazine
Mode of neuroendocrine action	Atrazine and its chlorometabolites modify hypothalamic catecholamine function and regulation, leading to alterations in pituitary LH and prolactin secretion
Mode of carcinogenic action	The postulated mode of carcinogenic action in female Sprague-Dawley rats involves acceleration of the reproductive ageing process (suppression of LH surge, subsequent estrous cycle disruption), which is not relevant to humans
Direct estrogenic activity	Atrazine has no intrinsic estrogenic activity
Aromatase expression	No effect on aromatase expression in rats
Effects on sexual development	Evidence of delayed sexual development in male and/or female rats by atrazine, DEA, DIA and DACT
Effects on neuronal development	Evidence of impaired postnatal CNS development (and subsequent alterations in prolactin regulation)
Immunotoxicity	Evidence for immune system modulation at doses greater than LOAELs for neuroendocrine disruption or reproductive and developmental effects
Medical data	
	No evidence of atrazine causing effects in manufacturing plant personnel. Epidemiology studies do not support a causal association between exposure to atrazine and cancer in humans.

Summary

Atrazine

	Value	Study	Safety factor
Group ADI[a]	0–0.02 mg/kg bw	Sprague-Dawley rats; 6-month study of LH surge/estrous cycle disruption	100
Group ARfD[a]	0.1 mg/kg bw	Rat; special 4-day study of prolactin release, supported by studies of developmental toxicity in rats and rabbits	100
Hydroxyatrazine			
ADI	0–0.04 mg/kg bw	Sprague-Dawley rats; 2-year study	25
ARfD	Unnecessary	—	—

[a] Group ADI or ARfD for atrazine, deethyl-atrazine (DEA), deisopropyl-atrazine (DIA) and diaminochlorotriazine (DACT).

CNS, central nervous sytem; GnRH, gonadotrophin-releasing hormone; LH, luteinizing hormone.

Appendix 1

Application of the IPCS Conceptual Framework for Cancer Risk Assessment (IPCS Framework for Analysing the Relevance of a Cancer Mode of Action for Humans): consideration of mammary gland tumours in female Sprague-Dawley rats exposed to atrazine

This framework, developed by an International Programme on Chemical safety (IPCS) working group, provides a generic approach to the principles commonly used in evaluating a postulated mode of action (MOA) for tumour induction by a chemical (Boobis et al., 2006). The framework was used by the 2007 JMPR to provide a structured approach to the assessment of the overall weight-of-evidence for the postulated MOA for the increased incidence of mammary tumours (adenomas, carcinomas, fibroadenomas) in rats that was observed after long-term administration of atrazine.

The first stage of the framework is to determine whether it is possible to establish an MOA. This process comprises a series of key events along the causal pathway to cancer, identified using a weight-of-evidence approach on the basis of the Bradford-Hill criteria. The key events are then compared first qualitatively and then quantitatively between the experimental animals and humans. Finally, a statement of confidence, analysis, and implications is provided.

Mammary gland tumours associated with atrazine exposure in female Sprague-Dawley rats

I. Is the weight of evidence sufficient to establish an MOA in animals?

1. Introduction

In seven studies of carcinogenicity in Sprague-Dawley rats fed diets containing atrazine at concentrations of up to 1000 ppm (equal to about 42 and 65 mg/kg bw per day in males and females, respectively), an increased incidence and/or an earlier appearance of spontaneously occurring mammary tumours (adenomas, carcinomas, fibroadenomas) was observed in four studies, while in two studies, there was an earlier appearance of mammary tumours, without any increase in their overall lifetime incidence (see monograph section 2.3).

2. Postulated MOA (theory of the case)

The postulated MOA for mammary tumours induced by atrazine in female Sprague-Dawley rats involves disruption of the hypothalamic–pituitary–ovary axis. Atrazine modifies catecholamine function and the regulation of gonadotropin-releasing hormone (GnRH) pulsatility in the rat hypothalamus, with the consequence that the pulse of luteinizing hormone (LH) released from the pituitary gland is of insufficient amplitude or duration to trigger the ovulation. The failure to ovulate results in persistent secretion of estrogen, which provides a feedback to the pituitary leading to increased secretion of prolactin. The persistent stimulation of the mammary gland by estrogen and prolactin translates into a proliferative response characterized by an earlier appearance and/or a higher incidence of adenocarcinomas (high estrogen, moderate prolactin levels) or fibroadenomas (high prolactin with a background of estrogen) (McConnell, 1989; O'Connor et al., 2000; Simpkins, 2000; Simpkins et al., 2000; Cooper et al., 2007).

3. Key events

The sequence of key events in the mode of carcinogenic action of atrazine in the mammary gland of female Sprague-Dawley rats includes:

- Effect on the hypothalamus, leading to a modification of catecholamine function and the regulation of GnRH pulsatility;
- In consequence, the pulse of LH released from the pituitary gland is of insufficient amplitude or duration to trigger the ovulation;

- The failure to ovulate results in persistent secretion of estrogen, which provides a feedback to the pituitary leading to increased secretion of prolactin;
- The prolonged exposure to estrogen and prolactin causes a hyperstimulation of the mammary gland and an earlier appearance and/or a higher incidence of mammary tumours.

The key events as described above include effects on the hypothalamic control of pituitary function, leading to disruption of the estrous cycle and a persistent stimulation of the mammary gland by endogenous estrogen and prolactin. These effects have been investigated and observed in female Sprague-Dawley rats in short-term and/or mechanistic studies, and at interim and terminal kills in a long-term study. The dose–response relationship and temporal analyses of the key events and tumour response are presented below.

4. Concordance of dose–response relationships

The no-observed-adverse-effect levels (NOAELs) and lowest-observed-adverse-effect levels (LOAELs) for the key effects in the MOA of atrazine in the mammary gland are provided in Table A1.

In ovariectomized rats, a biologically significant suppression of the serum LH surge was observed after 3 days or 3 weeks of treatment with atrazine at doses of 50 mg/kg bw per day and

Table A1. Summary of dose–response relationship in rats receiving atrazine

Effect	NOAEL/LOAEL	Reference
Hypothalamus		
Increase in GnRH and dopamine	< 25/25 mg/kg bw per day (4-day mechanistic study)	Cooper et al., (2007)
Pituitary		
Attenuation of LH surge	< 300/300 mg/kg bw per day (1-day mechanistic study)	Cooper et al. (2000)
	< 50/50 mg/kg bw per day (3-day mechanistic study)	Cooper et al. (2000)
	< 6.25/6.25 mg/kg bw per day (4-day mechanistic study)	Cooper et al. (2007)
	1.8/3.65 mg/kg bw per day (26-week mechanistic study)	Morseth (1996c)
Ovary		
Disruption of estrous cycle	< 300/300 mg/kg bw per day (1-day mechanistic study)	Cooper et al. (2000)
	< 75/75 mg/kg bw per day (21-day mechanistic study)	Cooper et al. (1996)
	1.8/3.65 mg/kg bw per day (26-week mechanistic study)	Morseth (1996c)
Mammary gland		
Increase in acinar-lobular development, secretory activity and galactocoele formation	3.5/20 mg/kg bw per day (2-year study, 1-year interim kill)	McConnell (1995)
Increase in incidence of palpable mammary masses	3.5/20 mg/kg bw per day (2-year study, 1-year interim kill)	Thakur (1991a, 1992a)
Increase in incidence of mammary tumours	0.5/3.5 mg/kg bw per day (2-year study)	Mayhew (1986)
	1.5/3.1 mg/kg bw per day (2-year study)	Morseth (1996d, 1998)

GnRH, gonadotrophin-releasing hormone; LH, luteinizing hormone.

greater, and in a 26-week study with atrazine at doses of 3.65 mg/kg bw per day and greater. Consistent with attenuation of the LH surge, disruption of the estrous cycle was seen in a 3-week study with atrazine at doses of 75 mg/kg bw per day and greater, and in a 26-week study at doses of 3.65 mg/kg bw per day and greater. Owing to the failure to ovulate and the subsequent persistent exposure to endogenous estrogen and prolactin, hyperstimulation of the mammary gland (increase in acinar-lobular development indicative of increased exposure to estrogen; increase in secretory activity and galactocoele formation indicative of increased exposure to prolactin) was observed with atrazine at doses of 20 mg/kg bw per day and greater, while increased incidences of mammary tumours were found at doses of 3.1 mg/kg bw per day and greater.

Generally, there was a good correlation between the doses causing attenuation of the LH surge and those causing an earlier onset and/or an increased incidence of mammary tumours.

5. Temporal association

The key events, such as attenuation of the LH surge and disruption of the estrous cycle were observed after a single high dose of atrazine at 300 mg/kg bw, after a 3-day or 3-week exposure at doses of 50 or 75 mg/kg bw per day and greater, respectively, and after a 26-week exposure at doses of 3.65 mg/kg bw per day and greater. In 2-year studies in rats, the onset-time for initial palpation of mammary masses was decreased when compared with controls (14 weeks at approximately 20 mg/kg bw per day vs 29 weeks in controls), while the incidence of palpable mammary masses was increased at interim kill (weeks 52–54) after exposure to atrazine at approximately 20 mg/kg bw per day. Thus, there is a logical temporal response with all key events preceding tumour formation.

6. Strength, consistency and specificity of association of tumour response with key events

The key events were observed consistently in a number of studies with differing experimental designs. On the basis of information from the studies described in the monograph, there is sufficient weight of evidence that the key events (attenuation of the LH surge, disruption of the estrous cycle) are linked to the morphological changes in the mammary gland indicative of stimulation of estrogen and prolactin (increase in acinar-lobular development, increase in secretory activity and galactocoele formation) which precede the occurrence of tumours. In addition, there is a substantial independent literature on the role of estrogen and prolactin in the pathogenesis of mammary tumours in rats. There are no significant contradictory data.

7. Biological plausibility and coherence

The relationship between sustained perturbation of the hypothalamic–pituitary axis (change of the regulation of GnRH pulsatility, attenuation of the LH surge), disruption of the estrous cycle, persistent secretion of estrogen and prolactin, prolonged stimulation of the mammary gland by endogenous estrogen and prolactin, and the development of mammary gland tumours is considered to be biologically plausible and has been shown in several studies in laboratory rats.

In long-term bioassays with natural and synthetic estrogens, it has been established that prolonged stimulation of the mammary gland with estrogen leads to development of adenocarcinomas. In contrast, high-level stimulation of the mammary gland with prolactin has been shown to be linked to the development of fibroadenoma.

The tumour response elicited by atrazine is typical of a rodent mammary-gland carcinogen in that mammary tumours are found in female Sprague-Dawley rats but not in male rats or mice. Rats tend to be more sensitive to mammary-gland carcinogenesis than mice, and female rats are frequently found to be more sensitive than male rats with respect to the proportion of chemicals that induce mammary tumours. Consistent with this, concentrations of estrogen and prolactin are typically higher in female rats than in males.

The relationship between increased concentrations of estradiol (E2) and prolactin and tumours of the mammary gland in female rats is further supported by the fact that tumours of the mammary gland tumours do not occur in female Sprague-Dawley rats ovariectomized at an early age and treated with atrazine, as concentrations of E2 and prolactin are both minimal in these females.

8. Other modes of action

Genotoxicity is always one possible MOA to consider, but in a large range of in studies of genotoxicity in vitro and in vivo with atrazine, mostly negative results were obtained in tests using standard methods. The weight of evidence suggests that it is unlikely that atrazine is genotoxic. This conclusion is also supported by the fact that atrazine did not induce a tumorigenic response in ovariectomized rats, while genotoxic carcinogens like dimethylbenz[a]anthracene and *N*-methyl-*N*-nitrosourea are capable of inducing mammary tumours in ovariectomized rats.

In addition, a possible intrinsic estrogenic activity of atrazine was considered. However, the majority of in-vitro studies reported that atrazine did not competitively bind to the rat cytosolic estrogen or progesterone receptors. Also, exposure to atrazine in vivo did not induce uterine growth nor increase the number of progesterone receptors in ovariectomized rats. Thus, the development of tumours of the mammary gland in female rats exposed to atrazine does not seem to be related to any intrinsic estrogenic activity of the compound.

Furthermore, studies in rats indicated no evidence for any atrazine-induced upregulation of aromatase expression in the brain, testes, or mammary gland.

9. Uncertainties, inconsistencies, and data gaps

No inconsistencies were identified in the database for atrazine with regard to the postulated MOA for mammary-gland tumours in female rats.

However, the precise mechanism by which atrazine disrupts the neuronal control of hypothalamic GnRH secretion in the rat remains to be determined.

10. Assessment of postulated MOA

There is sufficient experimental evidence that atrazine disrupts the neuroendocrine control of ovarian function in Sprague-Dawley rats, which leads to premature reproductive senescence (i.e. constant estrus) and a hormonal milieu conducive to the development of mammary gland tumours. The strength, consistency, and specificity of the available MOA information for female Sprague-Dawley rats is further confirmed by data showing that Fischer 344 rats, which have a different reproductive senescence, do not develop atrazine-related mammary tumours. Alternative hypotheses have been ruled out, such as genotoxicity, estrogenic activity or upregulation of aromatase expression, further showing the strength and specificity of the proposed MOA

II. Can human relevance of the MOA be reasonably excluded on the basis of fundamental, qualitative differences in key events between experimental animals and humans?

The MOA for the formation of mammary tumours in female Sprague-Dawley rats after exposure to atrazine depends on the rat-specific nature of the reproductive cycle and reproductive senescence. Because of the fundamental differences between female Sprague-Dawley rats and humans with regard to both the normal regulation of the pre-ovulatory LH surge and the reproductive senescence, the mammary tumorigenic effects of atrazine in female Sprague-Dawley rats are not expected to occur in humans.

1. Comparison of the reproductive cycles in rodents and humans

In the rodent, the estrous cycle is short and the pre-ovulatory LH surge is brief, timed by the light cycle and dependent on the brain (Simpkins, 2000; Goldman et al., 2007). The brain plays a deterministic role in the LH surge in rodents. Every afternoon during a critical period, a brain signal for LH secretion occurs that is driven by the increased activity of noradrenergic neurons (Wise et al., 1997). As such, selective blockage of this increased activity in noradrenergic neurons during this brief period blocks the pre-ovulatory LH surge (Ordog et al., 1998).

The human menstrual cycle is long, exhibits protracted pre-ovulatory LH surge and ends with menses due to the death of the corpus luteum and the resulting decline in estrogens and progestins. The driving force for the pre-ovulatory LH surge in women is ovarian estrogen secretion (Ordog et al., 1998; Simpkins, 2000). The role of brain regulation of GnRH is that the pre-ovulatory LH surge is permissive in women and other primates. Indeed, the entire menstrual cycle can be recapitulated in Rhesus monkeys in which the source of GnRH has been destroyed, by exogenous administration of pulses of GnRH (Pohl & Knobil, 1982; Ordog et al., 1998). In contrast to the observations in rodents, inhibitors of NE neurotransmission do not affect the pre-ovulatory LH surge in women or other primates (Knobil, 1974; Weiss et al., 1977).

2. Comparison of reproductive ageing in rodents and humans

Reproductive ageing in rodents and women is distinctively different. In female Sprague-Dawley rats, reproductive senescence is a result of a breakdown of the brain regulation of the LH surge, while the ovaries are functional very late into life. The decline in reproductive function is primarily a result of the inability of brain NE neurons to transmit the estrogen signal to GnRH neurons (Wise et al., 1997). The inability to stimulate a pre-ovulatory LH surge results in the maintenance of ovarian follicles and the persistent secretion of estrogens. Sequentially, the increased secretion of estrogens causes a persistent state of hyperprolactinaemia (Welsch et al., 1970; Sarkar et al., 1982; O'Connor et al., 2000). Thus in the Sprague-Dawley rat, reproductive senescence is characterized by persistent hyperestrogenaemia and hyperprolactinaemia with low concentrations of LH and follicle-stimulating hormone (FSH) (Simpkins, 2000).

Advancing age in the female Sprague-Dawley rat is associated with increasing numbers of days spent in estrus with eventual entry into a constant estrous state associated with elevated estrogen and prolactin concentrations and a high incidence of mammary tumours. Other rat strains (such as Fischer 344) are more likely to show age-related increases in the number of days spent in diestrus followed by a constant diestrus or pseudopregnant-like condition. This reproductive state is associated with the development of a number of corpora lutea that are stimulated to secrete progesterone by prolactin. Rats in this reproductive state show a much lower incidence of mammary tumours.

In women, reproductive ageing is characterized by exhaustion of ovarian follicles and the resulting menopause (Taylor, 1998). During menopause, the ability to induce a pre-ovulatory LH surge is normal, but estrogens, the driving force for the cycle, are absent. The menopause is characterized by low concentrations of estrogen and high concentrations of LH and FSH, while prolactin concentrations are usually unchanged or slightly reduced secondary to decreased estrogen (Simpkins, 2000). The major differences between the parameters of reproductive senescence in female Sprague-Dawley rats, female Fischer 344 rats and women are summarized in Table A2.

3. Similarities with reproductive pathologies in women

Polycystic ovarian syndrome (PCOS) is an anovulatory state in women that is characterized by the presence of a cystic ovary associated with elevated concentrations of LH, increased production of estrogen, elevated concentrations of androgens and marked hirsutism. It is associated with obesity,

Table A2. Comparison of parameters of reproductive senescence in female Sprague-Dawley rats, female Fischer 344 rats and humans

Parameter	Sprague-Dawley rats	Fischer 344 rats	Humans
Start of senescence (% of normal lifespan)	30–40%	60–70%	60–70%
Principle cause of senescence	Hypothalamic failure to stimulate LH/FSH	Hypothalamic failure to control prolactin surges	Depletion of ovarian follicle content
LH-surge capability	Lost	Maintained	Maintained
Predominant cycle pattern	Persistent estrus	Pseudopregnancy episodes	Menopause
Estrogen/progesterone ratio	Elevated/prolonged	Reduced	Reduced
Prolactin secretion	Persistently elevated	Episodically elevated	Reduced
Spontaneous incidence of mammary tumours (lifetime)	30–40%	2–5%	8–10%
Principal known factors that increase risk of mammary tumour s	Prolactin, estrogen, chemical mutagens	Prolactin, estrogen, chemical mutagens	Family history, parity, diet, and body weight
Prolactin dependence	High	Median	None

From Simpkins et al. (2000)
LH, luteinizing hormone; FSH, follicle-stimulating hormone.

insulin resistance and altered metabolism of carbohydrate and lipid. An association between the occurrence of PCOS and endometrial hyperplasia and/or cancer has been reported and is biologically plausible on the basis of estrogenic stimulation of the endometrium unopposed by progesterone. Epidemiological evaluation of the association between the occurrence of PCOS and ovarian and breast cancer is not compelling (Eldridge & Delzell, 2000a).

In contrast to women with PCOS, female Sprague-Dawley rats treated with atrazine at high doses display decreased LH, decreased or unaltered concentrations of androgen, weight loss and no association with endometrial or ovarian cancer. The earlier appearance and/or elevated incidence of mammary tumours observed in female Sprague-Dawley rats is attributed to persistent exposure to endogenous estrogen and prolactin resulting from an earlier failure of the neuro-endocrinological control mechanisms that regulate the estrous cycle in this strain of rat.

Hypothalamic amenorrhoea (HA) is characterized by decreased activity in the hypothalamic–pituitary ovarian axis and low-level exposure to endogenous estrogen (Eldridge & Delzell, 2000b). Amenorrhoea may occur as a result of a diverse number of conditions, including stress, anorexia-induced weight loss, exercise-induced weight loss or failure to gain weight, and the occurrence of lactation, and results in failure to have a normal menstrual cycle. Amenorrhoea, from whatever cause, is generally associated with a decreased risk for developing endocrine-mediated tumours in the breast, ovary and uterus. Exposure to atrazine would not lead to a state of persistently increased concentrations of estrogen in women and would not cause an increased incidence of estrogen-mediated tumours in women.

In contrast to hypothalamic amenorrhoea in women, female Sprague-Dawley rats given atrazine at high doses experience prolonged periods of high-level stimulation of estrogen-sensitive tissues by endogenous estrogens (endocrine ageing), which leads to the earlier appearance of mammary tumours.

III. Can human relevance of the MOA be reasonably excluded on the basis of quantitative differences in either kinetic of dynamic factors between experimental animals and humans?

It is not necessary to consider this section in this case because the postulated MOA for the carcinogenesis in the mammary gland of female Sprague-Dawley rats is unique to this strain and, thus, not relevant for risk assessment in humans.

IV. Statement on confidence, analysis, and implications

The postulated MOA is supported by data showing that Sprague-Dawley rats treated with atrazine maintain constant estrus as a result of reproductive senescence. The strength, consistency, and specificity of the available information on MOA for Sprague-Dawley females is further confirmed by data showing that Fischer 344 rats, which have a different reproductive senescence, do not form atrazine-related mammary tumours. A concordance analysis comparing key events in the animal MOA with related reproductive processes in the human female shows distinct differences. Alternative hypotheses have been ruled out, such as genotoxicity or estrogenic activity, further showing the strength and specificity of the proposed MOA. Because hypothalamic dysfunction leading to cessation of ovulation as the MOA for tumour formation appears to be specific to the female Sprague-Dawley rat and does not appear to have a counterpart in the human female, atrazine-related mammary tumours formed by this MOA in the Sprague-Dawley rat are qualitatively not relevant for risk assessment in humans (Meek et al., 2003).

In addition, epidemiological studies indicate that there is no known association between atrazine and any cancer. Three types of studies (cohort studies, case–control studies, and ecological or correlational studies) have assessed possible relationships between atrazine or other triazine herbicides and cancer in humans. Each study has shortcomings that limit conclusions. For example, limitations of the ecological studies include: (a) atrazine and other exposures were not measured at the level of the individual subject, but rather at the level of geographic region; (b) exposures were measured virtually concurrently with cancer occurrence; (c) factors such as duration of exposure and time since first exposure could not be analysed; and (d) controls for confounding was not adequate. Nonetheless, the weight of evidence from these studies indicated that atrazine is unlikely to cause cancer in humans.

References

Adler, I.D. (1980) A review of the coordinated research effort on the comparison of test systems for the detection of mutagenic effects, sponsored by the E.E.C. *Mutat. Res.*, **74**, 77–93.

Alavanja, M.C.R., Samanic, C., Dosemeci, M., Lubin, J., Tarone, R., Lynch, C.F., Knott, C., Thomas, K., Hoppin, J.A., Barker, J., Coble, J., Sandler, D.P. & Blair, A. (2003) Use of agricultural pesticides and prostate cancer risk in the agricultural health study cohort. *Am. J. Epidemiol.,* **157**, 800–814.

Arni, P. (1978). Salmonella/mammalian-microsome mutagenicity test with G 30027. Unpublished report No. 782527 dated 18 July 1978, from Ciba-Geigy Ltd, Basle, Switzerland. Submitted to WHO by Syngenta Crop Protection AG.

Arthur, A.T. (1984). A teratology of atrazine technical in New Zealand White rabbits. Unpublished report No. 832110 dated 18 September 1984 from Ciba-Geigy Corp., Research Department, Pharmaceuticals Division, Summit, New Jersey, USA. Submitted to WHO by Syngenta Crop Protection AG.

Ashby, J., Tinwell, H., Stevens, J., Pastoor, T. & Breckenridge, C.B. (2002) The effects of atrazine on the sexual maturation of female rats. *Regul. Toxicol. Pharmacol.,* **35**, 468–473.

Bachmann, M. (1994) G 30027. 3-Month oral toxicity study in rats. Unpublished report No. 931063 dated 5 August 1994, from Ciba Geigy Ltd, Stein, Switzerland. Submitted to WHO by Syngenta Crop Protection AG.

Ballantine, L., Murphy, T. G. & Simoneaux, B.J. (1985) Metabolism of ^{14}C atrazine in orally dosed rats. Unpublished report No. ABR-85104 dated 6 February 1985, from Ciba-Geigy Corp., Greensboro, North Carolina, USA.. Submitted to WHO by Syngenta Crop Protection AG.

Basler, A. & Rohrborn, G. (1978) SCE test in vivo: chromosome aberrations in bone marrow cells. Unpublished report No. 175-77-1 ENV D from University of Düsseldorf, Germany.

Blondell, J. & Dellarco, V. (2003) Review of atrazine cancer epidemiology. Unpublished report No. D295200 dated 28 October 2004 from United States Environmental Protection Agency, Health Effects Division (7509C).

Boobis, A.R., Cohen, S.M., Dellarco, V., McGregor, D., Meek, M.E., Vickers, C., Willcocks, D., Farland, W. (2006) IPCS framework for analyzing the relevance of a cancer mode of action for humans. *Crit. Rev. Toxicol.*, **36**, 78–192.

Brusick, D.J. (1994) An assessment of the genetic toxicity of atrazine: relevance to human health and environmental effects. *Mutat. Res.*, **317**, 133–144.

Brusick, D. (2000) An assessment of the genetic toxicology database for atrazine 1994–2000. Unpublished report dated 21 June 2000, pp 1–15. Submitted to WHO by Syngenta Crop Protection AG.

Butler, M.A. & Hoagland, R.E. (1989) Genotoxicity assessment of atrazine and some major metabolites in the Ames test. *Bull. Environ. Contam. Toxicol.*, **43**, 797–804.

Catenacci, G., Barbieri, F., Bersani, M., Feriolo, A., Cottica, D. & Maroni, M. (199). Biological monitoring of human exposure to atrazine. *Toxicol. Lett.*, **69**, 217–222.

Ceresa, C. (1988a) G 34048 - micronucleus test, mouse. Unpublished report No. 871373 dated 31 August 1988, from Ciba-Geigy Ltd, Basle, Switzerland, Submitted to WHO by Syngenta Crop Protection AG.

Ceresa, C. (1988b) G 30027 - micronucleus test, mouse. Unpublished report No. 871546 dated 31 May 1988, from Ciba-Geigy Ltd, Basle, Switzerland. Submitted to WHO by Syngenta Crop Protection AG.

Chau, R.Y., McCormick, G.C. & Arthur, A.T. (1990) Hydroxyatrazine - 13-week feeding study in dogs. Unpublished report No. 892076 dated 20 March 1990, from Ciba-Geigy Corp., Research Department, Pharmaceuticals Division, Summit, New Jersey, USA, Submitted to WHO by Syngenta Crop Protection AG.

Chollet, M.C., Degraeve, N., Gilot-Delhalle, J., Colizzi, A., Moutschen, J. & Moutschen-Dahmen, M. (1982). Mutagenic efficiency of atrazine with and without metabolic activation. *Mutation Res.* **97**, 237–238. Abstract presented at the 1st Meeting of the Belgian Environmental Mutagen Society, Brussels, 24 October 1981.

Chow, E. & Hart, E. (1995) Two-year dietary chronic toxicity/oncogenicity study with G-34048 technical in rats. Unpublished report No. F-00125 dated 27 January 1995, from Environmental Health Center, Farmington, Connecticut, USA.. Submitted to WHO by Ciba-Geigy Ltd.

Connor, K., Howell, J., Chen, I., Liu, H., Berhane, K., Sciarretta, C., Safe, S. & Zacharewski, T. (1996) Failure of chloro-s-triazine-derived compounds to induce estrogen receptor-mediated responses in vivo and in vitro. *Fundam. Appl. Toxicol.,* **30**, 93–101.

Connor, K., Howell, J., Safe, S., Chen, I., Liu, H., Berhane, K., Sciarretta, C., & Zacharewski, T. (1998) Failure of chloro-s-triazine-derived compounds to induce estrogenic responses in vivo and in vitro. In: Ballantine, L.G., McFarland, J.E. & Hackett, D.S., eds, *Triazine herbicides: risk assessment.* Washington, DC, Oxford University Press,.pp 424–431.

Cooper, R.L., Stoker, T.E., Goldman, J.M., Parrish, M.B. & Tyrey, L. (1996) Effect of atrazine on ovarian function in the rat. *Reprod. Toxicol.*, **10**, 257–264.

Cooper, R.L., Laws, S.C., Das, P.C., Narotsky, M.G., Goldman, J.M., Tyrey, E.L. & Stoker, T.E. (2007) Atrazine and reproductive function: mode and mechanism of action studies. *Birth Defects Res. B Dev. Reprod. Toxicol.*, **80**, 98–112.

Cooper, R.L., Stoker, T.E., Tyrey, L., Goldman, J.M. & McElroy, W.K. (2000) Atrazine disrupts the hypothalamic control of pituitary-ovarian function. *Toxicol. Sci.,* **53**, 297–307.

Cummings, A.M., Rhodes, B.E. & Cooper, R. L. 2000. Effect of atrazine on implantation and early pregnancy in four strains of rats. *Toxicol. Sci.,* **58**, 135–143.

Das, P.C., McElroy, W.K. & Cooper, R.L. (2000) Differential modulation of catecholamines by chlorotriazine herbicides in pheochromocytoma (PC12) cells in vitro. *Toxicol. Sci.,* **56**, 324–331.

Das, P.C., McElroy, W.K. & Cooper, R.L. (2001) Alteration of catecholamines in pheochromocytoma (PC12) cells in vitro by the metabolites of chlorotriazine herbicide. *Toxicol. Sci.,* **59**, 127–137.

Das, P.C., McElroy, W.K. & Cooper, R.L. (2003) Potential mechanisms responsible for chlorotriazine-induced alterations in catecholamines in pheochromocytoma (PC12) cells. *Life Sci.,* **73**, 3123–3138.

Davidson, I.W.F. (1988) Metabolism and kinetics of atrazine in man. Unpublished report No. 101947 dated 1989, from Bowman Gray School of Medicine, Department of Pysiology/ Pharmacology, North Carolina, USA. Submitted to WHO by Syngenta Crop Protection AG.

De Roos, A.J., Zahm, S.H., Cantor, K.P., Weisenburger, D.D., Holmes, F.F., Burmeister, L.F. & Blair, A. (2003) Integrative assessment of multiple pesticides as risk factors for non-Hodgkin's lymphoma among men. *Occup. Environ. Med.,* **60**, E11.

De Veer, I., Moriske, H.J., & Ruden, H. (1994) Photochemical decomposition of organic compounds in water after UV-irradiation: investigation of positive mutagenic effects. *Toxicol. Lett.,* **72**, 113–119.

Della Croce, C., Morichetti, E., Intorre, L., Soldani, G., Bertini, S. & Bronzetti, G. (1996) Biochemical and genetic interactions of two commercial pesticides with the monooxygenase ystem and chlorophyllin. *J. Environ. Pathol. Toxicol. Oncol.*, **15**, 21–28.

Deparade, E. (1986) G-30027 - Salmonella/mammalian microsome mutagenicity test. Unpublished report No. 861172 dated 5 December 1986, from Ciba-Geigy Ltd, Basle, Switzerland. Submitted to WHO by Syngenta Crop Protection AG.

Deparade, E. (1987) G 28273 - Salmonella/mammalian-microsome mutagenicity test. Unpublished report No. 871372 dated 10 November 1987 from Ciba-Geigy Ltd, Basle, Switzerland. Submitted to WHO by Syngenta Crop Protection AG.

Deparade, E. (1988) G 34048 - Salmonella/mammalian-microsome mutagenicity test. Unpublished report No. 871376 dated 15 February 1988 from Ciba-Geigy Ltd, Basle, Switzerland.. Submitted to WHO by Syngenta Crop Protection AG.

Deparade, E. (1989) G 30033 - Salmonella and Escherichia/liver-microsome test. Unpublished report No. 891236 dated 18 December 1989 (amended 21 December 1993) from Ciba-Geigy Ltd, Basle, Switzerland. Submitted to WHO by Syngenta Crop Protection AG.

Deparade, E. (1990) G 28279 - Salmonella and *Escherichia*/liver-microsome test. Unpublished report No. 891243 dated 18 January 1990 (amended 21 December 1993) from Ciba-Geigy Ltd, Basle, Switzerland. Submitted to WHO by Ciba-Geigy Ltd.

Devos, S., De Bosscher, K., Staels, B., Bauer, E., Roels, F., Berghe, W., Haegeman, G., Hooghe, R. & Hooghe-Peters, E.L. (2003) Inhibition of cytokine production by the herbicide atrazine. Search for nuclear receptor targets. *Biochem. Pharmacol.,* **65**, 303–308.

Devos, S., Van Den Heuvel, R., Hooghe, R. & Hooghe-Peters, E.L. (2004) Limited effect of selected organic pollutants on cytokine production by peripheral blood leukocytes. *Eur.Cytokine Netw.*, **15**, 145–151.

Dooley, G.P., Prenni, J.E., Prentiss, P.L., Cranmer, B.K., Andersen, M.E. & Tessari, J.D. (2006) Identification of a novel hemoglobin adduct in Sprague Dawley rats exposed to atrazine. *Chem. Res. Toxicol.,* **19**, 692–700.

Duchosal, F., Vogel, O., Chevalier, H.J., Luetkemeier, H., & Biedermann, K. (1990a) 28-Day oral toxicity (feeding) study with G 28279 tech. in the rat. Unpublished report No. RCC 252090 dated 21 September 1990 from RCC AG, Itingen, Switzerland.

Duchosal, F., Vogel, O., Wilson, J.T., Luetkemeier, H., & Biedermann, K. (1990b) 28-Day oral toxicity (feeding) study with G 30033 tech in the rat. Unpublished report No. RCC 252088 / CG 891235 dated 21 September 1990 from RCC AG, Itingen, Switzerland.

Dunkelberg, H., Fuchs, J., Hengstler, J.G., Klein, E., Oesch, F. & Struder, K. (1994) Genotoxic effects of the herbicides alachlor, atrazine, pendimethaline, and simazine in mammalian cells. *Bull. Environ. Contam. Toxicol.,* **52**, 498–504.

Ehling, U.H. 1979. Micronucleus test in mouse bone marrow. Progress report (March. 1979–December. 1979). Unpublished report No. 136-77-1 ENV D GSF from Gesellschaft für Strahlenschutz und Umweltforschung (GSF), Neuherberg, Germany.

Eldridge, J.C. & Delzell, E. (2000a) An evaluation of the association between polycystic ovarian syndrome in women and the occurrence of ovarian, breast, and endometrial cancer and an assessment of the relevance of finding in female Sprague-Dawley rats administered high doses of atrazine to PCOS in women. Unpublished report from Department of Physiology and Pharmacology, Wake Forest University School of Medicine, and Department of Epidemiology, University of Alabama School of Public Health. Submitted to WHO by Syngenta Crop Protection AG.

Eldridge, J.C. & Delzell, E. (2000b) Endocrinological characterization of hypothalic amenorrhea (HA) in women and an evaluation of the relevance of the endocrine effects observed in female Sprague-Dawley rats treated with high doses of atrazine to HA in women. Unpublished report from Department of Physiology and Pharmacology, Wake Forest University School of Medicine, and Department of Epidemiology, University of Alabama School of Public Health. Submitted to WHO by Syngenta Crop Protection AG.

Eldridge, J.C., Tennant, M.K., Wetzel, L.T., Breckenridge, C.B. & Stevens, J.T. (1994) Factors affecting mammary tumor incidence in chlorotriazine-treated female rats: hormonal properties, dosage, and animal strain. *Environ. Health Perspect.*, **102**, 29–36.

Eldridge, J.C., McConnell, R.F., Wetzel, L.T. & Tisdel, M.O. (1998) Appearance of mammary tumors in atrazine-treated female rats: probable mode of action involving strain-related control of ovulation and estrous cycling. In: Ballantine, L.G., McFarland, J.E. &. Hackett, D.S., eds, *Triazine herbicides: risk assessment,* Washington, DC: Oxford University Press, pp. 414–423.

Eldridge, J.C., Wetzel, L.T., Stevens, J.T. & Simpkins, J.W. (1999a) The mammary tumor response in triazine-treated female rats: a threshold-mediated interaction with strain and species-specific reproductive senescence. *Steroids*, **64**, 672–678.

Eldridge, J.C., Wetzel, L.T., Tisdel, M.O. & Luempert, L.G. (1993a) Revised supplement to final report: Determination of hormone levels in Sprague-Dawley rats treated with atrazine technical. Unpublished report dated 8 April 1993 from Bowman Gray School of Medicine, Winston-Salem, North Carolina; HWA study No. 483-278. Submitted to WHO by Syngenta Crop Protection AG.

Eldridge, J.C., Wetzel, L.T., Tisdel, M.O. & Luempert, L.G. (1993b) Revised supplement to final report: Determination of hormone levels in Fischer-344 rats treated with atrazine technical. Unpublished report dated 8 April 1993 from Bowman Gray School of Medicine, Winston-Salem, North Carolina; HWA study No. 483-279. Submitted to WHO by Syngenta Crop Protection AG.

Eldridge, J.C., Wetzel, L.T. & Tyrey, L. (1999b) Estrous cycle patterns of Sprague-Dawley rats during acute and chronic atrazine administration. *Reprod. Toxicol.*, **13**, 491–499.

Eldridge, J.C., Minnema, D., Breckenridge, C.B., McFarland, J. & Stevens, J.T. (2001) The effect of 6 months of feeding of atrazine or hydroxyatrazine on the leutinizing hormone surge in female Sprague-Dawley and Fischer 344 rats. *Soc. Toxicol. 28* March, Abstract No. 1525.

Engel, L.S., Hill, D.A., Hoppin, J.A., Lubin, J.H., Lynch, C.F., Pierce, J., Samanic, C., Sandler, D.P., Blair, A. & Alavanja, M.C. (2005) Pesticide use and breast cancer risk among farmers' wives in the agricultural health study. *Am. J. Epidemiol.*, **161**, 121–135.

Enoch, R.R., Stanko, J.P., Greiner, S.N., Youngblood, G.L., Rayner, J.L. & Fenton, S.E. (2007) Mammary gland development as a sensitive end point after acute prenatal exposure to an atrazine metabolite mixture in female Long-Evans rats. *Environ. Health Perspect.*, **115**, 541–547.

Fan, W., Yanase, T., Morinaga, H., Gondo, S., Okabe, T., Nomura, M., Komatsu, T., Morohashi, K.-I., Hayes, T.B., Takayanagi, R. & Nawata, H. (2007) Atrazine-induced aromatase expression is SF-1 dependent:

implications for endocrine disruption in wildlife and reproductive cancers in humans. *Environ. Health Perspect.,* **115**, 720–727.

Filipov, N.M., Pinchuk, L.M., Boyd, B.L. & Crittenden, P.L. (2005) Immunotoxic effects of short-term atrazine exposure in young male C57BL/6 mice. Toxicol. Sci., 86, 324–332.

Fournier, M., Friborg, J., Girard, D., Mansour, S. & Krzystyniak, K. (1992) Limited immunotoxic potential of technical formulation of the herbicide atrazine (AAtrex) in mice. *Toxicol. Lett.*, **60**, 263–274.

Gebel, T., Kevekordes, S., Pav, K., Edenharder, R. & Dunkelberg, H. (1997) In vivo genotoxicity of selected herbicides in the mouse bone-marrow micronucleus test. *Arch. Toxicol.,* **71**, 193–197.

Geleick, D. (1991a) G 30033 – autoradiographic DNA repair test on rat hepatocytes in vitro. Unpublished report No. 901310 dated 26 April 1991 (amended 21 December 1993) from Ciba-Geigy Ltd, Basle, Switzerland. Submitted to WHO by Syngenta Crop Protection AG.

Geleick, D. (1991b) G 28279 - autoradiographic DNA repair test on rat hepatocytes in vitro. Unpublished report No. 901308 dated 12 April 1991 (amended 21 December 1993) from Ciba-Geigy Ltd, Basle, Switzerland. Submitted to WHO by Ciba-Geigy Ltd.

Gerspach, R. (1991) G 30033 - 3-month oral toxicity study in rats (administration in food). Unpublished report No. 901264 dated 22 October 1991 from Ciba-Geigy Ltd, Stein, Switzerland, Submitted to WHO by Syngenta Crop Protection AG.

Giknis, M.L.A. (1989) Atrazine technical - a teratology (segment II) study in rats. Unpublished report No. 882049 dated 12 February 1989 from Ciba-Geigy Corp., Research Department, Pharmaceuticals Division, Summit, New Jersey, USA. Submitted to WHO by Syngenta Crop Protection AG.

Giurgea, R. & Koszta, M. (1979) Action of herbicide "Atrazine" on thymus and bursa in chickens. *Arch. Exp. Veterinarmed.,* **33**, 703–707.

Giurgea, R., Borsa, M. & Bucur, N. (1981) Immunological reactions of Wistar rats following administration of atrazine and prometryne. *Arch. Exp. Veterinarmed.,* **35**, 811–815.

Goldman, J.M., Murr, A.S. & Cooper, R.L. (2007) The rodent estrous cycle: characterization of vaginal cytology and its utility in toxicological studies. *Birth Defects Res. B Dev. Reprod. Toxicol.*, **80**, 84–97.

Guigas, C., Pool-Zobel, B.L. & Diehl, J.F. (1993) The combination effects of quercetin with the herbicides atrazine, cyanazine and gesamprim in mutagenicity tests. *Z Ernahrungswiss.,* **32**, 131–138.

Hamboeck, H., Fischer, R.W., Di Iorio, E.E. & Winterhalter, K.H. (1981) The binding of s-triazine metabolites to rodent haemoglobins appears irrelevant to other species. *Mol. Pharmacol.,* **20**, 579–584.

Hartmann, H.R. (1989) G 30027 - acute inhalation toxicity in the rat. Unpublished report No.891162 dated 24 August 1989 from Ciba-Geigy Ltd, Stein, Switzerland. Submitted to WHO by Syngenta Crop Protection AG.

Hartmann, H.R. (1993) G 30027 - acute dermal toxicity study of G 30027 in rats. Unpublished report No. 931184 dated 2 December 1993 from Ciba-Geigy Ltd, Stein, Switzerland. Submitted to WHO by Syngenta Crop Protection AG.

Hazelette, J.R. & Green, J. (1987) G 30027 - oncogenicity study in mice. Unpublished report No. 842120 dated 30 October 1987 from Ciba-Geigy Corp., Research Department, Pharmaceuticals Division, Summit, New Jersey, USA. Submitted to WHO by Syngenta Crop Protection AG.

Hazelette, J.R. & Arthur, A.T. (1989a) Hydroxy-atrazine - pilot 4-week oral feeding toxicity study in rats. Unpublished report No. 872282 dated 22 March 1989 from Ciba-Geigy Corp., Research Department, Pharmaceutical Division, Summit, New Jersey, USA.

Hazelette, J.R. & Arthur, A.T. (1989b) Hydroxy-atrazine - pilot 4-week oral feeding toxicity study in dogs. Unpublished report No. 872156 dated 22 March 1989 from Ciba-Geigy Corp., Research Department, Pharmaceutical Division, Summit, New Jersey, USA.

Heneweer, M., van den Berg, M. & Sanderson, J.T. (2004) A comparison of human H295R and rat R2C cell lines as in vitro screening tools for effects on aromatase. *Toxicol. Lett.*, **146**, 183–194.

Hertner, Th. (1988a) G 34048 - autoradiographic DNA repair test on rat hepatocytes. Unpublished report No. 871374 dated 22 January 1988 from Ciba-Geigy Ltd, Basle, Switzerland. Submitted to WHO by Syngenta Crop Protection AG.

Hertner, Th. (1988b) G 28273 - autoradiographic DNA repair test on rat hepatocytes. Unpublished report No. 871370 dated 10 March 1988 from Ciba-Geigy Ltd, Basle, Switzerland.. Submitted to WHO by Syngenta Crop Protection AG.

Hertner, Th. (1992) G 30027 - autoradiographic DNA repair test on rat hepatocytes. Unpublished report No. 911246 dated 14 April 1992 from Ciba-Geigy Ltd, Basle, Switzerland. Submitted to WHO by Syngenta Crop Protection AG.

Hertner, Th. (1993) G 30027 - dominant lethal test, mouse, 8-weeks. Unpublished report No. 911247 dated 7 January 1993 from Ciba-Geigy Ltd, Basle, Switzerland. Submitted to WHO by Syngenta Crop Protection AG.

Hessel, P.A., Kalmes, R., Smith, T.J., Lau, E., Mink, P.J. & Mandel, J. (2004) A nested case–control study of prostate cancer and atrazine exposure. *J. Occup. Environ. Med.*, **46**, 379–385.

Holbert, M. S. (1991) G 30027 - Acute inhalation toxicity study in rats. Unpublished report No. 8079-91 dated 18 July 1991 from Stillmeadow, Inc., Houston, Texas, USA. Submitted to WHO by Syngenta Crop Protection AG.

Hooghe, R.J., Devos, S. & Hooghe-Peters, E.L. (2000) Effects of selected herbicides on cytokine production in vitro. *Life Sci.*, **66**, 2519–2525.

Hool, G. (1981a) G 30027 - nucleus anomaly test in somatic interphase nuclei of Chinese hamster. Unpublished report No. 783027 dated 20 January 1981 (supplemented 6 October 1986) from Ciba-Geigy Ltd, Basle, Switzerland. Submitted to WHO by Syngenta Crop Protection AG.

Hool, G. (1981b) G 30027 - chromosome studies in male germinal epithelium in mouse (test for mutagenic effects on spermatogonia). Unpublished report No. 800209 dated 25 January 1981 (supplemented 6 October 1986) from Ciba-Geigy Ltd, Basle, Switzerland. Submitted to WHO by Syngenta Crop Protection AG.

Hool, G. (1981c) G 30027 - chromosome studies in male germinal epithelium in mouse (test for mutagenic effects on spermatocytes). Unpublished report No. 800210 dated 28 January 1981 (supplemented 6 October 1986) from Ciba-Geigy Ltd, Basle, Switzerland. Submitted to WHO by Syngenta Crop Protection AG.

Hool, G., (1981d) G 30027 – dominant lethal test in mouse. Unpublished report No. 801380 dated 8 August 1981 from Ciba-Geigy Ltd, Basle, Switzerland. Submitted to WHO by Syngenta Crop Protection AG.

Huber, K.R., Batastini, G., & Arthur, A.T. (1989) G 30027 - 21-day dermal toxicity study in the rabbit. Unpublished report No. 882035 dated 1 December 1989 from Ciba-Geigy Corp., Research Department, Pharmaceuticals Division, Summit, New Jersey, USA. Submitted to WHO by Syngenta Crop Protection AG.

Hui, X., Wester, R., Maibach, H.I., Gilman, S.D., Gee, S.J., Hammock, B.D., Simoneaux, B., Breckenridge, C. & Kahr, R. (1996a) Disposition of atrazine in Rhesus monkey following intravenous administration. Unpublished report of Study Nos UCSF 95SU04 (UCSF), BDH-081-1 (UC Davis), ABR 96066, 96073 dated 3 September 1996 from Ciba Crop Protection. Submitted to WHO by Syngenta Crop Protection AG.

Hui, X., Wester, R., Maibach, H.I., Simoneaux, B. & Breckenridge, C. (1996b) Disposition of atrazine in Rhesus monkey following oral administration. Unpublished report of study Nos UCSF 96SU01 (UCSF) ABR-96094; Ciba Study Number 306-96, dated 30 October 1996, from Ciba Crop Protection. Submitted to WHO by Syngenta Crop Protection AG.

Hummel, H., Youreneff, M., Giknis, M. & Yau, E.T. (1989) G 28273 - a teratology (segment II) study in rats. Unpublished report No. 872177 dated 15 August 1989 from Ciba-Geigy Corp., Research Department, Pharmaceuticals Division, Summit, New Jersey, USA. Submitted to WHO by Syngenta Crop Protection AG.

IARC (1999) Atrazine. In: *Some chemicals that cause tumours of the kidney or urinary bladder in rodents and some other substances*. Vol. 73, IARC Monographs on the Evaluation of Carcinogenic Risks to Humans. IARCPress, Lyon.

Infurna, R.N. (1984) Atrazine technical, a teratology study (segment II) in rats. Unpublished report No. 832109 dated 18 September 1984 from Ciba-Geigy Corp., Research Department, Pharmaceuticals Division, Summit, New Jersey, USA. Submitted to WHO by Syngenta Crop Protection AG.

Innes, J.R.M., Ulland, B.M., Valerio, M.G., Pertrucelli, L., Fishbein, L., Hart, E.R., Pallota, A.J., Bates, R.R., Falk, H.L., Gart, J.J., Klein, M., Mitchell, I. & Peters, J. (1969) Bioassay of pesticides and industrial chemicals for tumorogenicity in mice: a preliminary note. *J. Natl. Cancer Inst.*, **42**, 1101–1114.

Kahrs, R. & Thede, B. (1987) Study of ^{14}C-atrazine dose/response relationship in the rat. Unpublished report No. ABR-87087 dated 23 October 1987 form Ciba-Geigy Corp., Greensboro, North Carolina, USA. Submitted to WHO by Syngenta Crop Protection AG.

Karrow, N.A., McCay, J.A., Brown, R.D., Musgrove, D.L., Guo, T.L., Germolec, D.R. & White, K.L., Jr. (2005) Oral exposure to atrazine modulates cell-mediated immune function and decreases host resistance to the B16F10 tumor model in female B6C3F1 mice. *Toxicol.,* **209**, 15–28.

Kligerman, A.D., Doerr, C.L., Tennant, A.H. & Peng, B. (2000a) Cytogenetic studies of three triazine herbicides. In vivo micronucleus studies in mouse bone marrow. *Mutat. Res.,* **471**, 107–112.

Kligerman, A.D., Doerr, C.L., Tennant, A.H. & Zucker, R.M. (2000b) Cytogenetic studies of three triazine herbicides. In vitro studies. *Mutat. Res.,* **465**, 53–59.

Knobil, E. (1974) On the control of gonadotropin secretion in the Rhesus monkey. *Recent Prog. Hormone Res.,* **30**, 1–36.

Kuhn, J.O. (1988) G 30027 - acute oral toxicity study (mouse). Unpublished report No. 5421-88 dated 27 July 1988 from Stillmeadow Inc., Houston, Texas, USA. Submitted to WHO by Syngenta Crop Protection AG.

Kuhn, J.O. (1991a) G 30027 - acute oral toxicity study in rats. Unpublished report No. 7800-91 dated 18 March 1991 from Stillmeadow Inc., Sugar Land, Texas, USA. Submitted to WHO by Syngenta Crop Protection AG.

Kuhn, J.O. (1991b) DACT (G 28273) - technical ground FL-871776 - acute oral toxicitity study in rats. Unpublished report No. 7801-91 dated 21 March 1991 from Stillmeadow Inc., Sugar Land, Texas, USA. Submitted to WHO by Syngenta Crop Protection AG.

Kuhn, J.O. (1991c) G 30033 - acute oral toxicity study in rats. Unpublished report No. 7802-91 dated 22 March 1991 from Stillmeadow Inc., Sugar Land, Texas, USA. Submitted to WHO by Syngenta Crop Protection AG.

Kuhn, J.O. (1991d) Deisopropyl-atrazine (G 28279) - acute oral toxicity study in rats. Unpublished report No. 7803-91 dated 25 March 1991 from Stillmeadow Inc., Sugar Land, Texas, USA. Submitted to WHO by Syngenta Crop Protection AG.

Kuhn, J.O. (1991e) Hydroxyatrazine (G 34048) - technical FL-870869 - acute oral toxicity study in rats. Unpublished report No. 7938-91 dated 3 April 1991 from Stillmeadow Inc., Sugar Land, Texas, USA. Submitted to WHO by Syngenta Crop Protection AG.

Laws, S.C., Ferrell, J.M., Stoker, T.E., Schmid, J. & Cooper, R.L. (2000) The effects of atrazine on female Wistar rats: an evaluation of the protocol for assessing pubertal development and thyroid function. *Toxicol. Sci.*, **58**, 366–376.

Laws, S.C., Ferrell, J.M., Stoker, T.E. & Cooper, R.L. (2003) Pubertal development in female Wistar rats following exposure to propazine and atrazine biotransformation by-products, diamino-*S*-chlorotriazine and hydroxyatrazine. *Toxicol. Sci.*, **76**, 190–200.

Laws, S.C., Ferrell, J., Best, D., Buckalew, A., Murr, A. & Cooper, R.L. (2006) Atrazine stimulates the release of serum ACTH and adrenal steroids in male Wistar rats. Unpublished report (abstract) from Reprod. Tox. Division, NHEERL, ORD, U.S. EPA, Research Triangle Park, NC, USA.

Lindsay, L.A., Wimbert, K.V., Giknis, M.L.A. & Yau, E.T. (1989) Hydroxyatrazine (G 34048) - a teratology (segment II) study in rats. Unpublished report No. 872202 dated 14 February 1989 (supplemented 2 November 1990 and 14 May 1993) from Ciba-Geigy Corp., Research Department, Pharmaceuticals Division, Summit, New Jersey, USA. Submitted to WHO by Syngenta Crop Protection AG.

Lioi, M.B., Scarfi, M.R., Santoro, A., Barbieri, R., Zeni, O., Salvemini, F., Di Berardino, D. & Ursini, M.V. (1998) Cytogenetic damage and induction of pro-oxidant state in human lymphocytes exposed in vitro to glyphosate, vinclozolin, atrazine, and DPX-E9636. *Environ. Mol. Mutagen.*, **32**, 39–46.

Liu, C.Y. & Thakur, A.K. (1999) Statistical report for survival and mammary tumor analyses from the Fischer 344/Lati rat study (Pinter, 1990). Unpublished report No. 6117-998 dated 2 September 1999 from Covance Laboratories Inc., Vienna, Virginia, USA. Submitted to WHO by Syngenta Crop Protection AG.

Luebke, R.W., Chen, D.H., Dietert, R., Yang, Y., King, M. & Luster, M.I. (2006) The comparative immunotoxicity of five selected compounds following developmental or adult exposure. *J. Toxicol. Environ. Health B Crit. Rev.*, **9**, 1–26.

Lynch, S.M., Rusiecki, J.A., Blair, A., Dosemeci, M., Lubin, J., Sandler, D., Hoppin, J.A., Lynch, C.F. & Alavanja, M.C.R. (2006) Cancer incidence among pesticide applicators exposed to cyanazine in the Agricultural Health Study. *Environ. Health Perspect.*, **114**, 1248–1252.

MacLennan, P.A., Delzell, E., Sathiakumar, N., Myers, S.L., Cheng, H., Grizzle, W., Chen, V.W. & Wu, X.C. (2002) Cancer incidence among triazine herbicide manufacturing workers. *J. Occup. Environ. Med.,* **44**, 1048–1058.

MacLennan, P.A., Delzell, E., Sathiakumar, N. & Myers, S.L. (2003) Mortality among triazine herbicide manufacturing workers. *J. Toxicol. Environ. Health A.,* **66**, 501–517.

Mainiero, J., Youreneff, M. & Giknis, M. (1987) Two-generation reproduction study in rats: atrazine technical. Final report. Unpublished report No. MIN 852063 dated 2 September 1987 from Safety Evaluation Facility. Summit, New Jersey, USA. Submitted to WHO by Syngenta Crop Protection AG.

Marty, J.H. (1992a) G 28279 - developmental toxicity (teratogenicity) study in rats (oral administration). Unpublished report No. 901262 dated 1 June 1992 from Ciba-Geigy Ltd, Stein, Switzerland. Submitted to WHO by Ciba-Geigy Ltd.

Marty, J.H. (1992b) Developmental toxicity (teratogenicity) study in rats with G 30033 technical (oral administration). Unpublished report No. 901265 dated 1 June 1992 from Ciba-Geigy Ltd, Stein, Switzerland. Submitted to WHO by Syngenta Crop Protection AG.

Maurer, Th. (1983) G 30027 - skin sensitization effects in guinea pigs (optimization test). Unpublished report No. 830644 dated 29 November 1983 from Ciba-Geigy Ltd, Basle, Switzerland.. Submitted to WHO by Syngenta Crop Protection AG.

Mayhew, D.A. (1986) Two-year chronic feeding/oncogenicity study in rats administered atrazine. Unpublished report No. 410-1102 dated 29 April 1986 from American Biogenics Corporation, Decatur, IL, USA. Submitted to WHO by Syngenta Crop Protection AG.

McConnell, R.F. (1989) Comparative aspects of contraceptive steroids: effects observed in rats. *Toxicol. Pathol.,* **17**, 385–388.

McConnell, R.F. (1995) A histomorphologic reevaluation of the ovaries, uterus, vagina, mammary gland, and pituitary gland from Sprague-Dawley and Fischer 344 female rats treated with atrazine. Unpublished report Nos 483-278 & 483-279 dated 10 March 1995 from Hazleton Laboratories, Vienna, Virginia, USA. Submitted to WHO by Syngenta Crop Protection AG.

McElroy, J.A., Gangnon, R.E., Newcomb, P.A., Kanarek, M.S., Anderson, H.A., Vanden Brook, J., Trentham-Dietz, A.M. & Remington, P.L. (2007) Risk of breast cancer for women living in rural areas from adult exposure to atrazine from well water in Wisconsin. *J. Expo. Sci. Environ. Epidemiol.*, **17**, 207–214.

McMullin, T.S., Andersen, M.E., Nagahara, A., Lund, T.D., Pak, T., Handa, R.J. & Hanneman, W.H. (2004) Evidence that atrazine and diaminochlorotriazine inhibit the estrogen/progesterone induced surge of luteinizing hormone in female Sprague-Dawley rats without changing estrogen receptor action. *Toxicol. Sci.,* **79**, 278–286.

Meek, M.E., Bucher, J.R., Cohen, S.M., Dellarco, V., Hill, R.N., Lehman-McKeeman, L.D., Longfellow, D.G., Pastoor, T., Seed, J. & Patton, D.E. (2003) A framework for human relevance analysis of information on carcinogenic modes of action. *Crit. Rev. Toxicol.*, **33**, 591–653.

Meisner, L.F., Belluck, D.A. & Roloff, B.D. (1992) Cytogenetic effects of alachlor and/or atrazine in vivo and in vitro. *Environ. Mol. Mutagen.*, **19**, 77–82.

Mencoboni, M., Lerza, R., Bogliolo, G., Flego, G. & Pannacciulli, I. (1992) Effect of atrazine on hemopoietic system. *In Vivo.*, **6**, 41–44.

Mehta, C.S. (1980a) 2,4-Diamino-6-chloro-s-triazine (DACT, G 28273) FL-781538 ARS 1857 - rat acute oral toxicity. Unpublished report No. 1290A-79 dated 6 August 1980 from Stillmeadow Inc., Houston, Texas, USA. Submitted to WHO by Syngenta Crop Protection AG.

Mehta, C. S. (1980b) 2,4-Diamino-6-chloro-s-triazine (DACT, G 28273) FL-781538 ARS 1857 - rat acute oral toxicity. Unpublished report No. 1290B-79 dated 7 August 1980 from Stillmeadow Inc., Houston, Texas, USA. Submitted to WHO by Syngenta Crop Protection AG. Submitted by: Syngenta Crop Protection AG.

Meyer, A. (1987) G 28273 - autoradiographic DNA repair test on human fibroblasts. Unpublished report No. 871371 dated 20 November 1987 from Ciba-Geigy Ltd, Basle, Switzerland. Submitted to WHO by Syngenta Crop Protection AG.

Meyer, A. (1988) G 34048 - autoradiographic DNA repair test on human fibroblasts. Unpublished report No. 871375 dated 11 January 1988 from Ciba-Geigy Ltd, Basle, Switzerland. Submitted to WHO by Syngenta Crop Protection AG.

Mills, P.K. (1998). Correlation analysis of pesticide use data and cancer incidence rates in California counties. *Arch. Environ. Health.*, **53**, 410–413.

Mills, P.K. & Yang, R. (2003) Prostate cancer risk in California farm workers. *J. Occup. Environ. Med.*, **45**, 249–258.

Mills, P.K. & Yang, R. (2006) Regression analysis of pesticide use and breast cancer incidence in California Latinas. *J. Environ. Health.*, **68**, 15–22.

Minnema, D. (2001) Comparison of the LH surge in female rats administered atrazine, simazine or DACT via oral gavage for one month. Final report. Unpublished report No. 1198-98 dated 21 March 2001 from Covance Laboratories Inc., Vienna, Virginia, USA. Submitted to WHO by Syngenta Crop Protection AG.

Minnema, D. (2002) 52-Week toxicity study of simazine, atrazine, and DACT administered in the diet to female rats. Unpublished report No. 2214-01 dated 21 February 2002 from Covance Laboratories Inc., Vienna, Virginia, USA. Submitted to WHO by Syngenta Crop Protection AG.

Modic, W., Ferrell, J., Wood, C., Laskey, J., Cooper, R., & Laws, S. (2004) Atrazine alters steroidogenesis in male Wistar rats. *The Toxicologist.* Abstract. 568. Society of Toxicologists Annual Meeting 2004.

Modic, W.M. (2004) The role of testicular aromatase in the atrazine mediated changes of estrone and estradiol the male Wistar rat. Thesis submitted to Department of Molecular & Structural Biochemistry, North Carolina State University, Raleigh, NC, USA. pp. 1–61.

Moon, H.J, Han, S.Y., Shin, J.H., Kang, I.H., Kim, T.S., Hong, J.H., Kim, S.H .& Fenton, S.E. (2007) Gestational exposure to nonylphenol causes precocious mammary gland development in female rat offspring. *J. Reprod. Dev.*, **53**, 333–344.

Morseth, S.L. (1996a) Evaluation of the luteinizing hormone (LH) in female Sprague-Dawley rats - pilot study: final report. Unpublished report No. 2386-109:6791 dated 18 January 1996 from Corning Hazleton, Inc., Vienna, Virginia, USA. Submitted to WHO by Syngenta Crop Protection AG.

Morseth, S.L. (1996b) Evaluation of the luteinizing hormone (LH) surge in female Sprague-Dawley rats-method validation: final report. Unpublished report No. 2386-110 dated 18 January 1996 from Corning Hazleton, Inc., Vienna, Virginia, USA. Submitted to WHO by Syngenta Crop Protection AG.

Morseth, S.L. (1996c) Evaluation of the luteinizing hormone (LH) surge in atrazine-exposed female Sprague-Dawley rats: 6-month report. Unpublished report No. 2386-111 dated 25 October 1996 from Corning Hazleton, Inc., Vienna, Virginia, USA. Submitted to WHO by Syngenta Crop Protection AG.

Morseth, S.L. (1996d) Chronic (12–24 months) study in rats with atrazine technical: final 12-month report. Unpublished report No. 2386-108 dated 24 October 1996 from Corning Hazleton, Inc., Vienna, Virginia, USA. Submitted to WHO by Syngenta Crop Protection AG.

Morseth, S.L. (1998) Chronic (12–24 months) study in rats with atrazine technical. Unpublished report No. 2386-108 dated 15 April 1998 from Corning Hazleton, Inc., Vienna, Virginia., USA. Submitted to WHO by Syngenta Crop Protection AG.

Narotsky, M.G., Best, D.S., Guidici, D.L. & Cooper, R.L. (2001) Strain comparisons of atrazine-induced pregnancy loss in the rat. *Reprod. Toxicol.,* **15**, 61–69.

Neuberger, J.S. (1996) Atrazine and/or triazine herbicides exposure and cancer: an epidemiologic review. *J. Agromedicine,* **3**, 9–30.

O'Connor, D.J., McCormick, G.C. & Green, J.D. (1987) G 30027 – 52-week oral feeding study in dogs. Unpublished report No. MIN 852008 dated 27 October 1987 from Ciba-Geigy Corp., Research Department, Pharmaceuticals Division, Summit, New Jersey, USA. Submitted to WHO by Syngenta Crop Protection AG.

O'Connor, J.C., Plowchalk, D.R., Van Pelt, C.S., Davis, L.G. & Cook, J.C. (2000) Role of prolactin in chloro-*S*-triazine rat mammary tumorigenesis. *Drug Chem. Toxicol.,* **23**, 575–601.

Ogorek, B. (1991a) G 28279 - micronucleus test, mouse. Unpublished report No. 901307 dated 23 February 1991 (amended 21 December 1993) from Ciba-Geigy Ltd, Basle, Switzerland. Submitted to WHO by Ciba-Geigy Ltd.

Ogorek, B. (1991b) G 30033 - Micronucleus test, mouse. Unpublished report No. 901309 dated 25 March 1991 (amended 21 December 1993) from Ciba-Geigy Ltd, Basle, Switzerland. Submitted to WHO by Syngenta Crop Protection AG.

Oh, S.M., Shim, S.H., & Chung, K.H. (2003) Antiestrogenic action of atrazine and its major metabolites in vivo. *J. Health Sci.*, **49**, 65–71.

Ohno, K., Araki, N., Yanase, T., Nawata, H. & Iida, H. (2004) A novel nonradioactive method for measuring aromatase activity using a human ovarian granulosa-like tumor cell line and an estrone ELISA. *Toxicol. Sci.*, **82**, 443–450.

Ordog T., Goldsmith J.R., Chen M.D., Connaughton M.A., Hotchkiss J. & Knobil E. (1998) On the mechanism of the positive feedback action of estradiol on luteinizing hormone secretion in the Rhesus monkey. *J. Clin. Endocrinol. Metab.*, **83**, 4047–4053

Orr, G.R., Simoneaux, B. & Davidson, I.W.F. (1987) Disposition of atrazine in the rat. Unpublished report No. ABR-87048 dated 23 October 1987 from Ciba-Geigy Corp., Greensboro, North Carolina, USA. Submitted to WHO by Syngenta Crop Protection AG.

Osterburg, I., & Breckenridge, C. (2004) Oral (gavage) study on the effect of atrazine on pituitary hormone secretion of ovariectomized, estrogen-replaced female Rhesus monkeys. Unpublished report No. 1893-002 dated 14 July 2004 from Covance Laboratories, Munich, Germany. Submitted to WHO by Syngenta Crop Protection AG.

Osterloh, J., Letz, G., Pond, S. & Becker, C. (1983) An assessment of the potential testicular toxicity of 10 pesticides using the mouse-sperm morphology assay. *Mutat. Res.*, **116**, 407–415.

Paul, H.J., Dunsire, J.P., & Hedley, D. (1993) The absorption, distribution, degradation and excretion of U-^{14}C triazine G 30027 in the rat. Unpublished report No. IRI 153138 dated 8 July 1993 from Inveresk Research International Ltd, Scotland. Submitted to WHO by Syngenta Crop Protection AG.

Petterson, J.A. & Turnier, J. (1995) 1-Year chronic toxicity study with atrazine technical in rats: final report. Unpublished report No. F-00171 dated 8 December 1995. Submitted to WHO by Syngenta Crop Protection AG.

Pettersen, J.C., Richter, A.D. & Gilles, P.A. (1991) G 28273 - 90-day oral toxicity study in rats. Unpublished report No. F-00006 dated 5 November 1991 from Ciba-Geigy Corp., Environmental Health Center, Farmington, Connecticut, USA. Submitted to WHO by Syngenta Crop Protection AG.

Pino, A., Maura, A. & Grillo, P. (1988) DNA damage in stomach, kidney, liver and lung of rats treated with atrazine. *Mutat. Res.*, **209**,145–147.

Pintér, A., Török, G., Börzsönyl, M., Surján, A., Csík, M., Kelecsényi, Z., & Kocsis, Z. (1990) Long-term carcinogenicity bioassay of the herbicide atrazine in F 344 rats. *Neoplasma,* **37**, 533–544.

Pistl, J., Kovalkovicova, N., Holovska, V., Legath, J. & Mikula, I. (2003) Determination of the immunotoxic potential of pesticides on functional activity of sheep leukocytes in vitro. *Toxicology.* **188**, 73–81.

Pohl, C.R. & Knobil, E. (1982) The role of the central nervous system in the control of ovarian function in higher primates. *Annu. Rev. Physiol.*, **44**, 583–593.

Porter, W.P., Jaeger, J.W. & Carlson, I.H. (1999) Endocrine, immune, and behavioral effects of aldicarb (carbamate), atrazine (triazine) and nitrate (fertilizer) mixtures at groundwater concentrations. *Toxicol. Ind. Health.,* **15**, 133–150.

Pruett, S.B., Fan, R., Zheng, Q., Myers, L.P. & Hebert P. (2003) Modeling and predicting immunological effects of chemical stressors: characterization of a quantitative biomarker for immunological changes caused by atrazine and ethanol. *Toxicol. Sci.,* **75**, 343–354.

Puri, E. (1984a) G 30027 - autoradiographic DNA repair test on rat hepatocytes. Unpublished report No. 831171 dated 16 May 1984 (supplemented 14 October 1986) from Ciba-Geigy Ltd, Basle, Switzerland. Submitted to WHO by Syngenta Crop Protection AG.

Puri, E. (1984b) G 30027 - autoradiographic DNA repair test on human fibroblasts. Unpublished report No. 831172 dated 16 May 1984 from Ciba-Geigy Ltd, Basle, Switzerland. Submitted to WHO by Syngenta Crop Protection AG.

Rayner, J.L., Wood, C. & Fenton, S.E. (2004) Exposure parameters necessary for delayed puberty and mammary gland development in Long-Evans rats exposed in utero to atrazine. *Toxicol. Appl. Pharmacol.,* **195**, 23–34.

Rayner, J.L., Enoch, R.R. & Fenton, S.E. (2005) Adverse effects of prenatal exposure to atrazine during a critical period of mammary gland growth. *Toxicol. Sci.,* **87**, 255–266.

Ribas, G., Frenzilli, G., Barale, R. & Marcos, R. (1995) Herbicide-induced DNA damage in human lymphocytes evaluated by the single-cell gel electrophoresis (SCGE) assay. *Mutat. Res.*, **344**, 41–54.

Ribas, G., Surrallés, J., Carbonell, E., Creus, A., Xamena, N. & Marcos, R. (1998) Lack of genotoxicity of the herbicide atrazine in cultured human lymphocytes. *Mutat. Res.,* **416**, 93–99.

Roberge, M., Hakk, H. & Larsen, G. (2004) Atrazine is a competitive inhibitor of phosphodiesterase but does not affect the estrogen receptor. *Toxicol. Lett,* **154**, 61–68.

Roberge, M., Hakk, H., & Larsen, G. (2006) Cytosolic and localized inhibition of phosphodiesterase by atrazine in swine tissue homogenates. *Food Chem. Toxicol.*, **44**, 885–890.

Rodriguez, V.M., Thiruchelvam, M. & Cory-Slechta, D.A. (2005) Sustained exposure to the widely Used herbicide atrazine: altered function and loss of neurons in brain monoamine systems. *Environ. Health Perspect.,* **113**, 708–715.

Rooney, A.A., Matulka, R.A., Luebke, R.W. (2003) Developmental atrazine exposure suppresses immune function in male, but not female Sprague-Dawley rats. *Toxicol. Sci.,* **76**, 366–375.

Rowe, A.M., Brundage, K.M., Schafer, R. & Barnett, J.B. (2006) Immunodulatory effects of maternal atrazine exposure on male Balb/c mice. *Toxicol. Appl. Pharmacol.*, **214**, 69–77.

Rudzki, M.W., McCormick, G.C. & Arthur, A.T. (1989) Hydroxy-atrazine (G 34038) - 90-day oral toxicity study in rats. Unpublished report No. 882146 dated 25 October 1989 from Ciba-Geigy Corp., Research Department, Pharmaceuticals Division, Summit, New Jersey, USA. Submitted to WHO by Syngenta Crop Protection AG.

Rudzki, M.W., McCormick, G.C. & Arthur, A.T. (1991) Atrazine technical: chronic toxicity study in rats. Unpublished report No. 852214 dated 28 January 1991 from Ciba-Geigy Corp., Research Department, Pharmaceuticals Division, Summit, New Jersey, USA. Submitted to WHO by Syngenta Crop Protection AG.

Rudzki, M.W., Batastina, G., & Arthur, A.T. (1992) G-30033: 13-week feeding study in dogs. Unpublished report No. 902187 dated 16 April 1992 from Ciba-Geigy Corp., Research Department, Pharmaceuticals Division, Summit, New Jersey, USA. Submitted to WHO by Syngenta Crop Protection AG.

Ruiz, M.J. & Marzin, D. (1997) Genotoxicity of six pesticides by Salmonella mutagenicity test and SOS chromotest. *Mutat. Res.,* **390**, 245–255.

Rusiecki, J.A., De Roos, A., Lee, W.J., Dosemeci, M., Lubin, J.H., Hoppin, J.A., Blair, A. & Alavanja, M.C.R. (2004) Cancer incidence among pesticide applicators exposed to atrazine in the Agricultural Health Study. *J. Natl. Cancer Inst.* **96**, 1375–1382.

Sachsse, K. & Bathe, R. (1975a) Acute intraperitoneal LD_{50} of technical atrazine in the rat. Unpublished report No. Siss 4569 dated 10 March 1975 from Ciba-Geigy Ltd, Stein, Switzerland. Submitted to WHO by Syngenta Crop Protection AG.

Sachsse, K. & Bathe, R. (1975b) Acute oral LD_{50} of atrazine in the rat. Unpublished report No. Siss 4569 dated 10 March 1975 from Ciba-Geigy Ltd, Stein, Switzerland. Submitted to WHO by Syngenta Crop Protection AG.

Sachsse, K. & Bathe, R. (1975c) Acute oral LD_{50} of technical atrazine in the mouse. Unpublished report No. Siss 4569 dated 7 April 1975 from Ciba-Geigy Ltd, Stein, Switzerland. Submitted to WHO by Syngenta Crop Protection AG.

Sachsse, K. & Bathe, R. (1976) Acute dermal LD_{50} in the rat. Unpublished report No. Siss 5663 dated 6 December 1976 from Ciba-Geigy Ltd, Stein, Switzerland. Submitted to WHO by Syngenta Crop Protection AG.

Sanderson, J.T., Seinen, W., Giesy, J.P. & van den Berg, M. (2000) 2-Chloro-s-triazine herbicides induce aromatase (CYP19) activity in H295R human adrenocortical carcinoma cells: a novel mechanism for estrogenicity? *Toxicol. Sci.,* **54**, 121–127.

Sanderson, J.T., Letcher, R.J., Heneweer, M., Giesy, J.P. & van den Berg, M. (2001) Effects of chloro-s-triazine herbicides and metabolites on aromatase activity in various human cell lines and on vitellogenin production in male carp hepatocytes. *Environ. Health Perspect.,* **109**, 1027–1031.

Sanderson, J.T., Boerma, J., Lansbergen, G.W. & van den Berg, M. (2002) Induction and inhibition of aromatase (CYP19) activity by various classes of pesticides in H295R human adrenocortical carcinoma cells. *Toxicol. Appl. Pharmacol.,* **182**, 44–54.

Sarkar, D.K., Gottschall, P.E. & Meites, J. (1982) Damage to hypothalamic dopaminergic neurons is associated with development of prolactin-secreting pituitary tumors. *Science,* **218**, 684–686.

Sathiakumar, N. & Delzell, E. (1997) A review of epidemiologic studies of triazine herbicides and cancer. *Crit. Rev. Toxicol.,* **27**, 599–612.

Schneider, M. (1992) G 28279 - 3-month oral toxicity study in rats (administration in food). Unpublished report No. 901261 dated 8 May 1992 from Ciba-Geigy Ltd, Stein, Switzerland. Submitted to WHO by Syngenta Crop Protection AG.

Schoch, M. (1985) G 30027 - skin sensitization test in the guinea pig (maximization test). Unpublished report No. 841072 dated 4 June 1985 from Ciba-Geigy Ltd, Stein, Switzerland. Submitted to WHO by Syngenta Crop Protection AG.

Schwab, C.L., Fan, R., Zheng, Q., Myers, L.P., Herbert, P. & Pruett, S.B. (2005) Modeling and predicting stress-induced immunosuppression in mice using blood parameters. *Toxicol. Sci.,* **83**, 101–113.

Shafer, T.J., Ward, T.R., Meacham, C.A. & Cooper, R.L. (1999) Effects of the chlorotriazine herbicide, cyanazine on $GABA_A$ receptors in cortical tissue from rat brain. *Toxicology,* **142**, 57–68.

Shelanski, M.V. & Gittes, H.R. (1965) Atrazine 80 W - repeated insult patch test (in humans). Unpublished report No. IBL 2430 dated 21 January 1965 from Industrial Biology Laboratories Inc. Submitted to WHO by Syngenta Crop Protection AG.

Sielken, R.L. & Holden, L. (2002) Comparison of the LH surge in female rats administered atrazine, simazine, or DACT for six months. Statistical analysis of the LH surge: supplemental analysis. Unpublished report

No. 2214-01 dated 4 March 2002 from Sielken & Associates Consulting, Inc., Bryan, Texas, USA. Submitted to WHO by Syngenta Crop Protection AG.

Simoneaux, B.J. (2001). Comparative metabolism of two s-triazine herbicides: atrazine and ametryn in the rat. Unpublished report No. ABR-97077 dated 11 October 2001 from Norvartis Crop Protection, Inc., Greensboro, North Carolina, USA. Submitted to WHO by Syngenta Crop Protection AG.

Simpkins, J.W. (2000) Relevance of the female Sprague-Dawley (SD) rat for human risk assessment of chloro-s-triazines. A report to Novartis Crop Protection. Unpublished report dated 20 June 2000 frm Center for the Neurobiology of Aging, University of Florida, Gainesville, Florida, USA. Submitted to WHO by Syngenta Crop Protection AG.

Simpkins, J. & Eldridge, J.C. (2004) Evaluation of the potential effects of atrazine on pituitary hormones in estrogen replaced ovariectomized female Rhesus monkey. Unpublished report No. T020244-04 dated 15 September 2004 from Syngenta Crop Protection, Inc, Greensboro, North Carolina, USA. Submitted to WHO by Syngenta Crop Protection AG.

Simpkins, J.W., Andersen, M.E., Brusick, D., Eldridge, J.C., Delzell, E., Lamb, J.C., McConnell, R.F., Safe, S., Tyrey, L. & Wilkinson, C. (2000) Evaluation of hormonal mechanism for mammary tumorigenesis of the chloro-s-triazine herbicides: Fourth Consensus Panel report. Unpublished report dated 13 January 2000. Submitted to WHO by Syngenta Crop Protection AG.

Simpson, E.R., Clyne, C., Rubin, G., Boon, W.C., Robertson, K., Britt, K., Speed, C. & Jones, M. (2002) Aromatase – a brief overview. *Annu. Rev. Physiol.*, **64**, 93–127

Spindler, M. & Sumner, D.D. (1981) Two-year chronic toxicity study with technical atrazine in albino rats. Unpublished report No. 622-06769 dated 15 January 1981 from Industrial Bio-Test Laboratories, Northbrook, IL, USA. Submitted to WHO by Syngenta Crop Protection AG.

Stoker, T.E., Robinette, C.L. & Cooper, R.L. (1999) Maternal exposure to atrazine during lactation suppresses suckling-induced prolactin release and results in prostatitis in the adult offspring. *Toxicol. Sci.,* **52**, 68–79.

Stoker, T.E., Laws, S.C., Guidici, D.L. & Cooper, R.L. (2000) The effect of atrazine on puberty in male Wistar rats: an evaluation in the protocol for the assessment of pubertal development and thyroid function. *Toxicol. Sci.,* **58**, 50–59.

Stoker, T.E., Guidici, D.L., Laws, S.C. & Cooper, R.L. (2002) The effects of atrazine metabolites on puberty and thyroid function in the male Wistar rat. *Toxicol. Sci.,* **67**, 198–206.

Strasser, F. (1988) G 28273 - micronucleus test, mouse. Unpublished report No. 871369 30 March 1988 from Ciba-Geigy Ltd, Basle, Switzerland. Submitted to WHO by Syngenta Crop Protection AG.

Sumner, D.D. (1981) Carcinogenicity study with atrazine technical in albino mice. Unpublished report No. IBT 8580-8906 dated 20 June 1981 from Industrial Bio-Test Laboratories, Inc., Wedge's Creek Research Facility, Neillsville, Wisconsin, USA. Submitted to WHO by Syngenta Crop Protection AG.

Surrallés, J., Xamena, N., Creus, A. & Marcos, R. (1995) The suitability of the micronucleus assay in human lymphocytes as a new biomarker of excision repair. *Mutat. Res.*, **342**, 43–59.

Sutou, S., Kimura, Y., Yamamoto, K. & Ichihara, A. (1979) In-vitro microbial assays for mutagenicity testing of atrazine.Unpublished report No. NRI 79-2884 dated August 1979 from Nomura Research Institute. Submitted to WHO by Syngenta Crop Protection AG.

Taylor, A.E. (1998) Polycystic ovary syndrome. *Endocrinol. Metab. Clin. North Am.*, **27**, 877–902

Tennant, M.K., Hill, D.S., Eldridge, J.C., Wetzel, L.T., Breckenridge, C.B. & Stevens, J.T. (1994a) Possible antiestrogenic properties of chloro-s-triazines in rat uterus. *J. Toxicol. Environ. Health.,* **43**, 183–196.

Tennant, M.K., Hill, D.S., Eldridge, J.C., Wetzel, L.T., Breckenridge, C.B. & Stevens, J.T. (1994b) Chloro-s-triazines antagonism of estrogen action: interaction with estrogen receptor binding. *J. Toxicol. Environ. Health.,* **43**, 197–211.

Tennant, A.H., Peng, B. & Kligerman, A.D. (2001) Genotoxicity studies of three triazine herbicides: in vivo studies using the alkaline single cell gel (SCG) assay. *Mutat. Res.*, **493**, 1–10.

Terranova, P. (1991) F-00006: 90-day subchronic dietary toxicity study with G-28273 in rats. Report addendum: effects of G-28273 technical administration on estrous cycle parameters in female Sprague-Dawley rats. Unpublished report (addendum to report No. F-00006) dated 27 March 1991 from Department of Physiology, University of Kansas School of Medicine, Kansas City, Kansas, USA. Submitted to WHO by Syngenta Crop Protection AG.

Thakur, A.K. (1991a) Determination of hormone levels in Sprague-Dawley rats treated with atrazine technical. Unpublished report No. HWA 483-278 dated 17 October 1991 from Hazleton Laboratories, Vienna, Virginia, USA. Submitted to WHO by Syngenta Crop Protection AG.

Thakur, A.K. (1991b) Determination of hormone levels in Fisher-344 rats treated with atrazine technical. Unpublished report No. HWA 483-279 dated 8 November 1991 from Hazleton Laboratories, Vienna, Virginia, USA. Submitted to WHO by Syngenta Crop Protection AG.

Thakur, A.K. (1992a) Oncogenicity study in Sprague-Dawley rats with atrazine technical. Unpublished report No. HWA 483-275 dated 27 January 1992 from Hazleton Laboratories, Vienna, Virginia, USA. Submitted to WHO by Syngenta Crop Protection AG.

Thakur, A.K. (1992b) Oncogenicity study in Fisher-344 rats with atrazine technical. Unpublished reort No. 483-277 dated 18 February 1992 from Hazleton Laboratories, Vienna, Virginia, USA. Submitted to WHO by Syngenta Crop Protection AG.

Thede, B. (1988) Comparative metabolism of atrazine by mammalian hepatocytes: progress report. Unpublished report No. ABR-88139 dated 13 November 1988 from Ciba-Geigy Corp., Greensboro, North Carolina, USA. Submitted to WHO by Syngenta Crop Protection AG.

Thompson, S.S., Batastini, G.G. & Arthur A.T. (1990) G 28273 - 13/52-week oral toxicity study in dogs. Unpublished report No. 872151 dated 17 January 1990 from Ciba-Geigy Corp., Research Department, Pharmaceuticals Division, Summit, New Jersey, USA. Submitted to WHO by Syngenta Crop Protection AG.

Thompson, S.S., Batastini, G. & Arthur A.T. (1992) G 28279 - 13-week feeding study in dogs. Unpublished report No. 912021 dated 22 April 1992 from Ciba-Geigy Corp., Research Department, Pharmaceuticals Division, Summit, New Jersey, USA. Submitted to WHO by Ciba-Geigy Ltd.

Timchalk, C., Dryzga, M.D., Langvardt, P.W., Kastl, P.E. & Osborne, D.W. (1990) Determination of the effect of tridiphane on the pharmacokinetics of [^{14}C]-atrazine following oral administration to male Fischer 344 rats. *Toxicology*. **61**, 27–40.

Trentacoste, S.V., Friedmann, A.S., Youker, R.T., Breckenridge, C.B. & Zirkin, B.R. (2001) Atrazine effects on testosterone levels and androgen-dependent reproductive organs in peripubertal male rats. *J. Androl.*, **22**, 142–148.

Ullmann, L. (1976a) Eye irritation in the rabbit after single application of G 30027. Unpublished report No. SISS-5663 dated 24 November 1976 from Ciba-Geigy Ltd, Stein, Switzerland. Submitted to WHO by Syngenta Crop Protection AG.

Ullmann, L. (1976b) Skin irritation in the rabbit after single application of G 30027. Unpublished report No. SISS-5663 dated 24 November 1976 from Ciba-Geigy Ltd, Stein, Switzerland. Submitted to WHO by Syngenta Crop Protection AG.

Welsch, C.W., Jenkins, T.W. & Meites, J. (1970) Increased incidence of mammary tumors in the female rat grafted with multiple pituitaries. *Cancer Res.*, **30**, 1024–1029.

Whalen, M.M., Loganathan, B.G., Yamashita, N. & Saito, T. (2003) Immunomodulation of human natural killer cell cytotoxic function by triazine and carbamate pesticides. *Chem. Biol. Interact.*, **145**, 311–319.

Weiss, G., Schmidt, C., Kleinberg, D.L. & Ganguly, M. (1977) Positive feedback effects of estrogen on LH secretion in women in neuroleptic drugs. *Clin. Endocrinol.* **7**, 423–427.

Wise, P.M., Kashon, M.L., Krajnak, K.M., Rosewell, K.L., Cai, A., Scarbrough, K., Harney, J.P., McShane, T., Lloyd, J.M. & Weiland, N.G. (1997) Aging of the female reproductive system: a window into brain aging. *Recent Prog. Horm. Res.* **52,** 279–303.

Young, H.A., Mills, P.K., Riordan, D.G. & Cress, R.D. (2005) Triazine herbicides and epithelial ovarian cancer risk in central California. *J. Occup. Environ. Med.,* **47**, 1148–1156.

AZINPHOS-METHYL

First draft prepared by
U. Mueller[1] *and A. Moretto*[2]

[1]*Food Standards Australia New Zealand, Canberra, ACT, Australia; and*
[2]*Department of Occupational Medicine and Public Health, University of Milan, ICPS Ospedale Sacco, Milan, Italy*

Explanation

Azinphos-methyl is the International Organization for Standardization (ISO) approved common name for *S*-3,4-dihydro-4-oxo-1,2,3-benzotriazin-3-ylmethyl *O,O*-dimethyl phosphorodithioate (International Union of Pure and Applied Chemistry, IUPAC) or *O,O*-dimethyl *S*-[(4-oxo-1,2,3-benzotriazin-3(4*H*)-yl)methyl] phosphorodithioate (Chemical Abstracts Service, CAS; CAS No. 86-50-0). It is a broad-spectrum organophosphorus pesticide. Its toxicity was first evaluated by the 1965 JMPR, when an acceptable daily intake (ADI) of 0–0.0025 mg/kg bw was established based on a no-observed-adverse-effect level (NOAEL) of 0.25 mg/kg bw per day for inhibition of cholinesterase activity in serum and erythrocytes in a repeat-dose study in rats. The 1968 JMPR considered a number of additional studies that were not available to the Meeting in 1965. The ADI established in 1965 was confirmed. The 1973 JMPR considered new studies that involved human volunteers but, owing to the absence of sufficient information on the conduct of these studies in humans, the existing ADI was re-affirmed. In 1991, the ADI was increased to 0–0.005 mg/kg bw on the basis of reduced body-weight gain and fertility observed in the multigeneration study in rats. Azinphos-methyl was reviewed by the present Meeting within the periodic review programme of the Codex Committee on Pesticide Residues (CCPR). All studies previously submitted to JMPR were available for consideration by the present Meeting. Several new studies, including two double-blind clinical studies in human volunteers, were also considered by the present Meeting.

Most studies, excluding some described in previous JMPR monographs, were certified as having been performed in compliance with good laboratory practice (GLP) and in accordance with the relevant Organization for Economic Co-operation and Development (OECD) test guidelines. The studies in humans were conducted in accordance with the principles of good clinical practice and the Declaration of Helsinki, or equivalent statements prepared for use by national and/or multinational authorities. As these guidelines specify the tissues normally examined and the clinical pathology tests normally performed, only significant exceptions to these guidelines are reported here, to avoid repetitive listing of study parameters.

Evaluation for acceptable daily intake

1. Biochemical aspects

1.1 Absorption, distribution and excretion

The pharmacokinetic behaviour of carbonyl-^{14}C-labelled azinphos-methyl was investigated in male Sprague-Dawley rats. The material was almost completely absorbed from the digestive tract, and irrespective of dose and route of administration, 60–70% was eliminated in the urine and 25–35% in the faeces within 48 h. Less than 0.1% of the administered radiolabel was eliminated with the respiratory air within 24 h after dosing, and in rats with biliary fistulas about 30% of the intravenously administered radiolabel was eliminated in the bile within 24 h after dosing. Two days after dosing, the total amount of radioactivity in the animal (excluding digestive tract) was less than 5% of the administered dose; by 4 days this had declined to 2% and by 16 days to 1%. Six hours after dosing, the highest concentrations of radioactivity were found in the organs of elimination (liver and kidney) with relatively high concentrations found in blood. The activity concentrations decayed rapidly in all organs up to 2 days after dosing, but thereafter the activity was more slowly eliminated. At 16 days after dosing, the highest concentration was found in the erythrocytes. In studies in vitro, in which whole blood was incubated with labelled parent compound, there was no evidence for accumulation of radioactivity in the blood constituents (Patzschke et al., 1976; Annex 1, reference *64*).

The pharmacokinetic behaviour of benzazimide was investigated in rats using the ^{14}C-ring-labelled compound. After oral administration, the ^{14}C was almost completely absorbed from the gastrointestinal tract. Elimination of the radiolabel took place quickly, 24 h after administration only 1.3% of the amount administered was present in the animal not including the gastrointestinal tract. More than 99% of the amount administered was eliminated within 48 h (54–66% in the urine and 33–45% via the faeces) (Weber et al., 1980; Annex 1, reference *64*).

1.2 Biotransformation

The metabolism of azinphos-methyl was investigated by administration of ring-UL-^{14}C-labelled azinphos-methyl to male and female Sprague-Dawley rats. The proposed metabolic pathway of azinphos-methyl in rats is given in Figure 1. Upon absorption, azinphos-methyl is rapidly metabolized by mixed function oxidases and glutathione transferases in the liver and other tissues, which results in the formation of azinphos-methyl oxygen analogue, mercaptomethylbenzazimide, glutathionyl methylbenzazimide and desmethyl isoazinphos-methyl. Further hydrolysis, methylation and oxidation of mercaptomethyl-benzazimide forms benzazimide, methylthiomethylbenzazimide and its corresponding oxidized metabolites. Hydrolysis of glutathionyl methylbenzazimide may result in the formation of cysteinylmethyl-benzazimide. Subsequent oxidation of cysteinyl-methylbenzazimide forms its corresponding sulfoxide and sulfone (Kao, 1988; Annex 1, reference *64*).

The rate of disappearance of azinphos-methyl via a hepatic oxidative desulfurating system and a demethylating system was investigated in liver homogenates from four different species (rat, guinea-pig, chicken and monkey). Azinphos-methyl was metabolized by both systems and homogenates from all species were uniformly active (Rao & McKinley, 1969; Annex 1, reference *64*).

1.3 Effect on enzymes and other biochemical parameters

The acute oral toxicity of azinphos-methyl, dissolved in propylene glycol, was investigated in groups of female mice and the effect of the oxime antidote, toxogonin (80 mg/kg bw intraperitoneally, 15 min before oral dosing), was determined. Antidote treatment reduced the toxicity of azinphos-methyl by increasing the median lethal dose (LD_{50}) by a factor of 2 (Sterri et al., 1979; Annex 1, reference *64*).

2. Toxicological studies

2.1 Acute oral toxicity

The results of studies to establish the acute toxicity of azinphos-methyl are summarized in Table 1.

The acute oral toxicity of azinphos-methyl technical in mammals was high. Consistent with the cholinergic effects observed with other organophosphorus compounds, signs of acute intoxication with azinphos-methyl included diarrhoea, salivation, lacrimation and vomiting (muscarinic effects), muscular tremors and paralysis (nicotinic effects), and restlessness, ataxia and convulsions (central nervous system effects).

2.2 Short-term studies of toxicity

Rats

In a dose-range finding study with emphasis on the most sensitive end-point, namely inhibition of cholinesterase activity, groups of five male and five female SPF BOR:WISW (SPF/Cpb) rats were fed diets containing azinphos-methyl (purity, 93.3%) at a concentration of 0, 5, 20, or 50 ppm for 28 days. The average doses ingested at 5, 20, or 50 ppm were 0.35, 1.30, and 3.37 mg/kg bw per day for males and 0.46, 1.54, and 3.96 mg/kg bw per day for females, respectively. There were no effects on appearance, behaviour, mortality, body-weight gain, food consumption, and gross pathology. A statistically significant reduction in plasma cholinesterase activity of 44–66%, relative to the control group, was seen in females at 50 ppm. Appreciably less inhibition of plasma cholinesterase activity was observed in males and this was consistent across the entire sampling period. Erythrocyte acetylcholinesterase activity was unaffected in males at 20 ppm, but was significantly reduced (17–22%) in females. Brain cholinesterase activity was significantly in females at 50 ppm (Table 2).

The NOAEL was 20 ppm, equal to 1.30 mg/kg bw per day in males and 1.54 mg/kg bw per day in females, on the basis of inhibition of cholinesterase activity in the brain at 50 ppm (Eiben et al., 1983).

Figure 1. Proposed metabolic pathway for azinphos-methyl in rats

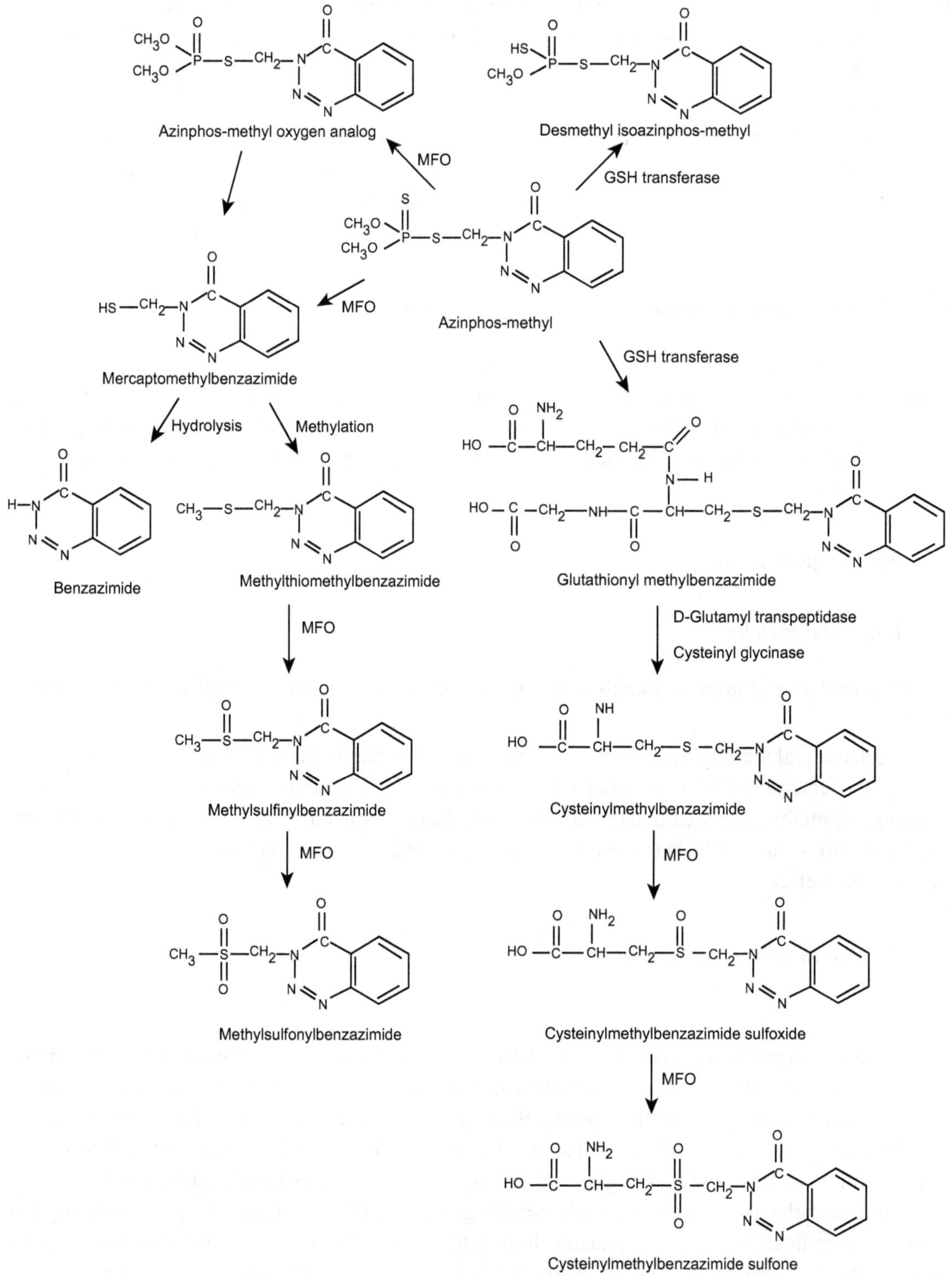

GSH, glutathione; MFO, mixed function oxidases.

Table 1. Acute oral toxicity of azinphos-methyl and its metabolites

Compound	Species	Strain	Sex *(n)*	Vehicle	Purity (%)	LD_{50} *(mg/kg bw)*	Reference
Azinphos-methyl	Mouse	ICR/SIM	Males (20)	Corn oil	NS	15	Simmon (1978)
	Rat	SD	Males and females (4)	DMSO	99	5.6 (fasted males) 6.4 (fasted females)	Crawford & Anderson (1974); Annex 1, reference *64*
	Rat	SD	Females (25)	Ethanol and propylene glycol	97	16.4	DuBois et al. (1955); DuBois et al. (1957)
	Rat	NS	Males (10)	Water and cremophor EL	NS	25.4	Flucke (1979); Annex 1, reference *64*
	Rat	Sherman	Males and females (NS)	Peanut oil	NS	13 (males) 11 (females)	Gaines (1969); Annex 1, reference *64*
	Rat	NS	Males (10)	Water and cremophor EL	NS	9.1 (fasted) 17.25 (non-fasted)	Heimann (1981); Annex 1, reference *64*
	Rat	Wistar	Males (10 or 20)	Water and cremophor EL	88.9	6.7 (fasted) 12.8 (non-fasted)	Heimann (1982); Annex 1, reference *64*
	Rat	NS	Males (5)	Water and cremophor EL	NS	7.1 (fasted)	Heimann (1987); Annex 1, reference *64*
	Rat	SD	Females (4)	CMC	99	16 (fasted) 10 (non-fasted)	Lamb and Anderson (1974); Annex 1, reference *64*
	Rat	Wistar	Males and females (15)	Water and cremophor EL	91.6	4.6 (fasted males) 4.4 (fasted females)	Mihail (1978); Annex 1, reference *64*
	Rat	SD	Females (4)	Panasol AN-2	NS (two samples)	12.2 15	Nelson (1968)
	Rat	CD	Males and females (2)	Methylene chloride + 10% arabic gum in Tween 80	> 95	26 (males) 24 (females)	Pasquet et al. (1976); Annex 1, reference *64*
	Rat	NS	Males (10)	Water and cremophor EL	92.7	15.5	Thyssen (1976a); Annex 1, reference *64*
	Rat	SD	Males and females (5)	Polyethylene glycol	NS	9.0 (males) 6.7 (females)	Crown & Nyska (1987)
	Guinea-pig	NS	Males (NS)	Ethanol and propylene glycol	97	80	DuBois et al. (1957); DuBois et al. (1955)
	Dog	Beagle	Males (1 or 2)	Water and Cremophor EL	91.6	> 10 (0/2 deaths), fasted	Mihail (1978); Annex 1, reference *64*

Benzazimide (azinphos-methyl metabolite in rats)	Rat	SD	Males and females (4)	DMSO	98.1	412 (fasted males) 269 (fasted females)	Crawford & Anderson (1974); Annex 1, reference *64*[a]
	Rat		Males and females (NS)	CMC	NS	576 (fasted males) 368 (fasted females) 576 (non-fasted males) 487 (non-fasted females)	Lamb & Anderson, (1974); Annex 1, reference *64*[b]
Methyl benzazimide (azinphos-methyl metabolite in rats)	Rat	SD	Males and females (4)	DMSO	NS	330 (fasted males and females)	Crawford & Anderson (1974); Annex 1, reference *64*
	Rat		Males and females (NS)	CMC	NS	412 (fasted males) 390 (fasted females) 524 (non-fasted males) 460 (non-fasted females)	Lamb & Anderson (1974); Annex 1, reference *64*[b]

CMC, carboxymethyl cellulose; DMSO, dimethyl sulfoxide; NS, not stated; NZW, New Zealand White; SD, Sprague-Dawley.

[a] Modified with reference to original data.

[b] Original data not provided.

Groups of 10 male and 10 female CD rats were given azinphos-methyl (purity, 93%; in corn oil) at a dose of 0, 0.22, 0.86, 3.44 mg/kg bw per day by gavage for 13 weeks. All rats were examined at least three times per day for mortality and clinical signs. Food consumption and body weights were recorded weekly, and water consumption was recorded daily. Food conversion ratios were calculated as the amount of food consumed per unit body weight gained. Ophthalmoscopic examinations were performed on all rats before the commencement of treatment and on rats in the control and the highest dose groups after 12 weeks of treatment. Haematology and clinical chemistry parameters (including cholinesterase activity) were determined at termination, after which all animals were given a macroscopic examination, and selected organs were removed and weighed. Histopathological examination was performed on selected tissues harvested from rats in the control group and in the group at the highest dose, while kidneys, liver and lungs were examined in all rats.

There were no treatment-related deaths. Salivation was observed in males at and above 0.86 mg/kg bw per day from approximately week 3, with the majority of rats being affected from week 8 at 0.86 (up to 8 out of 10) and 3.44 mg/kg bw per day (up to 10 out of 10). Salivation was observed shortly after dosing in the majority of males, but occasionally it was seen before dosing or up to an hour after dosing in some animals. Salivation was observed infrequently in females with the exception of 2 out of 20 rats at the intermediate dose only during week 7, and 4 out of 10 rats at the highest dose only during week 9. There were no treatment-related effects on ophthalmoscopy, body-weight gain, or food and water consumption. However, there was a clear treatment-related effect on the inhibition of plasma butyryl, plasma acetyl, erythrocyte and brain cholinesterase activities in males and females (Table 3).

Table 2. Cholinesterase activity in rats fed diets containing azinphos-methyl for 28 days

Time-point	Cholinesterase activity (% inhibition relative to control values)					
	Dietary concentration (ppm)					
	5		20		50	
	Males	Females	Males	Females	Males	Females
Plasma						
Day 1	0	0	8	0	10	15
4	0	0	21/0	0	25	44*
14	0	0	21	0	33*	53**
28	0	0	26*	0	26*	61**
Erythrocytes						
Day 1	0	0	0	0	0	0
4	0	0	0	+15	0	0
14	0	0	0	17*	22**	34**
28	0	0	0	22**	14**	35**
Brain						
Day 28	0	0	0	0	9	53**

From Eiben et al. (1983)
* $p < 0.05$; ** $p < 0.01$.

There were no treatment-related effects on absolute organ weights or histopathological abnormalities. The NOAEL was 0.86 mg/kg bw per day on the basis of salivation in male rats and significant inhibition of brain cholinesterase activity at 3.44 mg/kg bw per day (Broadmeadow et al., 1987).

Dogs

In a 52-week study of toxicity, groups of four male and four female beagle dogs were given diets containing azinphos-methyl (purity, 91.9%) at a concentration of 0 (control), 5, 25 or 125 ppm (equal to 0.16, 0.74, or 4.09 mg/kg bw per day). Clinical signs of reaction to treatment were confined to a higher incidence of diarrhoea in dogs at 125 ppm. Two males at 125 ppm failed to gain weight during the course of the study, but food intake remained unaffected by treatment. Haematology and urine analysis revealed no indication of any reaction to treatment. Clinical biochemistry tests revealed a depression of cholinesterase activity in plasma and erythrocytes in dogs at 25 and 125 ppm and in brain at termination of dogs at 125 ppm (Table 4).

There was also a very slight increase, compared with controls, in liver cytochrome P450 and *N*-demethylase activity at the highest dose and a reduction in albumin levels. Pathological investigations (macroscopic examination, organ weight analysis and histopathology) revealed no evidence of any reaction to treatment with azinphos-methyl. The NOAEL was 25 ppm (0.74 mg/kg bw per day) on the basis of reduced body-weight gain and inhibition of cholinesterase activity in brain (modified with reference to the original data, Allen, et al., 1990; Annex 1, reference *64*).

2.3 *Long-term studies of toxicity and carcinogenicity*

Mice

A bioassay of azinphos-methyl (purity, 90%; from manufacturing specification) for possible carcinogenicity was conducted by the National Cancer Institute (NCI). Osborne-Mendel rats and

Table 3. Cholinesterase activity in rats given azinphos-methyl by gavage for 13 weeks

Cholinesterase	Cholinesterase activity (% inhibition relative to control values)					
	Dose (mg/kg bw per day)					
	0.22		0.86		3.44	
	Males	Females	Males	Females	Males	Females
Plasma butyryl cholinesterase	18***	3	16**	10	30***	47***
Plasma acetyl cholinesterase	18**	−3	14*	7	42***	49***
Erythrocyte acetylcholinesterase	8*	9***	17***	29***	77***	78***
Brain cholinesterase	3	4	9*	3	67***	64***

From Broadmeadow et al. (1987)
* $p < 0.05$; ** $p < 0.01$; *** $p < 0.001$.

Table 4. Cholinesterase activity in dogs[a] fed diets containing azinphos-methyl for 52 weeks

Cholinesterase	Cholinesterase activity (% inhibition relative to control values)							
	Dietary concentration (ppm)							
	0		5		25		125	
	Males	Females	Males	Females	Males	Females	Males	Females
Plasma cholinesterase								
Week 4	0	0	11	−18	12	14	37	52*
Week 13	0	0	13	−2	15	17	53**	58**
Week 26	0	0	14	−10	12	33	58**	57**
Week 52	0	0	11	12	12	30	53*	53**
Erythrocyte acetylcholinesterase								
Week 4	0	0	−9	11	22	2	66**	86**
Week 13	0	0	8	16	40**	43**	87**	92**
Week 26	0	0	8	21	32	38**	88**	91**
Week 52	0	0	−5	15	27	35*	86**	86**
Brain cholinesterase	0	0	1	1	10	1	27**	20*

From Allen et al. (1990), Annex 1, reference *64*, modified with reference to the original data.
[a] $n = 4$
* $p < 0.05$; ** $p < 0.01$.

B6C3F$_1$ mice were fed diets containing azinphos-methyl. Groups of 50 male and 50 female Osborne-Mendel rats were given diets containing azinphos-methyl for 80 weeks, and were then observed for 34 or 35 weeks. Males received azinphos-methyl at a time-weighted average dietary concentration of 78 or 156 ppm (equivalent to 6.25 and 12.5 mg/kg bw per day respectively), while females received 62.5 or 125 ppm (equivalent to 3.12 and 6.25 mg/kg bw per day respectively). Matched controls consisted of 10 male and 10 female untreated rats; pooled controls consisted of matched controls combined with 95 male and 95 female untreated rats from similar bioassays of 10 other chemicals. The study in mice was of similar design; groups of 50 mice were treated for 80 weeks, then observed for 12 or 13 weeks, males received azinphos-methyl at a dietary concentration of 31.3 or 62.5 ppm (equivalent to 4.7 and 9.4 mg/kg bw per day respectively) and females received 62.5 or 125 ppm (equivalent to 9.4 and 18.75 mg/kg bw per day respectively), matched controls consisted of 10 males and 10 females, pooled controls comprised 130 males and 120 females. Typical signs of organophosphate intoxication (hyperactivity, tremors and dyspnoea) were observed in a few animals of both species at the highest dose. In both species, body-weight gain for treated males and in females at the highest dose was lower than in animals in the control groups. In rats there was some evidence of decreased survival at the highest dose compared with controls, but this was not seen in mice. In both sexes at all doses, survival to termination was adequate for assessment of effects on late-appearance tumours. The report concluded that the incidence of tumours of the pancreatic islets, and of follicular cells in the thyroid in males suggested, but did not clearly implicate, azinphos-methyl as a carcinogen in rats. There was no similar evidence in female rats, and in male and female mice sex there was no increased incidence of tumours that could be related to the administration of azinphos-methyl (National Cancer Institute, 1978; Annex 1, reference *64*, amended with reference to original data).

In a study of carcinogenicity in mice, groups of 50 male and 50 female CD-1 mice received azinphos-methyl (purity, 88.6%) at a dietary concentration of 0 (control), 5, 20, or 40/80 ppm (equal to 0.79, 3.49, or 11.33 mg/kg bw per day in males and 0.98, 4.12, or 14.30 mg/kg bw per day in females) for 2 years. The study was initially used 80 ppm as the highest dietary concentration, but this was reduced to 40 ppm after 1 week, owing to severe reaction (including mortality) to treatment at 80 ppm. After the reduction in the highest dietary concentration, there were no clinical signs and mortality remained unaffected by treatment. Weight gain and food intake remained unaffected by treatment at dietary concentrations up to and including 40 ppm. Haematological investigations revealed no indication of any reaction to treatment. Measurement of cholinesterase activity revealed that at 5 ppm, plasma, erythrocyte and brain cholinesterase activities remained comparable with control values. At 20 and 40 ppm there was a dose-related inhibition of cholinesterase activity in plasma and erythrocytes. A similar effect was noted in brain, except that males were only affected at 40 ppm, while females exhibited a depression of brain cholinesterase activity at 20 and 40 ppm (Table 5).

Pathological investigations revealed no evidence of any reaction to treatment, in particular there was no evidence of any carcinogenic effect of azinphos-methyl at any dose. The NOAEL was 5 ppm (equal to 0.79 and 0.98 mg/kg bw per day in males and females, respectively) on the basis of inhibition of cholinesterase activity in plasma, erythrocytes and brain at 20 ppm (Hayes, 1985; Annex 1, reference 64, amended with reference to original data).

Rats

In a combined long-term study of toxicity and carcinogenicity in rats, groups of 60 male and 60 female Wistar rats received diets containing azinphos-methyl (purity, 87.2%) at a concentration of 0 (control), 5, 15 or 45 ppm (equal to 0.25, 0.75, or 2.33 mg/kg bw per day for males and 0.31, 0.96, or 3.11 mg/kg bw per day for females, respectively). From each group, 10 males and 10 females were killed after 1 year, while all survivors were killed after 2 years of continuous treatment. There were no clinical signs of reaction to treatment and survival was unaffected by azinphos-methyl. Body-weight gain of males at the highest dose was slightly less than that of the controls (7%), but growth in other groups was not affected by treatment. Clinical biochemistry (apart from investigations of acetylcholinesterase activity), haematology and urine analysis revealed no indication of any reaction to treatment.

Determinations of cholinesterase activity in erythrocytes, plasma and brain revealed a marked inhibition, relative to controls, in males and females from the group at the highest dose (erythrocytes, plasma and brain) and a less marked effect at 15 ppm (males, erythrocytes; females, erythrocytes and plasma). Cholinesterase activity in brain from rats at 15 ppm and in erythrocytes, plasma and brain from rats at 5 ppm remained unaffected by treatment with azinphos-methyl (Table 6). Pathological examinations (including gross examination, organ-weight analysis and histological examination of tissues) revealed no evidence of any reaction to treatment; in particular, there was no evidence of any carcinogenic effect. The NOAEL was 15 ppm, equal to 0.75 and 0.96 mg/kg bw per day in males and females, respectively, on the basis of effects on body-weight gain and brain acetylcholinesterase activity (Schmidt, 1987; Annex 1, reference *64*, modified with reference to the original data).

Table 5. Cholinesterase activity in mice fed diets containing azinphos-methyl for 104 weeks

Time-point	Cholinesterase activity[a]							
	Dietary concentration (ppm)							
	0		5		20		40	
	Males	Females	Males	Females	Males	Females	Males	Females
Plasma (µmol/ml per min):								
6 months	3.11	5.76	3.33	5.45	2.57	3.08	1.62	1.48
1 year	3.88	6.51	4.81	5.44	2.63	3.27	1.32	1.50
2 years	4.33	4.98	3.95	4.93	2.97	3.86	1.89	1.65
Erythrocytes (µmol/ml per min):								
6 months	1.33	1.16	1.11	1.03	0.88	0.67	0.67	0.63
1 year	1.04	0.87	0.99	0.81	0.45	0.39	0.20	0.20
2 year	0.95	0.79	0.80	0.62	0.54	0.40	0.35	0.32
Brain (µmol/g per min):								
2 years	14.7/	14.4	12.9/	13.6	12.3	10.6	5.4	4.7

From Hayes (1985); Annex 1, reference *64*, amended with reference to original data.
[a] Results of statistical analysis not specified in the report.

2.4 *Genotoxicity*

The results of studies of genotoxicity with azinphos-methyl are summarized in Table 7.

An effort was made to evaluate the genotoxicity of a variety of pesticides, with the specific objectives of comparing different assays in vivo and in vitro, examining the spectrum of genetic activity displayed by the selected pesticides and examining the test results in relation to other biological and chemical features of the pesticides. In this research programme, azinphos-methyl was used in 14 teats for mutagenicity, examining point or gene mutations, DNA damage and chromosomal effects. Positive results for azinphos-methyl were seen in only two tests: an assay for forward mutation in mouse lymphoma L5178Y cells (only in the presence of metabolic activation) and an assay for mitotic recombination in *Saccharomyces cerevisiae* strain D3. Azinphos-methyl gave negative results in tests for point/gene mutation and DNA damage in prokaryotes and showed no positive results in tests for chromosomal effects (Waters et al., 1982; Annex 1, reference *64*).

2.4 *Reproductive toxicity*

(a) Multigeneration studies

Rats

In a two-generation (two litters per generation) study of reproduction, groups of 12 male and 24 female Wistar rats were given diets containing azinphos-methyl (purity, 87.2%) at a concentration of 0 (control), 5, 15 or 45 ppm (equal to 0, 0.33–0.42, 1.02–1.22 and 3.46–7.37 mg/kg bw per day in males and 0, 0.48–0.67, 1.48–2.02 and 4.84–10.27 mg/kg bw per day in females). At 45 ppm,

Table 6. Cholinesterase activity in rats fed diets containing azinphos-methyl for 2 years

Tissue	Cholinesterase activity (% of values for controls)					
	Dietary concentration (ppm)					
	5		15		45	
	Males	Females	Males	Females	Males	Females
Plasma						
1 month	93	88	95	88	62**	44**
3 months	91	88	102	65**	60**	35**
6 months	88	92	95	71*	57**	34**
1 year	84	90	87	65**	54**	33**
1.5 years	87	100	90	74*	55**	46**
2 years	113	102	88	81	51**	38**
Erythrocytes						
1 month	97	105	84**	87	76**	74*
3 months	101	110**	88**	88**	77**	72**
6 months	97	109*	90	86**	80**	77**
1 year	102	101	82*	81**	73**	69**
1.5 years	96	94	83**	80**	73**	73**
2 years	88**	98	78**	84**	63**	71**
Brain						
1 year	130**	112	137**	90	109	50**
2 years	117	102	112	79**	68**	45**

From Schmidt (1987); Annex 1, reference *64*, modified with reference to the original data.
* $p < 0.05$; ** $p < 0.01$.

there was a decrease in fertility of F_0 and F_{1b} dams and the total number of delivered pups (Table 8). The viability index was significantly reduced in F_{1a}, F_{1b} and F_{2a} litters at 45 ppm, while the effect was equivocal at 15 ppm (Table 9). At 45 ppm, there was an increased mortality of dams in the F_0 generation and reduced mean pup viability from days 5 to day 28 of lactation in the F_{1a} and F_{1b} litters. As a consequence of these effects, at 45 ppm only five females were available for mating in the F_{1b} generation. During mating of the F_{1b} generation, fertility was again slightly reduced at 45 ppm but not to as great an extent as it was for the F_0 generation.

At all stages of the study, there was no evidence of treatment-induced malformations and food intake remained unaffected. Clinical signs of reaction to treatment, including cholinergic signs, and reduced body-weight gain were observed at the highest dose tested (45 ppm). The NOAEL was 15 ppm, equal to 1.5 mg/kg bw per day, on the basis of reduced body-weight gain in F_0 and F_{1b} parental animals and decreased pup body weight and viability at 45 ppm (Eiben & Janda, 1987; Annex 1, reference *64*, modified by reference to original data).

Another study was conducted to investigate the effects on reproductive performance noted in the study described above. The objectives of this additional, one-generation study were to investigate whether the slight effect on fertility and pup viability observed at 15 ppm could be confirmed, and, if reproducible, to determine whether the effect was attributable to treatment of the male or the female and to determine whether reproductive effects were associated with treatment-induced inhibition of cholinesterase activity. Groups of 18 male and 46 female Wistar rats were given diets containing azinphos-methyl (purity, 92.0%) at a concentration of 0 (control), 5, 15, or 45 ppm (equal to 0, 0.43, 1.30,

Table 7. Results of studies of genotoxicity with azinphos-methyl

End-point	Test object	Concentration	Purity (%)	Results	Reference
In vitro					
Reverse mutation (Ames)	*S. typhimurium* TA98, TA100, TA1535, TA1537, TA1538	0–160 μg/plate	NS	Negative	Evanchik (1987)
Reverse mutation (Ames)	*S. typhimurium* TA98, TA100, TA1535, TA1537	4–2500 μg/plate	92.3	Negative	Herbold (1978)
Reverse mutation (Ames)	*S. typhimurium* TA98, TA100, TA1535, TA1537	75–9600 μg/plate	92.5	Negative	Herbold (1988)
Reverse mutation (Ames)	*S. typhimurium* TA98, TA100, TA1535, TA1537, TA1538	33–4000 μg/plate	88.8	Negative	Lawlor (1987)
Reverse gene mutation	*S. cerevisiae* S138 and S211α	33.3–10000 μg/ml	91.1	Negative	Hoorn (1983)
DNA damage	*E. coli* W3110, *E. coli* p3478	625–10000 μg/plate	91.1	Negative	Herbold (1984)
Unscheduled DNA synthesis	Rat hepatocytes	0.25–50.3 μg/ml, in DMSO	91.1	Negative	Myhr (1983)
Clastogenicity	Chinese hamster ovary cells-(K1)	60–120 μg/ml	Approx. 90	Dose-related chromatid breaks and exchanges	Alam et al. (1974)
	Human lymphocytes	5–500 μg/ml (+S9) 1–100 μg/ml (–S9) in DMSO	91.9	Clastogenicity observed at cytotoxic concentrations +S9. No effect −S9	Herbold (1986)
Sister chromatid exchange	Chinese hamster lung cells (V79)	5–25 μg/ml	98.7	Negative	Chen et al. (1982a)
		2.5–20 μg/ml			Chen et al. (1982b)
In vivo					
Micronucleus formation	Mouse	2 × 2.5 and 2 × 5 mg/kg bw single oral dose in Cremophor/water	92.3	Negative	Herbold (1979a)
		5 mg/kg bw, single IP dose in Cremophor/water	92.2	Negative	Herbold (1995)
Chromosomal aberration	Rat, femoral bone marrow	6.28 mg/kg bw, single oral dose in methyl-cellulose	92	Negative	Henderson et al. (1988)
Dominant lethal mutation	Mouse	0.125 or 0.25 mg/kg bw, single IP dose in corn oil	NS	Negative	Arnold (1971)
		1 × 4 mg/kg bw, single oral dose in Cremophor/water	92.3	Negative	Herbold (1979b)

DMSO, dimethyl sulfoxide; IP, intraperitoneal; NS, not specified; S9, 9000 × *g* supernatant from livers of rats.

and 3.73 mg/kg bw per day in males and 0.55, 1.54, and 4.87 mg/kg bw per day in females). Treated males and females were paired, and dams allowed to rear litters to weaning. Additional treated males were paired with untreated females.

Table 8. Fertility of F_0 and F_{1b} rats fed diets containing azinphos-methyl

Parameter	Dietary concentration (ppm)			
	0	5	15	45
	Mating 1/2	Mating 1/2	Mating 1/2	Mating 1/2
F_0				
Mated females (No.)	24/24	24/24	24/23	23/19
Insemination index (%)	100/100	95.8/100	91.6/91.3	95.6/94.7
Fertility index (%)	91.7/91.7	95.7/95.8	90.0/85.7	86.4/83.3
Gestation index (%)	100/100	100/100	100/100	100/93.3
Duration of gestation (days)	22.5/22.3	22.5/22.7	22.5/22.6	22.8/22.8
F_{1b}				
Mated females (No.)	24/24	24/24	24/23	5/5
Insemination index (%)	100/100	100/100	100/100	100/100
Fertility index (%)	91.7/87.5	100/91.7	87.5/95.7	100/80.0
Gestation index (%)	100/95.2	100/100	100/95.5	100/75.0
Duration of gestation (days)	22.1/22.2	22.3/22.6	22.6/22.2	22.4/22.3

From Eiben & Janda (1987); Annex 1, reference *64*, modified by reference to original data.

At 45 ppm, two females that were sacrificed in extremis exhibited clinical signs consistent with cholinergic effects, such as poor general condition, bloody noses, inertia and a stumbling gait. Males at the same dietary concentration showed no clinical signs. There were no treatment-related effects on fertility parameters (insemination, fertility and gestation indices, and duration of gestation) (Table 10). At 15 ppm, when males and females were treated, the viability index was slightly reduced but it was in the absence of any corresponding reduction in pup body-weight gain. In contrast, dams at 45 ppm delivered pups that had significantly reduced viability and body-weight gain by postnatal day 5. After treatment of male parental animals only, reproductive parameters remained unaffected, even at 45 ppm (Table 11). Investigations of cholinesterase activity in parental animals revealed a depression in activity in plasma and erythrocytes at all dietary concentrations, and a depression in activity in brain at 45 ppm in males and at 15 and 45 ppm in females (Table 12).

There were no treatment-related clinical signs observed in F_1 pups at birth or during the 4-week rearing period. Similarly, no treatment-related pathological changes were observed.

At 45 ppm, brain cholinesterase activity in pups was also depressed. The NOAEL was 5 ppm, equal to 0.55 mg/kg bw per day, on the basis of depression of brain cholinesterase activity at 15 ppm (Holzum, 1990; Annex 1, reference *64*, modified by reference to original data).

(b) Developmental toxicity

Mice and rats

The effects of azinphos-methyl (purity, 90.6%) on development in rats and mice were investigated in a series of experiments. On the basis of preliminary studies of toxicity, doses of 0, 1.25, 2.5 or 5.0 mg/kg bw per day were selected for two-phase studies of developmental toxicity in both species. During the first phase, pregnant rats and mice were treated for 10 days, starting on day 6 of gestation. During the second phase, pregnant rats were treated from day 6 of gestation to postnatal day 21. In the first phase, maternal toxicity was seen only in rats receiving the highest dose. When dams and fetuses were examined (day 18 of gestation for mice, day 20 for rats) there was no

Table 9. Pup parameters in rats fed diets containing azinphos-methyl

Parameter	Dietary concentration (ppm)			
	0	5	15	45
F_{1a}				
Live /dead pups at birth (No.)	252/1	247/0	204/8	197/9
Pups, males/females (%)	52/48	53/47	49/51	52/48
Litter size at day 0/day 5 (No.)	11.5/11.1	11.2/10.5	10.1/8.7*	10.1/3.9**
Viability index (%)	96.8	93.9	86.6**	38.7**
Pups at day 5[a]/week 4 (No.)	175/169	167/156	139/134	62/17
Lactation index (%)	96.6	93.4	96.4	27.4**
Body weight at day 0/week 3 (g)	5.8/36.7	5.7/37.5	5.9/35.9	5.4/25.8**
F_{1b}				
Live/dead pups at birth (No.)	235/1	236/11	175/1	133/0
Pups, males/females (%)	52/48	50/50	51/49	55/45
Litter size at day 0/day 5 (No.)	10.6/10.5	9.8/9.5	9.7/9.7	8.9/2.8**
Viability index (%)	98.3	97.3	98.9	31.6**
Pups at day 5[a]/week 4 (No.)	165/161	164/162	128/117	39/18
Lactation index (%)	97.6	98.8	91.4*	46.2**
Body weight at day 0/week 3 (g)	5.7/39.9	5.8/39.2	5.9/37.8	5.2**/27.2**
F_{2a}				
Live/dead pups at birth (No.)	259/3	270/0	230/0	43/0
Pups, males/females (%)	55/45	55/45	54/46	51/49
Litter size at day 0/day 5 (No.)	11.7/11.5	11.2/10.8	11.0/10.7	8.6*/7.0*
Viability index (%)	98.1	95.9	97.8	81.4**
Pups at day 5[a]/week 4 (No.)	176/173	185/174	152/134	29/21
Lactation index (%)	98.3	94.1	88.7**	72.4**
Body weight at day 0/week 3 (g)	5.7/37.3	5.7/35.6	5.7/36.0	5.4/22.4**
F_{2b}				
Live/dead pups at birth (No.)	223/1	244/2	214/3	25/0
Pups, males/females (%)	51/49	55/45	49/51	56/44
Litter size at day 0/day 5 (No.)	10.6/10.1	11.0/10.0	9.6/8.5	6.2/6.2
Viability index (%)	95.5	90.1*	88.6*	100
Pups at day 5[a]/week 4 (No.)	143/133	165/138	137/123	22/20
Lactation index (%)	93.0	83.6*	89.8	90.9
Body weight at day 0/week 3 (g)	5.8/40.2	5.9/39.6	5.6/37.8	5.8/27.0**

From Eiben & Janda (1987); Annex 1, reference *64*, modified by reference to original data.
[a] After reduction.
* $p < 0.05$; ** $p < 0.01$

dose-related increase in anomalies or malformation in rats or mice. In the second phase, dams in the group at the highest dose were more sensitive to azinphos-methyl in the latter stages of gestation and signs of anticholinesterase intoxication, including mortality, were observed. As a result, only 1 litter out of 13 survived to weaning in this group. The authors concluded that azinphos-methyl had little primary effect on development in rats and mice (Short et al., 1978; Short et al., 1980; Annex 1, reference *64*, modified by reference to original data).

Table 10. Fertility parameters in rats fed diets containing azinphos-methyl

Parameter	Dietary concentration (ppm)			
	Males, females	Males, females	Males, females	Males, females
	0, 0	5, 5 / 5, 0	45, 45 / 45, 0	15, 15 / 15, 0
Mated females (No.)	36	36/20	36/20	31/20
Insemination index (%)	100	100/100	100/100	100/100
Fertility index (%)	97.2	94.4/100	100/95.0	96.8/100
Gestation index (%)	88.5	100/100	100/100	95.5/95.0
Duration of gestation (days)	22.8	22.4/22.6	22.6/22.7	22.7/22.5

From Holzum (1990); Annex 1, reference *64*, modified by reference to original data.

Table 11. F_1 pup parameters in rats fed diets containing azinphos-methyl

Parameter	Dietary concentration (ppm)			
	Males, females	Males, females	Males, females	Males, females
	0, 0	5, 5 / 5, 0	15, 15 / 15, 0	45, 45 / 45, 0
Pups at birth (No.)	240	293 / 216	275 / 196	214 / 211
Pups dead at birth (No.)	13	4 / 2	3 / 1	3 / 3
Live birth index (%)	94.6	98.6* / 99.1*	98.9** / 99.5**	98.6* / 98.6*
Male pups (%)	48.8 / 52.3	48.7 / 48.5	49.5 / 65.4	45
Litter size (No.)	9.9	11.1 / 10.7	10.5 / 10.3	10.0 / 10.9
Viability index (%)	93.4	92.4 / 98.1*	86.0* / 90.8	48.3** / 95.7
Lactation index (%)	62.1	74.2 / —	69.8 / —	57.7 / —
Body weight (g):				
Day 0	5.9	5.7 / 5.8	5.8 / 5.9	5.9 / 5.7
Day 5	9.2	8.7 / 8.9	9.0 / 9.3	7.8* / 8.7
Day 14	24.4	22.5 / —	19.9 / —	21.4 / —
Day 28	52.4	52.7 / —	49.5 / —	46.3 / —

From Holzum (1990); Annex 1, reference *64*, modified by reference to original data.
* $p < 0.05$; ** $p < 0.01$.

Rats

Groups of 33 mated Crl:CD BR CD rats were given azinphos-methyl (purity, 87.7%) at a dose of 0 (control), 0.5, 1.0 or 2.0 mg/kg bw per day in 6% Emulphor® (EL-719) by gavage on days 6 to 15 of gestation (presence of spermatozoa was counted as day 0 of gestation). From each group, five dams were killed on day 16 of gestation and the remaining dams were killed on day 20. Appearance, behaviour, feed consumption, and body-weight gain of the dams were not adversely affected by treatment at any of the doses tested. On day 16 of gestation, cholinesterase activities in plasma, erythrocytes and brain were depressed, relative to values for controls, in dams at the highest dose only (fetal tissues were not examined). By day 20 of gestation, there was indication of recovery in cholinesterase activity in all previously affected tissues and fetal brain cholinesterase activity was comparable with control values (Table 13).

Azinphos-methyl did not affect any maternal reproductive parameters and there was no indication of treatment-related embryotoxicity, fetotoxicity or teratogenicity at any dose. The NOAEL for

Table 12. Inhibition of plasma, erythrocyte and brain cholinesterase activity in F_0 and in F_1 pups of rats fed diets containing azinphos-methyl

Dietary concentration (ppm)	Cholinesterase activity (% of values for controls)						
	F_0 males	F_0 females				F_1 pups	
	End of mating	End of pre-treatment	Post-coital day 11	Postnatal day 5	Postnatal day 28	Postnatal day 5	Postnatal day 28
Plasma							
5	2.3	ND	ND	26*	5		
15	14***	25*	18	46***	39**	ND	ND
45	43***	62***	60***	66***	63***		
Erythrocytes							
5	19**	ND	ND	25**	47***		
15	69***	46***	52***	75***	84***	ND	ND
45	94***	71***	81***	91***	89***		
Brain							
5	1	10*	ND	ND	12	ND	ND
15	ND	4	21**	38**	48***	1	14
45	19**	55***	69***	66***	68***	17*	46***

From Holzum (1990); Annex 1, reference 64, modified by reference to original data.
* $p < 0.05$; ** $p < 0.01$; *** $p < 0.001$.
ND, not determined.

Table 13. Cholinesterase activity in dams (and their fetuses) given azinphos-methyl by gavage on days 6–15 of gestation

Tissue	Cholinesterase activity			
	Dose (mg/kg bw per day)			
	0	0.5	1.0	2.0
Day 16				
Plasma, kU/l (% of control)	1.73 (100%)	1.56 (90%)	1.64 (95%)	1.08 (63%)
Erythrocytes, kU/l (% of control)	0.38 (100%)	0.34 (90%)	0.34 (90%)	0.08* (21%)
Brain, U/g (% of control)	2.53 (100%)	2.70 (107%)	2.55 (101%)	1.55* (61%)
Day 20				
Plasma, kU/l (% of control)	1.44 (100%)	1.47 (102%)	1.40 (97%)	1.32 (92%)
Erythrocytes, kU/l (% of control)	0.58 (100%)	0.59 (103%)	0.61 (106%)	0.44 (77%)
Brain, U/g (% of control)	2.67 (100%)	2.72 (102%)	2.46 (92%)	1.91* (72%)
Fetal brain, U/g (% of control)	1.04 (100%)	0.96 (91%)	1.04 (100%)	1.00 (96%)

From Kowalski et al. (1987a; 1987b); Annex 1, reference *64*, modified by reference to original data; Astroff & Young (1998).
* $p < 0.05$.

developmental toxicity was 2.0 mg/kg bw per day, on the basis of the absence of any effects at the highest tested dose. The NOAEL for maternal toxicity was 1.0 mg/kg bw per day, on the basis of the inhibition of brain cholinesterase activity seen on day 16 of gestation at 2.0 mg/kg bw per day (Kowalski et al., 1987a, 1987b; Annex 1, reference *64*, modified by reference to original data; Astroff & Young, 1998).

Groups of 22 mated Sprague-Dawley-derived CD rats were given azinphos-methyl (purity, 92.7%) at a dose of 0.4, 1.2, or 3.6 mg/kg bw per day by gavage in maize oil daily from day 6 to day 15 of gestation. All rats were observed daily for clinical signs and body weights were recorded on days 0, 3, 6-15, 17 and 20 of gestation. Food consumption was measured twice weekly. On day 20, the animals were killed and necropsied. No deaths or treatment-related clinical signs were observed during the study. Food consumption during gestation was similar in control and treated females. The group mean body weight of females at 3.6 mg/kg bw per day was slightly lower than controls during the period before treatment, but body weight and body-weight gain were unaffected by treatment. There were no treatment-related changes in gravid uterine weights, the number or sex of live fetuses, or the number of resorptions. Skeletal observations revealed retarded ossification in fetuses at 3.6 mg/kg bw per day, with reduced or absent ossification seen in the supraoccipital, pubic and hyoid bones. An increased incidence of supernumerary ribs (14th, lumbar) was also observed at 3.6 mg/kg bw per day (4.7%; $p < 0.01$), and the incidence of this finding was outside the range for historical controls in the testing laboratory (0–3.1%). In the absence of any maternal effects, the NOAEL for maternal toxicity was 3.6 mg/kg bw per day. The NOAEL for developmental toxicity was 1.2 mg/kg bw per day on the basis of increases in the incidence of supernumerary ribs (14th, lumbar) and delayed ossification (pubic, hyoid, and supraoccipital bones) in fetuses at 3.6 mg/kg bw per day (Rubin & Nyska, 1988).

Rabbits

In a teratology study in Himalayan rabbits, groups of 11 or 12 pregnant animals received azinphos-methyl (purity, 92.4%) as a daily oral dose at 0 (control), 0.3, 1.0 or 3.0 mg/kg bw per day from day 6 to day 18 of gestation (day of insemination was counted as day 0). Caesarean section was carried out on day 29 of gestation. Azinphos-methyl induced no evidence of maternal toxicity at any dose and there were no detectable effects on embryonic or fetal development (Machemer, 1975; Annex 1, reference *64*, modified with reference to the original data).

In a further study of teratology in American Dutch rabbits, groups of 20 inseminated does received azinphos-methyl (purity, 87.7%) as a daily oral dose at 0 (control), 1, 2.5 or 6 mg/kg bw per day from day 6 to day 18 of gestation (day of insemination was counted as day 0). Plasma and erythrocyte cholinesterase activities were determined on days 19 and 28 of gestation, and in brain on day 28 of gestation. Ataxia in four does at the highest dose and tremors in two of these represented clinical signs of reaction to treatment. At the intermediate and highest dose, plasma and erythrocyte cholinesterase activity on day 19 of gestation was depressed compared with controls. By day 28 of gestation there was clear evidence of recovery in plasma and erythrocyte cholinesterase activity, although activity in brain was depressed, compared with controls, at the highest dose. Azinphos-methyl did not affect any maternal reproductive parameters and there was no evidence of any treatment-related effect on embryotoxicity, fetotoxicity or teratogenicity at any dose. The NOAEL for maternal toxicity was 2.5 mg/kg bw per day on the basis of the inhibition of brain cholinesterase activity seen on day 28 of gestation at 6 mg/kg bw per day (Clemens et al., 1988; Annex 1, reference *64*, modified with reference to the original data).

In a study of developmental toxicity, groups of mated New Zealand White rabbits were given azinphos-methyl (purity, 92.7%) at a dose of 0 (vehicle control), 1.5, 4.75, or 15 mg/kg bw per day by oral gavage in maize oil from day 7 to 19 of gestation. To ensure that at least 12 pregnant rabbits per group were available at terminal sacrifice, the group sizes were adjusted to 16, 18, 15, and 18, at doses of 0, 1.5, 4.75 and 15 mg/kg bw per day, respectively. Rabbits were observed daily for clinical signs and weighed on days 0, 3, 7–19, 22, 25 and 29 of gestation. Food consumption was measured twice per week. Cholinesterase activity in erythrocytes and plasma was measured before pairing and again after 11 days of dosing.

Deaths occurred in all groups, including controls (one rabbit in the control group, two at 1.5 mg/kg bw per day, one at 4.75 mg/kg bw per day, and three at 15 mg/kg bw per day). Death in the control group and in the group at the lowest and intermediate dose was attributed to intercurrent infections, while at the highest dose it was attributed to accidental lung dosing. No treatment-related clinical signs or changes in body-weight gain were observed. After 11 days of dosing, erythrocyte acetylcholinesterase activity was significantly reduced ($p < 0.05$) by 27% at 15 mg/kg bw per day. Significant reductions in plasma cholinesterase activity were seen at all doses after 11 days of treatment. However, there was no consistent dose–response relationship and the magnitude of the inhibition was 22%, 29%, and 26%, at 1.5, 4.75 and 15 mg/kg bw per day, respectively. Erythrocyte acetylcholinesterase activity was only significantly reduced (27%) at 15 mg/kg bw per day.

No treatment-related effects were observed in gravid uterine weight, number of live fetuses, number of resorptions, placental weight or crown–rump length. However, the number of fetuses with a bodyweight of less than 30 g was significantly increased at 15 mg/kg bw per day. Skeletal examination revealed a statistically-significant increase in the incidence of reduced ossification of long bone epiphyses in fetuses at 4.75 and 15 mg/kg bw per day. The incidence of reduced ossification (42.4% and 44.3%, respectively) was also outside the range of mean values for historical controls (10.7–35.9%). The NOAEL for developmental effects was 1.5 mg/kg bw per day on the basis of retarded ossification of the long bones at 4.75 and 15 mg/kg bw per day and reduced fetal body weight at 15 mg/kg bw per day. The NOAEL for maternal toxicity was 4.75 mg/kg bw per day on the basis of significant inhibition of erythrocyte acetylcholinesterase activity at 15 mg/kg bw per day (Gal et al., 1988).

2.5 *Special studies: neurotoxicity*

Hens

In a published report of experiments designed to investigate the potential relationship between delayed neurotoxicity and copper concentration in the serum of hens, it was reported that azinphos-methyl failed to produce symptoms of neurotoxicity after either single or repeated doses (Kimmerle & Loser, 1974; Annex 1, reference *64*).

In a test for acute delayed neurotoxicity, a group of 30 white Leghorn hens was given two doses of azinphos-methyl (purity, 85%) at a dose of 330 mg/kg bw, with an interval of 21 days between doses. Hens in the untreated-control, vehicle-control and positive-control (triorthocresyl phosphate, TOCP at 600 mg/kg bw) groups, each comprising 10 animals, were also included. To protect against sudden death, atropine (15 mg/kg bw) was administered intramuscularly 15 min before dosing hens with TOCP or azinphos-methyl. A total of 11 hens treated with azinphos-methyl survived until sacrifice. These animals appeared normal during the last 12 or 13 days of the study, but exhibited varying degrees of impaired locomotor activity soon after dosing. Histopathological examinations indicated that azinphos-methyl did not increase the incidence or severity of lesions in the nerve tissue compared with untreated and vehicle controls. Investigations of neuropathy target esterase activity were not included in the study (Glaza, 1988; Annex 1, reference *64*, modified with reference to the original data).

Groups of eight HNL hens (age 18–20 months) were given diets containing azinphos-methyl (purity not stated; prepared as an 80% premix in Silkasil S) at a dose of 0, 75, 150, 300 or 600 ppm (equivalent to approximately 0, 9.4, 18.7, 37.5 and 75.0 mg/kg bw per day) for 30 days. This was followed by a post-treatment recovery period of 4 weeks during which the birds were examined for signs of neurotoxicity (leg weakness, limping and inability to walk). Blood (site or method of blood collection was not specified) cholinesterase activity was determined before the commencement of the study, after completion of the feeding trial, and at post-treatment observation periods using an unspecified method. Two birds per group were killed at the conclusion of dosing and the remaining birds were killed 1 day after the completion of the post-treatment observation period. One bird at 9.4 mg/kg bw day died during the post-treatment observation period as a result of suspected pneumonia, but no details of necropsy were provided. The group mean body weight of birds at 75 mg/kg bw per day was reduced to about 13% of that of the controls by day 14 on study and remained unchanged up until the conclusion of the treatment period. The hens, however, regained weight thereafter and body weights were comparable to those of the controls during the post-treatment observation period. Food consumption was unaffected by treatment. No signs of neurotoxicity or any behavioural changes were noted in any of the treated birds. No intergroup differences in blood cholinesterase activity was seen before treatment or at day 1 of the post-observation period. However, at day 1 after treatment, dose-related depressions in blood cholinesterase activity of about 15% and 27% were observed in birds at 37.5 and 75.0 mg/kg bw per day respectively relative to the controls (Kimmerle, 1964).

In a follow up study with essentially the same experimental protocol as described in an earlier study (Kimmerle, 1964), the administered dietary concentrations of azinphos-methyl (purity unspecified) were increased to 900, 1200, 1500, or 1800 ppm (equivalent to approximately 112.5, 150, 187 and 225 mg/kg bw per day respectively). One bird at 150 mg/kg bw per day died during the last week of the 4-week treatment period. The cause of the death was not specified and necropsy details were not provided. Food consumption in treatment groups was reduced by about 27%, 43%, 40% and 44% (at 112.5, 150.0, 187.0 and 225.0 mg/kg bw per day, respectively) relative to the controls during the treatment period. All birds including the controls lost body weight during the treatment period. Hens in the control group and hens at 112.5 and 150.0 mg/kg bw per day were about 15%, 23% and 21% respectively lighter relative to the corresponding pre-treatment body weights at the conclusion of the 4-week treatment period. The groups receiving the test substance at 187.0 and 225.0 mg/kg bw per day showed a body-weight loss of about 27% at the same observation time. The body-weight data appeared to be consistent with the observations on food consumption. The birds had not achieved their pre-treatment body weights by the completion of the recovery period, though there was some recovery of body weights in all groups. Blood cholinesterase activity was unaffected by treatment. No signs of neurotoxicity or any behavioural changes related to treatment were noted. Similarly, no histopathological findings were made (Grundman, 1965; Kimmerle, 1965).

Rats

In a study to investigate acute neurotoxicity, groups of 18 male and 18 female fasted Fischer 344 CDF/BR rats were given azinphos-methyl (purity, 92.8%) at a dose of 0 (vehicle), 2, 6 or 12 mg/kg bw for males, and 0, 1, 3 or 6 mg/kg bw for females by oral gavage in 0.5% (w/v) methylcellulose and 0.4% (w/v) Tween 80. Of the treated rats, six males and six females in satellite groups were used for determinations of cholinesterase activity and the remainder were involved in testing for changes in neurobehaviour.

Five males (one in the main study and four in the satellite group) and fifteen females (nine in the main study and all in the satellite group) in the group at the highest dose died. Treatment-related clinical signs in males were muscle fasciculations (2 out of 12 and 8 out of 11 at 6 and 12 mg/kg bw respectively), tremors (1 out of 11 at 12 mg/kg bw), uncoordinated gait (1 out of 11 at 12 mg/kg bw), oral stain (3 out of 12 and 10 out of 11 at 6 and 12 mg/kg bw, respectively), and urine stain (1 out of 12, 2 out of 12 and 10 out of 11 at 2, 6 and 12 mg/kg bw, respectively) and red nasal stain (2 out of 12, 2 out of 12 and 7 out of 11 at 2, 6 and 12 mg/kg bw, respectively). One of the three surviving females at 6 mg/kg bw had oral and urine staining for about 2 days after treatment. All clinical signs that did not involve staining were only observed about 30 min after treatment on the day of dosing. Staining was typically observed from between 1 to 3 days after dosing. There were no treatment related differences in group mean body weights of the surviving animals.

Treatment-related functional observational battery (FOB) findings were recorded at the time of maximal inhibition of cholinesterase activity (i.e. about 90 min after dosing) at doses of 1 mg/kg bw or greater on day 0, but not at any dose on day 7 or 14 after treatment. A summary of FOB findings is presented in Table 14. In addition, males at 6 and 12 mg/kg bw and females at 6 mg/kg bw had significantly lower ($p \leq 0.05$) body temperature (by about 3.4% and 3.9%, and 2.8% for males and females, respectively) after treatment on day 0, and appeared to be treatment-related. Males and females at the highest dose had significantly reduced forelimb grip performance (by about 31% and 27%, respectively). Similarly, significant reductions ($p \leq 0.05$) in hindlimb grip performance were noted only in males at 6 and 12 mg/kg bw (reduced by 18% and 30%, respectively). These manifestations were attributed to acute cholinergic effects of the test substance. Foot splay was unaffected by treatment.

In males, the response profile of remaining reflex/physiological observations such as touch (5 out of 12 and 6 out of 12), righting (slightly uncoordinated, 6 out of 12 and 3 out of 12) and righting (land on side, 2 out of 12 and 5 out of 12) for the groups at 6 and 12 mg/kg bw, respectively, was significantly different ($p \leq 0.05$) from the controls and appeared to be treatment-related, although a dose–response relationship was not always observed.

Maze activity in treated rats was lower on day 0 (by about 64%, 60%, 77% and 79% for males at 0, 2, 6 and 12 mg/kg and 71%, 78%, 79% and 85% for females at 0, 1, 3 and 6 mg/kg bw, respectively) compared with the corresponding values reported before treatment and appeared to be treatment-related. In comparison to the concurrent controls, dose-related reductions in motor activity were seen in males at 6 and 12 mg/kg bw (about 36% and 43%, respectively), and in females (about 19% for the groups at 1 and 3 mg/kg bw and about 44% at 6 mg/kg bw). The

Table 14. Functional observational battery findings after dosing (day 0) in rats given azinphos-methyl by gavage

Parameter	Males			Females		
	Dose (mg/kg bw)					
	2	6	12	1	3	6
Number of animals tested	12	12	11	12	12	3
Autonomic effects						
Lacrimation	0	3	1	0	0	1*
Salivation	0	4	4	0	0	1*
Neuromuscular effects						
Incoordination (home cage)	0	1	3	0	0	1*
Incoordination (open field)	0	6*	7*	0	0	1*
Repetitive chewing (home cage)	0	3	7*	0	0	1*
Repetitive chewing (open field)	0	8*	10*	0	0	1*
Posture	0	3*	6*	0	1	1*
Minimal movement	1	3	6	0	0	1
Central nervous system effects						
Muscle fasciculations (home cage)	0	8*	12*	0	0	1*
Muscle fasciculations (open field)	0	12*	12*	0	0	1*
Tremors (home cage)	0	3	9*	0	0	1*
Tremors (open field)	0	6*	9*	0	0	1*

From Sheets & Hamilton (1994)
* Significantly different from corresponding controls ($p \leq 0.05$).

reduced motor activity in males and females of the control group noted on day 0 was attributed to overnight fasting. In males, the motor activity was depressed by about 14% at 12 mg/kg bw relative to that of the controls, with partial recovery on day 7, and complete recovery of activity was seen by day 14. In females, motor activity appeared to be slower to recover, showing deficits of about 20%, 10% and 8% at 1, 3 and 6 mg/kg bw, respectively, relative to values for the corresponding controls on day 14.

Dose-related changes in locomotor activity were also noted in males on day 0, with about 47% and 77% depressions at 6 and 12 mg/kg bw respectively, relative to values for the corresponding controls. In females, the reductions in locomotor activity were about 26%, 15% and 63% at 1, 3 and 6 mg/kg bw respectively, in comparison to those for the parallel controls. Complete recovery of locomotor activity was seen in all males and females, except for males at 12 mg/kg bw that showed a deficit of 20% compared with the controls by day 7. Complete recovery was seen in all animals by day 14. Treatment-related reductions were also seen in interval motor and locomotor activities on day 0 after treatment, but recovery was complete by day 7. Habituation was not affected by treatment. Data on cholinesterase activities are shown in Table 15.

No intergroup differences were noted in terminal body weights, brain weight, gross pathology and histopathology of treated rats of either sex. The NOAEL was 2 mg/kg bw per day on the basis of a significant reduction in brain cholinesterase activity at higher doses (Sheets & Hamilton, 1994).

In a short-term study of neurotoxicity, groups of 18 male and 18 female Fischer 344 rats were given diets containing azinphos-methyl (purity, 92.2%) at a nominal dose of 0, 15, 45, or 120 ppm for males (mean intake, 0.91, 2.81, or 7.87 mg/kg bw per day) and 0, 15, 45, or 90 ppm for females (mean intake, 1.05, 3.23, or 6.99 mg/kg bw per day) for 13 weeks. Twelve of the rats in each group were used for behavioural testing with half of these being used for neuropathology. The remaining six males and six females per group were tested for cholinesterase activity. There were no deaths before terminal sacrifice. Treatment-related clinical signs (perianal stain, red lacrimation, increased reactivity, uncoordinated gait, tremor) were evident by cage-side observations

Table 15. Cholinesterase activity (relative to the corresponding controls) in rats given azinphos-methyl by gavage

Dose (mg/kg bw)	Cholinesterase activity (% relative to corresponding controls)		
	Plasma[a]	Erythrocytes[a]	Brain[a]
Males			
2	32*	33*	15
6	57*	67*	74*
12[b]	50*	63*	88*
Females			
1	11	17	5
3	36*	65*	51*
6[c]	—	—	—

From Sheets & Hamilton (1994)
[a] Six males and six females.
[b] Four animals died after treatment on day 0.
[c] All six animals in the group died after treatment on day 0.
[*] Significantly different from the corresponding controls ($p \leq 0.05$)

in males at 120 ppm and in females at 45 and 90 ppm. Body weight (reduced by 9–10%) and food consumption were affected by treatment at the highest dose for males (120 ppm) and females (90 ppm). The results from FOB revealed treatment-related effects in males at the highest dose (perianal stain), in females at the intermediate dose (urine stain) and in females at the highest dose (urine and perianal stains, increased reactivity, abnormal righting reflex, tremor, and decreased forelimb-grip strength).

Signs of toxicity persisted with continued exposure, but there was no evidence of cumulative toxicity beyond week 4. Treatment-related decreases in motor and locomotor activity were observed in males at the highest dose during weeks 4, 8, and 13 of exposure, with no evidence of cumulative toxicity beyond week 4, and in females at the highest dose during week 4 only. Greater than 20% inhibition of erythrocyte and plasma cholinesterase activity was observed at all doses, as was the inhibition of brain cholinesterase activity at the intermediate and highest doses (Table 16).

Gross lesions were not evident at terminal sacrifice. Brain weight was not affected by treatment in either sex. There were no treatment-related ophthalmic findings or microscopic lesions within neural tissues or skeletal muscle. The NOAEL for inhibition of brain cholinesterase activity was 15 ppm (equal to 1 mg/kg bw per day), while a NOAEL for inhibition of acetylcholinesterase activity in erythrocytes could not be identified in this study. The statistically significant inhibition of brain cholinesterase activity at 15 ppm in both sexes was discounted because it was less than 20% and other studies of a similar duration in which more rats per group were tested, inhibition of cholinesterase activity was not shown to be statistically significant (Sheets & Hamilton, 1995; Sheets et al., 1997).

3. Observations in humans

A randomized double-blind study with ascending single oral doses was reported in which a formal statement was provided to indicate that it complied with the principles of good clinical practice (CPMP/ICH/135/95) and the OECD principles of GLP. The study was conducted in accordance with the guidelines established in the Declaration of Helsinki 1964, as amended by the 29th World Medical Assembly in Tokyo 1975, the 35th World Medical Assembly in Venice 1983, the 41st World Medical Assembly in Hong Kong 1989 and the 48th General Assembly, Somerset West, Republic of South Africa October 1996. The study sought to determine the safety of azinphos-methyl (purity, 89.2%) and to establish a NOAEL for inhibition of cholinesterase activity in humans and compare

Table 16. Cholinesterase activity in ratsa given diets containing azinphos-methyl

Tissue	Cholinesterase activity (% inhibition relative to values for controls)		
	Dietary concentration (ppm)		
	15/15 (males/females)	45/45 (males/females)	120/90 (males/females)
Week 4			
Plasma	7/14*	42*/59*	75*/83*
Erythrocyte	37*/41*	88*/88*	98*/91*
Week 13			
Plasma	15*/13	44*/60*	69*/81*
Erythrocyte	37*/38*	84*/78*	95*/95*
Brain	8*/16*	46*/72*	82*/85*

From Sheets & Hamilton (1995) and Sheets et al. (1997)
[a] n = 6.
* $p < 0.05$

sensitivity with that of laboratory animal species. In this placebo-controlled study in healthy volunteers, 40 males and 10 females were given capsules containing azinphos-methyl at a dose of 0.25, 0.5, 0.75, or 1 mg/kg bw for males and 0.25, 0.5, or 0.75 mg/kg bw for females, according to the dosing schedule described in Table 17. Selection of doses was based on a study of neurotoxicity in rats given repeated doses, in which a NOAEL of 1 mg/kg bw was observed (Sheets & Hamilton, 1995), and a study in humans given repeated doses, in which the NOAEL was 0.29 mg/kg bw (Rider et al., 1971). Azinphos methyl was administered in capsules.

The mean age and weight of the male subjects was 32.7 ± 9.3 years and 75.52 ± 8.36 kg respectively, and the mean female age and weight were 31.0 ± 4.6 years and 63.83 ± 6.71 kg respectively. All subjects were selected from a panel of volunteers who were screened less than 3 weeks before study initiation. For inclusion in the study, all volunteers met the following criteria: age 18–50 years; with no clinically relevant physiological findings; no clinically relevant serum or blood findings (including cholinesterase activity); normal electrocardiogram (ECG); normal arterial pressure and heart rate; body weight between 50 and 100 kg; and within ± 15% of ideal body weight; able to communicate well with the investigator and to comply with the requirements of the entire study; and provision of written informed consent to participate. The study protocol was reviewed and approved by an independent research ethical committee.

Before the start of the study, a screening examination was performed which consisted of the following: medical history; complete physical examination and vital signs (pulse rate, respiratory rate and blood pressure); ECG recording; haematology, clinical chemistry, plasma and erythrocyte cholinesterase activity and urine analysis; hepatitis B, C and human immunodeficiency virus (HIV) status; drug screening; pregnancy status (females). Subjects could be withdrawn from the study in any of the following events: serious adverse effects; major violations to the protocol; withdrawal of consent; termination of the study by the sponsor. The study showed that azinphos-methyl was well tolerated at all doses. No clinical relevant reductions in plasma or erythrocyte cholinesterase values occurred. The NOAEL for azinphos-methyl when administered as a single oral dose was 1.0 mg/kg bw for males and 0.75 mg/kg bw for females (McFarlane & Freestone, 1999a).

Two male subjects each received 16 mg of azinphos-methyl orally each day for 30 days (body weight not stated, but assuming a 70 kg bw the administered dose would be 0.23 mg/kg bw per day). The daily urinary output of azinphos-methyl-related compounds, estimated by a method that converted them to anthranilic acid, was determined before, during and after cessation of treatment. The amount of anthranilic acid excreted in the urine was increased on the day after the start of treatment and remained at a high level during treatment with azinphos-methyl; it returned to a normal level

Table 17. Dosing schedule for a study in humans given capsules containing azinphos-methyl

Group	No. of subjects				
	Dose (mg/kg bw)				
	0	0.25	0.5	0.75	1.0
Session 1; males	1	1	0	0	0
Session 2; males	2	6	1	0	0
Session 3; males	3	0	6	1	0
Session 4; males	3	0	0	6	0
Session 5; females	3	0	0	7	0
Session 6; males	3	0	0	0	7

From McFarlane & Freestone (1999a)

on the day after cessation of treatment. Neither subject showed inhibition of blood cholinesterase activity during treatment (Thornton, 1971; Annex 1, reference *20*).

Groups of five male volunteers received 10, 12, 14 or 16 mg of azinphos-methyl by mouth each day for 30 days. Plasma and erythrocyte cholinesterase activity was measured before and during treatment. No significant inhibition of enzyme activity was found (Rider et al., 1971; Annex 1, reference *20*).

Additional groups of five male volunteers were given capsules containing azinphos-methyl (corn oil vehicle) at a dose of 18 or 20 mg/day (equivalent to 0.26 and 0.29 mg/kg bw per day respectively for a person of 70 kg bw) for 30 days. Cholinesterase activity was measured twice per week during the exposure period. There were no clinical signs or effects on clinical signs, hematology, prothrombin time, and urine analysis. No inhibition of plasma or erythrocyte cholinesterase activity was observed at doses of up to 20 mg/day. The NOAEL was considered to be 20 mg/day (0.29 mg/kg bw per day) on the basis of the absence of any effects on plasma and erythrocyte cholinesterase activity (Rider et al., 1972; Annex 1, reference *20*).

McFarlane & Freestone (1999b) conducted another study in humans in which eight men (age, 29.3 ± 9.6 years; body weight, 69.3 ± 5.62 kg) received a gelatin capsule containing azinphos-methyl (purity, 89.2%) at a dose of 0.25 mg/kg bw per day for 28 days. A formal statement was provided which indicated that the study complied with the principles of good clinical practice and GLP. The study was conducted in accordance with the guidelines established in the Declaration of Helsinki 1964, as amended by the 29th World Medical Assembly in Tokyo 1975, the 35th World Medical Assembly in Venice 1983, the 41st World Medical Assembly in Hong Kong 1989 and the 48th General Assembly, Somerset West, Republic of South Africa October, 1996.

Two healthy males (age, 35.3 ± 9.5 years; body-weight, 77.7 ± 12.51 kg) were given a placebo. The objective of the study was to establish a NOAEL for inhibition of erythrocyte acetylcholinesterase activity. The men resided in the clinic during the entire study under constant medical supervision and received a standardized diet. Volunteers had their blood pressure and heart rate monitored daily. ECGs and samples of blood and urine were obtained before dosing on days 1, 7, 14, 21 and 28. All observed or reported adverse events were recorded including duration and severity. An assessment of underlying cause and treatment were recorded. Baseline values for plasma and erythrocyte cholinesterase activity from eight time-points (days −14, −12, −10, −8, −6, −4, −2, and −1) were averaged for each individual to estimate the percentage change from baseline. There were no clinically significant changes in vital signs, ECG, haematology, clinical chemistry, urine analysis or physical examination observed for any subject receiving azinphos-methyl. Similarly, there was no treatment-related effect on plasma or erythrocyte cholinesterase activity. On the basis of the absence of any effects, the NOAEL was 0.25 mg/kg bw per day; the highest dose tested.

Medical surveillance of manufacturing plant personnel

Employees working in the formulation of products containing azinphos-methyl have been subjected to regular medical examinations and no general impairment of health has been observed. In one isolated case it was considered probable that contact with azinphos-methyl was the cause of generalized dermatosis resulting in sensitive dry skin (Faul, 1981; Annex 1, reference *64*).

A published report presented the findings of a scientific officer of the Division of Occupational Health, New South Wales Department of Public Health, following a visit to a guthion (azinphos-methyl) and azinphos-ethyl formulation plant. It was stated that the manufacturing plant attempted to formulate these chemicals under "primitive conditions". Further, according to the study author, safety measures to prevent worker exposure "did not exist" in the plant. Whole-blood cholinesterase activity for 15 operators was assayed randomly over a period of 4 months. The concentration of azinphos-methyl in the operator's breathing zone ranged from 0.5 to 1.0 mg/m^3. This concentration was found to be well above the tentative limit of 0.3 mg/m^3, set by the Division of Occupational Health of New South Wales Department of Public Health at that time. The concentration of azinphos-methyl found on drum tops was about 2.3 mg/cm^2. During the processing season (unspecified), 13 operators (details not provided) were removed from contact as their cholinesterase activity fell to below 60 units or 60% of the pre-exposure level. Two operators showed signs of organophosphate poisoning (onset, types and duration unspecified), and one of them had to be hospitalized. The cholinesterase levels of these two persons were 25 and 30 units respectively, and that of the remaining 13 persons ranged from 32 to 110 units. A line graph presented showed the recovery of whole-blood cholinesterase activity to about 80–90 units in about 15 days after completely removing these workers from the formulating plant. However, a comparable recovery of the enzyme activity in workers who remained in the plant doing other jobs was noticed only after about 35 days (Simpson, 1965).

Published reports from the pesticide incident monitoring system in the United States of America (USA) and additional data from the state of California in the USA have been reviewed. Between 1982 and 1988, a small number of incidents were reported annually (involving 5–12 persons each year) that have been definitely, probably or possibly associated with exposure to azinphos-methyl either alone or in combination with other pesticides. In addition, two incidents occurred in 1987, one involving 26 people, the other involving 32 people. The first involved spray drift in adverse weather conditions. The second involved workers who experienced symptoms including headache, nausea, weakness and vomiting upon entry to a field to pick peaches 3 days after methomyl was applied to the crop and about 6 weeks after an application of azinphos-methyl (United States Environment Protection Agency, 1981; Mahler, 1991; Annex 1, reference *64*).

Comments

Biochemical aspects

After oral administration of radiolabelled azinphos-methyl, the radiolabel was rapidly and completely (90–100% of the administered dose) absorbed from the mammalian intestinal tract, widely distributed throughout the organs, and eliminated in the urine (60–70% of the administered dose) and faeces via bile (25–35% of the administered dose) within 48 h. The maximum concentration in blood was reached within 2–3 h after administration. Azinphos-methyl was rapidly cleared from the blood and tissues, and consequently there is negligible potential for accumulation. In rats, the initial steps of metabolism involved the formation of the highly reactive oxon metabolite and mercaptomethylbenzazimide by cytochrome P450. Glutathionyl methylbenzazimide and desmethyl isoazinphos-methyl were formed via glutathione transferase. Subsequent hydrolysis of glutathionyl methyl-benzazimide resulted in the formation of cysteinyl-methylbenzazimide, which was then oxidized to form its corresponding sulfoxide and sulfone.

Toxicological data

Azinphos-methyl was highly acutely toxic (LD_{50} range, 4.4–26 mg/kg bw) when administered orally in an aqueous or non-aqueous vehicle to rats, and its profile of clinical signs was similar to those of other cholinesterase-inhibiting organophosphorus pesticides. Clinical signs observed in experimental animals after acute exposure were salivation, lacrimation, vomiting, diarrhoea, anorexia, reduced locomotor activity, piloerection, staggering gait and muscular tremors. These signs were generally evident within 5–20 min after dosing. There was very little difference in the sensitivity of male and female rats to the acute effects of azinphos-methyl.

The main toxicological findings in repeat-dose studies in rodents and dogs were inhibition of cholinesterase activity and, at higher doses, reduced body-weight gain and signs of neurotoxicity. In short-term studies of toxicity of less than 12 months duration, the NOAEL for inhibition of erythrocyte acetylcholinesterase activity was 0.2 mg/kg bw per day in rats and dogs. The NOAEL for inhibition of brain cholinesterase activity was 0.9 mg/kg bw per day in rats, and 0.7 mg/kg bw per day in dogs. Toxicity observed in rats and dogs was limited to the characteristic muscarinic signs (diarrhoea, salivation) and reduced body-weight gain. The effect doses for these clinical signs in short-term studies correlated with the high levels of inhibition of brain cholinesterase activity (> 80%) in rats and dogs.

Azinphos-methyl was tested in an adequate range of studies of genotoxicity in vitro and in vivo and showed no evidence of genotoxicity. The Meeting concluded that azinphos-methyl is unlikely to be genotoxic.

In long-term studies of toxicity, inhibition of cholinesterase activity was again the main toxicological finding in mice and rats. In mice, erythrocyte and brain cholinesterase activities were inhibited at 3.8 mg/kg bw per day, with a NOAEL of 0.9 mg/kg bw per day. Reduced body-weight gain and clinical signs involving hyperactivity and convulsions were observed in mice at higher doses (6.25 mg/kg bw per day). At equivalent doses in rats, body tremors and deaths were reported, although reduced body-weight gain was observed at 2.7 mg/kg bw per day. In rats, the NOAEL for inhibition of erythrocyte acetylcholinesterase activity was 0.3 mg/kg bw per day, while for brain cholinesterase activity it was 0.9 mg/kg bw per day and the NOAEL for a reduction in body-weight gain was 0.9 mg/kg bw per day. There was no evidence of carcinogenicity with azinphos-methyl at dietary concentrations of up to 40 ppm (equal to 12.8 mg/kg bw per day) in mice and up to 45 ppm (equal to 2.7 mg/kg bw per day) in rats; these were the highest doses tested.

In the absence of any carcinogenic potential in rodents and the lack of genotoxic potential in vitro and in vivo, the Meeting concluded that azinphos-methyl is unlikely to pose a carcinogenic risk to humans.

In multigeneration studies of reproductive toxicity in rats, the treatment-related effects of azinphos-methyl were cholinergic signs at high doses, reductions in body-weight gain and inhibition of cholinesterase activity. These effects were consistent with those seen in short- and long-term studies of toxicity. However, there was also evidence of reduced pup viability at 4.8 mg/kg bw per day. The NOAEL for inhibition of brain cholinesterase activity in dams was 5 ppm, equal to 0.5 mg/kg bw per day. The NOAEL for inhibition of brain cholinesterase activity in pups was 15 ppm, equal to 1.5 mg/kg bw per day.

In studies of developmental toxicity with azinphos-methyl in mice, rats and rabbits, teratogenicity was not observed at doses of up to 2, 3.6 and 15 mg/kg bw per day respectively. The only developmental effect noted in any of these studies was delayed ossification in rat and rabbit fetuses at doses that also caused inhibition of brain and erythrocyte cholinesterase activity in the dams. The NOAEL for developmental effects in fetuses was 2 mg/kg bw per day in rats and 1.5 mg/kg bw per day in rabbits. Inhibition of brain cholinesterase activity was not observed in rats at doses of 1 mg/kg bw per day or in rabbits at doses of 2.5 mg/kg bw per day.

In studies of delayed neurotoxicity, azinphos-methyl was administered to chickens either as a single dose at up to 330 mg/kg bw or as repeated doses of up to 225 mg/kg bw per day in the feed for 30 days; there was no evidence of delayed neuropathy.

In rats given azinphos-methyl as a single dose at up to 12 mg/kg bw by gavage or as repeated doses of up to 7.4 mg/kg bw per day in the diet for 13 weeks, cholinergic signs and significant inhibition of erythrocyte and brain cholinesterase activity were seen at a number of doses. In these studies, which included a FOB, clinical signs of intoxication (perianal stain, red lacrimation, increased reactivity, uncoordinated gait, tremor) were observed. However, cholinergic signs were observed only when brain cholinesterase activity was inhibited by more than 70% or when erythrocyte acetylcholinesterase activity was inhibited by approximately 65–80%. This occurred after repeated doses in excess of 3.2 mg/kg bw per day or after a single dose of 6 mg/kg bw. The NOAEL for inhibition of cholinesterase activity in the brain after a single dose was 2 mg/kg bw or 1 mg/kg bw per day after repeat dosing.

In a randomized double-blind study in human volunteers (seven of each sex) given ascending single oral doses, azinphos-methyl did not induce cholinergic signs or changes in acetylcholinesterase activity in erythrocytes at doses of up to 1 mg/kg bw in males and up to 0.75 mg/kg bw in females; these were the highest doses tested.

When eight male volunteers were given azinphos-methyl as a daily oral dose at 0.25 mg/kg bw per day for 28 days, there were no cholinergic signs and erythrocyte acetylcholinesterase activity was not significantly inhibited. In another study, two groups of five male volunteers were given azinphos-methyl at a dose of 0.26 or 0.29 mg/kg bw per day for 30 days did not induce cholinergic signs or changes in cholinesterase activity in erythrocytes or plasma. In a third study, a similar outcome was reported when two male volunteers were given azinphos-methyl orally at a dose of 0.23 mg/kg bw per day for 30 days.

Regular medical examinations of workers involved in formulating products containing azinphos-methyl had revealed no effects, except for one case of possible dermatosis resulting in sensitive dry skin.

The Meeting concluded that the existing database on azinphos-methyl was adequate to characterize the potential hazards to fetuses, infants, and children.

Toxicological evaluation

The Meeting established an ADI of 0–0.03 mg/kg bw per day based on a NOAEL of 0.29 mg/kg bw per day for the absence of inhibition of erythrocyte acetylcholinesterase activity in a 30-day study of toxicity in male volunteers and a safety factor of 10. Since the database indicated that rodents and dogs of each sex and humans had similar NOAEL values for the most sensitive end-point, namely inhibition of acetylcholinesterase activity in erythrocytes, the NOAELs identified in the studies in humans were considered to be protective for the entire population. The Meeting also considered the ADI to be protective for other, non-neurotoxic effects of azinphos-methyl observed in short- and long-term studies with repeated doses, and in studies of reproductive and developmental toxicity, where the use of a safety factor of 10 would be appropriate. The effect of azinphos-methyl on body-weight gain and fertility in dams at 15 ppm (0.5 mg/kg bw per day) in multigeneration studies of reproduction in rats was reconsidered. The Meeting concluded that it was a marginal effect that could not be directly attributed to treatment.

The Meeting established an ARfD of 0.1 mg/kg bw based on the NOAEL of 1 mg/kg bw and using a safety factor of 10. The NOAEL observed in a study of single doses in volunteers was the highest tested dose in males. Although the maximum dose given to females was only 0.75 mg/kg bw, there was no apparent observed difference in sensitivity between the sexes, so the NOAEL observed in males was also considered to be protective of effects in females. In a study of acute neurotoxicity in rats, the NOAEL was 2 mg/kg bw on the basis of inhibition of cholinesterase activity in the brain. At a dose of 2 mg/kg bw, significant inhibition of acetylcholinesterase activity in erythrocytes of male rats was observed, but not at 1 mg/kg bw in female rats. In rats, pup deaths as a result of inhibition of cholinesterase activity were observed at 6 mg/kg bw in females and at 12 mg/kg bw in males, suggesting a steep dose–response effect. Based on the median oral LD_{50} value of 13 mg/kg bw (range, 4.4–26 mg/kg bw) in all available studies in rats, there would be about a 130-fold margin between the ARfD and the LD_{50} in rats.

Levels relevant for risk assessment

Species	Study	Effect	NOAEL	LOAEL
Rat	Acute neurotoxicity[a]	Inhibition of brain cholinesterase activity	2.0 mg/kg bw	3.0 mg/kg bw
Human	Single dose[b,d]	No adverse effects	1.0 mg/kg bw per day	—
Mouse	Two-year study of toxicity and carcinogenicity[c]	Inhibition of erythrocyte and brain cholinesterase activity	5 ppm, equal to 0.9 mg/kg bw per day	20 ppm, equal to 3.8 mg/kg bw per day
		Carcinogenicity[d]	40 ppm, equal to 12.8 mg/kg bw per day	—
Rat	Three-month study of toxicity[c]	Inhibition of brain cholinesterase activity	0.9 mg/kg bw per day	3.4 mg/kg bw per day
	Two-year study of toxicity and carcinogenicity[c]	Reduced body-weight gain and inhibition of brain cholinesterase activity	15 ppm, equal to 0.9 mg/kg bw per day	45 ppm, equal to 2.7 mg/kg bw per day
		Carcinogenicity[d]	45 ppm, equal to 2.7 mg/kg bw per day	—
	Multigeneration study of reproductive toxicity[c,e]	Parental toxicity	5 ppm, equal to 0.5 mg/kg bw per day	15 ppm, equal to 1.5 mg/kg bw per day
		Offspring toxicity	15 ppm, equal to 1.5 mg/kg bw per day	45 ppm equal to 4.8 mg/kg bw per day
	Developmental toxicity[a,e]	Maternal toxicity	1.0 mg/kg bw per day	2.0 mg/kg bw per day
		Embryo/fetotoxicity	2.0 mg/kg bw per day	3.6 mg/kg bw per day
	Three-month study of neurotoxicity[c]	Inhibition of brain cholinesterase activity	15 ppm, equal to 1 mg/kg bw per day	45 ppm, equal to 3 mg/kg bw per day
Rabbit	Developmental toxicity[b,c]	Maternal toxicity	2.5 mg/kg bw per day	6.0 mg/kg bw per day
		Embryo/fetotoxicity	1.5 mg/kg bw per day	4.75 mg/kg bw per day
Dog	One-year study of toxicity[a]	Reduced body-weight gain and inhibition of brain cholinesterase activity	25 ppm, equal to 0.7 mg/kg bw per day	125 ppm, equal to 4.1 mg/kg bw per day
Human	Clinical 30-day study[b,d,e]	No adverse effects	0.29 mg/kg bw per day	—

[a] Gavage administration.
[b] Capsule administration.
[c] Dietary administration.
[d] Highest tested dose.
[e] Two or more studies combined.

Estimate of acceptable daily intake for humans

0–0.03 mg/kg bw

Estimate of acute reference dose

0.1 mg/kg bw

Information that would be useful for the continued evaluation of the compound

Results from epidemiological, occupational health and other such observational studies of human exposures

Critical end-points for setting guidance values of exposure for azinphos-methyl

Absorption, distribution, excretion and metabolism in mammals	
Rate and extent of oral absorption	Almost complete absorption. Maximum plasma concentration 2–3 h after dosing.
Dermal absorption	See previous azinphos-methyl monographs
Distribution	Extensive
Potential for accumulation	Low, no evidence of accumulation
Rate and extent of excretion	Largely complete within 48 h; approximately 95% excreted in urine and bile.
Metabolism in animals	Extensive; two major urinary metabolites and six other products. Five faecal metabolites (10–12% of the administered dose).
Toxicologically significant compounds in animals, plants and the environment	Azinphos methyl,azinphos-methyl oxon
Acute toxicity	
Rat, LD_{50}, oral	4.4–26 mg/kg bw
Rat, LD_{50}, dermal	See previous azinphos-methyl monographs
Rat, LC_{50}, inhalation	See previous azinphos-methyl monographs
Skin sensitization (test method used)	See previous azinphos-methyl monographs
Short-term studies of toxicity	
Target/critical effect	Inhibition of brain cholinesterase activity
Lowest relevant oral NOAEL	0.7 mg/kg bw per day (dog)
Lowest relevant dermal NOAEL	See previous azinphos-methyl monographs
Lowest relevant inhalation NOAEC	See previous azinphos-methyl monographs
Genotoxicity	
	Unlikely to pose a genotoxic risk in vivo
Long-term studies of toxicity and carcinogenicity	
Target/critical effect	Inhibition of brain cholinesterase activity
Lowest relevant NOAEL	0.9 mg/kg bw per day (rat)
Carcinogenicity	Not carcinogenic in rats and mice
Reproductive toxicity	
Reproduction target/critical effect	Inhibition of brain cholinesterase activity in dams
Lowest relevant reproductive NOAEL	0.5 mg/kg bw per day (rats)
Developmental target/critical effect	Delayed ossification at maternally toxic doses (rats, rabbits)
Lowest relevant developmental NOAEL	1.0 mg/kg bw per day (rats); 2.5 mg/kg bw per day (rabbits)
Neurotoxicity/delayed neurotoxicity	No evidence of delayed neuropathy observed in hens
NOAEL: 1 mg/kg bw in a repeat-dose study of neurotoxicity in rats	
Medical data	
	Medical examinations of workers involved in formulating azinphos-methyl products revealed no effects, except for one case of possible dermatosis resulting in sensitive dry skin.

Summary

	Value	Study	Safety factor
ADI	0–0.03 mg/kg bw per day	Humans, 30-day oral-dosing study	10
ARfD	0.1 mg/kg bw	Humans, study of acute toxicity	10

References

Alam, M.T., Corbeil, M., Chagnon, A., & Kasatiya, S.S. (1974). Chromosomal anomalies induced by the organic phosphate pesticide guthion in Chinese hamster cells. *Chromosoma (Berl.)*, **49**, 77–86.

Allen, T.R., Frei, T., Janiak, T., Luetkemeier, H., Vogel, O., Biedermann, K., & Wilson, J. (1990) 52-week oral toxicity (feeding) study with azinphos-methyl in the dog. Unpublished report from Research and Consulting Company AG. Submitted to WHO by Bayer AG, Leverkusen, Germany.

Arnold, D. (1971) Mutagenic study with Guthion in albino mice. Unpublished report No. E8921 from Industrial Bio-Test Laboratories Inc., Northbrook, IL, USA. Submitted to WHO by Chemagro Corp., Kansas City, MO, USA.

Astroff, A.B. &Young, A.D. (1998) The relationship between maternal and fetal effects following maternal organophosphate exposure during gestation in the rat. *Toxicol. Ind. Health*, **14**, 869–889.

Broadmeadow, A. (1987) Cotnion technical: Toxicity study by oral (gavage) administration to CD rats for 13 weeks. Unpublished Report no. 86/MAK057/342 from Life Sciences Research Ltd., Eye, Suffolk, England. Submitted to WHO by Makhteshim Chemical Works Ltd., Beer-Sheva, Israel.

Chen, H.H., Sirianni, S.R., & Huang, C.C. (1982a) Sister chromatid exchanges and cell cycle delay in Chinese hamster V79 cells treated with 9 organophosphorus compounds (8 pesticides and 1 defoliant). *Mutat. Res.*, **103**, 307–313.

Chen, H.H., Sirianni, S.R., & Huang, C.C. (1982b) Sister chromatid exchanges in Chinese hamster cells treated with seventeen organophosphorus compounds in the presence of a metabolic activation system. *Environ. Mutagen.*, **4**, 621–624.

Clemens, G.R., Bare, J.J., & Hartnagel, R.E. (1988) A teratology study in the rabbit with azinphos-methyl. Unpublished report from Miles Inc., Toxicology Department, Elkhart, IN, USA. Submitted to WHO by Bayer AG, Leverkusen, Germany.

Crawford, C.R. & Anderson, R.H. (1974) The acute oral toxicity of ®GUTHION technical, benzazimide and methyl benzazimide to rats. Unpublished report from Chemagro. Submitted to WHO by Bayer AG, Leverkusen, Germany.

Crown, S. & Nyska, A. (1987) Cotnion-M. Acute oral toxicity study in the rat. Unpublished report No. MAK/116/AZM from Life Science Research Israel Ltd, Ness Ziona, Israel. Submitted to WHO by Makhteshim Chemical Works Ltd., Beer-Sheva, Israel.

DuBois, K.P., Thursh, D.R. & Murphy, S.D. (1955) The acute mammalian toxicity and mechanism of action of Bayer 17147. Unpublished report from Department of Pharmacology, University of Chicago, Chicago, Illinois, USA. Submitted to WHO by Bayer AG, Leverkusen, Germany.

DuBois, K.P., Thursh, D.R. & Murphy, S.D. (1957) Studies on the toxicity and pharmacologic actions of the dimethoxy ester of benzotriazine dithiophosphoric acid (DBD, Guthion). *J. Pharmacol. Exp. Ther.*, **119**, 208–218.

Eiben, R. & Janda, B. (1987) R1582 (common name: azinphos-methyl, the active ingredient of Guthion): Two-generation study on rats. Unpublished study No. T 6006415 from the Institute of Toxicology, Industrial Chemicals, Fachbereich Toxikologie, Bayer AG, Wuppertal, Germany. Submitted to WHO by Bayer AG, Leverkusen, Germany.

Eiben, R., Schmidt, W., & Loeser, E. (1983) R1582 (common name: azinphos-methyl, the active ingredient of ®Guthion). Toxicity study on rats with particular attention to ChE activity (28-day feeding study as a range-finding test for a 2-year study. Unpublished study No: T 7011708 from the Institute of Toxicology Bayer AG, Wuppertal, Germany. Submitted to WHO by Bayer AG, Leverkusen, Germany.

Evanchik, Z. (1987) Cotnion: assessment of mutagenic potential in histadine auxotrophs in *Salmonella typhimurium*. Unpublished Report No. MAK/147/AZN from Life Sciences Research Ltd., Israel. Submitted to WHO by Makhteshim Chemical Works Ltd, Beer-Sheva, Israel.

Faul, J. (1981) Information on effects in man/occupational experience. Company letter of 2 June 1981, from Bayer AG medical department. Submitted to WHO by Bayer AG, Leverkusen, Germany.

Flucke, W. (1979) Determination of acute toxicity (LD_{50}) of azinphos-methyl. Unpublished report from Bayer AG, Leverkusen, Germany. Submitted to WHO by Bayer AG, Leverkusen, Germany.

Gaines, T.B. (1969) Acute toxicity of pesticides. *Toxicol. Appl. Pharmacol.*, **14**, 515–534.

Gal, N., Rubin, Y., Nyska, A. & Waner, T. (1988) Cotnion – M. Teratogenicity study in the rabbit. Unpublished report No. R4935 (Project No. MAK/126/AZM) from Life Science Research Israel Ltd, Ness Zion Israel. Submitted to WHO by Makhteshim Chemical Works Ltd., Beer-Sheva, Israel.

Glaza, S.M. (1988) Azinphos-methyl: acute delayed neurotoxicity study in the domestic fowl. Unpublished report from Hazleton Laboratories America Inc. Submitted to WHO by Bayer AG, Leverkusen, Germany.

Grundman, E. (1965) Histological findings: Gusation. Unpublished report from the Institute of Experimental Pathology, Wuppertal-Elberfeld, Bayer AG, Germany. Submitted to WHO by Bayer AG, Leverkusen, Germany.

Hayes, R.H. (1985) Oncogenicity study of azinphos-methyl in mice. Unpublished report from Mobay Corporation. Submitted to WHO by Bayer AG, Leverkusen, Germany.

Heimann, K.G. (1981) Determination of acute toxicity (LD_{50}) of azinphos-methyl. Unpublished report from Bayer AG, Leverkusen, Germany. Submitted to WHO by Bayer AG, Leverkusen, Germany.

Heimann, K.G. (1982) Azinphos-methyl study of the acute oral and dermal toxicity to rats. Unpublished report from Bayer AG, Leverkusen, Germany. Submitted to WHO by Bayer AG, Leverkusen, Germany.

Heimann, K.G. (1987) Determination of acute toxicity (LD_{50}) of azinphos methyl. Unpublished report from Bayer AG, Leverkusen, Germany. Submitted to WHO by Bayer AG, Leverkusen, Germany.

Henderson, L.M., Proudlock, R.J., & Gray, V.M. (1988) Analysis of metaphase chromosomes obtained from bone marrow of rats treated with azinphos-methyl. Unpublished report No. MBS 23/871337 from Huntingdon Research Centre Ltd, Huntingdon, Cambridgeshire, England. Submitted to WHO by Makhteshim Chemical Works Ltd., Beer-Sheva, Israel.

Herbold, B.A. (1978) Azinphos-methyl: Salmonella/microsome test to evaluate for point mutation. Unpublished report from Bayer AG, Leverkusen, Germany. Submitted to WHO by Bayer AG, Leverkusen, Germany.

Herbold, B.A. (1979a) Azinphos-methyl: micronucleus test on mice to evaluate for possible mutagenic effects. Unpublished report from Bayer AG, Leverkusen, Germany. Submitted to WHO by Bayer AG, Leverkusen, Germany.

Herbold, B.A. (1979b) Azinphos-methyl: dominant lethal study on male mouse to test for mutagenic effects. Unpublished report from Bayer AG, Leverkusen, Germany.

Herbold, B.A. (1984) Azinphos-methyl, Pol test on E. coli to evaluate for potential DNA damage. Unpublished report from Bayer AG, Leverkusen, Germany. Submitted to WHO by Bayer AG, Leverkusen, Germany.

Herbold, B.A. (1986) Azinphos-methyl: cytogenic study with human lymphocyte cultures in vitro to evaluate for harmful effect on chromosomes. Unpublished report from Bayer AG, Leverkusen, Germany. Submitted to WHO by Bayer AG, Leverkusen, Germany.

Herbold, B.A. (1988) Azinphos-methyl: Salmonella/microsome test to evaluate for point mutation. Unpublished report from Bayer AG, Leverkusen, Germany. Submitted to WHO by Bayer AG, Leverkusen, Germany.

Herbold, B.A. (1995) E1582 - Micronucleus test on the mouse. Unpublished report No 24015 from Bayer AG, Institute of Carcinogenecity and Genotoxicity, Wuppertal-Elberfeld, Germany. Submitted to WHO by Bayer AG, Leverkusen, Germany. Submitted to WHO by Bayer AG, Leverkusen, Germany.

Holzum, B.A. (1990) Azinphos-methyl: investigation of inhibition of cholinesterase activity in plasma, erythrocytes and brain, in a one-generation study. Unpublished report from Bayer AG, Leverkusen, Germany. Submitted to WHO by Bayer AG, Leverkusen, Germany.

Hoorn, A.J.W. (1983) Mutagenicity evaluation of azinphos-methyl in the reverse mutation induction assay with Saccharomyces cerevisiae strains S138 and S211. Unpublished report from Litton Bionetics, Holland. Submitted to WHO by Bayer AG, Leverkusen, Germany.

Kao, L.R.M. (1988) Disposition and metabolism of ring-UL-^{14}C azinphos-methyl in rats. Unpublished report from Mobay Corporation, submitted by Bayer AG, Leverkusen, Germany.

Kowalski, R.L., Clemens, G.R., Bare, J.J., & Hartnagel, R.E. (1987a) A teratology study with azinphos-methyl in the rat. Unpublished report from Miles Inc., Toxicology Department, Elkhart, IN, USA. Submitted to WHO by Bayer AG, Leverkusen, Germany.

Kowalski, R.L., Clemens, G.R., Bare, J.J., & Hartnagel, Jr. R.E. (1987b) Addendum: a teratology study with azinphos-methyl (Guthion® technical) in the rat. Unpublished report No. 94987 from Miles Inc., Toxicology Department, Elkhart, IN, USA. Sponsored by Mobay Chemical Corp., Kansas City, MO, USA. Submitted to WHO by Bayer AG, Leverkusen, Germany.

Kimmerle, G. (1964) Neurotoxicity study with Gusathion active ingredient. Unpublished study from the Institute of Toxicology, Farbenfabriken Bayer AG, Wuppertal-Elberfeld, Germany. Submitted to WHO by Bayer AG, Leverkusen, Germany.

Kimmerle, G. (1965) Neurotoxicity study with Gusathion active ingredient. Unpublished study from the Institute of Toxicology, Farbenfabriken Bayer AG, Wuppertal-Elberfeld, Germany. Submitted to WHO by Bayer AG, Leverkusen, Germany.

Kimmerle, G. & Loser, E. (1974) Delayed neurotoxicity of organophosphorus compounds and copper concentration in the serum of hens. *Environ. Qual. Safety*, **3**, 173–178.

Lamb, D.W. & Anderson, R.H. (1974) The acute oral toxicity of azinphos-methyl, benzazimide and methyl benzazimide to fasted and non-fasted rats using CMC as the excipient. Unpublished report from Chemagro. Submitted by Bayer AG, Leverkusen, Germany.

Lawlor, T.E. (1987) Azinphos-methyl: Salmonella/microsome plate incorporation mutagenicity assay. Unpublished report from Microbiological Associates Inc. Submitted to WHO by Bayer AG, Leverkusen, Germany.

Machemer, L. (1975) Azinphos-methyl: studies for embryonic and teratogenic effects on rabbits following oral administration. Unpublished report from Bayer AG, Leverkusen, Germany. Submitted to WHO by Bayer AG, Leverkusen, Germany.

Mahler, L. (1991) Program Director, Pesticide Illness Surveillance Program, Worker Health and Safety Branch, California Department of Agriculture. Data provided to Mobay Corporation. Submitted to WHO by Bayer AG, Leverkusen, Germany.

McFarlane, P. & Freestone, S. (1999a) A randomised double blind ascending single oral dose study with azinphos-methyl to determine the no effect level on plasma and RBC cholinesterase activity. Unpublished report No. 17067 from Inveresk Research Ltd, Edinburgh, Scotland. Submitted to WHO by Bayer AG, Leverkusen, Germany.

McFarlane, P. & Freestone, S. (1999b) A randomised double blind placebo controlled study with azinphos-methyl to determine the no effect level on plasma and RBC cholinesterase activity after repeated doses. Unpublished report No. 17360 from Inveresk Research Ltd, Edinburgh, Scotland. Submitted to WHO by Bayer AG, Leverkusen, Germany.

Mihail, F. (1978) Azinphos-methyl acute toxicity studies. Unpublished report from Bayer AG, Leverkusen, Germany. Submitted to WHO by Bayer AG, Leverkusen, Germany.

Myhr, B.C. (1983) Evaluation of azinphos-methyl in the primary rat hepatocyte unscheduled DNA synthesis assay. Unpublished report from Litton Bionetics USA. Submitted to WHO by Bayer AG, Leverkusen, Germany.

National Cancer Institute (1978) Bioassay of azinphos-methyl for possible carcinogenicity. Technical report series No. 69. Submitted to WHO by Bayer AG, Leverkusen, Germany.

Nelson, D.L. (1968) The acute mammalian toxicity of two samples of Guthion technical to adult female rats. Unpublished report from the Research and Development Dept., Chemagro Corporation, Kansas City, Missouri, USA. Submitted to WHO by Bayer AG, Leverkusen, Germany.

Pasquet, J., Mazuret, A., Fournel, J., & Koenig, F.H. (1976) Acute oral and percutaneous toxicity of phosalone in the rat, in comparison with azinphos-methyl and parathion. *Toxicol. Appl. Pharm.*, **37**, 85–92.

Patzschke, K., Wegner, L.A., & Weber, H. (1976) Biokinetic investigations of carbonyl-^{14}C azinphos methyl in rats. Unpublished report from Bayer AG, Leverkusen, Germany. Submitted to WHO by Bayer AG, Leverkusen, Germany.

Rao, S.L.N. & McKinley, W.P. (1969) Metabolism of organophosphorus insecticides by liver homogenates from different species. Can. *J. Biochem.*, **47**, 1155–1159

Rubin, Y. & Nyska, A. (1988) Cotnion – M. Teratogenicity study in the rat. Unpublished report No. R4678 (Project No. MAK/124/AZM) from Life Science Research Israel Ltd, Ness Zion Israel. Submitted to WHO by Makhteshim Chemical Works Ltd., Beer-Sheva, Israel.

Schmidt, W.M. (1987) Azinphos-methyl: study of chronic toxicity and carcinogenicity to rats. Unpublished report No. 16290 from Bayer AG, Leverkusen, Germany. Submitted to WHO by Bayer AG, Germany.

Sheets, L.P. & Hamilton, B.F. (1994) An acute oral neurotoxicity screening study with technical grade azinphos-methyl (Guthion) in Fischer 344 rats. Unpublished report No. 93-412-UM from Miles Inc., Agriculture Division, Toxicology, South Metcalf, Stilwell, KS, USA. Submitted to WHO by Bayer AG, Germany.

Sheets, L.P. & Hamilton, B.F. (1995) A subchronic dietary neurotoxicity screening study with technical grade azinphos-methyl (Guthion®) in Fischer 344 rats. Unpublished report No. 93-472-VJ from Miles Inc., Agriculture Division, Toxicology, Stilwell, KS, USA, Submitted to WHO by Bayer AG, Germany.

Sheets, L.P., Hamilton, B.F., Sangha, G.K. & Thyssen, J.H. (1997) Subchronic neurotoxicity screening studies with six organophosphate insecticides: an assessment of behavior and morphology relative to cholinesterase inhibition. *Fundam. Appl. Toxicol.*, **35**, 101–19.

Short, R.D., Minor, J.L., Unger, T.M., & Lee, C.C. (1978) Teratology of azinphos-methyl. Unpublished report from Mid West Research Institute, USA. Submitted to WHO by Bayer AG, Leverkusen, Germany.

Short, R.D., Minor, J.L., Lee, C.C., Chernoff, N., & Baron, R.L. (1980) Developmental toxicity of azinphos-methyl in rats and mice. *Arch. Toxicol.*, **43**, 177–186.

Simmon, V.F. (1978) In vivo and in vitro mutagenicity assays of selected pesticides. In: Hart, R.W., Kraybill, H.F. & De Serres, F.J (eds). *A rational evaluation of pesticidal vs. mutagenic/carcinogenic action*. Bethesda, Maryland, United States Department of Health, Education, and Welfare. DHEW Publication No. (NIH) 78-1306.

Simpson, G.R. (1965) Exposure to guthion during formulation. *Arch. Environ. Health*, **10**, 53–54.

Sterri, S.H., Rognerud, B., Fiskum, S.E., & Lyngaas, S. (1979) Effect of toxogonin and P2S on the toxicity of carbamates and organophosphorus compounds. *Acta Pharmacol. et Toxicol.*, **45**, 9–15

Thornton, J.S. (1971) Analysis of urine samples from human subjects treated orally with Guthion. Unpublished study No. 30201 from the Chemagro Corporation, Research and Development Department. Submitted to WHO by Bayer AG Leverkusen, Germany.

Thyssen, J. (1976a) Determination of acute toxicity (LD_{50}) of azinphos-methyl. Unpublished report from Bayer AG, Leverkusen, Germany. Submitted to WHO by Bayer AG, Leverkusen, Germany.

United States Environment Protection Agency (1981) Summary of reported pesticide incidents involving azinphos-methyl pesticide incident monitoring system. Report No. 469. Submitted to WHO by Bayer AG, Leverkusen, Germany.

Waters, M.D., Sandhu, S.S., Simon, V.F., Mortelmans, K.E., Mitchell, A.A., Jorgenson, T.A., Jones, D.C.L., Valencia, R., & Garrett, N.E. (1982) Study of pesticide genotoxicity. In: Fleck, R.A. & Holleander, A., eds, *Genetic Toxicology - an agricultural perspective*. New York, Plenum Press, pp 275–326.

Weber, H., Patzschke, K., & Wegner, L.A. (1980) Biokinetic investigation of ring-UL-^{14}C benzazimide in rats. Unpublished report from Mobay Corporation. Submitted to WHO by Bayer AG, Leverkusen, Germany.

LAMBDA-CYHALOTHRIN

First draft prepared by
G. Wolterink[1] & D. Ray[2]

[1] Centre for Substances and Integrated Risk Assessment,
National Institute for Public Health and the Environment,
Bilthoven, Netherlands; and
[2] Medical Research Council Applied Neuroscience Group, Biomedical Sciences,
University of Nottingham, Queens Medical Centre
Nottingham, England

Explanation

Lambda-cyhalothrin, the International Organization for Standardization (ISO) approved common name for (*R*)-cyano(3-phenoxyphenyl)methyl (1*S*,3*S*)-rel-3-[(1*Z*)-2-chloro-3,3,3-trifluoro-1-propenyl]-2,2-dimethylcyclopropanecarboxylate is a synthetic cyano-containing type II pyrethroid insecticide (Chemical Abstracts Service, CAS No. 91465-08-6).

Cyhalothrin (CAS No. 68085-85-8) was evaluated by the JMPR in 1984, when an acceptable daily intake (ADI) of 0–0.02 mg/kg bw was established based on a no-observed-adverse-effect level (NOAEL) of 20 ppm, equal to 2 mg/kg bw per day, identified on the basis of clinical signs in a 2-year study in mice; a NOAEL of 30 ppm, equal to 1.5 mg/kg bw per day, identified on the basis of decreased body-weight gain in a three-generation study in rats; and a NOAEL of 2.5 mg/kg bw per day, identified on the basis of neurotoxicity in a 6-month study in dogs, and using a safety factor of 100.

At its meeting in 2000, the Joint FAO/WHO Expert Committee on Food Additives (JECFA) established a temporary ADI of 0–0.002 mg/kg bw based on a lowest-observed-effect level (LOEL) of 1 mg/kg bw per day for induction of liquid faeces in dogs in a 26-week study, and using a safety factor of 500. The high safety factor was used to compensate for the absence of a no-observed-effect level (NOEL) in this study.

At its meeting in 2004, JECFA concluded that the toxicity of cyhalothrin is similar in rats and dogs. The Committee decided that the temporary ADI could be replaced by an ADI of 0–0.005 mg/kg bw, which was determined by dividing the LOEL of 1 mg/kg bw per day in dogs (also the NOEL for rats) by a safety factor of 200. The safety factor incorporated a factor of 2 to compensate for the absence of a NOEL in dogs.

Lambda-cyhalothrin consists of two of the four enantiomers (i.e. the *cis* 1*R*α*S* and *cis* 1*S*α*S* enantiomeric pair) of cyhalothrin. One of the two enantiomers of lambda-cyhalothrin is the insecticidally active gamma-cyhalothrin (CAS No. 76703-62-3). Cyhalothrin comprises about 50% lambda-cyhalothrin.

Lambda-cyhalothrin was evaluated by the present Meeting within the periodic review programme of the Codex Committee on Pesticide Residues (CCPR). For the present re-evaluation, studies with cyhalothrin and lambda-cyhalothrin were available.

For lambda-cyhalothrin, specifications were established by the FAO/WHO Joint Meeting on Pesticide Specifications (JMPS) and published as *WHO specifications and evaluations for public health pesticides: lambda-cyhalothrin*[1] (technical material, 2003). For other formulations, specifications also exist.

All pivotal studies with cyhalothrin and lambda-cyhalothrin were certified as being compliant with good laboratory practice (GLP).

Evaluation for acceptable daily intake

1. Biochemical aspects

1.1 Absorption, distribution and excretion

Rats

Groups of six male and six female Alderley Park Wistar (Alpk/Ap) (SPF) rats were given a single dose of [^{14}C]cyhalothrin (purity, >99%) at 1 or 25 mg/kg bw in corn oil by gavage. Cyhalothrin was labelled either at the 3-phenoxybenzyl side-chain (i.e. the carbon to which the cyanide moiety is attached: ^{14}CHCN, or at the 1 position of the cyclopropyl moiety (see Figure 1). In a separate experiment, groups of three or four rats with bile-duct cannulae were given an oral dose of ^{14}C-benzyl-labelled cyhalothrin at 1 mg/kg bw. An additional group of five males and six females received ^{14}C-benzyl-labelled cyhalothrin as a subcutaneous dose at 1 mg/kg bw. In a repeated-dose experiment to study the excretion and tissue accumulation of cyhalothrin given at a dose of 1 mg/kg per day by gavage, groups of six male and six female rats received daily doses of ^{14}C-benzyl-labelled or ^{14}C-cyclopropyl-labelled cyhalothrin for 14 days. Excreta (urine, faeces and, in selected rats, expired air) were collected for up to 7 days after dosing and analysed for total radioactivity and metabolites by liquid scintillation counting (LSC) and thin-layer chromatography (TLC). Blood samples were also collected at various times up to 48 h and analysed for total radioactivity and unchanged cyhalothrin. Statements of adherence to quality assurance (QA) and GLP were included.

[1] Available from: http://www.who.int/whopes/quality/en/Lambda-cyhalothrin_eval_specs_WHO_2003.pdf

Figure 1. Chemical structures of the two pairs of enantiomeric pairs of isomers comprising lambda-cyhalothrin

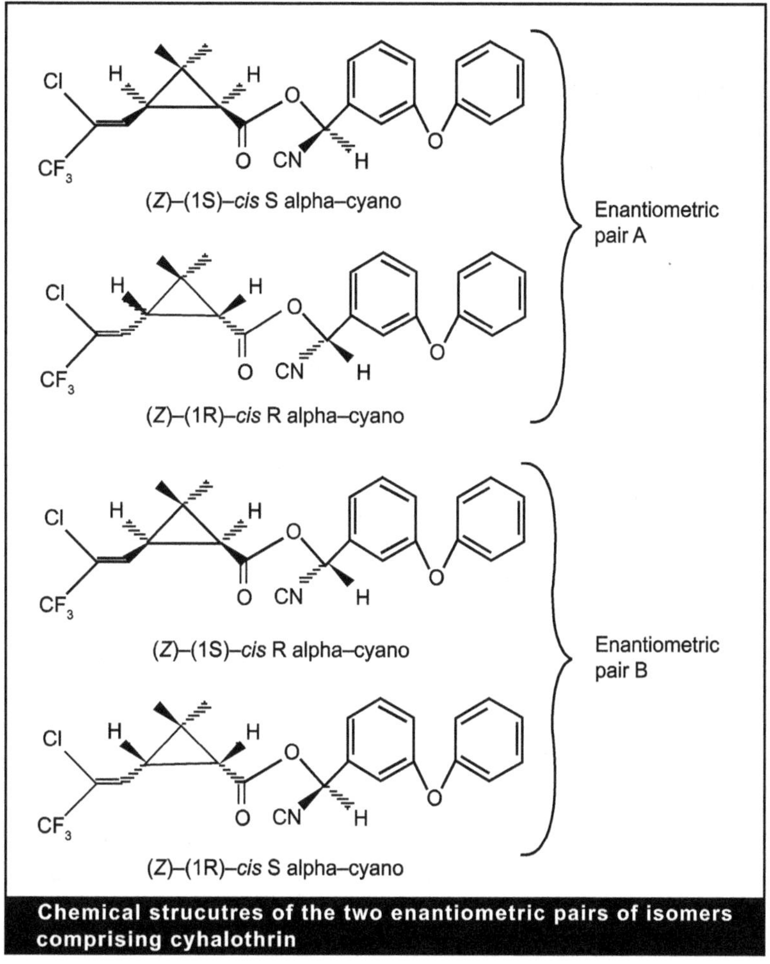

From IPCS (1990)

Recovery of radioactivity was 81–91%. Approximately 30–40% and 40–60% of the orally administered ^{14}C-benzyl-labelled material was recovered in urine and faeces respectively. No $^{14}CO_2$ was detected in expired air. The proportions absorbed and the rate of excretion were similar at both doses, although excretion in faeces tended to be slower at the higher dose. Peak blood concentrations (approximately 0.6 and 6 µg/kg of cyhalothrin equivalents at 1 and 25 mg/kg bw, respectively) were reached 7 h after dosing. By 48 h, the blood concentrations had depleted to less than 10% of the peak value. In bile-duct cannulated rats, biliary excretion accounted for 4.8% of the total excretion in males and 8.9% in females. The extent of absorption in rats equipped with a biliary fistula was less than in intact rats, however absorption was increased in rats (four males) given replacement bile, indicating that cyhalothrin is absorbed along with the fats of the oil formulation. In the males treated with replacement bile, biliary excretion was 11.2%.

At the lower dose, 70% of the administered material was recovered within 24 h, and only 2–3% of the administered radiolabel remained in the carcass after 7 days. At this time-point, the radiolabel was distributed mainly in fat (11.5 µg/g at a dose of 25 mg/kg bw).

Elimination of radiolabel was slower after subcutaneous administration of ^{14}C-benzyl-labelled cyhalothrin than after oral dosing; 58% of the radiolabel was still present in the carcass 7 days after dosing. This was attributed to retention of the oil formulation in subcutaneous fat.

With cyclopropyl-labelled cyhalothrin, smaller proportions of the administered radiolabel were recovered in the urine (19–30%). Peak blood concentrations were achieved at 4–7 h, but these had declined to 10% of the peak values by 48 h. Only 1–3% of the administered dose remained in the carcass after 7 days.

In the repeated-dosing experiment, more than 90% of the cumulative dose was recovered in urine and faeces within 7 days of the last dose. The daily excretion reached a constant after the second dose. Comparison with the single-dose experiment showed that radioactivity accumulated in fat (Harrison & Case, 1981; Harrison, 1984a, 1984b).

A study was undertaken to explore the retention, in fat, of cyhalothrin and lambda-cyhalothrin in rats. Male rats (Alpk/AP) received ^{14}C-cyclopropyl-labelled cyhalothrin (purity, 92.2%) as daily oral doses at 1 mg/kg per day, for up to 119 days. Rats in the control group received vehicle (corn oil). At several time-points (approximately weekly), groups of three rats were killed 24 h after the last dosing, and the concentrations of radioactivity in the liver, kidney, fat, and blood were determined. In addition, the concentration in fat of lambda-cyhalothrin and its opposite enantiomer pair (enantiomer pair A) was measured by high-performance liquid chromatography (HPLC). Furthermore, the elimination of radioactivity from blood, liver, kidneys and fat was assessed for 7 weeks after cessation of treatment. Statements of adherence to QA and GLP were included.

Levels of radioactivity in the blood remained fairly constant and low (approximately 0.2 μg cyhalothrin equivalents per gram) throughout the dosing period. In the liver and kidney, the radioactivity reached a plateau after approximately 70 days, at a level corresponding to cyhalothrin equivalents of approximately 2.5 μg/g liver and 1.2 μg/g kidney. The concentration of radioactivity in fat continued to increase with time, corresponding to cyhalothrin levels of 10 μg/g fat at the end of the treatment period. After cessation of dosing, concentrations of radioactivity in the kidney and blood declined rapidly, and were barely detectable after 5 weeks. In fat, the levels declined more slowly with an elimination half-life of 30 days. Concentrations of radioactivity in the liver initially declined rapidly, while subsequent elimination paralleled that of the fat. The radioactive material in fat was unchanged cyhalothrin; the ratio of enantiomeric pairs, one of which was lambda-cyhalothrin, was not significantly different from that in the dosing solution, indicating that the rate of metabolism of lambda-cyhalothrin was the same as that of cyhalothrin and that there was no preferential accumulation of lambda-cyhalothrin (Prout, 1984).

In a tissue-distribution time-course study, groups of 15 male and 15 female Alpk:AP_fSD Wistar-derived rats received [^{14}C]benzyl labelled cyhalothrin (radiochemical purity, > 97.5%) as a single oral dose at 1 or 25 mg/kg bw in corn oil (volume, 4 ml/kg). In the group receiving the lower dose, three males and three females were killed at intervals of 6.5, 11 (13 females), 24, 48 and 96 h after dosing. At the higher dose, groups of three male rats were killed at 10 h (time of peak plasma concentration), 17 h (half-life of elimination of radioactivity from blood) and at 24, 48 and 96 h. Also at this dose, groups of three female rats were killed at 7 h (time of peak plasma concentration), 21 h (half-life of elimination of radioactivity from blood) and at 30, 48 and 96 h. At these time-points, plasma, whole blood, liver, kidneys, lungs, spleen, bone, brain, heart, muscle, gonads, white and brown fat from each rat were analysed for radioactive content. Statements of adherence to QA and GLP were included.

Brown fat contained the highest concentrations of radioactivity (0.89–1.45 μg eq/g in rats at the lower dose and about 15 μg eq/g in rats at the higher dose). The concentration of radioactivity declined relatively quickly during the study in all tissues except white fat. The peak concentration of radioactivity in white fat increased in proportion to the dose. No marked sex differences in tissue distribution were noted. (Jones, 1989a, 1989b).

In a comparative study, the absorption, distribution, excretion, and metabolism of single oral doses of lambda-cyhalothrin (purity, 99.0%) and cyhalothrin (a 50 : 50 mixture of lambda-cyhalothrin and the opposite enantiomer pair A (see Figure 1); purity, 97.4%) was investigated in groups of four male rats (Alpk/AP). One group was given [^{14}C]cyclopropyl-labelled lambda-cyhalothrin (1 mg/kg bw), a second group was given [^{14}C]cyclopropyl-labelled lambda-cyhalothrin (1 mg/kg bw) plus the unlabelled enantiomeric pair A (1 mg/kg bw), and a third group was given [^{14}C]cyclopropyl-labelled cyhalothrin (1 mg/kg bw). The urinary and faecal excretion of radioactivity was monitored for 3 days and the residual radioactivity was then determined in selected tissues. The metabolite profile of the excreta was determined by TLC. Statements of adherence to QA and GLP were included.

The results of this study indicated that coadministration of enantiomer pair A with lambda-cyhalothrin had little or no effect upon the absorption, distribution, or tissue retention of radioactivity, and there was no effect upon the metabolite profile of lambda-cyhalothrin. Similarly, the absorption, distribution, excretion, and metabolism of cyhalothrin was indistinguishable from that of lambda-cyhalothrin (Prout & Howard, 1985).

Dogs

Groups of three male and three female beagle dogs received gelatin capsules containing either [^{14}C]cyclopropyl-labelled cyhalothrin or [^{14}C]benzyl-labelled cyhalothrin (purity, > 97%) as a single oral dose at 1 and 10 mg/kg bw dissolved in corn oil. The same dogs also received [^{14}C]benzyl- or [^{14}C]cyclopropyl-labelled material as a single intravenous injection at 0.1 mg/kg bw (in ethanol : 0.9% saline in a 3.2 : 1.2 ratio). There was an interval of at least 3 weeks between each dose. Blood samples and excreta were collected over 7 days after dosing and were analysed for radiolabel by liquid scintillation counting (LSC). Metabolites were measured by TLC, and the identities of selected metabolites were confirmed by mass spectrometry. Statements of adherence to QA and GLP were included.

Overall recovery ranged from 82% to 93%. After a single oral administration at a dose of 1 or 10 mg/kg bw, approximately 30% and 50% of the [^{14}C]benzyl-label was excreted in the urine and faeces respectively. In dogs that were dosed intravenously, 37% and 42% of [^{14}C]benzyl-label was excreted in the urine and faeces respectively. Excretion of radioactivity after both oral and intravenous dosing was initially rapid, with most of the administered radioactivity being excreted in the first 48 h after dosing. When [^{14}C]cyclopropyl-cyhalothrin was given orally, the proportion of residue in the urine was slightly lower (19%) and the amount in the faeces was higher (68%). The proportions of residue in the urine and in the faeces were approximately equal when the two radiolabelled forms of cyhalothrin were given intravenously. A large proportion (46–87%) of the faecal residue was in the form of unchanged cyhalothrin after oral dosing, but only small amounts of parent substance (1.4–1.5% of faecal radiolabel) were found in faeces after intravenous injection, suggesting that the oral dose of cyhalothrin was only partially absorbed from the gut lumen, leaving unabsorbed cyhalothrin in the faeces. No marked sex differences were observed (Harrison, 1984c).

1.2 *Biotransformation*

(a) In vivo

Rats

Groups of six male and six female Alderley Park Wistar (Alpk/Ap) (SPF) rats were given [^{14}C] cyhalothrin (purity, > 99%) as a single dose at 1 or 25 mg/kg bw in corn oil by gavage. Cyhalothrin was labelled either at the 3-phenoxybenzyl side-chain (i.e. the carbon to which the cyanide moiety is attached: ^{14}CHCN) or at the 1 position of the cyclopropyl moiety (for structure see Figure 1). Urine and faeces were collected for up to 7 days after dosing and analysed for metabolites by TLC. Bile samples were also collected at various times up to 48 h and analysed for metabolites.

The patterns of metabolites derived from the two ^{14}C-labelled forms of cyhalothrin were totally different, suggesting that metabolism involves initial cleavage of the ester bond. Two major urinary metabolites (not further identified), representing 75% and 17% of radioactivity in the urine, were derived from [^{14}C]benzyl-labelled cyhalothrin. Both were polar and resistant to glucuronidase. Of the urinary material derived from cyclopropyl-labelled cyhalothrin, 60% could be hydrolysed by glucuronidase. Three biliary metabolites (not further identified) were derived from [^{14}C]benzyl-labelled cyhalothrin, two of which (representing 12% and 67% of the radioactivity in bile) were different from the urinary metabolites. Urine and bile did not the contain parent compound. Faeces, on the other hand, contained mainly unchanged cyhalothrin, which was probably nonabsorbed material. In addition, the faeces contained three relatively non-polar metabolites which could be hydrolysis products of conjugates excreted in bile. Residues in fat probably represent unchanged cyhalothrin (Harrison & Case, 1981; Harrison, 1984a, 1984b).

In a study designed to identify the major pathways of cyhalothrin metabolism, groups of six male and six female Alpk:AP_fSD Wistar-derived rats were given [^{14}C]benzyl-labelled cyhalothrin at a dose of 12.5 mg/kg bw per day for 8 days until each rat had received approximately 25 mg of cyhalothrin. Urine and faeces were collected daily, after dosing and for up to 3 days after the last dose. A further six male and six female rats were each given fourteen consecutive daily doses of [^{14}C]cyclopropyl-labelled cyhalothrin at 1 mg/kg bw per day, and the urine was collected and combined. Metabolic profiles were determined by TLC, both before and after enzymic hydrolyses using aryl sulfatase and β-glucuronidase enzymes. Individual metabolites were purified by reverse-phase HPLC before analysis by gas chromatography-mass spectrometry (GC-MS), probe MS, fast atom-bombardment mass spectrometry (FAB-MS) and/or carbon nuclear magnetic resonance spectroscopy (^{13}C-NMR). Statements of adherence to QA and GLP were included.

The major radioactive component in urine of rats dosed with [^{14}C]benzyl-labelled cyhalothrin was the sulfate conjugate of 3-(4'-hydroxyphenoxy) benzoic acid (compound XXIII). Unconjugated compound XXIII and 3-phenoxybenzoic acid (compound V) were minor metabolites. The major radioactive component in urine of rats dosed with [^{14}C]cyclopropyl labelled cyhalothrin was the glucuronide conjugate of (1*RS*)-*cis*-3-(2-chloro-3,3,3-trifluoropropenyl)-2,2-dimethylcyclopropanoic acid (compound Ia). One of the minor metabolites also appeared to be a glucuronide conjugate, which was tentatively identified as a hydroxylated analogue of acid (compound XI) (Harrison & Case, 1983).

Dogs

Groups of three male and three female beagle dogs received gelatin capsules containing either [^{14}C]cyclopropyl-labelled or [^{14}C]benzyl-labelled cyhalothrin (purity, > 97%) as a single oral dose at 1 or 10 mg/kg bw dissolved in corn oil. The same dogs received a single intravenous injection of [^{14}C]cyclopropyl-labelled or [^{14}C]benzyl-labelled cyhalothrin at a dose of 0.1 mg/kg bw (in ethanol : 0.9% saline in a 3.2 : 1.2 ratio). The doses were separated by an interval of at least 3 weeks. Blood samples and excreta were collected for 7 days after dosing and were analysed for radiolabel by LSC. Metabolites were measured by TLC, and the identities of selected metabolites were confirmed by MS. Statements of adherence to QA and GLP were included.

Metabolism occurred initially by cleavage of the ester bond to give a phenoxybenzyl moiety and a cyclopropanoic acid moiety. Further metabolism of the phenoxybenzyl moiety produced *N*-(3-phenoxybenzoyl) glycine, 3-(4'-hydroxyphenoxy)benzoic acid (compound XXIII) and its sulfate conjugate, 3-phenoxybenzoyl glucuronide, small amounts of various other conjugates of these compounds, and a little free phenoxybenzoic acid. The cyclopropanoic acid moiety was extensively metabolized to at least 11 further metabolites, including 45% as the glucuronide of the cyclopropanoic acid and up to 23% as the free cyclopropanoic acid (i.e. 3-(*Z*-2-chloro-3,3,3-trifluoropropenyl)-2,2-dimethylcyclopropanoic acid; see Fig. 2) (Harrison, 1984c).

Figure 2. Main pathways of biotransformation of cyhalothrin in mammals

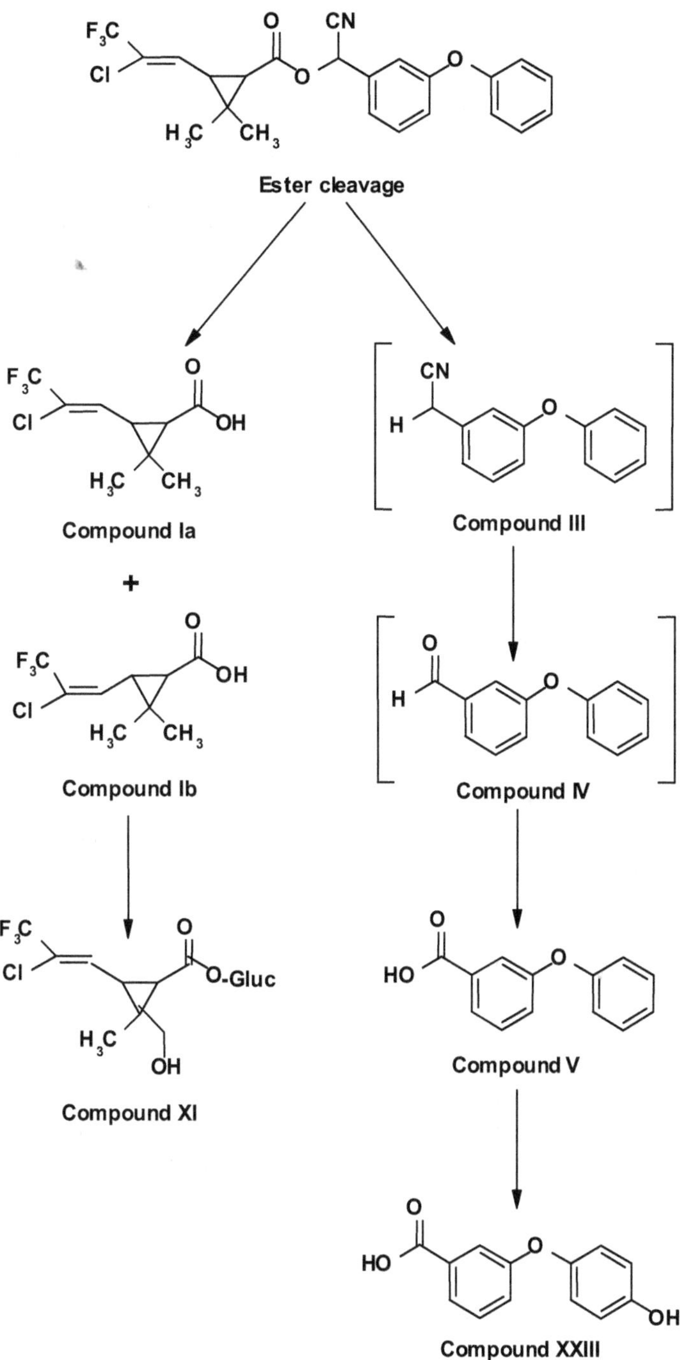

Compound Ia	(1*RS*)-*cis*-3-(*Z*-2-chloro-3,3,3-trifluoroprop-1-enyl)-2,2-dimethylcyclopropanoic acid
Compound Ib	(1*RS*)-*trans*-3-(*Z*-2-chloro-3,3,3-trifluoroprop-1-enyl)-2,2-dimethylcyclopropanoic acid
Compound III	(*RS*)-α-cyano-3-phenoxybenzyl alcohol,Mixture of R and S isomers
Compound IV	3-phenoxybenzaldehyde
Compound V	3-phenoxy benzoic acid
Compound XI	*EZ*)-1-(*RS*)-2-*trans*-3-*cis*-3-(2-chloro-3,3,3-trifluoromethylpropenyl)-2-hydroxymethyl-2-methylcyclopropanoic acid
Compound XXIII	3-(4'-hydroxyphenoxy) benzoic acid

2. Toxicological studies

2.1 Acute toxicity

(i) Lethal doses

The results of studies of acute toxicity with lambda-cyhalothrin are summarized in Table 1. It is noteworthy that lambda-cyhalothrin and cyhalothrin are unusual among pyrethroids in that these compounds cause lethality via the dermal route.

The observed clinical signs (ataxia, decreased activity, tiptoe gait, splayed gait, loss of stability, dehydration, urinary incontinence, hunched posture, piloerection, salivation, ungroomed appearance and pinched-in sides) were typical of this class of pyrethroids.

(ii) Dermal irritation

In a study of dermal irritation in female New Zealand White (NZW) rabbits, six rabbits were exposed to cyhalothrin (purity, 92.9%) or lambda-cyhalothrin (purity, 96.5%) for 4 h on the intact skin. Irritation was scored at 1, 20, 44 and 68 h, and 5, 7, 9 and 14 days after the removal of the occlusive dressing. Statements of adherence to QA and GLP were included.

In the rabbits treated with cyhalothrin, no erythema was observed. Very slight oedema was observed at 1 and 20 h, but not at other time-points after application. In the rabbits treated with lambda-cyhalothrin, no erythema was observed. Very slight oedema was observed only at 1 h, but not at other time-points after application (Pritchard, 1985a).

(iii) Ocular irritation

In a study of ocular irritation, the eyes of six male NZW rabbits were exposed to lambda-cyhalothrin (purity, 96.5%). The eyes were examined at 1–2 h and at 1, 2, 3, 4 and 7 days in all rabbits, and on days10, 11, 13, 14 and 17 in one to three rabbits. Statements of adherence to QA and GLP were included.

Lambda-cyhalothrin induced slight to moderate redness, slight or mild chemosis of the conjunctivae, and slight or severe discharge at 1 h after application. At day 2 after application, slight or moderate redness, slight chemosis and slight discharge were observed. In four rabbits, the effects had completely resolved within 4 days, while the other two rabbits had recovered at day 10. All rabbits appeared very agitated, flicking their heads, blinking and pawing at the treated eye during the first 24 h. This was attributed to paresthaesia effects. Also, additional irritation effects observed during the study may have been caused by rubbing of the eyes as a result of paresthaesia (Pritchard, 1985b).

Table 1. Results of studies of acute toxicity with lambda-cyhalothrin

Species	Strain	Sex	Route	Vehicle	Purity (%)	LD_{50} (mg/kg bw)	LC_{50} (mg/l)	Reference
Mouse	NS	Male Female	Oral	Corn oil	96.5	19.9 19.9	—	Southwood (1984)[a]
Rat	NS	Male Female	Oral	Corn oil	92.6-96	79 56	—	Southwood (1985)[a]
Rat	NS	Male Female	Dermal (24-h)	Propylene glycol	92.6	632 696	—	Barber (1985)[a]
Rat	Wistar	Male & Female	Inhalation (4-h)	NS	NS	—	0.06	Hext (1987)[a]

NS, not stated.

[a] Statements of adherence to QA were included.

(iv) Dermal sensitization

In a study of dermal sensitization using the Magnusson & Kligman maximization test, lambda-cyhalothrin (purity, 96.5%) was tested on 20 Dunkin Hartley guinea-pigs. The control group consisted of 10 guinea-pigs. In the induction phase, guinea-pigs received three intracutaneous injections: lambda-cyhalothrin (5% w/v in corn oil), Freund complete adjuvant and corn oil (1 : 1), and lambda cyhalothrin (5% w/v in Freund complete adjuvant and corn oil (1 : 1). One week later, this was followed by a topical application (48 h) of lambda-cyhalothrin (1% w/v) in corn oil. Two weeks after the topical application the guinea-pigs received the challenge dose of lambda-cyhalothrin (1% w/v in corn oil, 0.05 ml) for 24 h under occlusion. Skin reactions were examined immediately and 1 and 2 days after removal of the occlusive dressing. Statements of adherence to QA and GLP were included.

No response to the challenge dose was observed (Pritchard, 1984).

In deviation from what is required in the OECD 406 testing guideline, no skin irritation was observed in the induction phase. The skin was not treated with sodium lauryl sulfate before the topical application in the induction phase. Therefore the test was not been properly performed and no conclusion can be drawn as to the skin sensitizing properties of lambda-cyhalothrin.

In the IPCS Environmental Health Criteria 99 (IPCS, 1990) document on cyhalothrin it is reported that cyhalothrin induced skin sensitization in a Buehler test in guinea-pigs. In guinea-pigs previously induced with undiluted technical cyhalothrin, a moderate sensitization response was elicited in a Magnusson & Kligman maximization test.

2.2 *Short-term studies of toxicity*

Rats

In a limited 28-day study in Wistar rats, which included a recovery period, the toxicity of cyhalothrin for the liver was investigated. Groups of 32 male rats received diets containing cyhalothrin (purity, 89.2%) at a concentration of 0 or 250 ppm (equivalent to 12.5 mg/kg bw per day). Rats were checked daily for clinical signs. Body weights were measured at the start of treatment and weekly thereafter. After 28 days, eight rats from each group were killed. The remaining rats were maintained on a control diet and eight rats per group were killed after 7, 14 or 28 days. Livers from all rats were weighed and examined histologically and byelectron microscopy and hepatic aminopyrine-*N*-demethylase activity was measured. A statement of adherence to quality assurance (QA) was included.

The body-weight gain of the rats fed cyhalothrin decreased during treatment and remained decreased until the day of sacrifice. Although there was a tendency for a slight reduction of the absolute liver weights of the treated rats, relative liver weights were not affected. At the end of the 28-day treatment period, electron microscopy showed significant proliferation of smooth endoplasmic reticulum in only five rats of the group exposed to cyhalothrin. Such proliferation was no longer apparent in rats allowed a 7-day recovery. Hepatic aminopyrine-*N*-demethylase activity was elevated by 66% after 28 days feeding at 250 ppm, but had reverted to control levels 7 days after cessation of exposure (Lindsay et al., 1982).

In a 90-day study, groups of 20 male and 20 female Wistar derived Alderley Park rats were fed diets containing technical-grade cyhalothrin (purity, 89.2%; total pyrethroid content, 92.2%, of which 96.8% was cyhalothrin) at a concentration of 0, 10, 50, or 250 ppm (equal to 0, 0.56, 2.6, and 14 mg/kg bw per day for males and 0, 0.57, 3.2, and 15 mg/kg bw per day for females). Rats were checked daily for clinical signs. Detailed clinical examinations and body weights were assessed at the start of treatment and weekly thereafter. Food consumption was measured weekly. Haematology was performed on 10 males and 10 females per group at the start of the study and after 1 and 3 months of treatment. At the same time-points, clinical chemistry and urine analysis was performed on the 10 males and 10 females per group not designated for haematology. In addition, bone-marrow smears taken at 3 months were examined. During the last week of treatment, ophthalomoscopy was performed on all rats in the control group and those at the highest dose. At termination, the rats were killed and necropsied. Selected organs were weighed. An extensive range of tissues from rats in the control group and those at the highest dose, and the testes, epididymes, prostate, seminal vesicles, liver, kidneys, heart, spleen and abnormal tissues of all rats were histologically examined. Livers from six males and six females per group were examined by electron microscopy. Hepatic aminopyrine-*N*-demethylase activity was measured in liver samples from 10 males and 10 females per group. Statements of adherence to QA and GLP were included.

No effects of treatment on mortality, clinical signs, ophthalmoscopy and relative organ weights were observed. Minor effects on haematological parameters were considered not to be toxicologically relevant. The body-weight gain of males at the highest dose was consistently reduced (±10%) throughout the study. Food consumption was reduced in males at 50 and 250 ppm by 12% and 13%, respectively. At 13 weeks, plasma triglyceride concentrations were reduced (42% and 59%, respectively). Urinary glucose concentration was increased (49% and 69%) in males at 50 and 250 ppm, respectively. However, since the increases were small and similar increases in urine glucose concentrations were observed before treatment (54% and 79% at 50 and 250 ppm, respectively) this effect was not considered to be treatment-related. The results of gross examinations were not reported. No treatment-related changes were revealed by optical microscopy of organs. Electron microscopy of the liver showed mild proliferation of smooth endoplasmic reticulum in the hepatocytes of three males in the groups at 50 ppm and 250 ppm. Hepatic aminopyrine-*N*-demethylase activity was increased in males at 50 or 250 ppm by 34% and 68%, respectively, and in females at 250 ppm by 46%.

On the basis of reduced body-weight gain and food consumption in males at 250 ppm, the NOAEL was 50 ppm, equal to 2.6 mg/kg bw per day (Lindsay et al., 1981). The Meeting noted that the 1984 JMPR had stated that a NOEL could not be established owing to haematological effects in all treatment groups. However, the effects were very small (generally < 3%, and 6% for mean cell volume at 250 ppm).

In a 3-month feeding study, groups of 20 male and 20 female Alderley Park (Alpk/AP) rats were fed diets containing lambda-cyhalothrin (P321; purity, 96.5%) at a concentration of 0, 10, 50, or 250 ppm (equivalent to 0, 0.5, 2.5 and 12.5 mg/kg bw per day). The rats were checked daily for clinical signs. Detailed clinical examinations and body weights were assessed at the start of treatment and weekly thereafter. Food consumption was measured weekly. Haematology was performed on 10 males and 10 females per group at the start, and at 1 and 3 months of treatment. At the same time-points, clinical chemistry and urine analysis was performed on the 10 males and 10 females per group not designated for haematology. During the last week of treatment, ophthalomoscopy was performed on all rats in the control group and at the highest dose. After 13 weeks, the rats were killed and necropsied. Selected organs were weighed. A selection of tissues from rats in the control group and at the highest dose, and the liver, kidneys, lungs and abnormal tissues of all rats were examined histologically. Hepatic aminopyrine-*N*-demethylase activity was measured in liver samples of six males and six females per group. Statements of adherence to QA and GLP were included.

No toxicologically relevant effects on clinical signs, ophthalmoscopy, haematology, clinical chemistry and urine analysis were observed. Body-weight gain and food consumption were reduced in the group at the highest dose (males, 11%; females, 7–9%). Blood triglyceride concentrations, measured after 1 and 3 months of treatment, were reduced by 28–32% in males at the highest dose and by 15–17% in males at the intermediate dose. A slight increase in relative liver weight (8%) was observed in rats at the highest dose. Slight increases in hepatic aminopyrine-*N*-demethylase activity were observed in males (17%) and females (27%) at the highest dose. These were not accompanied by histopathological changes in the liver.

On the basis of the reduction in body-weight gain and food consumption in males at the highest dose, the NOAEL was 50 ppm, equivalent to 2.5 mg/kg bw per day (Hart, 1985).

Dogs

In a pilot study, groups of one male and one female beagle dog received gelatin capsules contaning lambda-cyhalothrin (purity, 87.7%; as a solution in corn oil) at a dose of 0, 0.75, 1.5, 3.0 or 4.0 mg/kg bw per day for 6 weeks. The test compound was given 2 h after feeding. the dogs were checked daily for clinical signs. Food consumption was measured daily and the dogs were weighed weekly. Ophthalmoscopy was performed before the start of treatment and before termination. Haematology and clinical chemistry was performed before the start of treatment and during weeks 1, 3 and 6. At the end of the treatment period, the dogs were killed, examined macroscopically and histologically and selected organs were weighed. Statements of adherence to QA and GLP were included.

Both dogs at 4.0 mg/kg bw per day were killed for humane reasons on day 15, after inappetence and body-weight loss that were observed from the start of dosing. Before this, occasionally slightly decreased activity, thin appearance, fluid faeces and regurgitation/vomiting was observed. In the groups at 1.5 and 3.0 mg/kg bw per day, occasional transient signs of minor neurological disturbance (e.g. decreased activity, slight tremors and unsteady gait) vomit and/or regurgitation were seen. Fluid faeces were observed in all treatment groups (Horner, 1996).

Groups of six male and six female beagle dogs received gelatin capsules containing cyhalothrin (purity unknown; as a solution in corn oil) at 0, 1.0, 2.5 or 10 mg/kg bw per day for 26 weeks. The test compound was given 1 h before feeding.

Clinical signs, body weight and food consumption were assessed daily. Water consumption was recorded before dosing and during weeks 1–3, 5–7, 9–11, 13–15, 17–19 and 21–24. Ophthalmoscopy was performed before the start of treatment and during weeks 6, 12 and 24. A neurological examination, investigating the cranial nerves, the segmental reflexes and the postural reactions, was carried out on all dogs in the control group and on dogs at 10 mg/kg bw per day before the start of treatment and during week 6 at approximately 1 h after administration. Haematological and clinical chemistry was performed before dosing started and during weeks 4, 8, 12, 16, 20 and 25. Urine analysis was performed before dosing and during weeks 8, 16, and 25.

On the day before autopsy, bone marrow was obtained by sternal puncture and a smear was prepared for examination. At the end of the treatment period, the dogs were killed and macroscopied, selected organs were weighed and a range of tissues microscopically examined. Statements of adherence to QA and GLP were included.

In the group at the highest dose, clinical signs indicative of a disturbance of the nervous system (e.g. unsteadiness, marked lack of coordination, collapse and occasionally muscular spasms) and vomiting were seen, from the first week of treatment onwards. The signs in general appeared within a few hours after administration. These observations were not accompanied by histological changes in nervous tissue. From the first week of treatment onwards, there was a dose-related increase in the passage of liquid faeces, observed in all treatment groups. Other pyrethroids produce this effect, which may represent a consequence of the gastrointestinal equivalent of paresthaesia in the skin (Ray & Fry, 2006). No compound-related changes in body weight were seen apart from a transient loss in weight for one male at 10.0 mg/kg bw per day that also showed transient reductions in food consumption. In the group at the highest dose, a transient increase in water consumption during the first half of the dosing period and a reduction in food consumption may have been treatment-related. No other treatment-related effects were observed.

On the basis of the neurotoxic effects observed at 10 mg/kg bw per day, the NOAEL for systemic effects was 2.5 mg/kg bw per day. On the basis of the increased incidence of liquid faeces the lowest-observed-adverse-effect level (LOAEL) for local gastrointestinal effects was 1.0 mg/kg bw per day (Chesterman, 1981).

Groups of six male and six female beagle dogs received capsules containing lambda cyhalothrin (purity, 96.5%; as a solution in corn oil) at a dose of 0, 0.1, 0.5 or 3.5 mg/kg bw per day, for 52 weeks. The test compound was given by capsule 1 h before feeding. Clinical signs and food consumption were recorded daily. Body weight was measured weekly. Haematology and clinical chemistry were performed before treatment and at weeks 4, 13, 26, 39 and 52. At week 52 the dogs were killed and examined macrosopically and histologically. A selection of organs was weighed. Statements of adherence to QA and GLP were included.

At the highest dose, ataxia and muscle tremors with the occasional instance of convulsions, as well as vomiting were observed on several occasions throughout the study. The signs were observed from the first week of treatment onwards and usually occurred within 3–7 h after dosing. In general, males showed a higher frequency and severity of observations and one dog in this group was killed for humane reasons during week 46. The clinical signs were not accompanied by histopathological lesions of the nervous system. From the start of treatment, a dose-dependent increased incidence in liquid faeces was observed in the groups at 0.5 and 3.5 mg/kg bw per day. The passing of fluid faeces was not accompanied by histopathological lesions of the alimentary tract. Other pyrethroids produce this effect, which may represent a consequence of the gastrointestinal equivalent of paraesthesia in the skin (Ray & Fry, 2006). Body weight and food consumption were not affected by treatment. There were no toxicologically significant effects on haematological and clinical chemistry parameters. Slight increases in relative liver weight (up to 17%) at the highest dose were not accompanied by increases in liver enzyme activity in plasma or histopathological changes and were not considered to be toxicologically significant.

On the basis of the neurotoxic effects observed at 3.5 mg/kg bw per day, the NOAEL for systemic effects was 0.5 mg/kg bw per day. On the basis of the increased incidence of liquid faeces the NOAEL for local gastrointestinal effects was 0.1 mg/kg bw per day (Hext et al. 1986).

2.3 Long-term studies of toxicity and carcinogenicity

Mice

In a 2-year dietary study, 52 male and 52 female CD-1 mice received cyhalothrin (purity unspecified) at a concentration of 0, 20, 100, or 500 ppm (equal to 0, 1.8, 9.2, and 53 mg/kg bw per day for males and 0, 2.0, 11, and 51 mg/kg bw per day for females). Satellite groups of 12 males and 12 females were given the same diets for 52 weeks and then killed. All mice were checked daily for mortality. Clinical observations were carried out daily for the first 4 weeks and then weekly for the remainder of the study. All mice were palpated weekly for masses. Food consumption per cage of mice (four per cage) and body weights were recorded weekly. Before the 1-year interim and terminal kills, samples of blood and urine were obtained for haematology, clinical chemistry and urine analysis.

At the termination of the study, all surviving mice were killed and macroscopied. Adrenals, brain, heart, kidneys, liver, lungs, ovaries, spleen and testes were weighed and a range of tissues, including any abnormal tissues, were examined histologically. Statements of adherence to QA and GLP were included.

Mortality was not affected by treatment. In both sexes at the highest dose and in males at 100 ppm, piloerection and hunched posture were observed. Males at 500 ppm also showed emaciation, pallor, hyperactivity, increased fighting activity, reduced food efficiency and reduced body-weight gain, especially during the first 13 weeks. Water consumption was also slightly increased in males at 500 ppm. Changes in haematological parameters were not consistent or dose-related and therefore considered not toxicologically relevant. Urine analysis revealed no significant differences from control values. The serum glucose concentration was slightly reduced (9–12%) in mice at 500 ppm. Increases (41–70%) in aspartate aminotransferase (AST) and alanine aminotransferase (ALT) activity were observed in males and females at 100 or 500 ppm at week 100, but not at week 50. Since the magnitude of the increases in AST were not dose-related, and not accompanied by histological changes in the liver, they are considered not toxicologically significant.

Macroscopic and histological evaluation and organ weights of the interim group revealed no treatment-related effects. In the main group, at the end of the study a number of findings, e.g. thickening of the forestomach, decreased ovarian weight and non-neoplastic lesions were not dose-related and were considered to be not related to treatment.

An increase in the incidence of mammary adenocarcinomas was seen in females in the main group (1 out of 52 in controls, 0 out of 52 at 20 ppm, 7 out of 52 (13%) at 100 ppm and 6 out of 52 (12%) at 500ppm). Data for historical controls from 17 studies from the same laboratory for 1980–1982 ranged from 2% to 12%. No preneoplastic changes were found in the mammary glands of mice in the main or satellite groups. Since the highest incidence of mammary adenocarcinoma in any group was only slightly above the range of historical controls and as there was no clear dose–response relationship, it was considered to be unlikely that these tumours were caused by treatment with cyhalothrin. On the basis of the clinical signs (piloerection and hunched posture) in males at 100 ppm, the NOAEL was 20 ppm, equal to 1.8 mg/kg bw per day (Colley et al., 1984).

Rats

Groups of 62 male and 62 female Alpk/AP rats were fed diets containing cyhalothrin (purity, 98.2%) at a concentration of 0, 10, 50 or 250 ppm (equal to about 0, 0.47, 2.3 and 12 mg/kg bw per day for males and 0, 0.55, 2.7, and 14 mg/kg bw per day for females) for 2 years. Additional groups of 10 males and 10 females were designated for interim sacrifice after 52 weeks. The rats were examined daily for clinical signs. A detailed clinical examination was carried out weekly. Body weights were recorded weekly for the first 12 weeks and every 2 weeks thereafter. Food consumption per cage of four rats was recorded weekly for the first 12 weeks and every fourth week thereafter. Haematology was performed on 10 males and 10 females per group before testing, at 4 and 13 weeks and subsequently at 13 week intervals. At termination, femoral bone-marrow smears were prepared from all rats submitted for haematological examination. Clinical chemistry and urine analysis were performed on a different group of 12 males and 12 females per dose before testing, at 4 and 13 weeks and subsequently at 13-week intervals. At week 52 and before termination, the eyes of 20 males and 20 females from the control group and the group at 250 ppm were examined.

The rats designated to the interim-kill group and the rats surviving to termination were killed and macroscopically examined. Gonads, spleen, heart, kidneys, adrenals, liver, lungs (with trachea) and brain were weighed. A range of tissues was examined histologically. Statements of adherence to QA and GLP were included.

There were no treatment-related effects on mortality and clinical signs. Compared with controls, body weights in the group at 250 ppm were decreased throughout treatment (up to 11%). The decreased body-weight gain was accompanied by a small decrease in food consumption during the first 12 weeks of the study. There were minor changes in blood biochemistry at this dose, consistent with reduced growth rate, i.e. reduced concentrations of plasma glucose (up to 9%), cholesterol (up to 26%) and triglycerides (up to 39%). No toxicologically relevant changes in haematology and ophthalmoscopy were observed.

Increased relative liver weight (up to 17%) was seen in male and female rats at 250 ppm at the interim sacrifice, but not at termination. Since these effects were not accompanied by histological lesions, these effects were considered to be not toxicologically significant. An unusual incidence of palatine fistulation and rhinitis observed from week 64 onwards was unrelated to treatment with cyhalothrin and appeared to be caused by long pointed fibres of cereal origin in the food as a consequence of a change in diet formulation. This may have contributed to a relatively high mortality rate in this study. Survival was highest in the group at 250 ppm. No significant treatment-related increase in the incidence of any tumour type was observed.

On the basis of the reduction in body-weight gain, the NOAEL was 50 ppm, equal to 2.3 mg/kg bw per day (Pigott et al., 1984).

2.4 *Reproductive toxicity*

(a) Multigeneration studies

Rats

In a dietary study of reproductive toxicity, groups of 15 male and 30 female Alpk/AP Wistar-derived rats were given diets containing cyhalothrin (purity, 89.2%) at a concentration of 0, 10, 30 or 100 ppm (equivalent to 0, 0.67, 2.0 and 6.7 mg/kg bw per day) for three successive generations. The parental rats produced two litters (F_{1a}, F_{1b}). The breeding programme was repeated with F_1 parents selected from the F_{1b} offspring and F_2 parents selected from the F_{2b} offspring. The rats were observed daily for clinical signs. Detailed examination and body-weight measurements were recorded After the premating period, the male rats were weighed at approximately 4-week intervals until termination and the females were weighed on days 1, 8, 15 and 22 of pregnancy (day 1 of pregnancy determined by the presence of sperm in a vaginal smear). Food consumption was recorded weekly throughout the premating period.

Litters were examined at least once daily and dead or grossly abnormal pups removed for soft tissue examination. The body weight, sex of the pups and any clinical abnormalities were recorded within 24 h of parturition and on days 5, 11, 22 and 29 after birth. All grossly abnormal pups, and those found dead at up to and including age 18 days were removed and examined by free-hand sectioning. Any pup aged more than 18 days found dead or moribund was removed for pathological examination and a range of tissues was taken and submitted for histopathological examination. On day 36, pups from 'a' litters were killed and approximately half of these were examined macroscopically and any abnormal tissues were histologically examined. All pups from 'b' litters except those selected to be parents of the following generation were killed on day 36 after birth and approximately five males and five females per group from the F_{1b} and F_{2b} litters and 10 males and 10 females per group from the F_{3b} litters were examined post mortem and a range of tissues taken and submitted for histopathological examination. Approximately half of the remaining rats were examined post mortem and only abnormal tissue submitted for examination. The F_0, F_1 and F_2 parents were killed and examined macroscopically and a range of tissues was taken and submitted for histopathological examination. Statements of adherence to QA and GLP were included.

A mild reduction in body-weight gain (up to 9%) was consistently found in parental rats at the highest dose, particularly males. A slight reduction in body-weight gain (7%) at 30 ppm cyhalothrin seen only in females of the F_1 generation, was considered to be not toxicologically significant. There were no treatment-related effects on food consumption, fertility rate, duration of pre-coital period or duration of gestation. No neurological effects were seen in parents or offspring. Small reductions in litter size in the group at 100 ppm were considered to be not toxicologically significant. In the group at 100 ppm, reductions in pup weight gain during lactation (up to –17%) were found in all generations except for the F_2 a generation. At autopsy, no treatment-related effects were found on pathological or histopathological examination.

The NOAEL for parental toxicity was 30 ppm, equivalent to 2.0 mg/kg bw per day, on the basis of a reduction in body-weight gain. The NOAEL for offspring toxicity was 30 ppm, equivalent to 2.0 mg/kg bw per day, on the basis of a reduced body-weight gain during lactation. The NOAEL for reproductive toxicity was 100 ppm, equivalent to 6.7 mg/kg bw per day, i.e. the highest dose tested (Milburn et al., 1984).

(b) Developmental toxicity

Rats

Groups of 24 time-mated (two females per male) CD rats were given cyhalothrin (purity, 98.2%) at a dose of 0, 5, 10 or 15 mg/kg bw per day by gavage on days 6–15 of gestation. The rats were checked daily for clinical signs. Body weight of the dams was recorded on days 0, 6–15, 18 and 20 of gestation. Food intake was recorded on days 3, 6, 9, 12, 15, 18 and 20. On day 20 of gestation, the rats were killed and examined macroscopically. The ovaries and uterus were removed and the fetuses were weighed and examined for visceral and skeletal abnormalities. A statement of adherence to QA was included.

No mortality was observed. At 15 mg/kg bw per day, two rats showed loss of limb coordination. At the highest dose, the dams initially lost weight and overall body-weight gain on days 6–20 was reduced by 12%. The effect on body weight was accompanied by a 7% reduction in food intake. There was no effect of treatment on the incidence of pregnancy, number, size, weight and sex of the fetuses and pre- and postimplantation loss. Gravid uterus weights were comparable between the groups. There was no treatment-related effect on external, visceral and skeletal development. Abnormalities in one litter at 10 mg/kg bw per day, in which 5 out of 17 fetuses had major defects (four bilateral agenesis of the kidneys; three skeletal malformations of the vertebral centrae, sternebrae, and/or metacarpals) were considered to be incidental and not related to treatment.

On the basis of reduction in body weight and the loss of limb coordination, the NOAEL for maternal toxicity was 10 mg/kg bw per day. The NOAEL for developmental toxicity was 15 mg/kg bw per day, i.e. the highest dose tested (Killick, 1981a).

Rabbits

Groups of 18–22 time-mated female New Zealand White rabbits received cyhalothrin (purity, 89.2%) at a dose of 0, 3, 10 or 30 mg/kg bw per day by gavage from day 6 to day 18 of gestation. All rabbits were examined daily for clinical signs. Body weight was recorded on days 0, 6–19, 24 and 28 of gestation. Food intake was recorded on days 3, 6, 9, 12, 15, 18, 21, 24 and 28. On day 28 of gestation, the rats were killed and examined macroscopically. The ovaries and uterus were removed and the fetuses were weighed and examined for visceral and skeletal abnormalities. A statement of adherence to QA was included.

The incidence of deaths was 1, 2, 6 and 2 in the groups at 0, 3, 10 and 30 mg/kg bw per day respectively. In the majority of rabbits, the deaths (three of which occurred before treatment had started) were attributed to pulmonary disorders, and were considered to be not related to treatment. In the surviving does, no treatment-related clinical signs and no macroscopic changes were observed.

At the highest dose, the does showed body-weight loss and reduced food consumption from days 6–9 of gestation. After that, body-weight gain of the dose at the highest dose was comparable to that of the other groups. No toxicologically relevant effects of treatment on incidence of pregnancy, gravid uterus weights, pre and postimplantation losses, number and sex of fetuses, litter weight, fetal weight or fetal crown/rump length were observed. Treatment with cyhalothrin did not affect the incidences of skeletal or visceral malformations or variations.

On the basis of the reduced body weight and food intake at the start of treatment, the NOAEL for maternal toxicity was 10 mg/kg bew per day. The NOAEL for developmental toxicity was 30 mg/kg bw per day, i.e. the highest dose tested (Killick, 1981b)

2.5 Genotoxicity

Lambda-cyhalothrin was tested for genotoxicity in a range of guideline-compliant assays, both in vitro and in vivo. No evidence for genotoxicity was observed in any test. A number of published studies, largely from the same laboratory, have reported significant increases in DNA damage in vitro (comet assay) and chromosomal aberrations in vitro and in vivo (Celik et al., 2003; Celik et al., 2005a, 2005b; Naravaneni & Jamil, 2005). Since the materials tested in these studies were either commercial formulations of unknown composition or were inadequately described, these studies were not further considered.

The results of tests for genotoxicity are summarized in Table 2. is the Meeting concluded that lambda-cyhalothrin is unlikely to be genotoxic.

2.6 Special studies: neurotoxicity

In a study of acute neurotoxicity that was performed according to OECD guideline 424, groups of 10 male and 10 female Alpk:AP$_f$SD (Wistar-derived) rats were given lambda-cyhalothrin (purity, 87.7%) at a dose of 0, 2.5, 10 or 35 mg/kg bw by gavage in corn oil. The rats were observed daily for clinical signs. Detailed clinical observations, including landing foot splay, sensory perception and muscle weakness and locomotor activity were recorded before treatment and on days 1, 8 and 15. Body weight and food consumption were measured weekly. Two weeks after dosing, five males

Table 2. Results of studies of genotoxicity with lambda-cyhalothrin

End-point	Test object	Concentration[a]	Lambda-cyhalothrin content (%)	Result	Reference
In vitro					
Reverse mutation	*S. typhimurium.* strains TA98, TA100, TA1535, TA1537 and TA1538	1.6–5000 µg/plate (±S9)	96.5	Negative	Callander (1984)[b]
Gene mutation	Mouse lymphoma L5178Y*Tk*$^{+/-}$	125-4000 µg/ml (±S9)	96.6	Negative	Cross (1985)[b]
Chromosomal aberration	Human lymphocytes	10^{-9}–10^{-2} mol/l (±S9)	96.5	Negative	Sheldon et al. (1985)[b]
Unscheduled DNA synthesis	Rat primary hepato-cytes	17–5000 µg/ml	96.6	Negative	Trueman (1989)[b]
In vivo					
Micronucleus formation	Mouse bone marrow	22 and 35 mg/kg bw (intraperitoneal)	96.5	Negative	Sheldon et al. (1984)[b,c]

[a]Positive and negative (solvent) controls were included in all studies.

[b] Statements of adherence to GLP and QA were included

[c] Doses were 44% and 70% of the median lethal dose at 7 days (based on mortality observed within 7 days after a singleintraperitoneal injection of lambda cyhalothrin). A reduction in the polychromatic erythrocytes-normochromatic erythrocytes (PCE/NCE) ratio indicated that lambda-cyhalothrin had reached the bone marrow.

and five females per group were killed and examined macroscopically. Selected nervous tissues were removed, processed and examined microscopically. Statements of adherence to QA and GLP were included.

In the group at the highest dose, clinical signs, including decreased activity, ataxia, increased breathing rate, reduced stability and/or tiptoe gait, upward curvature of the spine, piloerection, occurred about 7 h after dosing, and were also observed in some rats on days 2–4. There was a statistically significant reduction in landing-foot splay in males and a statistically significant increase in tail-flick response in females on day 1. Motor activity was reduced in both sexes on day 1. Increased breathing rate in five females and urinary incontinence, salivation and/or reduced response to sound in one to two rats were observed at 10 mg/kg bw. At this dose, no changes were seen in the functional observational battery (FOB). A slightly reduced hindlimb-grip strength, observed in the groups at the lowest and highest dose, showed no dose–response relationship and was attributed to slightly higher values for the controls. At the highest dose, food consumption (both sexes) and body weights (males only) were reduced during the first week of the study. At termination, no treatment-related macroscopic and histological changes were observed.

On the basis of signs of neurotoxicity in the groups at the intermediate and highest dose, the NOAEL was 2.5 mg/kg bw (Brammer, 1999).

In a study of acute neurotoxicity from published literature, groups of male Long Evans rats received lambda-cyhalothrin (purity, 87.7%), and a number of other pyrethroids, by gavage in corn oil to male Long Evans rats. It was reported that 8–18 rats per group, and 6–11 doses per compound were tested (not further specified for lambda-cyhalothrin).

Motor activity was assessed in an automated figure-eight maze. The data were analysed using a nonlinear exponential threshold model. The model was used to determine the dose associated with a 30% decrease in motor activity (ED_{30}) and the threshold dose (estimate of highest no-effect dose at which the rats would not display any decrease in motor activity) and the 95% confidence limits (CL).

Lambda-cyhalothrin induced a decrease in motor activity with an ED_{30} of 1.32 ± 0.13 mg/kg bw (lower 95% CL, 1.06 mg/kg bw) and a threshold dose of 0.52 ± 0.13 mg/kg bw (lower 95% CL, 0.28 mg/kg bw). The NOAEL was 0.5 mg/kg bw (Wolansky et al., 2006)

In a 90-day study of neurotoxicity that was performed according to OECD guideline 424, groups of 12 male and 12 female Alpk:APfSD (Wistar-derived) rats were fed diets containing lambda-cyhalothrin (purity, 87.7%) at a concentration of 0, 25, 60 or 150 ppm (equal to 0, 2.0, 4.6 and 11.4 mg/kg bw per day for males and 0, 2.2, 5.2 and 12.5 mg/kg bw per day for females). All rats were observed daily for clinical signs. Detailed clinical observations, including quantitative assessments of landing-foot splay, sensory perception (tail-flick test) and muscle weakness (fore-and hindlimb grip strength), and tests for locomotor activity were performed in weeks −1, 2, 4, 9 and 14. Body weights and food consumption were measured weekly. The eyes of rats at the highest dose and in the control groups were examined pre-study and in week 12. At the end of the 13 weeks, five males and five females were killed, brains were weighed and selected nervous system tissues were removed, processed and examined microscopically. Statements of adherence to QA and GLP were included.

No treatment-related clinical signs were reported. At 150 ppm, body-weight gain was slightly reduced in males (5–7%). At this dose, food consumption was decreased by 14–17% during the first week of treatment, and remained less than that of controls during the first half of the study. Food consumption was also decreased in the other treatment groups during the first week of treatment. This was considered to be due to the reduced palatability of the food. Ophthalmoscopy and the comprehensive battery of neurobehavioural and neuropathological examinations revealed no effects of treatment with lambda-cyhalothrin up to the highest dose tested. A statistically significant reduction in hindlimb grip strength in males at the highest dose during week 9 only was considered to be incidental and not related to treatment.

The reduction in body-weight gain at 150 ppm was slight, not progressive and occurred only in the first week of exposure and hence it is reasonable to attribute it to unpalatability of the diet. Therefore, the NOAEL was 150 ppm, equal to 11 mg/kg bw per day, i.e. the highest dose tested (Brammer, 2001).

In a study of developmental neurotoxicity, groups of 30 time-mated female Alpk:AP_fSD (Wistar-derived) rats received diets containing lambda cyhalothrin (purity, 87.7%) at a concentration of 0, 25, 60 or 150 ppm (equal to 0, 2.1, 4.9 and 11.4 mg/kg bw per day during gestation and 0, 4.6, 10.7 and 26.3 mg/kg bw per day during lactation) from day 7 of gestation through parturition and lactation to day 23 after birth. Pups were allocated to the F_1 phase of the study on postnatal day 5, separated from the dam on day 29 and allowed to grow to adulthood.

The dams were checked daily for clinical signs. Body weight and detailed clinical observations were recorded before administration on day 7, on days 15 and 22 of gestation and on days 1, 5, 8, 12, 15 and 22 after birth and on the day of termination. Food consumption was recorded weekly. On days 10 and 17 of gestation and on days 2 and 9 of lactation, the females were subjected to FOB. Each litter was examined as soon as possible after parturition on day one. On postnatal days 1 and, the sex, weight and clinical condition of each pup was recorded. Litters were checked daily for dead or abnormal pups. On postnatal day 5, litters were standardized to eight (four males and four females) randomly selected pups where possible. Litters of seven or eight pups with at least three of each sex were used for selection of the F_1 generation. F_1 rats were examined daily for mortality and clinical signs of toxicity. From postnatal day 5, detailed clinical observations were recorded at the same time as the rats were weighed which was on days 5, 12, 18, 22, 29, 36, 43, 50 and 57 and before termination on day 63. The F_1 females were examined daily from day 29 to determine the day of vaginal opening. The selected F_1 males were examined daily from day 41 to determine the day of preputial separation. Body weight was recorded on the day the landmark was achieved. For the FOB, at least 10 males and females per group were observed on days 5, 12, 22, 36, 46 and 61. One male and one female from each litter were assigned to tests for auditory startle reflex (days 23 and 61), locomotor activity (days 14, 18, 22 and 60) or a learning and memory (Y-shaped water maze, on days 21 or 59 followed 3 days later by retesting).

Parent females were killed on postnatal day 29 and were not examined. Any intercurrent death F_1 rats and those not selected and killed on postnatal day 5 were not examined. At scheduled termination on days 12 and 63, one male and one female from each litter was killed, and brains were weighed. On day 63, a further 10 males and females per group were killed, brains were weighed and selected nervous system tissues were collected. The brains and other selected nervous tissues of the rats in the control group and at the highest dose were examined microscopically. Statements of adherence to QA and GLP were included.

There were no treatment-related effects in parental females on clinical observations or in the FOB. During days 7–15 of gestation, body-weight gain of females at the highest dose was reduced by 51% when compared with that of controls. From day 15 of gestation, body-weight gain of the group at the highest dose was comparable to that of the other groups, although the body weights of these rats remained lower throughout the rest of the study. The reduction in body-weight gain was accompanied by a reduction in food consumption.

There were no treatment-related effects on reproduction parameters. In the group at 150 ppm, survival to day 5 was slightly lower than that of controls. Pup weights at birth were similar to those of controls. Individual pup weights and total litter weights in the group at 150 ppm were lower (up to

10–12%) than those of controls from days 5–22 after birth. There were no treatment-related clinical observations in F_1 rats.

In F_1 males at 150 ppm, the mean age of preputial separation was slightly but statistically significantly greater than that in controls (45.8 days vs 45 days), and accompanied by a slightly lower body weight. This probably reflects the poorer pup growth in this group. Time of vaginal opening was not affected by treatment.

There were no treatment-related effects on motor activity, auditory startle response or learning and memory. Also, no treatment-related macroscopic findings, effects on brain weight, changes in brain morphometry or microscopic findings were observed. There was an increased incidence of demyelination in the proximal sciatic and distal tibial nerves of males at 150 ppm on day 63. This is a common spontaneous finding in this age and strain of rat, and the increased incidences were within the ranges for historical controls and therefore was considered to be incidental.

On the basis of the reduction in body-weight gain during gestation, the NOAEL for maternal toxicity was 60 ppm, equal to 4.9 mg/kg bw per day. On the basis of the reduction in body-weight gain during lactation the NOAEL for offspring toxicity was 60 ppm, equal to 10.7 mg/kg bw per day (based on maternal lambda-cyhalothrin intake). No evidence for developmental neurotoxicity was found at doses up to and including 150 ppm, equal to 11.4 mg/kg bw per day (Milburn, 2004).

3. Observations in humans

Six male volunteers were each given a single oral dose of 5 mg of lambda-cyhalothrin as a solution in corn oil (25 mg/g) in a gelatin capsule, followed by 150 ml of water. A group of five volunteers received a dermal dose of 20 mg/800 cm^2. Blood samples were taken from an in-dwelling cannula before dosing and at 0.5, 1, 2 ,3, 4, 5, 6, 8, 10 and 12 h after dosing and by venepuncture at approximately 24, 31 and 48 h. Complete urine collections were made at 2-h intervals up to 14 h, then 14–24 h followed by 12-h periods up to 120 h. Faeces were collected daily (oral dose only).

Samples of blood and excreta, including hydrolysed samples, were extracted with solvent that was analysed for the test substance and for three specific metabolites, i.e. 3-(2-chloro-3,3,3,-trifluoroprop-1-enyl)-2,2-dimethylcyclopropanoic acid (TFMCVA, also known as compound Ia, see Fig. 1), 3-phenoxybenzoic acid (3PBA, also known as compound V) and 3-(4-hydroxyphenoxy)benzoic acid (4-OH-3PBA, also known as compound XXIII). These metabolites were analysed as pentafluoropropionyl derivatives by GC-MS. Statements of adherence to QA and GLP were included.

Approximately equal amounts of TFMCVA (56.7% of administered dose), and 3PBA (25.1% of administered dose) + 4-OH-3PBA (25.2% of administered dose) were excreted in the urine, with peak excretion rates of between 2 h and 14 h after dosing. No parent compound was detected in the urine. Based upon TFMCVA measurements in urine, the estimated amount of test substance absorbed ranged from 50% to 64%.

The presence of intact test substance in plasma showed that lambda-cyhalothrin can be absorbed unchanged; however, the observation that the highest plasma concentrations of TFMCVA and 3PBA occurred soon after dosing indicates possible pre-systemic hydrolysis and/or rapid hydrolysis of the test substance in the liver and blood. Unabsorbed test substance and TFMCVA were detected in the faeces, but, with the exception of one subject, accounted for less than 1.5% of the dose. It was stated in the report that other metabolites in faeces could not be estimated due to matrix interference.

After dermal administration, concentrations of metabolites in the urine were very low. On the basis of TFMCVA excretion in urine it was estimated that 0.12% of the administered dose of lambda-cyhalothrin was absorbed through the skin. Total recovery of radioactivity ranged from 56% to 90%. Radioactivity was still excreted in urine at the end of the 120-h sampling period. In view of the amount of radioactivity that was not recovered and the urinary excretion of radioactivity at the end of the study, dermal absorption was likely to be greater than 0.12%, albeit still low (Marsh et al, 1994).

No other data on human observations were available for the present evaluation. Observations in humans were described by the World Health Organization (WHO) in 1999 and by JECFA in 2000. The text below is copied from the JECFA 2000 evaluation.

> No clinical or haematological effects were observed in six volunteers given a single dose of 5 mg of lambda-cyhalothrin in corn oil (equal to 0.05–0.07 mg/kg bw) (European Medicines Evaluation Agency, 1999). The route of administration was not reported, but it seems likely to have been oral.
>
> In the study of Pakistani pesticide workers described in section 2.2, the average exposure of the workers to lambda-cyhalothrin was estimated to be 54 μg/person per day (extrapolated from measured metabolites in urine). Transient signs of toxicity, lasting up to 24 h, were reported by the workers, which included skin paraesthesia, feeling hot, feeling cold, numbness, irritation of the skin, red eyes, coughing, and sneezing. Medical examination revealed one case of face rash that lasted 2 days (Chester et al., 1992).
>
> In a study carried out in a village in the United Republic of Tanzania, a lambda-cyhalothrin-based insecticide was sprayed inside houses and shelters at a coverage of approximately 25 mg/m^2. The insecticide was supplied as a water-dispersible powder in a soluble sachet. Every day for 6 days, 12 spraymen and 3 squad leaders were interviewed about symptoms. Each sprayman used up to 62 g of lambda-cyhalothrin over 2.7–5.1 h each day. The spraymen wore personal protective equipment (rubber boots and gloves, cotton overalls, caps, and gauze nose-mouth masks) which left much of the face exposed. One sprayman also used a face shield for 3 days.
>
> All the spraymen complained at least once of symptoms related to exposure to lambda-cyhalothrin. The commonest symptoms were itching and burning of the face and nose and throat irritation, frequently accompanied by sneezing or coughing. Facial symptoms occurred only on unprotected areas, and the worker who wore a face shield was free of facial symptoms. All the symptoms had disappeared by the morning after the spraying. The number of subjects affected and the duration of facial symptoms were proportional to the amount of compound sprayed. These parameters were not affected by use of lambda-cyhalothrin in the previous 6 months. A sample of occupants was interviewed 1 and 5–6 days after their houses had been sprayed. One woman who entered her house 30 min after the end of spraying complained of periorbicular itching, but this lasted only a few minutes. The other inhabitants of sprayed houses reported no other insecticide-related adverse effects. Furthermore, a squad leader who entered almost every house a few minutes after spraying reported no symptoms (Moretto, 1991)
>
> As part of a field trial conducted in South India, electrophysiological tests were conducted on 15 spraymen aged 19–48 years, before and after exposure to lambda-cyhalothrin (formulation and purity unspecified). The tests performed on the subjects comprised conduction of the right median, common peroneal, and facial motor nerves; conduction of the right median and sural sensory nerves; blink response with stimulation of the right supra-orbital nerve and recording of R1 and R2 responses from the right orbicularis oculi muscle with a pair of surface electrodes; concentric needle electromyography of the tibialis anterior; repetitive stimulation of the right median nerve at the wrist at 3 and 20 Hz and recording of the responses from the abductor pollicis brevis; and multi-modality visual, brainstem, auditory and somatosensory evoked potentials. The evoked potentials were measured in only six of the subjects, but the other measurements were made in all 15 subjects.
>
> Clinical observation revealed no changes, and facial nerve conduction, blink response, responses to repetitive stimulation, and visual, auditory, and somatosensory evoked potentials were all normal. Six of the 15 subjects had mild changes in peripheral nerve conduction parameters (paired t test: $p < 0.05$), but comparison of the mean values for the various nerve conduction parameters before and after exposure showed no significant difference except for prolongation of distal motor latency of the median nerve. Studies of nerve conduction 12–16 months later in three subjects who had shown abnormalities immediately after exposure showed normal rates. The authors concluded that occupational exposure to lambda-cyhalothrin can produce transient, subclinical electrophysiological changes in the nerves of the upper limbs (Arunodaya et al., 1997).

Comments

Biochemical aspects

Oral doses of cyhalothrin were readily but incompletely absorbed (30–40% of radiolabel was recovered in urine) in rats and dogs. Peak blood concentrations were reached after 4–7 h. In male rats treated with replacement bile obtained from treatment-naive rats, biliary excretion was about 11%. At a low dose, most (70%) of the administered material was excreted in the faeces and urine within 24 h. After 7 days, 2–3% of the cyhalothrin administered persisted as unchanged residue in fat. Metabolism in rats and dogs was similar, involving initial cleavage of the molecule at the ester bond. In rats dosed with cyhalothrin, major metabolites identified in urine were the sulfate conjugate of 3-(4'-hydroxyphenoxy) benzoic acid (compound XXIII) glucuronide conjugate of (1*RS*)-*cis*-3-(2-chloro-3,3,3-trifluoropropenyl)-2,2-dimethylcyclopropanoic acid (i.e. the compound 1a glucuronide). Minor metabolites identified were unconjugated compound XXIII and 3-phenoxybenzoic acid (compound V).

In volunteers given a single dose of lambda-cyhalothrin in capsules, serum and urine contained the metabolites compound XXIII, compound V and compound 1a ((1*RS*)-*cis*-3-(*Z*-2-chloro-3,3,3-trifluoroprop-1-enyl)-2,2-dimethylcyclopropanoic acid, TMFVCA). Their presence suggests that the initial metabolism of this compound in humans is similar to that in rats and dogs.

Toxicological data

The acute oral LD_{50} of lambda-cyhalothrin in rats was 79 mg/kg bw in males and 56 mg/kg bw in females. The observed clinical signs (ataxia, decreased activity, tiptoe gait, splayed gait, loss of stability, dehydration, urinary incontinence, hunched posture, piloerection, salivation, ungroomed appearance and pinched-in sides) were typical of this class of pyrethroids.

In studies with lambda-cyhalothrin in rats, the inhalation LC_{50} value was 60 mg/m^3 (0.06 mg/l), and the dermal LD_{50} was 632 mg/kg bw in males and 696 mg/kg bw in females. Lambda-cyhalothrin was not irritating to the skin and only slightly irritating to the eyes. With respect to dermal sensitization, the results of a maximization test with lambda-cyhalothrin in guinea-pigs were inconclusive. Technical-grade cyhalothrin has been reported to cause skin sensitization in a Buehler test and a maximization test in guinea-pigs.

In a 90-day feeding study in rats given cyhalothrin, the NOAEL was 50 ppm, equal to 2.6 mg/kg bw per day, on the basis of reduced body-weight gain and food consumption. In a 90-day feeding study in rats given lambda-cyhalothrin, the NOAEL was 50 ppm, equivalent to 2.5 mg/kg bw per day, on the basis of reduced body-weight gain and food consumption. In a 26-week study in dogs fed capsules containing cyhalothrin and a 1-year study in dogs fed capsules containing lambda-cyhalothrin, increased incidences of liquid faeces was observed, with an overall NOAEL of 0.1 mg/kg bw per day. The increased incidences of liquid faeces were observed from the first week of treatment. Other pyrethroids produce this effect, which may be the consequence of the local gastrointestinal equivalent of paraesthesia in the skin. In the two studies in dogs, signs of systemic neurotoxicity (ataxia, tremors, and occasionally convulsions) were observed, with an overall NOAEL of 0.5 mg/kg bw per day. Signs of systemic neurotoxicity were observed from the first week and generally occurred within a few hours after treatment.

In a 2-year dietary study with cyhalothrin in mice, the NOAEL was 20 ppm, equal to 1.8 mg/kg bw per day, on the basis of clinical signs (piloerection and hunched posture) in males. An increase in the incidence of mammary adenocarcinomas in the groups receiving the intermediate or highest dose was at the upper limit of the range for historical controls and was not dose-related. The Meeting therefore considered that it was unlikely that these tumours were caused by treatment with cyhalothrin.

In a 2-year dietary study with cyhalothrin in rats, the NOAEL was 50 ppm, equal to 2.3 mg/kg bw per day, on the basis of a reduction in body-weight gain. No treatment-related changes in tumour incidence were observed in this study.

The Meeting concluded that cyhalothrin is not carcinogenic in rodents.

Lambda-cyhalothrin was tested for genotoxicity in an adequate range of assays, both in vitro and in vivo. No evidence for genotoxicity was observed in any test. A number of published studies, largely from the same laboratory, have reported significant increases in DNA damage in vitro (Comet assay) and chromosomal aberrations in vitro and in vivo. The materials tested in these studies were either commercial formulations of unknown composition or were inadequately described. In view of the uniform finding of a lack of genotoxicity in those studies in which lambda-cyhalothrin was adequately characterized, the Meeting concluded that lambda-cyhalothrin is unlikely to be genotoxic.

In view of the lack of genotoxicity of lambda-cyhalothrin and the absence of carcinogenicity shown by cyhalothrin in mice and rats, the Meeting concluded that lambda-cyhalothrin is unlikely to pose a carcinogenic risk to humans.

In a multigeneration dietary study with cyhalothrin in rats, the NOAEL for parental toxicity was 30 ppm, equivalent to 2.0 mg/kg bw per day, on the basis of a reduction in body-weight gain. The NOAEL for offspring toxicity was 30 ppm, equivalent to 2 mg/kg bw per day, on the basis of reduced body-weight gain during lactation. The NOAEL for reproductive toxicity was 100 ppm, equivalent to 6.7 mg/kg bw per day, i.e. the highest dose tested.

The effect of oral exposure to cyhalothrin on prenatal development was investigated in rats and rabbits. In a study of developmental toxicity in rats treated by gavage, the NOAEL for maternal toxicity was 10 mg/kg bw per day on the basis of a reduction in body weight and loss of limb coordination. The NOAEL for fetal toxicity was 15 mg/kg bw per day, i.e. the highest dose tested. In a study of developmental toxicity in rabbits treated by gavage, the NOAEL for maternal toxicity was 10 mg/kg bw per day on the basis of reduced body-weight gain and food consumption. The NOAEL for fetotoxicity was 30 mg/kg bw per day, i.e. the highest dose tested.

In a study of acute neurotoxicity in rats given lambda-cyhalothrin by gavage, the NOAEL was 2.5 mg/kg bw per day on the basis of signs of neurotoxicity (increased breathing rate, urinary incontinence, salivation, reduced response to sound).

In a comparative study on the acute effects of pyrethroids in rats treated by oral gavage, in which the data were analysed using a nonlinear exponential threshold model, lambda-cyhalothrin showed decreased motor activity with a benchmark threshold dose (estimate of the highest no-effect level at which the rats would not display any decrease in motor activity) of 0.5 mg/kg bw. In a 90-day dietary study, the NOAEL was 150 ppm (equal to 11 mg/kg bw per day), i.e. the highest dose tested.

In a study of developmental neurotoxicity in rats, the NOAEL for maternal toxicity was 60 ppm, equal to 4.9 mg/kg bw per day, on the basis of reduced body-weight gain during gestation. The NOAEL for offspring toxicity was 60 ppm, equal to 10.7 mg/kg bw per day, based on maternal lambda-cyhalothrin intake, on the basis of reduced body-weight gain during lactation. No evidence for developmental neurotoxicity was observed.

In case reports in humans, no systemic effects were reported. In most cases exposure was by the dermal and inhalation routes. Predominant signs were skin paraesthesia, numbness, irritation of the skin, red eyes, coughing and sneezing.

No toxicological studies on metabolites of cyhalothrin were available. However, the Meeting considered it likely that the metabolites would be less neurotoxic than cyhalothrin, as none contains an intact pyrethroid structure.

The Meeting concluded that the existing database on lambda-cyhalothrin was adequate to characterize the potential hazards to fetuses, infants and children.

Toxicological evaluation

Although increased incidences of liquid faeces were observed in dogs given lambda-cyhalothrin/cyhalothrin, which may represent a consequence of a local gastrointestinal equivalent of paraesthesia in the skin, the Meeting considered that it was not appropriate to base the ADI and acute reference dose (ARfD) on local effects on the gastrointestinal tract, observed after bolus administration.

The most sensitive systemic effect of lambda-cyhalothrin/cyhalothrin was neurotoxicity (decreased motor activity), which was observed in a study of acute toxicity in rats given lambda-cyhalothrin orally, with a threshold dose of 0.5 mg/kg bw, and in repeat-dose studies with cyhalothrin and lambda-cyhalothrin in dogs treated orally (ataxia, tremors, occasionally convulsions) with a NOAEL of 0.5 mg/kg bw per day. On the basis of these effects, the Meeting established a group ADI for cyhalothrin and lambda cyhalothrin of 0–0.02 mg/kg bw, using a safety factor of 25. Because lambda-cyhalothrin is relatively rapidly absorbed and excreted and the neurotoxic effects are rapidly reversible and dependent on C_{max}, the Meeting considered it appropriate to adjust the safety factor for the reduced variability in C_{max} compared with AUC. The Meeting considered that the ADI of 0.02 mg/kg bw is adequately protective against the other, non-neurotoxic effects of lambda-cyhalothrin/cyhalothrin observed in short- and long-term studies with repeated doses, and in studies of reproductive and developmental toxicity, where the use of a safety factor of 100 would be appropriate.

The Meeting established a group ARfD for cyhalothrin and lambda-cyhalothrin of 0.02 mg/kg bw on the basis of systemic neurotoxicity (decreased motor activity) observed in a study of acute toxicity in rats given lambda-cyhalothrin orally with a threshold dose of 0.5 mg/kg bw per day, and in repeat-dose studies with cyhalothrin and lambda-cyhalothrin in dogs treated orally, in which neurotoxic effects (ataxia, tremors, occasionally convulsions) occurred during the first week, within a few hours after treatment, with an overall NOAEL of 0.5 mg/kg bw per day, and using a safety factor of 25. For the same reasons as described above, the Meeting considered it appropriate to adjust the safety factor for the reduced variability in C_{max} compared with AUC.

Levels relevant for risk assessment

(a) Cyhalothrin

Species	Study	Effect	NOAEL	LOAEL
Mouse	Two-year study of toxicity and carcinogenicity[a]	Toxicity	20 ppm, equal to 1.8 mg/kg bw per day	100 ppm, equal to 9.2 mg/kg bw per day
		Carcinogenicity	500 ppm, equal to 51 mg/kg bw per day[c]	—
Rat	Ninety-day study of toxicity[a]	Toxicity	50 ppm, equal to 2.6 mg/kg bw per day	250 ppm, equal to 14 mg/kg bw per day
	Two-year study of toxicity and carcinogenicity[a]	Toxicity	50 ppm, equal to 2.3 mg/kg bw per day	250 ppm, equal to 12 mg/kg bw per day
		Carcinogenicity	250 ppm, equal to 12 mg/kg bw per day[c]	—
	Two-generation study of reproductive toxicity[a]	Parental toxicity	30 ppm, equivalent to 2.0 mg/kg bw per day	100 ppm, equivalent to 6.7 mg/kg bw per day[d]
		Offspring toxicity	30 ppm, equivalent to 2.0 mg/kg bw per day	100 ppm, equivalent to 6.7 mg/kg bw per day[d]
		Reproductive toxicity	100 ppm, equivalent to 6.7 mg/kg bw per day[c]	—

	Developmental toxicity[b]	Maternal toxicity	10 mg/kg bw per day	15 mg/kg bw per day
		Fetotoxicity	15 mg/kg bw per day[c]	—
Rabbit	Developmental toxicity[b]	Maternal toxicity	10 mg/kg bw per day	30 mg/kg bw per day
		Fetotoxicity	30 mg/kg bw per day[c]	—
Dog	Twenty-six-week study[b]	Toxicity	2.5 mg/kg bw per day	10 mg/kg bw per day

[a] Dietary administration.
[b] Gavage administration.
[c] Highest dose tested.

(b) Lambda-cyhalothrin

Species	Study	Effect	NOAEL	LOAEL
Rat	Ninety-day study of toxicity[a]	Toxicity	50 ppm, equivalent to 2.5 mg/kg bw per day	250 ppm, equivalent to 12.5 mg/kg bw per day
	Acute neurotoxicity[b]	Neurotoxicity	0.5 mg/kg bw[e]	1.3 mg/kg bw[f]
	Ninety-day study of neurotoxicity[a]	Neurotoxicity	150 ppm, equal to 11 mg/kg bwper day[c]	—
	Developmental neurotoxicity[a]	Maternal toxicity	60 ppm, equal to 4.9 mg/kg bw per day	150 ppm, equal to 11.4 mg/kg bw per day
		Offspring toxicity	60 ppm, equivalent to 10.7 mg/kg bw per day[d]	150 ppm, equivalent to 26.3 mg/kg bw per day[d]
		Developmental (neuro)-toxicity	150 ppm, equivalent to 11.4 mg/kg bw per day[c]	—
Dog	One-year study[b]	(Neuro)toxicity	0.5 mg/kg bw per day	3.5 mg/kg bw per day

[a] Dietary administration.
[b] Gavage administration.
[c] Highest dose tested.
[d] Based on maternal intake of lambda-cyhalothrin during lactation.
[e] Threshold dose obtained using a nonlinear exponential threshold model.
[f] ED_{30} (dose associated with a 30% decrease in motor activity) obtained using a nonlinear exponential threshold model.

Estimate of acceptable daily intake for humans

0–0.02 mg/kg bw

Estimate of acute reference dose

0.02 mg/kg bw

Information that would be useful for the continued evaluation of the compound

Results from epidemiological, occupational health and other such observational studies of human exposures

Critical end-points for setting guidance values for exposure to cyhalothrin/lambda-cyhalothrin

Absorption, distribution, excretion and metabolism in animals	
Rate and extent of absorption	Rapid, incomplete absorption (about 40–50% in rats)
Distribution	Highest concentrations in fat, followed by liver and kidney (rats)
Potential for accumulation	Low
Rate and extent of excretion	Rapid (70% in faeces and urine within 24 h in rats)

Metabolism in animals	Sulfate conjugate of 3-(4'-hydroxyphenoxy) benzoic acid (compound XXIII) and glucuronide conjugate of (1RS)-cis-3-(2-chloro-3,3,3-trifluoropropenyl)-2,2-dimethylcyclopropane carboxylic acid. Unconjugated compound XXIII and 3-phenoxybenzoic acid (compound V) were minor metabolites.
Toxicologically significant compounds in animals, plants and the environment	Cyhalothrin, lambda-cyhalothrin
Acute toxicity	
Rat, LD_{50}, oral	56 mg/kg bw
Rat, LD_{50}, dermal	632 mg/kg bw
Rat, LC_{50}, inhalation	0.060 mg/l
Rabbit, skin irritation	Not an irritant (cyhalothrin)
Rabbit, eye irritation	Slightly irritating (lambda-cyhalothrin)
Guinea-pig, skin sensitization	Sensitizing (cyhalothrin, Buehler test and Magnusson & Kligman)
Short-term studies of toxicity	
Target/critical effect	Neurotoxicity, i.e. ataxia, tremors, occasionally convulsions (dogs)
Lowest relevant oral NOAEL	0.5 mg/kg bw per day (lambda-cyhalothrin, dogs)
Lowest relevant dermal NOAEL	No data
Lowest relevant inhalation NOAEC	No data
Long-term studies of toxicity and carcinogenicity	
Target/critical effect	Decreased body-weight gain (rats)
Lowest relevant NOAEL	50 ppm, equal to 2.3 mg/kg bw per day (cyhalothrin, rats)
Carcinogenicity	Not carcinogenic (cyhalothrin, mice, rats)
Genotoxicity	
	Not genotoxic (lambda-cyhalothrin)
Reproductive toxicity	
Reproduction target/critical effect	No reproductive effects (rats)
Lowest relevant reproductive NOAEL	100 ppm, equal to 6.7 mg/kg bw per day, i.e. highest dose tested (cyhalothrin, rats)
Developmental target	No developmental effects (rabbits)
Lowest relevant developmental NOAEL	30 mg/kg bw per day (lambda-cyhalothrin, rabbits)
Neurotoxicity/delayed neurotoxicity	
Neurotoxicity	Type II pyrethroid toxicity (choreoathetosis/salivation syndrome)
Lowest relevant oral NOAEL	0.5 mg/kg bw (lambda-cyhalothrin, rats, dogs)
Other toxicological studies	
	No data
Medical data	
	No systemic poisoning reported. Skin paraesthesia, numbness, irritation of the skin, red eyes, coughing and sneezing.

Summary for cyhalothrin and lambda-cyhalothrin

	Value	Study	Safety factor
Group ADI	0–0.02 mg/kg bw	Rat, acute neurotoxicity, lambda-cyhalothrin;[a] dog, 1-year, lambda-cyhalothrin	25
Group ARfD	0.02 mg/kg bw	Rat, acute neurotoxicity, lambda-cyhalothrin; dog, 1-year, lambda-cyhalothrin[b]	25

[a] The lowest NOAEL for the primary action of the chemical and considered to be protective of other non-neurotoxic effects from studies of repeated doses.

[b] Neurotoxicity occurred a few hours after dosing during the first week of treatment.

References

Barber, J.E. (1985) PP321: acute dermal toxicity study. Unpublished report No. CTL/P/1149 from Central Toxicology Laboratory, Macclesfield, England. Submitted to WHO by Syngenta Crop Protection AG.

Brammer, A. (1999) Lambda-cyhalothrin: acute neurotoxicity study in rats. Unpublished report No. CTL/P/6151 from Central Toxicology Laboratory, Macclesfield, England. Submitted to WHO by Syngenta Crop Protection AG.

Brammer, A. (2001) Lambda-cyhalothrin: subchronic neurotoxicity study in rats. Unpublished report No. CTL/PR1125 from Central Toxicology Laboratory, Macclesfield, England. Submitted to WHO by Syngenta Crop Protection AG.

Callander, R.D. (1984) PP321 - an evaluation in the Salmonella mutagenicity assay. Unpublished report No. CTL/P/1000 from Central Toxicology Laboratory, Macclesfield, England. Submitted to WHO by Syngenta Crop Protection AG.

Celik, A., Mazmanci, B., Camlica, Y., Aşkin, A., & Cömelekoglu, U. (2003) Cytogenetic effects of lambda-cyhalothrin on Wistar rat bone marrow. *Mutation Res.*, **539**, 91–97.

Celik, A., Mazmanci, B., Camlica, Y., Cömelekoglu, U. & Aşkin, A. (2005) Evaluation of cytogenetic effects of lambda-cyhalothrin on Wistar rat bone marrow by gavage administration. *Ecotoxicol. Environ. Saf.*, **61**, 12–133.

Celik, A., Mazmanci, B., Camlica, Y., Aşkin, A., & Cömelekoglu, U. (2005) Induction of micronuclei by lambda-cyhalothrin in Wistar rat bone marrow and gut epithelial cells. *Mutagenesis*, **20**, 125–129.

Chesterman, H., Heywood, R., Allen, T.R., Street, A.E., Kelly, D.F., Gopinath, C. & Prentice, D.E. (1981) Cyhalothrin oral toxicity study in beagle dogs (repeated daily dosing for 26 weeks). Unpublished report No. CTL/C/1093 from Central Toxicology Laboratory, Macclesfield, England. Submitted to WHO by Syngenta Crop Protection AG.

Colley, J., Dawe, S., Heywood, R., Almond, R., Gibson, W.A., Gregson, R. & Gopinath, C. (1984) Cyhalothrin: potential tumorigenic and toxic effects in prolonged dietary administration to mice. Unpublished report No CTL/C/1260 from Central Toxicology Laboratory, Macclesfield, England. Submitted to WHO by Syngenta Crop Protection AG.

Cross, M. (1985) PP321: Assessment of mutagenic potential using L5178Y mouse lymphoma cells. Unpublished report No. CTL/P/1340 from Central Toxicology Laboratory, Macclesfield, England. Submitted to WHO by Syngenta Crop Protection AG.

Harrison, M.P. & Case, D.E. (1981) Cyhalothrin: the disposition and metabolism of ^{14}C-ICI 146814 in rats. Unpublished report No. 8/HA/003990 from ICI Pharmaceuticals Division. Submitted to WHO by Syngenta Crop Protection AG. Submitted to WHO by Syngenta Crop Protection AG.

Harrison, M.P. & Case, D.E. (1983) Cyhalothrin: the metabolism and disposition of ICI 146814 in the rat: part IV. Unpublished report No. 6/HC/005683 from Central Toxicology Laboratory, Macclesfield, England. Submitted to WHO by Syngenta Crop Protection AG.

Harrison, M.P. (1984a) Cyhalothrin: the metabolism and disposition of ICI 146,814 in rats; part II. Tissue residues derived from [^{14}C-benzyl] or [^{14}C-cyclopropyl] ICI 146,814, after a single oral dose of 1 or 25 mg/kg. Unpublished report No. CTL/C/1279B from Central Toxicology Laboratory, Macclesfield, England. Submitted to WHO by Syngenta Crop Protection AG.

Harrison, M.P. (1984b) Cyhalothrin: the metabolism and disposition of [^{14}C]-ICI 146,814 in rats; part III. Studies to determine the radioactive residues in the rat following 14 days repeated oral administration. Unpublished report No. CTL/C/1279C from Central Toxicology Laboratory, Macclesfield, England. Submitted to WHO by Syngenta Crop Protection AG.

Harrison, M.P. (1984c) Cyhalothrin (ICI 146, 814): the disposition and metabolism of [^{14}C]-ICI 146,814 in the dog. Unpublished report No. CTL/C/1277 from Central Toxicology Laboratory, Macclesfield, England. Submitted to WHO by Syngenta Crop Protection AG.

Hart, D., Banham, P.B., Chart, I.S., Evans, D.P., Gore, C.W., Stonard, M.D., Moreland, S., Godley, M.J. & Robinson, M. (1985) PP321: 90-day feeding study in rats. Unpublished report No. CTL/P/1045 from Central Toxicology Laboratory, Macclesfield, England. Submitted to WHO by Syngenta Crop Protection AG.

Hext, P.M. (1987) PP321: 4-hour acute inhalation toxicity study in the rat. Unpublished report No. CTL/P/1683 from Central Toxicology Laboratory, Macclesfield, England. Submitted to WHO by Syngenta Crop Protection AG.

Hext, P.M., Brammer, A., Chalmers, D.T., Chart, I.S., Gore, C.W., Pate, I. & Banham, P.B. (1986) PP321: 1-year oral dosing in dogs. Unpublished report No. CTL/P/1316 from Central Toxicology Laboratory, Macclesfield, England. Submitted to WHO by Syngenta Crop Protection AG.

Horner, S. (1996) Lambda-cyhalothrin: 6-week oral toxicity study in dogs. Unpublished report No. CTL/P/5256 from Central Toxicology Laboratory, Macclesfield, England. Submitted to WHO by Syngenta Crop Protection AG.

JECFA (2000) Cyhalothrin. In: Toxicological evaluation of certain veterinary drug residues in food. WHO Food Additives Series 45.Geneva, World Health Organization.

JECFA (2005) Cyhalothrin addendum. In: Toxicological evaluation of certain veterinary drug residues in food WHO Food Additives series 53. Geneva, World Health Organization.

Jones, B.K. (1989a) Cyhalothrin: tissue distribution and elimination following a single oral dose (1 mg/kg) in the rat. Unpublished report No. CTL/P/2489 from Central Toxicology Laboratory, Macclesfield, England. Submitted to WHO by Syngenta Crop Protection AG.

Jones, B.K. (1989b) Cyhalothrin: tissue distribution and elimination following a single oral dose (25 mg/kg) in the rat. Unpublished report No. CTL/P/2490 from Central Toxicology Laboratory, Macclesfield, England. Submitted to WHO by Syngenta Crop Protection AG.

Killick, M.E. (1981a) Cyhalothrin: oral (gavage) teratology study in the rat. Unpublished report No. CTL/C/1075 from Central Toxicology Laboratory, Macclesfield, England. Submitted to WHO by Syngenta Crop Protection AG.

Killick, M.E. (1981b) Cyhalothrin: oral (gavage) teratology study in the New Zealand White rabbit. Unpublished report No. CTL/C/1072 from Central Toxicology Laboratory, Macclesfield, England. Submitted to WHO by Syngenta Crop Protection AG.

Lindsay, S., Chart, I.S., Godley, M.J., Gore, C.W., Hall, M., Pratt, I., Robinson, M. & Stonard, M. (1981) Cyhalothrin: 90-day feeding study in rats. Unpublished report No. CTL/P/629 from Central Toxicology Laboratory, Macclesfield, England. Submitted to WHO by Syngenta Crop Protection AG.

Lindsay, S., Doe, J.E., Godley, M.J., Hall, M., Pratt, I., Robinson, M. & Stonard, M.D. (1982) Cyhalothrin induced liver changes: reversibility study in male rats. Unpublished report No. CTL/P/668 from Central Toxicology Laboratory, Macclesfield, England. Submitted to WHO by Syngenta Crop Protection AG.

Marsh, J.R., Woollen, B.H. & Wilks, M.F. (1994) The metabolism and pharmacokinetics of lambda-cyhalothrin in man. Unpublished report No. CTL/P/4208 from Central Toxicology Laboratory, Macclesfield, England. Submitted to WHO by Syngenta Crop Protection AG.

Milburn, G.M. (2004) Lambda-cyhalothrin: developmental neurotoxicity study in rats. Unpublished report No. CTL/RR0969 from Central Toxicology Laboratory, Macclesfield, England. Submitted to WHO by Syngenta Crop Protection AG.

Milburn, G.M., Banham, P., Godley, M.J., Pigott, G. & Robinson, M. (1984) Cyhalothrin: three generation reproduction study in the rat. Unpublished report No. CTL/P/906 from Central Toxicology Laboratory, Macclesfield, England. Submitted to WHO by Syngenta Crop Protection AG.

Naravaneni, R. & Jamil, K. (2005) Evaluation of cytogenetic effects of lambda-cyhalothrin on human lymphocytes. *J. Biochem. Mol. Toxicology*, **19**, 304–310.

Pigott, G.H., Chart, I.S., Godley, M.J., Gore, C.W., Hollis, K.J., Robinson, M., Taylor, K. & Tinston, D.J. (1984) Cyhalothrin: two year feeding study in rats. Unpublished report No. CTL/P/980 from Central Toxicology Laboratory, Macclesfield, England. Submitted to WHO by Syngenta Crop Protection AG.

Pritchard, V.K. (1984) PP321: skin sensitisation study. Unpublished report No. CTL/P/1054 from Central Toxicology Laboratory, Macclesfield, England. Submitted to WHO by Syngenta Crop Protection AG.

Pritchard, V.K. (1985a) PP321 and cyhalothrin: skin irritation study. Unpublished report No. CTL/P/1139 from Central Toxicology Laboratory, Macclesfield, England.

Pritchard, V.K. (1985b) PP321: eye irritation study. Unpublished report No: CTL/P/1207 from Central Toxicology Laboratory, Macclesfield, England. Submitted to WHO by Syngenta Crop Protection AG.

Prout, M.S. (1984) Cyhalothrin: bioaccumulation in the rat. Unpublished report No. CTL/P/1014 from Central Toxicology Laboratory, Macclesfield, England. Submitted to WHO by Syngenta Crop Protection AG.

Prout, M.S. & Howard, E.F. (1985) PP321: comparative absorption study in the rat (1 mg/kg) Unpublished report No. CTL/P/1214 from Central Toxicology Laboratory, Macclesfield, England. Submitted to WHO by Syngenta Crop Protection AG.

Ray, D.E. & Fry, J.R. (2006) A reassessment of the neurotoxicity of pyrethroid insecticides. *Pharmacol. Therapeut.* **111**, 174–193.

Sheldon, T., Richardson, C.R., Shaw, J. & Barber, G. (1984) An evaluation of PP321 in the mouse micronucleus test. Unpublished report No. CTL/P/1090 from Central Toxicology Laboratory, Macclesfield, England. Submitted to WHO by Syngenta Crop Protection AG.

Sheldon, T., Howard, C.A. & Richardson, C.R. (1985) PP321: a cytogenetic study in human lymphocytes in vitro. Unpublished report No. CTL/P/1333 from Central Toxicology Laboratory, Macclesfield, England. Submitted to WHO by Syngenta Crop Protection AG.

Southwood, J. (1984) PP321: acute oral toxicity to the mouse. Unpublished report No. CTL/P/1066 from Central Toxicology Laboratory, Macclesfield, England. Submitted to WHO by Syngenta Crop Protection AG.

Southwood, J. (1985) PP321: acute oral toxicity studies. Unpublished report No. CTL/P/1102 from Central Toxicology Laboratory, Macclesfield, England. Submitted to WHO by Syngenta Crop Protection AG.

Trueman, R.W. (1989) Lambda-cyhalothrin: Assessment for the Induction of Unscheduled DNA Synthesis in Primary Rat Hepatocyte Cultures. Unpublished report No CTL/P/2707 from Central Toxicology Laboratory, Macclesfield, England. Submitted to WHO by Syngenta Crop Protection AG.

WHO (1990) *Environmental Health Criteria 99: cyhalothrin*. Geneva, WHO.

Wolansky, M.J., Gennings, C. & Crofton, K.M. (2006) Relative potencies for acute effects of pyrethroids on motor function in rats. *Toxicol. Sciences*, **89**, 271–277.

DIFENOCONAZOLE

First draft prepared by
D.B. McGregor[1] and Roland Solecki[2]

[1]Toxicity Evaluation Consultants, Aberdour, Scotland; and
[2]Federal Institute for Risk Assessment, Berlin, Germany

Explanation

Difenoconazole (Chemical Abstracts Service, CAS name, 1-[[2-[2-chloro-4-(4-chlorophenoxy) phenyl]-4-methyl-1,3-dioxolan-2-yl]methyl]-1*H*-1,2,4-triazole, CAS No. 119446-68-3) is a systemic triazole fungicide that controls a broad spectrum of foliar, seed and soil-borne diseases caused by *Ascomycetes, Basidiomycetes* and *Deuteromycetes* in cereals, soya, rice, grapes, pome fruit, stone fruit, potatoes, sugar beet and several vegetable and ornamental crops. It is applied by foliar spray or seed treatment and acts by interference with the synthesis of ergosterol in the target fungi by inhibition of the 14α-demethylation of sterols, which leads to morphological and functional changes in the fungal cell membrane. Difenoconazole is being reviewed for the first time by the present Meeting at the request of the Codex Committee on Pesticide Residues (CCPR). All critical studies complied with good laboratory practice (GLP).

Evaluation for acceptable daily intake

1. Biochemical aspects

1.1 Absorption, distribution and excretion

The absorption, distribution and excretion characteristics of difenoconazole were studied extensively in male and female rats (Esumi, 1992; Craine, 1987a, 1987b). The structure and positions of the radiolabel are shown in Figure 1. A single oral dose of [^{14}C-phenyl]difenoconazole at 0.5 mg/kg bw was rapidly and almost completely absorbed by male and female rats. A single oral dose of [^{14}C-phenyl]difenoconazole at 300 mg/kg bw was less extensively absorbed since bile-duct cannulated male and female rats excreted 17% and 22%, respectively, of the dose in faeces. Maximum blood concentrations were reached within 2 h, followed by a rapid decline for the lower dose and after approximately 4 h, with an initial slow decline up to 24 h, for the higher dose, after which the rate of decline was similar to the lower dose.. In females the difference in area under the blood concentration–time curve (AUC) measurements between the lower and higher doses matched the dose differential, while in males the AUC differential was 400-fold.

The systemic dose was eliminated predominantly via bile where it accounted for 73% of the administered dose (0.5 mg/kg bw) in males and 76% in females. Urinary excretion by bile-duct cannulated rats accounted for 14% of the administered dose in males and 9% in females, with faecal excretion representing less than 4% of the dose and thereby confirming the almost complete absorption at the lower dose. When bile from male rats at 0.5 mg/kg bw was administered intraduodenally to other bile-duct cannulated rats, 80% of the dose was re-eliminated via the bile and just 4% in the urine, thereby demonstrating enterohepatic recirculation. Bile was also the major route of eliminryation at 300 mg/kg bw, accounting for 56% of the administered dose in males and 39% in females, while urinary excretion represented only 1% of the dose. In non-cannulated male and female rats given a single oral dose of [^{14}C-phenyl]difenoconazole or [^{14}C-triazole]difenoconazole at 0.5 mg/kg bw, 13–22% of the administered dose was excreted in the urine and 81–87% in the faeces. In non-cannulated male and female rats given a single oral dose of [^{14}C-phenyl]difenoconazole or [^{14}C-triazole]difenoconazole at 300 mg/kg bw, 8–15% of the administered dose was excreted in the urine and 85–95% in the faeces. There was no pronounced difference in excretion profiles, either between the sexes or between the two radiolabelled forms at either dose. When a similar dose of [^{14}C-phenyl]difenoconazole or [^{14}C-triazole]difenoconazole was given to rats pre-treated with 14 daily oral doses of unlabelled difenoconazole at 0.5 mg/kg bw, there was neither

Figure 1. Position of the radiolabels on [^{14}C-phenyl]difenoconazole and [^{14}C-triazole] difenoconazole

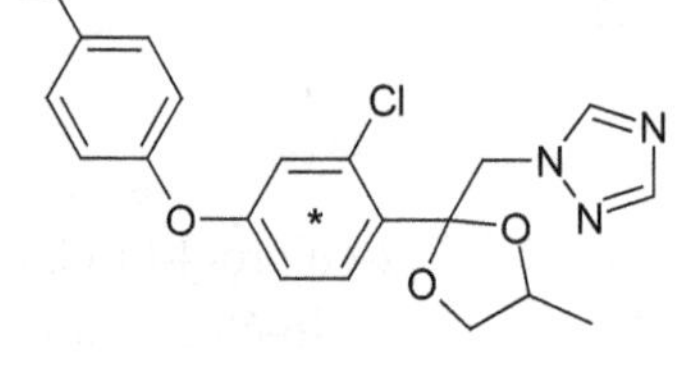

[^{14}C-phenyl]difenoconazole

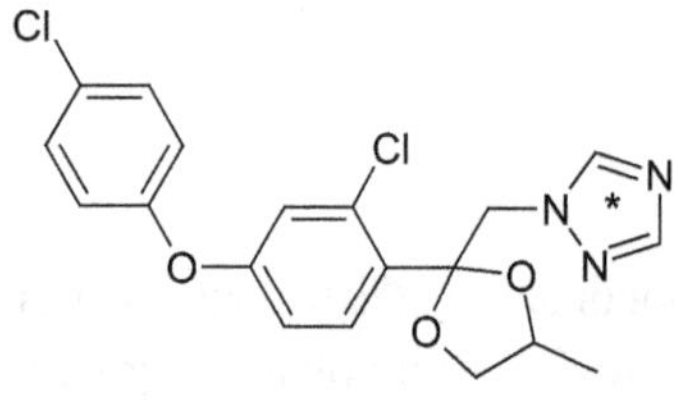

[^{14}C-triazole]difenoconazole

a sex difference nor any marked difference in excretion profiles from the rats with no pre-treatment. The excretion data for rats at 0.5 mg/kg bw showed that while enterohepatic recirculation was evident, biliary metabolites were largely excreted in the faeces. At the lower dose, the half-life of excretion was approximately 20 h. At 300 mg/kg bw, more of the administered dose w as excreted in the urine by non-cannulated rats than by bile-duct cannulated rats, apparently after reabsorption and further metabolism of some biliary metabolites; however, the predominant route of excretion of biliary radioactivity was again in faeces, as observed at the lower dose. At the higher dose, the half-life of excretion was approximately 33–48 h. At both doses, elimination kinetics were thus independent of sex and radiolabel position.

Tissue distribution

The concentration of radiolabel in the blood reached a maximum concentration (C_{max}) at 2 h after oral administration in male rats at 0.5 mg/kg bw and declined rapidly thereafter. The AUC up to 168 h after administration was 6.19 µg equivalent/h per ml. T_{max} was shorter in females than in males and C_{max} and AUC in females reached only about 50% of the respective values in males. The disappearance of radioactivity in female rats was slightly faster than in male rats (Figure 2, Table 1).

Figure 2. Concentrations of difenoconazole equivalents in the blood of rats given a single oral dose of radiobelled difenoconazole at (a) 0.5 mg/kg bw or (b) 300 mg/kg bw

(a) 0.5 mg/kg bw

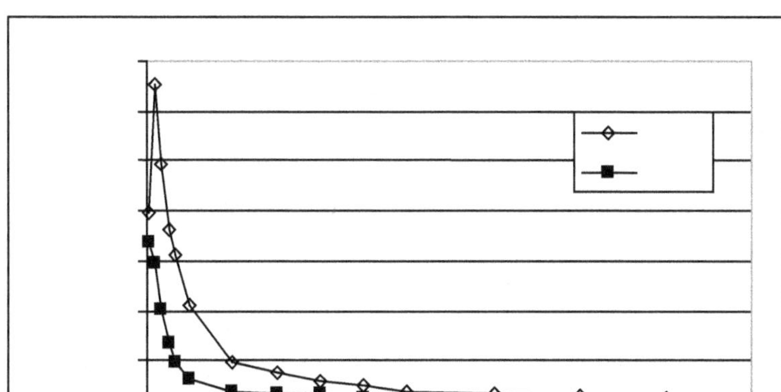

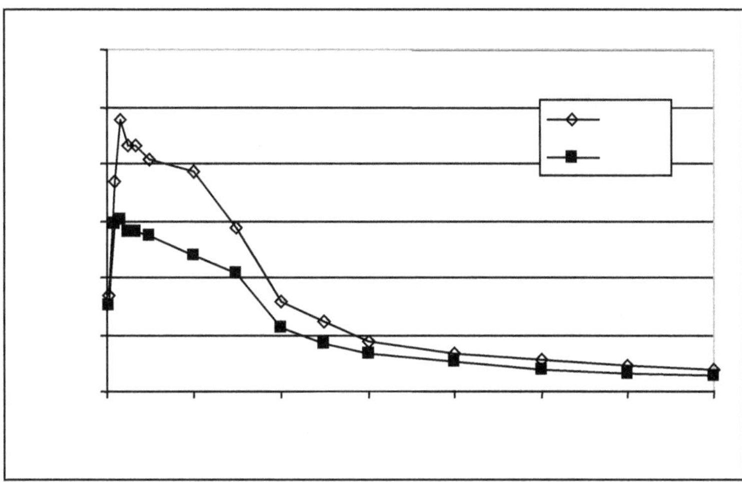

From Esumi (1992)

Table 1. Key parameters of blood kinetics in rats given a single oral dose of radiolabelled difenoconazole

Parameter	0.5 mg/kg bw		300 mg/kg bw	
	Males	Females	Males	Females
C_{max} (ppm)	0.327	0.169	47.89	30.02
T_{max} (h)	2	0.5	4	4
$T_{1/2}$ 1st phase (h)	6.2	4.4	22	24
$T_{1/2}$ 2nd phase (days)	2.8	3.7	3.8	3.4
AUC (0–168 h) (µg equivalents/h·per ml)	6.19	2.78	2460	1710

From Esumi (1992)

Tissue depletion results showed that at 2 h and 24 h after a dose of [^{14}C-phenyl]difenoconazole at 0.5 mg/kg bw, only in the liver and kidney were concentrations consistently higher than those in plasma for both sexes. Consistent results were observed in whole-body autoradiography sections of similarly-dosed male rats, since at 2 h and 24 h after dosing most of the radioactivity was present in the gastrointestinal tract contents and in bile, with lower concentrations in the liver and kidneys. Other tissues with concentrations initially higher than in plasma were adrenal glands in both sexes and Harderian glands and adipose tissue in females; however, these concentrations declined rapidly. After 168 h, [^{14}C-phenyl]difenoconazole residues in tissue were very low, with only fat showing concentrations comparable to those present in plasma, while all tissue residues of [^{14}C-triazole]difenoconazole were either below the limit of detection or below the limit of quantification. For both radiolabelled forms of difenoconazole, residues in female tissues also tended to be slightly lower than those in males and pre-treatment with unlabelled test substance had no effect on tissue distribution.

Tissue depletion results showed that at 4 h after a dose of [^{14}C-phenyl]difenoconazole at 300 mg/kg bw, most tissue concentrations were similar to or higher than those in plasma in both sexes. The highest tissue concentrations were present in fat in both sexes, with progressively lower concentrations in the liver, Harderian glands, adrenal glands, kidney and pancreas. In all other tissues that initially showed concentrations higher than those in plasma concentrations declined rapidly by 48 h after dosing and by 168 h all tissue residues of [^{14}C-phenyl] difenoconazole had declined markedly, only fat showing residues that were higher than in plasma. Tissue residues of [^{14}C-triazole]difenoconazole were significantly lower than those of [^{14}C-phenyl]difenoconazole residues and by 168 h were measurable only in the liver. Measurements in the contents of the gastrointestinal tract were consistent with the observed absorption and elimination profiles.

After 14 days of [4-chloro-phenoxy-U-^{14}C]difenoconazole at a dose of at 0.5 mg/kg bw per day (Hassler, 2003a), the absorbed radiolabel was rapidly and almost completely excreted, predominantly via the faeces. More than 90% of the total administered radiolabel was excreted within 24 h after the last dose and 98.5% of the total radiolabel was recovered within 7 days. At that time, less than 0.5% of the administered dose remained in the tissues and carcass. Metabolite profiles in the urine and faeces were qualitatively similar at each time-point, although small quantitative differences were observed after the administration of single and multiple oral doses. Concentrations of radioactive residues reached a plateau in most tissues after 7 days. Residue concentrations increased with continued dosing in liver, kidneys, fat and pancreas and did not reach a plateau during the dosing period; however, it was estimated that residue concentrations would reach a plateau within 3 weeks. The depletion of radioactive residues from tissues was moderately fast. Assuming first-order kinetics and monophasic depletion kinetics for the depuration of radiolabel from tissues, the half-lives ranged typically from 4–6 days. Depletion was more rapid in the liver, kidneys and pancreas, with half-lives of 1–3 days, and slower in fat, with a half-life of 9 days. Experiments with position-specific radiolabelled compounds (Capps et al., 1988) demonstrated that the highest tissue concentrations were found in the liver for the triazole label, and in fat and plasma for the phenyl label. Residues from the triazole-labelled compound were significantly less than those from the phenyl-labelled compound. Tissue residues in females were slightly less than in males. Multiple pre-treatment with unlabelled difenoconazole had no effect on tissue distribution.

Dermal absorption

Studies of dermal penetration in rats given difenoconazole as an application to the skin in vivo and studies with rat and human skin in vitro were conducted using non-radiolabelled difenoconazole (purity, 99.3%) and [triazole-U-^{14}C]difenoconazole.

The absorption, distribution and excretion of radioactivity was studied in male HanBrlWIST (SPF) rats after a single dermal administration of ^{14}C-labelled difenoconazole mixed with the non-labelled difenoconazole formulated as SCORE 250 EC. At the highest dose, the specific activity was 54 kBq/mg (1.5 µCi/mg).

For the highest dose, difenoconazole was dissolved in blank formulation at a concentration of 250 g/l, representing the undiluted product. For the intermediate and lowest dose, this formulated material was mixed with water in a ratio of 1 : 200 (w/v) and 1 : 5000 (w/v), respectively. The applied nominal doses were 0.5, 13 and 2600 µg/cm^2 and these were left in place for 6 h. The determination of dermal absorption at the highest dose was repeated in a separate experiment, using an application of 2400 µg/cm^2. In each experiment, groups of 16 rats were exposed for 6 h, after which the skin was washed, and four rats per group were killed at 6, 24, 48 and 72 h after dose application. The amount of radiolabelled difenoconazole remaining in the skin was determined at each sampling time after washing and skin-stripping to separate the stratum corneum from the epidermis. Urine and faeces were collected up to 72 h and blood residue concentrations were measured at 72 h.

At the lowest, intermediate and highest doses, respectively, systemic absorption within 6 h was 15.3%, 7.5% and 7.1% and the penetration rates was 0.013, 0.162 and 30.4 µg/cm^2 per h. The ratios of these penetration rates, i.e. 1 : 12 : 2400 were proportional to the application concentrations, i.e. 1 : 26 : 5100. There was, however, a substantial variation in the data that was attributed to the irritancy of the formulation vehicle, SCORE 250 EC; this resulted in a broad range of individual values for dermal absorption of up to 45% of the administered dose. For the study in rats in vivo, in a worst-case scenario the mean dermal absorption values at the lowest intermediate and highest doses, respectively, were 37.6%, 14.6% and 10.6%. Nevertheless, concentrations of residues in the blood were generally very low: the highest concentrations (as difenoconazole equivalents) were 0.01 µg/ml and 0.26 µg/ml at the intermediate and highest doses, respectively, 6–8 h after appliciation (Hassler, 2003c).

The penetration of non-radiolabelled difenoconazole (purity, 99.3%) and [triazole-U-^{14}C] difenoconazole through isolated rat and human epidermal membrane preparations in vitro after stripping the stratum corneum from skin was determined after an exposure of 24 h to difenoconazole at a concentration of 0.05, 1.28 or 250 mg/ml, achieving applications of 0.5, 12 or 2345 µg/cm^2. Before dose application, the integrity of the epidermal membranes and their suitability for use in the study was assessed by measuring the penetration of tritiated water applied to the membranes and comparison of the results with exclusion criteria. Rat skin membranes with permeability coefficients (Kp) of $> 3.5 \times 10^{-3}$ cm·h^{-1} and human skin membranes with Kp $> 2.5 \times 10^{-3}$ cm·h^{-1}were excluded from the subsequent experiment. After this membrane integrity check and before dose application, the receptor chamber was refilled with ethanol : water (1 : 1 v/v).

Within 24 h, the proportions of radiolabel penetrating the membranes at the lowest, intermediate and highest concentrations, respectively, were 71%, 64% and 23% for the preparations of rat skin and 7.6%, 7.0% and 0.7% for the preparations of human skin. Although the studies with human skin in vitro indicated that dermal absorption was approximately 8% (7.6%) for the lowest concentration, if the amount retained in skin is considered as potentially absorbable it increases to 15%. However, the main objective of the comparative studies of rat and human skin in vitro was to evaluate the differences in the percentage actually absorbed, permitting the estimation of an appropriate species ratio. The flux (i.e. rate of penetration under steady-state conditions) at the lowest, intermediate and highest concentrations, respectively, was 0.020, 0.455 and 26.0 μg/cm^2 h^{-1} for the rat skin and 0.002, 0.037 and 0.82 μg/cm^2 h^{-1} for the human skin. The flux ratios rat : human were therefore about 1 : 10, 1 : 12 and 1 : 32 at the lowest, intermediate and highest concentrations, respectively (Hassler, 2003c). Comparison of the penetration rates at the different doses (concentration ratio, 1 : 25 : 5000), revealed an increase in the penetration rates of 1 : 23 : 1300 for rat epidermal membranes, while the penetration rates for human epidermal membranes revealed a ratio of only 1 : 24 : 500 (Hassler, 2003b). These data acquired in vitro are used to arrive at an estimate of dermal absorption partly because the inter-rat variation observed in the study in vivo was very large (attributed to the irritancy of the formulation vehicle and for this reason was not considered reliable).

Metabolism

Metabolites were isolated from the urine and faeces of male and female rats given [^{14}C-phenyl] difenoconazole or [^{14}C-triazole]difenoconazole as single oral dose at 0.5 or 300 mg/kg bw (Figure 1a and 1b), or 0.5 mg/kg bw after pre-treatment with 14 daily oral doses of unlabelled difenoconazole at 0.5 mg/kg bw (Craine, 1987a, 1987b).

Balance data for the phenyl and triazole ring labels (Craine 1987a, 1987b) summarized in a report by Capps et al. (1988) showed that > 97% of the radiolabel in all cases was excreted with > 78% in all cases eliminated in the faeces. Three main metabolites, A, B and C, were isolated from faeces and together accounted for an average of 68% of the administered dose. Metabolite B (hydroxy-CGA 169374) was hydroxylated in the outer phenyl ring and spectral analysis showed that that it comprised two isomers, one with a rearrangement of the chlorine on the outer phenyl ring, attributed to a mechanism analogous to an NIH shift.1 Metabolite C (CGA 205375) was the hydroxyl product of cleavage of the dioxolane ring from the difenoconazole molecule and was present only in the faeces of rats given the higher dose. Metabolite A (hydroxy-CGA 205375) was the outer phenyl-ring hydroxylated product of metabolite C and comprised two diastereoisomers, as described for metabolite B. The profile of urinary metabolites was more complex and showed more variability between the two radiolabelled forms. Free 1,2,4-triazole was identified in male urine, accounting for less than 10% of the administered dose and represented cleavage of the alkyl bridge between the ring systems. Other urinary metabolites included metabolite C, its sulfate conjugate, ring-hydroxylated metabolite C and its sulfate conjugate, as well as an hydroxyacetic acid metabolite of the chlorophenoxy-chlorophenyl moiety, all of which were present in minor quantities, each representing less than 3% of the administered dose. One metabolite CGA 189138 (chlorophenoxy-chlorobenzoic acid) was also isolated from the liver. The difenoconazole molecule was therefore extensively metabolized, although with limited cleavage of the triazole and dioxolane rings. The extensive biliary elimination observed was consistent with the major metabolites being of relatively high molecular mass.

The major steps of the metabolism of difenoconazole in the rat were deduced to involve hydrolysis of the ketal in difenoconazole, resulting in CGA 205375 (1-[2-chloro-4-(4-chloro-phenoxy)-phenyl]-2-(1,2,4-triazol)-1-yl-ethanol), with the ketone CGA 205374 (1-(2-chloro-4-(4-chloro-phenoxy)-phenyl)-2-(1,2,4-triazol)-1-yl-ethanone) as a postulated but not identified intermediate, and hydroxylation on the outer phenyl ring of the parent (hydroxy-CGA 169374) and in CGA 205375 (hydroxy-CGA 205375). As a minor process, cleavage of the alkyl chain between the triazole and the inner phenyl ring occurs, resulting in a hydroxyacetic acid (NOA 448731) or CGA 189138 (2-chloro-4-(4-chlorophenoxy)-benzoic acid) and free triazole. Sulfate conjugates were identified for CGA 205375 and for hydroxy-CGA 205375. The proposed metabolic pathway in mammals is shown in Figure 3 (Capps et al., 1990).

[1] An NIH shift is a chemical rearrangement (first described in studies from the United States National Institutes of Health, NIH) in which a hydrogen atom on an aromatic ring undergoes an intramolecular migration primarily during a hydroxylation reaction.

Figure 3. Proposed metabolic pathway of difenoconazole in mammals

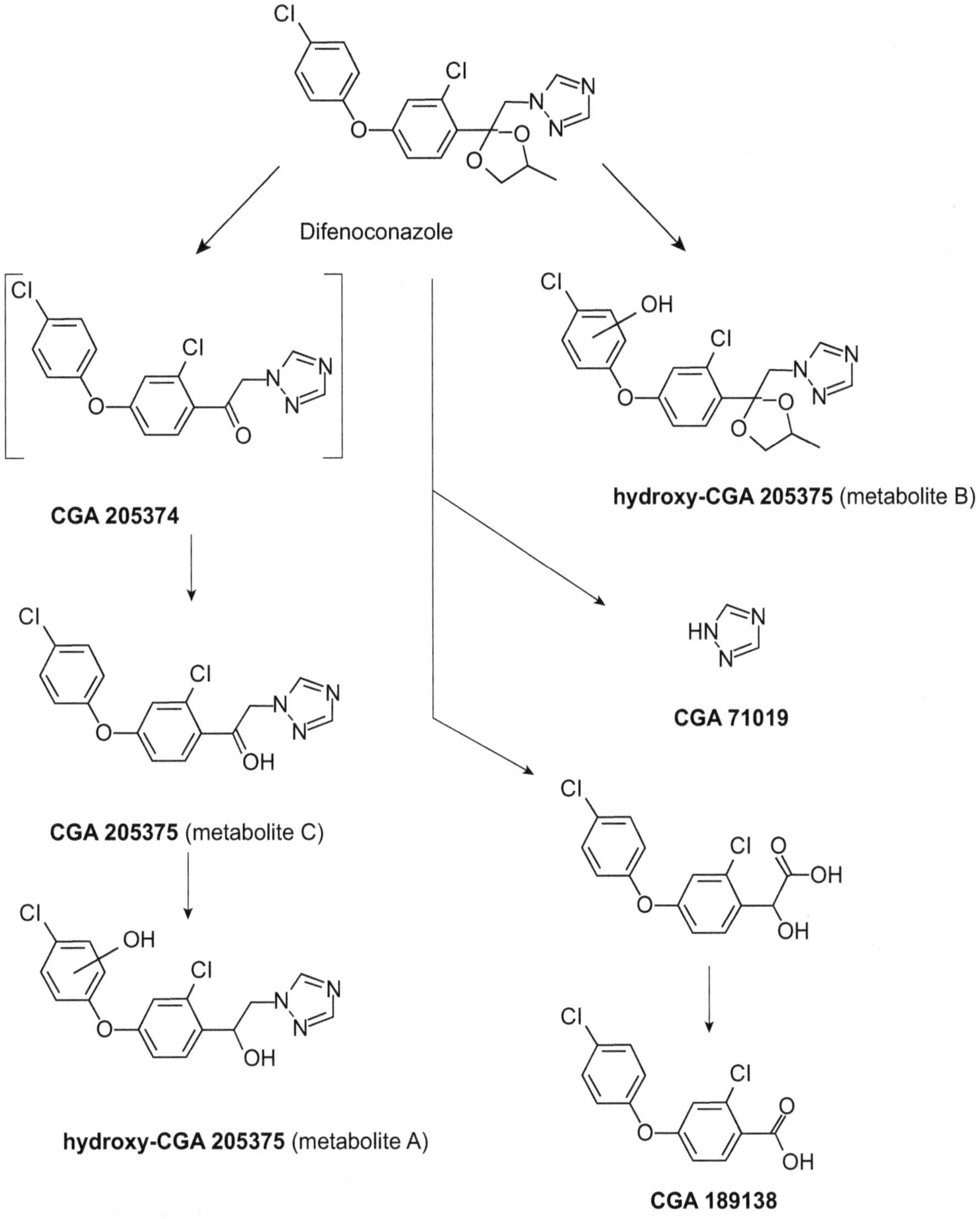

Difenoconazole

CGA 205374

hydroxy-CGA 205375 (metabolite B)

CGA 71019

CGA 205375 (metabolite C)

hydroxy-CGA 205375 (metabolite A)

CGA 189138

CGA 205374: 1-(2-chloro-4-(4-chloro-phenoxy)-phenyl)-2-(1,2,4-triazol)-1-yl-ethanone;
CGA 205375 (metabolite C): 1-[2-chloro-4-(4-chloro-phenoxy)-phenyl]-2-(1,2,4-triazol)-1-yl-ethanol;
CGA 71019: 1,2,4-triazole.
CGA 189138: 2-chloro-4-(4-chlorophenoxy)-benzoic acid;
From Capps et al. (1990)

2. Toxicological studies

2.1 *Acute toxicity*

(a) *Oral administration*

The acute oral toxicity of difenoconazole was evaluated in groups of five male and female Sprague-Dawley rats given difenoconazole technical (purity not specified) as a single dose at 0, 1000, 2000 or 3000 mg/kg bw in 3% cornstarch and 1% Tween 80 in water in a dose volume of 50 ml/kg bw administered orally by gavage. The rats were then observed frequently after dosing on test day 1 and then twice per day on days 2–14. Body weights were recorded twice during the period before testing and on test days 1, 8 and 15. Necropsy was performed on all rats.

Signs of toxicity in male and female rats in all treatment groups were observed from 0.5 h after dosing with no clear differences between the groups at 1000, 2000 and 3000 mg/kg bw. These signs included hypoactivity, stains around the mouth, perineal staining, ataxia, lacrimation, soft faeces, hypothermia, prostration, chromodacryorrhoea, chromorhinorrhoea, spasms, salivation, unkempt appearance, rhinorrhoea, hypopnoea, or ptosis. Mortality of 40%, 40%, and 100% occurred at doses of 1000, 2000, and 3000 mg/kg bw, respectively. The oral median lethal dose (LD_{50}) was calculated to be approximately 1453 mg/kg bw for both sexes. In rats that were found dead on test days 3–5 after receiving a dose of 2000 mg/kg bw, pronounced stomach lesions were observed. No other significant findings on necropsy were noted for rats at other doses (Argus et al., 1987).

The acute oral toxicity of difenoconazole was evaluated in groups of five male and female Tif:MAG f (SPF) mice given difenoconazole technical (purity not specified) as a single dose at 1000 (females only) or 2000 mg/kg bw in arachis oil and administered orally by gavage in a dose volume of 10 ml/kg bw. Mortality and clinical symptoms were recorded 1, 3 and 5 h after treatment and at daily intervals thereafter for a total of 14 days. Body weights were recorded immediately before treatment and on test days 7 and 14 or at death. Autopsy was performed on all mice.

Piloerection, abnormal body positions, and dyspnoea were observed, these being common symptoms in tests for acute toxicity. Additionally, reduced locomotor activity and ataxia was observed in the mice of both dose groups from 5 h after dosing. The group of mice at 2000 mg/kg bw experienced tonic spasms. At autopsy, no deviations from normal morphology were found. On the basis of limited mortality, the acute oral LD_{50} was estimated to be > 2000 mg/kg bw in male and female mice (Hartmann, 1990).

(b) *Dermal administration*

The acute percutaneous (dermal) toxicity of difenoconazole technical (purity not specified) was investigated after administering the test material at a dose of 2010 mg/kg bw as a 50% ethanol preparation to the shaved flank skin of groups of five New Zealand White rabbits. The treated area represented about 10% of the body surface. The treated area was covered with a gauze dressing attached by tape. Exposure lasted 24 h; the gauze was then removed and the skin was washed with ethanol and dried with paper towels. Rabbits were observed regularly after dosing on test day 1 and at least daily thereafter until day 14. Body weights were recorded before treatment on day 1 and on days 8 and 15. Autopsy was performed on all rabbits.

No mortality occurred after dermal application of difenoconazole at a dose of 2010 mg/kg bw for 24 h. Clinical observations that were performed over 14 days after treatment were unremarkable for all rabbits; however, slight erythema (Draize grade 1) at the treatment site of three rabbits was observed at the end of the dosing period and fissuring or desquamation of the skin was noted in rabbits of both sexes, probably as a result of exposure to the ethanol vehicle. Body-weight data indicated that all rabbits gained weight during the study and gross visceral evaluations performed on all rabbits at the termination of the study were unremarkable. On the basis of these data, difenoconazole was considered to be essentially non-toxic when applied topically under occlusion to intact rabbit skin (Mastrocco et al., 1987a).

(c) Inhalation

A test for the acute toxicity of difenoconazole technical (batch No. P 012060, purity, 96.2%) was conducted in a group of Tif RAI f (SPF) rats (five of each sex) exposed by inhalation. Exposure was in a head/nose inhalation system for 4 h to a mean analysed atmospheric concentration of 3967/3285 (nominal/gravimetric) mg/m^3 air (technically highest achievable concentration). In order to maintain a constant aerosol concentration with acceptable particle size distribution, it was necessary to blend difenoconazole with 5% Sipernat 50S, an inert silica. The aerosol was generated from the solid test material by means of a brush-feed micronizing jet mill. The aerosol from the dust generator was diluted with filtered humidified air. Owing to the properties of the test material, it was not possible to generate higher concentrations of aerosol. A control group of equal size was exposed to purified air under the same conditions as the test group. The rats were examined for clinical signs of toxicity and mortality during and after exposure and daily thereafter. Individual body weights were recorded immediately before exposure, on day 7 and day 14. The study was terminated 14 days after dosing. Autopsy was performed on all rats. Particular attention was given to the respiratory tract.

No rats died during the study. Symptoms observed in rats of both sexes and to a similar extent were dyspnoea during and after exposure and piloerection, hunched posture and reduced locomotion after exposure. Reduced locomotion was not observed after the day of exposure, while hunched posture was also observed on the day after exposure, dyspnoea persisted to include day 4 and piloerection persisted to include day 6. All symptoms disappeared within 7 days. In comparison with rats in the control group, the males showed a lower body-weight gain in the first week and the females showed a higher body-weight gain in the second week after exposure. No treatment-related macroscopic findings were observed. The median lethal concentration (LC_{50}) for difenoconazole suspended in air was > 3300 mg/m^3 (3.3 mg/l) for male and female rats (Hartmann, 1991).

(d) Dermal irritation

Difenoconazole technical (purity, 91.5%) was evaluated for acute dermal irritation potential in three male and three female Hra:(NZW)SPF rabbits. Difenoconazole (0.5 g) was moistened with 0.9% saline and applied to a 6.25 cm^2 gauze pad that was placed on the shaved dorsal area. The gauze patch was covered with plastic, secured with tape and left for 4 h. Thereafter, patches were removed and the test sites were washed using tap water and paper towels. Adjacent areas of untreated skin of each rabbit served as controls for the test. Signs of skin erythema or oedema were scored after 30 min, 24 h, 48 h, and 72 h after dosing. Dermal irritation was scored and recorded according to the Draize technique

One female was found with grade 1 erythema after 30 min. After 24 h, this reaction had cleared. No other skin reactions were noted with any rabbit (Glaza, 1991a). The Meeting concluded that difenocoazole was very slightly and transiently irritating to the skin of rabbits.

(e) Ocular irritation

Difenoconazole technical (purity, 91.5%) was evaluated for acute eye irritation potential in New Zealand White (Hra:(NZW)SPF) rabbits (initial body weight, 2442–2696 g). Three males and three females in group 1 and two males and one female in group 2 were used. Each rabbit received 0.1 ml of difenoconazole placed into the everted lower lid of the right eye. The upper and lower lids were gently held together for 1 s to prevent loss of material and then released. In group 1 (three males and three females), the eyes of the rabbits were not washed, while in group 2 (two males and one female), the eyes of the rabbits were washed with lukewarm tap water for 60 s starting 30 s after instillation of difenoconazole. The eyes were examined at 1, 24, 48 72 and 96 h after dosing. Eye irritation was scored and recorded according to the Draize technique.

Signs of irritation were observed in the cornea, iris and conjunctiva of unwashed eyes (Table 2). All reactions had cleared by day 4 after treatment. Reactions in the washed eye were generally weaker and of shorter duration (Glaza, 1991b).

The Meeting concluded that difenoconazole was moderately and transiently irritating to the eyes of rabbits.

(f) Sensitization

The potential of difenoconazole technical (purity not specified) to produce delayed contact hypersensitivity was evaluated in female Hartley guinea-pigs (initial body weight, 278–385 g) via a modified Buehler method. Ten guinea-pigs were used for each control and treatment group. The positive-control substance was 1-chloro-2,4-dinitrobenzene (DNCB). Difenoconazole was applied undiluted; DNCB was applied as a solution in ethanol/water (20 : 80).

On the basis of a preliminary test for irritation it was established that 0.5 g of undiluted difenoconazole and 0.05% DNCB could be applied for topical induction and topical challenge phases of the test.

Guinea-pigs in the treatment groups were dosed on test days 1, 3, 6, 8, 10, 13, 15, 17, 20 and 22. Difenoconazole or DNCB were applied to a gauze patch, which was placed on the hair-clipped left flank of the appropriate animals. Patches were secured with occlusive tape and left in place on the guinea-pigs for 6 h on each day. At the end of each exposure period, tape and patch were removed and the test site was wiped with 95% ethanol and water. The guinea-pigs were challenged on test day 36 on the hair-clipped right flank by exposure for 6 h as described for the induction phase. At 24

Table 2. Ocular irritation in rabbits after instillation of difenoconazole

Draize score	Time after patch removal[a]					
	1 h	24 h	48 h	72 h	96 h	Mean score (24–72 h)
Group 1 (eyes unwashed)						
Cornea opacity	0/0/0/0/0/0	1/1/1/2/0/0	1/0/0/1/0/0	0/0/0/1/0/0	0/0/0/0/0/0	—
Mean score	0	0.83	0.33	0.17	0	0.44
Iris lesions	0/1/0/0/1/1	1/1/1/1/1/0	1/1/0/1/0/0	0/0/0/0/0/0	0/0/0/0/0/0	—
Mean score	0.5	0.83	0.5	0	0	0.44
Conjunctivae, redness	2/2/2/2/2/2	2/3/3/3/2/2	2/3/2/3/1/2	2/2/1/2/0/1	0/0/0/0/0/0	—
Mean score	2.0	2.5	2.16	1.33	0	2.0
Conjunctivae, chemosis	1/1/1/1/1/1	2/1/1/2/1/1	1/1/1/2/0/1	0/0/0/1/0/0	0/0/0/0/0/0	—
Mean score	1.0	1.33	1.0	0.17	0	0.83
Group 2 (eyes washed)						
Cornea opacity	0/0/0	0/0/0	0/0/0	0/0/0	0/0/0	—
Mean score	0	0	0	0	0	0
Iris lesions	1/0/1	1/0/0	0/0/0	0/0/0	0/0/0	—
Mean score	0.67	0.33	0	0	0	0.11
Conjunctivae, redness	2/2/2	2/2/2	1/2/2	1/1/1	0/0/0	—
Mean score	2.0	2.0	1.67	1.0	0	1.56
Conjunctivae, chemosis	1/1/1	1/1/0	1/0/0	0/0/0	0/0/0	—
Mean score	1.0	0.67	0.33	0	0	0.33

From Glaza (1991b

[a] Numbers are the individual scores for each rabbit tested.

Table 3. Results of studies of acute toxicity with difenoconazole

Species	Strain	Sex	Route	LD_{50} (mg/kg bw)	LC_{50} (mg/l air)	Other results, clinical signs	Reference
Rat	Sprague-Dawley	Males and females	Oral	1453	—	Hypoactivity, stains around the mouth, perineal staining, ataxia, lacrimation, soft faeces, hypothermia, prostration, chromodacryorrhoea, chromorhinorrhoea, spasms, salivation, unkempt appearance, rhinorrhoea, hypopnoea, or ptosis.	Argus et al. (1987)
Mouse	Tif: MAG f (SPF)	Males and females	Oral	> 2000	—	Piloerection, abnormal body positions, and dyspnoea, reduced locomotor activity and ataxia, tonic spasms	Hartmann (1990)
Rabbit	NZW	Males and females	Dermal	> 2010	—	Clinical observations were unremarkable for all rabbits	Mastrocco et al. (1987a)
Rat	Tif: RAI f (SPF)	Males and females	Inhalation	—	3.3	Piloerection, hunched posture, dyspnoea, and reduced locomotion.	Hartmann (1991)
Rabbit	NZW	Males and females	Dermal irritation	—	—	Very slightly and transiently irritating	Glaza (1991a)
Rabbit	NZW	Males and females	Ocular irritation	—	—	Moderately and transiently irritating	Glaza (1991b)
Guinea- pig	Hartley	Female	Dermal sensitization (modified Buehler)	—	—	Non-sensitizing	Mastrocco et al. (1987b)

NZW, New Zealand White.

and 48 h after challenge applications dermal reactions were assessed and graded. Body weights were recorded at weekly intervals.

Epidermal induction caused no skin reactions in any of the guinea-pigs given undiluted difenoconazole. No positive reactions were observed at the application site in any of these animals 24 and 48 h after challenge application and there were no signs of toxicity. The positive control, DNCB, produced significant skin reactions in six out of nine guinea-pigs after 24 h and in four out of nine guinea-pigs after 48 h. One guinea-pig in the positive-control group died during the induction phase (Mastrocco et al., 1987b).

The Meeting concluded that difenoconazole was not sensitizing in this modified Buehler test.

The results of these studies of acute toxicity are summarized in Table 3.

2.2 *Short-term studies of toxicity*

Mice

Groups of 15 male and 15 female CD-1®(ICR) mice were given diets containing difenoconazole technical (purity, 94.5%) at a concentration of 0, 20, 200, 2500, 7500 or 15 000 ppm, which, in the groups receiving the four lower dose was equal to 0, 3.3, 34.2 and 440 mg/kg bw per day in males and 0, 4.6, 45.2 and 639 mg/kg bw per day in females, for 13 weeks. Because of mortalities occurring at a very early stage in this study, the intake of difenoconazole was not calculated for the two groups at the highest dose. An additional five males and five females were included in the control group. Mortality was checked twice per day, clinical signs were recorded once per day and detailed physical examinations were performed each week. Food consumption and body weight were determined once per week. Eye examinations were conducted on all mice before treatment and at the end of treatment in week 13. Haematology investigations were conducted on 10 males and 10 females per group at the end of the treatment period.

Ten males and ten females per group were killed in week 13. All these mice were subjected to complete gross examinations, and weights of selected organs were determined. Autopsies were also carried out on mice that were found dead or killed in extremis. Microscopic examinations were conducted on all organs from mice in the control group and the group at the highest dose and on lung, liver, kidneys and all gross lesions from mice from all dose groups. Specified organs of mice scheduled to be killed were weighed and all specified tissues were collected, fixed and examined microscopically. The stability of difenoconazole in the diet was verified and the homogeneity of the dietary mixtures was verified before the start of the study.

Dietary administration of difenoconazole at doses of 7500 and 15 000 ppm for up to 3 months resulted in very high mortality by week 3. All female mice receiving difenoconazole at 7500 or 15 000 ppm died or were found in a moribund condition and were killed on study days 3–7. Males survived a few days longer, but all males at 15 000 ppm were dead by day 11 and 13 out of 15 males at 7500 ppm were dead by day 18. The remaining two males at 7500 ppm were removed from the study by day 21. One female at 2500 ppm was also found dead during week 3. In addition, one male in the control group, one male at 20 ppm and one female at 200 ppm died on day 16. A female at 200 ppm was considered to have died accidentally after blood collection during week 13.

Clinical signs in the groups at 2500 ppm or less were noted at low frequencies and there was no dose–response pattern, although polypnoea was seen in most females at 2500 ppm during the first study week. Ophthalmoscopy revealed no treatment-related findings. Unilateral diffuse retinochoroidal degeneration was seen in one male at 2500 ppm and in one female at 2500 ppm, but owing to the low incidence these findings were considered to be incidental to treatment. Body weights of the group of mice at 2500 ppm were somewhat lower than those of the mice in the control group throughout the study, but food consumption was not affected by treatment. Dark areas in the stomach were noted at necrospy in 5 out of 13 males at 7500 ppm, 14 out of 15 males at 15 000 ppm, 15 out of 15 females at 7500 ppm and 14 out of 15 females at 15 000 ppm. These findings corresponded to microscopic findings of erosion/ulceration of the glandular stomach and/or hyperkeratosis in the non-glandular stomach in most of these mice. Haematology parameters were not different among the groups. There was a statistically significant positive trend in platelet counts in males, but there was no trend in the females and no other findings that would indicate this was a meaningful effect. None of the other parameters measured were remarkable. Absolute and relative liver weights were increased in mice at 2500 ppm and relative liver weights were increased in males at 200 ppm. The only other statistically significant organ-weight changes were a 25% reduction in absolute ovary weight and a 7.7% reduction in absolute heart weight in females in the group at 2500 ppm. Liver enlargement and prominent reticular patterns were seen at an increased frequency in the group of mice at 2500 ppm. There was an increased incidence of diffuse hepatocellular enlargement in all male and female mice of this same group and some minimal centrilobular hepatocellular enlargement in males at 200 ppm. Hepatic vacuolization also occurred with greater frequency and severity in the group of mice at 2500 ppm. Since minimal centrilobular hepatocellular enlargement only occurred in males of the group at 200 ppm it probably represented a less severe manifestation of the diffuse response to difenoconazole at higher doses seen in this organ. This can be considered to be an adaptive change at this dose (Cox, 1987a).

The no-observed-adverse-effect level (NOAEL) was 200 ppm, equal to 34.2 mg/kg bw per day, on the basis of clinical signs, reduced body weight and ovary weight and changes in liver weight and histology at 2500 ppm, equal to 440 mg/kg bw per day.

Rats

Groups of 10 male and 10 female Wistar rats (outbred KFM-Han, SPF) were given diets containing difenoconazole technical (purity, ≥ 95%) at a concentration of 0, 250, 1500 or 10 000 ppm, equal to 0, 27, 156 and 914 mg/kg bw per day for male rats and 0, 27, 166 and 841 mg/kg bw per day for female rats, for 33 days. The stability of difenoconazole in the diet and the homogeneity of the dietary mixtures were verified before the start of the study. Analyses for correct concentration were performed before the study start. Food consumption and body weight were determined once each week. The state of health was checked twice per day. Eye examinations were carried out before the start and at the end of the dosing period. Blood samples were taken from all rats for haematology and blood chemistry examination at the end of the dosing period. Urine was analysed at the end of the exposure period. All rats were subjected to complete gross examinations, and weights of selected organs were determined. Microscopic examinations were conducted on liver and all gross lesions in rats from all groups receiving difenoconazole.

There were no mortalities in the study and difenoconazole did not induce any remarkable clinical signs of toxicity in any of the groups receiving difenoconazole. Body weights and body-weight gain in the group of rats at 10 000 ppm were significantly ($p < 0.01$) lower than those in the control group throughout the study. No body-weight losses were recorded, but body weights of males and females at 10 000 ppm were 42% and 36% lower, respectively, than those of the males and females in the control group after 26 days. Body weights of males at 250 ppm were also significantly lower than those of males in the control group at weeks 2 and 3, but these reductions were not considered to be treatment-related owing to the absence of effects in the males at 1500 ppm. Food consumption by the group of rats at 10 000 ppm was also markedly reduced throughout the study, most notably during the first study week (reductions of 75% and 71% for males and females, respectively). Food consumption by the group of rats at 10 000 ppm increased, relative to the controls, in the following weeks, but remained low. No statistical tests were performed on food consumption data owing to the low number of observations ($n = 2$ cages per dose group).

Ophthalmoscopy revealed no treatment-related findings. Treatment-related differences in haematological parameters were observed between the control group and the group at 10 000 ppm. These consisted of slightly decreased concentration of haemoglobin, erythrocyte volume fraction, mean corpuscular volume, mean corpuscular haemoglobin and slightly shorter thromboplastin times in males and females and slightly decreased thrombocyte count and slightly increased reticulocyte count in females only. Slightly shorter thromboplastin times in the group of males at 1500 ppm were also considered to be related to treatment. All other statistically significant differences (primarily in the group of females at 250 ppm) were considered to be incidental and within the normal range of biological variation.

Treatment-related, statistically significant changes in blood chemistry parameters for rats at 10 000 ppm included moderately increased cholesterol concentrations, slightly decreased sodium concentrations, slightly increased alkaline phosphatase and γ-glutamyl transferase activities in males and females, slightly increased aspartate aminotransferase activity for males only, and slightly increased phosphate concentrations for females only. The more notable changes were the increase in total cholesterol concentration (230% and 240% in males and females, respectively) and in gamma-glutamyl transferase activity (75% and 73% in males and females, respectively). In addition, protein electrophoresis revealed treatment-related, statistically significant effects at all doses: slightly increased albumin concentrations in males at all doses and in females of the group at 10 000 ppm; slightly decreased α1-globulin fractions in males at all doses and in females of the group at 10 000 ppm; slightly decreased β-globulin fractions in males at all doses; slightly increased albumin/globulin ratios in males at all doses and in females of the group at 10 000 ppm. In the absence of morphological changes, many of these differences might be considered to be adaptive in nature, but the marked increases in total cholesterol concentration and in γ-glutamyl transferase activity at 10 000 ppm are nevertheless adverse. Since these increases also occur in the absence of morphological changes, it was not clear whether the other blood chemistry changes observed at 10 000 ppm were toxicologically relevant. All other statistically significant differences were considered to be incidental and within the normal range of biological variation.

No toxicologically significant effects were observed in the quantitative and qualitative analyses of urine. There was a slight to moderate increase in ketone body formation for males at 250 ppm, 1500 ppm and 10 000 ppm, as well as females at 10 000 ppm. The increases in the rats at 10 000 ppm were attributed to reduced food consumption and/or prolonged fasting. The few other statistically significant differences were considered to be incidental and within the range of normal biological variation. Almost all organ weights were significantly reduced in the group of rats at 10 000 ppm, but because of the reduced carcass weights, the organ-to-body weight ratios were actually increased in this group. There were statistically significant reductions in liver weight of males at 250 ppm and 10 000 ppm, while there was a statistically significant increase in liver weight in the group at 1500 ppm. This lack of a dose–response relationship paralleled the fluctuations in male body weight described earlier. In female rats, there were no significant changes in absolute liver weight at any dose; however, liver weights relative to body weights were significantly increased in males and females in the groups at 1500 and 10 000 ppm. In the group at 1500 ppm, kidney weights relative to body weights increased in males, while absolute kidney weights were decreased in females. Decreased spleen weights were seen in males of the group at 250 ppm, but not in the group at 1500 ppm or in females at either of these doses. No treatment-related observations were recorded at autopsy or upon microscopic examination. Thus, on the basis of organ-weight changes, the liver appeared to be the target organ for toxicity (Suter, 1986a).

The NOAEL in rats given diets containing difenoconazole for 33 days was 250 ppm, equal to 27 mg/kg bw per day, on the basis of liver-weight changes at 1500 ppm, equal to 156 mg/kg bw per day. Changes in liver weight observed at 250 ppm were considered to be adaptive responses. The maximum tolerated dose (MTD) was clearly exceeded at 10 000 ppm, equal to 841 mg/kg bw per day.

Groups of 10 male and 10 female Wistar rats (outbred KFM-Han, SPF) were given diets containing difenoconazole technical (purity, 94.5%) at a concentration of of 0, 40, 250 or 1500 ppm, equal to 0, 3.3, 20 and 121 mg/kg bw per day in males and 0, 3.5, 21 and 129 mg/kg bw per day in females, for 13 weeks. An additional 10 males and 10 females were included in the control group and the group at 1500 ppm for the 13 weeks and then continued on control diet for a 4-week recovery period. The doses were selected on the basis of the results of a 28-day study (Suter, 1986a), in which treatment-related effects were seen at 1500 ppm and the MTD was exceeded at 10 000 ppm. The stability of difenoconazole in the diet had been verified for the 33-day study (see above) that was conducted within 2 months of this study. The homogeneity of the dietary mixtures was verified before the start of the study. Analyses for correct concentration also were performed before the start of the study and were satisfactory.

Food consumption and body weight were determined once per week. The state of health was checked twice per day. Hearing tests and eye examinations were carried out before the start and at the end of dosing and recovery periods. Blood samples were taken from all rats for haematology and blood chemistry examination at the end of the dosing period. Urine was analysed at the end of the exposure period. All rats were subjected to complete gross examinations, and weights of selected organs were determined. Microscopic examinations were conducted routinely on selected tissues and all gross lesions.

There were no mortalities in the study and difenoconazole did not induce clinical signs of toxicity at any dose. Body weights of the group of rats at 1500 ppm were significantly and/or clearly lower than those of rats in the control group throughout the study. In addition to occasional body-weight losses, body weights of the groups of males and females at 1500 ppm were 13% and 10% lower, respectively, than their controls after 13 weeks on test. During the 4-week recovery period, body-weight gains of the rats at 1500 ppm were slightly greater than those among rats in the control group, resulting in differences of only −9% for males and −4% for females by week 17. Body weights of the groups of rats at 40 and 250 ppm were not affected by treatment. Hearing tests and ophthalmoscopy revealed no treatment-related findings. There were no treatment-related differences between the control and treated rats in the haematology parameters and there were no clearly treatment-related changes in blood chemistry parameters. At the end of the treatment period, there was a slight increase in alkaline phosphatase activity for males and females at 1500 ppm and a slight decrease in total protein concentrations among males at 1500 ppm, but these values returned to the normal range by the end of recovery. All other statistically significant differences were considered to be incidental and within normal biological variation. Decreased glucose concentrations in males at 250 ppm and increased chloride values in males at 40, 250 and 1500 ppm did not follow dose–response patterns. Increased values for urea, gamma-glutamyl transferase activity and phosphate concentration in males at 1500 ppm and potassium and phosphate in females at 1500 ppm, as well as decreased sodium in females at 1500 ppm, were well within the range for historical controls. The increased gamme-glutamyl transferase activity noted in males at 1500 ppm at 13 weeks was lower than the control value at the end of the 4-week recovery period. A similar result was obtained for cholesterol concentrations in female rats, suggesting that variations in these parameters, although statistically significant, were most probably random variations. Also, significant reductions from control values in lactate dehydrogenase (LDH) activity in males at 1500 ppm were less than the reductions in LDH in control groups at the end of the recovery period compared with the control group value at 13 weeks. The lower creatinine values for females at 1500 ppm at 13 weeks were due to a few exceptionally low values and the means were not different from the control animals after recovery. Slight variations in albumin and globulin fractions were not of sufficient magnitude to be meaningful. The quantitative and qualitative analyses of urine revealed no toxicologically significant effects. The few statistically significant differences were considered to be incidental and within the range for normal biological variation. Increased urobilinogen scores in males of the groups at 250 and 1500 ppm did not follow a dose-related pattern.

Increased absolute liver weights (18% in male and 22% in female rats) and liver-to-body weight ratios were observed in male and female rats at 1500 ppm. Liver-to-body weight ratios were also increased among males and females at 250 ppm. These differences were not apparent among the rats killed at the end of the recovery period. The few other statistically significant differences were not considered to be toxicologically relevant and could, at least for the group at 1500 ppm, be attributed to the significantly reduced carcass weights in this group and to random variation (female kidney weights) in the group at 250 ppm. No differences were seen between the group of rats at 40 ppm and their controls. No treatment-related observations were recorded at autopsy or upon microscopic evaluation of tissues. Observations that were made were consistent with those expected in rats of this strain and age.

The NOAEL in rats fed diets containing difenoconazole for 13 weeks was 250 ppm, equal to 20 mg/kg bw per day, on the basis of increased liver weight at 1500 ppm, equal to 121 mg/kg bw per day (Suter, 1986b).

Groups of 15 male and 15 female CRL:CD(SD)® rats were given diets containing difenoconazole (purity, 94.5%) at a concentration of 0, 20, 200, 750, 1500 or 3000 ppm, equal to 0, 1.3, 13, 51, 105 and 214 mg/kg bw per day in males and 0, 1.7, 17, 66, 131 and 275 mg/kg bw per day in females, for 3 months. An additional five males and five females were included in the control group. Analyses for correct concentrations were performed on all batches of diet, which were prepared every 2 weeks.

The stability of difenoconazole in the diet for at least 16 days was verified and the homogeneity of the dietary mixtures was also verified. The analyses were satisfactory.

Food consumption and body weight were determined once per week. Mortality was checked twice per day, clinical signs were recorded once per day and detailed physical examinations were performed each week. Ophthalmology was carried out on all rats before the start and at the end of dosing. Blood samples for haematology and blood chemistry examination at the end of the dosing period were taken from 10 males and 10 females per group. Urine was analysed at the end of the exposure period. All rats were subjected to complete gross examinations, and weights of selected organs were determined. Ten males and ten females per group were killed in week 13. The remaining rats were removed from the study in week 17. Autopsies were conducted on all rats that were killed at the scheduled times or in extremis and selected organs were weighed. Microscopic examinations were conducted on selected organs from all rats from the scheduled kill at week 13 and on all gross lesions.

There were no treatment-related mortalities or clinical signs. Body-weight reductions were observed in the male group at 3000 ppm and in group of femaless receiving 200, 750, 1500 and 3000 ppm throughout the study, although no statistical analyses were performed. At the end of the study and in comparison with the controls, the mean body weights were reduced to 90% in males at 3000 ppm and, in females of the groups receiving 200, 750, 1500 and 3000 ppm, to 94%, 93%, 89% and 80%, respectively. Body weights of groups of males at 200, 750 and 1500 ppm and females at 20 ppm were not affected by treatment. Food consumption in the group at 3000 ppm was 3.5% lower in males and 6.9% lower in females compared with controls; and while there was no significant trend in males there was a significant ($p < 0.05$) negative trend in females.

Ophthalmoscopy revealed no treatment-related findings. Such findings that were made were considered to be incidental. They consisted of bilateral diffuse posterior subcapsular cataract in one male at 3000 ppm and a focal posterior subcapsular cataract in one female at 200 ppm.

Examination of blood indicated slight, statistically significant decreases in erythrocyte counts and erythrocyte volume fraction in groups of males at 750, 1500 and 3000 ppm and females at 1500 and 3000 ppm; haemoglobin concentration was also reduced in these two groups of females. There were no observed differences in the groups of males or females at 20 and 200 ppm. Blood chemistry analysis showed some statistically significant changes that included a negative trend for glucose concentration in males, a positive trend for blood urea nitrogen (BUN) in males and higher BUN concentrations in groups of males at 1500 and 3000 ppm, lower total bilirubin values for males at 3000 ppm and for females at 1500 and 3000 ppm, a positive trend for cholesterol in females and a negative trend in alanine aminotransferase (ALT) activity in females. The biological significance of these differences was considered to be negligible. It is notable that there were no significant changes in serum cholesterol concentrations in males or females of this study, although the mean value for females of the group receiving 3000 ppm was 30% higher than the control value. It is also notable that gamma-glutamyl transferase activity was below the level of detection in all groups, in contrast to the findings at the much higher dose in the 33-day study described above.

Urine analyses revealed an increase in the incidence and severity of ketones in the male group at 3000 ppm, which was attributed by the sponsor to reduced nutritional status; however, this suggestion does not seem to be supported by the data on food consumption. Absolute liver weights were increased in the groups of males at 750, 1500 and 3000 ppm by 25.7%, 30.9% and 30.2%, respectively and in females by 24.5%, 26.2% and 40.2%, respectively. Liver-to-body weight ratios were increased in these same groups and among females at 200 ppm. The few other statistically significant differences were not considered to be toxicologically relevant and could be largely attributed to reduced carcass weights in the higher dose groups. No organ-weight differences were seen in the group of rats at 20 ppm. No treatment-related effects were observed at autopsy. Microscopic, treatment-related changes were observed only in the liver and consisted of increased incidence and severity of diffuse hepatocellular enlargement. In the group at 3000 ppm, this affected all males and females and was of minimal (two males and three females) or slight (eight males and seven females) severity. Diffuse hepatocellular enlargement also affected all males at 1500 ppm (minimal severity, 9 out of 10; slight severity, 1 out of 10) and four females at 1500 ppm (minimal severity, 4 out of 10), compared with one male and no females in the control groups. There was no indication of a treatment-related association for the other histopathological changes in liver or in other organs (Cox, 1987b). The slight effects observed as liver weight increase and minimal diffuse hypertrophy were regarded as adaptive responses to treatment.

The NOAEL in rats given diets containing difenoconazole for 3 months was 200 ppm equal to 13 mg/kg bw per day, on the basis of a reduction in body-weight gain, an increase in liver weight and haematological effects at 750 ppm equal to 51 mg/kg bw per day and above.

Difenoconazole technical (purity, 91.8%) suspended in 1% carboxymethyl cellulose (CMC) in 0.1% Tween 80 was applied to the skin of 10 male and 10 female HanIbm:WIST (SPF) rats at doses of 0, 10, 100 or 1000 mg/kg bw for 28 days. Applications were for 6 h per day, 5 days per week for the first 3 weeks and every day thereafter. Analysis of test article dosing suspensions before testing (three samples per dose) indicated that the suspensions were close to nominal concentrations, homogeneous and stable. Mean concentrations of the dosing suspensions used during the test were found to be 109%, 116% and 111% of the nominal concentrations at 10, 100 and 1000 mg/kg bw, respectively. Samples of suspensions taken during the test were also found to be homogeneous and stable for 7 h at room temperature.

The state of health of the rats was checked twice per day and clinical signs were recorded daily. Eye examinations were conducted on all rats on day −4 before dosing and on all surviving rats in the control group and those at 1000 mg/kg bw on day 23. Body weights were measured daily for dosing purposes and were recorded weekly. Food consumption was recorded weekly. Detailed clinical examinations (in the home cage, in a standard arena, and during handling) were performed before the dosing period and once per week thereafter. Neurological examinations were performed after the detailed clinical examinations, and included tests for sensorimotor function (approach, touch, vision, audition, pain, vestibular). Laboratory investigations were carried out on all rats at the end of the treatment period. Rats were killed on study day 29 and were subjected to a detailed gross pathological examination. Selected organs were collected, weighed, and examined microscopically.

There was no treatment-related mortality or clinical signs indicative of a treatment-related effect. Also, there were no abnormal findings in the neurological examinations and no signs of irritation at the skin application site. There were no effects on body weights or body-weight gains at any dose. No significant differences or treatment-related trends were observed in any of the haematology parameters, although one male at 1000 mg/kg bw had a high leukocyte count with associated neutrophilia, lymphopenia, eosinopenia and high monocyte and large unstained-cell counts. Treatment-related changes in blood chemistry parameters were slightly lower values for globulin with an associated increase in the albumin to globulin ratio and minimally lower plasma bilirubin and calcium concentrations in males at 1000 mg/kg bw. An elevated mean value for ALT activity among females at 10 mg/kg bw was due to high individual values for two rats that, clearly, were not dose-related. Mean absolute liver weights of groups of male and female rats at 1000 mg/kg bw were 12% and 9% higher than their respective controls, and relative liver weights were similarly 16% and 11% higher. All other organ weights were comparable among the control and treated rats. No treatment-related observations were noted at autopsy. However, microscopic examination revealed treatment-related changes in the skin application site, liver and thyroid gland (Table 4). The skin changes were statistically significant minimal to slight increases in the number of epidermal cell rows, as well as in thickness of the retained lamellar keratin layer (hyperkeratosis) in male and female rats at 1000 mg/kg bw. There was an increased incidence of minimal centrilobular hepatocellular hypertrophy in males and

Table 4. Treatment-related histopathological findings in rats treated topically with difenoconazole for 28 days

Site/finding	No. of rats or weighted grade[a]	Dose (mg/kg bw)							
		Males				Females			
		0	10	100	1000	0	10	100	1000
Skin application site	No. examined	10	9	10	10	10	10	10	10
Hyperkeratosis	Incidence	2	4	2	6	4	7	6	10
	Weighted grade[a]	1.0	1.0	1.0	1.1	1.3	1.0	1.7	1.2
Liver	No. examined	10	10	10	10	10	10	10	10
Hepatocellular hypertrophy	Incidence	2	1	2	7	1	1	1	7
	Weighted grade[a]	1.0	1.0	1.0	1.0	1.0	1.0	1.0	1.0
Thyroid gland	No. examined	10	10	10	10	10	10	10	10
Hypertrophy of follicular epithelium	Incidence	8	6	9	8	7	6	7	9
	Weighted grade[a]	1.6	1.5	1.7	2.0	1.3	1.3	1.4	1.7

From Gerspach (2000)

[a] Weighted grade is Σ (severity grade × No. of rats with that grade)/total examined.

females at 1000 mg/kg bw. In the thyroid, the incidence of minimal to moderate severity grades of hypertrophy of the follicular epithelium was slightly increased in the group of rats at 1000 mg/kg bw.

The observations in liver were regarded as adaptive responses and therefore not adverse. Although the thyroid follicular-cell responses could also be considered adaptive, the reason for stimulation of the thyroid to occur was not so obvious, i.e. whether it was the result of a direct effect of difenoconazole delivery by a route where liver is not the first pass, or it was an indirect effect secondary to a hepatic response. There were no other treatment-related findings.

The NOAEL in rats exposed topically to difenoconazole for 28 days was 100 mg/kg bw per day on the basis of marginal effects on thyroid and the site of skin application at 1000 mg/kg bw per day (Gerspach, 2000).

Dogs

Groups of three male and three female pure-bred beagle dogs were given diets containing difenoconazole technical (purity, 96.1%) at a concentration of of 0, 100, 1000, 3000 or 6000 ppm, equal to 0, 3.6, 31.3, 96.6 and 157.8 mg/kg bw for males and 0, 3.4, 34.8, 110.6 and 203.7 mg/kg bw for females, for 28 weeks. The dogs were aged 5–6 months at the start of dosing. Dietary analyses showed that the difenoconazole was homogeneously distributed in the diet and was present at the targeted concentrations throughout the study. Analyses performed before feeding was started indicated that the diets were stable for at least 15 days at room temperature. The state of health of the dogs was checked and recorded daily. Daily measurements were made of food consumption and body weights before the dosing period and weekly thereafter. Physical examinations, including heart rate, rectal temperature and hearing tests, were performed before the dosing period and during weeks 13 and 28. Eye examinations were conducted on all dogs before the dosing period and every 2–3 weeks thereafter. Haematology, blood chemistry and urine analyses were carried out on all dogs before the dosing period and during weeks 17 and 28. All dogs were subjected to gross pathological assessment after at least 28 weeks, followed by microscopic examinations of selected tissues and organs.

There were no mortalities in the study and difenoconazole did not induce clinical signs of toxicity at any dose. Body-weight losses occurred in all dogs during the first week of the study, except for one male and one female at 100 ppm.. The losses were marked in the group of dogs at 6000 ppm, i.e. males lost 15.4% and females lost 17.9% of their body weight. The pattern of weight loss suggested that palatability was reduced at this, the highest dose. Weight losses continued in all dogs at 6000 ppm into week 2 and some into week 3. By week 4, the dogs at 6000 ppm were gaining weight at rates similar to those for the dogs in the other groups, but the initial loss was never recovered. At study termination, body weights of groups of males and females at 6000 ppm were approximately 30% lower than those of the respective controls. Body-weight gains of the group of dogs at 3000 ppm and of males but not females at 1000 ppm were also lower relative to those of the dogs in the control group. The mean body-weight gain of males at 1000 ppm was 16% lower than the that of the controls. Body weights of the group of dogs at 100 ppm were not affected by treatment. Food consumption by the dogs at 6000 ppm was also reduced throughout the study in a pattern that was consistent with reduced palatability. All dogs in all groups consumed less feed during the first study week than during the last week before testing. For the group of dogs at 6000 ppm, the reduction was marked: during week 1, three out of six dogs in this group ate less than 10% of the amount of food they consumed the previous week. Food consumption increased in the second and third study weeks in all groups, but at a slower rate among the group of dogs at 6000 ppm. From week 4 through study termination, the males at 6000 ppm consumed about 35–45% less feed than did the males in the control group and the females at 6000 ppm consumed about 30–40% less than did the females in the control group. Slight reductions were also noted for males at 1000 and 3000 ppm, but not females. Food consumption by males and females at 100 ppm was not affected by treatment. Thus, body-weight effects appeared to correlate in some groups with reduced food consumption that, in turn, was probably due to reduced palatability.

Physical examinations, including heart rate, rectal temperature and hearing tests, revealed no treatment-related findings. Ophthalmoscopic examinations performed on dogs in week 11 revealed bilateral subcapsular, equatorial, anterior cortical and posterior cortical lenticular aberrations (cataracts) in all dogs at 6000 ppm, and in one male and all females at 3000 ppm. Subsequent examinations revealed slight to marked progression of the lenticular aberrations. The changes varied from an increase in vacuoles, opacities and feathering of the lens suture to mature cataracts. In addition, iridic changes were manifested as irregular pupillary margins and miosis, considered to have occurred from lens-induced uveitis associated with rapidly developing cataracts. No changes were seen in the other dogs at 3000 ppm or in any dog in the control group or in groups at 100 or 1000 ppm.

Examination of blood revealed slight, non-significant treatment-related differences in the group of dogs at 6000 ppm at weeks 17 and 28. These consisted of reductions in erythrocyte count in females of 19.6% and 18.7%, respectively, but with no consistent effect in males; reductions in haemoglobin concentration of 6.9% and 9.6%, respectively, in males and 18.4% and 16.0% in females and reductions in erythrocyte volume fraction of 5.7% and 7.1% in males and 16.8% and 14.0% in females. The effects in females were larger than in males and were consistent, but not progressive. Platelet counts were elevated in the group of dogs at 6000 ppm at both 17 and 28 weeks by 37.7% and 120.7%, respectively in males and 47.5% and 31.4% in females. There were no notable changes in platelet counts at other doses and no dose-related changes in other haematological parameters. Blood chemistry analysis showed significantly increased alkaline phosphatase activity at weeks 17 and 28 for males at 6000 ppm and for females at 3000 and 6000 ppm, although the data for females, while consistent over time, did not follow a clear dose–response relationship. Decreased serum total protein and calcium concentrations were also noted in females at 6000 ppm. These changes were attributed by the sponsor to the moderate to severe reductions in food intake in these dogs; however, there was no such change in serum protein concentration in the males, where there was also a reduction in food intake.

Urine analysis showed that in males at 6000 ppm there was a small reduction in pH, a dose-related increase in bilirubin trend score and a higher severity score for casts. In females there was a reduction in urinary pH at 6000 ppm, but no other notable effects.

Absolute liver weights and liver-to-body weight ratios were increased in females at 3000 ppm and 6000 ppm. There was also a tendency towards higher liver to body-weight ratio in males receiving

3000 ppm and 6000 ppm, but these changes were not statistically significant. Kidney weight relative to body weight was also increased in the female group at 6000 ppm. Reductions were observed in absolute weights of heart, prostate and salivary gland in the male group at 6000 ppm. No organ-weight differences were seen among the groups of males and females at 100 or 1000 ppm or in the group of males at 3000 ppm. At autopsy, treatment-related bilateral ocular opacity was observed in one female at 3000 ppm, one male at 6000 ppm and one female at 6000 ppm. No other treatment-related observations were made. Microscopic examination did not reveal any treatment-related changes other than to the eyes, which confirmed the observations made in vivo. Moderate to severe cataracts were seen in the left lens of all animals at 6000 ppm and minimal to severe cataracts in the right lens. Cataracts were noted for one male at 3000 ppm (bilateral) and two females at 3000 ppm (both unilateral) which were not seen in the ophthalmoscopic examinations. Differences in observations between the left and right lenses were attributed to the use of 10% neutral buffered formalin as the fixative for the right eye and Russell's fixative for the left eye.

The NOAEL in dogs given diets containing difenoconazole for 6 months was 1000 ppm, equal to 31.3 mg/kg bw per day, on the basis of decreased body-weight gain in males, cataracts in males and females and significantly increased serum alkaline phosphatase activity in females at 3000 ppm, equal to 96.6 mg/kg bw per day (O'Connor et al., 1987).

Groups of four male and four female pure-bred beagle dogs were given diets containing difenoconazole technical (purity, 96.1%) at a concentration of of 0, 20, 100, 500 or 1500 ppm, equal to 0, 0.7, 3.4, 16.4 and 51.2 mg/kg bw for males and 0, 0.6, 3.7, 19.4 and 44.3 mg/kg bw for females, for 52 weeks. The dogs were aged approximately 5 months at the start of dosing. Dietary analyses showed that the difenoconazole was homogeneously distributed in the diet and was present at the targeted concentrations throughout the study. Analyses performed before feeding was started indicated that the diets were stable for at least 46 days at room temperature. The state of health of the dogs was checked and recorded daily. Measurements were made of food consumption daily and body weights before the dosing period, weekly for the first 13 weeks and monthly thereafter. Physical examinations, including heart rate, rectal temperature and hearing tests, were performed before the dosing period and during weeks 14, 25, 39 and 52. Eye examinations were conducted on all dogs before the dosing period and during weeks 11, 27, 39 and 51. The eyes of dogs in the control group and dogs at 1500 ppm were also examined at nine additional times during the study. Haematology, blood chemistry and urine analysis were carried out on all dogs before the dosing period and during weeks 13, 26 and 52–53. All dogs were subjected to gross pathological assessment on days 365–367 (during week 52), followed by microscopic examinations on selected tissues and organs.

There were no mortalities in the study and difenoconazole did not induce clinical signs of toxicity in any of the dose groups. Absolute body weights were not significantly different among males or females of any the groups during the study and body-weight gains of male dogs were not affected by any of the treatments, but while body-weight gains among the females receiving 100, 500 and 1500 ppm were not significantly affected by treatment, they tended to be lower relative to values for females in the control group at some intervals during the study. There was no effect on body-weight gain in females of the group at 20 ppm. During the first 3 months of the study, percentage body-weight gains of the females at 100, 500 and 1500 ppm were slightly lower than those of females in the control group. During the intermediate 6 months, this reduction continued for the group of femaless at 500 and 1500 ppm, but during the last 3 months, body weights were relatively stable or decreased in all groups, particularly in the females at 500 and 1500 ppm. The terminal body-weight gains for females in the groups at 0, 20, 100, 500 and 1500 ppm, respectively, were 42%, 45%, 43%, 25% and 29%.

Food consumption by the group of females at 1500 ppm was reduced throughout the study. Across all measurement intervals, the females at 1500 ppm consumed about 20% less feed than did the females in the control group; statistically significant reductions were recorded on days 7, 35, 70 and 357. This reduced food consumption corresponded to reduced body-weight gains among the group of females at 1500 ppm. Food consumption was not affected in females at 20, 100 and 500 ppm or in any of the groups of males.

Physical examinations, including heart rate, rectal temperature, hearing tests and ophthalmoscopic examinations, revealed no treatment-related findings. A single incidence of ocular change (corneal opacity) was seen in one female in the control group.

There were no treatment-related differences among the groups in haematology parameters. A statistically significant reduction in reticulocyte count from 0.8% to 0.3% in females at 1500 ppm was observed at week 52, but this difference was considered to be spurious, there being no accompanying reductions in erythrocyte parameters that could indicate a response to toxicity; furthermore, the results showed little consistency in that the reticulocyte count in females at 1500 ppm at 25 weeks increased to 0.8% from a control mean of 0.5%. Differentials were also unremarkable; the only apparent difference among the groups was a statistically significant decrease in relative eosinophils among females at 100 ppm at week 13, which was not considered to be related to treatment. Blood chemistry analysis showed significantly higher alkaline phosphatase activity in the groups of males at 1500 ppm at all measurement times and in males at 500 ppm at week 52. All other statistical differences (reduced BUN in the groups of males receiving 100 and 1500 ppm, increased total bilirubin concentration in males at 100 ppm, increased sodium concentration in females at 1500 ppm) were considered to be incidental and within the range for normal biological variation. Measurements at weeks 13 and 26 were not noticeably different from those at week 52.

No significant or remarkable changes in urine analysis were observed.

There were no treatment-related differences in organ weights among the groups. The few differences noted were not considered to be toxicologically relevant. Decreased adrenal weights in males at 20 and 500 ppm and increased thyroid weights in females at 20 ppm were statistically different from the respective control animals but not meaningful due to the absence of a dose–response relationship and corresponding histopathology. No treatment-related observations were recorded at autopsy or upon microscopic examination of the organs and tissues.

The NOAEL in dogs given diets containing difenoconazole for 52 weeks was 100 ppm, equal to 3.7 mg/kg bw per day, on the basis of reduced body-weight gain in females at 500 ppm, equal to 19.4 mg/kg bw per day (Rudzki et al., 1988).

2.3 Long-term studies of toxicity and carcinogenicity

Mice

Groups of 60 male and 60 female CD-1®(ICR) mice were given diets containing difenoconazole technical—first batch, of 94.5% purity, was given during weeks 1–20 and the second batch, of 95% purity, was given in weeks 21–80—at a concentration of of 0, 10, 30, 300, 3000 or 4500 ppm for 18 months. The diet containing difenoconazole at 3000 ppm was exchanged after the first 3 weeks for one containing difenoconazole at 2500 ppm and the group of females at 4500 ppm was discontinued because of excessive toxicity very early in the study. The doses had been selected on the basis of the results of a 3-month study (Cox, 1987a). These dietary exposures were equal to 0, 1.5, 4.7, 46.3, 423 and 819 mg/kg bw per day for males and 0, 1.9, 5.6, 57.8 and 513 mg/kg bw per day for females in the groups at 0, 10, 30, 300, 2500 ppm (and 4500 ppm in males), respectively. Additional groups of 10 male and 10 female mice were fed diets containing difenoconazole at 0, 3000 or 4500 ppm then allowed a recovery phase after 12 months of treatment. However, all of the females at 4500 ppm and 16 of the females at 3000 ppm died or were killed because of their moribund condition within the first 2 weeks.

Food consumption and body weights were recorded weekly for weeks 1–16 and every 4 weeks thereafter. Mortality was checked twice per day, clinical signs were recorded once per day and detailed physical examinations (including palpation) were performed weekly. Eye examinations were conducted on all mice during the period before dosing and on all mice in the control group and those at the highest dose at 6-month intervals thereafter. Haematology, consisting of differential counts and cell morphology, was conducted on: (a) 10 males and 10 females before dosing; (b) 10 males and 10 females in the control group and at the highest dose at 52 and 78 weeks; (c) all mice in the recovery group at week 57; and (d) all mice killed because of their moribund condition. Blood chemistry investigations and urine analysis were also performed on: (a) 10 males and females that were not assigned to the study before dosing; (b) 10 males and females in the control group and at the highest dose at 52 and 78 weeks; and (c) all mice in the recovery group at week 57.

Ten males and ten females per group were killed during week 52. Ten additional mice from the groups of females at 2500 ppm and the groups of males at 0, 2500 and 4500 ppm were placed on control diet in weeks 53–56 and killed during week 57. The remaining surviving mice were killed during week 79 or 80. All mice that were killed were subjected to a detailed gross pathological examination. Mice that were found dead or killed in extremis were also autopsied. Specified organs of mice killed at scheduled times were weighed and specified tissues were collected, fixed and examined microscopically.

Dietary analyses performed on each diet batch (prepared every 2 weeks) revealed that difenoconazole was present at the targeted concentrations. Diet mixes were an average of 101.0%, 99.8%, 100.2%, 97.3% and 99.4% of nominal concentrations at 10, 30, 300, 2500 and 4500 ppm, respectively. The homogeneous distribution of difenoconazole in the diet mixes and its stability for at least 16 days at room temperature was also determined.

Table 5. Mortality during an 18-month study in mice fed diets containing difenoconazole

Parameter	Dietary concentration (ppm)											
	Males						Females					
	0	10	30	300	2500[a]	4500	0	10	30	300	2500[a]	4500
No. at initiation	70	60	60	60	70	70	70	60	60	60	70	70
Died/moribund, killed. days 1–9	0	0	0	0	0	3	0	0	0	0	16	52
Moribund, killed day 10	0	0	0	0	0	1	0	0	0	0	0	18
Re-assignment day 10							−10				+10	
Died/moribund, killed. days 11–21	0	0	0	0	0	7	0	0	0	0	3	0
Died/moribund, killed. weeks 4–52	2	2	2	4	0	3	3	3	4	4	2	0
Interim kill, week 53	10	10	10	10	10	10	10	10	10	10	10	0
Post-recovery kill, week 57	9[b]	0	0	0	10	10	0	0	0	0	10	0
Died/moribund, killed weeks 53–68	9	8	8	6	2	7	10	5	6	6	1	0
Died/moribund, killed weeks 69–80	9	8	13	16	14	13	13	7	11	5	9	0
Terminal kill, weeks 79–80	31	32	27	24	34	16	24	35	29	35	29	0
Percentage survival to termination[c]	62	64	54	40	68	32	48	70	58	70	36	0

From Cox et al. (1989a)

[a] Received diet containing difenoconazole at 3000 ppm until day 21, when the dietary concentration was reduced to 2500 ppm.

[b] One mouse in the recovery group died during recovery.

[c] Excluding interim and post-recovery kills.

There was very high mortality at the beginning of the study, with 52 out of 70 of the group of female mice at 4500 ppm dying or being killed in a moribund condition within the first 9 days (Table 5). The remaining 18 females in the group at 4500 ppm were killed in a moribund condition on day 10, and 10 of the females at 0 ppm were transferred to the 3000/2500 ppm diet on the same day. Four of the males at 4500 ppm died or were killed in a moribund condition during the same time period and an additional seven died during week 3. Primarily due to these early deaths, survival to termination was significantly reduced among the males at 4500 ppm.

The early deaths were accompanied by clinical signs of thinness, hunched posture and rough hair coat. The female mice at 4500 ppm that survived to the first measurement of body weight had lost approximately 25% of their initial weight. At autopsy, findings were recorded for: liver, males at 4500 ppm, 7 out of 11, females at 4500 ppm, 22 out of 70 and females at 2500 ppm, 2 out of 9; and, stomach, males at 4500 ppm, 5 out of 11, females at 4500 ppm, 31 out of 70 and females at 2500 ppm, 3 out of 9. Tissues of females in the group at 4500 ppm were not examined microscopically, but the liver findings in the group of males at 4500 ppm and females at 2500 ppm that died early tended to be correlated with individual cell necrosis and hepatocellular hypertrophy. For the most part, the stomach findings were not confirmed microscopically.

Thinness, hunched posture and rough hair coat were also observed at a higher incidence in the group of females at 2500 ppm and males at 4500 ppm that survived the first three study weeks than in the other treated and control groups. In addition, reduced motor activity was noted for the males at 4500 ppm. No other clinical signs that were noted showed dose-related incidences and all were common for studies of this duration.

Body weights of the groups of males at 4500 ppm and 2500 ppm and group of females at 2500 ppm were lower than those in mice at 0 ppm throughout the study. Nearly all of the males at 2500 ppm and all of the females at 2500 ppm and males at 4500 ppm lost weight during the first study week. These weight losses were approximately 18% of initial weight in the latter two groups. Body-weight gains then began to recover and approach values in the group at 0 ppm, but body weights in these groups remained lower through to the end of the study. Body-weight gains of the male mice at 300 ppm were also somewhat reduced relative to those of the controls in the first 36 weeks of the study, but recovered thereafter. Body weights of the mice at 10 and 30 ppm were not affected by treatment.

There were no clear differences in weekly food consumption among the groups.

Ophthalmoscopy at weeks 27, 53 and 78 revealed no treatment-related findings. Corneal calcification (corneal dystrophy) was observed in about 20% of the mice examined after 53 weeks, but the incidence was similar among mice in the control group and mice treated with difenoconazole. Posterior subcapsular or complete cataracts were observed in about 50% of the mice at 0 ppm and 25% of the males at 4500 ppm and females at 2500 ppm at the 78-week examination. These findings were considered to be incidental to treatment.

Haematology revealed no treatment-related effects. The percentage segmented neutrophil count was increased and percentage of lymphocytes was decreased in females at 2500 ppm at week 79, but the biological significance of this change was unclear. Decreases in the percentage of eosinophils in males at 4500 ppm and females at 2500 ppm at week 79 relative to the group of mice at 0 ppm were too small to be meaningful. No treatment-related changes were seen in the differential counts on blood from mice that were killed while moribund. Blood chemistry analysis at 53 weeks revealed significant increases in ALT activity in the group of males at 2500 and 4500 ppm and sorbitol dehydrogenase activity in males at 300, 2500 and 4500 ppm; but these values decreased towards values for the control group after the 4-week recovery period (week 57). At the end of the study, ALT activities were elevated in males at 4500 ppm and females at 2500 ppm; sorbitol dehydrogenase activities were elevated in groups of males at 4500 ppm and 2500 ppm and females at 2500 ppm and alkaline phosphatase activities were increased in males at 4500 ppm. The significantly lower values for sorbitol dehydrogenase activity in group of females at 10 and 30 ppm at week 53 were considered not to be toxicologically meaningful.

Absolute liver weights and liver-to-body weight ratios were increased in males at 4500 ppm and 2500 ppm and in females at 2500 ppm and 300 ppm at weeks 53 and 79–80. However, mean liver weights from mice in the recovery group were markedly lower than the weights for the mice killed at 53 weeks. The few other statistically significant organ weight differences were considered not to be toxicologically relevant. Increased relative adrenal, heart and pituitary weights of males at 4500 ppm and/or 2500 ppm at week 57 (after recovery) can be attributed to significantly reduced carcass weights. Dose–response patterns were not followed for the observed increased kidney weights in males at 10 ppm, decreased relative pituitary weights in males at 30 ppm at weeks 79–80 and decreased absolute testes weight in males at 300 ppm at week 53. Decreased absolute brain weights in males at 4500 ppm at week 53 were not seen at other times (week 57 or weeks 79–80) and were not corroborated by autopsy or microscopic findings.

Autopsies of the mice killed at 53, 57 and 79–80 weeks (interim, recovery and terminal kills), as well as the unscheduled deaths, revealed treatment-related findings only in the liver. The most prominent findings were enlargement, pale areas and masses. The overall incidence of findings in the liver of cases of unscheduled death was increased over that in the control group in the groups of males at 4500 ppm and 2500 ppm and females at 2500 ppm. Enlarged livers or liver masses were seen in about half of the males in the group at 4500 ppm and one quarter of the groups of males and females at 2500 ppm. All other findings were considered to be spontaneous and commonly encountered in mice of this strain and age.

Microscopic examination also revealed treatment-related changes only in the liver (Table 6). Effects were seen in males at 300, 2500 and 4500 ppm and females at 2500 ppm. Non-neoplastic changes in the liver included various combinations of necrosis (primarily single-cell but also focal/multifocal), hypertrophy, fatty change and bile stasis. For the most part, these changes were observed at the interim kill as well as in the mice that died during the second study year or were killed at termination. The incidences of necrosis, hypertrophy and fatty change were lower in the mice in the recovery group than in the mice at interim kill, indicating partial recovery. In terms of neoplastic changes, the overall incidence of hepatocellular adenomas and/or carcinomas was increased in males at 2500 ppm, females at 2500 ppm and males at 4500 ppm; the trends were statistically significant. The incidence of adenomas and/or carcinomas was already somewhat elevated in males at 4500 ppm at the interim and recovery-group kill. The other liver findings were considered to be incidental. There was no observable neoplastic response at 300 ppm.[2]

With the exception of the hepatocellular adenomas and carcinomas noted above, none of the neoplastic findings were present at incidences that would suggest a relationship to treatment. The other histopathological changes that were noted are commonly seen in mice of this strain and age and the incidence, distribution and morphological appearance of these findings did not give any indication of a treatment-related association.

The NOAEL in mice fed diets containing difenoconazole for 18 months was 30 ppm, equal to 4.7 mg/kg bw per day, on the basis of reduced body-weight gain and increased liver weight in females and increased incidence of hepatocellular hypertrophy in males at 300 ppm, equal to 46.3 mg/kg bw per day. The incidences of hepatocellular adenomas and carcinomas were increased in male and female mice of the group at 2500 ppm , equal to 423 mg/kg bw per day for male mice and 513 mg/kg bw per day for female mice. Clearly, the MTD was exceeded at this level in females, body-weight gain in the survivors of the early mortality in this group being 67–77% of that of mice in the control group. Subsequent studies of the mode of action (see below) have shown that difenoconazole induces a profile of hepatic enzymes that is similar to that of phenobarbital (Cox et al., 1989a).

[2] The United States Environment Protection Agency (EPA) Benchmark software was applied to the combined incidences of hepatocellular adenoma and carcinoma in mice investigated between weeks 53 and 79. The 10% benchmark-dose lower 95% confidence level (BMDL10) was 673 ppm for male mice and 355 ppm for female mice, using a gamma multi-hit model.

Table 6. Treatment-related histopathology findings in a study of carinogenicity in mice fed diets containing difenoconazole

Finding	Death/ scheduled termina-tion	Dietary concentration (ppm)																							
		Males												Females											
		0		10		30		300		2500[a]		4500		0		10		30		300		2500[a]			
		N	E	N	E	N	E	N	E	N	E	N	E	N	E	N	E	N	E	N	E	N	E		
Liver																									
Hepatocellular adenoma	Early[b]	0	2	0	2	0	2	0	4	0	0	0	14	0	3	0	3	0	4	0	4	0	11[g]		
	Interim[c]	0	10	1	10	2	10	0	10	1	10	2	10	0	10	0	10	0	10	0	10	1	10		
	Recovery[d]	0	9	—	—	—	—	—	—	0	10	3	10	—	—	—	—	—	—	—	—	0	10		
	Late[e]	0	18	3	16	1	21	2	22	9	16	6	20	0	26	0	14	0	21	0	15	5	10		
	Terminal[f]	4	31	6	32	5	27	7	24	3	34	9	16	0	24	0	35	0	29	1	35	10	29		
Hepatocellular carcinoma	Early	0	2	0	2	0	2	0	4	0	0	0	14	0	3	0	3	0	4	0	4	0	11		
	Interim	0	10	0	10	0	10	0	10	0	10	2	10	0	10	0	10	0	10	0	10	0	10		
	Recovery	0	9	—	—	—	—	—	—	1	10	1	10	—	—	—	—	—	—	—	—	0	10		
	Late	0	18	0	16	1	21	0	22	1	16	4	20	0	26	0	14	0	21	0	15	2	10		
	Terminal	1	31	0	32	0	27	0	24	3	34	6	16	0	24	0	35	1	29	0	35	2	29		
Single-cell necrosis	Early	0	2	0	2	0	2	1	4	0	0	14	14	0	3	0	3	0	4	0	4	9	11		
	Interim	0	10	1	10	1	10	3	10	10	10	10	10	0	10	0	10	0	10	3	10	10	10		
	Recovery	1	9	—	—	—	—	—	—	6	10	3	10	—	—	—	—	—	—	—	—	0	10		
	Late	0	18	1	16	0	21	2	22	11	16	17	20	0	26	0	14	0	21	1	15	0	10		
	Terminal	4	31	3	32	1	27	7	24	25	34	9	16	3	24	0	35	0	29	2	35	8	29		
Focal/multifo-cal necrosis	Early	0	2	0	2	1	2	1	4	0	0	6	14	0	3	0	3	0	4	0	4	1	11		
	Interim	0	10	0	10	0	10	2	10	1	10	0	10	0	10	0	10	0	10	1	10	0	10		
	Recovery	0	9	—	—	—	—	—	—	0	10	0	10	—	—	—	—	—	—	—	—	0	10		
	Late	1	18	1	16	3	21	2	22	5	16	5	20	1	26	1	14	0	21	0	15	2	10		
	Terminal	3	31	1	32	0	27	1	24	5	34	5	16	3	24	1	35	0	29	6	35	3	29		

Hepatocellular hypertrophy	Early	0	2	0	2	0	2	1	4	0	0	14	14	0	3	0	3	0	4	0	4	6	11
	Interim	0	10	1	10	0	10	1	10	10	10	10	10	0	10	0	10	0	10	0	10	10	10
	Recovery	1	9	—	—	—	—	—	—	4	10	2	10	—	—	—	—	—	—	—	—	1	10
	Late	4	18	9	16	7	21	13	22	15	16	6	34	1	26	1	14	1	21	3	15	8	10
	Terminal	12	31	6	32	8	27	11	24	32	34	11	16	1	24	6	35	1	29	4	35	28	29
Fatty change	Early	0	2	0	2	0	2	1	4	0	0	2	14	0	3	0	3	0	4	0	4	1	11
	Interim	2	10	1	10	0	10	2	10	5	10	9	10	0	10	0	10	2	10	1	10	4	10
	Recovery	0	9	—	—	—	—	—	—	0	10	1	10	—	—	—	—	—	—	—	—	0	10
	Late	0	18	0	16	0	21	0	22	1	16	13	20	0	26	0	14	0	21	0	15	0	10
	Terminal	0	31	0	32	0	27	1	24	7	34	7	16	0	24	1	35	0	29	3	35	4	29
Biliary stasis	Early	0	2	0	2	0	2	0	4	0	0	1	14	0	3	0	3	0	4	0	4	0	11
	Interim	0	10	0	10	0	10	0	10	3	10	10	10	0	10	0	10	0	10	0	10	6	10
	Recovery	0	9	—	—	—	—	—	—	9	10	6	10	—	—	—	—	—	—	—	—	9	10
	Late	0	18	0	16	0	21	0	22	12	16	20	20	0	26	0	14	0	21	3	15	6	10
	Terminal	1	31	0	32	0	27	3	24	32	34	13	16	0	24	0	35	0	29	0	35	29	29

From Cox et al. (1989a)

N, No. of mice with finding; E, No. of mice examined.

[a] Fed diet containing difenoconazole at 3000 ppm until day 21.

[b] Early unscheduled deaths in weeks 1–52.

[c] Interim kill in week 53.

[d] Recovery group, killed in week 57.

[e] Late unscheduled deaths in weeks 53–79.

[f] Terminal kill in weeks 79–80.

[g] No histopathology on 10 early deaths (mice ‘replaced’ by mice from control group).

Rats

Groups of 80 CRL:CD(SD)® rats were given diets containing difenoconazole technical of one of two batches (weeks 1–20: batch 1, purity, 94.5%; and weeks 21–106: batch 2, purity, 95%) at a concentration of of 0, 10, 20, 500 or 2500 ppm (90 animals in the control group and group at 2500 ppm), equal to 0, 0.5, 1.0, 24.1 and 124 mg/kg bw per day for males and 0, 0.6, 1.3, 32.8 and 170 mg/kg bw per day for females for 2 years. The doses were selected on the basis of the results of two 3-month feeding studies in rats (Suter, 1986b; Cox, 1987b). Ten male and ten female rats per group were killed at 52 weeks and ten males and ten females from the control group and the group at 2500 ppm were placed on basal diet from weeks 53–56 and then killed. Mortality was checked twice per day, clinical signs were recorded once daily and detailed physical examinations were performed weekly. Individual body weights and food consumption were recorded weekly for the first 4 months and monthly thereafter. Eye examinations were conducted on all rats before the dosing period and on the groups of rats at 0 and 2500 ppm at 6-month intervals. Laboratory investigations were carried out on 20 males and 20 females per group at 27, 52, 78 and 104 weeks, as well as on 10 males and 10 females before the dosing period. Surviving rats were killed after 104 weeks and subjected to a detailed autopsy. All rats killed at 52 weeks and after the recovery period and all rats that were found dead or killed in extremis were also examined. Specified organs of the rats were weighed and specified tissues were collected, fixed and examined microscopically. Dietary analyses performed on each diet batch (prepared every 2 weeks) revealed that the test item was present at the targeted concentrations with the exception of the week 7–8 mix at 500 ppm (44% of target) and week 47–48 mix at 10 ppm (77% of target). Homogeneous distribution in the diet mixes was demonstrated for the first set of diet mixes and stability for at least 16 days at room temperature was determined on samples containing difenoconazole at 10 ppm.

There were no treatment-related differences between the groups in terms of survival to study termination and all clinical signs noted were considered to be incidental to treatment and without marked differences between the groups. Body weights and body-weight gains in males and females at 500 and 2500 ppm were lower than in the group at 0 ppm in the first year of the study, a trend that continued throughout the study in the rats at 2500 ppm, although the differences were only occasionally statistically significant. By study termination, the mean body-weight gains of the males and females at 2500 ppm were 11% and 37% lower than those of rats at 0 ppm. Body-weight gains of the males and females at 500 ppm were reduced at week 52 by 6% and 10% respectively. Reductions in body-weight gain were also recorded for this group at weeks 13 and 24. Body weights of the rats at 10 and 20 ppm were not affected by treatment. Weekly food consumption was somewhat reduced among the rats at 2500 ppm throughout the study; only three weekly measures were analysed statistically. Food spillage was high among the females at 2500 ppm during the first study month, but was similar to the other groups by week 5. Food consumption among the rats at 10, 20 and 500 ppm was not affected by treatment.

Ophthalmoscopy of the rats at 0 and 2500 ppm at weeks 28, 52, 78 and 104 revealed no treatment-related findings. All abnormalities were present at low incidences (maximum, three rats in any group). Treatment-related differences in haematology between rats in the control group and treated rats were decreases in erythrocyte counts, haemoglobin concentration and erythrocyte volume fraction in the group of females at 2500 ppm, especially early in the study. Mean corpuscular volume (MCV) and mean corpuscular haemoglobin concentrations (MCHC) were also reduced in this group. MCV tended to be lower and mean cell haemoglobin (MCH) and mean corpuscular haemoglobin concentration (MCHC) tended to be higher among males at 2500 ppm, but these changes were not accompanied by changes in other erythrocyte parameters. Platelet counts were decreased for males at 500 ppm at weeks 28 and 53 and for males at 2500 ppm throughout the study. Leukocyte counts were depressed for males and females at 2500 ppm at week 105, resulting from lower absolute segmented neutrophil and lymphocyte counts. None of the other parameters measured showed notable differences, nor were any differences in haematology seen in the groups at 10 and 20 ppm or in females of the group at 500 ppm.

Blood chemistry measurements showed an increase in albumin and decrease in globulin concentrations in males of the group at 2500 ppm, resulting in increased albumin : globulin ratios throughout the study. Serum albumin concentrations were increased in females at 2500 ppm only in week 28. There was a transient decrease in glucose and a transient increase in total cholesterol concentrations in rats at 2500 ppm in week 28. Decreases in total bilirubin in males at 2500 ppm in week 28 and females at 2500 ppm in weeks 28, 53 and 79 were considered to be too small to be meaningful. An increase in BUN in females at 2500 ppm in week 53 was considered to be incidental. ALT activities were increased in males receiving 500 and 2500 ppm at week 53 but decreased in females at 500 and 2500 ppm in week 28 and females at 2500 ppm in week 53. The inconsistency of these changes suggests that there was no relationship to treatment. All other differences were considered to be spurious owing to the low magnitude of the change, inconsistency across study intervals or the lack of a dose–response relationship.

Urine analyses showed a slight increase in ketone bodies and decrease in pH in males at 2500 ppm at week 28, suggesting accelerated metabolism of lipids, which would be consistent with the diminished nutritional status of these rats. Values for specific gravities, concentrations of protein, urobilinogen and occult blood were comparable among the groups; bilirubin was not detected in any samples and traces of glucose were found in only five rats. There were no apparent differences among the groups in terms of microscopic examination of sediment.

In the females at 2500 ppm, absolute liver weights were 15.7% higher than those of rats in the control group at 53 weeks, while relative liver weights of both males and females of the group at 2500 ppm were significantly increased by 14.2% and 48.2% as a result of significantly lower carcass weights. Liver weights of the rats at 2500 ppm killed after 4 weeks of recovery were not different from those of rats in the control group. At termination, absolute liver weights of the males and females at 2500 ppm were 19.6% and 6% higher than those of their respective controls. The few other statistically significant differences were considered not to be toxicologically relevant. These consisted of increased relative brain and kidney weights in females at 2500 ppm at week 53 and relative brain weight at week 105 that could be attributed to lower carcass weights; decreased absolute adrenal weights at week 53 in males at 2500 ppm and spleen weights at week 57 in females at 2500 ppm that were not seen at other times, and an increase in ovary weights at week 105 (500 and 2500 ppm) that were attributable to one rat in each group with massive ovarian cysts. No differences were seen between the groups of rats at 10, 20 or 500 ppm and the group at 0 ppm.

Autopsy did not reveal any treatment-related findings after 105 weeks. The most common findings in the liver (target organ) were pale or dark areas and enlargement, but without any dose–response relationship.

Microscopic examinations revealed an increased incidence and severity of hepatocellular hypertrophy in males and females at 500 and 2500 ppm at 105 weeks (Table 7). These changes were not evident in those rats killed at week 53, although ALT activities were increased in males at this time. In contrast, however, ALT activities were significantly reduced in female rats at 28 weeks in the groups at 500 and 2500 ppm and at 53 weeks in the group at 2500 ppm. The other liver findings, including neoplasia, were considered to be incidental, owing to the lack of a dose–response relationship and/or the low incidence.

The other histopathological changes that were noted are commonly seen in rats of this strain and age and the incidence, distribution and morphological appearance of these findings did not give any indication of a treatment-related association. There was no evidence for carcinogenicity or oncogenicity in rats.

The NOAEL was 20 ppm, equal to 1.0 mg/kg bw per day, on the basis of reduced body-weight gain during the first year, reduced platelet counts and increased incidence and severity of hepatocellular hypertrophy at 500 ppm, equal to 24.1 mg/kg bw per day (Cox et al., 1989b).

Table 7. Histopathology findings in the liver in a long-term combined study of toxicity/carcinogenicity study in rats fed diets containing difenoconazole

Finding	Deaths/ kill	Dietary concentration (ppm)																			
		Males										Females									
		0		10		20		500		2500		0		10		20		500		2500	
		N	E	N	E	N	E	N	E	N	E	N	E	N	E	N	E	N	E	N	E
Hepatocellular hypertrophy	Unscheduled	0	30	0	39	0	30	1	27	0	26	0	38	0	35	0	44	3	29	4	32
	Interim kill, week 53	0	10	0	10	0	10	0	10	0	10	0	10	0	10	0	10	0	10	0	10
	Recovery group, week 57	0	10	—	—	—	—	—	—	0	10	0	10	—	—	—	—	—	—	0	10
	Terminal kill, week 79–80	7	40	5	31	8	40	28	43	39	44	4	32	0	35	0	26	14	41	32	38
Severity of finding at terminal kill:	Minimal	7	4	5	24	7	4	—	—	10	18										
	Slight	0	1	3	3	27	0	—	—	4	8										
	Moderate	0	0	0	1	5	0	—	—	0	6										

From Cox et al. (1989b)
N, No. of rats with finding; E, No. of rats examined.
U = unscheduled deaths I = interim kill week 53 R = recovery kill week 57 T = terminal week 79–80
[a] Only findings seen in more than one rat in any group.

(a) Discussion of a possible mode of carcinogenic action of difenoconazole.

A. Postulated mode of action.

In mice and rats, treatment with difenoconazole caused an increasee in the incidences of hepatocellular adenomas and carcinomas in male and female mice given diets containing difenoconazole at 2500 ppm, equal to 423 and 513 mg/kg bw per day, respectively, for 18 months (Cox et al., 1989a), but not in male or female rats given the same diet, equal to 124 and 170 mg/kg bw per day, respectively, for 24 months (Cox et al., 1989b). Difenoconazole has also been shown to induce various hepatic enzymes of xenobiotic metabolism and it is not genotoxic. These general properties suggest that difenoconazole could have a mode of carcinogenic action (MOA) similar to that of phenobarbital. The generally proposed MOA for phenobarbital is tumour promotion resulting from a long-standing hepatomegaly, hepatocellular hypertrophy and hyperplasia of the liver. Treatment with phenobarbital alone produces hepatocellular adenomas and carcinomas in mice and, when administered after a short treatment with a known liver carcinogen, the incidences of tumours are increased. The potency of CYP induction in mice and rats correlates with the degree of tumour promotion. In addition, phenobarbital has mitogenic effects in liver and has been shown in many studies to inhibit intercellular communication (reviewed in IARC, 2001). Phenobarbital has been studied to a far greater degree than difenoconazole and so it is not expected that there should be complete concordance of the data in order to conclude that the MOAs are most likely the same; however, clear discord should not be demonstrated.

B. Key events.

Difenoconazole is expected to induce hepatic CYP enzymes. A study on such an effect was performed in male TIF:Magf (SPF) mice given difenoconazole at a dose of 0, 1, 10, 100 or 400 mg/kg bw per day for 14 days. Reversibility was studied in the same experiment in which groups were given difenoconazole at 0 or 400 mg/kg bw per day for 14 days followed by 28 days without treatment. Other groups were given phenobarbital, 3-methylcholanthrene or nafenopin (Thomas, 1992). Difenoconazole increased liver weight, hepatic cytochrome P450 protein and a number of hepatic CYP enzyme and other microsomal (UDP-glucuronosyltranferase, lauric acid 11- and 12-hydroxylase, epoxide hydrolase) enzyme activities, as well as microsomal cytochrome P450-dependent testosterone hydroxylation in a way that was generally very similar to that observed in the same strain of male mouse when phenobarbital was administered. Most effects were observed in only the groups at 100 and 400 mg/kg bw per day, although certain enzyme activities were increased even with difenoconazole at 1 mg/kg bw per day. All effects were dose-dependent and regressed within the 28-day recovery period. Usually, regression was complete. The pattern of enzyme induction by difenoconazole was very similar to that of phenobarbital for 7-ethoxyresorufin *O*-deethylase (EROD) and 7-pentoxyresorufin *O*-depentylase (PROD) activities in particular. This similarity is important because it is quite different from the induction pattern observed after treatment with either 2-methylcholanthrene or nafenopin.

Clearly, other key events must follow, but there was no experimental evidence with difenoconazole to support their participation in an MOA.

C. Dose–response relationships.

All effects on enzyme induction in the 14-day study and the tumour response in the 18-month study were dose-dependent. Liver-weight increases were dose-dependent in a 3-month study in mice. At daily doses of 200 and 2500 ppm, equal to 34 and 440 mg/kg bw in males and 45 and 639 mg/kg bw in females, the absolute liver-weight increases were 9% and 82% in males and 11% and 71% in females. In rats, on the other hand, the responses were less clear. At daily doses of 750, 1500 and 3000 ppm equal to 51, 105 and 214 mg/kg bw in males and 66, 131 and 275 mg/kg bw in females, the absolute liver-weight increases were 25.7%, 30.9% and 30.2% in males and 24.5%, 26.2% and 40.2% in females.

D. Temporal association.

The few data that were available reflected a coherent temporal relationship in that enzyme induction was observed within a few days and tumours were observed much later. This is a particularly weak piece of evidence, since there has been no suggestion as to the intervening steps and when they occurred.

E. Strength, consistency and specificity of association of tumour response with key events.

Although a phenobarbital-like MOA has been proposed, there has been only one key event identified in the case of difenoconazole: induction of hepatic enzymes. While this shows consistency, it is nevertheless considered weak in view of the absence of evidence for intervening steps.

F. Biological plausibility.

There has been no suggestion as to why induction of hepatic enzymes per se should have resulted in a neoplastic response.

G. Possible alternative modes of action.

A genotoxic mechanism should always be considered. In the case of phenobarbital, there were some data (usually weak and inconsistent) suggesting that it might have some genotoxic activity. Nevertheless, a non-genotoxic MOA is generally accepted. In the case of difenoconazole, there is no evidence for genotoxicity of difenoconazole or any of its metabolites. This MOA would therefore appear to be ruled out. No other MOA has been proposed.

H. Uncertainties, inconsistencies and data gaps.

Within the dataset reported, there are uncertainties because so much relies upon a single experiment on enzyme induction and a single experiment on carcinogenicity. On the other hand, the results of several other studies are consistent with the conclusion that enzyme induction has occurred because they show that treatment with difenoconazole produces hepatomegaly and hepatocellular hypertrophy. An uncertainty, which applies even to the experiment with male mice, is that although the dose and dose route were similar to the effective dose used in the study of carcinogenicity, the dose rate would have been entirely different, since gavage administration was used in the 14-day study and dietary administration in the study of carcinogenicity. The data available showed no inconsistencies; however, large data gaps are apparent. Thus, there is no information to describe what happens between the induction of liver enzyme activity and the emergence of tumours. An important data gap regards the pivotal mechanistic study in male mice, which did not investigate whether similar results would be found in female mice or how rats might respond to a similar treatment. In the case of phenobarbital, there is evidence for a correlation between the magnitude of enzyme induction and the tumour response in different rodent species and strains (IARC, 2001).

I. Assessment of postulated mode of acton.

The postulated MOA is weak because of the total lack of a description of events between enzyme induction and the discovery of tumours.

2.4 Genotoxicity

Difenoconazole was tested for genotoxicity in a range of assays, both in vitro and in vivo (Table 8). There was no evidence for induction of gene mutation in *Salmonella typhimurium* or *Escherichi coli* in vitro (Ogorek, 1990). The dose range used extended into a range that was bacteriotoxic for this compound, particularly for *S. typhimurium* TA1537 and TA98. A confirmatory test with these strains used a highest dose of 1362 μg/plate. No significant response was observed in a single study for mutations at the *Tk* locus in mouse lymphoma L5178Y cells in vitro, after exposure for 4 h to difenoconazole at concentrations of up to 150 μg/ml or 50 μg/ml in the absence or presence of an exogenous metabolic activation system, respectively (Dollenmeier, 1986a). Thus, difenoconazole did not induce gene mutations in either bacterial cells or cultured mammalian cells.

Four studies on chromosomal aberrations were conducted, two in Chinese hamster ovary (CHO) cells and two in human lymphocytes. The studies with CHO cells gave inconclusive results, in that they were judged to be significant in one experiment in each study, but the result could not be confirmed in independent experiments (Lloyd, 2001; Ogorek, 2001). On the other hand, the two studies on chromosomal aberrations in human lymphocytes gave unambiguous negative results (Fox, 2001; Strasser, 1985).

No genotoxic activity was observed in a study for the induction of unscheduled DNA synthesis (UDS) in primary cultures of rat hepatocytes exposed for 18–20 h to difenoconazole at concentrations of up to 50 μg/ml. Higher concentrations were toxic (Hertner, 1992).

In a test for clastogenicity and aneugenicity, micronucleus formation was examined in bone-marrow cells of Tif:MAGf (SPF) mice. Groups of five male and five female mice were given difenoconazole at a dose of 0, or 1600 mg/kg bw orally by gavage on a single occasion and bone-marrow cells were sampled after 16, 24 and 48 h. In a second experiment, groups of eight male and eight female mice were given difenoconazole at a dos at 0, 400, 800 or 1600 mg/kg bw, and marrow cells were sampled after 24 h. No increases in the frequency of micronucleated polychromatic erythrocytes were observed in any group dosed with difenoconazole. Large increases in the frequency of micronucleus formation were observed in both of the positive-control groups (Ogorek, 1991).

Table 8. Results of studies of genotoxicity with difenoconazole

End-point	Test object	Concentration/ dose[a] (LED/HID)	Purity (%)	Result	Reference
In vitro					
Gene mutation	*S. typhimurium strains* TA100, TA1535, TA1537, TA98; *E.coli* WP2 *uvrA* (± S9), standard plate test	5447 μg/plate	91.8	Negative	Ogorek (1990)
Gene mutation	Mouse lymphoma L5178Y *Tk* locus	150 μg/ml −S9 50 μg/ml +S9	94.5	Negative	Dollenmeier (1986a)
Chromosomal aberration	Chinese hamster ovary cells	34 μg/ml −S9 67 μg/ml +S9	94.3	EQ EQ	Lloyd (2001)
Chromosomal aberration	Chinese hamster ovary	59 μg/ml −S9 18 μg/ml +S9	94.3	Negative EQ	Ogorek (2001)
Chromosomal aberration	Human lymphocytes	40 μg/ml −S9 40 μg/ml +S9	94.5	Negative	Strasser (1985)
Chromosomal aberration	Human lymphocytes	75 μg/ml −S9, 3 h 10 μg/ml −S9, 20 h 62 μg/ml +S9	94.3	Negative	Fox (2001)
Unscheduled DNA synthesis	Male Wistar rat, liver cells	50 μg/ml, 18–20 h	91.8	Negative	Hertner (1992)
In vivo					
Micronucleus formation	Male and female Tif:MAGf (SPF) mice, bone-marrow cells 16, 24 and 48 h after dosing	1600 mg/kg bw	91.8	Negative	Ogorek (1991)

HID, highest ineffective dose; EQ, equivocal response; LED, lowest effective dose; S9, 9000 × g supernatant from rat livers.

2.5 *Reproductive toxicity*

(a) Multigeneration studies

Rats

Groups of 30 male and 30 female sexually immature albino rats were fed diets containing difenoconazole technical (purity, 97.4%) at a concentrations of 0, 25, 250 or 2500 ppm in a two-generation study of reproduction. These diets delivered difenoconazole at the doses given in Table 9. The grand mean daily consumption of difenoconazole (mean of all calculated means) was 0, 1.75, 17.3 and 178.0 mg/kg bw per day at 0 25, 250 and 2500 ppm, respectively.

After 11 weeks, the rats were mated and allowed to rear the ensuing F_1 litters to weaning. Litters were culled to four males and four females on postnatal day 4. F_0 parental males and females were killed and autopsied after weaning of the last litter. Paired testes and ovaries of the parental rats were weighed. Microscopic examinations were performed on specified sex organs and the pituitary from the rats at 0 ppm and 2500 ppm, as well as any gross lesions found in any rat. After weaning (age, 31–48 days), 30 males and 30 females per group were selected as the F_1 generation; pups not selected for mating were killed and autopsies performed. Feeding and mating of the F_1 generation followed the same procedures as the F_0 generation, except that the period before mating was 14 weeks. Clinical signs, body weights, food consumption, mating, gestation and delivery parameters, pup survival and development were recorded. F_1 parental males and females were killed and autopsies and microscopic examinations performed as for the F_0 generation. Analysis of the diet for difenoconazole content was performed on seven of the diet mixes prepared throughout the study. The concentrations ranged from −10% to +7% of the target concentrations. The grand means of the analytical concentrations were 24.0, 246 and 2524 ppm. Pre-test mixes at 25 and 2500 ppm were also analysed for stability and found to be stable at room temperature for at least 30 days. Two sets of mixes were analysed for homogeneity and the concentrations were found to vary between −2% and +2% of the mean concentrations, demonstrating homogeneous distribution in the diet.

F_0 females received difenoconazole at the same dietary concentrations during gestation and lactation. The breeding programme was repeated with the F_1 parents (selected from the F_1 offspring)

Table 9. Mean consumption of difenoconazole by parental animals given diets containing difenoconazole

Generation	Mean consumption (mg/kg bw per day)					
	25 ppm		250 ppm		2500 ppm	
	Start	End	Start	End	Start	End
F_0 generation						
Males:						
Before mating	2.75	1.31	27.4	13.3	247.5	132.1
Females:						
Before mating	2.65	1.65	26.3	16.2	227.6	166.3
Gestation	1.82	1.52	17.5	15.5	173.2	168.2
F_1 generation						
Males:						
Before mating	2.56	1.15	27.6	11.5	300.9	122.7
Females:						
Before mating	2.54	1.41	25.6	13.7	269.7	149.3
Gestation	1.62	1.56	15.2	14.1	164.5	158.0

From Giknis (1988)

after an exposure period of at least 10 weeks. The ensuing F_2 litters were reared to weaning. The diets containing difenoconazole were fed continuously throughout the study.

There were no adverse effects of difenoconazole on survival in the parental groups of the F_0 or the F_1 generation and there was no evidence of treatment-related clinical changes. Compared with the males at 0 ppm, mean body weights of F_0 males in the group at 2500 ppm were significantly lower throughout the treatment period (Table 10), beginning with significantly reduced body-weight gains during the first study week and continuing significantly lower during most study weeks. Their body weights were approximately 8% lower by the end of the pre-mating phase and continued at about that level until they were killed. Mean body weights of the F_0 females at 2500 ppm were significantly lower than those of the females at 0 ppm throughout pre-mating, gestation and lactation. While the body-weight gains of the females at 2500 ppm were significantly lower than those of the females at 0 ppm during pre-mating weeks 1, 3, 7, 8 and 11, during gestation and lactation, the deficits were relatively small. Body weights of the groups of females at 25 ppm and 250 ppm were not affected by treatment. Food consumption by rats of both sexes at 2500 ppm was consistently lower than that of rats in the control group throughout the pre-mating period. For males, the difference was statistically significant at every pre-mating week and was an average of about 9% lower. For females, the difference was statistically significant in every pre-mating week except weeks 2 and 3 and averaged about 12% lower. Food consumption by the group of females at 2500 ppm was also lower (about 16%) than that of the group of females at 0 ppm during the first 2 weeks of gestation. Food consumption by the males and females at 25 and 250 ppm was not affected by treatment.

No treatment-related changes were observed at autopsy of the F_0 parental rats. There were no differences in absolute organ weights between treated and control rats. Significantly increased relative

Table 10. Body-weight development of F_0 rats in a multigeneration study of difenoconazole

Time-point/phase	Study day(s)	Dietary concentration (ppm)							
		Males				Females			
		0	25	250	2500	0	25	250	2500
Body weight (g)									
Start of treatment	0	175.3	175.6	176.7	178.7	144.3	145.3	143.8	146.4
	7	233.1	230.0	233.1	222.8*	173.4	173.4	171.6	166.0*
	42	437.3	435.2	435.0	402.7*	259.2	255.5	255.7	232.2*
End of pre-mating	77	542.6	540.7	549.7	500.7*	303.9	295.2	297.0	258.3*
Beginning of gestation[a]	79/98	—	—	—	—	297.9	295.7	298.6	260.1*
End of mating	98	568.7	567.7	577.9	520.2*	—	—	—	—
End of gestation [a]	101/118	—	—	—	—	430.6	427.3	422.3	379.7*
Beginning of lactation [a]	102/119	—	—	—	—	337.6	335.4	336.5	294.3*
End of lactation [a]	123/140	—	—	—	—	346.5	342.7	341.4	300.8*
Termination	141	610.5	614.3	618.1	554.4*	324.7	317.4	313.6	284.6*
Body-weight change (g)									
Premating	0–77	367.3	365.1	373.0	322.0*				
Gestation	Approx. 79–118	—	—	—	—	132.7	131.6	123.7	119.6
Lactation	Approx. 102–140	—	—	—	—	8.9	7.3	4.9	6.6
Full duration	0–141	434.9	438.6	441.4	345.7*	—	—	—	—

From Giknis (1988)

[a] Days variable based on when each dam started/ended the phase.

* $p \leq 0.05$, ANOVA + Dunnett test.

testes weights in the males at 2500 ppm could be attributed to the significantly lower carcass weights, because the absolute weights were essentially equal among the groups. No treatment-related microscopic changes were observed in the reproductive organs or in gross lesions from F0 males and females of the control group and group at 2500 ppm and the few rats at 25 and 250 ppm with macroscopic findings. Tissue masses in one male at 0 ppm and one at 2500 ppm were found to be spermatic granulomas. A tissue mass in a female at 2500 ppm was diagnosed as an adenocarcinoma. Changes seen microscopically for which no evidence was noted at autopsy included pituitary cysts or congestion in 5 of the males at 0 ppm, 4 of the females at 0 ppm and 1 female at 2500 ppm, hydrometria and lipofuscin pigment in the uterus in 7 of the females at 0 ppm and 13 of the females at 2500 ppm and lipofuscin pigment in the uterus ligament in 17 of the group of females at 0 ppm and 15 of those at 2500 ppm. These findings were not unusual for rats of this strain and age and were not present in a dose-related pattern.

There were no treatment-related differences among the groups in the number of rats mating (0 ppm, 29 out of 30; 2500 ppm, 30 out of 30), the number of females becoming pregnant (0 ppm, 26 out of 30; 2500 ppm, 30 out of 30) or on the mean pre-coital time (0 ppm, 2.5 days; 2500 ppm, 2.9 days). Only two males (one in the control group and one at 25 ppm) failed to mate. The mean duration of gestation was approximately 23 days in all groups. A total of 26, 25, 29 and 29 females in the groups at 0, 25, 250 and 2500 ppm, respectively, were pregnant and gave birth to live young.

There were no differences in litter size among the groups at birth, the percentage of the pups born alive, or sex ratios (Table 11)

There was a slight, incidental decrease in survival from days 0 to 4 among the male pups at 2500 ppm and lower body weights among the male and female pups at 2500 ppm. Body weights of

Table 11. Litter data for the F_1 generation rats in a multigeneration study of difenoconazole

Parameter	Dietary concentration (ppm)							
	0		25		250		2500	
No. of litters	26		25		29		29	
Total pups born	346		342		389		365	
Total stillbirths (pups/litters)	5/5		9/6		11/7		4/2	
Mean litter size day 0 (live births)	13.12		13.32		13.03		12.45	
Live birth index[a]	98.6		97.4		97.2		98.9	
Sex ratio (% males, day 0)	51.3		48.6		50.5		47.6	
	Males	Females	Males	Females	Males	Females	Males	Females
Viability index (day 0–4)[b]	98.7	97.6	100.0	98.2	99.6	98.8	95.2#	95.3
Viability index (day 4–21)[c]	99.0	100.0	98.2	100.0	100.0	99.1	97.3	100.0
Mean pup weight:								
Day 0	6.58	6.15	6.71	6.46	6.35	6.06	6.20*	5.82
Day 4, pre-cull	10.00	9.38	9.83	9.55	9.46	9.11	8.71*	8.33*
Day 4, post-cull	10.07	9.42	9.84	9.60	9.52	9.10	8.71*	8.37*
Day 7	16.54	15.43	16.30	15.68	15.48	14.73	12.81*	12.28*
Day 14	32.49	30.67	32.07	31.30	30.93	29.82	23.56*	22.65*
Day 21	52.44	49.00	51.54	49.45	48.82*	46.76	36.85*	35.00*

From Giknis (1988)
[a] Percent of total pups born which were alive
[b] Percentage of live-born pups that survived until day 4
[c] Percentage of post-cull pups (day 4) that survived until day 21
$p \leq 0.05$, Mantel test for trend; * $p \leq 0.05$, Healy analysis (weighted for variation in litter size).

male pups at 250 ppm were significantly lower than those of male pups in the control group on day 21, but this difference was considered to be incidental in view of the absence of any effect in the F_2 generation (see Table 13). However, there was no measurement of food intake at this age and no detailed behavioural observations. There were no treatment-related clinical signs or ophthalmologic findings and no treatment-related macroscopic findings were noted at autopsy of the F_1 pups.

There were no treatment-related clinical signs, ophthalmological findings or treatment-related mortalities among the F_1 parental rats. One female at 25 ppm was found dead on day 2 of lactation, one female at 2500 ppm was found dead on day 1 of lactation and a second female at 2500 ppm was found dead on day 5 lactation. The timing of these deaths and the clinical signs associated with them suggested that they were due to non-treatment-related difficulties in labour and delivery.

Body weights of the selected F1 parental rats were lower than those of rats at 0 ppm in the selected males at 250 ppm and 2500 ppm and females at the beginning of dietary exposure (Table 12). Males in the group at 2500 ppm tended to have lower body-weight gains throughout the treatment period. The males at 250 ppm began the treatment period with body weights that were 8% lower than those of males in the control group during the first week, resulting from a slight body-weight decrease after lactation, but body-weight gains thereafter were sporadically higher or lower than those of males in the control group. During the pre-mating phase, females in the group at 2500 ppm had significantly lower mean body weights than did females in the control group and lower body-weight gains during most intervals. Absolute body weights of the females at 25 and 250 ppm were not different from those of females in the control group during pre-mating, although body-weight gains were occasionally significantly higher or lower. Absolute body weights for the females at 2500 ppm

Table 12. Body-weight development of F1 generation in a multigeneration study of rats fed diets containing difenoconazole

Time-point/phase	Study day(s)	Dietary concentration (ppm)							
		Males				Females			
		0	25	250	2500	0	25	250	2500
Body weight (g)									
Start of treatment	0	237.0	235.5	217.6*	169.0*	178.2	177.3	166.5	138.2*
	7	304.3	298.7	280.8*	225.9*	206.9	208.3	194.0	163.2*
	49	525.4	517.6	504.0	432.8*	298.6	289.1	288.5	241.3*
End of pre-mating	98	641.6	619.5	611.7	534.3*	346.2	329.1	333.1	268.8*
Beginning of gestation[a]	101/121	—	—	—	—	331.7	325.1	321.1	263.5*
End of mating	119	665.0	639.2	631.5	560.4*	—	—	—	—
End of gestation[a]	121/143	—	—	—	—	452.6	439.3	446.8	375.6*
Beginning of lactation[a]	122/144	—	—	—	—	371.3	365.0	358.3	290.3*
End of lactationa	144/165	—	—	—	—	360.1	358.5	341.6	294.4*
Termination	145/169 (male/female)	683.1	653.0	647.6	572.6*	375.2	354.8	361.4	291.3*
Body weight change (g)									
Pre-mating	0–98	404.6	384.1	394.1	365.3*	—	—	—	—
Gestation	Approx. 110–132	—	—	—	—	120.9	114.2	125.7	112.2
Lactation	Approx. 132–155	—	—	—	—	−11.2	−7.3	−16.7	1.7
Full duration	0–145	446.1	417.6	430.0	403.6*	—	—	—	—

From Giknis (1988)

[a] Days variable based on when each dam started/ended the phase.

* $p \leq 0.05$, ANOVA + Dunnett test.

continued to be lower than those of the females in the control group during gestation and lactation; body-weight gain of the females at 2500 ppm was also lower during the first week of gestation and body-weight losses during the last week of lactation were smaller.

Food consumption by males and females at 2500 ppm was significantly lower than that of rats in the control group throughout the pre-mating phase, approximately 11% for the males and 17% for the females. During gestation, the females at 2500 ppm continued to consume about 17% less feed than did the females in the control group. Food consumption by the females at 250 ppm was significantly lower than that of females in the control group during the second week of gestation, but this difference was considered to be incidental. Food consumption by rats at 25 and 250 ppm was not affected by treatment.

Absolute weights of testes and ovary were not different among the groups. As observed in the F_0 generation, differences in relative weights of these organs between rats at 0 ppm and 2500 ppm were attributed to reduced carcass weights at 2500 ppm rather than specific toxic effects.

No treatment-related changes were observed at autopsy of the F_1 parental rats. Minor changes in the epididymides, seminal vesicles and/or testis were seen in two males in the control group, two males at 25 ppm, two at 250 ppm and five at 2500 ppm but none of these was considered to be related to treatment. There were macroscopic findings in two females in the control group (mammary mass and ovarian cyst), a female at 25 ppm that was found dead (fluid contents in lungs), two females at 250 ppm (uterine dilation and cyst) and one of the females at 2500 ppm that was found dead (reddish kidney and enlarged, reddish adrenals). These findings were not considered to be unusual for rats of this strain and were considered to be incidental to treatment.

Table 13. Litter data for the F_2 generation in a multigeneration studying rats fed diets containing difenoconazole

Parameter	Dietary concentration (ppm)							
	0		25		250		2500	
No. of litters	19		25		21		23	
Total pups born	234		254		297		281	
Total stillbirths (pups/litters)	5 / 5		30 / 9		9 / 6		29 / 9	
Mean litter size day 0 (live births)	12.05		8.96		13.71		11.39	
Live birth index[a]	97.9		88.2		97.0		93.2	
Sex ratio (% males day 0)	44.1		56.3		47.9		51.9	
	Males	Females	Males	Females	Males	Females	Males	Females
Viability index (days 0–4)[b]	98.3	94.2	91.4	96.0	97.4	100.0	93.8	93.9
Viability index (days 4–21)[c]	98.6	98.6	100.0	98.8	90.5	89.3	95.5	95.5
Mean pup weight:								
Day 0	6.61	6.25	6.77	6.46	6.42	6.15	6.07*	5.79*
Day 4 pre-cull	9.99	9.47	10.23	9.70	9.58	9.29	8.58*	8.23*
Day 4 post-cull	10.07	9.51	10.22	9.71	9.57	9.36	8.61*	8.19*
Day 7	16.07	15.21	15.66	15.12	15.46	15.06	12.77*	12.18*
Day 14	31.76	30.34	30.75	30.15	30.80	30.02	23.45*	22.58*
Day 21	50.99	48.35	50.11	49.02	52.00	50.63	34.22*	32.99*

From Giknis (1988)

[a] Percentage of total pups born that were born alive.

[b] Percentage of live-born pups that survived until day 4.

[c] Percentage of pups remaining after cull (day 4) that survived until day 21.

* $p \leq 0.05$, Healy analysis (weighted for variation in litter size).

There were also no treatment-related microscopic changes of the reproductive organs from F_1 rats of the groups at 0 ppm and 2500 ppm or in the few rats with macroscopic findings. A mammary mass in a female at 0 ppm was diagnosed as a fibroadenoma. The few other findings that were noted were considered to be of no toxicological relevance.

There were no treatment-related effects on the number of rats mating (0 ppm, 25 out of 30; 2500 ppm, 30 out of 30) the number of females becoming pregnant (0 ppm, 20 out of 25; 2500 ppm, 24 out of 30), or the mean number of days until evidence of mating (0 ppm, 3.4 days; 2500 ppm, 2.4 days). The mean duration of gestation was approximately 23 days in all groups. A total of 19, 25, 21 and 23 females in the groups at 0, 25, 250 and 2500 ppm, respectively, were pregnant and gave birth to live young. One female in the control group and two at 250 ppm did not deliver but were subsequently found to be pregnant; one female at 2500 ppm delivered a litter with all pups stillborn.

The mean litter size, the number of live pups at birth and the sex ratios were similar in all groups (Table 13). Survival of pups from days 0 to 4 and days 4 to 21 was not different among the groups. Similar to the parental animals, body weights of the F_2 pups at 2500 ppm were significantly different from those of pups in the control group from birth onwards, but body weights of the pups at 25 and 250 ppm were not affected by treatment.

There were no treatment-related clinical signs for the F_2 pups during lactation and no treatment-related macroscopic findings were noted at necropsy of the F_2 pups. Histopathological examination of five male and five females pups per group did not reveal any treatment-related findings.

The NOAEL for toxicity in adult rats was 250 ppm, equal to 11.5 mg/kg bw per day in males and 13.3 mg/kg bw per day in females, on the basis of reduced body-weight gain during the pre-mating period in F0 and F1 generations at 2500 ppm, equal to 122.7 mg/kg bw per day in males and 149.3 mg/kg bw per day in females. The NOAEL for offspring toxicity was 250 ppm, equal to 14.1 mg/kg bw per day in females, on the basis of reduced pup weight at birth and pup-weight gain at 2500 ppm in F_1 and F_2 generations, equal to 158.0 mg/kg bw per day in females. There were no effects on reproductive indices at doses up to and including 2500 ppm, equal to 132.1 mg/kg bw per day in males and 158.0 mg/kg bw per day in females, the highest dose tested. The Meeting concluded that difenoconazole is not a reproductive toxicant in rats (Giknis, 1988).

(b) Developmental toxicity

Rats

In a study of developmental toxicity, groups of 30 female Crl:COBS®CD®(SD)BR rats were paired with males of the same strain for a maximum of 5 days. The day that successful mating was established (presence of a vaginal plug or sperm in a vaginal smear) was designated as day 0. Difenoconazole technical (purity, 95.7%) was prepared for dosing by first combining it with HiSil using acetone as a solvent to reduce to a powder and then mixing it in the vehicle consisting of 0.5% carboxymethyl cellulose in purified water. This preparation was administered at a dose of 0, 2, 20, 100 or 200 mg/kg bw per day by gavage on days 6–15 of presumed gestation. The control group of rats received vehicle only. A standard dose volume of 10 ml/kg bw was used. Analysis of the dosing suspensions used in the study (two sets of preparations, each sampled on the first and last days of use) indicated that the concentrations of suspensions were consistently lower than the nominal concentrations and somewhat variable under the conditions of the test. Actual concentrations were 71%, 78%, 85% and 86% of the target concentrations for the doses of 2, 20, 100 and 200 mg/kg bw, respectively. Concentrations of the samples taken on the last day of use were within −11% to +20% of the comparable samples taken on the first day of use.

Clinical signs were recorded daily before and after dosing and several times daily during days 6–15. Body weights were recorded five times before mating, on day 0 and daily on days 6–20. Food consumption was recorded on days 0–6 and then daily on days 6–20. The dams were killed on day 20.

Corpora lutea in each ovary were identified and counted. Each uterus was examined for pregnancy, number and placement of implantations, early and late resorptions and live and dead fetuses. After removal from the uterus, fetuses were weighed, numbered, examined externally and sexed. Live fetuses were killed. Approximately 50% of the fetuses in each litter were examined for visceral alterations using a modification of the Wilson sectioning technique. The remaining fetuses in each litter were eviscerated, cleared, stained with alizarin red S and examined for skeletal alterations.

There were no premature deaths in the study. Clinical observations possibly related to treatment included excess salivation in 14 out of 23 and 19 out of 25 of the dams at 100 and 200 mg/kg bw, respectively, compared with none in the groups at 0, 2, or 20 mg/kg bw and red vaginal exudate in 3 out of 23 dams at 100 mg/kg bw and 3 out of 25 dams at 200 mg/kg bw, compared with 0 out of 25 rats in the control group, 1 out of 25 dams at 2 mg/kg bw and 1 out of 25 dams at 20 mg/kg bw. The excess salivation was first observed in most rats on days 7 or 8. Nearly all dams at 100 and 200 mg/kg bw lost body weight during the first few days of treatment. Also, the body-weight gains of the dams at 100 and 200 mg/kg bw were significantly lower than those of rats in the control group throughout the dosing period. Mean body-weight gains in the rats at 100 and 200 mg/kg bw after cessation of dosing were greater than in controls. Body weights of the dams at 2 and 20 mg/kg bw were not affected by treatment. Relative to the control group, food consumption was significantly reduced in the groups at 100 and 200 mg/kg bw during the treatment period (days 6–15) by about 10% and 22%, respectively. Food consumption by the dams in these groups was greater than that of rats in the control group after the end of dosing. Food consumption was not affected in the groups at 2 and 20 mg/kg bw.

One dam at 20 mg/kg bw and two at 100 mg/kg bw showed signs of having mated, but were not pregnant. Except for one dam at 200 mg/kg bw, all the pregnant rats were killed as scheduled and all were found to have viable fetuses, so there were 25, 25, 24, 23 and 24 litters for evaluation in the groups at 0, 2, 20, 100 and 200 mg/kg bw, respectively. Data from caesarean sections indicated that the mean numbers of corpora lutea and implantation sites were comparable between the groups. There were no dead fetuses and there was only one dam (at 200 mg/kg bw) with total implant loss. The number of resorptions per litter (mainly early) was slightly higher and litter size was slightly lower among the rats at 200 mg/kg bw, but these variations were not statistically significant.

The weights of male and female fetuses were not significantly different between the groups, although the means of the fetuses at 200 mg/kg bw were slightly less than those for fetuses in the control group. Fetal sex ratios were not affected by treatment. Maternal autopsy examinations did not reveal any treatment-related findings.

Data derived from fetal examinations were classified as either malformations (irreversible changes) or developmental changes (reversible accelerations or delays of development). Examinations did not reveal any treatment-related effects on the incidences of external or visceral findings. One fetus at 100 mg/kg bw was found during external examination to have a depressed left-eye bulge and one fetus at 200 mg/kg bw had an umbilical hernia. These irreversible changes were considered to be incidental. The only irreversible visceral finding was agenesis of the diaphragmatic lobe of the lung in one fetus at 100 mg/kg bw. Delays in visceral development were seen in renal pelvic dilation in 0, 1, 4, 1, and 1 fetuses from the control litters and litters at, 2, 20, 100 and 200 mg/kg bw, respectively, and slight dilation of the lateral and/or third ventricle of the brain in two fetuses in the control group (two litters) and one fetus at 200 mg/kg bw. These findings were not considered to be treatment related.

Skeletal alterations were seen in approximately 7% of the fetuses and 30% of the litters. Overall, more alterations were found in fetuses in the control group and litters than in any of the treated groups. Wavy and/or incompletely ossified ribs, incompletely ossified or non-ossified sternebra(e) and incompletely ossified pelvis (ischia or pubis) were seen in significantly more control fetuses than in the fetuses from the treated dams. Fetal incidence of bifid and unilaterally ossified thoracic vertebral centres was significantly higher in the fetuses at 200 mg/kg bw than in the controls, but the litter incidence was not statistically different.

Table 14. Incidence of skeletal observations in fetuses in a developmental study in rats given difenoconazole by gavage

Observation	Dose (mg/kg bw per day)																			
	0				2				20				100				200			
	Fetus		Litter		Fetus		Litter		Fetus		Litter		Fetus		Litter		Fetus		Litter	
	(*n* = 182)		(*n* = 25)		(*n* = 176)		(*n* = 25)		(*n* = 172)		(*n* = 24)		(*n* = 168)		(*n* = 23)		(*n* = 160)		(*n* = 24)	
	No.	%	No.	%	No.	%	No.	%	No.	%	No.	%	No.	%	No.	%	No.	%	No.	%
Any skeletal findings	24	13	13	52	4	2	3	12	10	6	6	25	8	5	6	26	18	11	10	42
Vertebrae, cervical																				
C7, rib present	2	1	2	8	—	—	—	—	1	1	1	4	1	1	1	4	1	1	1	4
Vertebrae, thoracic																				
T10 or T11, bifid centra	—	—	—	—	1	1	1	4	—	—	—	—	2	1	2	9	4**	3	3	13
T10 + T12, bifid centra	—	—	—	—	—	—	—	—	—	—	—	—	—	—	—	—	1**	1	1	4
T2 or T8 or T1 centra, unilateral ossification	—	—	—	—	—	—	—	—	—	—	—	—	—	—	—	—	3**	2	3	13
T5 + T8 + T10 centra, unilateral ossification	—	—	—	—	—	—	—	—	—	—	—	—	—	—	—	—	1**	1	1	4
T3 centrum, not ossified	—	—	—	—	—	—	—	—	—	—	—	—	—	—	—	—	1	1	1	4
Vertebrae, lumbar																				
V5 or V6, arch(es) incompletely ossified	1	1	1	4	—	—	—	—	—	—	—	—	—	—	—	—	1	1	1	4
V6 + V7, arches incompletely ossified	1	1	1	4	—	—	—	—	—	—	—	—	—	—	—	—	—	—	—	—
Ribs																				
Wavy	5**	3	3	12	—	—	—	—	1	1	1	4	—	—	—	—	—	—	—	—
Incompletely ossified/hypoplastic	6**	3	3	12	—	—	—	—	—	—	—	—	—	—	—	—	—	—	—	—
Fused (3 pairs)	—	—	—	—	—	—	—	—	—	—	—	—	—	—	—	—	1	1	1	4
Manubrium																				
Not ossified	—	—	—	—	—	—	—	—	1	1	1	4	—	—	—	—	—	—	—	—
Incompletely ossified	2	1	2	8	—	—	—	—	—	—	—	—	—	—	—	—	1	1	1	4
Sternebrae																				
S1 or S2, not ossified	2**	1	2	8	—	—	—	—	2	1	2	8	2	1	2	9	5	3	3	13
S1 or S2, incompletely. ossified	8**	4	6	24	—	—	—	—	2	1	1	4	3	2	2	9	5	3	5	21

S1 + S2, not ossified	—	—	—	—	—	—	—	—	1	1	1	4	—	—	—	—	—	—	—	—
S1 + S2, incompletely. ossified	3**	2	3	12	—	—	—	—	—	—	—	—	—	—	—	—	—	—	—	—
Forelimb																				
Polydactyly	1	1	1	4	—	—	—	—	—	—	—	—	—	—	—	—	—	—	—	—
Pelvis/pubes																				
Not ossified	—	—	—	—	—	—	—	—	2	1	2	8	—	—	—	—	1	1	1	4
Incompletely ossified	8**	4	6	24	4	2	3	12	4	2	2	8	1	1	1	4	1	1	1	4
Pelvis/ischia																				
Incompletely ossified	6**	3	4	16	—	—	—	—	—	—	—	—	—	—	—	—	—	—	—	—

From Lochry (1987)
** $p \leq 0.01$, binomial test.

Analyses of the sites of fetal ossification revealed a significant increase in the average number of ribs in the fetuses from the dams at 200 mg/kg bw, with a related significant increase in the number of thoracic vertebrae and decrease in the average number of lumbar vertebrae. Increases in the average number of ossified hyoid and caudal vertebrae and decreases in the average number of sternal centres and hind-paw phalanges also occurred for fetuses at 200 mg/kg bw. All other incidences were within the expected range for fetuses of this strain and age and were not significantly different among the groups. The differences in skeletal observations that were recorded as statistically significant are summarized in Table 14. As may be seen, the changes were extremely small and unlikely to have produced any deleterious effects.

TheNOAEL for maternal toxicity was 20 mg/kg bw per day on the basis of reduced body-weight gain at 100 mg/kg bw per day. The NOAEL for fetal toxicity was 100 mg/kg bw per day on the basis of a low incidence of minor deviations in skeletal ossification found at 200 mg/kg bw per day. There were no indications of embryotoxicity or teratogenicity at any dose up to 200 mg/kg bw (Lochry, 1987).

Rabbit

In a study of developmental toxicity, groups of 20 artificially inseminated New Zealand White rabbits were given difenoconazole technical (purity, 95.7%) at a dose of 0, 1, 25 or 75 mg/kg bw per day in an aqueous vehicle (3% aqueous corn starch containing 1.0% Tween 80) administered by stomach tube on days 7–19 after insemination. A standard dose volume of 5 ml/kg bw was used. The control group was dosed with the vehicle only. Food consumption and body weights were recorded regularly throughout the study period. The state of health of the rabbits was checked each day. Eye examinations were performed before dosing and at the end of the dosing period on all rabbits. On day 29 after insemination, all surviving females were killed and assessed by gross pathology (including weights of the unopened uterus and the placentae). For each dam, corpora lutea were counted and number and distribution of implantation sites (differentiated as resorptions, live and dead fetuses) were determined. The fetuses were removed from the uterus, sexed, weighed and further investigated for any external, soft tissue and skeletal findings. The stability of the test substance was proven by reanalysis. The stability and homogeneity of the test substance preparation was analytically verified over at least a 21-day period. The correctness of the concentrations was analytically demonstrated to be 96%, 96% and 97% of the target concentrations of 1, 25 and 75 mg/kg bw.

Premature mortality occurred in three study groups: one control doe and one doe at 1 mg/kg bw per day were found dead on days 15 and 16, respectively, apparently due to dosing accidents and one doe at 75 mg/kg bw per day was found dead on day 18, after a period of apparently treatment-related anorexia. Clinical observations possibly related to treatment included stool variations in 7 out of 19 and 12 out of 19 of the groups of does at 25 and 75 mg/kg bw per day, respectively, compared with two or three in the groups at 0 and 1 mg/kg bw per day, although this was thought to be secondary to variations in food consumption. There were two abortions in the group at 75 mg/kg bw per day, which were attributed to treatment. Other signs, including alopecia, anogenital stains and blood in pan were considered incidental. All does at 75 mg/kg bw per day lost body weight during the first few days of treatment and body-weight gains of the does at 75 mg/kg bw per day were significantly lower than rabbits in the control group for most of the dosing period. This effect of treatment was consistent with reduced food consumption and indicative of toxicity. Body-weight gains of the does at 25 mg/kg bw per day were also somewhat lower than those of the controls, but not significantly so. Body-weight gains of the does at 1 mg/kg bw per day were not affected by treatment. Body-weight changes over the full duration of the study (and including uterus and its contents) were 0.61, 0.54, 0.48 and 0.40 kg in the groups at 0, 1, 25 and 75 mg/kg bw per day, respectively. Relative to the control group, food consumption was significantly reduced in the group at 75 mg/kg bw per day during the treatment period (days 7–19), with mean values being about 50% lower than those for the controls during the first week of dosing. Food consumption increased in all groups during the second week of dosing, most prominently among the rabbits at 75 mg/kg bw per day. Food consumption by the does in the control group and in the groups at 1 and 25 mg/kg bw per day tended to decrease after the end of dosing, while the rabbits at 75 mg/kg bw per day continued to increase their food consumption.

Fifteen does that were inseminated were not pregnant at caesarean section on day 29: three in the control group, five at 1 mg/kg bw per day, three at 25 mg/kg bw per day and four at 75 mg/kg bw per day. Three does died (one each in the control group and the groups at 1 mg/kg bw per day and 75 mg/kg bw per day) and two does at 75 mg/kg bw per day aborted, so there were 15, 13, 16 and 12 litters for evaluation in the control group and the groups receiving 1, 25 and 75 mg/kg bw per day, respectively. Data from caesarean sections indicated that the mean numbers of corpora lutea, implantation sites and live fetuses were comparable between the groups. There were no dead fetuses. The numbers of resorptions per pregnancy were only slightly and not significantly higher among the rabbits at 75 mg/kg bw per day.

The weights of male and female fetuses were not significantly different between the groups, although some statistical differences were described in the original report. It was subsequently determined that the statistical routine used for generating the report tables ran incorrectly, i.e., post hoc comparisons were inappropriately performed although the overall F statistics were not significant ($p = 0.072$ for males; $p = 0.104$ for females). It was also suggested in the original report that variations in litter size could account for apparent differences in fetal weights. To this end, the data were subsequently analysed by analysis of covariance with litter size as the covariate. These analyses were also not significant at the $p \leq 0.05$ level for males and females. Moreover, it was stated in the original report that the weights for fetuses from the control group (males, 47.3 ± 1.6 g; females, 45.6 ± 1.5 g) were exceptionally high in comparison with the mean values for historical controls of 43.4 g for male fetuses and 42.2 g for female fetuses. The weights for the fetuses in the treated groups were closer to these values than were those for the control group in this study. Fetal sex ratios were not affected by treatment. Maternal autopsy examinations did not reveal any treatment-related findings

Fetal examinations did not reveal any treatment-related effects on the incidences of external, visceral or skeletal variations or malformations. One fetus in the group at 25 mg/kg bw per day was found during external examination to have a raised, discoloured area on the ventral thoracic region that may have resulted from technical manipulation and was considered to be incidental. The only visceral findings were a horseshoe kidney in one fetus at 75 mg/kg bw per day, microencephaly in one fetus at 25 mg/kg bw per day and partial cryptophthalmos in one fetus at 75 mg/kg bw per day. Skeletal alterations were seen in approximately 80–90% of the fetuses and all of the litters. There were no statistically significant or biologically meaningful differences among the groups (by fetus or by litter) in the incidence of any finding.

The NOAEL for maternal toxicity was 25 mg/kg bw per day on the basis of reduced body-weight gain at 75 mg/kg bw per day. The NOAEL for fetal toxicity was 75 mg/kg bw per day, the highest dose tested. There were no indications of embryotoxicity, fetotoxicity or teratogenicity at a dose up to 75 mg/kg bw per day (Hummel et al., 1987).

2.6 *Special studies*

(a) Studies on neurotoxic potential

In a study of neurotoxicity, groups of 10 male and 10 female Alpk:AP$_f$SD (Wistar-derived) rats were given single doses of difenoconazole technical (purity, 94.3 %) at a dose of 0, 25, 200 or 2000 mg/kg bw in 1% carboxymethyl cellulose andthen observed for 14 days.

All rats were observed before the study start and daily throughout the study for any changes in clinical condition. In addition, a functional observation battery (FOB), including quantitative assessments of landing foot splay, sensory perception and muscle weakness, was performed in week −1, and on days 1 (at the time of anticipated maximum effect), 8 and 15. Locomotor activity was also monitored in week −1, and on days 1, 8 and 15. Body weights and food consumption were measured weekly throughout the study. At the end of the scheduled period, five male and five females per group were perfused in situ. Selected nervous system tissues were removed, processed and examined microscopically.

There were no deaths; this is a remarkable difference between this study, in which cage-side observation is an important feature, and an earlier single-dose study (Argus, 1987) in which there was 40% mortality and significant clinical signs of toxicity at 2000 mg/kg bw. Evidence of a toxic effect at 2000 mg/kg bw per day on day 1, at the time of peak effect, was reflected in a number of adverse clinical abnormalities, with females being more affected than males. The more common signs were upward curvature of the spine, tip-toe gait and decreased activity. All treatment-related clinical signs seen on day 1 showed complete recovery by day 5 (males) and day 7 (females). In addition on day 1, body weights at 2000 mg/kg bw per day were lower than those of males and females in the control group (approximately 7%). Body weight remained lower for males throughout the study (4% by day 15) with food consumption lower for males in week 1. At 200 mg/kg bw per day, body weights were statistically significantly lower (2%) than those of the controls on day 1 for both sexes.

On day 1, forelimb grip strength was lower than that of controls for males at 200 and 2000 mg/kg bw per day. Forelimb grip strength was statistically significantly higher than that of controls on day 15 for females at 2000 mg/kg bw. The difference was seen only in females at one time-point; in the absence of any other findings at this time-point, this small difference from control values was considered to be incidental to treatment. There were no other differences from control values for forelimb grip strength. At the 200 mg/kg bw, the lower forelimb grip strength observed on day 1 in males was not associated with perturbations in any of the other functional assessments or microscopic pathology. There were no treatment-related effects on hindlimb grip strength, landing-foot splay measurements, or time to tail-flick at any dose.

At 2000 mg/kg bw, motor activity was lower than that for controls for females on day 1, with evidence of some recovery on day 8 and complete recovery by day 15. This lower motor activity in females was not associated with perturbations in any of the other functional assessments or microscopic pathology, but may have reflected the adverse general clinical observations that had not recovered until day 7. On day 1, at the time of anticipated maximum effect, males at 200 or 2000 mg/kg bw had slightly higher activities than did males in the control group. The differences were small, the response was not proportional to dose and was in the opposite direction from females and inconsistent with the general clinical condition of the rats (e.g. decreased activity). Therefore, these small differences from control values were considered to be incidental to treatment. There were no treatment-related effects on body weight, food consumption, FOB or motor activity in rats at 25 mg/kg bw.

There were no effects on brain weight, nor any treatment-related microscopic findings in the tissues of the nervous system after single doses of difenoconazole of up to 2000 mg/kg bw.

The NOAEL for neurotoxicity in rats treated with difenoconazole was 25 mg/kg bw on the basis of reduced forelimb grip strength in males at 200 mg/kg bw. The Meeting considered that this effect was not necessarily an expression of neurotoxicity, although it was an adverse effect of treatment, because it was not observed in females as well as males, even at a dose that was 10 times higher, and in males it was not supported by any of the multiple other possible end-points for neurotoxicity examined (Pinto, 2006a).

In a multiple-dose study of neurotoxicity, groups of 12 male and 12 female Alpk:AP_fSD (Wistar-derived) rats were given diets containing difenconazole (purity, 94.3 %) at a concentration of of 0, 40, 250 or 1500 ppm, equal to 0, 2.8, 17.3 and 107.0 mg/kg bw per day in males and 0, 3.2, 19.5 and 120.2 mg/kg bw per day in females, for at least 90 days.

Food and water consumption and body weights were measured once each week throughout the study. A check of the general state of health of the rats was made before the study began and daily throughout the study. Detailed observations (during which each rat was removed from its cage and physically examined for changes in general health status) and quantitative assessments of landing foot splay, muscle weakness (fore- and hindlimb grip strength) and sensory perception (tail-flick test) were made in week −1, and in weeks 2, 5, 9 and 14. The observations were made by one observer who was "blind" with respect to the treatment of the rats, and recorded on a computer system by personnel not directly involved in the clinical observations. The examination proceeded from the least to the most interactive test with observations recorded as follows:

(i) *Assessment in the home cage*: bizarre behaviour (circling, head-flicking, head-searching, walking backwards, rolling-over sideways, paw-flicking), vocalization.

(ii) *Removal from the cage*: approach response, response to touch (increased, decreased), vocalization.

(iii) *In the standard arena*: activity (increased, decreased), comatosed, prostration, hunched posture, bizarre behaviour (circling, head flicking, head searching, walking backwards, rolling over sideways, paw flicking), convulsions (tonic, clonic), vocalization, ataxia, tremors, reduced stability, abnormal gait, splayed gait, tiptoe gait, reduced limb function (fore-, hind-), upward curvature of the spine, downward curvature of the spine, piloerection, sides pinched in, ungroomed appearance, urinary incontinence, diarrhoea.

(iv) *Handling the rat*: response to touch, convulsions, vocalization, tremors, piloerection, skin colour (pale, hyperaemia, cyanosis), ungroomed appearance, hyperthermia, hypothermia, chromodacryorrhoea, lachrimation, ptosis, endophthalmus, exophthalmus, miosis, mydriasis, stains around the mouth, stains around the nose, salivation, respiratory abnormalities (breathing rate, breathing depth, laboured breathing, gasping, irregular breathing, whistling, wheezing, croaking), thin appearance, sides pinched in, dehydration, abdominal tone (increased, decreased), urinary incontinence, diarrhoea.

(v) *Reflex tests*: righting reflex (from supine position), response to sound (to finger click/clap), splay reflex (degree of splay when rat was lifted by base of tail), visual placing response (rat was lifted by base of tail and slowly moved downwards towards the edge of arena), pupil response to light (after eye had been held closed for 10 s), palpebral membrane reflex (palpebral membrane touched with bristle and blink response observed), corneal reflex (hair was touched against cornea and blink reflex observed (only performed if palpebral reflex was absent), pinna reflex (bristle poked into ear canal), foot-withdrawal reflex (to toe pinch).

(vi) *Quantitative measures*: forelimb and hindlimb grip strength, landing-foot splay, time to tail-flick.

Locomotor activity was monitored by an automated activity recording apparatus (Coulbourn Lab Linc Infra-red Motion Activity System), which records small and large movements as an activity count. Rats were placed individually in stainless steel cages with an infra-red sensor attached and the recording session started immediately. All rats were tested in weeks −1, 2, 5, 9 and 14 of the study. Each observation period was divided into ten scans of 5 min duration, during which the rats did not have access to food, water or items of environmental enrichment. Treatment groups were counterbalanced across test times and across devices, and when the trials were repeated each rat was returned to the same activity monitor at approximately the same time. At the end of the dosing period, five males and five females per group were killed, then fixed by perfusion in situ and subjected to neuropathological examinations. The liver was removed and weighed from non-perfused rats from the control group and the group at 2000 mg/kg bw per day. The remaining rats were killed under carbon dioxide anesthaesia without any further examinations. The stability, homogeneity and correctness of the concentrations of difenoconazole in the diet were analytically verified.

There were no difenoconazole-related adverse effects at any dose. Body weight was lower than that of controls for both sexes at 1500 ppm for all or most of the study (maximum difference, 10% and 6% respectively). There was no effect of treatment on food consumption at any dose. Food utilization for males at 1500 ppm was lower than that of controls throughout the study. Hindlimb grip strength was lower for males at 1500 ppm in weeks 9 and 14. Since this pattern was observed at the

Table 15. Scores for hindlimb grip in male rats fed diets containing difenoconazole for 90 days

Time-point	Dietary concentration (ppm)			
	0	40	250	1500
Week −1	571	573	600	481
Week 2	971	767**	844	752**
Week 5	833	856	831	806
Week 9	960	846	829	788*
Week 14	1131	1100	902*	823**

From Pinto (2006b)
* $p \leq 0.05$; ** $p \leq 0.01$.

highest dose tested at two consecutive time-points, it was considered to be a potentially treatment-related effect at 1500 ppm in males only. Hindlimb grip strength was also significantly lower in the group at 250 ppm in week 14, but in no other week (Table 15). It may be argued that this observation was incidental to treatment or that it was a late development caused by treatment.

Hindlimb grip strength was significantly lower at some other observation times, but these observations were considered to be incidental variations. In week 2, hindlimb grip strength was statistically different from control values for males at 40 ppm and 1500 ppm. There was no difference from control values at the intermediate dose of 250 ppm and, in addition, there were no effects in any group at the next time-point (week 5). There were no treatment-related clinical abnormalities observed in the FOB, with no treatment-related effects on landing-foot splay, time to tail-flick, fore- or hindlimb grip strength (with the exception of males at 250 and 1500 ppm), or motor activity.

Liver weight for rats at 1500 ppm was higher than that of controls. Brain weight was unaffected by treatment. Microscopic examination of the central and peripheral nervous system showed no effects of treatment with difenoconazole at dietary concentrations of up to 1500 ppm in males or females.

The NOAEL was 40 ppm, equal to 2.8 mg/kg bw per day, on the basis of reduced hindlimb grip strength in male rats at 250 ppm, equal to 17.3 mg/kg bw per day. It should be noted that no other effects on neurobehavioral parameters (forelimb grip strength, motor activity, time to tail-flick, FOB) were observed at any time point for males at 250 ppm, which argues for the reduction in grip strength not being an expression of neurotoxicity, although it should remain to be considered an effect of treatment. Furthermore, there were no effects considered to be caused by treatment in males at 40 ppm, equal to 2.8 mg/kg bw per day, or in females at any dose, including 1500 ppm, equal to 120.2 mg/kg bw per day. Thus, the observed grip weaknesses in both of these studies of neurotoxicity were considered to be non-specific effects of difenoconazole, there being a lack of consistency (hindlimb versus forelimb effects), a complete absence of effects at other end-points for neurotoxicity and no neuropathology (Pinto, 2006b).

(b) Toxicology of metabolites.

The following are plant metabolites of difenoconazole:

- CGA 169374: difenoconazole or 1-[[2-[2-chloro-4-(4-chlorophenoxy)phenyl]-4-methyl-1,3-dioxolan-2-yl]methyl]-1*H*-1,2,4-triazole;
- CGA 205374 (1-(2-chloro-4-(4-chloro-phenoxy)-phenyl)-2-(1,2,4-triazol)-1-yl-ethanone);
- CGA 205375 (1-[2-chloro-4-(4-chloro-phenoxy)-phenyl]-2-(1,2,4-triazol)-1-yl-ethanol);
- CGA 189138 (2-chloro-4-(4-chlorophenoxy)-benzoic acid);

Table 16. Results of studies of acute toxicity and genotoxicity with metabolites of difenoconazole found in plants and rats

Metabolite	Assay	HID/ concentration	Result	Reference
CGA 189138	Reverse mutation	2000 µg/plate, ±S9	Negative	Nakajima (1991a)
CGA 205374	Study of acute oral toxicity in mice	0, 5000 mg/kg bw	LD_{50} > 5000 mg/kg bw	Ohba (1991a)
	Reverse mutation	5000 µg/plate, ±S9	Negative	Nakajima (1991b)
CGA 205375	Study of acute oral toxicity in mice	0, 1000, 1300, 1600, 2000, 2500 mg/kg bw	LD_{50} > 2309mg/kg bw (male mice)	Ohba (1991b)
	Reverse mutation	320 µg/plate, ±S9	Negative	Nakajima (1991c)

HID, highest ineffective dose, in the bacterial mutation tests; S9, 9000 × g supernatant from rat livers.

- CGA 71019, (1,2,4-triazole);
- 1,2,4-triazolyl alanine; and
- 1,2,4-triazolyl acetic acid.
- All these metabolites, except for 1,2,4-triazolyl alanine and 1,2,4-triazolyl acetic acid, are also found in the metabolism of rats.

The acute oral LD_{50} values for metabolites CGA 205374 and CGA 205375 in mice were in excess of 2000 mg/kg bw. In addition, none of these metabolites was mutagenic in *S. typhimurium* strains TA100, TA98, TA1535 and TA1537 and *E. coli* WP2 *uvrA* (Table 16)

(i) CGA 71019 (1,2,4-triazole)

In three studies of metabolism and toxicokinetics (Lai & Simoneaux, 1986a; Weber et al., 1978; Ecker, 1980), 1,2,4-triazole was shown to be rapidly absorbed and readily eliminated, mainly via the urine. In the study of biokinetics (Weber et al., 1978), elimination was shown to be proportional to dose and independent of the route of administration. The bioavailability of the oral dose was virtually 100%, and the bulk of the amount excreted via bile was reabsorbed. In the study of biotransformation (Ecker, 1980), 1,2,4-triazole was rapidly eliminated via the urine, predominantly in an unchanged form.

1,2,4-Triazole is of low acute oral and dermal toxicity in rats (oral LD_{50}, 1650 and 1648 mg/kg bw for males and females, respectively; dermal LD_{50}, 4200 and 3129 mg/kg bw for males and females, respectively). No dermal irritation and only moderate ocular irritation were noted after application to the skin and to the eye of rabbits. The studies of acute toxicity after inhalation in rats and mice were deficient in that exposure to the test substance was not demonstrated, and classification was not possible. In a Magnusson & Kligman maximization test, no signs of allergic skin reactions were observed. There was no evidence of genotoxicity in two assays for reverse mutation in bacteria.

The results from the short-term studies of toxicity are summarized in Table 17.

In an older 90-day feeding study in rats, 1,2,4-triazole caused retarded body-weight development, temporary slight effects on the central nervous sytem (convulsions), slight microcytic hypochromic anaemia (males only), and accumulation of hepatocellular fat (males only) when administered at a dietary concentration of 2500 ppm, equivalent to an average test substance intake of 212 and 267 mg/kg bw per day for males and females, respectively. Comparable results were obtained in a more recent short-term study in rats that included additional specific neurotoxicological investigations. In this study, a retardation of body-weight gain and clinical findings, such as tremor, muscle fasciculations and uncoordinated gait, were seen at 3000 ppm (equivalent to 183 mg/kg bw per day in

Table 17. Results of studies of acute toxicity and irritation with CGA 71019 (1,2,4-triazole)

Study	Species	Results	Reference
Acute oral LD_{50}	Rats	500 < LD_{50} < 5000 mg/kg bw	Procopio & Hamilton (1992)
		LD_{50} = 1650 and 1648 mg/kg bw for males and females, respectively.	Thyssen & Kimmerle (1976)
Acute dermal LD_{50}	Rats	LD_{50} = 4200 and 3129 mg/kg bw for males and females, respectively	Thyssen & Kimmerle (1976)
	Rabbits	200 < LD_{50} < 2000 mg/kg bw	Procopio & Hamilton (1992)
Acute inhalation LC_{50} (4 h)	Rats	Exposure to the test article not demonstrated	Thyssen & Kimmerle (1976)
Acute inhalation LC_{50} (6 h)	Mice	Exposure to the test article not demonstrated	Thyssen & Kimmerle (1976)
Acute skin irritation	Rabbits	Not irritating	Procopio & Hamilton (1992); Thyssen & Kimmerle (1976)
Acute eye irritation	Rabbits	Irritating	Procopio & Hamilton (1992); Thyssen & Kimmerle (1976)
Skin sensitization (maximization test)	Guinea-pigs	Not ssensitizing	Frosch (1998)

males and 234 mg/kg bw per day in females) and at 1000/4000 ppm (210 mg/kg bw per day in males; 275 mg/kg bw per day in females). At these doses, degenerative lesions were seen in the cerebellum, the lumbar dorsal root ganglion and other peripheral nerves. The brain lesions were limited to the anterior, dorsal cerebellum and were coded overall as an increased incidence of cellular degenerations and necrosis. Findings were characterized by extensive loss of Purkinje cells, variable white matter degeneration and gliosis. Subtle atrophy of the molecular layer, primarily at the cerebellar surface, or loss of granule cells was occasionally present. The effects tended to be slightly more prominent in males. Individual nerve-fibre degeneration, which is frequently seen in healthy rats of this age, was increased in incidence and severity in the peripheral nerves (sciatic, tibial, sural). Additionally, minimal to mild neuronal chromatolysis and vacuolation of the lumbar dorsal root ganglia was observed at ≥ 3000 ppm, but no similar change was seen in the cervical dorsal root ganglia. The NOAEL for general effects and effects on the nervous system in rats was 500 ppm (33 mg/kg per day in males and 41 mg/kg per day in females).

In comparison to rats, mice proved to be less sensitive to 1,2,4-triazole. In a 13-week dietary study in mice, clinical symptoms were found only at 6000 ppm (988 mg/kg bw per day in males and 1346 mg/kg bw per day in females). Body-weight gain was decreased at this dose in males and females and at 3000 ppm in males (487 mg/kg bw per day). In mice treated for 13 weeks, the central nervous system was established as a target organ for 1,2,4-triazole, albeit only near or above the limit dose of 1000 mg/kg bw per day (6000 ppm). In histopathology, a minimal to mild loss of Purkinje cells was seen in the cerebellum at the end of the 13-week exposure period in mice at 6000 ppm, with no effects on the peripheral nervous system. Lower absolute brain weights were observed in males at 3000 ppm and males and females at 6000 ppm; however, no histopathological effects on the brain were observed at 3000 ppm. In a separate 28-day dietary study in mice, no effects on the nervous system were observed at doses of up to 2000 ppm (356 mg/kg bw per day in males and 479 mg/kg bw per day in females).

In the 28-day study in mice, effects on the testes and epididymides were observed at the highest dose only (2000 ppm). Similar effects on the testes were observed in a dose–responsive manner at 3000 and 6000 ppm (487 and 988 mg/kg bw per day) in the 13-week study in mice. This effect on the testes was related to increased incidence and severity (minimal to slight) of certain findings that also can occur in untreated mice, including apoptotic-like bodies, degeneration, depletion and asynchrony of spermatids, and tubular atrophy. Secondary to these testes effects, exfoliated germ cells and debris were found in the epididymides at increased incidences only at the highest doses of 2000 ppm (28-day study) or 6000 ppm (13-week study). The NOAEL for testes effects in mice after 13 weeks of treatment was 1000 ppm, equivalent to 161 mg/kg bw per day.

No data were available to directly evaluate the potential for carcinogenicity of 1,2,4-triazole. Available data on mutagenicity data were limited (*Salmonella* assays and a report of a study of chromosomal aberration in rat bone-marrow cells), but they do not indicate any activity. A large number of parent triazole-derivative pesticides had been classified as carcinogens (most also are non-mutagenic), but the relevance of that finding to expected effects of free triazole may be limited. The types of tumours associated with exposure to the parent chemicals were most commonly hepatocellular adenomas/carcinomas in mice. None of the tumour types were clearly associated with the proportion of free triazole formed in the available studies of metabolism in rats and 1,2,4-triazole showed little or no effects in the livers of rats and mice in short-term studies.

In a two-generation study of reproductive toxicity in rats, retardation of body-weight gain was the most sensitive end-point in parent animals. At 250 ppm (equivalent to 16 mg/kg bw per day), body weights of the F1 males were slightly but significantly lower than those of rats in the concurrent control group. In some respects this is in contrast to the results of the two short-term studies of toxicity in rats, where 500 ppm was a clear NOAEL with regard to body-weight reduction in males. In females, body weights were lower only at 3000 ppm. Absolute brain weights were decreased in males and females of the parental generation receiving 3000 ppm. At this dose also, microscopic lesions such as mild to moderate degeneration and necrosis were evident in the cerebellum. The lesions were identical to that observed in the short-term study of combined toxicity and neurotoxicity in rats and are characterized by variable loss of Purkinje cells, white matter degeneration and gliosis. The white matter tract degeneration presented as nerve fibre (axonal) swelling or fragmentation often with digestion chambers containing debris and macrophages. No similar changes were noted in the rats at 500 ppm.

In the two-generation study of reproduction, at a feed concentration of 500 ppm, all parameters of reproduction were similar to those of rats in the control group through two generations. Fertility was significantly decreased only at the highest feed concentration of 3000 ppm (equivalent to 218 mg/kg bw per day in females). Two out of 30 females delivered viable offspring (one pup each); there were only three implantation sites compared with 265 in the control group. No effects were observed on the length or the number of the estrous cycles. No treatment-related effects on sperm parameters were observed at any dose, including the males at 3000 ppm. In parental females at 3000 ppm, the number of total corpora lutea and the ovary weights were increased, and a dilatation of uterine horns was found in some rats.

Two studies of developmental toxicity in rats (Renhof, 1988a, 1988b) demonstrated maternal toxicity (retarded weight gain) at 100 mg/kg bw or higher, developmental toxicity (increased incidence of runts, lower fetal and placental weights, and a higher incidence of minor skeletal deviations) at 100 mg/kg bw or higher, and an increased incidence of malformations at 200 mg/kg bw. The maternal and the developmental NOAEL in rats was 30 mg/kg bw per day.

In the study in rabbits (Hoberman, 2004), lower body-weight gain and clinical signs of systemic toxicity such as excess salivation, hyperpnoea and ptosis were evident at 45 mg/kg bw per day. Five out of 25 dams were sacrificed in a moribund condition at this dose. Developmental effects included lower weights of fetuses at 45 mg/kg bw per day, and there were a few alterations of the urogenital system which occurred in several fetuses at the maternally toxic dose of 45 mg/kg bw per day. Similar to rats, the maternal and the developmental NOAEL was 30 mg/kg bw per day in rabbits.

Table 18. Studies of toxicity after repeated doses, reproductive toxicity and genotoxicity with 1,2,4-triazole

Study	Species or test object	Dietary concentration (ppm)	Test concentration	NOAEL (mg/kg per day)	Result	Adverse effects	Reference
Short-term studies of toxicity							
Ninety-day, feeding	Rat	0, 100, 500, 2500	—	37.9 in males 54.2 in females	—	Retarded body-weight development (males and females), temporary slight CNS effects, lower erythrocyte parameters (males only), hepatocellular fat accumulation (males only)	Bomhard et al. (1979)
Ninety-day, feeding, neurotoxicity	Rat	0, 250, 500, 3000, 1000/4000	—	33 in males 41 in females	—	Clinical symptoms, body-weight decrease; triglycerides decrease (males), corpora-lutea increase, absolute brain-weight decrease, degenerative effects in brain and peripheral nerves	Wahle & Sheets (2004)
Twenty-eight-day, feeding,	Mouse	0, 50, 250, 500, 2000	—	90 in males 479 in females	—	Testes: increased incidence and severity of background lesions (spermatid degeneration, depletion and asynchrony, tubular atrophy); epididymis: exfoliated germ cells and debris	Wahle (2004a)
Ninety-day, feeding	Mouse	0, 500, 1000, 3000, 6000	—	161 in males 663 in females	—	Males: body-weight decrease, absolute brain-weight increase; testes: apoptotic-like bodies increase, spermatid degeneration, depletion and asynchrony, tubular atrophy. Females: clinical symptoms, body-weight decrease; absolute brain-weight decrease; brain/cerebellum: loss of Purkinje cells	Wahle (2004b)
Genotoxicity							
Gene mutation	*S. typhimurium* TA98, TA100, TA1535 and TA1537	—	10–5000 µg/plate	—	Non-mutagenic with and without metabolic activation in all test systems	—	Poth (1989)
Gene mutation	*S. typhimurium* TA98, TA100, TA1535 and TA1537	—	100–7500 µg/plate	—	Non-mutagenic with and without metabolic activation in all test systems	—	Melly & Lohse (1982)

Study of reproductive toxicity in vitro							
Embryotoxicity	Rat	500, 1500, 2500, 5000 µmol/l	—	1500 µmol/l	—	Visceral yolk sac anaemia and slight developmental retardation	Menegola et al. (2001)
Studies of reproductive toxicity in vivo							
Two-generation, feeding	Rat	0, 250, 500, 3000	—	Paternal toxicity: < 16	—	Paternal toxicity: body-weight decrease (F_1 males).	Young & Sheets (2005)
				Maternal toxicity: 36.2 Reproductive toxicity: 34.4 Developmental toxicity: > 35.8	—	Maternal: body-weight decrease, ovary-weight increase, corpora-lutea increase, uterine-horn dilatation, cerebellar degeneration and necrosis Reproductive toxicity: fertility decrease Developmental toxicity: no effects at 500 ppm in F_1 and F_2 pups.	

1,2,4-Triazole does not modulate ovarian estrogen biosynthesis in vitro. Rat embryos (aged 9.5 days) exposed in vitro to 1,2,4-triazole at concentrations of up to 5000 μmol/l showed a markedly anaemic visceral yolk sac and visibly paler erythrocytes at 2500 or 5000 μmol/l.

The results from these studies are summarized in Table 18.

(ii) Toxicology of 1,2,4-triazolyl alanine

1,2,4-Triazolyl alanine acid (Figure 5) is a significant residue in plants treated with difenoconazole.

An initial study of absorption, distribution, metabolism and excretion and subsequent studies of metabolism were conducted. Hamboeck (1983a) showed that the compound was rapidly absorbed and readily eliminated, mainly via the urine, and that up to 86% of the administered dose was excreted unchanged. There was no difference between the sexes with respect to excretion, the pattern of tissue residues, or the qualitative pattern of urinary metabolites.

In the second study of metabolism (Hambock, 1983b), 1,2,4-triazolyl alanine and one metabolite were identified: DL-triazolyl alanine was isolated and identified by nuclear magnetic resonance (NMR) and mass spectrometry (MS) and represented 69–86% of the administered dose in the urine and 1–2% of the dose in the faeces, while *N*-acetyl-DL-triazolyl alanine represented 8–19% of the administered dose in urine and < 1% of the dose in faeces. Thus, absorbed DL-triazolyl alanine is very rapidly and completely eliminated via the urine, predominantly in unchanged form. Approximately 15% of the administered dose was excreted as *N*-acetyl-DL-triazolyl alanine.

Experiments on acute oral toxicity were performed in rats and mice. A test for acute intraperitoneal toxicity was performed in rats. 1,2,4-Triazolyl alanine was found to be of very low acute toxicity.

One 28-day study in rats was conducted in order to identify potential target organs after administration of high doses of 1,2,4-triazolyl alanine. There were no treatment-related findings and the NOAEL was the highest dose tested (400 mg/kg bw per day).

Ninety-day studies of toxicity were performed in rats and dogs. In the study in rats, the NOAEL was 5000 ppm, equivalent to 500 mg/kg bw per day, on the basis of body-weight effects at 20 000 ppm, equivalent to 2000 mg/kg bw per day, the highest dose tested. There were no other treatment-related effects. In the study in dogs, the NOAEL was 8000 ppm, equivalent to 200 mg/kg bw per day, on the basis of reduced body-weight gain at 20 000 ppm, equivalent to 500 mg/kg bw per day, the highest dose tested. There were no other treatment-related findings (Table 19).

1,2,4-Triazolyl alanine produced no evidence of genotoxic potential in assays for reverse mutation/DNA repair in bacteria and in assays for forward mutation/DNA repair/cell transformation in mammalian cells in vitro. No clastogenic or aneugenic effects were observed in tests for micronucleus formation in mice and hamsters in vivo (Table 20).

No studies of carcinogenicity with 1,2,4-triazolyl alanine have been identified

A preliminary study of reproduction in rats (Birtley, 1983) identified a NOAEL of 10 000 ppm for maternal toxicity and a NOAEL for fetal toxicity as 2500 ppm on the basis of slight reductions in neonatal body weights of male and female pups at 10 000 ppm. The follow-up main study also identified the NOAEL for maternal toxicity as 10 000 ppm, equivalent to an average intake of

Figure 5. 1,2,4-Triazolyl alanine acid or 2-amino-3-(1,2,4]triazol)-1-yl-propionic acid

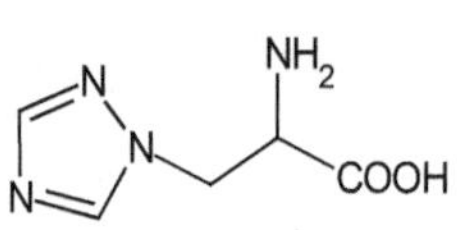

N N N NH$_2$ COOH

Table 19. Results of studies of acute toxicity and short-term studies of toxicity with 1,2,4-triazolyl alanine

Study	Species	Dose	LD_{50} (mg/kg bw)	NOAEL (mg/kg bw per day)	Adverse effect	Reference
Acute toxicity						
Oral	Rat, mouse	—	> 5000	—	—	Mihail (1982)
Oral	Rat	—	> 2000	—	—	Henderson & Parkinson (1980)
Intraperitoneal	Rat	—	> 5000	—	—	Mihail (1982)
Short-term studies of toxicity						
Twenty-eight-day, gavage	Rat	0, 25, 100 or 400 mg/kg bw per day	—	400 mg/kg bw per day	None observed	Mihail & Vogel (1983)
Ninety-day, feeding	Rat	0, 1250, 5 000 and 20 000 ppm	—	5000 ppm, equivalent to 500 mg/kg bw per day	Reduced body-weight gains in males at 20 000 ppm	Maruhn & Bomhard (1984)
Ninety-day, feeding	Dog	0, 3 200, 8 000 and 20 000 ppm	—	8000 ppm equivalent to 200 mg/kg bw per day	Reduced food intake and body-weight gains in females at 20 000 ppm	von Keutz & Gröning (1984)

500 mg/kg bw per day, and the NOAEL for fetal toxicity of 2000 ppm, equivalent to 100 mg/kg bw per day, on the basis of reduced birth weights at 10 000 ppm, equivalent to 500 mg/kg bw per day (Table 21).

(iii) Toxicology of 1,2,4-triazolyl acetic acid

1,2,4-Triazolyl acetic acid (Figure 6) is a significant residue in plants treated with difenoconazole.

The metabolism and pharmacokinetics of 1,2,4-triazolyl acetic acid have been evaluated in rats (Lai & Simoneax; 1986b, 1986c). In a repeat-dose study, oral administration of radiolabelled 1,2,4-triazolyl acetic acid resulted in negligible tissue residue with no unusual accumulation in any tissue or organ. Recovery of the administered dose was high and elimination was rapid, with 66–75% of the administered dose being excreted via the urine and 11–32% via the faeces. In all cases, only unchanged 1,2,4-triazolyl acetic acid was excreted.

A balance study to further refine the uptake and rates of excretion of 1,2,4-triazolyl acetic acid after oral administration again confirmed that the primary route of excretion was via the urine. No significant differences were observed in excretion patterns between males and females. Excretion of radioactivity during the first 48 h after administration accounted for > 90% of the total activity at all three doses used, indicating that clearance was not saturable at the doses tested.

The toxicity of 1,2,4-triazolyl acetic acid has been studied much less extensively than that of 1,2,4-triazolyl alanine. At a dose of 5000 mg/kg bw, no mortality was observed in rats given 1,2,4-triazolyl acetic acid as a single oral dose and after a 14-day observation period. The LD50 was > 5000 mg/kg bw (Thevenaz, 1984).

In a 14-day dietary study, male and female rats were given diets containing 1,2,4-triazolyl acetic acid at a concentration of 0, 100, 1000 or 8000 ppm, equal to 0, 11, 103 and 788 mg/kg bw per day in males and 0, 10, 97 and 704 mg/kg bw per day in females. No mortality occurred in the study, nor were there any clinical signs of systemic toxicity. Ophthalmic and auditory examinations revealed no treatment-related reactions. The mean body-weight values and mean food and water

Table 20. Results of studies of genotoxicity with 1,2,4-triazolyl alanine

End-point	Test system	Concentration or dose tested	Result	Reference
In vitro				
Gene mutation	S. typhimurium TA98, TA100, TA102, TA1535 and TA1537	20–5 000 μg/plate	Not mutagenic ±S9 in all test systems	Deparade (1986)
Gene mutation	S. typhimurium TA98, TA100, TA1535 and TA1537; E. coli strain WP2 uvrA	312.5–5 000 μg/plate	Not mutagenic ±S9 in all test systems	Hertner (1993)
Gene mutation	S. typhimurium TA98, TA100, TA1535, TA1537 and TA1538	20–12 500 μg/plate	Not mutagenic ±S9 in all test systems	Herbold (1983a)
Gene mutation	Chinese hamster V79 cells	500–10 000 μg/ml	Not mutagenic ±S9	Dollenmeier (1986b)
DNA repair	E. coli (K 12) p 3478 (pol A1-) and E. coli W3110 (pol A+).	62.5–1 000 μg/plate	No induction of DNA damage ±S9	Herbold (1983b)
	Bacillus subtilis (rec+) strain H17 and (rec-) strain M15	20–1 000 μg / plate	No induction of DNA damage ±S9	Watanabe (1993)
DNA repair	Rat hepatocytes	80–10 000 μg/ml	No induction of DNA damage	Puri (1986)
Cell transformation	Balb/3T3 mouse fibroblasts	62.5–10 000 μg/ml	No induction of cell transformation ±S9	Beilstein (1984)
Cell transformation	Baby hamster kidney (BHK 21 C13) cells	500–8 000 μg/ml –S9; 1 000–16 000 μg/ml +S9	Induction of cell transformation suggested. Method inadequate. Invalid result.	Richold et al. (1981)
In vivo				
Micronucleus formation	Mouse (NMRI)	8 000 mg/kg, oral	Not clastogenic or aneugenic	Herbold (1983c)
Micronucleus formation	Mouse (CBC F1)	2 500 and 5 000 mg/kg, intraperitoneal	Not clastogenic or aneugenic	Watkins (1982)
Micronucleus formation	Chinese hamsters	5 000 mg/kg bw, oral	Not clastogenic or aneugenic	Strasser (1986)

S9, 9000 × g supernatant from rat livers.

Figure 6. 1,2,4-Triazolyl acetic acid

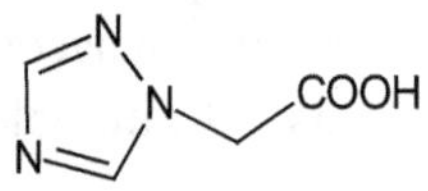

consumption values of all treated rats were similar to those of the respective controls during the entire treatment period. No differences were observed in clinical chemistry evaluations and analysis of organ weights and organ weight ratios revealed no treatment-related effects. Autopsy revealed no evidence of a reaction to treatment and there were no microscopic lesions or changes that could be attributed to treatment. The NOAEL was 8000 ppm, equal to 788 mg/kg bw per day, the highest dose tested (Thevenaz, 1986).

Table 21. Results of studies of reproductive toxicity with 1,2,4-triazolyl alanine

Study	Species	Dose	NOAEL	Adverse effects	Reference
Two-generation, feeding	Rat	0, 500, 2 000 and 10 000 ppm	Parental: 10 000 ppm, equivalent to 500 mg/kg bw per day Reproduction: 2 000 ppm, equivalent to 100 mg/kg bw per day	Reduced neonatal weights	Milburn et al. (1986)
Teratogenicity, gavage	Rat	0, 100, 300 and 1 000 mg/kg bw per day	Maternal: 1 000 mg/kg bw per day Developmental: 100 mg/kg bw per day	Retarded ossification	Clapp et al. (1983)

1,2,4-Triazolyl acetic acid was tested for mutagenic effects at doses of up to 5120 µg/plate in *S. typhimurium* strains TA98, TA100, TA1535 and TA1537 in the presence and absence of an exogenous metabolic activation system. There were no increases in the numbers of mutants per plate (Deparade, 1984).

(c) Supplementary studies with difenoconazole.

Supplementary studies conducted with difenoconazole consisted of a mechanistic study, examining biochemical and morphological changes in the mouse liver, to help elucidate the species-specificity of the incidence of tumours seen in the long-term studies, and two studies (one in hens and one in dogs) to help understand the finding of cataracts in a 6-month study in dogs.

A study with difenoconazole was conducted to examine the effects on selected biochemical and morphological parameters in the liver, including activities of various drug-metabolizing enzymes, after repeated doses in male mice.

Groups of nine male TIF:Magf (SPF) mice were given difenoconazole technical (purity, 91.8%) at a dose of 0, 1, 10, 100 or 400 mg/kg bw per day by gavage in carboxymethyl cellulose for 14 days. Additional groups at 0 and 400 mg/kg bw per day were dosed for 14 days and then permitted to live without treatment for another 28 days (recovery groups) (Thomas, 1992). Groups of male mice were given phenobarbitone, 3-methylcholanthrene (3-MC) or nafenopin (NAF) as reference substances. Mortality and clinical signs were checked twice per day during treatment and twice per week during recovery. Body weights were recorded daily during treatment and twice per week during recovery. From each group, six mice were killed for biochemical investigations of the liver and three were killed for electron microscopy of the liver. The following parameters were measured: protein content of supernatant, microsomal and cytosolic fractions; microsomal cytochrome P450 content; monoclonal antibodies vs cytochrome P450 isoenzymes (immunoblot analyses); microsomal EROD and PROD activities; region- and stereoselective microsomal testosterone hydroxylation; cyanide-insensitive peroxisomal β-oxidation; microsomal hydroxylation of lauric acid; microsomal epoxide hydrolase activity; microsomal UDP-glucronosyltransferase activity; cytosolic glutathione S-transferase activity; spectral interaction of difenoconazole with microsome P450.

There was a marked increase in liver weights (+79%) in the mice treated with difenoconazole at 400 mg/kg bw per day, relative to the controls. However, after a 28-day recovery period, liver weights of mice in the control group and mice at 400 mg/kg bw per day were similar (1.41 g vs 1.40 g, respectively). Liver weights of the mice treated with 3-MC and NAF were also significantly higher than those of mice in the control group, although they were treated for much shorter periods of time.

The protein contents of the liver supernatant fractions were not affected by treatment, either with difenoconazole or the reference substances. There was a moderate, dose-dependent increase in the protein content of the microsomal liver fractions in mice treated with difenoconazole at 100 and 400 mg/kg bw per day; these increases were not apparent in animals which had a subsequent 28-day recovery period. The microsomal protein content was elevated in the mice treated with the three reference substances. The protein content of the cytosolic liver fractions was not different between the control and treated groups.

Microsomal cytochrome P450 contents were significantly elevated in mice treated with difenoconazole at 100 and 400 mg/kg bw per day. The values returned to the control levels after a 28-day recovery period. Cytochrome P450 contents were also markedly elevated in mice treated with the three reference substances.

Examination of the cytochrome P450 gene families detected by cross-reaction with different monoclonal antibodies showed that the P450 families CYP1A and CYP3A were detectable at enhanced concentrations in mice treated with difenoconazole at 100 or 400 mg/kg bw per day. There appeared to be a dose-dependent decrease in the expression of the CYP2B isoenzymes. A similar profile was obtained with phenobarbitone, while 3-MC caused a 45-fold increase in CYP1A and NAF caused a 9-fold increase in CYP4A. Microsomal epoxide hydrolase activity was markedly elevated in mice treated with difenoconazole at 100 and 400 mg/kg bw per day relative to mice in the control group, by about 50% and 250%, respectively. No differences were seen after the 28-day recovery period. Substantial increases were also noted for phenobarbitone and NAF, but not for 3-MC. Activities of morphine UDP-glucuronosyltransferase were slightly increased in a dose-related manner in the mice treated with difenoconazole; this increase was reversed after the 28-day recovery period. Similar increases were seen in mice treated with phenobarbitone, 3-MC and NAF. Microsomal 1-naphthol UDP-glucuronosyltransferase activities were essentially unchanged by treatment with difenoconazole, while significant increases were seen after treatment with phenobarbitone, 3-MC and NAF. EROD and PROD activities were increased approximately 3- and 30-fold, respectively, in the mice treated with difenoconazole at 400 mg/kg bw; these increases were comparable in magnitude to the increases produced by phenobarbitone. Changes were also induced by 3-MC, but much greater changes were noted in EROD than PROD (opposite of difenoconazole). NAF only marginally induced PROD activities. All changes were reversible after the 28-day recovery period. The activity of cytosolic glutathione S-transferase in mice treated with difenoconazole at 500 mg/kg bw was increased about 50% over that for mice in the control group; a similar level of induction was seen with phenobarbitone, and to a lesser extent with NAF.

Total testosterone hydroxylation, dependent on cytochrome P450, was induced in a dose-related manner to about sixfold the control value in mice treated with difenoconazole at 400 mg/kg bw. Testosterone hydroxylation rates were also significantly increased with phenobarbitone, 3-MC and NAF. Except for 7α-hydroxy-testosterone, all testosterone metabolites were increased by between 3- and 120-fold in mice treated with difenoconazole at 400 mg/kg bw compared with mice in the control group. The increases of 6β- and 15β-hydroxy-testosterone indicate either a strong barbiturate- or steroid-type induction. The extent of 6α-hydroxylation also suggests difenoconazole to be a phenobarbitone-type inducer. Testosterone metabolite profiles of animals treated with 3-MC and NAF were clearly different in magnitude and distribution, indicating that difenoconazole is not likely to be a 3-MC- or NAF-type inducer.

Treatment with difenoconazole at 100 and 400 mg/kg bw per day resulted in approximately 2-fold increases in lauric acid 11-hydroxylation, while lauric acid 12-hydroxylation decreased at all doses. Mice treated with phenobarbitone had increases in lauric acid 11-hydroxylation that were comparable to those in mice treated with difenoconazole and 12-hydroxylation remained unchanged. On the other hand, 3-MC did not increase 11-hydroxylation and significantly reduced 12-hydroxylation, while NAF increased both 11- and 12-hydroxylation markedly.

Cyanide-insensitive peroxisomal fatty acid β-oxidation appeared to be slightly and dose-dependently decreased by treatment with difenoconazole, although this change was not statistically significant and was reversible after 28 days of recovery. Neither phenobarbitone nor 3-MC affected peroxisomal β-oxidation, but NAF resulted in a more than fourfold increase in this process.

Titration of liver microsomal suspensions from mice treated with difenoconazole, phenobarbitone, 3-MC or NAF revealed type II binding spectra, which are indicative of an inhibitory action of the test article on microsomal cytochrome P450. All spectra indicated low- and high-affinity microsomal binding sites for difenoconazole, with approximate spectral dissocation constants K_{S1} and K_{S2} between 0.22 and 0.43 μmol/l and 1.73 and 5.27 μmol/l, respectively. A unique binding site for difenoconazole (K_{S1} = 1.99 μmol/l) was identified in mice treated at 400 mg/kg bw per day.

The ultramorphological analysis of liver sections from the mice treated with difenoconazole at 400 mg/kg bw revealed a distinct proliferation of smooth and rough endoplasmic reticulum membranes and a disorganization of rough endoplasmic reticulum membranes, leading to a mixture of smooth and rough endoplasmic reticulum elements, mostly in vesicular form. After the 28-day recovery period, hepatocytes from mice in the control group and mice treated with difenoconazole were essentially the same (Thomas, 1992).

Difenoconazole is considered to be a reversible barbiturate-type inducer of metabolizing enzymes in the mouse liver. No peroxisome proliferation was observed. The dose of 10 mg/kg bw per day was a no-observed-effect level (NOEL), without an inductive effect on metabolizing enzymes and other parameters in the mouse liver.

(d) Cataract induction

Chickens

In a study of cataract induction, a group of five male and five female Hisex chickens were given diets containing difenoconazole technical (purity ≥ 95%) at a concentration of 5000 ppm for 8 weeks. An additional three male and three female chickens received the basal diet only and three male and three female chickens received the diet admixed with 2,4-dinitrophenol at 2500 ppm as a positive control. Mortality was checked and clinical signs were recorded daily. Eye examinations were conducted on all chickens twice per week. Individual body weights were recorded weekly and food consumption was recorded daily. Surviving chickens were killed on study day 57. The eyes were removed and fixed in buffered 10% neutral formalin and subsequently processed for histopathological examination.

There was no mortality among the mice in the negative- or positive-control groups or among the male treated with difenoconazole. One female receiving difenoconazole died on study day 36. The clinical sign of ruffled feathers was seen in all test groups and all hickens in the positive-control group from day 7 until study termination; reduced locomotor activity and ventral recumbency were also noted for the female that died. There were no clinical signs among the chickens in the control group. Body weights of the males receiving difenoconazole were significantly ($p < 0.05$) lower than those of the chickens in the control group from day 14 onwards; body weights for females receiving difenoconazole were significantly lower from day 7 onwards. Mean body weights of males and females in the positive-control group were also consistently lower than those of chickens in the control group and slightly higher than those of the chickens receiving difenoconazole. Food consumption by the difenoconazole-treated chickens was markedly less than that of the chickens in the control group throughout the study. Food consumption by chickens in the positive-control was also lower than that of the controls, but higher than that of chickens receiving difenoconazole.

Ophthalmoscopic examinations revealed lens alterations in all males treated with difenoconazole, two out of five females treated with difenoconazole, all chickens in the positive-control group and one in the negative-control group. Slight lens alterations were seen between days 24 and 35 in three males and one female treated with difenoconazole. Slight to moderate but irreversible lens alterations developed between days 45 and 56 in one female and four males treated with difenoconazole. In contrast, all chickens treated with 2,4-dinitrophenol showed marked lens opacities on day 3, which diminished in severity on day 7 and persisted as slight alterations through study termination (except for one female for which there were no findings after day 38). Yellow/grey discoloration was also observed in the eyes of chickens in the positive-control group from day 31 onwards. The slight, transient alterations seen in the one male in the negative-control group were considered to be spurious.

Microscopic examinations revealed initial changes in the lens, indicative of cataract, in three out of five males and one out of five females treated with difenoconazole. The lesions comprised slight swelling of the epithelial cells either at the equator or anteriorly, and/or necrosis of the lens fibres posteriorly, under the capsule or in the outer cortex. Among the chickens in the positive-control group, two out of three males developed changes indicative of cataract (necrosis of the lens fibres, posterior under the capsule); one female in the positive-control group showed a slight swelling of the lens epithelium at the equator. No changes were seen in the lenses of the chickens in the negative-control group (Schoch & Schneider, 1989). The sponsor considered that, in the absence of dose–response information, this high-dose phenomenon should be considered of no human relevance; however, the experimental design was such that a dose–response relationship could never have been observed.

Dogs

In a study of cataract induction, groups of three male and three female pure-bred beagle dogs (KFM Kleintierfarm Madoerin, Füllinsdorf, Switzerland) were given diets containing difenoconazole technical (purity 95%) at a concentration of of 0, 3000, 4000 or 6000 ppm for up to 18 weeks according to a complex dosing schedule (Table 22).

Mortality was checked and clinical signs were recorded daily. Individual body weights were recorded weekly and food consumption was recorded daily. Eye examinations were conducted on all dogs before exposure and every 2 weeks thereafter. Haematology and blood chemistry investigations were carried out on all dogs before the exposure period and during weeks 3, 13 and 19. Surviving dogs were killed in week 19. All dogs were subjected to a detailed autopsy. Specified organs were weighed. Eyes and tissues with macroscopic findings were examined microscopically.

There were no deaths in the test. The only clinical signs noted were vomiting in the male in the reference group and a female in the recovery group during the first study week and faecal changes in the female in the reference group during weeks 6 and 9, a male in the recovery group during week 10 and a female in the recovery group during week 14.

All dogs lost body weight during the first study week, 13–14% from initial values in the males and 11–15% in the females. After the dietary concentration was reduced to 3000 ppm, all dogs began to gain weight. The males in the reference and recovery groups gained weight at approximately the same rate, but the female in the reference group gained weight at a slower rate than those in the recovery group and lost weight again when the dietary concentration was increased to 4000 ppm.

Food consumption was markedly depressed in all dogs during the first study week, being between 72 and 85% lower than the last value recorded before dosing for males and 81–88% lower for the females. Food consumption increased for all dogs when the concentration was lowered to 3000 ppm. By week 3, all males and the females in the recovery group were consuming all of the diet offered. The female in the reference group consistently consumed less than the other dogs.

Table 22. Dosing schedule used in a special study of cataract induction in dogs fed diets containing difenoconazole

Time-point	Dietary concentration (ppm)					
	Group 1 (reference)		Group 2 (recovery)			
	Male No. 1	Female No. 1	Male No. 2	Female No. 2	Male No. 3	Female No. 3
Week 1	6000 ppm	6000 ppm	6000 ppm	6000 ppm	6000 ppm	6000 ppm
Weeks 2–3	3000 ppm	3000 ppm	3000 ppm	3000 ppm	3000 ppm	3000 ppm
Weeks 4–9	3000 ppm	3000 ppm	0 ppm	0 ppm	0 ppm	0 ppm
Weeks 10–18	4000 ppm	4000 ppm	0 ppm	0 ppm	0 ppm	0 ppm

Ophthalmoscopic examinations performed every 2 weeks did not reveal any alterations in the lens in any dog.

There were no clear differences between the dogs in any of the haematological parameters measured. Although the male in the reference group had lower erythrocyte, haemoglobin and erythrocyte volume fraction values at termination than did the males in the recovery group (as seen in the 6-month feeding study in dogs; O'Connor et al., 1987), lower values were also evident before the exposure period began. The values for leukocytes in the male in the reference group and erythrocytes in the female in the reference group were clearly above the 95% upper limit of the range for historical controls, but these variations were not considered to be related to treatment.

There were also no clear differences in clinical chemistry parameters between dogs in the reference and recovery groups. There was an apparent increase in alkaline phosphatase activity in the male and female in the reference groups, which is consistent with findings in the 6-month and 12-month feeding studies in dogs (O'Connor et al., 1987; Rudzki et al., 1988); the value for the male at 19 weeks was also clearly above the 95% upper limit of the range for historical controls, while the value for the female was within the range.

Analysis of organ weights revealed increased absolute and relative liver weights in the male but not in the female in the reference group. The few other differences observed were considered not to be related to treatment.

Macroscopic examinations did not reveal any treatment-related findings. The few findings noted (congestion in the spleen, worms in the small intestine, foci in the lungs, discoloration in the cervical lymph node, cysts in the ovaries and thickening in the vagina) were considered to be spontaneous findings usually encountered in dogs of this strain and age.

There were no treatment-related histopathology findings and there were no histological alterations of any sort in the eye. All findings that were seen in dogs in the reference groups were also seen in those in the recovery groups, and were not unusual in young beagle dogs (Janiak et al., 1989). The Meeting concluded that treatment of single male and female dogs with difenoconazole in this small study did not result in the formation of cataracts.

3. Observations in humans

3.1 Medical surveillance on personnel at manufacturing sites

Manufacturing employees are medically examined by company physicians at the beginning of their employment and then routinely once per year. In Switzerland, routine medical examinations according to the criteria of the Swiss Accident Insurance Institution (SUVA) include: anamnesis; physical examination including blood pressure measurement; blood analysis: haemoglobin, erythrocytes, leukocytes, thrombocytes, complete blood count, blood sedimentation rate, blood sugar, blood pressure, cholesterol, triglycerides, ALT, aspartate aminotrasferase, alkaline phosphatase, bilirubin, creatinine and urate; and urine analysis.

Difenoconazole has been manufactured in production plants at Monthey in Switzerland and formulated at various sites around the world. Data was sought from all sites handling either the technical active ingredient or formulated product. A request for data on adverse health events covering the years 1992 to 2002, carried out in December 2002, provided the information that questionnaires were sent to those responsible for the production sites and company physicians:

Monthey, Switzerland (production); reporting period, 1992 until December 2002

Difenoconazole had been handled since 1992 in three buildings, with an estimated 5300 tonnes being produced over the decade. In 2002, about two thirds of the product was handled and packed as powder. The remaining one third was used in two premixes, which were either formulated in Monthey or sent in bulk to Aigues Vives, France. The exact number of people potentially in contact with the material was unknown, but it was estimated that "several hundred" temporary workers had been involved with difenaconazole over the 10-year period. No adverse health effects had been recorded within the three buildings concerned or within the medical service based at Monthey.

Roosendaal, the Netherlands (formulation); reporting period, 1992 until December 2002

Difenoconazole had been handled at this site since 1997, with 60–70 tonnes being handled annually. Up to 16 people in quality-control roles and 34 people in the formulation plant and warehouse had been involved in handling difenaconazole. No adverse health effects had been reported.

Aigues Vives, France (formulation); reporting period, early 1990s until December 2002

Difenoconazole had been handled since the early 1990s, with 132 tonnes being handled in the years 1998–2003. Up to 19 people had been involved with the formulation and packing activities. One case of allergic reaction to a formulated product, "Score", was reported in 1995.

Iksan, Korea (formulation); reporting period, early 1990s until December 2002

Difenoconazole technical material and formulated products ("Bogard" WG – approximately 30 tonnes produced annually in 2001 and 2002) were handled at the Iksan site without any reports of adverse health effects.

Pendle Hill, Australia (formulation); reporting period, 1995 until December 2002

The site first handled difenaconazole in 1995. No adverse health effects had been reported.

Since December 2002, all the above sites have been required to report any adverse reactions or occupational illnesses related to chemical exposure. A review of the database revealed that no such adverse reactions or occupational illnesses had been reported.

3.2 Direct observations (e.g. clinical cases and poisoning incidents)

No cases of poisoning had been reported to the sponsor and no cases of sensitization had been observed. There were no published cases of human poisoning with difenoconazole or its formulations. No epidemiological study had been performed by the company. No reports from the open medical literature were on record.

Comments

Biochemical aspects

In rats given [^{14}C]difenoconazole labelled in either the triazole or phenyl rings as a single oral dose at 0.5 or 300 mg/kg bw, the radiolabel was rapidly and almost completely absorbed from the gastrointestinal tract, widely distributed, and eliminated from the body with excretion half-lives of about 20 h at the lower dose and about 33–48 h at the higher dose. At doses of 0.5 and 300 mg/kg bw approximately 13–22% and 8–15%, respectively, of the administered radioactivity was excreted via the urine in rats. Excretion via the faeces accounted for 81–87% (lower dose) to 85–95% (higher dose) of the radiolabel; however, approximately 73% (males) to 76% (females) of the lower dose and approximately 39% (females) to 56% (males) of the higher dose was eliminated from the body via the bile in bile-duct cannulated rats. Thus, bioavailability decreased with increasing dose. When bile from rats given difenoconazole was introduced into the duodenum of other bile-duct cannulated rats, 80% of the radioactivity was re-eliminated in the bile and 4% in the urine, indicating that there was extensive enterohepatic recirculation. The greater quantities of radioactivity were distributed to the gastrointestinal tract, liver and kidneys. The initial plasma half-life was approximately 4–6 h at the lower dose and 22–24 h at the higher dose and the terminal half-life was approximately 3–4 days at both doses. In spite of these long terminal half-lives, there was no evidence for bioaccumulation of difenoconazole.

After oral administration to male and female rats, the systemically available portion of [^{14}C-phenyl]difenoconazole was rapidly and extensively metabolized to a number of biotransformation products. Three main metabolites isolated from the faeces accounted for approximately 68% of the administered doses. These were: the hydroxylated product of cleavage of the dioxolane ring; its further metabolite resulting from hydroxylation of the chlorophenoxy ring; and the product of the hydroxylation of difenoconazole, occurring directly in the chlorophenoxy ring. A minor metabolic pathway involves cleavage of the alkyl chain between the triazole and the inner phenyl ring, resulting in a hydroxy acetic acid, 2-chloro-4-(4-chlorophenoxy)-benzoic acid and free 1,2,4-triazole. Sulfate conjugates were identified for some hydroxylated metabolites.

Toxicological data

The acute toxicity of difenoconazole was low, with values for the oral LD_{50} being 1453 mg/kg bw in rats and > 2000 mg/kg bw in mice. The dermal LD_{50} in rabbits was > 2010 mg/kg bw and the inhalation LC_{50} in rats was > 3.3 mg/l. Difenoconazole is very slightly and transiently irritating to the skin and moderately and transiently irritating to the eyes of rabbits and is non-sensitizing in a modified Buehler test in guinea-pigs.

Overall, in short-term studies with orally-administered difenoconazole, the signs of toxicity observed in mice, rats and dogs were similar, with reduced body-weight gain and increased liver weights being common features. Histopathology confirmed the liver as a target organ with observation of diffuse or centrilobular hypertrophy of hepatocytes in rats and mice, although this can also be indicative of an adaptive response. Cataracts were found in dogs fed diets containing difenoconazole at a concentration of ≥ 3000 ppm, equal to 96.6 mg/kg bw per day, for 6 months, with a NOAEL of 1000 ppm, equal to 31.3 mg/kg bw per day; however, cataracts were not induced in a second study in dogs given diets containing difenoconazole at up to 1500 ppm, equal to 51.2 mg/kg bw per day, for 1 year. Increased activity of alkaline phosphatase was observed in two studies in rats and in one in dogs. No other blood chemistry changes were consistently observed, although reduced concentrations of blood protein were observed in dogs given diets containing difenoconazole at 6000 ppm, equal to 157.8 mg/kg bw per day. Also in dogs, a reduction in erythrocyte count of almost 20% was observed in females at this high level of exposure.

For the short-term dietary studies, the NOAELs were: in studies of up to 90 days in rats, 200 ppm (equal to 17 mg/kg bw per day) on the basis of increased hepatocellular hypertrophy and liver weight; in a 90-day dietary study in mice, 200 ppm (equal to 34.2 mg/kg bw per day) on the basis of clinical signs of toxicity and changes in liver weight and increased incidence of centrilobular hepatocellular hypertrophy; in a 28-week study in dogs, 1000 ppm (equal to 31.3 mg/kg bw per day) on the basis of cataracts and liver-weight changes; in a 12-month study in dogs, 100 ppm (equal to 3.6 mg/kg bw per day) on the basis of reduced body-weight gain; in a 4-week study of dermal toxicity with difenoconazole in rats, 100 mg/kg bw per day, on the basis of minimal centrilobular hepatocellular hypertrophy, minimal to moderate thyroid follicular cell hypertrophy and skin lesions at the site of application.

Long-term feeding studies in rats and mice fed with difenoconazole confirmed that the primary target organ was the liver. There was no evidence for any carcinogenic potential in rats, in which hepatic effects were increases in liver weight and hepatocellular hypertrophy in male and female rats. In addition, there were reductions in erythrocyte parameters in female rats at the highest dose, 2500 ppm, equal to 170 mg/kg bw per day.

In mice, there was very high, treatment-related mortality at the beginning of the 18-month study. In groups of 70 mice, there were 52 deaths among females receiving difenoconazole at 4500 ppm and

16 deaths among females at 3000 ppm (reduced to 2500 ppm after 1 week) within the first 2 weeks, while, among the male mice there were 11 deaths in the group at 4500 ppm within the first 3 weeks of the study. There was an increased incidence of hepatocellular adenomas and carcinomas in the group of male and female mice fed diet containing difenoconazole at 2500 ppm, equal to 423 and 513 mg/kg bw per day, respectively, and males at 4500 ppm, equal to 819 mg/kg bw per day. No increase in the incidence of tumours was observed at 300 ppm, equal to 46.3 and 57.8 mg/kg bw per day in males and females, respectively. However, the neoplastic responses occurred at highly toxic doses that also caused the death of substantial proportions of the groups of mice. Among the survivors, biliary stasis and hepatic single-cell necrosis as well as hepatocellular hypertrophy were significantly increased in male and female mice at the tumorigenic doses. On the basis of a study of enzyme activities in male mice, difenoconazole is considered to be a reversible barbiturate-type inducer of metabolizing enzymes in the mouse liver. No peroxisome proliferation was observed. The NOEL was 10 mg/kg, there being no inductive effect on metabolizing enzymes and other parameters in the mouse liver.

The NOAEL in long-term studies in rats was 20 ppm, equal to 1.0 mg/kg bw per day, on the basis of reduced body-weight gains during the first year in males and females, reduced platelet counts in males and hepatic centrilobular hypertrophy in males and females at 500 ppm, equal to 24 mg/kg bw per day. In long-term studies in mice, the NOAEL was 30 ppm, equal to 4.7 mg/kg bw per day, on the basis of decreased body-weight gain in males, increased liver weight in females and hepatocellular hypertrophy in males at 300 ppm, equal to 46.3 mg/kg bw per day.

Difenoconazole was tested for genotoxicity in an adequate range of assays, both in vitro and in vivo. No evidence for genotoxicity was observed in any test.

The Meeting concluded that difenoconazole is unlikely to be genotoxic.

The Meeting concluded that difenoconazole caused an increase in the incidence of hepatocellular adenomas and carcinomas in mice (but not in rats) by a non-genotoxic mode of action, the nature of which has not been established but which resembles that for phenobarbital in its liver enzyme-inducing characteristics. It is therefore unlikely to pose a carcinogenic risk to humans at exposure levels that do not cause changes in the liver.

The reproductive toxicity of difenoconazole was investigated in a two-generation study of reproduction in rats and in studies of developmental toxicity in rats and rabbits.

Reproductive function was not affected in rats in the two-generation study and the NOAEL for reproductive function was 2500 ppm, equal to 132.1 mg/kg bw per day, the highest dose tested. The NOAEL for systemic toxicity in the parental animals was 250 ppm, equal to 11.5 mg/kg bw per day on the basis of reduced body-weight gain during the pre-mating period in F_0 and F_1 generations at 2500 ppm, equal to 122.7 mg/kg bw per day during these periods. In pups, lower birth weight and subsequent decreased body-weight gain at 2500 ppm, equal to 158.0 mg/kg bw per day, were the effects that defined the NOAEL for offspring toxicity at 250 ppm, equal to 14.1 mg/kg bw per day in the females.

In a study of developmental toxicity in which rats were given difenoconazole by gavage on days 6–15 of gestation, the NOAEL for maternal toxicity was 20 mg/kg bw per day on the basis of reduced body-weight gain and excess salivation first observed on day 2 of dosing in 14 out of 23 dams at 100 mg/kg bw per day. There was a statistically significant, but small increased incidence of changes in thoracic vertebral ossification centres in fetuses at 200 mg/kg bw. The NOAEL for fetal toxicity was 100 mg/kg bw per day on the basis of these skeletal anomalies.

In a study of developmental toxicity in which rabbits were given difenoconazole by gavage on days 7–19 of gestation, the NOAEL for maternal toxicity was 25 mg/kg bw per day on the basis of reduced body-weight gain in the first few days of dosing at 75 mg/kg bw per day. Examination of the fetuses did not reveal any treatment-related effects in soft or ossified tissues. The NOAEL for developmental toxicity was 75 mg/kg bw per day, the highest dose tested.

In a single-dose study of neurotoxicity in rats treated by gavage, the NOAEL was 25 mg/kg bw on the basis of reduced fore-limb grip strength interpreted as a non-specific response at 200 mg/kg bw. In a 90-day study of neurotoxicity in rats given diets containing difenoconazole, the NOAEL was 40 ppm, equal to 2.8 mg/kg bw per day, on the basis of reduced hind-limb grip strength in males during the final week of the study at 250 ppm, equal to 17.3 mg/kg bw per day. These responses were considered to be non-specific effects of difenoconazole because of the absence of any changes in the multiple end-points of neurotoxicity that were measured and the absence of neuropathological findings.

The Meeting concluded that difenoconazole is unlikely to cause neurotoxicity in humans.

There were no indications of immunotoxicity in general studies of toxicity in dogs, rats and mice.

Some aspects of the toxicology of three plant metabolites of difenoconazole that are also found in rats given difenoconazole were investigated. These metabolites were: 1-[2-chloro-4-(4-chloro-phenoxy)-phenyl]-2-(1,2,4-triazol)-1-yl-ethanone, 1-[2-chloro-4-(4-chloro-phenoxy)-phenyl]-2-(1,2,4-triazol)-1-yl-ethanol and 2-chloro-4-(4-chlorophenoxy)-benzoic acid. The LD_{50} value for each of these metabolites was > 2000 mg/kg bw and none of them showed any alerts for mutagenic activity. In addition, there are three metabolites which are common to all parent triazoles: 1,2,4-triazole, 1,2,4-triazolyl alanine and 1,2,4-triazolyl acetic acid. 1,2,4-Triazole is a soil metabolite that also appears to a small extent (< 10%) in some studies in plants and is found as at least 10% of the total metabolites in rats administered difenoconazole. 1,2,4-Triazolyl alanine and 1,2,4-triazolyl acetic acid are plant-specific metabolites. The acute toxicity (LD_{50}) values for 1,2,4-triazole, 1,2,4-triazolyl alanine and 1,2,4-triazolyl acetic acid were similar to or higher than the LD_{50} values for difenoconazole. Some recent studies conducted with 1,2,4-triazole have shown certain effects at high doses, i.e. testicular atrophy in mice at ≥ 487 mg/kg bw per day; effects on the central nervous system and pathology in rats at ≥ 183 mg/kg bw per day; and a reduction in fertility in a two-generation study of reproduction in rats at 218 mg/kg bw per day. The NOAELs for 1,2,4-triazole for these effects were 90 mg/kg bw per day for testicular atrophy in mice, 33 mg/kg bw per day for neurotoxicity in rats and 16 mg/kg bw per day for effects on fertility in rats.

Repeat-dose studies in which 1,2,4-triazolyl alanine was administered for up to 90 days did not show any significant effects other than reduced body-weight gains at 20 000 ppm, equivalent to 2000 mg/kg bw per day in rats and 500 mg/kg bw per day in dogs. The NOAELs in these studies were 5000 ppm, equivalent to 500 mg/kg bw per day in rats and 8000 ppm, equivalent to 200 mg/kg bw per day, in dogs. In a two-generation study with 1,2,4-triazolyl alanine in rats, the NOAEL for parental toxicity was 10 000 ppm, equivalent to 500 mg/kg bw per day, the highest dose tested, and the NOAEL for reproductive toxicity was 2000 ppm, equivalent to 100 mg/kg bw per day, on the basis of reduced birth weights at 10 000 ppm. In a study of teratogenicity in rats, the NOAEL for maternal toxicity was 1000 mg/kg bw per day, the highest dose tested, and the NOAEL for developmental toxicity was 100 mg/kg bw per day on the basis of retarded ossification at 1000 mg/kg bw per day.

1,2,4-Triazolyl acetic acid has been tested in a 14-day dietary study in which the NOAEL was 8000 ppm, equal to 704 mg/kg bw per day, the highest dose tested.

None of these metabolites has been tested for carcinogenicity. The extent to which 1,2,4-triazole, 1,2,4-triazolyl alanine and 1,2,4-triazolyl acetic acid had been tested for mutagenicity varied, but no significant responses were obtained in any of the studies.

Medical surveillance of personnel has been conducted since the early 1990s at sites in several countries where difenoconazole is manufactured. Reports were available up to the end of 2002. There have been no reports of toxicity in workers during manufacture and there was only one case of allergic reaction to a formulated product. There were no reports of poisoning and no reports of sensitization during use of the formulated product.

The Meeting concluded that the existing database was adequate to characterize the potential hazards to fetuses, infants and children.

Toxicological evaluation

An acceptable daily intake (ADI) of 0–0.01 mg/kg bw was established for difenoconazole based on the NOAEL of 1.0 mg/kg bw per day in rats, identified on the basis of reduced body-weight gains, reduced platelet counts and hepatic hypertrophy in a 24-month long-term dietary study of toxicity and carcinogenicity. A safety factor of 100 was applied. The increased incidence of hepato-cellular adenomas and carcinomas observed in mice at 423 mg/kg bw per day, the NOAEL being 4.7 mg/kg bw per day, in an 18-month long-term dietary study of toxicity and carcinogenicity, was likely to be due to a mode of action without human relevance at exposure levels that do not cause changes in the liver.

An acute reference dose (ARfD) of 0.3 mg/kg bw was established for difenoconazole. This was based on the NOAEL of 25.0 mg/kg bw in rats, identified on the basis of clinical signs in a single-dose study of neurotoxicity and using a safety factor of 100. This ARfD is supported by the NOAEL of 25 mg/kg bw per day for maternal toxicity in a study of developmental toxicity in rats and rabbits on the basis of excess salivation in rats at 100 mg/kg bw per day and body-weight loss in rabbits during the first few days of treatment at 75 mg/kg bw per day.

Levels relevant to risk assessment

Species	Study	Effect	NOAEL	LOAEL
Mouse	Eighteen-month study of toxicity and carcinogenicity	Toxicity	30 ppm, equal to 4.7 mg/kg bw per day	300 ppm, equal to 46.3 mg/kg bw per day
		Carcinogenicity	300 ppm, equal to 46.3 mg/kg bw per day	2500 ppma, equal to 423 mg/kg bw per day
Rat	Twenty-four-month studies of toxicity and carcinogenicity	Toxicity	20 ppm, equal to 1.0 mg/kg bw per day	500 ppm, equal to 24 mg/kg bw per day
		Carcinogenicity	2500 ppm,a equal to 124 mg/kg bw per day	—
	Two-generation study of reproductive toxicity[b]	Reproductive toxicity	2500 ppm,a equal to 132.1 mg/kg bw per day	—
		Parental toxicity	250 ppm, equal to 11.5 mg/kg bw per day	2500 ppm,a equal to 122.7 mg/kg bw per day
		Offspring toxicity	250 ppm, equal to 14.1 mg/kg bw per day	2500 ppm,a equivalent to 158 mg/kg bw per day
	Developmental toxicity[c]	Maternal toxicity	20 mg/kg bw per day	100 mg/kg bw per day
		Embryo and fetal toxicity	100 mg/kg bw per day	200 mg/kg bw per daya
	Acute neurotoxicity	Toxicity	25 mg/kg bw	200 mg/kg bw per daya
Rabbit	Developmental toxicity[c]	Maternal toxicity	25 mg/kg bw per day	75 mg/kg bw per daya
		Embryo and fetal toxicity	75 mg/kg bw per day[a]	—
Dog	One-year study of toxicity	Toxicity	100 ppm, equal to 3.6 mg/kg bw per day	500 ppm, equal to 16.4 mg/kg bw per day

a Highest dose tested.

b Measurements of intake of the compound are the mean for the pre-mating phases for females.

c Gavage administration.

Estimate of acceptable daily intake for humans

0–0.01 mg/kg bw

Estimate of acute reference dose

0.3 mg/kg bw

Information that would be useful for the continued evaluation of the compound

Results from epidemiological, occupational health and other such observational studies of human exposure

Critical end-points for setting guidance values for exposure to difenoconazole

Absorption, distribution, excretion and metabolism	
Rate and extent of oral absorption	High, but incomplete (rats)
Dermal absorption	Approximately 8% (rats)
Distribution	Distributed throughout the body; higher concentrations in liver and gastrointestinal tract
Potential for accumulation	Low, no evidence of accumulation
Rate and extent of excretion	High; essentially 100% excretion in bile and urine within 7 days
Metabolism in animals	Extensive; a small number (< 10) of metabolites; little parent compound remaining
Toxicologically significant compounds in animals, plants and environment	Parent, 1,2,4-triazole
Acute toxicity	
Rat, LD_{50}, oral	1453 mg/kg bw
Rat, LC_{50}, inhalation	3.3 mg/l (4 h)
Rabbit, LD_{50}, dermal	2010 mg/kg bw
Rabbit, skin irritation	Very slightly and transiently irritating
Rabbit, eye irritation	Moderately and transiently irritating
Guinea-pig, skin sensitization	Not sensitizing (modified Buehler test)
Short-term studies of toxicity	
Target/critical effect	Liver; body weight
Lowest relevant oral NOAEL	3.6 mg/kg bw per day (12-month study in dogs)
Lowest relevant dermal NOAEL	100 mg/kg bw per day (4-week study in rats)
Lowest relevant inhalation NOAEC	No data
Genotoxicity	
	Not genotoxic
Long-term studies of toxicity and carcinogenicity	
Target/critical effect	Liver; body weight
Lowest relevant NOAEL	1.0 mg/kg bw per day (24-month study in rats)
Carcinogenicity	Not carcinogenic at levels below those causing changes in the liver
Reproductive toxicity	
Reproductive target/critical effect	None
Lowest relevant reproductive NOAEL	132 mg/kg bw per day
Developmental target/critical effect	Not teratogenic; reduced fetal body weight, delayed ossifications in rats, but not in rabbits

Lowest relevant developmental NOAEL	100 mg/kg bw per day (rats)
Neurotoxicity/delayed neurotoxicity	
Single-dose study of neurotoxicity	No signs of neurotoxicity, NOAEL was 25 mg/kg bw (rats)
Ninety-day study of neurotoxicity	No signs of neurotoxicity, NOAEL was 2.3 mg/kg bw (rats)
Other toxicological studies	
	Induction of liver xenobiotic metabolizing enzymes
	Studies of mammalian and plant metabolites
Medical data	
	No reports of toxicity in workers exposed during manufacture or use

Summary

	Value	Study	Safety factor
ADI	0–0.01 mg/kg bw	Rat, 2-year study of toxicity and carcinogenicity	100
ARfD	0.3 mg/kg bw	Rat, single-dose study of neurotoxicity, supported by maternal effects in a study of developmental toxicity in rabbits	100

References

Argus, M.A.; Ricci, J.M.; Huber, K.R.; Schiavo, D.M.; Hazelette, JR. & Green, J.D. (1987) CGA 169374 tech.: acute oral toxicity study in rats. Unpublished report No. MIN 862119 from Novartis Crop Protection AG, Basel, Switzerland and Ciba-Geigy Corp., Summit, USA. Submitted to WHO by Syngenta Crop Protection AG, Basel, Switzerland.

Beilstein, P. 1984) CGA 131013 tech.: transformation/liver-microsome test. Unpublished report No. 840324 from Novartis Crop Protection AG, Basel, Switzerland Ciba-Geigy Ltd, Basel, Switzerland. Submitted to WHO by Syngenta Crop Protection AG, Basel, Switzerland.

Birtley, R.D.N. (1983) Triazole alanine: preliminary reproduction study in the rat. Unpublished report No. CTL/L/470 from Novartis Crop Protection AG, Basel, Switzerland and Central Toxicology Laboratory (CTL), Cheshire, England. Submitted to WHO by Syngenta Crop Protection AG, Basel, Switzerland.

Bomhard, E. (1982) THS 2212: preliminary subacute toxicity study on male rats, administration in the drinking water. Unpublished report No. 11253 from Novartis Crop Protection AG, Basel, Switzerland and Bayer AG Toxicological Institute, Wuppertal-Elberfeld, Germany. Submitted to WHO by Syngenta Crop Protection AG, Basel, Switzerland.

Bomhard, E., Loeser, E., Schilde, B. (1979) 1,2,4-triazole (metabolite of CGA 64250) - subchronic toxicological study with rats (feeding study over three months).Unpublished report No. 8667 from Syngenta Crop Protection AG, Basel, Switzerland and Bayer AG Toxicological Institute, Wuppertal-Elberfeld, Germany. Submitted to WHO by Syngenta Crop Protection AG, Basel, Switzerland.

Capps, T.M., McFarland, J.E. & Cassidy, J.E. (1988) Metabolism of triazole-^{14}C and phenyl-^{14}C-CGA 169374 in the rat: distribution of radioactivity.Unpublished report No. ABR-88043 from Novartis Crop Protection AG, Basel, Switzerland and Ciba-Geigy Corp., Greensboro, USA. Submitted to WHO by Syngenta Crop Protection AG, Basel, Switzerland.

Capps, T.M., Barr, H.P. & Carlin, T.J. (1990) Characterization and identification of major triazole-^{14}C and phenyl-^{14}C-CGA 169374 metabolites in rats. Unpublished report No. ABR-90019 from Novartis Crop Protection AG, Basel, Switzerland and Ciba-Geigy Corp., Greensboro, USA. Submitted to WHO by Syngenta Crop Protection AG, Basel, Switzerland.

Clapp, M.J.L., Killick, M.E., Hollis, K.J and Godley, M.J. (1983) Triazole alanine: teratogenicity study in the rat. Unpublished report No. CTL/P/875 from Novartis Crop Protection AG, Basel, Switzerland and Central Toxicology Laboratory (CTL), Cheshire, England. Submitted to WHO by Syngenta Crop Protection AG, Basel, Switzerland.

Cox, R.H. (1987a) CGA 169374 tech.: subchronic toxicity/metabolism study in mice. Unpublished report No. 483-241 from Novartis Crop Protection AG, Basel, Switzerland and Hazleton Laboratories America Inc., Vienna, USA. Submitted to WHO by Syngenta Crop Protection AG, Basel, Switzerland.

Cox, R.H. (1987b) CGA 169374 tech.: subchronic toxicity/metabolism study in rats. Unpublished report No. 483-242 from Novartis Crop Protection AG, Basel, Switzerland and Hazleton Laboratories America Inc., Vienna, USA. Submitted to WHO by Syngenta Crop Protection AG, Basel, Switzerland.

Cox, R.H. (1989a) CGA 169374 tech.: oncogenicity study in mice. Unpublished report No. 483-250 from Novartis Crop Protection AG, Basel, Switzerland and Hazleton Laboratories America Inc., Vienna, USA. Submitted to WHO by Syngenta Crop Protection AG, Basel, Switzerland.

Cox, R.H. (1989b) Combined chronic toxicity and oncogenicity study of CGA 169374 technical in rats. Unpublished report No. 483-249 from Novartis Crop Protection AG, Basel, Switzerland and Hazleton Laboratories America Inc., Vienna, USA. Submitted to WHO by Syngenta Crop Protection AG, Basel, Switzerland.

Craine, E.M. (1987a) Metabolism of triazole 14C-CGA 169374 in the rat. Unpublished report No. WIL-82014 from Novartis Crop Protection AG, Basel, Switzerland and WIL Research Lab. Inc., Ashland, USA. Submitted to WHO by Syngenta Crop Protection AG, Basel, Switzerland.

Craine, E.M. (1987b) Metabolism of phenyl 14C-CGA 169374 in the rat. Unpublished report No WIL-82013 from Novartis Crop Protection AG, Basel, Switzerland and WIL Research Lab. Inc., Ashland, USA. Submitted to WHO by Syngenta Crop Protection AG, Basel, Switzerland.

Deparade, E. (1984) CGA 142856; salmonella/mammalian-microsome mutagenicity test. Unpublished report No. 840864 from Novartis Crop Protection AG, Basel, Switzerland and Ciba-Geigy Ltd, Basel, Switzerland. Submitted to WHO by Syngenta.

Deparade, E. (1986) CGA 131013 tech.: Salmonella/mammalian-microsome mutagenicity test. Unpublished report No. 860187 from Novartis Crop Protection AG, Basel, Switzerland and Ciba-Geigy Ltd, Basel, Switzerland. Submitted to WHO by Syngenta Crop Protection AG, Basel, Switzerland.

Dollenmeier, P. (1986a) CGA 169374 tech.: L5178Y/TK+/- mouse lymphoma mutagenicity test. Unpublished report No. 850570 from Novartis Crop Protection AG, Basel, Switzerland and Ciba-Geigy Ltd, Basel, Switzerland.. Submitted to WHO by Syngenta Crop Protection AG, Basel, Switzerland.

Dollenmeier, P. (1986b) CGA 131013 tech.: point mutation test with Chinese hamster cells V79. Unpublished report No. 860258 from Novartis Crop Protection AG, Basel, Switzerland and Ciba-Geigy Ltd, Basel, Switzerland.. Submitted to WHO by Syngenta Crop Protection AG, Basel, Switzerland.

Ecker, W. (1980) Biotransformation of 1,2,4-(3(5)-14C) triazole in rats. Unpublished report No. PF 1471 from Novartis Crop Protection AG, Basel, Switzerland. Submitted to WHO by Syngenta Crop Protection AG, Basel, Switzerland.

Esumi, Y. (1992) Absorption, distribution and excretion of CGA 169374 in rats. Unpublished report No. AE-1488 from Novartis Crop Protection AG, Basel, Switzerland and Tokai Research Lab. Daiichi Pure Chemicals Co., Ltd, Muramatsu, Japan. Submitted to WHO by Syngenta Crop Protection AG, Basel, Switzerland.

Fox, V. (2001) CGA 169374 tech.: in vitro cytogenetic assay in human lymphocytes. Unpublished report No. SV1090 from Syngenta Crop Protection AG, Basel, Switzerland and Central Toxicology Laboratory (CTL), Cheshire, England. Submitted to WHO by Syngenta Crop Protection AG, Basel, Switzerland.

Frosch, I. (1998) Evaluation of skin sensitization by 1,2,4-triazole with guinea-pig maximization test. Unpublished report No. ToxLabs/1998/7050 SEN from SKW Stickstoffwerke Piesteritz GmbH, Cunnersdorf, Germany and Toxlabs Prüflabor GmbH, Greppin, Germany. Submitted to WHO by Syngenta Crop Protection AG, Basel, Switzerland.

Gerspach, R. (2000) CGA 169374 tech.: 28-day repeated dose dermal toxicity study in rats. Unpublished report No. 993072 from Novartis Crop Protection AG, Basel, Switzerland and Novartis Crop Protection AG, Stein, Switzerland. Submitted to WHO by Syngenta Crop Protection AG, Basel, Switzerland.

Giknis, M.L.A. (1988) CGA 169374 tech.: a two-generation reproductive study in albino rats. Unpublished report No. MIN 862091 from Novartis Crop Protection AG, Basel, Switzerland and Ciba-Geigy Corp., Summit, USA. Submitted to WHO by Syngenta Crop Protection AG, Basel, Switzerland.

Glaza, S.M. (1991a) Primary dermal irritation study of CGA 169374 technical in rabbits. Unpublished report No. HWI 10503687 from Novartis Crop Protection AG, Basel, Switzerland and Hazleton Laboratories, Madison, USA, Submitted to WHO by Syngenta Crop Protection AG, Basel, Switzerland.

Glaza, S.M. (1991b) Primary eye irritation study of CGA 169374 technical in rabbits. Unpublished report No. HWI 10503688 from Novartis Crop Protection AG, Basel, Switzerland and Hazleton Laboratories, Madison, USA. Submitted to WHO by Syngenta.

Hamboeck, H. (1983a) CGA 131013 tech.: distribution, degradation and excretion of D,L-2-amino-3-(1-H-1,2,4-triazol-1-yl)-propanoic acid (D,L-triazolylalanine) in the rat. Unpublished report No. 1/83 from Novartis Crop Protection AG, Basel, Switzerland and Ciba-Geigy Ltd, Basel, Switzerland. Submitted to WHO by Syngenta Crop Protection AG, Basel, Switzerland.

Hamboeck, H. (1983b) CGA 131013 tech.: the metabolism of D,L-2-amino-3-(1H-1,2,4-triazol-1-yl)-propanoic acid (D,L-triazolylalanine) in the rat. Unpublished report No. 11/83 from Novartis Crop Protection AG, Basel, Switzerland and Ciba-Geigy Ltd, Basel, Switzerland. Submitted to WHO by Syngenta Crop Protection AG, Basel, Switzerland.

Hartmann, H.R. (1990) CGA 169374 tech.: acute oral toxicity in the mouse. Unpublished report No. 891514 from Novartis Crop Protection AG, Basel, Switzerland and Ciba-Geigy Ltd, Stein, Switzerland. Submitted to WHO by Syngenta Crop Protection AG, Basel, Switzerland.

Hartmann, H.R. (1991) CGA 169374 tech.: acute inhalation toxicity in the rat. Unpublished report No. 901476 from Novartis Crop Protection AG, Basel, Switzerland and Ciba-Geigy Ltd, Stein, Switzerland. Submitted to WHO by Syngenta Crop Protection AG, Basel, Switzerland.

Hassler, S. (2003a) Disposition of [4-Chloro-phenoxy-U-^{14}C] CGA 169374 in the rat after multiple oral administrations. Unpublished report No. 051AM03 from Syngenta Crop Protection AG, Basel, Switzerland and Central Toxicology Laboratory (CTL), Cheshire, England. Submitted to WHO by Syngenta Crop Protection AG, Basel, Switzerland.

Hassler, S. (2003b) Dermal absorption of [triazole-U-^{14}C] CGA 169374 formulated as SCORE R 250 EC (A-7402 G) in the rat (in vivo). Unpublished report No. 051AM01 from Syngenta Crop Protection AG, Basel, Switzerland. Submitted to WHO by Syngenta Crop Protection AG, Basel, Switzerland.

Hassler, S. (2003c) The percutaneous penetration of [triazole-U-^{14}C] CGA 169374 formulated as SCORE R 250 EC (A-7402 G) through rat and human split-thickeness skin membranes (in vitro). Unpublished report No. 051AM02 from Syngenta Crop Protection AG, Basel, Switzerland. Submitted to WHO by Syngenta Crop Protection AG, Basel, Switzerland.

Henderson, C. & Parkinson, G.R. (1980) R152056: acute oral toxicity to rats. Unpublished report No. CTL/P/600 from Novartis Crop Protection AG, Basel, Switzerland and Central Toxicology Laboratory (CTL), Cheshire, England. Submitted to WHO by Syngenta Crop Protection AG, Basel, Switzerland.

Herbold, B. (1983a) THS 2212 triazolylalanine: salmonella/microsome test for point mutagenic effect. Unpublished report No. 11388 from Novartis Crop Protection AG, Basel, Switzerland and Bayer AG Toxicological Institute, Wuppertal-Elberfeld, Germany. Submitted to WHO by Syngenta Crop Protection AG, Basel, Switzerland.

Herbold, B. (1983b) THS 2212 triazolylalanine: Pol A1-test on E. coli during testing for effects harmful to DNA. Unpublished report No. 11390 from Novartis Crop Protection AG, Basel, Switzerland and Bayer AG Toxicological Institute, Wuppertal-Elberfeld, Germany. Submitted to WHO by Syngenta Crop Protection AG, Basel, Switzerland.

Herbold, B. (1983c) THS 2212 triazolylalanine: micronucleus test for mutagenic effect on mice. Unpublished report No 11054 from Novartis Crop Protection AG, Basel, Switzerland and Bayer AG Toxicological Institute, Wuppertal-Elberfeld, Germany. Submitted to WHO by Syngenta Crop Protection AG, Basel, Switzerland.

Hertner, T. (1992) CGA 169374 tech.: autoradiographic DNA repair test on rat hepatocytes in vitro. Unpublished report No. 923124 from Novartis Crop Protection AG, Basel, Switzerland and Ciba-Geigy Ltd, Basel, Switzerland. Submitted to WHO by Syngenta Crop Protection AG, Basel, Switzerland.

Hertner, T. (1993) CGA 131013 tech.: salmonella and escherichia/liver-microsome test. Unpublished report No. Report No 933002 from Novartis Crop Protection AG, Basel, Switzerland and Ciba-Geigy Ltd, Basel, Switzerland. Submitted to WHO by Syngenta Crop Protection AG, Basel, Switzerland.

Hoberman, A. (2004) Oral (stomach tube) developmental toxicity study of 1,2,4-triazole in rabbits. Unpublished report No. VCB00002 from Syngenta Crop Protection AG, Basel, Switzerland and Argus Research Lab. Inc., Horsham, USA. Submitted to WHO by Syngenta Crop Protection AG, Basel, Switzerland.

Hummel, H.E., Youreneff, M.A., Giknis, M.L.A. & Yau, E.T. (1987) CGA 169374 tech.: teratology study in rabbits. Unpublished report No. MIN 862107 from Novartis Crop Protection AG, Basel, Switzerland and Ciba-Geigy Corp., Summit, USA. Submitted to WHO by Syngenta Crop Protection AG, Basel, Switzerland.

Janiak, T., Frei, T., Luetkemeier, H., Vogel, O., Pappritz, G. & Mladenovic, P. (1989) CGA 169374 tech.: oral toxicity (feeding) study in dogs - assessment of cataractogenic potential. Unpublished report No. RCC 097132A from Novartis Crop Protection AG, Basel, Switzerland and RCC Ltd, Itingen, Switzerland. Submitted to WHO by Syngenta Crop Protection AG, Basel, Switzerland.

Lai, K. & Simoneaux, B. (1986a) Balance study of ^{14}C-triazole in orally dosed rats. Unpublished report No. ABR-86021 from Novartis Crop Protection AG, Basel, Switzerland and Ciba-Geigy Corp., Greensboro, USA. Submitted to WHO by Syngenta Crop Protection AG, Basel, Switzerland.

Lai, K. & Simoneaux, B. (1986b) Balance study of ^{14}C-triazole acetic acid in orally dosed rats. Unpublished report No. ABR-86022 from Novartis Crop Protection AG, Basel, Switzerland and Ciba-Geigy Corp., Greensboro, USA. Submitted to WHO by Syngenta Crop Protection AG, Basel, Switzerland.

Lai, K. & Simoneaux, B. (1986c) The metabolism of ^{14}C-triazole acetic acid in the rat. Unpublished report No. ABR-86028 from Novartis Crop Protection AG, Basel, Switzerland and Ciba-Geigy Corp., Greensboro, USA. Submitted to WHO by Syngenta Crop Protection AG, Basel, Switzerland.

Lloyd, M. (2001) CGA 169374 tech.: induction of chromosome aberrations in cultured Chinese hamster ovary (CHO) cells. Unpublished report No. 252/293-D6172 from Syngenta Crop Protection AG, Basel, Switzerland and Covance Laboratories, North Yorkshire, England., Submitted to WHO by Syngenta Crop Protection AG, Basel, Switzerland.

Lochry, E.A. (1987) Developmental toxicity study of CGA 169374 technical (FL-851406) administered orally via gavage to Crl:COBS CD (SD)BR presumed pregnant rats. Unpublished report No. No 203-005 from Novartis Crop Protection AG, Basel, Switzerland and Argus Research Lab. Inc., Horsham, USA. Submitted to WHO by Syngenta Crop Protection AG, Basel, Switzerland.

Maruhn, D. & Bomhard, E. (1984) Triazolylalanine (THS 2212): study for subchronic toxicity to rats. Unpublished report No. 12397 from Novartis Crop Protection AG, Basel, Switzerland and Bayer AG Toxicological Institute, Wuppertal-Elberfeld, Germany. Submitted to WHO by Syngenta Crop Protection AG, Basel, Switzerland.

Mastrocco, F., Ricci, J.M., Huber, K.R., Schiavo, D.M., Hazelette, J.R. & Green, J.D. (1987a) CGA 169374 tech.: acute dermal toxicity study in rabbits. Unpublished report No. MIN 862122 from Novartis Crop Protection AG, Basel, Switzerland and Ciba-Geigy Corp., Summit, USA. Submitted to WHO by Syngenta Crop Protection AG, Basel, Switzerland.

Mastrocco, F., Ricci, J.M., Huber, K.R., Schiavo, D.M., Hazelette, J.R and Green, J.D. (1987b) CGA 169374 tech.: dermal sensitization study in female guinea-pigs Unpublished report No. MIN 862076 from Novartis Crop Protection AG, Basel, Switzerland and Ciba-Geigy Corp., Summit, USA. Submitted to WHO by Syngenta Crop Protection AG, Basel, Switzerland.

Melly, J. & Lohse, K. (1982) Microbial mutagen test. Unpublished report No. 81R-252 from Syngenta Crop Protection AG, Basel, Switzerland and Rohm and Haas, Philadelphia, USA. Submitted to WHO by Syngenta Crop Protection AG, Basel, Switzerland.

Menegola, E., Broccia, M.L. Di Renzo, F. & Giavini, E. (2001) Antifungal triazoles induce malformations in vitro. *Reprod. Toxicol.*, 15, 421–427.

Mihail, F. (1982) Triazolylalanine (THS 2212): expanded version of Bayer report No. 11229, dated October 19, 1982. Unpublished report No. 11229A from Novartis Crop Protection AG, Basel, Switzerland Bayer AG Toxicological Institute, Wuppertal-Elberfeld, Germany. Submitted to WHO by Syngenta Crop Protection AG, Basel, Switzerland.

Mihail, F. & Vogel, O. (1983) Triazolylalanine (THS 2212): subacute oral toxicity study on rats. Unpublished report No. 11491 from Novartis Crop Protection AG, Basel, Switzerland and Bayer AG Toxicological Institute, Wuppertal-Elberfeld, Germany. Submitted to WHO by Syngenta Crop Protection AG, Basel, Switzerland.

Milburn, G., Birtley, R., Pate, I., Hollis, K. & Moreland, S. (1986) Triazole alanine: two-generation reproduction study in the rat. Unpublished report No. CTL/P/1168 (revised) from Novartis Crop Protection AG, Basel, Switzerland and Central Toxicology Laboratory (CTL), Cheshire, England. Submitted to WHO by Syngenta Crop Protection AG, Basel, Switzerland.

Nakajima, M. (1991a) Reverse mutation assay of CGA-189138. Unpublished report No. 1809 from Novartis Crop Protection AG, Basel, Switzerland and Biosafety Research Center of Japan, Japan. Submitted to WHO by Syngenta Crop Protection AG, Basel, Switzerland.

Nakajima, M. (1991b) Reverse mutation assay of CGA-205374. Unpublished report No. 1746 from Novartis Crop Protection AG, Basel, Switzerland and Biosafety Research Center of Japan, Japan. Submitted to WHO by Syngenta Crop Protection AG, Basel, Switzerland.

Nakajima, M. (1991c) Reverse mutation assay of CGA-205375. Unpublished report No. 1747 from Novartis Crop Protection AG, Basel, Switzerland and Biosafety Research Center of Japan, Japan. Submitted to WHO by Syngenta Crop Protection AG, Basel, Switzerland.

O'Connor, D.J., McCormick, G.C. & Green, J.D. (1987) CGA 169374 tech.: 26-week oral toxicity study in dogs. Unpublished report No. MIN 852197 from Novartis Crop Protection AG, Basel, Switzerland and Ciba-Geigy Corp., Summit, USA. Submitted to WHO by Syngenta Crop Protection AG, Basel, Switzerland

Ogorek, B. (1990) CGA 169374 tech.: salmonella and escherichia/liver-microsome test. Unpublished report No. 901061 from Novartis Crop Protection AG, Basel, Switzerland and Ciba-Geigy Ltd, Basel, Switzerland. Submitted to WHO by Syngenta Crop Protection AG, Basel, Switzerland.

Ogorek, B. (1991) CGA 169374 tech.: micronucleus test, mouse. Unpublished report No. 911041 from Novartis Crop Protection AG, Basel, Switzerland and Ciba-Geigy Ltd, Basel, Switzerland. Submitted to WHO by Syngenta Crop Protection AG, Basel, Switzerland.

Ogorek, B. (2001) CGA 169374 tech.: cytogenetic test on Chinese hamster cells in vitro. Unpublished report No. 20013013 from Syngenta Crop Protection AG, Basel, Switzerland and Syngenta Crop Protection AG, Stein, Switzerland. Submitted to WHO by Syngenta Crop Protection AG, Basel, Switzerland.

Ohba, K. (1991a) Acute oral toxicity study of CGA-205374 in mice. Unpublished report No. CG-1B250 from Ciba-Geigy Japan Ltd, Tokyo, Japan and Medical Research Laboratories, Saitama, Japan. Submitted to WHO by Syngenta Crop Protection AG, Basel, Switzerland.

Ohba, K. (1991b) Acute oral toxicity study of CGA-205375 in mice. Unpublished report No. CG-1B260 from Ciba-Geigy Japan Ltd, Tokyo, Japan and Medical Research Laboratories, Saitama, Japan. Submitted to WHO by Syngenta.

Pinto, P. (2006a) Difenoconazole technical (CGA169374) - acute neurotoxicity study in rats. Unpublished report No. AR7517-REG-R1T002709-05 from Syngenta Crop Protection AG, Basel, Switzerland and Central Toxicology Laboratory (CTL), Cheshire, England. Submitted to WHO by Syngenta Crop Protection AG, Basel, Switzerland.

Pinto, P. (2006b) Difenoconazole technical (CGA169374) - subchronic neurotoxicity study in rats. Unpublished report No. PR1330-REG-R1 from Syngenta Crop Protection AG, Basel, Switzerland and Central Toxicology Laboratory (CTL), Cheshire, England. Submitted to WHO by Syngenta Crop Protection AG, Basel, Switzerland.

Poth, A. (1989) *Salmonella typhimurium* reverse mutation assay with 1H-1,2,4-triazole. Unpublished report No. 158400 from Syngenta Crop Protection AG, Basel, Switzerland and Cytotest Cell Research GmbH & Co. KG, Rossdorf, Germany. Submitted to WHO by Syngenta Crop Protection AG, Basel, Switzerland.

Procopio, K. & Hamilton, J. (1992) Acute toxicity range-finding study. Unpublished report No. 81R 057A from Novartis Crop Protection AG, Basel, Switzerland and Rohm and Haas, Philadelphia, USA. Submitted to WHO by Syngenta Crop Protection AG, Basel, Switzerland.

Puri, E. (1986) CGA 131013 tech.: autoradiographic DNA repair test on rat hepatocytes. Unpublished report No. 860184 from Novartis Crop Protection AG, Basel, Switzerland and Ciba-Geigy Ltd, Basel, Switzerland. Submitted to WHO by Syngenta Crop Protection AG, Basel, Switzerland.

Renhof, M. (1988a) 1,2,4-Triazole - investigations into embryotoxic effects on rats after oral administration. Unpublished report No. 17401 from Novartis Crop Protection AG, Basel, Switzerland and Bayer AG Toxicological Institute, Wuppertal-Elberfeld, Germany. Submitted to WHO by Syngenta.

Renhof, M. (1988b) 1,2,4-Triazole - investigations into embryotoxic effects on rats after oral administration. Unpublished report No. 17402 from Novartis Crop Protection AG, Basel, Switzerland and Bayer AG Toxicological Institute, Wuppertal-Elberfeld, Germany. Submitted to WHO by Syngenta Crop Protection AG, Basel, Switzerland.

Richold, M., Allen, J.A., Williams, A. & Ransome, S.J. (1981) Cell transformation test for potential carcinogenicity of R152056. Unpublished report No. ICI 394A/81153 from Novartis Crop Protection AG, Basel, Switzerland and Huntingdon Research Centre Ltd, Huntingdon, England. Submitted to WHO by Syngenta Crop Protection AG, Basel, Switzerland.

Rudzki, M.W., McCormick, G.C. & Arthur, A.T. (1988) CGA 169374 tech.: chronic toxicity study in dogs. Unpublished report No. MIN 862010 from Novartis Crop Protection AG, Basel, Switzerland and Ciba-Geigy Corp., Summit, USA. Submitted to WHO by Syngenta Crop Protection AG, Basel, Switzerland.

Schoch, M. & Schneider, M. (1989) CGA 169374 tech.: 56-day feeding cataractogenicity in young chickens. Unpublished report No. 871210 from Novartis Crop Protection AG, Basel, Switzerland and Ciba-Geigy Ltd, Stein, Switzerland. Submitted to WHO by Syngenta Crop Protection AG, Basel, Switzerland.

Strasser, F. (1985) CGA 169374 tech.: chromosome studies on human lymphocytes in vitro. Unpublished report No. 850569 from Novartis Crop Protection AG, Basel, Switzerland and Ciba-Geigy Ltd, Basel, Switzerland. Submitted to WHO by Syngenta Crop Protection AG, Basel, Switzerland.

Strasser, F. (1986) CGA 131013 tech.: micronucleus test (Chinese hamster). Unpublished report No. 860185 from Novartis Crop Protection AG, Basel, Switzerland and Ciba-Geigy Ltd, Basel, Switzerland. Submitted to WHO by Syngenta Crop Protection AG, Basel, Switzerland.

Suter, P. (1986a) 28-Day cumulative oral toxicity (feeding) study with CGA 169374 in the rat. Unpublished report No. 040770 from Novartis Crop Protection AG, Basel, Switzerland and RCC Ltd, Itingen, Switzerland. Submitted to WHO by Syngenta Crop Protection AG, Basel, Switzerland.

Suter, P. (1986b) 13-Week oral toxicity (feeding) study with CGA 169374 in the rat. Unpublished report No. RCC 040757 from Novartis Crop Protection AG, Basel, Switzerland and RCC Ltd, Itingen, Switzerland. Submitted to WHO by Syngenta.

Thevenaz, P. (1984) CGA 142856: acute oral LD_{50} in the rat. Unpublished report No. 840887 from Novartis Crop Protection AG, Basel, Switzerland and Ciba-Geigy Ltd, Stein, Switzerland. Submitted to WHO by Syngenta Crop Protection AG, Basel, Switzerland.

Thevenaz, P. (1986) CGA 142856: 14-day subacute toxicity study in rats (dietary administration). Unpublished report No. 841140 from Novartis Crop Protection AG, Basel, Switzerland and Ciba-Geigy Ltd, Stein, Switzerland. Submitted to WHO by Syngenta Crop Protection AG, Basel, Switzerland.

Thomas, H. (1992) The effect of CGA 169374 tech. on selected biochemical and morphological liver parameters following subchronic administration to male mice. Unpublished report No. CB 91/15 from Novartis Crop Protection AG, Basel, Switzerland and Ciba-Geigy Ltd, Basel, Switzerland. Submitted to WHO by Syngenta Crop Protection AG, Basel, Switzerland.

Thyssen, J. & Kimmerli, G. (1976) 1,2,4-Triazole occupational toxicology report. Unpublished report No. 5926 from Bayer AG, Leverkusen, Germany and Bayer AG Toxicological Institute, Wuppertal-Elberfeld, Germany. Submitted to WHO by Syngenta Crop Protection AG, Basel, Switzerland.

von, Keutz E. & Groening, P. (1984) THS 2212 (triazolylalanine): subchronic toxicity to dogs on oral administration. Unpublished report No. 12562 from Novartis Crop Protection AG, Basel, Switzerland and Bayer AG Toxicological Institute, Wuppertal-Elberfeld, Germany. Submitted to WHO by Syngenta Crop Protection AG, Basel, Switzerland.

Wahle, B.S. (2004a) A subacute toxicity testing study in the CD-1 mouse with 1,2,4-triazole. Unpublished report No. 200808 from Syngenta Crop Protection AG, Basel, Switzerland and Bayer Crop Science LP, Stilwell, USA. Submitted to WHO by Syngenta Crop Protection AG, Basel, Switzerland.

Wahle, B.S. (2004b) A subchronic toxicity testing study in the CD-1 mouse with 1,2,4-triazole. Unpublished report No. 201052 from Syngenta Crop Protection AG, Basel, Switzerland and Bayer Crop Science LP, Stilwell, USA. Submitted to WHO by Syngenta Crop Protection AG, Basel, Switzerland.

Wahle, B.S. & Sheets, L.P. (2004) A combined subchronic toxicity/neurotoxicity screening study in the Wistar rat with 1,2,4-triazole. Unpublished report No. 201024 from Syngenta Crop Protection AG, Basel, Switzerland and Bayer Crop Science LP, Stilwell, USA. Submitted to WHO by Syngenta Crop Protection AG, Basel, Switzerland.

Watanabe, M. (1993) CGA 131013: DNA repair test (rec assay). Unpublished report No. IET 93-0010 from Novartis Crop Protection AG, Basel, Switzerland and the Institute of Environmental Toxicology, Tokyo, Japan. Submitted to WHO by Syngenta Crop Protection AG, Basel, Switzerland.

Watkins, P. (1982) R152056: 3-(1,2,4-triazol-l-yl) alanine (ICI 156,342) – micronucleus test in CBC F1 mice. Unpublished report No. CTL/C/1164 from Novartis Crop Protection AG, Basel, Switzerland and Central Toxicology Laboratory (CTL), Cheshire, England. Submitted to WHO by Syngenta Crop Protection AG, Basel, Switzerland.

Weber, H., Patzschke, K. & Wegner, L. (1978) 1,2,4-Triazole-14C - biokinetic studies on rats. Unpublished report No. 7920 from Brian Christen Companies, Inc., Minnetonka, USA. Submitted to WHO by Syngenta Crop Protection AG, Basel, Switzerland.

Young, A. & Sheets, L.P. (2005) A two-generation reproductive toxicity study in the Wistar rat with 1,2,4-triazole. Unpublished report No. 201220 from Syngenta Crop Protection AG, Basel, Switzerland and Bayer Crop Science LP, Stilwell, USA. Submitted to WHO by Syngenta Crop Protection AG, Basel, Switzerland.

DIMETHOMORPH

First draft prepared by
Jürg Zarn[1] and Maria Tasheva[2]

[1]Food Toxicology Section, Swiss Federal Office of Public Health, Zurich, Switzerland; and
[2]National Center of Public Health Protection, Sofia, Bulgaria

Explanation

Dimethomorph is a cinnamic acid derivative for which the chemical name is (*E,Z*)-4-[3-(4-chlorophenyl)-3-(3,4-dimethoxyphenyl)acryloyl]morpholine or (*EZ*)-4-[3-(4-chlorophenyl)-3-(3,4-dimethoxyphenyl)-1-oxo-2-propenyl]morpholine, according to International Union of Pure and Applied Chemistry (IUPAC) and Chemical Abstract Service (CAS) nomenclatures respectively (CAS No. 110488-70-5). Dimethomorph is a mixture of *E* and *Z* isomers in the ratio of approximately 1 : 1.

Dimethomorph is a fungicide that disrupts fungal cell-wall formation. Fungicidal activity is primarily associated with the *Z* isomer.

Dimethomorph has not been evaluated previously by the JMPR and was reviewed at the present Meeting at the request of the Codex Committee on Pesticide Residues (CCPR). All pivotal studies with dimethomorph were certified as complying with good laboratory practice (GLP).

Evaluation for acceptable daily intake

1. Biochemical aspects

1.1 Absorption, distribution, metabolism and excretion

Absorption, distribution, metabolism and excretion of [^{14}C]dimethomorph was studied in male and female Sprague-Dawley CD rats. [^{14}C]Dimethomorph labelled uniformly on the chlorophenyl ring (radiochemical purity, 98.5%) had a specific activity of 0.647 GBq/mmol and the *E* : *Z* isomer ratio was 44–47 : 56–53. Non-labelled dimethomorph had a purity of 99.2% and the isomer ratio was virtually 50 : 50.

Figure 1.* E- *and* Z-*isomers of dimethomorph

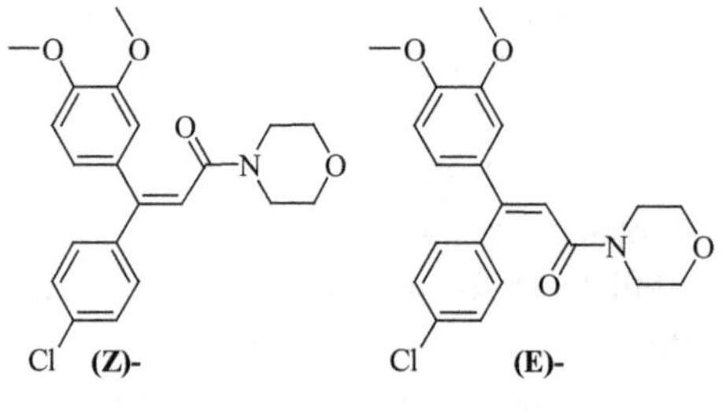

Cl (Z)-

Cl (E)-

Table 1. Dosing regimen for a study in rats treated with [^{14}C]dimethomorph

Group	No. of rats of each sex	Dose (mg/kg bw)	Remark
A	1	500	Single dose, expired air
B	5	10	Single dose
C	5	10	Repeated doses of non-labelled dimethomorph for 14 days, followed by a single radiolabelled dose
D	5	500	Single dose
E	5	10	Repeated doses of radiolabelled dimethomorph for 7 consecutive days

From Schluter (1990)

Five different treatment groups were investigated (Table 1). The study complied with GLP.

Within 24 h no radioactivity was found in the expired air of animals in group A.

In the group receiving a single lower dose (group B), the group receiving a single higher dose (group D), and the group receiving pre-treatment with non-radiolabelled dimethomorph (group C), nearly 90% of the radioactivity administered was excreted via the faeces within 3 days (Table 2). While only marginal differences in faecal excretion were observed in males and females, urinary excretion was 6–7% in males and 10–16% in females. Residual radioactivity in the carcass was low, ranging from 0.08% to 0.43% of the administered dose.

One male and one female in group E were sacrificed 1, 6, 24, 48 or 120 h after the last administration of radiolabelled dimethomorph and organs and tissues were analysed for radiolabel. Radioactivity at a concentration of greater than 0.01% of the administered dose after 1 day was only found in the liver, the muscles, fat and the gastrointestinal tract (GIT). Except for liver and GIT, residual radioactivity was less than 1% of the administered dose after 1 day. Therefore, only liver, kidney and fat of rats in groups B, D, and C were investigated for radioactivity on day 7 after treatment (Table 3). From the results for groups B, C, D and E, no potential for accumulation was evident.

Table 2. Excretion patterns in rats treated with [^{14}C]dimethomorph (groups B, C and D)

Group	Day	Excretion (% of administered dose)					
		Males			Females		
		Faeces	Urine	Carcass	Faeces	Urine	Carcass
B	1	75.09	4.93	—	72.41	12.66	—
	2	13.1	0.51	—	12.4	0.75	—
	3	1.43	0.07	—	1.16	0.13	—
	4	0.24	0.03	—	0.21	0.04	—
	Total (7 days)	90.07 ± 2.87	5.57 ± 0.81	0.1 ± 0.02	86.37 ± 4.43	13.62 ± 4.88	0.08 ± 0.03
D	1	58.91	4.57	—	33.14	6.15	—
	2	25.44	1.45	—	39.39	2.67	—
	3	1.98	0.13	—	13.62	1.37	—
	4	0.27	0.04	—	2.36	0.2	—
	Total (7 days)	86.79 ± 9.62	6.23 ± 2.52	0.13 ± 0.04	88.91 ± 8.36	10.44 ± 3.55	0.19 ± 0.03
C	1	65.64	5.83	—	52.35	12.53	—
	2	16.36	1.48	—	22.31	3.11	—
	3	6.32	0.33	—	5.12	0.51	—
	4	0.56	0.05	—	0.58	0.08	—
	Total (7 days)	89.17 ± 1.73	7.74 ± 1.71	0.43 ± 0.03	80.61 ± 16.31	16.31 ± 1.3	0.36 ± 0.03

From Schluter (1990)

Table 3. Recovery of radioactivity in selected organs of rats treated with [^{14}C]dimethomorph

Group	Radioactivity on day 7 after treatment (% of administered dose)					
	Males			Females		
	Liver	Kidney	Fat	Liver	Kidney	Fat
B	0.088	0.002	0.001	0.050	0.002	0.003
D	0.020	0.001	0.002	0.043	0.002	0.014
C	0.111	0.002	0.002	0.064	0.002	0.009

From Schluter (1990)

Profiles for urinary and faecal metabolites were not substantially different between sexes. In urine samples for group D, major metabolites were demethylated products in position 3 and 4 in the dimethoxyphenyl moiety and small amounts of unchanged dimethomorph as well as oxo metabolites of the morpholine ring. In the faeces of rats at the lower dose (groups B and C), 5–9% of the administered dose was recovered as demethylation products and another 5% was unchanged dimethomorph (see Figure 2). In rats receiving the higher dose (group D), dimethomorph accounted for up to 50% of the administered dose, with traces of demethylation products present (Schluter, 1990).

The absorption, distribution, metabolism and excretion of [^{14}C]dimethomorph via bile was studied in bile-duct cannulated male and female Sprague-Dawley CD rats. [^{14}C]Dimethomorph labelled uniformly at the chlorophenyl ring (radiochemical purity, 99.7%) had a specific activity of 0.338 GBq/mmol and the $E : Z$ isomer ratio was 57.9 : 42.1. Non-radiolabelled dimethomorph was of 99.2% purity and the isomer ratio $E : Z$ was 49.5 : 50.5. Rats were dosed by oral gavage as deszcribed in Table 4. The study complied with GLP.

In males and females at the lower dose, nearly 100% of the administered dose was absorbed and 92–95% was excreted within 48 h via the bile (Table 5). At the higher dose, absorption is saturated and, especially in females, large amounts of the dose (44.1%) remained in the GIT. Biliary excretion had a half-life of approximately 3 h at the lower dose and was saturated at the higher dose, with half-lives of 11 h for males and approximately 6 h for females.

In the bile of rats at the lower and higher doses, more than 10 metabolite fractions were identified, most of them being glucuronidated. Quantitatively, the most important metabolites in all treated groups were the demethylated compounds Z 67 and Z 69 (Figure 2) (Van Dijk, 1990).

In a supplementary study, dimethomorph metabolites in urine samples from rats of group B from the previous study (Schluter, 1990) were further characterized. The study complied with GLP.

As the major pathway, the demethylation of one methoxy group at either position 3 or 4 was identified with formation of conjugates (Figure 2). A second pathway included the oxidative degradation of the morpholine ring moiety (Schluter, 1991).

Table 4. Dosing regimen for bile-duct cannulated rats treated with [^{14}C]dimethomorph

Group	No. of males/females	Dose (mg/kg bw)
1	Four males	10
2	Three females	10
3	Three males	500
4	Three females	500

From Van Dijk (1990)

Table 5. Excretion and half-lives of biliary excretion in bile-duct cannulated rats treated with [^{14}C]dimethomorph

Dose (mg/kg bw)	Excretion after 48 h (% of administered dose)							
	Males				Females			
	Faeces	Urine	Bile	$t_{1/2}$ bile (h)	Faeces	Urine	Bile	$t_{1/2}$ bile (h)
10	7.6 ± 3.2	6.6 ± 4.9	95.1 ± 11.5	2.9	4.0 ± 1.0	6.9 ± 1.2	92.6 ± 2.6	3.4
500	21.8 ± 13.4	14.8 ± 3.0	49.1 ± 9.7	11.0	3.5 ± 4.2	8.6 ± 11.9	31.2 ± 15.7[a]	5.7

From Van Dijk (1990)

[a] 44.1% in the gastrointestinal tract and 10.1% in residual carcass.

To investigate the significance of the label position for the outcome of studies of excretion and metabolism, five male and five female Sprague-Dawley CD rats were given dimethomorph labelled uniformly on either the chlorophenyl ring or the morpholine ring as a single dose at 500 mg/kg bw by oral gavage. Chlorophenyl-ring labelled test compound (radiochemical purity, > 98%) had a specific activity of 0.647 GBq/mmol and the *E* : *Z* isomer ratio was 45.3% *E* and 54.7% *Z*. Morpholine-ring labeled test compound (radiochemical purity, > 98%) had a specific activity of 0.268 GBq/mmol and the *E* : *Z* isomer ratio was close to 50 : 50 (mixed from separated isomers). The study complied with GLP.

The excretion patterns of dimethomorph labelled at one of two different positions did not exhibit significant differences (Table 6). Most of the radioactivity was excreted within 3 days after administration, with faecal excretion being the major elimination pathway accounting for approximately 80% in females and approximately 90% in males. Urinary excretion was approximately 5% in males and approximately double (10%) in females. No difference in metabolite profiles between the two test compounds was identified, although an additional very minor metabolic pathway as cleavage of the amide bond was found with the chlorophenyl ring-labelled test compound (Schluter, 1993).

To further characterize the faecal metabolites, the metabolism and excretion of [^{14}C]dimethomorph was investigated in 10 male and 10 female Wistar rats (BRL-HAN) given a single dose at 50 mg/kg bw by gavage. Uniformly chlorophenyl-ring labelled [^{14}C]dimethomorph (radiochemical purity, 99%) had a specific activity of 0.327 GBq/mmol and the *E* : *Z* isomer ratio was 51 : 49. Non-radiolabelled dimethomorph was of 99.1% purity and the isomer ratio *E* : *Z* was 48 : 52. The study complied with GLP.

Again, faecal excretion was predominant while urinary excretion was less than 20% of the administered dose (Table 7). When compared with studies in Sprague-Dawley rats, rates of urinary excretion seemed to be higher in male Wistar rats (17.1% of the administered dose).

In metabolite analyses, apart from the known degradation products (see Figure 2) an additional metabolite (Z 98) was identified (Schluter & Grahl, 1994).

To investigate the pharmacokinetic features of dimethomorph, groups of four male and four female Sprague-Dawley rats (Crl:CD BR) were given [^{14}C]dimethomorph as single oral gavage doses at 10 mg/kg bw or 500 mg/kg bw, respectively. Test compound was uniformly chlorophenyl-ring labelled [^{14}C]dimethomorph (radiochemical purity, 96%) with a specific activity of 0.363 GBq/mmol

Table 6. Excretion patterns in rats given dimethomorph labelled with ^{14}C in two different positions

Position of radiolabel	Day	Excretion (% of administered dose)					
		Males			Females		
		Faeces	Urine	Total	Faeces	Urine	Total
Chlorophenyl ring	1	63.71	4.41	NR	34.08	6.13	NR
	2	25.87	1.30	NR	36.21	6.26	NR
	3	2.37	0.15	NR	11.63	0.87	NR
	Total	91.95 ± 3.62	5.86 ± 2.01	97.80 ± 1.91	81.92 ± 6.94	13.26 ± 4.73	95.2 ± 2.92
Morpholine ring	1	76.71	3.24	NR	44.35	6.01	NR
	2	13.89	0.86	NR	33.5	3.22	NR
	3	1.09	0.17	NR	4.49	0.34	NR
	Total	91.69 ± 1.50	4.27 ± 2.02	96.0 ± 1.54	82.34 ± 2.58	9.57 ± 1.10	91.9 ± 1.80

From Schluter (1993)
NR, not reported.

Table 7. Excretion pattern in rats treated with [^{14}C]dimethomorph labelled on the chlorophenyl ring

Day	Excretion (% of administered dose)			
	Males		Females	
	Faeces	Urine	Faeces	Urine
1	63.0	13.3	65.4	11.0
2	19.4	3.2	31.0	1.8
3	2.8	0.6	2.7	0.7
Total	85.2	17.1 ± 5.4	99.0	13.6 ± 3.7

From Schluter & Grahl (1994)

Figure 2. Proposed metabolic pathways for dimethomorph in the rat

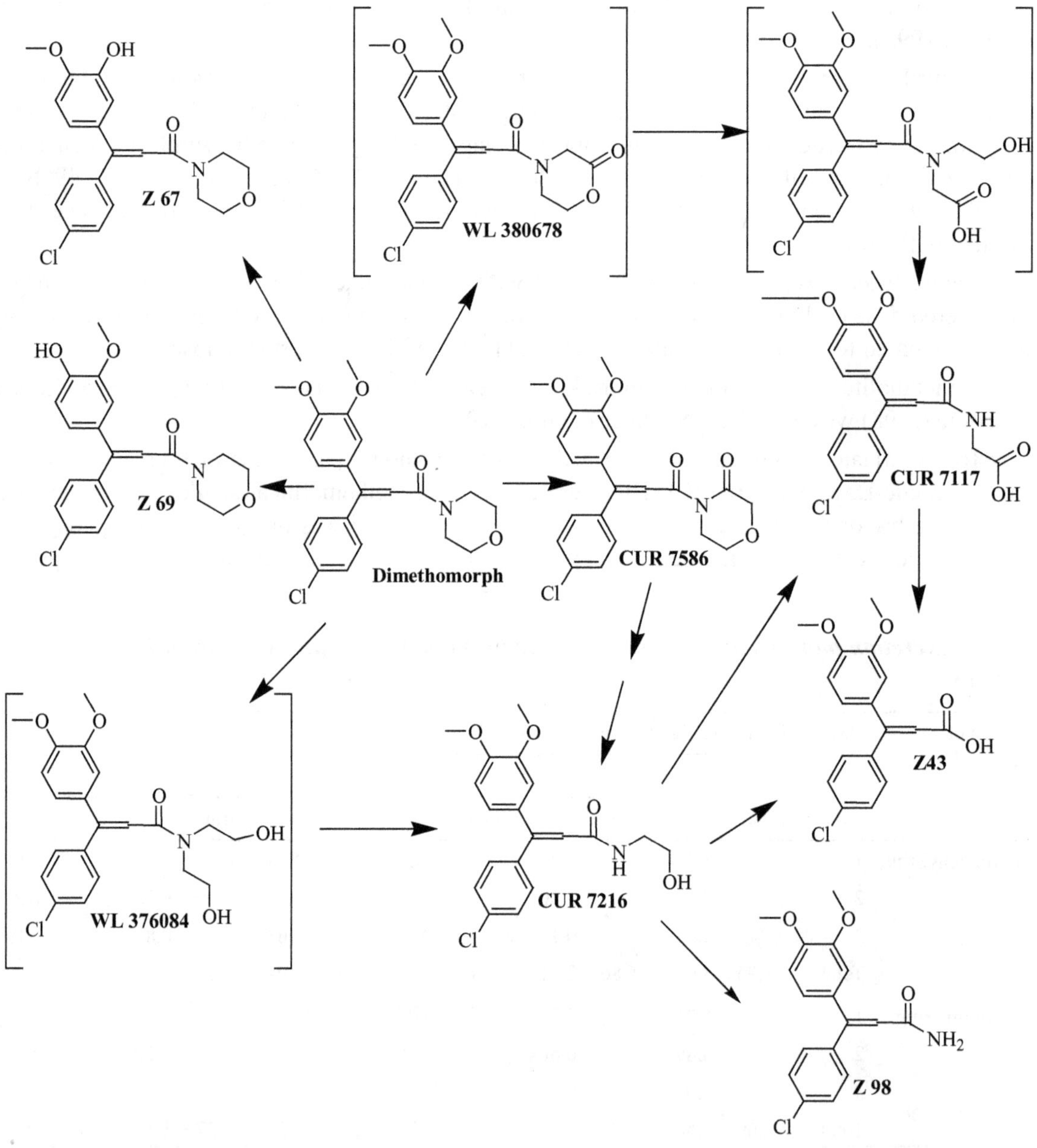

and the $E : Z$ isomer ratio was approximately 50 : 50 after dilution with non-radiolabelled dimethomorph. Non-labelled dimethomorph was 98.8% chemically pure. Another four males and four females served as a control group and were treated with vehicle only. Blood samples at different time intervals from 0.25 h to 144 h after dosing were collected and analysed for radioactivity. The study complied with GLP.

In the groups receiving the lower dose, values for both sexes were comparable; t_{max} was reached at < 3h, terminal $t_{1/2}$ was 59–68 h and $AUC_{0-\infty}$ at 10–15 µg × h/g (Table 8). In the groups receiving the higher dose, increased c_{max} and $AUC_{0-\infty}$ values were found in females and this correlated with a prolonged half-life.

Total radioactive residues (TRR) in plasma and erythrocytes 168 h after treatment were below the limit of detection (LOD) in both rats of groups at the lower dose and in the plasma of rats in groups receiving the higher dose (LOD at lower dose, 0.020–0.022 ppm; LOD at higher dose, 0.201–0.223 ppm). In the groups at the higher dose, TRR in erythrocytes were 0.75 ppm and 0.939 ppm in male and female rats, respectively (Afzal & Wu, 1995a).

To investigate the tissue distribution of dimethomorph, groups of nine male and nine female Sprague-Dawley rats (Crl:CD BR) were given [^{14}C]dimethomorph as single oral doses at 10 mg/kg bw or 500 mg/kg bw by gavage. The test compound was uniformly chlorophenyl-ring labelled [^{14}C] dimethomorph (radiochemical purity, 96%) and had a specific activity of 0.363 GBq/mmol and the $E : Z$ isomer ratio was approximately 50 : 50 after dilution with non-radiolabelled dimethomorph (purity, 98.8%). Another three males and three females served as a control group and were treated with vehicle only. At t_{max}, 24 h and 168 h after dosing, three males and three females from each group were killed and tissues collected for TRR analysis. T_{max} values were derived from a previous study (Afzal & Wu, 1995a) considering the lower bounds of t_{max} ranges in actual dosing groups; for females at the lower dose, t_{max} was set at 1.5 h; for males at the lower dose at 0.5 h; and for males and females at the higher dose at 8 h. The study complied with GLP.

No signs of toxicity were recorded in any group. Highest TRR levels in all dosed groups at t_{max} and 24 h after dosing were found in the GIT and contents, followed by the carcass and liver (Table 9 and Table 10). In liver, maximal TRR values were 1.05% of the administered dose. In all other organs, low levels of less than 0.1% were observed even at t_{max} and they were significantly depleted within 24 h in the groups at the lower dose. In the groups at the higher dose, a certain delay in depletion from organs within 24 h was observed when compared with the groups at the lower dose. Levels at t_{max} and 24 h were all less than 0.1% of the administered dose. At 168 h, liver contained the highest levels of residues, with a range of 0.04–0.1% TRR of the administered dose; in all other tissues, TRR was virtually negligible (Afzal & Wu, 1995b).

To investigate the dermal absorption of dimethomorph, groups of four male Sprague-Dawley CD rats were exposed dermally to [^{14}C]dimethomorph at a dose of 7.73 or 79.62 mg/kg bw for 8 h. The test compound was uniformly chlorophenyl-ring labelled [^{14}C]dimethomorph (radiochemical purity, 96%; purity, 98.8%) and had a specific activity of 0.363 GBq/mmol and the $E : Z$ isomer ratio was approximately 45 : 55. The chemical purity of dimethomorph before labelling was 97.6%. The study complied with GLP.

In the group at the lower dose, 4.75% of the applied dose was absorbed, based on the sum of radioactivity recovered 168 h after topical application in urine, faeces, cage wash, carcass, blood and untreated skin. In the group at the higher dose, 1.2% of the applied dose was absorbed (Bounds, 1995).

1.2 Effects on enzymes and other biochemical parameters

No information was available.

Table 9. Tissue distribution of radiolabel in groups of rats receiving [^{14}C]dimethomorph at a lower dose of 10 mg/kg bw

Tissue	Recovery (% of the total radioactive residues in the administered dose)					
	t_{max}		24 h		168 h	
	Males	Females	Males	Females	Males	Females
GIT contents	67.67	67.52	6.68	9.03	0.01	0.02
GIT	13.91	11.87	0.43	1.06	0.00	0.00
Carcass	7.47	6.89	0.33	0.26	0.04	0.01
Liver	5.36	3.29	0.09	0.75	0.10	0.07
Kidneys	0.25	0.20	0.02	0.02	0.00	0.00
Pancreas	0.10	0.07	0.00	0.00	0.00	0.00
Plasma	0.06	0.10	0.01	0.01	0.00	0.00
Erythrocytes	0.06	0.09	0.01	0.01	0.00	0.00
Heart	0.06	0.04	0.00	0.00	0.00	0.00
Lungs	0.06	0.06	0.00	0.00	0.00	0.00
Fat	0.04	0.09	0.00	0.00	0.00	0.00
Muscle	0.03	0.06	0.00	0.00	0.00	0.00
Spleen	0.02	0.02	0.00	0.00	0.00	0.00
Testes	0.02	—	0.00		0.00	—
Bone	0.01	0.02	0.00	0.00	0.00	0.00
Brain	0.01	0.02	0.00	0.00	0.00	0.00
Thymus	0.01	0.02	0.00	0.00	0.00	0.00
Adrenals	0.00	0.01	0.00	0.00	0.00	0.00
Bone marrow	0.00	0.00	0.00	0.00	0.00	0.00
Pituitary	0.00	0.00	0.00	0.00	0.00	0.00
Thyroid	0.00	0.00	0.00	0.00	0.00	0.00
Ovaries	—	0.01	—	0.00	—	0.00
Uterus	—	0.00	—	0.00	—	0.00
Total	95.16	90.39	6.37	11.14	0.15	0.10

From Bounds (1995)
GIT, gastrointestinal tract.

2. Toxicological studies

2.1 Acute toxicity

(a) Oral administration

Groups of five male and five female overnight fasted CD rats were given dimethomorph (purity not stated) as a single dose at 3200, 4000 or 5000 mg/kg bw in 0.1% Tween 80 by oral gavage. The study complied with GLP.

Mortalities occurred between 24 h and 48 h after dosing. One female died in the group at the lower dose, three males and four females died in the group at the intermediate dose, and three males and all females died in the group at the higher dose. Median lethality for males was at 4300 mg/kg bw, for females at 3500 mg/kg bw, and for both sexes combined, 3900 mg/kg bw. Animals found

Table 10. Tissue distribution of radiolabel in groups of rats receiving [^{14}C]dimethomorph at a higher dose of 10 mg/kg bw

Tissue	Recovery (% of the total radioactive residues in the administered dose)					
	t_{max}		24 h		168 h	
	Males	Females	Males	Females	Males	Females
GIT contents	74.40	72.33	22.16	21.76	0.01	0.01
GIT	5.30	3.87	2.19	3.35	0.00	0.00
Carcass	4.61	3.62	0.96	3.86	0.04	0.04
Liver	1.14	0.82	0.85	1.05	0.08	0.04
Kidneys	0.08	0.07	0.04	0.08	0.00	0.00
Erythrocytes	0.06	0.02	0.03	0.03	0.01	0.00
Plasma	0.05	0.03	0.04	0.04	0.00	0.00
Lungs	0.03	0.03	0.01	0.03	0.00	0.00
Fat	0.02	0.06	0.00	0.04	0.00	0.00
Heart	0.02	0.02	0.01	0.02	0.00	0.00
Pancreas	0.02	0.02	0.01	0.03	0.00	0.00
Testes	0.02	—	0.01	—	0.00	—
Bone	0.01	0.00	0.00	0.01	0.00	0.00
Muscle	0.01	0.00	0.00	0.02	0.00	0.00
Spleen	0.01	0.01	0.00	0.01	0.00	0.00
Thymus	0.01	0.01	0.00	0.01	0.00	0.00
Adrenals	0.00	0.00	0.00	0.00	0.00	0.00
Bone marrow	0.00	0.00	0.00	0.00	0.00	0.00
Brain	0.00	0.01	0.00	0.01	0.00	0.00
Pituitary	0.00	0.00	0.00	0.00	0.00	0.00
Thyroid	0.00	0.00	0.00	0.00	0.00	0.00
Ovaries	—	0.00	—	0.00	—	0.00
Uterus	—	0.02	—	0.01	—	0.00
Total	85.79	80.98	26.29	30.37	0.14	0.10

From Bounds (1995)
GIT, gastrointestinal tract.

dead had congested lungs and pallor of the liver, kidneys and spleen. Signs of toxicity were observed 30 min after dosing, including hunched posture, pilo-erection, abnormal gait, lethargy, decreased respiratory rates. Recovery was not observed before 10 days after dosing. Surviving animals showed no treatment-related pathological findings (Kynoch, 1985).

To investigate the possible differences in isomer-specific toxicity in Wistar rats, groups of five male and five female overnight fasted rats were given the dimethomorph *Z*-isomer (purity not stated) as a single doses at 0 or 5000 mg/kg bw in 0.1% Tween 80 by oral gavage. Rats were observed for 15 days. The study complied with GLP.

The only finding in the treated rats was pale faeces in the first 2 days after dosing. No signs of toxicity and no mortalities were recorded within the 2 weeks of observation. Therefore, the median lethal dose (LD_{50}) was greater than 5000 mg/kg bw for males and female (Heusener & Jacobs, 1987a).

In an additional study on possible isomer-specific acute toxicity in Wistar rats, groups of five male and five female overnight fasted rats were given dimethomorph *E*-isomer (purity not stated) as a single dose of 0, 4000 or 5000 mg/kg bw in 0.1% Tween 80 by oral gavage. Rats were observed for 15 days. The study complied with GLP.

In all dosed groups, signs of toxicity started within 1 h after dosing and included (among others) posture and locomotion disturbances, salivation, blood-crusted snouts and haemorrhagic lacrimation. Mortalities were observed between study day 2 and day 7; in the group at the lower dose, four females and no males died; in the group at the higher dose, two females and four males died. Findings in rats that died were restricted to the GIT and included erosions. In rats sacrificed at study termination, congested lungs and opaque eyes were found. The LD_{50} for the *E*-isomer was calculated to be 4715 mg/kg bw in males and 4754 mg/kg bw in females, in both sexes combined, 4472 mg/kg bw (Heusener & Jacobs, 1987b).

(b) Dermal application

To investigate the dermal toxicity of the dimethomorph *Z*-isomer in Wistar rats, groups of five males and five females were exposed dermally to the *Z*-isomer (purity, 98.7%) at a dose of 0 or 5000 mg/kg bw for 24 h and then observed for 15 days. The study complied with GLP.

No signs of toxicity and no mortalities were observed. Therefore, the LD_{50} was greater than 5000 mg/kg bw in males and females (Heusener & Eberstein, 1985).

Dermal toxicity was investigated in a second study in Fischer 344 rats. Groups of five males and five females exposed dermally to dimethomorph (purity, 98.5%; *E* : *Z*, 46 : 54) at a dose of 2000 mg/kg bw for 24 h and then observed for 14 days. The study complied with GLP.

No signs of toxicity and no mortalities were observed. Therefore, the LD_{50} was greater than 2000 mg/kg bw in males and females (Gardner, 1989)

(c) Exposure by inhalation

Groups of five male and five female Wistar rats were exposed by inhalation to dimethomorph *Z*-isomer (purity, 98.7%) at a concentration of 4.24 mg/l (the highest concentration attainable) in a whole-body chamber for 4 h. 56% of particles had a mean aerodynamic diameter of 5.5 µm or less. Animals were then observed for 14 consecutive days. The study complied with GLP.

All rats survived till study termination. During the exposure, the rats showed changes in posture and respiratory patterns. Noticeable breathing was resolved after 5 days. Within the first 2–3 days, reduced feed intake and reduced body-weight gain was found. Therefore, the median lethal concentration (LC_{50}) of the *Z*-isomer was greater than 4.24 mg/l for males and females (Jackson & Hardy, 1986).

(d) Dermal and ocular irritation

The dermal and ocular irritation potential of dimethomorph was investigated in New Zealand White (NZW) rabbits. The test compound was dimethomorph (purity, 98.5%; *E* : *Z*, 46 : 54). For dermal irritation, three rabbits (two males, one female) were exposed to 500 mg of test compound for 4 h under semi-occlusive conditions and observed for another 72 h after removal. For ocular irritation, three animals (one male, two females) were given 50 mg of test compound in a volume of 0.1 ml into the conjunctival sac of one eye. The eyes were not rinsed. The study complied with GLP.

In the test for dermal irritation, no signs of an irritating potential were recorded.

In the test for ocular irritation, all rabbits showed reddened conjunctivae and slight chemosis. These signs resolved within 4 h after dosing (Gardner, 1989).

(e) *Dermal sensitization*

The dermal sensitization potential of dimethomorph (purity, 98.0%) was investigated in male Crl (HA)BR guinea-pigs via the maximization test of Magnusson & Kligman. In two pre-test studies, four males received 0.1 ml of dimethomorph intradermally at a concentration of 1%, 5%, 10%, 15%, and 25% in mineral oil and another four males received dimethomorph topically at a concentration of 1%, 5%, 10%, and 25% in petrolatum occluded for 24 h by patches. All guinea-pigs in both groups received all doses and were evaluated for skin irritation 24 h and 48 h after application. In the group treated by intradermal injection, all doses induced dose-dependent skin irritations that were moderate-intense at 5%. In the group treated by topical application, no irritation was observed at any dose. For the main study, the concentrations of 5% for intradermal injection and 25% for topical application were selected.

Twenty males were induced intradermally with 5% dimethomorph in mineral oil and Freund complete adjuvant and then on day 9 topically with 25% dimethomorph in petrolatum. On day 23, the guinea-pigs were challenged by topical application of 25% dimethomorph in petrolatum. The negative-control group consisted of 20 males. The study complied with GLP.

No skin irritation was recorded in any of the challenged or control guinea-pigs and the Meeting concluded that dimethomorph has no skin sensitization potential (Glaza, 1997)

Table 11. Results of studies of acute toxicity with dimethomorph

Isomer	Species	Strain	Sex	Route	LD_{50} (mg/kg bw)	LC_{50} (mg/l)	Purity (%)	Reference
E : Z	Rat	CD	Male	Oral	4300	—	NS	Kynoch (1985)
E : *Z*		CD	Female	Oral	3500	—	NS	Kynoch (1985)
Z		Wistar	Males and females	Oral	> 5000	—	NS	Heusener & Jacobs (1987a)
E		Wistar	Male	Oral	4715	—	NS	Heusener & Jacobs (1987b)
E		Wistar	Female	Oral	4754	—	NS	Heusener & Jacobs (1987b)
E : *Z*		Wistar	Males and females	Dermal	> 5000	—	98.7	Heusener & Eberstein (1985)
E : *Z*		Fischer	Males and females	Dermal	> 2000	—	98.5	Gardner (1989)
E : *Z*		Wistar	Males and females	Inhalation	—	> 4.24	98.7	Jackson & Hardy (1986)
E : *Z*	Rabbit	NZW	Males and females	Dermal	Not irritating	—	98.5	Gardner (1989)
E : *Z*	Rabbit	NZW	Males and females	Ocular	Not irritating	—	98.5	Gardner (1989)
E : *Z*	Guinea-pig	Crl (HA)BR	Male	Dermal	No sensitization	—	98.0	Glaza (1997)

NS, not stated; NZW, New Zealand White.

2.2 *Short-term studies of toxicity*

Mouse

In a 6-week dose range-finding study, 10 male and 10 female CD-1 mice were initially fed diets containing dimethomorph (purity, 96.6%) at a concentration of 0, 300, 800, or 2000 ppm. Since there were no signs of toxicity in any groups, the doses were increased at treatment day 23 from 300 ppm to 10000 ppm in males and 8000 ppm in females, and from 2000 ppm to 5000 ppm in males and 4000 ppm in females for the remaining study period. The intended 4-week exposure was prolonged to 6 weeks. Clinical signs were recorded twice per day and body weight, feed and water consumption each week. At the end of the study, investigations of gross pathology were carried out and histopathology on livers was performed. The study complied with GLP.

There were no clinical signs of toxicity and body-weight development was unaffected in all groups. The only finding was a dose-related increase in relative liver weights, with no concomitant histological finding (Table 12).

On the basis of the pronounced liver weight increases in the groups at the intermediate and the higher dose, the NOAEL was 800 ppm, equal to 184 mg/kg bw per day (Perry, 1986).

Rats

In a 4-week dose range-finding feeding study, groups of five male and five female Sprague-Dawley rats were fed diets containing dimethomorph (purity, 99%) at a concentration of 0, 200, 1000, or 5000 ppm, equal to 0, 15.8, 80.9, and 305.9 mg/kg bw per day in males and 0, 17.5, 81.1, 283.5 mg/kg bw per day in females. Rats were examined twice per day for general health, moribundity and mortality, and detailed clinical examinations were performed daily. Feed and water consumption and body-weight development were recorded regularly. At the end of the study, haematology, clinical chemistry, and urine analysis were performed and organ weights and gross pathological changes were recorded. No statement on GLP compliance was provided.

In the groups receiving the highest dose, severe signs of toxicity were recorded as loose faeces, brown staining of the perinasal and perigenital regions, swollen abdomen, hunched posture, pilo-erection, emaciation, lethargy, bad grooming and unsteady gait. One male and two females died during the study and the remaining rats were killed before termination owing to severely impaired general health in week 4 after sampling for laboratory investigations. Males gained weight only minimally and females did not (Table 13). Statistical significant increases in neutrophils and platelets were recorded in males and females at the highest dose. In males, the only statistical significant change in clinical chemistry was an increase in urea nitrogen in the group at the highest dose. In

Table 12. Body-weight adjusted increases in liver weight in mice fed diets containing dimethomorph for 6 weeks

Sex	Body-weight increase (% of control)			
	Dietary concentration (ppm)			
	0	800	2000/5000 (males) 2000/4000 (females)	300/10000 (males) 300/8000(females)
Males	—	10	18**	33***
Females	—	15**	30***	40***

From Perry (1986)

** Significantly different from controls, $p < 0.01$

*** Significantly different from controls, $p < 0.001$

females at the highest dose, total protein and creatinine concentrations were decreased and urea nitrogen, potassium and phosphorus concentrations were increased. Urea nitrogen was also increased in females at the intermediate dose. In both sexes receiving the highest dose, urine volumes were significantly increased and protein concentrations were decreased in males—three out of five males did not excrete any protein in the urine. Similarly to rats found dead/killed in a moribund condition before termination that showed thickened and congested large and small intestines, rats at the highest dose at termination showed thickened stomachs and males also showed thickened ileums. Males at the highest dose had empty seminal vesicles. Absolute pituitary weights of males at the intermediate (9 ± 2.2 mg) and highest dose (9 ± 1.0 mg), and of females at the highest dose (7 ± 0.6 mg) were reduced relative to controls (controls: males, 14 ± 3.6 mg; females, 14 ± 2.1), as was the absolute uterus weight of females at the highest dose. In the group at the highest dose, absolute weights of adrenals were increased in males and relative liver weights were increased in females. No histopathology was performed on organs.

On the basis of decreases in absolute pituitary weight in males at 1000 ppm, the NOAEL was 200 ppm, equal to 15.8 mg/kg bw per day (Warren, 1985).

In a second 4-week dose range-finding feeding study, groups of 10 male and 10 female Sprague-Dawley rats were fed diets containing dimethomorph at a concentration of 0, 2000, 3000, or 4000 ppm, equal to approximately 0, 200, 290, and 370 mg/kg bw per day in males and 0, 220, 300, 420 mg/kg bw per day in females (purity, 96.6%). The rats were examined twice per day for general health, moribundity and mortality, and weekly detailed clinical examinations were performed. Feed and water consumption and body-weight development were recorded regularly. At the end of the study, organ weights were recorded and gross pathology performed. Additionally, histopathology on livers was performed. The study complied with GLP.

All rats survived until scheduled death. At the highest dose, males and females showed piloerection and swollen abdomens and perigenital staining were observed in females. In all dosed females and in males at the highest dose, statistically significant decreases in body-weight gains were recorded (Table 14). Feed consumption was reduced by 16% in males and 26% in females at the highest dose. Two males at the highest dose and two females at the intermediate dose and six females at the highest dose showed distended intestines, in some instances with thickened mucosa and often filled with fluids or gelatinous material. Body-weight adjusted organ weights were only significantly changed for livers of rats at 2000 ppm (males, 4%; females, 14%), at 3000 ppm (males, 10%; females, 25%) and at 4000 ppm (males, 13%; females, 36%). Increased dose-related incidences of liver hypertrophy were found in males at the intermediate and highest dose and in females at the lowest and intermediate dose and only in one female at the highest dose. In this group of females, at the highest dosehypertrophy may have been masked by vacuolation in the other females of this group.

On the basis of the body-weight and liver findings, the NOAEL was less than 2000 ppm, equal to approximately 200 mg/kg bw per day (Scott, 1986).

Table 13. Body-weight gain and feed consumption in rats fed diets containing dimethomorph for 4 weeks

Dietary concentration (ppm)	Dose (mg/kg bw)		Body-weight gain (g), weeks 0–4		Feed consumption (% of control)	
	Males	Females	Males	Females	Males	Females
0	0	0	176	51	—	—
200	15.8	17.5	157	60	99	100
1000	80.9	81.1	157	57	101	93
5000	305	283.2	23	0.3	62	53

From Warren (1985)

To investigate the contribution of the *E* and *Z* isomers to the toxicity of dimethomorph, two 4-week studies in rats treated by oral gavage were performed with both isomers. Gavage administration was chosen to reduce exposure of the test compounds to light, since photoisomerization is known to occur between the *E*- and *Z*-isomers of dimethomorph.

In the first 4-week study, groups of seven male and seven female Fischer 344 rats received dimethomorph *E*-isomer (purity, 99.5%) at a dose of 0, 10, 100, or 750 mg/kg bw per day. During the 4 weeks of the study, no changes in the composition of the test compound were observed. Clinical examinations were performed twice per day, feed consumption and body-weight development were recorded weekly. At the end of the study, haematology and clinical chemistry parameters were investigated and organ weights (brain, heart, liver, kidneys, spleen, adrenals, testes) were recorded and gross pathology performed. Histopathology was performed for rats in the control group and at the highest dose for adrenals, heart, intestines, liver (also for rats at the lowest and intermediate dose), kidneys, pituitary, spleen and stomach. The study complied with GLP.

There were no mortalities in the study. A minimal increase in body-weight gain was observed in males at the highest dose coinciding with a minimal increase in feed intake. In rats at the highest dose there were some statistical significant haematological changes, including reduced leukocyte counts in males, and mildly reduced haemoglobin and increased platelet counts and some changes in platelet-associated parameters in males and females. Statistical significant changes in clinical chemistry for males and females at the highest dose included increased concentrations of protein, urea, bilirubin, cholesterol, calcium, creatinine, and phosphate. In males at the intermediate and highest dose, there was also an increase in urea concentration, while females at the intermediate and at the highest dose showed an increased albumin concentration and increased activity of gamma-glutamyltransferase.

Significant decreases in spleen weight were observed in males at the highest dose, and in males at the intermediate and the highest dose liver and adrenal weights were also increased. In females at the highest dose, liver weights were increased, as were adrenal weights, although this was not statistically significant (Table 15). Macroscopic and histological changes found were caecal enlargements, liver enlargements, liver dark discolorations and liver fatty vacuolations in both sexes. In males, most findings were present in groups at the intermediate dose and the highest dose, but in females mostly only at the highest dose (Table 16).

Table 14. Body weight and body-weight gain in rats fed diets containing dimethomorph for 4 weeks

Dietary concentration (ppm)	Body weight (g)					Body-weight gain	
	Week 0	Week 1	Week 2	Week 3	Week 4	(g)	% of control
Males							
0	194	253	303	345	380	186	—
2000	194	242	285	324	357	163	88
3000	196	241*	286	323	353	157	84
4000	193	224***	264***	292***	312***	119	64
Females							
0	135	160	183	202	219	84	—
2000	133	150*	165**	179**	190**	57	68
3000	137	151	164**	170***	172***	35	42
4000	130	140***	151***	154***	155***	25	30

From Scott (1986)

* $p < 0.05$; ** $p < 0.01$; *** $p < 0.001$.

The NOAEL was 10 mg/kg bw per day on the basis of several changes in the liver in males and females at 100 mg/kg bw per day (Esdaile, 1991a).

In the second 4-week study, groups of seven male and seven female Fischer 344 rats received dimethomorph *Z*-isomer (purity, 95.6%) at a dose of 0, 10, 100, or 750 mg/kg bw per day. During the 4 weeks of the study, no changes in the composition of the test compound were observed. Clinical examinations were performed twice per day, feed consumption and body-weight development were recorded weekly. At the end of the study, haematology and clinical chemistry parameters were investigated and organ weights (brain, heart, liver, kidneys, spleen, adrenals, testes) were recorded and gross pathology performed. Histopathology was performed for rats in the control group and in the group at the highest dose, for the adrenals, brain, heart, intestines, liver (also for rats at the lowest and intermediate dose), kidneys, pituitary, spleen and stomach. The study complied with GLP.

All rats survived till study termination and body-weight gain was not affected. Minimal haematological changes in males and females in the groups at the intermediate and the highest dose showed no dose–response relationship and were therefore judged not to be treatment-related. Treatment-related changes in clinical chemistry in both sexes comprised increased plasma protein concentrations in the groups at the intermediate and highest dose, while bilirubin concentrations were increased in males at the intermediate and highest dose and in females only at the highest dose. In the

Table 15. Organ weights[a] in rats given dimethomorph* E*-isomer by gavage for 4 weeks

Organ	Organ weight (g)			
	Dose (mg/kg bw per day)			
	0	10	100	750
Males				
Spleen	0.4	0.42	0.41	0.36**
Liver	6.14	6.23	6.46*	8.14**
Adrenals	0.033	0.036	0.038*	0.044**
Females				
Liver	4.05	3.97	4.18	5.81**
Adrenals	0.045	0.043	0.045	0.049

From Esdaile (1991a)

[a] Organ weights were adjusted for terminal body weight.

* $p \leq 0.05$; **, $p \leq 0.01$.

Table 16. Macroscopic and histological changes in organs of rats given dimethomorph* E*-isomer by gavage for 4 weeks

Change	Dose (mg/kg bw per day)							
	Males				Females			
	0	10	100	750	0	10	100	750
Caecal enlargement	0	0	1	7	0	0	0	7
Liver enlargement	0	0	1	7	0	0	0	7
Liver, dark discoloration	0	0	2	5	0	0	0	1
Liver, patchy midzonal cytoplasmic lipid vacuolation	0	0	2	7	0	0	4	7

From Esdaile (1991a)

groups at the highest dose, females had increased cholesterol and males increased calcium and phosphate concentrations. Body-weight adjusted liver weights were increased in both sexes at 100 and 750 mg/kg bw (Table 17). As with the *E*-isomer, caecal and liver enlargements, liver discoloration and hepatocyte lipid vacuolation were observed, all effects starting at 100 mg/kg bw (Table 18).

The NOAEL was 10 mg/kg bw per day on the basis of several changes in the liver in males and females at 100 mg/kg bw per day (Esdaile, 1991b).

For *E*- and *Z*-isomers of dimethomorph, no obvious differences in toxicity profile or potency were evident.

In a 13-week feeding study , groups of 10 male and 10 female Sprague-Dawley rats were fed diets containing dimethomorph (purity, 98.7) at a concentration of 0, 40, 200, or 1000 ppm (equal to 0, 2.9, 14.2, 73 mg/kg bw per day in males and 0, 3.2, 15.8, 82 mg/kg bw per day in females). Additionally, recovery groups of 10 male and 10 female rats were fed diets containing dimethomorph at a concentration of 0 or 1000 ppm for 13 weeks and were then given control feed for an additional 4 weeks to evaluate the reversibility of any effects induced during treatment with dimethomorph. Daily cage-side observations, daily/weekly detailed clinical examinations were performed and ophthalmology at the end of the study in the control group and in the group at the highest dose. Body weights and feed consumption were recorded weekly and at the end of the study, haematology, clinical chemistry and urine-analysis parameters were analysed, and a gross necropsy with extensive histopathological examination of tissues and organ weights was conducted. The study complied with GLP.

There were no mortalities in the study and body-weight gain and feed consumption were comparable within all groups and sexes. Minimal changes in haematology parameters were noted in males and most of them were reversible in the recovery group (Table 19). Mildly decreased activity of alanine aminotransferase in males and minimally increased calcium and decreased chloride

Table 17. Liver weights[a] in rats given dimethomorph Z-isomer by gavage for 4 weeks

Sex	Liver weight (g)			
	Dose (mg/kg bw per day)			
	0	10	100	750
Males	6.59	6.65	7.2**	7.97**
Females	4.27	4.21	4.87**	5.81**

From Esdaile (1991b)

[a] Adjusted for terminal body weight.

** $p \leq 0.01$.

Table 18. Macroscopic and histological changes in organs of rats given dimethomorph Z-isomer by gavage for 4 weeks

Change	Dose (mg/kg bw per day)							
	Males				Females			
	0	10	100	750	0	10	100	750
Caecal enlargement	0	0	1	6	0	0	0	1
Liver enlargements	0	1	2	7	0	0	3	7
Liver, dark discoloration	0	0	1	4	0	1	3	7
Liver, patchy midzonal cytoplasmic lipid vacuolation	0	0	3	7	0	0	4	7

From Esdaile (1991b)

concentrations in females were found in rats at the highest dose only. Mild changes in a few body-weight adjusted organ weights did not follow a dose–response relationship and were therefore judged to be incidental. Only the relative liver weights in females at the highest dose were higher than in controls and showed a dose–response relationship (12.3 g at 1000 ppm and 10.5 g in the controls). Although less pronounced, the same was found in the recovery groups (11.6 g at 1000 ppm and 10.7 g in the controls). On histopathology in males at the highest dose, an increased incidence (six out of ten) of vacuolation of adrenal cells in the zona fasciculata was observed (two out of ten in controls, zero out of ten in other groups) and focal vascular congestion in the mucosa of ileum was found in both sexes (two out of ten in males and three out of ten in females, one out of ten in males in the control group and zero out of ten in all other groups).

The NOAEL was 200 ppm, equal to 14.2 mg/kg bw per day, on the basis of liver-weight increases in females at 1000 ppm (Ruckman, 1987).

Table 19. Changes in haematology parameters in male rats fed diets containing dimethomorph for 13 weeks

Parameter	Study groups				Recovery groups	
	Dietary concentration (ppm)					
	0	40	200	1000	0	1000
Packed cell volume[a] (%)	51	51	51	49*	48	45***
Erythrocytes × 10^6/mm^3	8.6	8.2	8.4	8.1*	8.0	7.7
Mean cell haemoglobin concentration (%)	30.1	29.8	30.6	31.2*	32.4	32.7
Lymphocytes × 10^3/mm^3	12.43	11.42	12.08	9.15**	11.04	9.83

From Ruckman (1987)
[a] Erythrocyte volume fraction.
* $p \leq 0.05$; ** $p \leq 0.01$; *** $p \leq 0.001$.

Table 20. Histological findings in the ileum of rats and mice fed diets containing dimethomorph

Finding	Dietary concentration (ppm)							
	Males (n = 10)				Females (n = 10)			
	0	2000	3000	4000	0	2000	3000	4000
Rats (Scott, 1986)	0	0	0	0	0	0	0	0
Dilatation	0	2	1	2	0	4	7	5
Inflammation	0	0	0	2	0	1	4	2
Serosa monocytes infiltrates	0	0	0	0	0	0	3	5
Mucosal hyperplasia	0	0	0	2	0	1	6	7
Fiber separation in muscle layer	0	0	0	0	0	1	5	6
Villous atrophy	0	2	2	1	1	0	1	1
	Dietary concentration (ppm)							
	Males (n = 10)				Females (n = 10)			
	0	800	2000/5000	300/10000	0	800	2000/4000	300/8000
Mice (Perry, 1986)								
Dilatation	0	0	0	1	0	0	1	2
Villous atrophy	2	0	3	4	1	1	1	4

From Perry (1987) and Scott (1986)

The ileums from all mice in the 6-week dose range-finding study (Perry, 1986) and all rats in the 4-week dose range-finding study (Scott, 1986) were examined histologically, this organ seemed to be a target of toxicity of dimethomorph in the rat. The study complied with GLP.

Table 20 presents histological findings in the ileum of rats and mice. For both species, the ileum is a target of toxicity attributable to dimethomorph. Rats were more sensitive to changes in the ileum than were mice, and female rats were more sensitive than male rats. This difference between the sexes was not evident in mice (Perry, 1987).

Dog

In the first part of a two-part study to determine the maximum tolerated dose in beagle dogs, one male and one female were fed diets containing dimethomorph (purity, 96.6%) at a concentration of 1000 ppm for the first 7 days, then at 750 ppm for 7 days, then at 900 ppm for 7 days and finally at 1200 ppm for 7 days. In the second part, one male and one female received diet containing dimethomorph at a concentration of 1200 ppm for 14 consecutive days. The study complied with GLP.

In the first part, clinical signs in the male were emesis, subdued behaviour and increased micturition on a few occasions. No signs of toxicity were seen in the female.

In the second part, the male showed occasionally body tremors.

In both parts of the study, no marked effects on body-weight gain were recorded and there wre no evident changes in clinical chemistry, urine analysis and macroscopic lesions in organs.

The maximum tolerated dose was greater than 1200 ppm (Greenough & Goburdhun, 1986a).

In a 13-week feeding study, groups of four male and four female beagle dogs were fed 400 g of diet containing dimethomorph (purity, 96.6%) at a concentration of 0, 150, 450, or 1350 ppm (equal to 0, 6, 15, 43 mg/kg bw per day) each day. The dogs were examined daily for general health, moribundity and mortality, and weekly detailed clinical examinations were performed. Feed consumption and body-weight development were recorded regularly. Before treatment and in the final week of treatment, detailed ophthalmic examinations were performed. At the beginning, at week 6 and at the end of the study, haematological and clinical chemistry and urine analysis parameters were analysed; at the end of the study, organ weights were recorded and gross pathology and histopathology were performed. The study complied with GLP.

There were no mortalities in the study and the only treatment-related clinical signs were increased incidences of lip-licking and subdued behaviour in males and females at 1350 ppm most frequently observed from week 4 on and usually within 2 h after feeding. A very slight body-weight reduction in females at 1350 ppm coincided with a minor decrease in feed intake in these dogs; body-weight development and feed intake in all other groups were not affected. There were no significant haematological changes in any group. In all dogs receiving dimethomorph, an increase in alkaline phosphatase activity was observed at 6 and 13 weeks and this became statistically significant for males at the highest dose only at 6 and 13 weeks. Alkaline phosphatase activity in these dogs was nearly doubled. No other biochemistry parameters were changed. In males at 1350 ppm, body-weight adjusted thymus weights were increased and prostate weights were decreased, and in females a significant increase in liver weights was recorded (Table 21). Histopathologically, only in the prostate were treatment-related findings recorded. All males at 1350 ppm had increased fibrosis in the prostate and prostatitis was found in two males. Prostatitis was also diagnosed in one male at 150 ppm.

The NOAEL was 450 ppm, equal to 15 mg/kg bw per day, on the basis of weight and histological changes in organs and behavioural findings at 1350 ppm (Greenough & Goburdhun, 1986b).

In a 52-week feeding study, groups of four male and four female beagle dogs were fed 400 g of diet containing dimethomorph (purity, 96.6%) at a concentration of 0, 150, 450, or 1350 ppm (equal to 0, 5, 15.2, and 44.8 mg/kg bw per day) each day. The dogs were examined daily for general health, moribundity and mortality, and weekly detailed clinical examinations were performed. Feed consumption and body-weight development were recorded regularly. Before treatment, at week 26 and in the final week of treatment, detailed ophthalmic examinations were performed. At the beginning, at weeks 13 and 26 and at the end of the study, haematological and clinical chemistry and urine analysis parameters were analysed. At the end of the study, organ weights were recorded and gross pathology and histopathology examinations were performed. The study complied with GLP.

There were no mortalities in the study and no dose-related changes in body-weight gain and feed consumption were recorded. In haematology and urine analysis, no significant changes were found at any dose. Alkaline phoshatase activity was increased slightly but statistically significantly in week 13 and then nearly doubled at week 26 and 51 in males at 1350 ppm; the same effect was somewhat more pronounced in females at 1350 ppm. There were no other treatment-related changes in biochemical parameters. Organ-weight analysis (Table 22) showed significant liver-weight increases in males and females at the highest dose and testes-weight increases in all groups receiving dimethomorph, these effects becoming statistically significant at the intermediate and the highest dose. In the testes, no histological changes were observed. Although statistically not significant, prostate weights were decreased in dogs at 1350 ppm. Histopathologically, three out of four males at 1350 ppm had mildly increased hepatic lipid while none did in the other groups; two out of four females at 1350 and 450 ppm showed this effect as well as one out of four in the control group and none in the group at 150 ppm. A very slight increase in severity of prostate interstitial fibrosis was observed at the highest dose. All other organs investigated histologically were not different from controls.

The NOAEL was 450 ppm, equal to 15.2 mg/kg bw per day, on the basis of changes in the weights of the liver, prostate and testes and prostate fibrosis at 1350 ppm (Greenough & Goburdhun, 1988).

2.3 *Long-term studies of toxicity and carcinogenicity*

Mouse

In a 24-month study of toxicity and carcinogenicity, groups of 50 male and 50 female CD-1 mice were fed diets containing dimethomorph (purity, 96.6%) adjusted to provide doses of 0, 10, 100 and 1000 mg/kg bw per day. Adjustment of dietary concentrations was weekly for the first 13 weeks and every 4 weeks thereafter. In a satellite study, four males and six females served as a control group and 15 males and 15 females were fed diets adjusted to provide dimethomorph at doses of 1000 mg/kg bw per day. In this satellite study, eight males and eight females at the highest dose and all mice in

Table 21. Mean relative organ weights of dogs fed diets containing dimethomorph for 13 weeks

Organ	Mean relative weight (% of body weight, ± SD)			
	Dietary concentration (ppm)			
	0	150	450	1350
Males				
Thymus	0.0692 ± 0.0192	0.0511 ± 0.0286	0.083 ± 0.0223	0.1215 ± 0.0485*
Prostate	0.0701 ± 0.0264	0.0598 ± 0.0079	0.057 ± 0.029	0.0271 ± 0.0104**
Females				
Liver	2.821 ± 0.209	3.297 ± 0.091	3.208 ± 0.247	3.585 ± 0.510**

From Greenough & Goburdhun (1986b)
* $p < 0.05$; ** $p < 0.01$
SD, standard deviation.

the control group were terminated at week 13, the remaining mice at the highest dose at week 52 of dosing. In both the main and the satellite study, mice were daily observed for mortalities and clinical signs; body weights, feed and water consumption and palpable masses were recorded weekly. Haematology data in the main study were collected at weeks 52, 78 and 104 of dosing and gross necropsy with extensive histopathology examination of tissues was done after terminal sacrifice. In the satellite study, blood samples were taken in week 14 and 52 for clinical chemistry analysis and at study termination a full necropsy, as in the main study, was performed with additional recording of liver weight and liver histology. The study complied with GLP.

The survival rate in males at 10 and 1000 mg/kg bw per day was higher than in the control group and no clinical signs of toxicity were seen throughout the study in any group nor were changes in feed and water consumption or in differential blood counts detectable. Absolute terminal body weight and body-weight gain was reduced significantly in males at 1000 mg/kg bw per day (Table 23). In the satellite group, males and females at the highest dose at week 13 showed slight increases in alkaline phosphatase activity (32% and 26%, respectively) were observed and at week 52 no obvious changes in clinical chemistry were evident, although no concurrent control was available. Liver weights in males were increased by 15% (relative, body-weight adjusted) in week 14 and 17% in week 52 (absolute, compared with laboratory historical controls); in females, 28% (relative, body-weight adjusted) in week 14 and 32% in week 52 (absolute, compared with laboratory historical controls). Histologically, there were no correlates to the liver weight increases at week 13 and 52 in both sexes but at week 13 moderate dilations of and villous atrophy in the ileums of one male and some females at 1000 mg/kg bw per day. At week 52, no treatment-related histological findings were recorded.

In the main study, an increase in the incidence of enlarged spleen was observed in both sexes at 1000 mg/kg bw per day (Table 24), but no histological changes were reported and organ weights not recorded. In females at the highest dose, the incidence of mammary adenocarcinoma of 6% was higher than in all other groups and at the upper bound of the range for historical controls (range,

Table 22. Absolute body and organ weights of dogs fed diets containing dimethomorph for 52 weeks

Body or organ	Absolute body or organ weight (body-weight adjusted group mean, ± SD)[a]			
	Dietary concentration (ppm)			
	0	150	450	1350
Males				
Body weight (kg)	13.1 ± 1.1	12.1 ± 1.0	11.4 ± 1.5	12.1 ± 1.1
Liver (g)	371.65 ± 33.52 (368.14)	327.71 ± 31.93 (328.01)	345.36 ± 23.96 (349.20)	433.44 ± 30.87* (433.82*)
Prostate (g)	8.25 ± 2.4 (7.4)	8.97 ± 2.52 (9.04)	5.55 ± 1.35 (6.23)	4.27 ± 1.86* (4.36)
Testes (g)	27.25 ± 2.52 (25.44)	27.41 ± 4.56 (27.56)	30.04 ± 3.12 (31.49*)	32.53 ± 4.37 (32.73*)
Females				
Body weight (kg)	10.7 ± 0.8	11.5 ± 0.6	10.1 ± 1.1	10.3 ± 1.0
Liver (g)	281.83 ± 17.56 (280.25)	297.15 ± 11.55 (264.79)	328.97 ± 105.06 (350.02)	399.36 ± 49.13** (412.26**)

From Greenough & Goburdhun (1988)

[a] In parentheses: body-weight adjusted group means determined by covariance analysis.

* $p < 0.05$; ** $p < 0.01$.

SD, standard deviation.

0–9%, mean, 3.5%; only one out of 12 control groups with an incidence of equal or higher incidence than 6%, control groups were from studies performed in 1997–2002). Although the historical control groups were not contemporary, no altered time trend in the background incidence of mammary adenocarcinomas between 1993 and 2003 was evident. In males at the highest dose, the incidence of lung adenoma was higher than in controls but lower than in the group receiving the next lower dose and, also considering the lack of any other histological changes in the lungs, a relationship with treatment is questionable.

The NOAEL was 100 mg/kg bw per day on the basis of reduced body-weight gain in males and increased incidence of mammary adenocarcinomas in females at 1000 mg/kg bw per day (Smith, 1991).

Rats

In a 24-month study of toxicity, groups of 20 male and 20 female Sprague-Dawley rats were fed diets containing dimethomorph (purity, 96.6%) at a concentration of 0, 200, 750 or 2000 ppm (equal to 0, 9.4, 36.3, 99.9 mg/kg bw per day in males and 0, 11.9, 57.7, 157.2 mg/kg bw in females). The rats were daily observed for mortalities, clinical signs and palpable masses; body weights, feed and water consumption were recorded weekly and ophthalmoscopy was performed pre-trial and at weeks 51 and 102. Data on haematology, clinical chemistry, and urine analysis were collected from

Table 23. Body-weight development in mice fed diets containing dimethomorph

Dietary concentration (ppm)	Body weight (g)				Body-weight gain		Terminal body weight
	Week 0	Week 17	Week 52	Week 104	(g)	% of control	% of control
Males							
0	30.7	38.8	43.8	43.0	12.3	—	—
10	30.6	37.3*	41.4*	43.4	12.8	104	101
100	30.4	37.7	42.4	42.0	11.6	94	98
1000	30.1	36.7**	40.5**	40.3*	10.2	83	94
Females							
0	23.2	30.9	34.9	38.8	15.6	—	—
10	23.7	31.9	36.0	38.1	14.4	92	98
100	23.3	31.4	34.7	37.3	14.0	89	96
1000	24.2	31.7	34.6	36.4	12.2	78	94

From Smith (1991)
* $p < 0.05$; ** $p < 0.01$.

Table 24. Gross pathology, non-neoplastic and neoplastic histological changes in mice fed diets containing dimethomorph

Gross histopathological change	Dose (mg/kg bw per day)							
	Males (n = 50)				Females (n = 50)			
	0	10	100	1000	0	10	100	1000
Spleen, enlarged	4	4	5	7	14	12	14	20
Lung, adenomas	13	14	23	17	9	5	12	5
Mammary, adenocarcinoma	—	—	—	—	1	0	1	3

From Smith (1991)

the groups at weeks 11 (only haematology), 25, 51, 77 and 102 of dosing. Organ weights and gross necropsy with extensive histopathology examination of tissues were done after terminal sacrifice. The study complied with GLP.

Male and female rats at the intermediate and highest dose showed a slight increased survival time and no treatment-related clinical signs were observed. Feed and water consumption was comparable in all groups, but body-weight gain was decreased in both sexes at 2000 ppm. The final body weight in males was 67% and in females 83% of that of the controls. In males, the effect became statistically significant from week 68 onwards and in females from the beginning of dosing. Starting at week 11, the erythrocyte system was affected at 2000 ppm in both sexes, this being more pronounced in females (Table 25). After 1 year, males at 750 ppm were also affected but at the end of the study, erythrocyte parameters were no longer changed statistically significantly in any group. Statistically significant, only body-weight adjusted kidneys weights in females at the highest dose were increased (Table 26). However, increases of relative liver weights and decreases of relative pituitary weights were recorded in both sexes at 2000 ppm. At terminal kill, macroscopically no treatment-related changes were evident. Histologically, in the liver of females at 2000 ppm hepatocyte hypertrophy and pigmentation was increased and in males more ground glass foci were recorded (Table 27). The effects did not attain statistical significance. In both groups at the highest dose, arteritis in the pancreas and the mesenterium was increased. In males, a dose-related increase in testes interstitial cell hyperplasia was found in all dosed groups and the incidence of adenomas was also increased. In both sexes at 2000 ppm, an increased severity in sternum marrow cellularity was recorded that might be a compensatory sign of anaemia at higher doses. In females at the highest dose, a significant decrease in the incidence of pituitary hyperplasias and tumours was found.

The NOAEL was 750 ppm, equal to 36.3 mg/kg bw per day, on the basis of body-weight decreases and histological changes in the liver of females at 2000 ppm (Everett, 1990).

In another 24-month study of carcinogenicity, groups of 50 male and 50 female Sprague-Dawley rats were fed diets containing dimethomorph (purity, 96.6%) at a concentration of 0, 200, 750 or 2000 ppm (equal to 0, 8.8, 33.9, 94.6 mg/kg bw per day in males and 0, 11.3, 46.3, 132.5 mg/kg bw

Table 25. Changes in erythrocyte parameters for rats fed diets containing dimethomorph for 24-months

Parameter	Dietary concentration (ppm)							
	Males				Females			
	0	200	750	2000	0	200	750	2000
Week 11								
Haemoglobin (g/dl)	15.6	15.6	15.4	15.2	15.1	15.1	15.0	14.5**
Erythrocytes (10^{12}/l)	7.63	7.64	7.5	7.32**	7.43	7.41	7.33	7.00***
Erythrocyte volume fraction (1/l)	0.433	0.439	0.429	0.425	0.428	0.429	0.425	0.414*
Week 51								
Haemoglobin (g/dl)	14.8	14.8	14.1**	14.3*	13.8	13.9	13.5	12.7**
Erythrocytes (10^{12}/l)	7.87	7.71*	7.49***	7.44***	7.02	7.05	6.87	6.35***
Erythrocyte volume fraction (1/l)	0.399	0.402	0.382**	0.390	0.380	0.384	0.375	0.354**
Week 102								
Haemoglobin (g/dl)	14.7	13.3	13.8	13.9	13.6	13.5	13.3	12.6
Erythrocytes (10^{12}/l)	7.54	6.71	7.08	7.15	6.63	6.37	6.28	5.86
Erythrocyte volume fraction (1/l)	0.419	0.387	0.393	0.396	0.391	0.389	0.380	0.366

From Everett (1990)
* $p < 0.05$; ** $p < 0.01$; *** $p < 0.001$.

Table 26. Organ weights adjusted to terminal body weight for rats fed diets containing dimethomorph for 24 months

Organ	Dietary concentration (ppm)							
	Males				Females			
	0	200	750	2000	0	200	750	2000
Liver	18.73 ± 0.97	17.24 ± 0.8	17.62 ± 0.77	19.93 ± 0.75	12.00 ± 0.78	12.94 ± 0.74	12.82 ± 0.63	15.08 ± 0.78
Pituitary	0.038 ± 0.02	0.061 ± 0.017	0.051 ± 0.016	0.022 ± 0.015	0.091 ± 0.024	0.061 ± 0.022	0.063 ± 0.019	0.043 ± 0.023
Kidney	4.44 ± 0.19	4.46 ± 0.16	4.45 ± 0.15	4.03 ± 0.15	2.68 ± 0.1	2.65 ± 0.09	2.82 ± 0.08	3.15 ± 0.10**

From Everett (1990)
** $p < 0.01$.

Table 27. Non-neoplastic and neoplastic histological changes in rats fed diets containing dimethomorph for 24 months

Histological change	Dietary concentration (ppm)							
	Males (n = 20)				Females (n = 20)			
	0	200	750	2000	0	200	750	2000
Liver, ground glass foci	7	7	10	14	1	4	6	6
Liver, periacinar pigmentation	0	0	0	0	0	0	3	16
Liver, periacinar hypertrophy	0	0	0	0	0	0	2	8
Pancreas, arteritis	1	0	4	5	0	0	1	3
Testes, interstitial cell hyperplasia	2	5	6	7	—	—	—	—
Testes, interstitial adenoma	2	5	4	6	—	—	—	—
Mesenterium, arteritis	1	0	2	8	0	0	1	3
Sternum, high-grade cellularity in marrow	5	0	2	7	5	2	3	13

From Everett (1990)

in females). This study was conducted in parallel with a similar long-term study of toxicity (Everett, 1990). Rats were daily observed for mortalities, clinical signs and palpable masses; body weights, feed and water consumption were recorded weekly. Haematology data were collected from the groups at weeks 52, 78 and 103 of dosing. Gross necropsy with extensive histopathology examination of tissues was done after terminal sacrifice. The study complied with GLP.

From the Kaplan-Meier survival curves, no significant differences between groups with regard to time to death and total number of pre-term deaths are evident. In females at the highest dose, feed intake was decreased by 7%, but not in other groups; terminal body weights in males at 2000 ppm were 14% lower and in females 9%, 23% and 38% lower at 200, 750 and 2000 ppm, respectively. In the first 12 weeks, five females at the highest dose had severely swollen hind limbs and were killed pre-term. The reason for this effect remains unclear. Haematology was not affected in any group. On gross pathology, at 2000 ppm increased incidences of enlarged adrenals and lymph node cysts in males were recorded and, in both sexes, swollen hind limbs, pancreas masses and dilated blood vessels, in females also ovary cysts (Table 28). Non-neoplastic histological findings in the liver at 750 and 2000 ppm were found in females and in males at 2000 ppm and arteritis was found in the pancreas for males and females at 2000 ppm (Table 29). In males, increased incidences of tumours in the adrenal medulla and the testes were found. The interstitial testes tumours coincided with increased interstitial

Table 28. Gross pathology in rats fed diets containing dimethomorph for 24 months

Gross pathology	Dietary concentration (ppm)							
	Males (n = 50)				Females (n = 50)			
	0	200	750	2000	0	200	750	2000
Adrenals, enlarged	1	2	5	5	0	0	0	0
Lymph node, cysts	2	4	4	9	0	0	2	0
Hind feet, hindlimbs, swollen	11	16	12	20	5	4	6	19
Pancreas, masses	1	3	1	5	0	2	1	8
Blood vessel, dilated	1	0	4	13	0	1	1	4
Ovary, cysts	—	—	—	—	8	2	9	15

From Everett (1991)

Table 29. Non-neoplastic and neoplastic histological changes in rats fed diets containing dimethomorph for 24 months

Histological change	Dietary concentration (ppm)							
	Males (n = 50)				Females (n = 50)			
	0	200	750	2000	0	200	750	2000
Liver, ground-glass foci	17	18	21	27	12	9	16	18
Liver, periacinar hepatocyte pigmentation	0	0	0	0	1	0	4	26
Liver, periacinar hypertrophy	0	0	0	0	0	0	1	15
Lymph node, mesenteric, reactive hyperplasia	0	0	0	3	0	0	0	1
Pancreas, arteritis	3	2	0	24	0	3	2	6
Testes, interstitial cell hyperplasia	6	6	10	10	—	—	—	—
Testes, interstitial adenoma	5	7	8	10	—	—	—	—
Adrenals, benign medulla tumours	4	5	4	8	2	3	1	2

From Everett (1991)

cell hyperplasia. The testes tumour incidences of 16% at 750 ppm and 20% at 2000 ppm are at the upper bound of the range for historical controls (range, 4–20%; mean, 10.4 ± 6.6%; control groups from studies performed 1986–1992) and statistically significant at the highest dose. The incidence of benign adrenal medulla tumours of 16% in males is at the upper bound of the range for historical controls (range, 1–17%; mean, 12%; control groups from studies performed 1997–2002).

The NOAEL was 750 ppm, equal to 33.9 mg/kg bw per day, on the basis of reduced body-weight gain in both sexes and histological changes in the liver of females at 2000 ppm (Everett, 1991).

2.4 *Genotoxicity*

Dimethomorph was tested for genotoxicity in assays for reverse mutation with *Salmonella typhimurium* and *E. coli*, an assay for HGPRT forward mutation assay with V79 cells, two assays for chromosomal aberration in vitro with V79 cells, one assay for chromosomal aberration in vitro with human peripheral lymphocytes, an assay for cell transformation, an assay for unschedulaed DNA synthesis in vitro and an assay for micronucleus formation in mouse bone marrow in vivo. All studies complied with GLP.

The summary of results is given Table 31. Except for the test for chromosomal aberration, all tests for genotoxicity gave clear negative results. In the first of two assays for chromosomal aberration in vitro with V79 cells (Heidemann & Miltenburger, 1986), at the first time-point of collection at 7 h an increase in the percentage of aberrant cells was observed at the highest doses with (8%) and

Table 30. Results of tests for chromosomal aberration with dimethomorph

Concentration (µg/ml)	Percentage of cells that are aberrant, including gaps (excluding gaps)								
	Harvest at 7 h			Harvest at 18 h			Harvest at 28 h		
	+S9	−S9	MI	+S9	−S9	MI	+S9	−S9	MI
Heidemann & Miltenburger (1986)									
0 (solvent)	3.0 (0.5)	2.0 (0.5)	100	2.5 (1.0)	7.0 (3.0)	100	2.5 (0.5)	3.5 (0.5)	100
12	—	—	—	7.5 (3.5)	3.0 (2.5)	157/159	—	—	—
60	—	—	—	1.0 (1.0)	2.5 (2.0)	177/138	—	—	—
160 (−S9); 170 (+S9)	8.0 (3.5)	8.0 (4.0)	26.2/47.6	1.5 (1.0)	5.5 (4.0)	243/212	4.5 (1.5)	4.5 (1.5)	74/97
Heidemann & Miltenburger (1987)									
0 (solvent)	3.75 (1.25)	4.25 (1.25)	100	—	4.0 (1.25)	100	—	—	—
12	—	—	—	—	4.0 (1.75)	—/75	—	—	—
60	—	—	—	—	5.0 (1.25)	—/64	—	—	—
160 (−S9); 170 (+S9)	9.5 (6.75)	6.25 (3.5)	73/76	—	7.0 (3.5)	—/77	—	—	—
van de Waart (1991a)									

	Harvest at 24 h			Harvest at 48 h		
	+S9	−S9	MI	+S9	−S9	MI
0 (solvent)	1.5 (1.0)	0 (0)	100	2 (0.5)	3 (3)	100
1	2 (1.5)	—	94/—	—	—	104/—
10	1.5 (1.5	0.5 (0.5)	86/79	—	—	105/70
100	—	2 (1)	62/66	—	—	71/65
333	2 (0)	2.5 (1)	61/53	—	1.5 (1)	83/74
422[a]	18 (16)	—	32/24	2 (1.5)	5.5 (3.25)	87/43

MI, mitotic index; S9, 9000 × *g* supernatant from rat livers.
[a] Slight precipitation observed.

without metabolic activation (8%) (Table 30). The negative controls showed incidences of aberrant cells of 3% and 2%, respectively. At later time-points, no dose-related increase in aberrant cells was observed. Based on this inconclusive result, a second assay for chromosomal aberration in vitro with V79 cells was performed (Heidemann & Miltenburger, 1987). Again, at 7 h an increase in the percentage of aberrant cells was observed at the highest doses with (9.5%) and without metabolic activation (6.25%). The negative controls showed incidences of aberrant cells of 3.75% and 4.25%, respectively. For time-point at 18 h, only tests without metabolic activation were performed with 4% aberrant cells at 12 µg/ml, 5% at 60 µg/ml and 7% at 160 µg/ml. In both studies, the mitotic indices were greater than 70%. In a similar study with human peripheral lymphocytes, increases in aberrant cells were observed only at 422 µg/ml without metabolic activation at 48 h and with metabolic activation at 24 h (statistical significant), a concentration level inducing slight precipitations and clearly reduced mitotic indices with metabolic activation at 24 h but not at 48 h and without metabolic activation at 24 h and 48 h (van de Waart, 1991a).

In cytogenic assays in vitro (V79 Chinese hamster cells and human lymphocytes), an increase in the frequency of chromosome aberrations was detected only at high concentrations, which were strongly cytotoxic and partly caused precipitation. No clastogenicity was found in an assay for

Table 31. Results of studies of genotoxicity with dimethomorph

End-point	Test object	Concentration	Purity (%)	Result	Reference
In vitro					
Reverse mutation	*Salmonella typhimurium* TA98, TA100, TA1535, TA1537, TA1538 and *E. coli* WP2*uvr*A	31.25–5000 µg/plate in DMSO ± S9	94	Negative	Brooks & Wiggins (1989)
Forward mutation	*Hgprt* locus in V79 cells	33–180 µg/ml –S9 133–333 µg/ml +S9	98.6	Negative	van de Waart (1991b)
Chromosomal aberration	V79 cells	12–160 µg/ml –S9 13–170 µg/ml +S9	NS	Increase in aberrant cells at 7 h ±S9, but not at later time-points	Heidemann & Miltenburger (1986)
Chromosomal aberration	V79 cells	160 µg/ml –S9 and 170 µg/ml +S9 for 7 h and 12–160 µg/ml for 18 h –S9	NS	Increase in aberrant cells at 7 h ±S9 and at 18 h –S9	Heidemann & Miltenburger (1987)
Chromosomal aberration	Peripheral human lymphocytes	10–333 µg/ml –S9 at 24 h and 333–422 µg/ml –S9 at 48 h; 1–422 µg/ml +S9 for 24 h and 422 µg/ml for 48 h +S9	98.6	Increase in aberrant cells at 422 µg/ml ±S9	van de Waart (1991a)
Cell transformation	Syrian Hamster cells	5–50 µg/ml –S9 25–265 µg/ml +S9	98.7	Negative	Poth & Miltenburger (1986)
Unscheduled DNA synthesis	Male rat hepatocytes	2.5–250 µg/ml	NS	Negative	Timm & Miltenburger (1986)
In vivo					
Micronucleus formation	Male and female mice	Single doses of 20, 100, 200 mg/kg bw, sampling at 24 h, 48 h and 72 h	98.6	Negative	van de Waart (1991c)

NS, not stated; S9, 9000 × *g* supernatant from rat livers.

unscheduled DNA synthesis, the test for micronucleus formation in the mouse and did not induce cell transformation in the Syrian hamster cell cultures. On the basis of all these results, the Meeting concluded that dimethomorph is neither mutagenic nor genotoxic.

2.5 *Reproductive toxicity*

(a) *Multigeneration studies*

Rats

Groups of 30 male and 30 female Sprague-Dawley rats were fed diets containing dimethomorph (purity, 96.6%) at a concentration of 0, 100, 300 and 1000 ppm for approximately 13 weeks before breeding, and continuing through breeding (3 weeks), gestation and lactation for each of two generations to evaluate the potential for reproductive toxicity and effects on neonatal growth and survival. F_1 consisted of 25 male and 25 female rats. An evaluation of mating performance, conception, gestation, parturition, lactation and weaning, as well as survival, growth and development of the offspring was done. In-life observations, body weights, feed consumption and litter data were recorded. In addition, a gross necropsy of the P and F_1 adults was conducted. The study complied with GLP.

Table 32. Compound intake and body-weight developments in female parent rats fed diets containing dimethomorph in a study of reproductive toxicity

Parameter	Period	Time-point	Dietary concentration (ppm)			
			0	100	300	1000
Compound intake (mg/kg bw per day)	Premating	Weeks 1–15	—	8.0	24.0	79.3
	Gestation	Days 1–20	—	6.8	20.8	71.4
	Lactation	Days 1–21	—	13.5	40.7	140.2
Body-weight development (g)	Premating	Body weight at week 15	292.3	291.2	280.8	267.2
		Body-weight gain week 1–15	138.0	137.5	128.3	117.7*
	Gestation	Body weight at day 20	390.9	391.6	384.2	377.0
		Body-weight gain day 1–20	108.9	114.2	110	116.3
	Lactation	Body weight at day 21	320.0	316.0	317.4	310.0
F_1						
Compound intake (mg/kg bw per day)	Premating	Weeks 1–15	—	8.9	27.0	89.2
	Gestation	Days 1–20	—	7.0	21.5	74.1
	Lactation	Days 1–21	—	14.6	46.2	151.8
Body-weight development (g)	Premating	Body weight at week 15	292.4	285.2	281.0	277.4
		Body-weight gain weeks 1–15	183.4	178.2	174.4	171.0
	Gestation	Body weight at day 20	400.5	400.2	388.5	393.5
		Body-weight gain day 1–20	117.9	118.8	109.3	116.5
	Lactation	Body weight at day 21	332.4	327.1	326.1	336.0

From Osterburg (1990)
* $p < 0.05$.

Clinical observations, feed consumption and body weight gain were comparable in all groups in parents of both generations with the exception of the premating period for P, where body weight gain was slightly decreased in the group at 1000 ppm (Table 32). Since the body-weight development thereafter was comparable with that of the other groups and fertility was not affected, this effect was judged not to be of toxicological relevance.

There were no treatment-related findings on mating performance, fertility and litter data in P and F_1. In F_1, F_{2a} and F_{2b} pups at 1000 ppm, there were signs of retarded development indicated by a decreased percentage of pups with erupted incisors when compared with the other groups (Table 34 and Figure 3). However, virtually all pups of the group at the highest dose had incisor eruption within postnatal days 8 and 11–12, as had the pups in the other groups. No other markers for postnatal physical development of the pups (eye opening, hair growth and pinna unfolding) were affected and behaviour was not changed.

The NOAEL for maternal toxicity and reproductive performance was 1000 ppm, equal to approximately 80 mg/kg bw per day, the highest dose tested. The NOAEL for pup development was 300 ppm, equal to approximately 20 mg/kg bw per day, on the basis of slight delays in incisor eruption in pups at 1000 ppm (Osterburg, 1990).

(b) Developmental toxicity

Rats

In a dose range-finding study, groups of eight mated female Sprague-Dawley rats received dimethomorph (purity, 98.7%) at a dose of 0, 50, 120, or 300 mg/kg bw per day by oral gavage suspension on days 6 to day 15 of gestation. Clinical examinations were done twice per day and body weights were recorded regularly. On day 20, dams were terminated and the fetuses delivered

Table 33. Duration of gestation in a study of reproductive toxicity in rats fed diets containing dimethomorph

Generation	Duration of gestation (days)			
	Dietary concentration (ppm)			
	0	100	300	1000
F_{1a}	22.0 ± 0.3	22.1 ± 0.3	21.9 ± 0.4	21.8 ± 0.4*
F_{2a}	21.9 ± 0.4	21.9 ± 0.3	22.1 ± 0.3	21.7 ± 0.5**
F_{2b}	21.9 ± 0.2	22.2 ± 0.5	21.9 ± 0.3	21.8 ± 0.4

From Osterburg (1990)
*Statistically different to group at 100 ppm, $p < 0.05$.
**Statistically different to group at 300 ppm, $p < 0.05$.

Table 34. Incisor eruption in a study of reproductive toxicity in rats fed diets containing dimethomorph

Generation	Group mean percentage of pups with incisor eruption on postnatal day 10			
	Dietary concentration (ppm)			
	0	100	300	1000
F_{1a}	82.6 ± 30.2	70.3 ± 35.9	63.7 ± 35.0	43.1 ± 37.3
F_{2a}	79.0 ± 24.0	68.4 ± 28.5	76.2 ± 24.8	48.0 ± 35.9
F_{2b}	78.8 ± 21.6	77.9 ± 27.8	71.1 ± 36.7	40.3 ± 27.0

From Osterburg (1990)

Figure 3. Cumulative percentage of F_{1a}, F_{2a}, F_{2b} pups with erupted incisors at postnatal days in a study of reproductive toxicity in rats fed dimethomorph

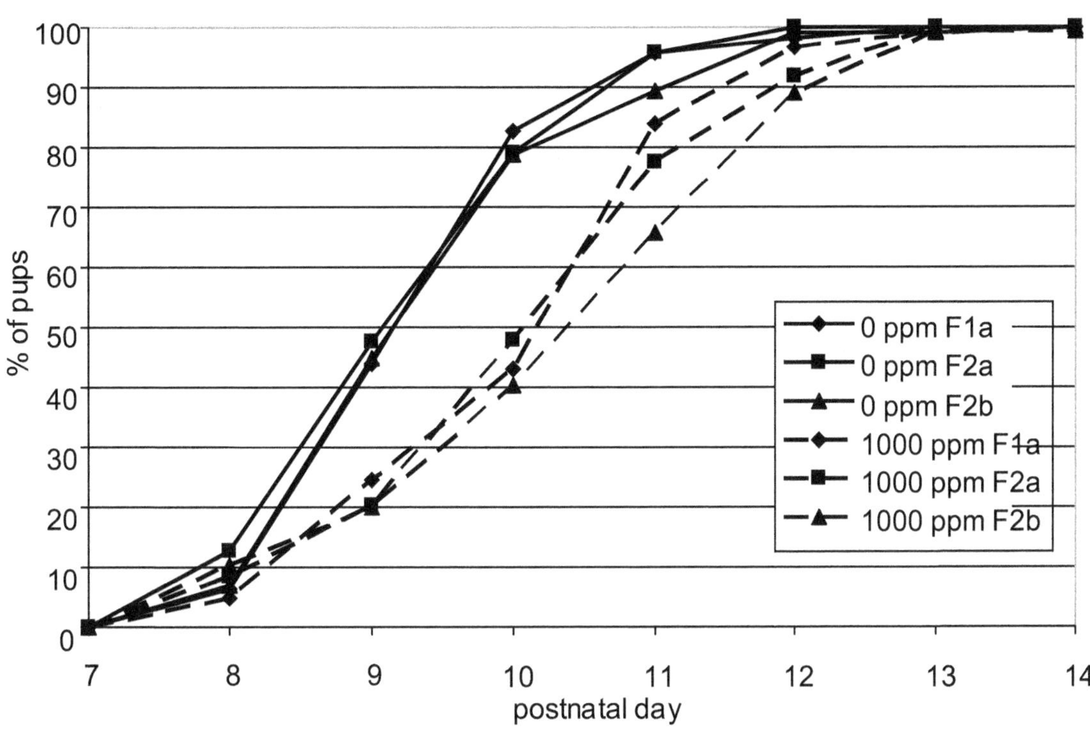

From Osterburg (1990)

by caesarian section. The numbers of corpora lutea, implantation sites, viable fetuses and resorptions were recorded. Fetuses were sexed, body weights measured and they were subjected to external examinations. The study complied with GLP.

At 300 mg/kg bw per day, one rats showed bloody urine for 2 days and another died from an emphysema in the lung; relationship to treatment is questionable. No treatment-related body-weight changes in dams were seen. Additionally, pregnancy and fetal parameters in the treated groups were not different from those of the control group.

The NOAEL was 300 mg/kg bw per day, the highest dose tested. No effect on dams and the investigated reproductive parameters was seen (Osterburg, 1986).

In a second, very poorly reported dose-range finding study, a group of four mated female Sprague-Dawley rats received dimethomorph (purity not stated) at a dose of 150 mg/kg bw per day by oral gavage as a suspension and another group of three females received dimethomorph at a dose of 300 mg/kg bw per day on day 6 until day 15 of gestation. In the group at 150 mg/kg bw per day, one female had a total litter loss and the other three had each one to three postimplantation losses. In the group at 300 mg/kg bw per day, two females had a total litter loss and the third had one postimplantation loss. Mean pup weight at 300 mg/kg bw per day was reduced (Müller, 1987b).

In the definitive study of developmental toxicity in rats, groups of 30 mated female Sprague-Dawley rats received dimethomorph (purity, 96.6%) at a dose of 0, 20, 60, or 160 mg/kg bw per day as a suspension at 10 ml/kg bw by oral gavage on day 6 until day 15 of gestation. Clinical examinations were done twice per day and feed consumption and body weights were recorded regularly. On day 20, dams were terminated and the fetuses delivered by caesarian section. The numbers of corpora lutea, implantation sites, viable fetuses and resorptions were recorded. Fetuses were sexed, body weights measured and they were subjected to external, visceral and skeletal examinations. The study complied with GLP.

The only pre-term death was due to an intubation error in the group at the highest dose. There were no treatment-related clinical signs in any group, but feed consumption was reduced in the group at the highest dose from day 6 to day 15 of gestation (Table 35). This resulted in reduced body-weight gain from day 6 to day 10 of gestation. Thereafter, body-weight gain was comparable to the other groups. At 160 mg/kg bw per day, there were two total litter losses and one at 60 mg/kg bw per day but none in the other groups (Table 36). These rats had decreases in feed intake and in body weight that were at the lower bound within their respective group and therefore contributed most to the mean group changes. If the intrauterine deaths of the two dams at the highest dose are included, postimplantation deaths were significantly increased to 15.6% at 160 mg/kg bw per day and this was beyond the upper limit of the range for historical controls (11.8%; 23 control groups, 4 out of 23 with one total loss). All postimplantation losses were described as early losses. There were no signs of developmental toxicity.

The NOAEL for maternal and for embryo toxicity was 60 mg/kg bw per day on the basis of body-weight gain decreases in dams and increased total litter losses at the next higher dose. The NOAEL for developmental toxicity was 160 mg/kg bw per day, the highest dose tested (Müller, 1989b).

Table 35. Body weight, body-weight gain and feed consumption in a study of developmental toxicity in rats given dimethomorph by gavage

Parameter	Day of gestation	Dose (mg/kg bw per day)			
		0	20	60	160
Body-weight gain (g)	0–6	33.4	30.8	33.3	32.0
	6–10	21.8	22.2	19.3	7.0
	10–15	30.0	28.5	30.0	30.3
	15–20	66.8	67.3	63.1	65.0
Body weight (g)	20	356.2	352.3	346.2	335.8**
Feed consumption (g/animal per day)	0–6	21.8	21.6	21.8	21.6
	6–10	23.2	23.	22.0	16.5**
	10–15	26.2	25.7	25.7	21.3**
	15–20	28.3	28.3	25.9	28.3

From Müller (1989b)
** $p < 0.01$.

Table 36. Postimplantation losses in a study of developmental toxicity in rats given dimethomorph by gavage

Parameter	Dose (mg/kg bw per day)			
	0	20	60	160
Pregnant females at terminal kill	25	20	22	22
Females with live fetuses	25	20	21	20
Females with total intrauterine deaths	0	0	1	2
% deaths in females with live fetuses	3.9 ± 6.1	6.3 ± 7.2	4.4 ± 5.9	7.2 ± 8.8
% deaths in all females	3.9 ± 6.1	6.3 ± 7.2	8.8 ± 21.2	15.6 ± 28.6

From Müller (1989b)

Rabbits

In a dose-range finding study, groups of eight time-mated female New Zealand White rabbits received dimethomorph (purity, 96.6%) at a dose of 0, 300, 600, or 1000 mg/kg bw per day by oral gavage as a suspension in a volume of 10 ml/kg, on day 6 until day 18 of gestation. In the group at the highest dose, nine animals were used. Dams were examined twice per day for clinical and behavioural signs of toxicity and body weights were recorded regularly. On day 28, dams were terminated and the fetuses removed and ovaries and uteri examined. The numbers of corpora lutea, implantation sites, viable fetuses and resorptions were recorded. Fetuses were sexed and body weights measured and fetuses were subjected to external examinations. The study complied with GLP.

One dam at 1000 mg/kg bw per day was killed on day 7 post coitum because of an injury to a hindlimb. Another six aborted and were terminated pre-term; for three of these, there were findings in the liver and/or spleen. After dosing, all rabbits at thee highest dose showed reduced feed/water consumption and two of them were severely emaciated at abortion. At 300 mg/kg bw per day, one dam aborted, but none did so at 600 mg/kg bw per day. In all dosed groups, body-weight gain was reduced during days 6–12 of gestation when compared with controls, an analysis of the group at the highest dose was not possible due to the abortions (Table 37). Of the two remaining dams at 1000 mg/kg bw per day, one had total postimplantation losses at necropsy. A slight increase of postimplantation losses was also evident at 600 mg/kg bw per day, but not at 300 mg/kg bw per day, numbers and sex of fetuses were not affected. At 600 mg/kg bw per day, mean fetal body weight was slightly reduced. No increased incidences in malformations were observed.

The NOAEL for maternal toxicity was 300 mg/kg bw on the basis of reduced maternal body-weight gain at 600 mg/kg bw. The NOAEL for developmental toxicity was 300 mg/kg bw per day on the basis of reduced fetal body weight and increased postimplantation losses at 600 mg/kg bw per day (Müller, 1987a).

In the definitive study, groups of 22 time-mated female New Zealand White rabbits received dimethomorph (purity, 96.6%) at a dose of 0, 135, 300, or 650 mg/kg bw per day by oral gavage as a suspension in a volume of 10 ml/kg, from day 6 until day 18 of gestation. Dams were examined twice per day for clinical and behavioural signs of toxicity and body weights and feed consumption were recorded regularly. On day 28, dams were terminated and the fetuses removed and ovaries and uteri examined. The numbers of corpora lutea, implantation sites, viable fetuses and resorptions were recorded. Fetuses were sexed and body weights measured and fetuses were subjected to external, visceral and skeletal examinations. The study complied with GLP.

In the control group and in the group at the lowest dose, one dam died and one was terminated because they aborted. In the group at the intermediate dose, three dams died for different not treatment-related reasons, no abortions occurred. In the group at the highest dose, four rabbits died, three

Table 37. Body-weight gain in a study of developmental toxicity studying rabbits given dimethomorph by gavage

Days of gestation	Body-weight gain (kg)			
	Dose (mg/kg bw per day)			
	0	300	600	1000
0–6	0.2	0.2	0.2	—
6–12	0.2	0.1	−0.1	—
12–18	0.1	0.2	0.2	—
18–24	0.1	0.1	0.2	—
0–28	0.7	0.7	0.7	—

From Müller (1987a)

of them with signs of improper dosing; three rabbits aborted and two of them had findings in the liver and/or spleen. Mortalities in all groups were not considered to be treatment-related, while the abortions at 650 mg/kg bw per day probably were treatment-related (Table 38). There were no additional clinical or necropsy findings considered to be treatment-related. The decreased body-weight gain during days 6–12 of gestation coincided with a statistically significant reduced feed intake in this period (Table 39). Postimplantation losses in females with live fetuses at necropsy were comparable in all groups. There were no malformations or increased incidences in variations that are considered to be treatment-related.

The NOAEL for maternal and developmental toxicity was 300 mg/kg bw per day on the basis of intermittently reduced body-weight gain in dams and abortions at 650 mg/kg bw per day. No treatment-related malformations were observed (Müller, 1989a).

2.6 *Special studies: pharmacology*

Thirteen studies on the pharmacological effects of dimethomorph in different species were performed. They all complied with GLP.

(a) Anticonvulsive activity in mice

In six male mice, dimethomorph at a dose of 100 mg/kg bw did not suppress leptazol- (pentylenetetrazole) induced convulsions and therefore the study authors concluded that dimethomorph has no central nervous system depressing activity (Ainsworth & Wright, 1991a).

Table 38. Postimplantation losses in a study of developmental toxicity in rabbits given dimethomorph by gavage

Parameter	Dose (mg/kg bw per day)			
	0	135	300	650
Inseminated females	22	22	22	22
Pregnant females	20	17	18	20
Females found dead	1	0	2	4
Abortions	1	1	0	3
Females with total intrauterine deaths	1	2	0	1
Females with live fetuses at necropsy	17	14	16	12

From Müller (1989a)

Table 39. Body weight gain in dams in a study of developmental toxicity in rabbits given dimethomorph by gavage

Days of gestation	Body-weight gain (kg)			
	Dose (mg/kg bw per day)			
	0	135	300	650
0–6	0.3	0.2	0.3	0.2
6–12	0.1	0.1	0.1	0
12–18	0.1	0.1	0.1	0.1
18–24	0.1	0.1	0.2	0.2
0–28	0.8	0.6	0.7	0.5

From Müller (1989a)

(b) Potentiation of hexobarbiton sleeping time in mice

In six male mice, dimethomorph at a dose of 100 mg/kg bw increased the barbital-induced mean sleeping time from 187.17 min to 257 min. According to the company, this might be explained by competition of dimethomorph with barbital for metabolism pathways (Ainsworth & Wright, 1991e).

(c) Alteration of body temperature in mice

In six male mice, dimethomorph at a dose of 100 mg/kg bw did not alter the body temperature (Ainsworth & Wright, 1991f).

(d) Spontaneus motor activity in mice

In six male mice, dimethomorph at a dose of 100 mg/kg bw did not statistically significant increase spontaneous locomotor activity. However, a very slight increase in grooming was observed (Ainsworth & Wright, 1991g).

(e) Tail flick test for analgesia

In six male mice, dimethomorph at a dose of 100 mg/kg bw did not change the reaction time (tail flicking) to pain. Therefore, dimethomorph has no analgesic activity (Ainsworth & Wright, 1991i).

(f) Effect on the Irwin test in mice

In groups of five male and five female mice given dimethomorph at a dose of 30, 100 or 300 mg/kg bw, the results of the Irwin test showed very mildly increased incidences of parameters that suggest a sedative effect (Ainsworth & Wright, 1991k).

(g) Charcoal meal transit times in the rat small intestines

In groups of five male and five female rats given dimethomorph at a dose of 30, 100 or 300 mg/kg bw, intestinal motility measured as charcoal-meal transit (migration distance at a certain time-point as percentage of the whole length of intestine) was not influenced in males but was increased in females in a dose-related manner (Ainsworth & Wright, 1991d).

(h) Effect on motility of uterine smooth muscle in rats

Smooth muscles from preparations of rat uterus were reproducibly induced to contract by acetylcholine and this contraction could be inhibited by adrenaline. Dimethomorph at concentrations of up to 30 µg/ml had no effect greater than that of the vehicle (dimethylformamide) on muscle contraction and did not interfere with the cholinergic and adrenergic receptor dependant reactions (Ainsworth & Wright, 1991j).

(i) Assessment of potential anti-inflammatory activity in rats

In groups of eight male and eight female rats given dimethomorph at a dose of 30, 100 or 300 mg/kg bw, the inflammation potential of dimethomorph was investigated by kappa carrageenan injection in the hind paw. Edema formation was inhibited at 300 mg/kg bw in males (Ainsworth & Wright, 1991b).

(j) Cardiovascular, respiratory and nictitating membrane alterations in cat

Three anaesthetized cats were surgically prepared to measure blood pressure, heart rates, electrocardiogram, respiration and electrical stimulation of nictitating membrane after intravenous application of dimethomorph at a dose of 0.01, 0.03 or 0.1 mg/kg bw. The only effect different from vehicle control injection (dimethylformamide) was a small increase in heart rates at 0.1 mg/kg bw (Ainsworth & Wright, 1991c).

(k) Effect on the spontaneous motility of isolated rabbit ileum

Acetylcholine-induced contraction forces of ileum preparations from rabbits were reproducibly increased; this effect could be inhibited by adrenaline. Dimethomorph at concentrations of up to 30 µg/ml had no significant direct effects on ileum contractions and did not interfere with the cholinergic and adrenergic receptor dependant reactions (Ainsworth & Wright, 1991l).

(l) Effect on the isolated guinea-pig ileum and action on four agonists

Acetylcholine, histamine, 5-hydroxytryptamine and $BaCl_2$ induced contractions of ileum preparations from guinea-pigs; this effect could be inhibited by atropine. Dimethomorph at concentrations of up to 30 µg/ml had no significant direct effects on ileum contractions but inhibited the effects of all four compounds almost completely. Unfortunately, dimethylformamide as vehicle also significantly inhibited the four contraction agonists (Ainsworth & Wright, 1991m).

(m) Surface anaesthetic activity in the guinea- pig

In male guinea-pigs, 1% dimethomorph as a dermal injection of 0.1 ml did not have a local anaesthetic effect (Ainsworth & Wright, 1991h).

3 Observations in humans

In a plant producing dimethomorph in Brazil, no health effects due to possible exposure to dimethomorph in 75 workers from the start of production on 1 October 1996 until 1 April 1999 were reported (Milanez, 1999; Milanez et al., 2000).

Comments

Biochemical aspects

In most studies, the batch of dimethomorph used consisted of mixtures of the *E* and *Z* isomers in approximately equal amounts. It was reported that the two isomers can interconvert on exposure to light.

In several studies, the absorption, distribution, metabolism and excretion of dimethomorph were investigated in rats treated orally. After single oral doses of 10 or 500 mg/kg bw administered by gavage to male and female rats, more than 90% of the lower dose was absorbed and excreted via bile and 7% via urine. At 500 mg/kg bw, absorption decreased to 65% in males and 40% in females. Pre-treatment of the animals with nonlabelled dimethomorph at the lower dose did not influence the pattern of excretion. At 10 mg/kg bw, t_{max} for total radioactivity was reached after 1.4–2.8 h and excretion was virtually complete after 48 h. After 24 h, up to 10% of the administered dose was found in the gastrointestinal tract (including contents, 0.4–1% in the gastrointestinal tract only). Less than 1% of the administered dose was found in the carcass and in liver, and

0.2% or less of the administered dose was found in the kidneys, plasma and erythrocytes. In all other organs, concentrations of radioactivity were no longer quantifiable. At 500 mg/kg bw, some delay in depletion from organs was observed; residual radioactivity in tissues as a percentage of administered dose was approximately threefold that at 10 mg/kg bw. After 168 h, the pattern of distribution was very similar. Dimethomorph is extensively metabolized by demethylation of one of the methoxy groups and formation of *O*-conjugate and degradation products of morpholine ring-opening were found.

Toxicological data

Dimethomorph is of low acute toxicity in rats; the oral LD_{50} was 350 mg/kg bw in females and 4300 mg/kg bw in males. The oral LD_{50}s for the *E* and *Z* isomers were similar. In studies with dimethomorph administered dermally or by inhalation, the LD_{50} was > 2000 mg/kg bw and the LC_{50} was > 4.24 mg/l, respectively. Dimethomorph was not a skin irritant and was not a skin sensitizer in the Magnusson & Kligman test. In a test for ocular irritation, all animals showed reddened conjunctivae and slight chemosis. These effects had resolved within 4 h after dosing.

In short- and long-term studies, the liver was consistently a target organ; increased organ weights were often accompanied by hepatocyte hypertrophy.

In short-term studies of toxicity in mice fed diets containing dimethomorph at up to 10 000 ppm, equal to 1145 mg/kg bw per day, dimethomorph was generally well tolerated, but increased liver weights were observed at all doses. In 4-week studies in rats given doses of 220 mg/kg bw per day and greater, decreased body-weight gains, liver-weight increases with increased hepatocyte hypertrophy and, at higher doses, histological changes in the intestine were observed. Generally, the same effects were also seen in two 4-week studies in rats given isolated *E* and Z isomers; NOAELs were 10 mg/kg bw per day. In a 13-week feeding study in rats allowed a recovery period after dosing, the NOAEL was 14 mg/kg bw per day on the basis of increased liver weights in females at higher doses; in males, an increase in vacuolation in adrenals was found and slightly changed haematology parameters at higher doses. In both sexes, vascular congestion of the ileum mucosa was also observed. The severity of most effects was reduced in the recovery period. In a later re-examination of histology slides, findings in the ileum of mice and rats, dilatation, mucosal hyperplasia and fiber separation in muscle layers were identified with rats being more sensitive than mice. In a 13-week feeding study in dogs, lip-licking and subdued behaviour were seen usually shortly after feeding at the highest dose of 43 mg/kg bw per day. At this dose, alkaline phosphatase activity in both sexes had nearly doubled, and males showed an increase in relative thymus weight and decreased prostate weight with increased fibrosis, while females showed an increase in liver weight. The NOAEL was 450 ppm, equal to 15 mg/kg bw per day. In the 52-week feeding study in dogs, liver weights in both sexes were increased at a dietary concentration of 450 ppm, equal to 15.2 mg/kg bw per day, and greater. At 1350 ppm, equal to 44.8 mg/kg bw per day, the highest dose, alkaline phosphatase activity was increased in males and females and, as in the 13-week study, prostate weights were decreased, with a very slightly increased severity of fibrosis. Testes weights were statistically significantly increased at the intermediate dose of 450 ppm and greater, but without a histological correlate and therefore this was not considered toxicologically relevant. The NOAEL was 450 ppm, equal to 15.2 mg/kg bw per day, on the basis of weight changes in the liver and prostate, and prostate fibrosis at 1350 ppm.

In a 24-month feeding study in mice, dimethomorph was well tolerated. Treatment-related findings were reduced absolute terminal body weight in males at 1000 mg/kg bw per day without decreased feed intake. In an interim sacrifice of animals at the highest dose at week 52, liver-weight increases and dilatations in the ileum were found; both effects were more pronounced in females. At study termination, more animals (males and females) with enlarged spleens were recorded, without a histological correlate. In females at the highest dose, an increase in the incidence of mammary adenocarcinomas was found that was within the range for historical controls. The Meeting concluded that the increased incidence in mammary adenocarcinoma was not due to a tumorigenic potential of dimethomorph. The NOAEL was 100 mg/kg bw per day on the basis of body-weight gain decreases at the highest dose.

There were two 24 month feeding studies in rats, one of toxicity and another one of carcinogenicity. In the long-term feeding study in rats, feed intake was not affected in any group. At the highest dose (2000 ppm, equal to 99.9 mg/kg bw per day), body weight was decreased in females and an increase in relative kidney weights in females and a statistically non-significant increase in liver weights were observed in both sexes. Incidences of histological changes in the liver included ground-glass foci, periacinar hypertrophy and pigmentation, and were increased statistically significantly at the highest dose of 2000 ppm. The NOAEL was 750 ppm, equal to 36.3 mg/kg bw per day in males and 57.7 mg/kg bw per day in females, on the basis of body-weight decreases and histological changes in the liver of females at 2000 ppm. Effects observed in the study of carcinogenicity in rats were similar to those seen in the study of toxicity. Feed intake was decreased only decreased by 6% in females at the highest dose only. At the lowest dose of 200 ppm, equal to 8.8 mg/kg bw per day, body-weight gain among females was reduced by 9%, with reductions of 23% and 38% at the next higher doses. Although body-weight gain in females at 750 ppm was statistically significantly decreased, the effect was not considered to be treatment-related because there was high variability in body-weight development in the control and the dosed groups between the two 2-year studies. In males, body-weight gain was only impaired in the group at the highest dose of 2000 ppm, equal to 94.6 mg/kg bw per day. In animals at the highest dose, an increase in swollen hind feet/limbs was seen, with unknown etiology. In males, the frequency of findings of lymph node cysts and dilated blood vessels was also increased. At dietary concentrations of 2000 ppm, statistically significantly increased incidences of histological changes in the liver of males and females were seen, increased pancreatitis being observed in males at the highest dose. As in the study of toxicity, males receiving dimethomorph showed more interstitial-cell hyperplasia and adenoma of the testes than did the controls. In both studies, the incidence of adenoma at 2000 ppm, equal to 94.6 mg/kg bw per day, the highest dose, were close to the upper limit of the range for historical controls. However, there was no clear dose–response relationship and therefore adenoma was considered as part of a continuum with interstitial-cell hyperplasia. Additionally, this type of tumour is usually secondary to hormonal perturbation, an effect that is not suggested by the current toxicology database for any species. In males at the highest dose, there was also an increase in the incidence of benign medulla tumours of the adrenals when compared with concurrent controls, but this was within the range for historical controls. The Meeting concluded that dimethomorph was not tumorigenic in rats. When evaluating the two 2-year studies together, the overall NOAEL was 750 ppm, equal to 36.3 mg/kg bw per day, on the basis of reduced body-weight gain in both sexes and histological changes in the liver of females at 2000 ppm.

Dimethomorph was not carcinogenic in mice or rats.

With the exception of three assays for chromosomal aberration in V79 cells and in human lymphocytes, dimethomorph gave negative results in a battery of appropriate tests for genotoxicity. A slight increase in the frequency of aberrant cells was found in V79 cells and in human lymphocytes at high doses, with reduced mitotic indices and slight precipitation of the compound.

The Meeting concluded that dimethomorph was unlikely to be genotoxic in vivo.

On the basis of the absence of carcinogenicity and genotoxicity, the Meeting concluded that dimethomorph is unlikely to pose a carcinogenic risk to humans.

In a two-generation study of reproductive toxicity studying rats, there was a reduction in body-weight gain in dams at the highest concentration of 1000 ppm, equal to 80 mg/kg bw per day, in the pre-mating period; this was compensated for thereafter and had no impact on reproductive performance. In F_1, F_{2a} and F_{2b} pups, the only finding was a slight delay in incisor eruption, which did not affect feeding capacity. Overall, pups developed normally; no other developmental markers, such as eye-opening, pinna unfolding or hair growth, were affected. Therefore, the NOAEL for maternal and reproductive toxicity was 1000 ppm, equal to 80 mg/kg bw per day, the highest dose tested.

In studies of developmental toxicity in rats, reduced feed consumption and reduced body-weight gain were recorded at the highest dose of 160 mg/kg bw per day on days 6–10 of gestation, but not thereafter. On day 20 of gestation, the difference in absolute body weight compared with that in control animals was marginal. At the highest dose, an increase in total litter losses and an increase in postimplantation losses in females with live fetuses at terminal sacrifice were observed. There were no other effects on fetal development. The NOAEL for developmental toxicity was 160 mg/kg bw per day, the highest dose tested. The NOAEL for maternal toxicity and embryotoxicity was 60 mg/kg bw per day, on the basis of intermittent decreases in body-weight gain in dams and postimplantation losses.

In two studies of developmental toxicity in rabbits, dams lost weight or did not show an increase in body weight during days 6–12 of gestation at 600 and 650 mg/kg bw per day, respectively. At this dose, the incidence of total litter losses was increased, but there were no increases in malformations or variations. The NOAEL for maternal and developmental toxicity was 300 mg/kg bw per day.

The Meeting concluded that dimethomorph is not teratogenic.

The Meeting considered that dimethomorph is not neurotoxic on the basis of the available data.

The Meeting concluded that the existing database on dimethomorph was adequate to characterize the potential hazards to fetuses, infants and children.

No health effects related to exposure to dimethomorph were reported in personnel working in a production plant.

Levels relevant to risk assessment

Species	Study	Effect	NOAEL	LOAEL
Mouse	Two-year study of toxicity and carcinogenicity[a]	Toxicity	100 mg/kg bw per day	1000 mg/kg bw per day
		Carcinogenicity	1000 mg/kg bw per day[c]	—
Rat	Two-year studies[d]	Toxicity	750 ppm, equal to 36.3 mg/kg bw per day	2000 ppm, equal to 99.9 mg/kg bw per day
		Carcinogenicity	2000 ppm, equal to 99.9 mg/kg bw per day[c]	—
	Two-generation study of reproductive toxicity[a]	Parental toxicity	1000 ppm, equal to 80 mg/kg bw per day[c]	—
		Offspring toxicity	1000 ppm, equal to 80 mg/kg bw per day[c]	—
	Developmental toxicity[b]	Maternal toxicity	60 mg/kg bw per day	160 mg/kg bw per day
		Embryo/fetotoxicity	60 mg/kg bw per day	160 mg/kg bw per day
Rabbit	Developmental toxicity[b]	Maternal toxicity	300 mg/kg bw per day	650 mg/kg bw per day
		Embryo/fetotoxicity	300 mg/kg bw per day	650 mg/kg bw per day
Dog	Thirteen-week and 1-year studies of toxicity[ad]	Toxicity	450 ppm, equal to 15.2 mg/kg bw per day	1350 ppm, equal to 44.8 mg/kg bw per day

[a] Dietary administration.
[b] Gavage administration.
[c] Highest dose tested.
[d] The results for two studies were combined.

Toxicological evaluation

The Meeting established an ADI of 0–0.2 mg/kg bw based on a NOAEL of 15.2 mg/kg bw per day identified on the basis of the liver weight and clinical chemistry changes and prostate weight changes and prostate fibrosis observed at higher doses in the 13-week and the1-year studies in dogs. A safety factor of 100 was applied.

The Meeting established an ARfD of 0.6 mg/kg bw based on a NOAEL of 60 mg/kg bw per day idenitified on the basis of postimplantation losses at higher doses in the study of developmental toxicity in rats. A safety factor of 100 was applied.

Estimate of acceptable daily intake for humans

0–0.2 mg/kg bw

Estimate of acute reference dose

0.6 mg/kg bw

Information that would be useful for the continued evaluation of the compound

Results from epidemiological, occupational health and other such observational studies of human exposures.

Critical end-points for setting guidance values for exposure to dimethomorph

Absorption, distribution, excretion and metabolism in mammals	
Rate and extent of oral absorption	Rapid, > 90% within 24 h
Dermal absorption	4.75% after application of single dose of 7.7 mg/kg bw for 8 h
Distribution	Extensive
Potential for accumulation	Low, no evidence of accumulation
Rate and extent of excretion	Rapid, close to 100% within 48 h, mainly via faeces
Metabolism in animals	Extensive, demethylation and morpholine ring-opening
Toxicologically significant compounds in animals, plants and the environment	Dimethomorph
Acute toxicity	
Rat, LD_{50}, oral	3900 mg/kg bw
	> 5000 mg/kg bw (*Z* isomer)
	4715 mg/kg bw (*E* isomer)
Rat, LD_{50}, dermal	> 2000 mg/kg bw; > 5000 mg/kg bw (*Z* isomer)
Rat, LC_{50}, inhalation	> 4.24 mg/l
Rabbit, skin irritation	Not irritating
Rabbit, eye irritation	Initially slightly irritating
Guinea-pig, skin sensitization (test method used)	Not a sensitizer (Magnusson & Kligman)
Short-term studies of toxicity	
Target/critical effect	Prostate and liver and clinical chemistry (dogs)
Lowest relevant oral NOAEL	15.2 mg/kg bw per day (dog)
Lowest relevant dermal NOAEL	No data

Lowest relevant inhalation NOAEC	No data
Genotoxicity	
	Unlikely to be genotoxic in vivo
Long-term studies of toxicity and carcinogenicity	
Target/critical effect	Liver histology and body-weight decrease (rats)
Lowest relevant NOAEL	36.3 mg/kg bw per day
Carcinogenicity	Not carcinogenic
Reproductive toxicity	
Reproduction target/critical effect	No reproductive effects
Lowest relevant reproductive NOAEL	80 mg/kg bw per day, the highest dose tested
Developmental target/critical effect	Increased incidence of total litter losses (rats and rabbits)
Lowest relevant developmental NOAEL	60 mg/kg bw per day (rats)
Neurotoxicity/delayed neurotoxicity	
	No evidence in conventional studies
Other toxicological studies	
	In pharmacological studies, evidence for an increase in phenobarbital sleeping time
Medical data	
	Medical surveillance of workers in a plant producing dimethomorph did not reveal any adverse health effects.

Summary

	Value	Study	Safety factor
ADI	0–0.2 mg/kg bw	Dog, 13-week and 1-year study	100
ARfD	0.6 mg/kg bw	Rat, study of developmental toxicity	100

References

Afzal, J. & Wu, D. (1995a) Dimethomorph (CL 336,379): blood pharmacokinetics of C-14 CL 336,379 derived residues in the rat. Unpublished report No. DK-452-008 from XenoBiotic Laboratories, Inc., North Branch, USA. Submitted to WHO by BASF, Ecully Cedex, France.

Afzal, J. & Wu, D. (1995b) Dimethomorph (CL 336,379): tissue distribution of C-14 CL 336,379 derived residues in the rat. Unpublished report No. DK-440-013 from XenoBiotic Laboratories Inc., Plainsboro NJ 08536, USA. Submitted to WHO by BASF, Ecully Cedex, France.

Ainsworth, G.A. & Wright, A. (1991a) Anticonvulsive activity in mice. Unpublished report No. DK-451-004 from Toxicol Laboratories Ltd, Ledbury Herefordshire, England. Submitted to WHO by BASF, Ecully Cedex, France.

Ainsworth, G.A. & Wright, A. (1991b) Assessment of potential anti-inflammatory activity using the carrageenan induced rat paw oedema model. Unpublished report No. DK-452-004 from Toxicol Laboratories Ltd, Ledbury Herefordshire, England. Submitted to WHO by BASF, Ecully Cedex, France.

Ainsworth, G.A. & Wright, A. (1991c) Cardiovascular, respiratory and nictitating membrane alterations produced by test compound in the anaesthetised cat. Unpublished report No. DK-452-006 from Toxicol Laboratories Ltd, Ledbury Herefordshire, England. Submitted to WHO by BASF, Ecully Cedex, France.

Ainsworth, G.A. & Wright, A. (1991d) Charcoal meal transit times in the rat small intestine. Unpublished report No. DK-452-003 from Toxicol Laboratories Ltd, Ledbury Herefordshire, England. Submitted to WHO by BASF, Ecully Cedex, France.

Ainsworth, G.A. & Wright, A. (1991e) Potentiation of hexobarbitone sleeping time in mice. Unpublished report No. DK-451-003 from Toxicol Laboratories Ltd, Ledbury Herefordshire, England. Submitted to WHO by BASF, Ecully Cedex, France.

Ainsworth, G.A. & Wright, A. (1991f) Safety pharmacology body temperature alteration in mice. Unpublished report No. DK-452-005 from Toxicol Laboratories Ltd, Ledbury Herefordshire, England. Submitted to WHO by BASF, Ecully Cedex, France.

Ainsworth, G.A. & Wright, A. (1991g) Spontaneous motor activity in mice. Unpublished report No. DK-451-002 from Toxicol Laboratories Ltd, Ledbury Herefordshire, England. Submitted to WHO by BASF, Ecully Cedex, France.

Ainsworth, G.A. & Wright, A. (1991h) Surface anaesthetic activity in the guinea-pig. Unpublished report No. DK-451-005 from Toxicol Laboratories Ltd, Ledbury Herefordshire, England. Submitted to WHO by BASF, Ecully Cedex, France.

Ainsworth, G.A. & Wright, A. (1991i) Tail flick test for analgesia. Unpublished report No. DK-451-001 from Toxicol Laboratories Ltd, Ledbury Herefordshire, England. Submitted to WHO by BASF, Ecully Cedex, France.

Ainsworth, G.A. & Wright, A. (1991j) The effect of test compound on rat uterine smooth muscle motility. Unpublished report No. DK-452-007 from Toxicol Laboratories Ltd, Ledbury Herefordshire, England. Submitted to WHO by BASF, Ecully Cedex, France.

Ainsworth, G.A. & Wright, A. (1991k) The effect of test compound on the Irwin test in mice. Unpublished report No. DK-451-006 from Toxicol Laboratories Ltd, Ledbury Herefordshire, England. Submitted to WHO by BASF, Ecully Cedex, France.

Ainsworth, G.A. & Wright, A. (1991l) The effect of test compound on the spontaneous motility of the isolated rabbit ileum. Unpublished report No. DK-452-001 from Toxicol Laboratories Ltd, Ledbury Herefordshire, England. Submitted to WHO by BASF, Ecully Cedex, France.

Ainsworth, G.A. & Wright, A. (1991m) The effect of the test compound on the isolated guinea-pig ileum and its actions on acetylcholine, histamine, 5-hydroxytryptamine and barium chloride. Unpublished report No. DK-452-002 from Toxicol Laboratories Ltd, Ledbury Herefordshire, England. Submitted to WHO by BASF, Ecully Cedex, France.

Bounds, S.V.J. (1995) CL 336,379 Absorption study in the male rat after topical application. Unpublished report No. DK-440-012 from Huntingdon Life Sciences Ltd, Huntingdon Cambridgeshire PE28 4HS, England. Submitted to WHO by BASF, Ecully Cedex, France.

Brooks, T. & Wiggins, D. (1989) Bacterial mutagenicity studies with CME-151. Unpublished report No. DK-435-018 from Sittingbourne Research Centre, Kent, England. Submitted to WHO by BASF, Ecully Cedex, France.

Esdaile, D.J. (1991a) SAG 151 E isomer: a 28 day oral toxicity study in rats. Unpublished report No. DK-470-015 from Sittingbourne Research Centre, Kent, England. Submitted to WHO by BASF, Ecully Cedex, France.

Esdaile, D.J. (1991b) SAG 151 Z isomer: a 28 day oral toxicity study in rats. Unpublished report No. DK-470-016 from Sittingbourne Research Centre, Kent ME9 8AG, England. Submitted to WHO by BASF, Ecully Cedex, France.

Everett, D. (1990) SAG 151 104-week dietary toxicity study in rats. Unpublished report No. DK-427-006 from Inveresk Research International Ltd, Tranent, Scotland. Submitted to WHO by BASF, Ecully Cedex, France.

Everett, D. (1991) SAG151 104 Week dietary carcinogenicity study in rats. Unpublished report No. DK-428-005 from Inveresk Research International Ltd, Tranent, Scotland. Submitted to WHO by BASF, Ecully Cedex, France.

Gardner, J.R. (1989) CME 151 technical material: Acute dermal toxicity, skin and eye irritancy. Unpublished report No. DK-412-002 from Sittingbourne Research Centre, Kent, England. Submitted to WHO by BASF, Ecully Cedex, France.

Glaza, S.M. (1997) Dermal sensitization study of AC 336379 in guinea pigs - maximization test (EC Guidelines). Unpublished report No. DK-416-003 from Corning Hazleton Inc, Vienna, USA. Submitted to WHO by BASF, Ecully Cedex, France.

Greenough, R.J. & Goburdhun, R. (1986a) CME 151 - dietary maximum tolerated dose study in dogs. Unpublished report No. DK-420-001 from Inveresk Research International, Musselburgh EH21 7UB, England. Submitted to WHO by BASF, Ecully Cedex, France.

Greenough, R.J. & Goburdhun, R. (1986b) CME 151: 13 Week dietary toxicity study in dogs. Unpublished report No. DK-425-002 from Inveresk Research International Ltd, Tranent, Scotland. Submitted to WHO by BASF, Ecully Cedex, France.

Greenough, R.J. & Goburdhun, R. (1988) SAG 151 52-week dietary toxicity study in dogs. Unpublished report No. DK-427-003 from Inveresk Research International Ltd, Musselburg, Scotland. Submitted to WHO by BASF, Ecully Cedex, France.

Heidemann, A. & Miltenburger, H.G. (1986) Chromosome aberrations in cells of Chinese hamster cell line V79. Unpublished report No. DK-435-004 from Laboratorium fur Mutagenitatsprufung, TH Darmstadt, Germany. Submitted to WHO by BASF, Ecully Cedex, France.

Heidemann, A. & Miltenburger, H.G. (1987) Chromosome aberrations in cells of Chinese hamster cell line V79. Unpublished report No. DK-435-006 from Laboratorium fur Mutagenitatsprufung, TH Darmstadt, Germany. Submitted to WHO by BASF, Ecully Cedex, France.

Heusener, A. & Eberstein, M. (1985) Acute toxicity study in rats after epicutaneous administration. Unpublished report No. DK-412-001 from E. Merck, 64271 Darmstadt, Germany. Submitted to WHO by BASF, Ecully Cedex, France.

Heusener, A. & Jacobs, M. (1987a) Acute toxicity study in rats after oral administration. Unpublished report No. DK-411-008 from E. Merck, 64271 Darmstadt, Germany. Submitted to WHO by BASF, Ecully Cedex, France.

Heusener, A. & Jacobs, M. (1987b) Acute toxicity study in rats after oral administration. Unpublished report No. DK-411-009 from E. Merck, 64271 Darmstadt, Germany. Submitted to WHO by BASF, Ecully Cedex, France.

Jackson, G.C. & Hardy, C.J. (1986) ZTH 236 Z50 Acute inhalation toxicity study in rats 4-hour exposure. Unpublished report No. DK-413-001 from Huntingdon Research Centre, Huntingdon, England. Submitted to WHO by BASF, Ecully Cedex, France.

Kynoch, S.R. (1985) Acute oral toxicity to rats of ZTH 236 Z50 and Report Amendment No. 1. Unpublished report No. DK-411-004 from Huntingdon Research Centre, Huntingdon, England. Submitted to WHO by BASF, Ecully Cedex, France.

Milanez, A. (1999) Health surveillance for dimethomorph at the Cyanamid Resende Plant, Resende, Brazil. Unpublished report No. DK-445-006 from Cyanamid Quimica Brazil. Submitted to WHO by BASF, Ecully Cedex, France.

Milanez, A., Almeida, J.M. & Nass, W. (2000) Dimethomorph: medical data information - status February 2000. Unpublished report No. 2000/1022687 from Cyanamid Agrar, Resende, Brazil. Submitted to WHO by BASF, Ecully Cedex, France.

Müller, W. (1987a) CME 151 - preliminary oral (gavage) embryotoxicity study in the rabbit. Unpublished report No. DK-432-003 from Hazleton Laboratories Deutschland GmbH, Kesselfeld 29, Germany. Submitted to WHO by BASF, Ecully Cedex, France.

Müller, W. (1987b) Treatment of pregnant Sprague-Dawley rats with CME 151 - Letter report. Unpublished report No. DK-432-006 from Hazleton Laboratories Deutschland GmbH, Kesselfeld 29, Germany. Submitted to WHO by BASF, Ecully Cedex, France.

Müller, W. (1989a) SAG 151 Oral (gavage) teratogenicity study in the rabbit. Includes Amendment No. 1 to Final Report. Unpublished report No. DK-432-004 from Hazleton Laboratories, Münster, Germany. Submitted to WHO by BASF, Ecully Cedex, France.

Müller, W. (1989b) SAG 151 Oral (gavage) teratogenicity study in the rat. Unpublished report No. DK-432-002 from Hazleton Laboratories, Münster, Germany. Submitted to WHO by BASF, Ecully Cedex, France.

Osterburg, I. (1986) ZTH 236Z50 - preliminary oral (gavage) embryotoxicity study in the rat. Unpublished report No. DK-432-001 from Hazleton Laboratories Deutschland GmbH, Kesselfeld 29, Germany. Submitted to WHO by BASF, Ecully Cedex, France.

Osterburg, I. (1990) SAG 151 Two generation oral (dietary administration) reproduction toxicity study in the rat (two litters in the F1 generation). Unpublished report No. DK-430-001 from Hazleton Laboratories, Münster, Germany. Submitted to WHO by BASF, Ecully Cedex, France.

Perry, C.J. (1986) CME 151 – 6-week dietary dose range finding study in mice. Unpublished report No. DK-420-003 from Inveresk Research International, Musselburgh, Scotland. Submitted to WHO by BASF, Ecully Cedex, France.

Perry, C.J. (1987) CME 151 - histopathological examination of ileum sections taken from the rat (IRI project No. 435025) and mouse (IRI project No. 435067) dose range finding studies. Unpublished report No. DK-420-004 from Inveresk Research International, Musselburgh, Scotland. Submitted to WHO by BASF, Ecully Cedex, France.

Poth, A. & Miltenburger, H.G. (1986) Cell transformation assay with Syrian hamster embryo (SHE) cells. Unpublished report No. DK-435-005 from Laboratorium fur Mutagenitaetspruefung, TH Darmstadt, Darmstadt, Germany. Submitted to WHO by BASF, Ecully Cedex, France.

Ruckman, S.A. (1987) CME 151 toxicity to rats by dietary admixture for 13 weeks with a 4-week withdrawal period (final report). Unpublished report No. DK-425-001 from Huntingdon Research Centre, Huntingdon, England. Submitted to WHO by BASF, Ecully Cedex, France.

Schluter, H. (1990) The biokinetics and metabolism of ^{14}C-dimethomorph in the rat. Unpublished report No. DK-440-001 from Shell Forschung GmbH, Schwabenheim, Germany. Submitted to WHO by BASF, Ecully Cedex, France.

Schluter, H. (1991) ^{14}C-Dimethomorph (CME 151): investigation on the nature of metabolites occurring in rats. Unpublished report No. DK-440-006 from Shell Forschung GmbH, Schwabenheim, Germany. Submitted to WHO by BASF, Ecully Cedex, France.

Schluter, H. (1993) ^{14}C-Dimethomorph (CME 151): investigation of the metabolite profiles in the rat following treatment with two different labels. Unpublished report No. DK-440-011 from Shell Forschung GmbH, Schwabenheim, Germany. Submitted to WHO by BASF, Ecully Cedex, France.

Schluter, H. & Grahl, U. (1994) ^{14}C-Dimethomorph: additional investigation on the nature of metabolites occurring in rats. Unpublished report No. DK-440-010 from Cyanamid Forschung GmbH, Schwabenheim, Germany. Submitted to WHO by BASF, Ecully Cedex, France.

Scott, K. (1986) CME 151 4-week dietary dose range finding study in rats. Unpublished report No. DK-420-002 from Inveresk Research International Ltd, Tranent, Scotland. Submitted to WHO by BASF, Ecully Cedex, France.

Smith, A. (1991) SAG 151 104-week dietary carcinogenicity study in mice. Unpublished report No. DK-428-004 from Inveresk Research International Ltd, Scotland. Submitted to WHO by BASF, Ecully Cedex, France.

Timm, A. & Miltenburger, H.G. (1986) Unscheduled DNA synthesis in hepatocytes of male rats in vitro (UDS test). Unpublished report No. DK-435-002 from Laboratorium fur Mutagenitatsprufung, TH Darmstadt, Germany. Submitted to WHO by BASF, Ecully Cedex, France.

van de Waart, E.J. (1991a) Evaluation of the ability of dimethomorph to induce chromosome aberrations in cultured peripheral human lymphocytes. Unpublished report No. DK-435-013 from RCC Notox B.V., Hertogenbosch, Netherlands. Submitted to WHO by BASF, Ecully Cedex, France.

van de Waart, E.J. (1991b) Evaluation of the mutagenic activity of dimethomorph in an in vitro mammalian gene mutation test with V79 Chinese hamster cells (with independent repeat). Unpublished report No. DK-435-014 from RCC Notox B.V., Hertogenbosch, Netherlands. Submitted to WHO by BASF, Ecully Cedex, France.

van de Waart, E.J. (1991c) Micronucleus test in bone marrow cells of the mouse with dimethomorph. Unpublished report No. DK-435-012 from RCC Notox B.V., Hertogenbosch, Netherlands. Submitted to WHO by BASF, Ecully Cedex, France.

Van Dijk, A. (1990) ^{14}C-Dimethomorph (CME 151): absorption, distribution, metabolism and excretion after bile cannulation and single oral administration to the rat. Unpublished report No. DK-440-002 from RCC Umweltchemie AG, Itingen, Switzerland. Submitted to WHO by BASF, Ecully Cedex, France.

Warren, S. (1985) ZTH 236 Z50 Preliminary assessment of toxicity to rats by dietary admixture for 4 weeks. Unpublished report No. DK-420-005 from Huntingdon Research Centre, Huntingdon, England. Submitted to WHO by BASF, Ecully Cedex, France.

FLUSILAZOLE

First draft prepared by
C. Adcock[1] and M. Tasheva[2]

[1]Health Evaluation Directorate, Pest Management Regulatory Agency, Health Canada, Canada; and
[2]National Center of Public Health Protection, Sofia, Bulgaria

Explanation

Flusilazole is the International Organization for Standardization (ISO) approved name for 1-[[bis(4-fluorophenyl)methyl]silyl]methyl]-1*H*-1,2,4-triazole (Chemical Abstracts Service, CAS No. 85509-19-9). It is a broad-spectrum fungicide that belongs to the triazole subclass of ergosterol biosynthesis inhibitors. Flusilazole was previously evaluated by the Joint Meeting in 1989 (Annex 5, references *56*, *58*) and in 1995. An acceptable daily intake (ADI) of 0–0.001 mg/kg bw was allocated in 1989, based on a no-observed-adverse-effect level (NOAEL) of 0.14 mg/kg bw per day (5 ppm) for liver toxicity in a 1-year feeding study in dogs. This was confirmed in 1995. The compound was re-examined by the present Meeting as part of the periodic re-evaluation programme of the Codex

Committee on Pesticide residues (CCPR). Three new studies were provided, two studies of developmental toxicity in rats (one of oral and one of dermal administration) and a 28-day mechanistic study in dogs.

Owing to the age of the database, some studies predate the establishment of good laboratory practice (GLP); however, all critical studies complied with GLP.

Evaluation for acceptable daily intake

1. Biochemical aspects

The toxicokinetics of technical-grade flusilazole have been studied after oral administration of radiolabelled test material in rats. Summaries of the relevant data are presented below.

1.1 Absorption, distribution and excretion

In a GLP-compliant study, groups of two male and two female Charles River CD rats were given [^{14}C]flusilazole (uniformly labelled on the phenyl ring) orally by one of the following three regimens: (1) a single low dose of at approximately 8 mg/kg bw; (2) a single low dose of approximately 8 mg/kg bw after pre-treatment for 21 days on a diet containing non-labelled flusilazole at 100 ppm; (3) a single high dose of approximately 200 mg/kg bw. By 96 h (regimen 1) or 168 h (regimen 2 or 3), a total of approximately 90% of the administered radioactivity was eliminated in the urine and faeces. The excretion half-life ($t_{1/2}$) was approximately 34 h. The amount of radioactivity recovered in the expired air (as CO_2 or volatile metabolites) was not significant. The main route of elimination was in the faeces and there was an apparent sex difference in the excretion pattern. In males, 87% and 8% of the administered dose was eliminated in faeces and urine, respectively, while in females, the corresponding figures were 59% and 23%. Pre-treatment with non-radiolabelled flusilazole did not affect excretion. Tissue retention of radioactivity was very low, with total residues accounting for < 2.5% of the administered dose. The highest amounts of radioactivity were found in the carcass, gastrointestinal tract and liver (mean, < 1% of administered dose, in each case). Tissue concentrations within the groups at the lower and higher dose were proportional to the amount of flusilazole administered. In the same study, an additional group of one male and one female rat was given a single oral dose of [^{14}C]flusilazole (triazole-3-labelled) at approximately 8 mg/kg bw. As with the phenyl-labelled flusilazole, 88% of the administered radioactivity was recovered in the urine and faeces within 96 h after dosing. However, urine was the predominant route of excretion. In males, urinary and faecal radioactivity accounted for 78% and 11% of the administered dose, respectively, while in females, the corresponding figures were 59% and 26%. Tissue retention of radioactivity was very low, with total residues accounting for approximately 3% of the administered dose. The highest concentrations of radioactivity were found in the carcass (approximately 2% of the administered dose), skin, gastrointestinal tract and liver (mean, < 0.5% of the administered dose, in each case) (Anderson et al., 1986).

In another GLP-compliant study, groups of five male and five female Charles River Crl:CD(SD) BR rats were given [^{14}C]flusilazole (triazole-3-labelled) orally by one of the following three regimens: (1) a single dose at approximately 8 mg/kg bw; (2) a single dose at approximately 8 mg/kg bw after pre-treatment with non-radiolabelled flusilazole at approximately 8 mg/kg bw per day by gavage for 14 days; (3) a single dose at approximately 224 mg/kg bw. At 96 h (regimen 1) or 120 h (regimen 2 or 3), total recovery of radioactivity was 92.6–99.2%, with approximately 90% of the radioactivity eliminated within the first 48 h after dosing. Urine was the primary route of excretion, accounting for approximately 72% of the administered dose, while faecal excretion accounted for approximately 17%. Sex or regimen differences were not observed in this study. Tissue retention of radioactivity was very low. Total residues excluding the carcass (which accounted for < 3% of the administered dose) were < 0.2% (Cheng, 1986).

1.2 Biotransformation

[^{14}C]Flusilazole was extensively metabolized when administered orally to Charles River CD rats. Recovered parent compound accounted for only 2–11% of the administered dose in all groups regardless of dose (low/high) or preconditioning (with/without), found predominantly in faeces; urinary levels were < 1% of the administered dose. After absorption, flusilazole was cleaved at the triazole ring. With phenyl-labelled test material, the major faecal metabolites identified were: (i) [bis(4-fluorophenyl)methyl] silanol (approximately 30% of the administered dose in males, approximately 19% in females); (ii) [bis(4-fluorophenyl)methylsilyl] methanol (approximately 9% of the administered dose in both sexes); (iii) the fatty acid conjugates of [bis(4-fluorophenyl)methylsilyl] methanol (19% of the administered dose in males, 10% in females); and (iv) disiloxane (approximately 11% in males, approximately 7% in females). Except for the fatty acid conjugates, the same metabolites were found in the urine. In males, all three urinary metabolites were < 1% of the administered dose; in females, [bis(4-fluorophenyl)methyl] silanol, [bis(4-fluorophenyl)methylsilyl] methanol and siloxane accounted for 7.5%, 2.2% and 1.9% of the administered dose, respectively. With triazole-labelled material, the main metabolite identified was 1*H*-1,2,4-triazole, which was found predominantly in the urine (63.8% of the administered dose in males, 51.6% in females). Faeces contained only a minor amount of the metabolite (4% of the administered dose in males, 17% in females). On the basis of these results, a metabolic pathway for flusilazole in the rat was proposed (Anderson, 1986).

2. Toxicological studies

2.1 Acute toxicity

The results of studies of acute toxicity with flusilazole technical are summarized in Table 1. All studies complied with GLP.

Flusilazole is slightly to moderately toxic in rats when given as a single dose via the oral route; and minimally toxic in rats and rabbits when administered acutely via the dermal or inhalation route. Symptoms of toxicity after oral administration included weight loss, weakness, lethargy and at higher doses, prostration, salivation, laboured breathing, convulsions and loss of righting reflex. Dermal administration resulted only in mild erythema at site of application. Inhalation effects were mainly laboured breathing and lung sounds.

Table 1. Results of studies of acute toxicity with flusilazole

Species	Strain	Sex	Route	LD_{50} (mg/kg bw)	LC_{50} (mg/l air)	Purity (%)	Reference
Rat	Crl:CD	Males	Oral	1110	—	> 92.7	Wylie et al. (1984a)
		Females		674			
	Crl:CD(SD)IGC BR	Males	Oral	1216	—	> 92.7	Finlay (2000)
		Females		672			
	Crl:CD(SD)BR	Males & females	Inhalation	—	6.8–7.7	> 92.7	Turner et al. (1985)
Rabbit	New Zealand White	Males & females	Dermal	> 2000	—	> 92.7	Gargus & Sutherland (1983)

(a) Ocular irritation

Rabbits

Two male New Zealand White rabbits were given undiluted flusilazole (purity, 90%) at a dose of 0.01 ml in the right conjunctival sac. The left eyes were left untreated to serve as negative controls. One rabbit had its right eye washed with 100 ml of water after 20 s of exposure. Ocular irritation was graded according to the Draize scale at 1, 4, 24, 48 and 72 h after instillation. This study did not comply with GLP, but was considered to be a conclusive study. Both rabbits showed mild conjunctival redness and chemosis at 1–4 h on day of instillation. The washed eye also showed some discharge at 1 h; but by 24 h these effects were no longer observed. The unwashed eye showed slight corneal opacity at 1–4 h and biomicroscopic examination revealed slight cloudiness at 24–48 h; the eye was normal at 72 h. Flusilazole was minimally irritating to the eyes of male New Zealand White rabbits, causing transient mild irritation. Flushing the eye with water after exposure eliminated the corneal but not the conjunctival effects (Hall et al., 1984).

Flusilazole was evaluated for potential acute eye irritation in a GLP-compliant study performed in compliance with test guideline OECD 405. Six young adult female New Zealand White rabbits were given approximately 46 mg of the test substance in a volume of approximately 0.1 ml in one eye. The eyes remained unwashed after treatment. The conjunctiva, iris, and cornea of each treated eye were evaluated and scored according to a numerical scale approximately 1, 24, 48, and 72 h after administration of the test substance. The test substance produced corneal opacity (score of 1), iritis (score of 1), conjunctival redness (score of 1 or 2), conjunctival chemosis (score of 1, 2, or 3), and discharge (score of 2 or 3) in the treated eyes of all six rabbits. The treated eye of one rabbit was normal by 48 h, and the treated eyes of the remaining five rabbits were normal by 72 h after instillation of the test substance. No adverse weight loss or clinical signs of toxicity occurred during the study. All the rabbits pawed their treated eye after instillation of the test substance. Mean values were calculated from numerical scores obtained from the quantitative evaluation of ocular response observed in all rabbits at 24, 48, and 72 h after treatment. The results of this study indicate that flusilazole is minimally irritating to the eye of rabbits (Finlay, 1998).

(b) Dermal irritation

Rabbits

In a GLP-compliant study of primary skin irritation in New Zealand White rabbits, six young males received flusilazole formulation (Nustar 20DF, containing 21% flusilazole technical and 79% inert ingredients that were not specified) at a concentration of approximately 0.5 g (undiluted, moistened with distilled water)applied topically to shaved skin sites on the back . The test site was covered with gauze patch and a sheet of rubber for 4 h, after which all coverings were removed and the site was washed gently with warm water. Each test site was evaluated for irritation potential according to the Draize scale at 4, 24, 28 and 72 h after initial dosing. At 4 h, erythema (grade 1) was observed in four out of six rabbits and oedema was found in two out of six rabbits (grade 1 or 2). By 24 h, only two rabbits had erythema and no oedema was observed. No signs of dermal irritation were evident in any rabbit at 72 h.

The primary irritation index was 0.167. On the basis of these results, flusilazole (as formulated for this test) was minimally irritating to the skin of rabbits (Brock, 1988).

Guinea-pigs

In young adult Hartley guinea-pigs, a range-finding GLP-compliant study indicated mild erythema 24 h after dermal application of flusilazole (purity, 90%) at a concentration of 100% (aliquots of 0.05 ml); no dermal irritation was observed at a concentration of 50%. In a study of primary skin irritation, flusilazole (purity, 90%) was applied topically to the skin at concentrations of 5% or 50% (w/v, solution in demethyl phthalate) and the test sites (two per animal) were scored for signs of irritation at 24 h and 48 h. No dermal irritation was evident at any time in any of the 10 animals exposed. Flusilazole was found to be non-irritating to the skin of guinea-pigs (Wylie et al., 1984b).

(c) Dermal sensitization

In a GLP-compliant study of sensitization in young adult Hartley guinea-pigs, intradermal injections of flusilazole (purity, 90%) (1% solution, w/v; in methyl phthalate) weekly for 4 weeks caused erythema and edema with necrotic centres at sites of injection at 24 h. Challenge with topical applications of 5 or 50% solutions to the skin did not demonstrate sensitization. Flusilazole was not a dermal sensitizer in the guinea pig (Wylie et al., 1984b)

In another GLP-compliant study of sensitization, 10 male and 10 female young adult Duncan Hartley albino guinea-pigs were given three weekly dermal applications of flusilazole (purity, 97.7%) of 0.4 ml (equivalent to 0.192 g). The test material (slightly moistened with dimethyl phthalate) was applied to the shaved intact skin at the back of the guinea-pig and covered with plastic wrap for 6 h. The dermal irritation response was scored at 24 and 48 h after treatment. No signs of dermal irritation were observed at any time in any of the guinea-pigs after any of the induction applications. Two weeks after the last induction treatment, the guinea-pigs were challenged with a single dermal application of flusilazole of 0.4 ml (0.192 g) on an untreated site that was then covered for 6 h. the irritation response was scored again at 24 h and 48 h. No signs of dermal irritation were observed at any time in any of the guinea-pigs tested. A positive-control group treated with 1-chloro,-2,4-dinitrobenzene (DNCB) showed severe erythema with necrosis at 24 h after the second and third induction applications and severe erythema after the challenge dose 2 weeks later. Flusilazole was not a dermal sensitizer in guinea-pigs under the conditions of the study (Brock et al., 1988)

2.2 *Short-term studies of toxicity*

(a) Oral administration

Mice

Groups of 20 male and 20 female Crl:CD-1 mice were given diets containing flusilazole technical (purity, 96.7%) at a concentration of 0, 25, 75, 225, 500 or 1000 ppm (equal to 0, 4, 12, 36, 82 and 164 mg/kg bw per day in males and 0, 5, 15, 43, 92 and 222 mg/kg bw per day in females) for up to 90 days. Ten males and ten females from each group were killed after 4 weeks of dosing; the remaining ten males and ten females from each group were killed at study termination (90 days). No histopathological examination was conducted on mice killed at 4 weeks. This study was not GLP-compliant, but was performed in compliance with the United States Environmental Protection Agency (EPA) guidelines and was considered to be a conclusive study. There were no clinical symptoms of toxicity and no treatment-related effects on body weight or food consumption in mice killed at either interval. The target organs of toxicity were liver and urinary bladder; females were slightly more sensitive to the test substance than males. No treatment-related effects were observed at the lowest dietary concentration of 25 ppm. At the next higher concentration of 75 ppm, increased absolute and relative liver weights and increased incidence (1 out of 10) of hepatocellular vacuolar cytoplasmic changes were observed in females, which was considered to be an adaptive response. At 225 ppm and/or higher, elevated liver weight, dose-related increases in the incidence of hepatocellular vacuolar cytoplasmic changes, hepatocellular hypertrophy and urinary bladder urothelial-cell hyperplasia were evident in both sexes. At the highest dose of 1000 ppm, absolute and relative kidney weights were decreased in males, but no treatment-related renal lesions were observed. In addition, slight decreases in erythroid parameters (haemoglobin, erythrocyte volume fraction and erythrocyte counts) were also noted in both sexes. The NOAEL was 75 ppm (equal to 12 mg/kg bw per day) (Pastoor et al., 1984).

In a GLP compliant study, groups of 16 male and 16 female Crl:CD-1(ICR)BR mice were fed diets containing flusilazole technical (purity, 94%) at a concentration of 0, 1000, 2500 or 5000 ppm (equal to 0, 161, 436 and 1004 mg/kg bw per day in males and 0, 239, 601 and 1414 mg/kg bw per day in females) for at least 90 days. Additional groups of six males and six females were assessed for cellular proliferation in the liver and urinary bladder, three males and three females being assessed each on days 14 and 106. At the lowest dose of 1000 ppm (equal to 161 mg/kg bw per day), treatment-related effects included: reduced mean body weight and food efficiency (males only); increased absolute and relative liver weights (both sexes); decreased absolute and relative kidney weights (females only); and hypertrophy/hyperplasia, cytoplasmic vacuolation and inflammation in the liver and urinary bladder (both sexes). At the intermediate dose of 2500 ppm, increased cellular proliferation of the urinary bladder (both sexes) was also observed. At the highest dose of 5000 ppm, severe body-weight loss occurred in males and reduced body weight and food efficiency were noted in females. The males at 5000 ppm were all killed on day 44 owing to excess mortality and moribund condition. No unscheduled death/killing occurred in females. No definitive NOAEL could be identified in this study (Keller, 1990).

Rat

In a non-GLP compliant, supplemental study, groups of six male Crl:CD rats were given flusilazole technical (purity, 95.5%) at a dose of 0 or 300 mg/kg bw per day orally by gavage (in corn oil) for 5 days per week for 2 weeks. Three rats per group were killed at the end of the treatment period; the remaining three rats per group were killed after a 2-week recovery period. All rats were examined histopathologically. One treated rat died after the fifth dose (on day 7). Clinical signs of toxicity (lower body weight, alopecia, diarrhoea, stained/wet perineal area, salivation and hypersensitivity) were observed in four out of six treated rats during the treatment period. Treatment-related histopathological changes were observed in liver, kidney, urinary bladder and testes; the lesions appeared to be less severe in rats killed after a 2-week recovery period. Histopathological findings included: hepatocellular vacuolation (six out of six treated rats); hyperplasia and vacuolation of the urinary bladder transitional epithelium (six out of six treated rats) and renal pelvis urothelium (two out of six treated rats); and necrosis and cellular degeneration of the seminiferous tubule germinal epithelium (two out of six treated rats) (Wylie et al., 1984c).

Four groups of 10 male and 10 female Charles River CD rats were fed daily diets containing flusilazole technical (purity, 96.7%) at a concentration of 0, 25, 125, 375 or 750 ppm (equal to 0, 2, 9, 27 and 55 mg/kg bw per day in males and 0, 2, 11, 31 and 70 mg/kg bw per day in females) for 90 days. This study was not GLP-compliant, but was considered to be conclusive. No treatment-related effects were oberved at the lower doses of 25 and 125 ppm. At the next higher dose of 375 ppm, an increased concentration of serum cholesterol (both sexes) and increased incidence of mild bladder urothelial hyperplasia in males (1 out of 10) and females (4 out of 10) were noted. At the highest dose of 750 ppm, bladder urothelial hyperplasia was evident in 5 out of 10 males and 8 out of 10 females. Additional treatment-related changes included decreased body weight in females; increased absolute and relative liver weights (both sexes); and histopathological findings (hepatocellular hypertrophy, mild fatty changes and hepatocytolysis) in the liver of 5 out of 10 males. The NOAEL was 125 ppm, equal to 9 mg/kg bw per day (Pastoor et al., 1983).

Dogs

Groups of four male and four female beagle dogs were fed diets containing flusilazole technical (purity, 93%) at a concentration of 0, 25, 125 or 750/500 ppm (equal to 0.9, 4.3 and 13.4 mg/kg bw per day in males and 0, 0.9, 4.3 and 14.2 mg/kg bw per day in females) for 3 months. The group receiving the highest dose (750/500 ppm) received diet containing flusilazole at 750 ppm for the first 3 weeks, then control diet for 1 week and diet containing flusilazole at 500 ppm for the remainder of the study period. The dose was reduced owing to marked body-weight loss and decreased food intake at 750 ppm. This study was not GLP-compliant, but was considered to be conclusive.

At the lowest dose of 25 ppm (equal to 0.9 mg/kg bw per day), there was a treatment-related increase in the incidence of hyperplasia of the lymphoid follicles in the pyloric glandular mucosa of the stomach in males (zero out of four, three out of four, three out of four, and four out of four at 0, 25, 125 and 750/500 ppm, respectively). At the next higher dose of 125 ppm, an increased incidence of pyloric granular mucosa hyperplasia was also observed in females (zero out of four, zero out of four, three out of four, and four out of four at 0, 25, 125 and 750/500 ppm, respectively). Higher levels (relative to control values) of alanine aminotransferase activity and an increased incidence of mild urinary-bladder mucosal hyperplasia were observed in males at 125 ppm and in both sexes at the highest dose of 750/500 ppm. At 750/500 ppm (13.4 mg/kg bw per day), additional treatment-related effects included: clinical signs of toxicity (weakness or tremors) in both sexes; reduced mean body-weight gain (males); body-weight loss (females) and decreased mean food consumption (both sexes); a slight increase in leukocyte and monocyte counts (males) and lower (relative to control values) plasma concentrations of cholesterol, total protein and albumin concentrations (both sexes); and increased absolute and relative liver weights (both sexes). No NOAEL was identified in this study (Rickard et al., 1983).

In a GLP-compliant study, groups of five male and five female beagle dogs were a daily diet containing flusilazole technical (purity, 95.8%) at a concentration of 0, 5, 20 or 75 ppm (equal to 0, 0.14, 0.7 and 2.4 mg/kg bw per day in males and 0, 0.14, 0.7 and 2.6 mg/kg bw per day in females) for 1 year. No treatment-related effects were observed at the lowest dose of 5 ppm. At the next higher dose of 20 ppm, serum albumin concentration was decreased (males) and there was a dose-related increase in the incidence of centrilobular hepatocellular hypertrophy in males (zero out of five, zero out of five, four out of five, and five out of five) and females (zero out of five, zero out of five, two out of five, and five out of five). An increased severity of gastric-mucosa lymphoid hyperplasia (from minimal/mild at 0–20 ppm to moderate at 75 ppm) was also noted in both sexes. At the highest dose of 75 ppm, additional treatment-related effects included: higher leukocyte counts (relative to control values, both sexes); elevated alkaline phosphatase activity and lower cholesterol, total protein and albumin concentrations (males only); and increases in relative liver (both sexes) and relative kidney weights (females). All dogs at the highest dose exhibited a greater degree of hepatic centrilobular inflammatory infiltration; distinct centrilobular hepatocellular vacuolation was observed in three out of five males. Lymphoid hyperplasia in the gastric mucosa was of moderate severity in males and females. The NOAEL was 20 ppm (equal to 0.7 mg /kg bw per day) (O'Neal et al., 1985)

(b) Dermal application

Rabbits

Groups of five male and five female New Zealand White rabbits were treated dermally with flusilazole technical (purity, 94.9%) at a concentration of 0, 1, 5, 25 or 200 mg/kg bw per day for 21 days. The study was performed in compliance with GLP and test guideline OECD 410. The test material was applied daily as a paste in distilled water; the exposure sites were occluded for 6 h and then washed with water. No evidence of any systemic toxicity was observed. The NOAEL for dermal toxicity was 5 mg/kg bw per day on the basis of diffuse epidermal hyperplasia/thickening (slight to mild) at 25 mg/kg bw per day and greater. At the highest dose of 200 mg/kg bw per day, mild erythema was also evident from days 6–12. The NOAEL for systemic toxicity was 200 mg/kg bw per day (Sarver et al., 1986).

2.3 *Long-term studies of toxicity/carcinogenicity*

Mice

In a GLP-compliant study, groups of 80 male and 80 female Crl:CD-1(ICR)BR mice were given diets containing flusilazole technical (purity, 96.5%) at a concentration of 0, 5, 25 or 200 ppm (equal to 0, 0.66, 3.4, and 27 mg/kg bw per day in males and 0, 0.92, 4.6 and 36 mg/kg bw per day in females) for 18 months. Each dose group contained a satellite group of 10 males and 10 females for interim killing at 6 months.

There were no treatment-related effects on mortality (terminal survival rates were 76–86% in males and 57–80% in females), body weight, food consumption, haematology or serum chemistry at any dose tested in the study. At the highest dose of 200 ppm, a toxicologically significant increase in incidence of hepatocellular fatty changes was observed (incidence at termination was 4 out of 80, 3 out of 80, 10 out of 80 and 40 out of 80 in males and 2 out of 80, 4 out of 80, 3 out of 80 and 24 out of 80 in females at 0, 5, 25 and 200 ppm, respectively). In addition, there was an increase in absolute and relative liver weights (both sexes), a decrease in absolute kidney weight (females) and an increase in lymphocytic infiltration in the lung and urinary bladder (males). There was no treatment-related increase in the incidence of any tumour type at any dose up to and including 200 ppm (equal to 27 mg/kg bw per day), the highest dose tested. Flusilazole technical was not oncogenic in the mouse under the conditions of the study. The NOAEL for systemic toxicity was 25 ppm, equal to 3.4 mg/kg bw per day (Brock et al., 1985).

Groups of 100 male and 100 female Crl:CD-1(ICR)BR mice were given daily diets containing flusilazole technical (purity, 94%) at a concentration of 0, 100, 500 or 1000 ppm in males (equal to 0, 14.3, 73.1 and 144 mg/kg bw per day) and 0, 100, 1000 or 2000 ppm in females (equal to 0, 19.4, 200 and 384 mg/kg bw per day) for 18 months. An additional group of 100 males and 100 females was fed a daily diet containing flusilazole technical at a concentration of 25 ppm (equal to 3.51 mg/kg bw per day in males and 4.38 mg/kg bw per day in females) and was used only for the study of cell proliferation; no other end-points of toxicity were measured in this group. Doses were selected on the basis of the results of a 90-day study in mice, which indicated a greater sensitivity of males to the test compound. The study was performed in compliance with GLP and test guideline OECD 451.

At the lowest dose tested (100 ppm, equal to 14.3 mg/kg bw per day), there was a slight decrease in absolute kidney weight, an increase in focal necrosis of the liver and urinary bladder hyperplasia in males. No remarkable changes were evident in females at 100 ppm. Females at higher doses (1000 and 2000 ppm) showed increased cellular hyperplasia in the urinary bladder and urethra, but no focal necrosis of the liver was observed. At the two higher doses (500 and 1000 ppm in males; or 1000 and 2000 ppm in females), additional treatment-related systemic effects included: increased absolute and relative liver weights; lower absolute and relative kidney weights; an increased number of foci of hepatocellular alteration and increased incidence of hepatic vesicular/vacuolar changes with cellular hypertrophy. Significantly higher mortality occurred at 1000 ppm in males and 2000 ppm in females. In addition, increased cellular proliferation in the urinary bladder (but not in the liver) was observed in the females at 2000 ppm. Flusilazole was found to be oncogenic in this study. At terminal sacrifice, an increased incidence of hepatocellular tumours (adenomas and carcinomas) was observed in males at all doses (13 out of 80, 23 out of 79, 20 out of 80, and 18 out of 78 at 0, 100, 500, and 1000 ppm,

Table 2. Incidence of liver tumours and associated histopathological lesions reported in combined studies in mice fed diets containing flusilazole

Finding	Study	Dietary concentration (ppm)									Range for historical controls (%)[c]
	Brock et al. (1985)	0	—	5	25	—	200	—	—	—	
	Keller (1992b)	—	0	—	—	100	—	500	1000	2000	
Males											
Dietary intake (mg/kg bw per day)		0	0	0.66	3.4	14.3	27.0	73.1	144.0	—	—
Hepatocellular tumours (%)[a]		20	16	16	15	29	30	25	23	—	7.5–21.8
Hepatocellular adenomas (%)		10	10	10	10	18	14	19	15	—	5.0–21.8
Hepatocellular carcinomas		10	6	6	4	10	11	6	6	—	0–6.8
Foci of cellular alteration (%)		4	4	5	3	5	4	12	29*	—	—
Degeneration, centrilobular (%)[b]		5	0	4	10	1.2	50*	11.2*	11.4*	—	—
Degeneration, focal (%)		4	3.8	3	8	12.5	3	13.8	15.0*	—	—
Relative liver weight (% body weight)		4.9	5.4	4.8	4.7	6.4*	5.5*	7.6*	9.2*	—	—
Females											
Dietary intake (mg/kg bw per day)		0	0	0.92	4.6	19.4	36.0	—	200	384	—
Hepatocellular tumours (%)[a]		1.3	1.3	2.5	0	3.8	1.2	—	13.8*	53.8*	0–2.6
Hepatocellular adenomas (%)		1.0	0	3.0	0	2.5	1.0	—	10.0*	14.0*	0–2.6
Hepatocellular carcinomas		0	1.3	0	0	1.2	0	—	1.3	26.0*	0–1.3
Foci of cellular alteration (%)		0	2.5	0	0	1.2	1.0	—	9.0*	34.0*	—
Degeneration, centrilobular (%)[b]		3.0	1.3	5.0	4.0	0	30.0*	—	1.2	0	—
Degeneration, focal (%)		10.0	12.7	9.0	5.0	7.5	5.0	—	10.0	12.5	—
Relative liver weight (% body weight)		5.4	5.2	5.0*	5.3	5.3	5.5	—	8.9*	12.7*	—

From Brock et al. (1985) and Keller (1992b)

[a] Mice with at least one liver tumour.

[b] Includes lesions described as fatty change.

[c] Historical controls or controls from laboratory carrying out the study.

* $p < 0.05$.

respectively; range for historical controls, 7.5–21.8%) and in females at the two higher doses (1 out of 79, 3 out of 80, 11 out of 77 and 43 out of 76 at 0, 100, 1000 and 2000 ppm, respectively; range for historical control, 0–2.6%). The increase in incidence of tumours was not dose-related nor statistically significant in males, but found to be dose-dependent and statistically significant in females. On the basis of these results, no NOAEL for systemic toxicity could be identified. The overall NOAEL for oncogenicity was 200 ppm (36 mg/kg bw per day) for females and 1000 ppm (144 mg/kg bw per day) for males (Keller, 1992b).

Rats

In a GLP-compliant study, groups of 70 male and 70 female rats of strain Crl:CD(SD)BR were given diets containing flusilazole technical (purity, 95.6%) at a concentration of 0, 10, 50 or 250 ppm (equal to 0, 0.4, 2.0, and 10 mg/kg bw per day in males and 0, 0.5, 2.6 and 13 mg/kg bw per day in females) for 2 years. Each group had two satellite groups of 10 males and 10 females for interim killing at 6 months and 1 year, respectively. At approximately 100 days after the start of treatment, 20 males and 20 females from each group were mated for the reproduction phase of the study. These rats were returned to the long-term study after weaning of the second litter. The rats killed at 6 months were examined for bladder lesions only. No treatment-related effect on mortality was observed; survival rate was > 50% in all groups until week 98. No signs of systemic toxicity were evident at the lowest dose of 10 ppm. At the next higher dose of 50 ppm, there was a dose-related increase in the incidence of pyelonephritis in females at 2 years (3 out of 66, 3 out of 62, 8 out of 67 and 10 out of 65 at 0, 10, 50, and 250 ppm, respectively). Increases in relative liver weight (females) and incidence of hydronephrosis (males) were also observed at interim kill at 1 year. These effects were not considered adverse. At the highest dose of 250 ppm, additional treatment-related changes included: increased incidence of hepatic lesions in females (centrilobular hepatocellular hypertrophy and polyploidy) and hydro-nephrosis in males at 1 year; and increased incidence of acidophilic foci of hepatocellular alteration (13 out of 65 compared with 3 out of 66 in controls) and hepatic diffuse fatty changes (23 out of 65 compared with 9 out of 66 in controls) in females at 2 years. No hepatic lesions were observed in males at any time during the study and no lesions of the urinary bladder were evident in either sex at either interim (6 or 12 months) or terminal kill. There was no treatment-related increase in incidence of any tumour type at any dose tested in the study. At 250 ppm, males showed a slightly higher incidence of squamous cell carcinomas of the oral and nasal cavities (0 out of 66, 1 out of 63, 0 out of 67 and 3 out of 64, at 0, 10, 50, and 250 ppm, respectively). On the basis of data on historical controls from six 2-year feeding studies in rats (five studies with no tumours of this type and one study with an incidence of 2 out of 60), the incidence of nasal tumours in this study was judged to be incidental. Flusilazole technical was, therefore, not oncogenic in the rat under the conditions of this study. The NOAEL for systemic toxicity was 50 ppm, equal to 2.0 mg/kg bw per day. The Meeting noted, however, that the systemic effects reported in this study were relatively mild and higher doses might have been tolerated (Pastoor et al., 1986).

In a more recent study, groups of 65 male and 65 female Crl:CD(SD)BR rats were fed diets containing flusilazole technical (purity, 95%) at a concentration of 0, 125, 375, or 750 ppm (equal to 0, 5.03, 14.8, and 30.8 mg/kg bw per day in males and 0, 6.83, 20.5 and 45.6 mg/kg bw per day in females) for 2 years. Each group included a satellite group of 10 males and 10 females destined for interim kill at 1 year. The study complied with GLP and test guideline OECD 451.

At the lowest dose of 125 ppm (equal to 5.03 mg/kg bw per day), dose-related increases in incidence of hepatocellular hypertrophy were observed in both sexes at either interim (males, 0 out of 10, 4 out of 10, 8 out of 10 and 8 out of 10; females, 0 out of 10, 7 out of 10, 10 out of 10 and 10 out of 10; at 0, 125, 375 and 750 ppm, respectively) or terminal kill (males, 2 out of 53, 2 out of 51, 10 out of 53 and 19 out of 53; females, 2 out of 56, 4 out of 55, 12 out of 53 and 25 out of 54; values exclude deaths that occurred before termination). There was an apparent sex difference in the type of hepatic lesions observed: periportal hepatocellular hypertrophy with lamellar bodies was seen in males, while females showed centrilobular hepatocellular hypertrophy with eosinophilic cytoplasm, but no lamellar bodies. At terminal kill, there was also an increased incidence of mixed foci of cellular alteration in males (6 out of 53, 14 out of 51, 17 out of 53 and 19 out of 53). At the next two higher doses of 375 and 750 ppm, there were also significant decreases in terminal body weight (females), increases in absolute and/or relative liver weights at interim and/or terminal kill (males and/or females) and an increase in the incidence of urinary-bladder transitional-cell hyperplasia at study termination (both sexes; 3 out of 46, 6 out of 45, 27 out of 47 and 42 out of 51 in males, and 5 out of 47, 3 out of 49, 15 out of 49 and 33 out of 53 in females). At 750 ppm, additional treatment-related changes included: increased incidence of hepatic fatty changes (males); and increased cellular proliferation in the liver and urinary bladder. Flusilazole technical was found to be oncogenic in this study. At the highest dose of 750 ppm, a treatment-related increase in incidence of urinary-bladder transitional-cell tumours (papillomas and carcinomas) was observed in males and females at termination (males, 0 out of 46, 0 out of 45, 1 out of 47 and 5 out of 51; and females, 0 out of 47, 1 out of 49, 0 out of 49 and 13 out of 53; at doses of 0, 125, 375 and 750 ppm, respectively). The incidence of testicular interstitial-cell (Leydig cell) tumours in the males at 750 ppm was also increased relative to that in the controls (2 out of 53, 4 out of 51, 2 out of 53 and 9 out of 53, at 0, 125, 375 and 750 ppm, respectively). On the basis of the results, no NOAEL for systemic toxicity could be identified; the NOAEL for oncogenicity was 375 ppm, equal to 14.8 mg/kg bw per day (Keller, 1992c).

Table 3. Incidence of tumours of the bladder and associated histopathological lesions in combined studies in rats fed diets containing flusilazole

Finding	Study	Dietary concentration (ppm)								Range for historical controls (%)[a]
	Pastoor et al. (1986)	0	—	10	50	—	250	—	—	
	Keller (1992a)		0	—	—	125	—	375	750	
Males										
Dietary intake (mg/kg bw per day)		0	0	0.4	2.0	5.0	10.0	14.8	30.8	—
Papilloma, transitional cell (%)		0	0	0	0	0	0	2.1	2.0	0–1.9/0–1.5
Carcinoma, transitional cell (%)		0	0	0	0	0	1.6	0	7.8*	0/0–2.0
Papillomas + carcinomas (%)		0	0	0	0	0	1.6	0	9.8*	—
Mucosal hyperplasia (%)		1.5	6.5	0	1.5	13.3	3.2	57.4*	82.4*	—
Females										
Dietary intake (mg/kg bw per day)		0	0	0.5	2.6	6.8	13.0	20.5	43.6	—
Papilloma, transitional cell (%)		0	0	0	0	0	0	0	3.8*	0–2.0/0–1.4
Carcinoma, transitional cell (%)		0	0	0	0	2.0	0	0	20.8*	0/0–1.4[b]
Papillomas + carcinomas (%)		0	0	0	0	2.0	0	0	24.5*	—
Mucosal hyperplasia (%)		1.6	10.6	0	0	6.1	0	30.6*	62.3*	—

From Pastoor et al. (1986) and Keller (1992a)

[a] Range for historical controls (%) in laboratories: DuPont Haskell laboratory / Charles River laboratory.

[b] Total number of transitional cell tumours rats (papilloma + carcinoma) in females.

* $p < 0.05$.

Table 4. Incidence of Leydig-cell tumours and associated histopathological lesions in combined studies in male rats fed diets containing flusilazole

Finding	Study	Dietary concentration (ppm)								Range for historical controls[a]
	Pastoor et al. (1986)	0	—	10	50	—	250	—	—	
	Keller (1992a)	—	0	—	—	125		375	750	
Dietary intake (mg/kg bw per day)		0	0	0.4	2.0	5.0	10.0	14.8	30.8	0
Leydig-cell adenomas (%)		4.5	3.8	1.6	4.4	7.8	3.1	3.8	17.0*	0–17.7 / 0–10.0
Interstitial-cell hyperplasia (%)		0	0	1.6	2.9	7.8	1.6	5.7	17.0*	—

From Pastoor et al. (1986) and Keller (1992a)

[a] Range for historical controls (%): DuPont Haskell laboratory/Charles River laboratory.

* $p < 0.05$.

2.4 *Genotoxicity*

A battery of studies of mutagenicity with flusilazole technical was conducted to assess potential for inducing gene mutation, chromosome aberration or unscheduled DNA synthesis. The study results (summarized in Table 5) were clearly negative. Flusilazole technical did not demonstrate any genotoxic potential under the conditions tested.

2.5 *Reproductive toxicity*

(a) Multigeneration studies

Rats

Groups of six male and six female Crl:CD(SD)BR rats from a 90-day feeding study were used in a one-generation study of reproduction. The rats were fed daily diets containing flusilazole technical (purity, 96.7%) at a concentration of 0, 25, 125, or 375 ppm (equal to 0, 2, 9, and 27 mg/kg bw per day in males and 0, 2, 11, and 31 mg/kg bw per day in females) for 90 days before mating. Males and females within the same dosing group were mated for 15 days; females were examined daily for evidence of a copulation plug. After the mating period, females were housed individually. The fertility index was low in all groups, especially the control group (67.7%); the number of pregnant females per group ranged from three to six. At the highest dose of 375 ppm (equal to a mean of 29 mg/kg bw per day), a lower (relative to controls) gestation index, lower percentage of liveborn pups and lower pup weight at day 4 were observed. The study was not GLP-compliant, and the small group sizes and absence of individual data on some parameters limited the usefulness of this study for evaluation of reproductive toxicity; the Meeting therefore considered the results of this study as supplementary information (Pastoor et al., 1983).

Groups of 20 male and 20 female Crl:CD(SD)BR rats from a 2-year study of toxicity/oncogenicity were used in a two-generation (two-litter) study of reproduction, which was GLP-compliant. The rats were fed daily diets containing flusilazole technical (purity, 95.6%) at a concentration of 0, 10, 50, or 250 ppm (equal to 0, 1, 3, and 18 mg/kg bw per day in males and 0, 1, 4, and 20 mg/kg bw per day in females; pre-mating intake data) for at least 100 days before mating. Males and females (F_0) within the same dose group were mated (1 : 1) for 15 days; the females were examined daily for evidence of a copulation plug. After the mating period, F_0 females were housed individually and allowed to give birth to F_{1a} litters. About 1 week after weaning the last F_{1a} litter, F_0 females were mated (with different F_0 males of the same group) to produce F_{1b} litters. At weaning of F_{1b} litters, 20 males and 20 females per

Table 5. Results of studies of genotoxicity with flusilazole

End-point	Test object	Concentration	Purity (%)	Results	GLP-compliant	Reference
In vitro						
Reverse mutation [a,b]	*S. typhimurium* TA98, TA100, TA1535, TA1537	1–250 µg/plate	90.0	Negative	No	Donovan & Irr (1982)
Reverse mutation [c,d]	*S. typhimurium* TA97, TA98, TA100, TA1535	5–250 µg/plate	97.7	Negative	Yes	Arce et al. (1988)
Reverse mutation [c,e]	*S. typhimurium* TA97, TA98, TA100, TA1535	10–300 µg/plate	95.0	Negative	Yes	Reynolds (1991)
Forward mutation [f,g]	Chinese hamster ovary cells (CHO-K1/BH4*)* *Hgprt* locus	0.04–0.275 mmol/l	95.5	Negative	No	McCooey et al. (1983)
Unscheduled DNA synthesis[h]	Rat primary hepatocyte cultures	$1 \times 10^{-5} - 1.1 \times 10^{2}$ mM	95.5	Negative	Yes	Chromey et al. (1983)
Chromosome aberration[f,i]	Human lymphocytes	1.7–100 µg/ml	94.9	Negative	Yes	Vlachos et al. (1986)
In vivo						
Chromosome aberration[j]	Rat (Crl:CD, 20 males and 20 females per dose) bone-marrow cells	50–500 mg/kg bw (single oral gavage dose in corn oil)	Not reported	Negative	Yes	Farrow et al. (1983)
Micronucleus formation[k]	Mouse (CD-1, 5 males and 5 females per dose) bone-marrow cells	375 mg/kg bw (single oral gavage dose in corn oil) ()	91.5	Negative	Yes	Sorg et al. (1984)

GLP, good laboratory practice; QA, quality assurance; S9, 9000 × *g* supernatant from rat livers.
[a] Positive and negative controls included, ± S9.
[b] Cytotoxicity (decrease in colony size) seen at 100 and 250 µg/plate.
[c] Positive and negative controls included, ± S9; GLP and QA statements included.
[d] Cytotoxicity observed at 500 µg/plate in TA98 ± S9.
[e] Cytotoxicity observed at 500 µg/plate and slight cytotoxicity at 250 µg/plate in TA100 ± S9.
[f] Test in duplicate; positive and negative controls included; ± S9.
[g] Cytotoxicity observed at ≥ 10 µg/plate −S9 and ≥ 15 µg/plate +S9.
[h] Test in duplicate; positive controls included.
[i] Cytotoxicity observed at 135 µg/ml.
[j] Aberrations analysed in 20 males and females at each dose; positive and negative controls included; only 50 cells were scored instead of 100 cells/animal as required by OECD; clinical signs at the highest dose: slightly depressed, soft faeces, urine stains, red stains on nose and/or ears and/or decreased body weight. One mortality.
[k] Analysed five males and five females per group; positive and negative controls included; clinical signs included: decreased body tone and activity at 4 and 24 h after administration. One mortality at 48 h. No clinical signs at 72 h.

group were selected as F_1 parents for the F_2 generation. These rats were maintained on the same diets as their F_0 parents for 90 days before mating to produce F_{2a} and F_{2b} litters. Sibling mating was avoided. Mean body weight of the F_{1b} males at 250 ppm were decreased during the pre-mating period. No other treatment-related systemic effects were observed in any F_0 or F_1 adult in the study. There was no evidence of a test material-related effect on reproductive parameters (mating and fertility). Treatment-related embryofetal- and litter toxicity were observed at the highest dose (250 ppm). An increased number of stillborn pups and decreased viability index (days 0–4) occurred in all litters (F_{1a}, F_{1b}, F_{2a}, F_{2b}). Litter survival after day 4 was similar in all groups in F_1 and F_2 generations. Mean weights of F_{1b} and F_{2a} pups (both sexes) at weaning were slightly reduced and absolute and relative liver weights of F_{2b} weanling pups (males only) were increased. An increased incidence of hydronephrosis (unilateral and/or bilateral) was noted in F_{2b} female weanlings (1 out of 10, 4 out of 10, 3 out of 10 and 5 out of 10, at 0, 10, 50 and 250 ppm, respectively). Since neither the severity nor the incidence of the lesions showed a dose–response relationship and considering the fact that hydronephrosis is a common lesion in weanling pups (range for historical controls, 0–30% in 13 in-house studies in rats), the finding was judged to be toxicologically insignificant. The NOAEL was 50 ppm, equal to 3 mg/kg bw per day, on the basis of embryo/fetotoxicity and decreased pup viability (Pastoor et al., 1986).

In a two-generation (one to two litters per generation) GLP-compliant study of reproduction, groups of 30 male and 30 female Crl:CD(SD)BR rats were fed diets containing flusilazole technical (purity, 94%) at a concentration of 0, 5, 50 or 250 ppm (equal to 0, 0.34, 3.46, and 17.3 mg/kg bw per day in males and 0, 0.40, 4.04, and 19.6 mg/kg bw per day in females; pre-mating intake data) for 73 days (F_0 rats) or 91 days (F_1 parents) before mating. Males and females within the same dosing group were randomly paired (1 : 1) for 15 days; females were examined daily for evidence of a copulation plug. One litter (F_{1a}) was produced in the F_0 generation and two litters (F_{2a} and F_{2b}) in the F_1 generation.

No treatment-related effects were evident in either generation (F_0 or F_1) at the lowest dose of 5 ppm. At the next higher dose of 50 ppm, increased smooth endoplasmic reticulum in the hepatocytes of males and centrilobular hepatocellular hypertrophy in females was observed, which was considered to be an adaptive response. At the highest dose of 250 ppm, F_1 females showed a slightly but consistently lower body weight. Reproductive toxicity was indicated by higher mortality during parturition and increased duration of gestation (mean values, 22.9–23.2 days compared with 22.4–22.6 days for the controls) of the F_0 and F_1 dams. No treatment-related effects on the mating and fertility indices were observed. Embryofetal and litter toxicity were also evident, with reduced numbers of pups/litter and an increased number of stillborn pups/litter in F_{1a}, F_{2a} and F_{2b} litters and decreased mean pup weights on days 14 and 21 of lactation (F_{2a} litters only) were noted at 250 ppm. The NOAEL for systemic toxicity was 50 ppm, equal to 4.04 mg/kg bw per day, on the basis of decreased body-weight gain at 250 ppm, and the NOAEL for reproductive and offspring toxicity was 50 ppm, equal to 4.04 mg/kg bw per day, on the basis of treatment-related mortality during parturition and increased duration of gestation, as well as embryofetal and litter toxicity at 250 ppm (Mullin, 1990).

(b) Developmental toxicity

Rats

In a range-finding stdy of developmental toxicity, groups of seven pregnant rats were given flusilazole technical (purity, 99%) at a concentration of 0, 100 or 300 mg/kg bw per day orally by gavage on days 7–16 of gestation. The results indicated that flusilazole caused maternal and embryo toxicity at 100 and 300 mg/kg bw per day, and a dose of 300 mg/kg bw per day resulted in cleft palates in approximately 51% of the fetuses in each litter.

In the main study of developmental toxicity, groups of 25 mated female Crl:CD(SD)BR rats were given flusilazole technical (purity, 95.6%) at a dose of 0, 10, 50, or 250 mg/kg bw per day orally by gavage (in corn oil) on days 7–16 of presumed gestation (the day a copulation plug was observed was designated as day 1 of gestation). On day 21 of gestation, all surviving dams were killed and necropsied. Fetuses were delivered by caesarean section and examined for external, visceral and skeletal abnormalities. This study was not GLP-compliant, but was considered conclusive.

No treatment-related signs of maternal toxicity were evident at the lowest dose of 10 mg/kg bw per day. At the next higher dose of 50 mg/kg bw per day, slight decreases in body-weight gain and food consumption during the dosing period, and a slight increase in relative liver weight of dams were observed. At the highest dose of 250 mg/kg bw per day, additional maternal toxic effects included: higher mortality (2 out of 25), clinical signs of toxicity (23 out of 25, effects including chromodacryorrhoea, chromorhinorrhoea, wet/stained perineal areas, red vaginal discharges/stains and focal alopecia). Fetal examinations revealed treatment-related increases in the incidence of skeletal anomalies (misaligned sternebra, extra ossification centres in ribs and delayed ossification in sternebra) at all doses including 10 mg/kg bw per day, the lowest dose tested in the study. At the next higher dose of 50 mg/kg bw per day, there was reduced mean number of liveborn fetuses per litter, a higher total number of stunted fetuses (0, 1, 4 and 3 at 0, 10, 50 and 250 mg/kg bw per day, respectively) and an increased incidence of rudimentary rib. At the highest dose of 250 mg/kg bw per day, additional signs of embryofetotoxicity were: increased mean incidence of resorptions, reduced mean fetal weight/litter and increased incidence of extra ribs in the fetuses. Flusilazole technical was found to be teratogenic in this study. An increased incidence of cleft palate (28 out of 241 compared with 0 out of 331 in controls) and absence of renal papilla (21 out of 155 compared with 0 out of 175 in controls) were observed in fetuses from dams at 250 mg/kg bw per day, but not at lower doses. An unusually high incidence of hydrocephalus and/or dilated lateral ventricles of the brain was also noted in all test groups including controls; the effect was not dose-related and not considered to be related to treatment. The NOAEL for maternal toxicity was 10 mg/kg bw per day; no NOAEL for embryo/fetotoxicity could be identified in this study; and the NOAEL for teratogenicity was 50 mg/kg bw per day (Lamontia et al., 1984a).

In a second study of developmental toxicity, groups of 24 mated female Crl:CD(SD)BR rats were given flusilazole technical (purity, 95.6%) at a dose of 0, 0.4, 2, 10, 50, or 250 (10 females only) mg/kg bw per day orally by gavage (in corn oil) on days 7–16 of gestation (the day a copulation plug was observed was designated as day 1 of gestation). On day 21 of gestation, all surviving dams were killed and necropsied. Fetuses were delivered by caesarean section and examined for external, visceral and skeletal abnormalities. This study was not GLP-compliant, but was considered to be conclusive.

No treatment-related signs of maternal toxicity were observed at doses of 10 mg/kg bw per day or less. Slightly reduced body-weight gain and food consumption during the dosing period, and increased relative liver weight of dams were evident at the next higher dose of 50 mg/kg bw per day. At the highest dose of 250 mg/kg bw per day, clinical signs of maternal toxicity (alopecia, brown stains on face/limbs and stained perineal area) were also observed. No treatment-related embryo/fetotoxic effects were evident at the lower doses of 0.4 and 2 mg/kg bw per day. At higher doses of 10 mg/kg bw per day and greater, increased number of dams with median/late resorptions, increased total number of stunted fetuses, higher incidence of visceral (large renal pelvis and small renal papilla) and skeletal (rib) anomalies, and an increased incidence of delayed ossification (sternebra and vertebral arch) were observed. Flusilazole technical was teratogenic in this study. An increased incidence of cleft palate (21 out of 116 compared with 0 out of 291 in controls) was observed in fetuses from dams at 250 mg/kg bw per day, but not at lower doses. Hydrocephalus was not observed in any group. The NOAEL for maternal toxicity was 10 mg/kg bw per day, the NOAEL for embryo/fetotoxicity was 2 mg/kg bw per day, and the NOAEL for teratogenicity was 50 mg/kg bw per day (Lamontia et al., 1984b).

In a dietary study of developmental toxicity, groups of 24 mated female rats, Crl:CD(SD)BR strain, received daily diets containing flusilazole technical (purity, 96.5%) at a concentration of 0, 50, 100, 300 or 900 ppm (equal to 0, 4.6, 9.0, 26.6, and 79.2 mg/kg bw per day) on days 7–16 of gestation (the day a copulation plug was observed was designated as day 1 of gestation). On day 21 gestation, all surviving dams were killed and necropsied. Fetuses were delivered by caesarean section; weighed, sexed and then examined for external, visceral and skeletal abnormalities. The study was performed in compliance with GLP and test guideline OECD 414.

No treatment-related signs of maternal toxicity were evident at doses of 100 ppm or less. Significant, dose-related reductions in body-weight gain and food consumption were observed during the dosing period at the two higher doses of 300 and 9000 ppm. Treatment-related embryo/fetal toxic effects were noted at 100 ppm and greater; these were an increased incidence of median/late resorptions, small litters (fewer than 10 fetuses/litter) and significant dose-related increases in skeletal variations with extra ossification of the sternebra. At the two higher doses (300 and 900 ppm), additional symptoms of fetal toxicity included an increased number of stunted fetuses; a higher incidence of rudimentary ribs, extra ossification in cervical ribs and delayed ossification in the cervical vertebral arches. Flusilazole was not teratogenic in this study. No fetuses showed cleft palate and there were no treatment-related malformations observed at any dose up to and including 900 ppm, the highest dose tested. The NOAEL for maternal toxicity was 100 ppm (9.0 mg/kg bw per day), the NOAEL for embryo/fetotoxicity was 50 ppm (4.6 mg/kg bw per day), and the NOAEL for teratogenicity was 900 ppm (79.2 mg/kg bw per day) (Alvarez et al., 1984).

In a GLP-compliant study of prenatal and postnatal toxicity in rats, groups of 24 (phase I, prenatal study) or 22 (phase II, postnatal study) mated female rats, strain Crl:CD(SD)BR, were given flusilazole technical (purity, 96.5%; in 0.5% aqueous methyl cellulose) at a dose of 0, 0.2, 0.4, 2, 10, or 100 mg/kg bw per day orally by gavage on days 7–16 of gestation (the day a copulation plug was observed was designated as day 1 of gestation). In phase I, dams were killed on day 21 of gestation for examination of the uterine contents. Rats in an additional control group and group at 100 mg/kg bw per day were killed on day 22 of gestation to determine whether the absence of renal papillae was a compound-related effect or an anomaly. In view of the fact that the first few dosing solutions used in phase I were found to be considerably below the nominal concentrations (1–19% of nominal) and the next analysis (75–110% of nominal) was done only on day 7, no definitive conclusions could be made from results of this part of the study. Maternal toxicity was evident at the highest dose of 100 mg/kg bw per day in the form of clinical signs (chin/perinasal staining and/or wet perineum); reduced body-weight gain and food consumption during the dosing period; and increased absolute and relative liver weights. Embryo/fetotoxicity was observed at 10 mg/kg bw per day and above, including increases in number of stunted fetuses and in dams with stunted fetuses; and a higher incidence of visceral anomalies (small renal papilla and distended ureter). At the highest dose of 100 mg/kg bw per day, additional embryo/fetotoxicity noted included an increased incidence of median/late resorptions and a decrease in the mean number of live fetuses per litter. Treatment-related malformations (absence of renal papilla) were observed in three fetuses (two litters) from dams in the group at 100 mg/kg bw per day.

In phase II, dams were permitted to deliver naturally and raise their litters to weaning. All dams and pups were killed on day 21 of lactation and subjected to gross necropsy. The dosing solutions were found to be adequately prepared in phase II. No treatment-related maternal toxicity was evident at doses of 10 mg/kg bw per day or less. At the next higher dose of 100 mg/kg bw per day, increased mortality (5 out of 22 compared with 0 out of 22 in controls), clinical signs of difficult parturition (pallor, hunching, weakness and/or dystocia during parturition and lactation in four dams), reduced body-weight gain and food consumption during the early part of dosing, and increased liver weight in dams were observed. No treatment-related embryo/fetotoxicity was noted at doses of 2 mg/kg bw per day or less. At 10 mg/kg bw per day and above, there were dose-related overt increases in the mean duration of gestation (22.8, 23.1, 22.9, 23.1, 23.5 and 24.7 days at 0, 0.2, 0.4, 2, 10 and 100 mg/kg bw per day, respectively), decreases in mean litter size and number of liveborn fetuses per litter, an increased number of small litters (with fewer than 10 fetuses per litter), and a higher incidence of dilated renal pelvic and/or ureter in pups at weaning. At the highest dose of 100 mg/kg bw per day, additional signs of treatment-related embryo/fetotoxicity included a reduced total number of live litters, an increased mean number of dead fetuses per litter, and a reduced viability index (days 0–4, 82% compared with

98–99% in the control group and at the lowest dose). Flusilazole technical was not teratogenic in this study. There was no evidence of any treatment-related malformations in any live pups examined. Of the 42 pups found dead (29 were in the group at 100 mg/kg bw per day), absence of renal papilla was observed in only 2 out of 42 and small papilla in 4 out of 42. On the basis of the results of phase II of the study, the NOAEL for maternal toxicity was 10 mg/kg bw per day, the NOAEL for embryo/fetotoxicity was 2 mg/kg bw per day, and the NOAEL for teratogenicity was 100 mg/kg bw per day, the highest dose tested (Alvarez et al., 1985a).

Five groups of 25 mated female Crl:CD®(SD)IGS BR rats were given flusilazole technical (purity, 94.85%; in aqueous methyl cellulose) at a dose of 0, 0.5, 2, 10 or 50 mg/kg bw per day by gavage on days 6–20 of gestation (the day a copulation plug was observed was designed as day of gestation). On day 21 of gestation, all surviving does were killed and necropsied. Two additional groups of rats were given flusilazole at a dose of 50 mg/kg bw per day on days 6–15 of gestation and killed on day 16 or 21 of gestation. A second control group was given a dose of 0 mg/kg bw per day (vehicle only) on days 6–15 of gestation and killed on day 16. On day 21 of gestation, all surviving dams were killed and necropsied. Fetuses were delivered by caesarean section; weighed, sexed and then examined for external, visceral and skeletal abnormalities. The study was performed in compliance with GLP and test guideline OECD 414.

No treatment-related signs of maternal toxicity were evident at 0.5 mg/kg bw per day. At doses of 2 mg/kg bw per day or greater, there was an increased incidence of red vaginal discharge during the latter part (days 13–21) of gestation (0, 0, 3, 12, and 22 at 0, 0.5, 2, 10 and 50 mg/kg bw per day, respectively) and an increase in placental weights. In addition, at doses of 10 mg/kg bw per day and greater, maternal body weight, body-weight gain and food consumption were reduced. Fetal examinations revealed a slight increase in the incidence of rudimentary cervical ribs at doses of 2 mg/kg bw per day and greater (3 out of 332, 4 out of 331, 9 out of 306, 27 out of 314 and 141 out of 302 at 0, 0.5, 2, 10 and 50 mg/kg bw per day, respectively) and an increase in patent ducts arteriosis at 50 mg/kg bw per day. At the highest dose tested (50 mg/kg bw per day), an increased incidence of the malformation naris atresia was observed in the two groups killed at day 21, but not in the group killed on day 16. The NOAEL for maternal toxicity, and embryo/fetotoxicity was 2.0 mg/kg bw per day in this study and the NOAEL for teratogenicity was 10 mg/kg bw per day (Munley, 2000).

In a study of developmental toxicity, groups of 25 Crl:CD®(SD)BR female rats were given flusilazole (purity, 94.85%) at a dose of 0, 2, 10, 50 or 250 mg/kg bw per day administered dermally on days 6–19 of gestation. The study was performed in compliance with GLP and test guideline OECD 414.

Placental changes consisting of enlargement with labyrinth angiectasis and trophoblast necrosis were observed starting at the lowest dose of 2 mg/kg bw per day. At doses of 10 mg/kg bw per day and greater, there was an increase in the number of total late resorptions as well as an increase in placental mineralization affecting the placental capillaries. An increase in the number of early and late resorptions and decreases in the total number of live fetuses and number of live fetuses per dam occurred at doses of 50 mg/kg bw per day or greater. There was an increase in the incidence of red vaginal discharge and/or red material found around the vagina or around the nose (250 mg/kg bw per day) or forelimbs at doses of 50 mg/kg bw per day and greater. Body-weight gains were decreased at 250 mg/kg bw per day at the end of the treatment period and food consumption was decreased sporadically in the group at 50 mg/kg bw per day and throughout the treatment period in the group at 250 mg/kg bw per day. Albumin, total protein and potassium concentrations were decreased at doses of 10 mg/kg bw per day or greater, globulin concentrations were decreased at doses of 50 mg/kg bw per day or greater and creatinine concentrations were decreased at 250 mg/kg bw per day. There were no treatment-related malformations. There were increases in the incidence of skeletal variations at 10 mg/kg bw per day, consisting of an increase in the incidence of rudimentary fourteenth ribs, and at

Table 6. Incidence of selected fetal malformations and variations in rats exposed prenatally to flusilazole

Study[a]	Dose (mg/kg bw per day)	Fetal malformation or variation			
		Cleft palate	Naris atresia	Absent renal papilla	Rudimentary cervical rib
		Total No. of fetuses (No. of litters) affected			
Alvarez et al., (1984)	0	—	—	1 (1)	3 (1)
	4.6	—	—	5 (2)	—
	9.0	—	—	1 (1)	—
	26.6	—	—	2 (2)	1 (1)
	79.2	—	—	1 (1)	—
Lamontia et al., (1984a)	0	—	—	—	—
	10	—	—	—	1 (1)
	50	—	—	—	13 (9)*[c]
	250	28 (10)*[b]	—	21 (12)*[b]	9 (7)*[c]
Lamontia et al., (1984b)	0	—	—	—	—
	0.4	—	—	5 (5)	—
	2	—	—	2 (2)	—
	10	—	—	5 (5)	—
	50	—	—	5 (4)	18 (8)*[c]
	250	21 (5)*[c]	—	4 (2)	15 (5)*[c]
Munley (2000)	0	—	—	—	3 (3)
	0.5	—	1 (1)	—	4 (4)
	2	—	—	—	9 (6)
	10	—	—	—	27 (15)*[c]
	50	—	3 (2), 3 (3)[d]	—	141 (22)[c]

—, no incidence.

[a] Purity of flusilazole used in each study: Alvarez et al. (1984), 96.5%; Lamontia et al. (1984a), 95.6%; Lamontia et al. (1984b), 95.6%; Munley (2000), 94.85%.

[b] Jonckheere test of statistical significance.

[c] Fischer exact test of statistical significance.

[d] Incidence in each of the two groups at 50 mg/kg bw per day.

* $p \leq 0.05$.

250 mg/kg bw per day, consisting of an increase in full fourteenth ribs and seventh cervical ribs and a decrease in ossification of the first cervical centra. The LOAEL for maternal and embryo/fetotoxicity was 2 mg/kg bw per day. No NOAEL could be identified; however, the study authors suggested that the data supported the fact that the lowest dose tested (2 mg/kg bw per day) was near the bottom end of the dose–response curve and was approaching a NOAEL (Schardein, 1998)

Rabbits

Four groups of 18 artificially inseminated New Zealand White female rabbits were given flusilazole technical (purity, 96.5%; in corn oil) at a nominal dose of 0, 2, 5, or 12 mg/kg bw per day (equal to 0, 1.9, 4.8, and 10.1 mg/kg bw per day based on analysis of dosing solutions) orally by gavage on days 7–19 of presumed gestation (the day of insemination was designated as day 0 of gestation). On day 29 of gestation, all surviving does were killed and necropsied. Fetuses were delivered by caesarean section and examined for external, visceral and skeletal abnormalities. There was no treatment-related mortality, clinical signs of maternal toxicity nor disturbances of the intra-uterine development of the conceptuses at any dose up to and including 12 mg/kg bw per day, the highest dose tested in the study. All fetuses delivered showed normal development. No evidence of fetotoxicity and no treatment-related increases in malformations were observed in fetuses at any dose tested. The high incidence of hydrocephalus observed in this study—1 (1), 2 (1), 4 (2), and 4 (3) fetuses (litters) at 0, 2, 5, and 12 mg/kg bw per day, respectively, which was not confirmed in three subsequent studies (hydrocephalus being found in only one fetus at 35 mg/kg bw per day and in none at any lower dose nor in the controls), was not considered to be treatment-related. The NOAEL for maternal toxicity, embryo/fetotoxicity and teratogenicity was 12 mg/kg bw per day (or 10.1 mg/kg bw per day by analysis). This study was not GLP compliant and was considered to provide supplementary information, as the doses administered were not high enough to elicit maternal toxicity (Solomon et al., 1984).

In a second study, three groups of 20 artificially inseminated New Zealand White female rabbits were given flusilazole technical (purity, 96.5%) at a nominal dose of 0, 12, or 35 mg/kg bw per day (equal to 0, 11.2 and 31.5 mg/kg bw per day based on analysis of dosing solutions) orally by gavage on days 7–19 of presumed gestation (the day of insemination was designated as day 0 of gestation). On day 29 of gestation, all surviving does were killed and necropsied. Fetuses were delivered by caesarean section and examined for gross pathology. This study was not GLP compliant, but was considered conclusive.

No evidence of maternal toxicity nor embryo/fetotoxicity was observed at the lowest dose of 12 mg/kg bw per day and there were no increases in the incidence of any malformations or fetal variations, compared with the controls. At the highest dose of 35 mg/kg bw per day, there was an increased incidence of red vaginal discharge and stained tail, and an increased incidence of periodic anorexia. Abortion occurred in 2 out of 13 does (compared with 0 out of 16 in the control group) and total early resorption occurred in 10 out of 13 does (compared with 1 out of 16 in the control group). Teratogenicity could not be assessed at this dose because only one live litter was produced. The NOAEL for maternal and embryo/fetotoxicity was 12 mg/kg bw per day (or 11.2 mg/kg bw per day by analysis), and for teratogenicity was 12 mg/kg bw per day (or 11.2 mg/ kg bw per day by analysis) and greater (Zellers et al., 1985).

In a range-finding dietary study of developmental toxicity in rabbits, four groups of seven artificially inseminated New Zealand White female rabbits were fed daily diets containing flusilazole technical (purity, 94.8%) at a concentration of 0, 500, 1000, or 2000 ppm on days 7–19 of gestation. The pregnancy rate (three out of seven per group) was low in groups at 500 and 1000 ppm. The incidence of mortality in utero was high at 2000 ppm (four out of seven females with total resorptions).

In the main study, groups of 20 artificially inseminated, New Zealand White female rabbits were given diets containing flusilazole technical (purity, 94.8%) at a concentration of 0, 300, 600, or 1200 ppm (equal to 0, 8.9, 21.2 an 37.8 mg/kg bw per day) on days 7–19 of presumed gestation (day of insemination was designated as day 0 of gestation). All surviving does were killed on day 29 of gestation and necropsied. Fetuses were delivered by caesarean section and examined for external, visceral and skeletal abnormalities. At the highest dose of 1200 ppm, maternal body weight and food consumption were decreased during dosing. Pregnancy rate was reduced in all treated groups (fertility, 9 out of 20, 10 out of 20, and 7 out of 20 at 300, 600, and 1200 ppm, respectively). The number of does with total resorption was increased in the groups at the intermediate (600 ppm) and highest (1200 ppm) dose. There were no treatment-related effects on mean number of live fetuses/litter, mean number of resorptions (in dams with liver fetuses) or fetal weight. The small number of litters available at 600 and 1200 ppm (three per dose) precluded any definitive assessment of fetotoxicity or teratogenic potential at these doses. There were no apparent treatment-related effects at the lowest dose of 300 ppm.

In a supplementary study, groups of 18 or 25 (300 ppm) inseminated, New Zealand White female rabbits were given flusilazole technical at a concentration of 0, 30, 100, or 300 ppm (equal to 0, 0.81, 2.84, and 8.32 mg/kg bw per day) on days 7–19 of presumed gestation. The pregnancy rate was again low in all groups, including the control group (fertility, 8 out of 18). Total resorption occurred at 0 and 300 ppm (25% and 29%, respectively, of the pregnant does), but not at the lowest (30 ppm) or intermediate (100 ppm) dose. Because of the low number of live litters available for examination, the data in this study did not allow adequate assessment of the embryo/fetotoxicity and teratogenic potential of the test compound. The NOAEL for maternal toxicity was 600 ppm (21.2 mg/kg bw per day) on the basis of decreased body weight and food consumption during dosing at 1200 ppm. No definitive NOAEL for embryo/fetotoxicity nor teratogenicity could be identified in this study. The study was not GLP compliant and was stated to, in general, conform to test guideline OECD 414, and was considered as supplemental information, owing to the lack of litters available for examination (Alvarez et al., 1985b).

In another study, groups of 18 artificially inseminated New Zealand White female rabbits were given flusilazole technical (purity, 93.8%; in 0.5% methylcellulose) at a dose of 0, 7, 15 or 30 mg/kg bw per day orally by gavage on days 7–19 of gestation. The study was performed in compliance with GLP and test guideline OECD 414.

The pregnancy rate was acceptable in all test groups (12 out of 18, 14 out of 18, 16 out of 18, and 16 out of 18 at 0, 7, 15 and 30 mg/kg bw per day, respectively). No treatment-related maternal nor embryo/fetal toxicity was evident at the lowest dose of 7 mg/kg bw per day. At the next two higher doses of 15 and 30 mg/kg bw per day, clinical signs of maternal toxicity (cageboard red discharge and brown/yellow-stained tail) and increased incidence of abortion (one dam/dose group) and total resorptions (4 out of 16 and 12 out of 16, respectively) were observed. At 30 mg/kg bw per day, food consumption was also decreased during the dosing period. The mean number of live fetuses per litter, mean number of dead fetuses (zero), mean fetal weight and mean male : female ratio were comparable between the control group and all treated groups. No treatment-related external, visceral or skeletal malformations/variations were observed in any of the fetuses from does at any dose, including those at 30 mg/kg bw per day, the highest dose tested in the study. However, it must be noted that assessment of fetotoxicity and teratogenic potential at 30 mg/kg bw per day was based on data from only three live litters available for that dose (compared with 11–12 live litters in the control group and at lower doses). The limited data points reduce confidence in the accuracy of any study conclusions made on the basis of observations at this dose. Therefore, the NOAEL for maternal and embryo/fetal toxicity was 7 mg/kg bw per day, and for teratogenicity was 15 mg/kg bw per day (Alvarez, 1990).

2.6 Special studies

(a) Studies on mechanisms by which Leydig-cell tumours are induced

Rats

In a GLP-compliant study in vivo, groups of 10 Crl:CD BR male rats were given flusilazole technical (purity, 94%; in corn oil) at a dose of 0, 20, 50, 150, or 250 mg/kg bw per day (given as two equal half-doses, twice per day) by subcutaneous injection for 14 days. The control group (0 mg/kg bw per day) and group at 250 mg/kg bw per day each contained an additional subgroup of 10 male rats that were treated with human chorionic gonadotropin (hCG) 1 h before killing. Ketoconazole (a known inhibitor of 17 α-hydroxylase) was used as the positive control. Groups of 10 male rats were treated with ketoconazole (in saline) at a dose of 0, 20, 50, 100 or 200 mg/kg bw

Table 7. Incidence of hydrocephaly in rabbits exposed prenatally to flusilazole

Study[a]	Dose (mg/kg bw per day)	Total No. of fetuses (litters) affected
Alvarez (1990)	0	—
	7	—
	15	1 (1)
	30	1 (1)
Alvarez et al. (1985b)	0	—
	8.9	—
	21.2	—
	37.8	—
Solomon et al. (1984)	0	1 (1)
	1.9	2 (1)
	4.8	5 (3)
	10.1	4 (3)
Zellers et al. (1985)	0	—
	11.2	—
	31.5	1 (1)

—, no incidence.

[a] Purity of flusilazole: Alvarez (1990), 93.8%; Alvarez et al. (1985b), 94.8%; Solomon et al. (1984), 96.5%; Zellers et al. (1985), 96.5%.

per day (given as two equal half-doses, twice per day) for 14 days. The control group (0 mg/kg bw per day) and group at 200 mg/kg bw per day each contained an additional subgroup of 10 male rats that were treated with hCG 1 h before killing. All surviving rats were killed on day 15 of treatment and necropsied. Samples of testicular interstitial fluid and serum were collected from rats that were not treated with hCG; interstitial fluid was analysed for testosterone and serum was analysed for testosterone, estradiol, luteinizing hormone (LH) and follicle-stimulating hormone (FSH). Serum samples collected from the rats treated with hCG were analysed for testosterone, androstenedione, 17 α-hydroxyprogesterone and progesterone.

At the lowest dose of 20 mg/kg bw per day, and greater, increased absolute and relative liver weights and a dose-related inhibition of serum testosterone (statistically significant at 150 mg/kg bw per day and greater) and estradiol (statistically significant at all doses) concentrations were observed. At the two higher doses (150 and 250 mg/kg bw per day), clinical symptoms of toxicity (sores, stained/wet fur, dehydration and diarrhoea), decreased body weight and body-weight gain, and reduced food consumption were evident. Increased mortality (8 out of 10) also occurred at the highest dose of 250 mg/kg bw per day. At 250 mg/kg bw per day in the rats treated with hCG, serum testosterone concentrations were significantly lower than in the controls; no other significant differences in hormone levels were noted in rats treated with flusilazole. In the positive-control group treated with ketoconazole, rats showed significantly lower concentrations of testosterone, androstenedione and 17-hydroxyprogesterone, and higher concentrations of progesterone, indicating inhibition of 17α-hydroxylase. On the basis of the results, no NOAEL for the portion of this study that was carried out in vivo could be identified.

In a GLP-compliant study in vitro, Leydig cells were collected from rat testes at termination of the study in vivo and cultured in microplate wells. The cells were then incubated with either flusilazole technical or ketoconazole at doses of 0.05–100 μmol/l (in 70% ethanol) for 2 h. Culture media were sampled and analysed for testosterone, androstenedione, 17 α-hydroxyprogesterone and progesterone. The concentration at which 50% inhibition was produced (IC_{50}) of testosterone was determined. Results of the study in vitro confirmed the hormonal changes observed in vivo. Incubation of testicular Leydig cells with flusilazole technical caused a dose-dependent lowering of testosterone and androstenedione concentrations, suggesting inhibition of enzymes involved in steroid biosynthesis. The IC_{50} for testosterone was 3.475–1.455 µmol/l without hCG and 2.774–0.646 µmol/l with hCG pre-treatment. In cells treated with ketoconazole (from rats in the positive-control group), the IC_{50} for testosterone was 0.97–0.83 µmol/l without hCG and 0.154–0.065 µmol/l with hCG (Cook, 1993).

In a GLP-compliant study, groups of 52 male and 52 female Crl:CD BR rats were fed daily diets containing flusilazole technical (purity, 95%) at a concentration of 0, 10, 125, 375, or 750 ppm (equal to 0, 0.58, 7.27, 22.1, and 44.7 mg/kg bw per day in males and 0, 0.74, 9.40, 27.6, and 59.0 mg/kg bw per day in females) for up to 91 days. Each group was divided into three subgroups in order to study possible mechanisms of action of toxicity in the liver and urinary bladder. In each group, subgroups of 20 males and 20 females were assigned for evaluation of the liver or urinary bladder; five males and five females per subgroup were killed on days 7 or 8, 14, 46 and 91 for cellular proliferation and histopathological assessment. An additional 10 males and 10 females per subgroup were used for evaluation of cytochrome P450 and peroxisome proliferation; five males and five females were killed on days 14 and 90 (males) or 15 and 91 (females). Serum concentrations of testosterone, estradiol and LH were determined in all males killed on days 14 and 90 ($n = 10$). No treatment-related adverse effects were observed at the lower doses of 10 and 125 ppm. At 125 ppm, absolute and relative liver weights were increased; in the absence of histopathological changes or any other symptoms of toxicity, the finding was not considered to be toxicologically significant. At the next two higher doses (375 and 750 ppm), liver hypertrophy was observed; the response was sex-specific: lamellar bodies and periportal hypertrophy occurred in males and centrilobular hypertrophy without lamellar bodies was evident in females. In the urinary bladder, transitional-cell necrosis, exfoliation and hyperplasia were noted. Hepatic cytochrome P450 levels were higher than control values, but there was no evidence of peroxisome proliferation in the liver of these treated rats. There was no increase in cellular proliferation: in the liver there was decreased proliferation in females at doses of 125 ppm and greater and in the urinary bladder of males at 750 ppm. No significant changes in the serum concentrations of testosterone, estradiol and LH were found in any of the males examined. The NOAEL was 125 ppm, equal to 7.27 mg/kg bw per day (Keller, 1992a)

(b) Study on the mechanism of liver effects in dogs

In order to distinguish an adaptive or toxic response at 20 ppm in a 1-year feeding study in dogs, groups of ten male, outbred, beagle dogs were fed diets that contained flusilazole (purity, 92.5%) at a concentrations of 0, 5, 20 or 75 ppm (equivalent to 0, 0.18, 0.68 and 2.65 mg/kg bw per day). Five dogs per group were killed after 28 days and the remaining five after a recovery period of 28 days. Daily observations were made for mortality and clinical signs; food consumption and body weights were measured weekly; haematology and clinical chemistry parameters were evaluated at study initiation, and at weeks 4 and 7. This study was GLP compliant and was performed in compliance with test guideline OECD 452.

There were no effects on survival, clinical signs or nutritional parameters. A slight increase in mean absolute and relative liver weight was observed after 4-weeks exposure in all treated groups. Aspartate aminotransferase (AST) activity was also minimally increased at 20 and 75 ppm, but was not clearly dose-related, and was not considered to be an adverse effect. The minimal liver-weight and AST changes had reversed during the 28-day recovery period. Cytochrome P450 isozymes were significantly induced at 20 and 75 ppm (i.e. total cytochrome P450 and CPYB1/2,

CYP3A and CYP4A isozymes). This was indicative of an adaptive response and is consistent with observations of increased liver weight in the absence of hepatotoxicity. It is also consistent with the suggestion of increased smooth endoplasmic reticulum in livers of dogs at 75 ppm. The observed effects were reversible, which further supports the conclusion that these are not adverse findings (Mertens, 1999).

Comments

Biochemical aspects

In rats, orally administered ^{14}C-labelled flusilazole was readily absorbed from the gastrointestinal tract and rapidly excreted in urine (72% of triazole label) and faeces (up to 87% of phenyl label), with little or no radioactivity recovered in the expired air. The excretion half-life was approximately 34 h and more than 90% of the administered dose was eliminated within 96 h. Tissue retention of radiolabelled material was low. Total tissue residues excluding the carcass (which accounted for approximately 2% of the administered dose) were less than 1% of the administered dose, therefore demonstrating no evidence of bioaccumulation.

[^{14}C]Flusilazole was extensively metabolized in rats. Recovered parent compound accounted for only 2–11% of the given dose, found predominantly in the faeces (urinary concentration, < 1%). After absorption, flusilazole was cleaved at the triazole ring. With phenyl-labelled test material, the major faecal metabolites identified were [bis(4-fluorophenyl)methyl] silanol, [bis(4-fluoro -phenyl)methylsilyl] methanol and its fatty acid conjugates, and disiloxane. Except for the fatty acid conjugates, the same metabolites were found in the urine. With triazole-labelled material, the main metabolite identified was 1*H*-1,2,4-triazole, which was found predominantly in the urine (63.8% of the administered dose in males, 51.6% in females); faeces contained only a small amount of the metabolite.

Toxicological data

Flusilazole is moderately to slightly toxic in rats when given as a single oral dose; and minimally toxic to rats and rabbits when administered as a single dose dermally or by inhalation. The oral LD_{50} in rats was 672–1216 mg/kg bw, the dermal LD_{50} in rabbits was > 2000 mg/kg bw and the inhalation LC_{50} in rats was 6.8–7.7 mg/l. Flusilazole was found to be minimally irritating to the eyes and the skin of New Zealand White rabbits. It was practically non-irritating to the skin and was not a dermal sensitizer in guinea-pigs in a Buehler test.

Short- and long-term studies of repeated oral doses of flusilazole in mice (90-day dietary study), rats (90-day studies of gavage and dietary administration) and dogs (90-day and 1-year dietary studies) resulted primarily in lesions of the liver (hepatocellular hypertrophy, fatty change, focal inflammation/necrosis (mouse only) and vacuolation) and urinary bladder (urothelial hyperplasia and vacuolation). In addition, the gastrointestinal tract was a target in dogs. Clinical chemistry was not assessed in the studies in mice, the only finding in the studies in rats was a decrease in cholesterol in both sexes in the 90-day study and increase in cholesterol in females only in the long-term studies. On the basis of the hepatic and/or urinary bladder histopathology, the NOAEL was 75 ppm (equal to 12 mg/kg bw per day) in mice, 125 ppm (equal to 9 mg/kg bw per day) in rats and 20 ppm (equal to 0.7 mg/kg bw per day) in dogs (1-year study). Lymphoid hyperplasia of the gastric mucosa was observed in all treated dogs in the 90-day study, but not in the controls. In the 1-year study, this finding was observed in all dogs, including controls, with severity increasing in a dose-related manner. Effects at the LOAEL in the 1-year study in dogs included hepatocellular hypertrophy, inflammatory infiltration and vacuolation (males only), decreased cholesterol, total protein and albumin, and increased alkaline phosphatase activity and leukocyte counts. A mechanistic study in male dogs at the doses used in the 1-year study indicated that after 28 days of exposure, the effects observed on the liver were adaptive and reversible (weight, increased AST activity and cytochrome P450). The dog appeared to be the most sensitive species in these studies, with a NOAEL of 0.7 mg/kg bw per day in the 1-year study, on the basis of histopathology changes in the liver and stomach and changes in clinical chemistry.

After repeated short-term (21-day) dermal application of flusilazole, there was no evidence of any treatment-related systemic toxicity in rabbits given doses of up to 200 mg/kg bw per day.

Flusilazole was tested for genotoxicity in an adequate range of assays in vitro and in vivo. It was not genotoxic in mammalian or microbial systems. The Meeting concluded that flusilazole was unlikely to be genotoxic.

Two 18-month dietary studies with flusilazole were conducted in mice. In the first study in which flusilazole was administered at concentrations of up to 200 ppm in the diet, the target organs identified were the liver (hepatocellular fatty changes), kidney (decreased weight), and urinary bladder (histopathological change). There was no evidence of carcinogenicity in this study. Concentrations from 100 to 2000 ppm were used in the second 18-month study. Systemic toxicity was observed at all doses. At doses of 500 and 1000 ppm in males (73.1 and 144 mg/kg bw per day, respectively) or 1000 and 2000 ppm in females (200 and 384 mg/kg bw per day, respectively), overt hepatic lesions (increased foci of hepatocellular alteration and hepatocellular hypertrophy with cytoplasmic vesiculation and/or vacuolation) and cellular hyperplasia in the urinary bladder were observed. Increased incidences of liver tumours (hepatocellular adenomas and carcinomas) were observed at concentrations of more than 1000 ppm. Liver tumours occurred at doses in excess of the maximum tolerated dose (MTD) and were preceded at lower concentrations by clear histopathological changes in the liver. The overall NOAEL for systemic toxicity was 25 ppm, equal to 3.4 mg/kg bw per day, on the basis of hepatotoxicity and urinary bladder hyperplasia at 100 ppm (14.3 mg/kg bw per day) in males and hepatocellular fatty changes at 200 ppm (27 mg/kg bw per day) in both sexes. The overall NOAEL for carcinogenicity was 200 ppm (equal to 36 mg/kg bw per day) in females and 1000 ppm (equal to 144 mg/kg bw per day) for males. The incidence of tumours at the NOAEL was within the range for historical controls.

The toxicity and carcinogenicity of flusilazole were investigated in two 2-year studies in rats. The target organs identified were the liver and bladder. The overall NOAEL for systemic toxicity was 50 ppm, equal to 2.0 mg/kg bw per day, on the basis of mild nephrotoxicity (pyelonephritis in females) and hepatotoxicity (hepatocellular hypertrophy in both sexes), acidophilic foci, and diffuse fatty change (females only). There was no treatment-related increase in the incidence of any tumour type AT up to 250 ppm (the highest dose tested in the first study). Concentrations of between 125 and 750 ppm, the latter exceeding the MTD, were used in the second study. Flusilazole was found to be tumorigenic at the highest dose of 750 ppm (30.8 mg/kg bw per day) causing bladder transitional cell neoplasia in both sexes and testicular Leydig-cell tumours in males. There was no evidence of any treatment-related increase in tumour incidence at a dietary concentration of 375 ppm. The overall NOAEL for carcinogenicity was 375 ppm (14.8 mg/kg bw per day)

A special 2-week study to investigate the possible mechanism for the induction of testicular Leydig-cell tumours was conducted in rats. The results demonstrated that flusilazole caused a dose-dependent lowering of estradiol concentrations at 20 mg/kg bw per day and above, and of serum and interstitial testosterone concentrations at 150 mg/kg bw per day in vivo after subcutaneous exposure (n = 10) and a dose-related decrease in testosterone and androstenedione production in testicular Leydig-cell cultures by inhibition of enzymes involved in steroid biosynthesis in vitro at less than 5 μmol/l. In the 90-day mechanistic study in rats given flusilazole at doses similar to those used in the second long-term study in rats (0, 10, 125, 375 or 750 ppm), there were no changes in serum concentrations of testosterone, estradiol or LH, which would be expected for this mode of action. However, there was appreciable inter-animal variability in the hormone measurements.

Overall, the data suggested that flusilazole may induce Leydig-cell tumours via an endocrine-related mechanism—inhibition of testosterone and estradiol biosynthesis could contribute to disruption of the hypothalamus–pituitary–testis axis, resulting in overstimulation of the testicular endocrine tissues. Exposure to flusilazole at doses not causing disruption of the hypothalamus–pituitary–testis axis would, therefore, be unlikely to induce an increase in Leydig-cell tumours. Although this mode of action is relevant to humans, there was good evidence to suggest that humans are less sensitive to chemically-induced Leydig-cell tumours than are rats, owing to differences in sensitivity to LH on the basis of number of Leydig-cell receptors and control of LH-receptor expression (e.g. by prolactin in rodents but not in humans) (Foster, 2007; Cook et al., 1999).

The Meeting concluded that the weight of evidence indicated that the mode of action for bladder tumours was via cell injury and regenerative hyperplasia (Cohen, 1998).

In view of the lack of genotoxicity and the finding of hepatocellular tumours in mice and testicular and bladder transitional cell tumours in rats only at doses at which marked toxicity was observed, the Meeting concluded that flusilazole is not likely to pose a carcinogenic risk to humans at dietary levels of exposure.

The effect of flusilazole on reproduction in rats was investigated in two two-generation studies. The first was a part of a 2-year feeding study. No parental toxicity was observed at doses of up to 250 ppm. The same doses were used in the second definitive two-generation study. The NOAEL for parental systemic toxicity was 50 ppm, equal to 4.04 mg/kg bw per day, on the basis of slightly lower body-weight gain in F_1 females. The main reproductive effects at 250 ppm included increased duration of gestation and increased maternal mortality during parturition. The NOAEL for reproductive toxicity was 50 ppm, equal to 3.46 mg/kg bw per day. Toxicity observed in offspring at 250 ppm included a reduced number of live pups per litter and decreased pup growth. The NOAEL for offspring toxicity was 50 ppm, equal to 4.04 mg/kg bw per day.

Nine studies of developmental toxicity were carried out with flusilazole administered orally, of which five were in rats (one dietary study and four with gavage administration) and four (one dietary study and three with gavage administration) in rabbits to characterize potential teratogenicity observed in some studies.

In most of the studies in rats, the NOAEL for maternal toxicity was 10 mg/kg bw per day on the basis of reduced body-weight gain and decreased food consumption. In one study, the NOAEL for maternal toxicity was 2 mg/kg bw per day on the basis of increased incidence of red vaginal discharge during the latter part of gestation and an increase in placental weights at 10 mg/kg bw per day, which was not assessed in the other studies. At maternally toxic doses, specific malformations noted were cleft palate, nares atresia and absent renal papillae. An increased incidence of anomalies (extra cervical ribs, patent ductus arteriosis) was also observed. The incidence of rudimentary cervical ribs was slightly, but not statistically significantly increased at 2 mg/kg bw per day (3 out of 3, 4 out of 4, 9 out of 6, 27 out of 15, and 141 out of 22 fetuses per litter in the groups at 0, 0.5, 2, 10, 50 mg/kg bw per day, respectively). The overall NOAEL for embryo/fetotoxicity was 2 mg/kg on the basis of a higher incidence of skeletal variations (extra cervical ribs) at 10 mg/kg bw per day. No malformations were found at doses of less than 50 mg/kg bw per day.

In four studies of developmental toxicity in rabbits, the NOAEL for maternal and embryo/fetal toxicity was 7 mg/kg bw per day on the basis of clinical signs of toxicity, increased incidence of abortion and total resorption at 15 mg/kg bw per day. There was no evidence for any teratogenic potential in rabbits given flusilazole at doses of up to 15 mg/kg bw per day, the maximum tolerated dose in this study.

A major metabolite identified was 1*H*-1,2,4-triazole, which was found predominantly in the urine (63.8% of the administered dose in males, 51.6% in females). Studies with this metabolite are summarized in the evaluation of difenoconazole in the present report.

No neurotoxic effects were seen during conventional repeat-dose studies with flusilazole.

There were no reports of adverse health effects in manufacturing plant personnel or in operators and workers exposed to flusilazole formulations during their use. Also, there was no evidence or data to support any findings in relation to poisoning with flusilazole.

The Meeting concluded that the existing database on flusilazole was adequate to characterize the potential hazards to fetuses, infants and children.

Toxicological evaluation

The Meeting established an ADI of 0–0.007 mg/kg bw based on the NOAEL of 0.7 mg/kg bw per day for lymphoid hyperplasia in the gastric mucosa, liver histopathology (hypertrophy, inflammatory infiltration in males and females, and vacuolation in males only), and clinical chemistry (decreased concentrations of cholesterol, total protein and albumin and increased alkaline phosphatase activity and leukocyte counts) in the 1-year dietary study in dogs and a safety factor of 100.

The Meeting established an acute reference dose (ARfD) of 0.02 mg/kg bw based on the NOAEL of 2 mg/kg bw per day for skeletal anomalies in the study of developmental toxicity in rats treated orally and a safety factor of 100.

Levels relevant to risk assessment

Species	Study	Effect	NOAEL	LOAEL
Mouse	Two-year studies of toxicity and carcinogenicity [a]	Toxicity	25 ppm, equal to 3.4 mg/kg bw per day	200 ppm, equal to 27 mg/kg bw per day
		Carcinogenicity[d]	200 ppm equal to 36 mg/kg bw per day (females)	1000 ppm equal to 384 mg/kg bw per day
Rat	Two-year studies of toxicity and carcinogenicity [a,c]	Toxicity	50 ppm, equal to 2 mg/kg bw per day	250 ppm, equal to 10 mg/kg bw per day
		Carcinogenicity	375 ppm, equal to 14.8 mg/kg bw per day	750 ppm, equal to 30.8 mg/kg bw per day
	Multigeneration reproductive toxicity[a,c]	Parental toxicity	50 ppm, equal to 4.04 mg/kg bw per day	250 ppm, equal to 19.6 mg/kg bw per day
		Offspring toxicity	50 ppm, equal to 4.04 mg/kg bw per day	250 ppm, equal to 19.6 mg/kg bw per day
		Reproduction	50 ppm, equal to 4.04 mg/kg bw per day	250 ppm, equal to 19.6 mg/kg bw per day
	Developmental toxicity [a,b,c]	Maternal toxicity	2 mg/kg bw per day	10 mg/kg bw per day
		Embryo/fetotoxicity	2 mg/kg bw per day	10 mg/kg bw per day
Rabbit	Developmental toxicity[a,b,c]	Maternal toxicity	7 mg/kg bw per day	15 mg/kg bw per day
		Embryo/fetotoxicity	7 mg/kg bw per day	15 mg/kg bw per day
Dog	One-year study of toxicity [a]	Toxicity	20 ppm, equal to 0.7 mg/kg bw per day	75 ppm, equal to 2.4 mg/kg bw per day

[a] Dietary administration.

[b] Gavage administration.

[c] Two or more studies combined.

[d] Greater than the maximum tolerated dose (MTD).

Estimate of acceptable daily intake for humans

0–0.007 mg/kg bw

Estimate of acute reference dose

0.02 mg/kg bw

Information that would be useful for the continued evaluation of the compound

Results from epidemiological, occupational health and other such observational studies of human exposures.

Critical end-points for setting guidance values for exposure to flusilazole

Absorption, distribution, excretion and metabolism in mammals	
Rate and extent of oral absorption	Rapid and extensive (up to 80%)
Dermal absorption	Data not available
Distribution	Widely
Potential for accumulation	Low
Rate and extent of excretion	Rapidly excreted
Metabolism in animals	Extensively metabolized
Toxicologically significant compounds in animals, plants and the environment	Parent compound,1,2,4-triazole
Acute toxicity	
Rat, LD_{50}, oral	674 mg/kg bw
Rat, LD_{50}, dermal	> 2000 mg/kg bw
Rat, LC_{50}, inhalation	2.7–3.7 mg/l, 4 h
Guinea-pig, skin sensitization (test method used)	Non-sensitizing (Buehler)
Short-term studies of toxicity	
Target/critical effect	Liver and urinary bladder
Lowest relevant oral NOAEL	0.7 mg/kg bw per day (1-year study in dogs)
Lowest relevant dermal NOAEL	5 mg/kg bw per day (21-day study in rabbits)
Lowest relevant inhalation NOAEC	No data presented
Genotoxicity	
	Not genotoxic
Long-term studies of toxicity and carcinogenicity	
Target/critical effect	Liver and bladder
Lowest relevant NOAEL	2.0 mg/kg bw per day (2-year study in rats)
Carcinogenicity	No carcinogenic concern at levels of dietary exposure
Reproductive toxicity	
Reproduction target/critical effect	Increased gestation length, reduced live born pups/litter and decreased pup growth

Lowest relevant reproductive NOAEL	50 ppm (4.04 mg/kg bw per day)
Developmental target/critical effect	Skeletal anomalies, malformations at higher doses
Lowest relevant developmental NOAEL	2 mg/kg bw per day (rats)
Neurotoxicity/delayed neurotoxicity	
	No indications of neurotoxicity in studies of acute toxicity or repeated doses
Other toxicological studies	Disruption of the hypothalamus–pituitary–testis axis
Mechanistic studies	Necrosis and hyperplasia in the rat bladder
Medical data	
	No occupational or accidental poisoning reported

Summary

	Value	Study	Safety factor
ADI	0–0.007 mg/kg bw	Dog, 1-year study	100
ARfD	0.02 mg/kg bw	Rat, study of developmental toxicity	100

References

Alvarez, L. (1990) Teratogenicity study of DPX-H6573-66 in rabbits. Unpublished report No. HLR 216-90 from Haskell Laboratory for Toxicology and Industrial Medicine, DE, USA. Submitted to WHO by E.I. du Pont de Nemours and Co., Inc., DE, USA.

Alvarez, L., Krauss, W.C. & Staples, R.E. (1984) Developmental toxicity study in rats given INH-6573-66 in the diet on days 7-16 of gestation. Unpublished report No. HLR 431-84 from Haskell Laboratory for Toxicology and Industrial Medicine, DE, USA. Submitted to WHO by E.I. du Pont de Nemours & Co., Inc., DE, USA.

Alvarez, L., Staples, R.E., & Kaplan, A.M. (1985a) INH-6573: prenatal and postnatal toxicity study in rats dosed by gavage on days 7-16 of gestation. Unpublished report No. HLR 654-85 from Haskell Laboratory for Toxicology and Industrial Medicine, DE, USA. Submitted to WHO by E.I. du Pont de Nemours & Co., Inc., DE, USA.

Alvarez, L., Staples, R.E., Driscoll, C.D. & Kaplan, A.M. (1985b) INH-6573: developmental toxicity study in rabbits treated by diet on days 7-19 of gestation. Unpublished report No. HLR 337-85 from Haskell Laboratory for Toxicology and Industrial Medicine, DE, USA. Submitted to WHO by E.I. du Pont de Nemours & Co., Inc., DE, USA.

Anderson, J.J., Stadalius, M.A. & Schlueter, D.D. (1986) Metabolism of ^{14}C-DPX-H6573 in rats. Unpublished report No. AMR-196-84 from E.I. du Pont de Nemours & Co., Inc., DE, USA. Submitted to WHO by E.I. du Pont de Nemours & Co., Inc., DE, USA.

Arce, G.T., Matarese, C.C. & Sarrif, A.M. (1988) Mutagenicity testing of INH-6573-21 in *Salmonella typhimurium* plate incorporation assay. Unpublished report No. HLR 59-88 from Haskell Laboratory for Toxicology and Industrial Medicine, DE, USA. Submitted to WHO by E.I. du Pont de Nemours & Co., Inc., DE, USA.

Brock, W.J. (1988) Primary dermal irritation study with INH-6573-106 in rabbits. Unpublished report No. HLR 484-88 from from Haskell Laboratory for Toxicology and Industrial Medicine, DE, USA. Submitted to WHO by E.I. du Pont de Nemours & Co., Inc., DE, USA.

Brock, W.J., Rickard, R.W., Kaplan, A.M. & Gibson, J.R. (1985) Long-term feeding study in mice with INH-6573. Unpublished report No. HLR 278-85 from Haskell Laboratory for Toxicology and Industrial Medicine, DE, USA. Submitted to WHO by E.I. du Pont de Nemours & Co., Inc., DE, USA.

Brock, W.J., Vick, D.A. & Chromey, N.C. (1988) Closed-patch repeated insult dermal sensitization study (Buehler method) with INH-6573-21 in guinea pigs. Unpublished report No. HLR 34-88 from Haskell Laboratory for Toxicology and Industrial Medicine, DE, USA. Submitted to WHO by E.I. du Pont de Nemours & Co., Inc., DE, USA.

Cheng, T. (1986) Rat metabolism study of [triazole-3-^{14}C] DPX-H6573. Unpublished report No. 6129-128 from Hazleton Laboratories America, Inc., VA, USA. Submitted to WHO by E.I. du Pont de Nemours & Co., Inc., DE, USA.

Chromey, N.C., Horst, A.L., McCooey, K.T. & Sarrif, A.M. (1983) Unscheduled DNA synthesis/rat hepatocytes in vitro. Unpublished report No. HLR 209-83 from Haskell Laboratory for Toxicology and Industrial Medicine, DE, USA. Submitted to WHO by E.I. du Pont de Nemours & Co., Inc., DE, USA.

Cohen, S.M. (1998) Urinary bladder carcinogenesis. *Toxicol. Pathol.*, **26**, 121–127.

Cook, J. (1993) Mechanisms of rat Leydig cell tumor induction by DPX-H6573-193 (flusilazole). Unpublished report No. HLR 410-93 from Haskell Laboratory for Toxicology and Industrial Medicine, DE, USA. Submitted to WHO by E.I. du Pont de Nemours & Co., Inc., DE, USA.

Cook, J.C., Klinefelter, G.R., Hardisty, J.F., Sharpe, R.M. & Foster, P.M.D. (1999) Rodent Leydig Cell tumorigenesis: a review of the physiology, pathology, mechanisms, and relevance to humans. *Crit. Rev. Toxicol.*, **29**, 169–261.

Donovan, S.M. & Irr, J.D. (1982) Mutagenicity evaluation in *Salmonella typhimurium.* Unpublished report No. HLR 611-82 from Haskell Laboratory for Toxicology and Industrial Medicine, DE, USA. Submitted to WHO by E.I. du Pont de Nemours & Co., Inc., DE, USA.

Farrow, M.G., Cortina, T. & Padilla-Nash, H. (1983) *In vivo* bone marrow chromosome study in rats - H # 14,728. Unpublished report No. 201-636 from Hazleton Laboratories America, Inc., VA, USA. Submitted to WHO by E.I. du Pont de Nemours & Co., Inc., DE, USA.

Finlay, C. (1998) Flusilazole technical: primary eye irritation study in rabbits. Unpublished report No. DuPont-1300 from Haskell Laboratory for Toxicology and Industrial Medicine, DE, USA. Submitted to WHO by E.I. du Pont de Nemours & Co., Inc., DE, USA.

Finlay, C. (2000) Flusilazole technical: acute oral toxicity study in male and female rats. Unpublished report No. DuPont-3749 from Haskell Laboratory for Toxicology and Industrial Medicine, DE, USA. Submitted to WHO by E.I. du Pont de Nemours & Co., Inc., DE, USA.

Foster, M.D. (2007) Induction of Leydig cell tumors by xenobiotics. In: Hardy, M. & Payne, A. (eds), *The Leydig cell in health and disease.* Humana Press, Totuwa, NJ, USA. Chapter 27, pp 383–392.

Gargus, J.L. & Sutherland, J.D. (1983) Acute skin absorption LD_{50} test on rabbits. Unpublished report No. 288-83 from Hazelton Laboratories America Inc., VA, USA. Submitted to WHO by E.I. du Pont de Nemours & Co., Inc., DE. USA.

Hall, J.A., Dashiell, O.L. & Kennedy, G.L. (1984) Eye irritation test in rabbits. Unpublished report No. HLR 582-82 from Haskell Laboratory for Toxicology and Industrial Medicine, DE, USA. Submitted to WHO by E.I. du Pont de Nemours & Co., Inc., Wilmington, DE. USA.

Keller, D.A. (1990) Subchronic oral toxicity: 90-day study with DPX-H6573-193 feeding study in mice. Unpublished report No. HLR 60-90 from Haskell Laboratory for Toxicology and Industrial Medicine, DE, USA. Submitted to WHO by E.I. du Pont de Nemours & Co., Inc., DE, USA.

Keller, D.A. (1992a) Mechanism of toxicity: 90-day feeding study in rats with DPX-H6573-194 (flusilazole). Unpublished report No. HLR 628-92 from Haskell Laboratory for Toxicology and Industrial Medicine, DE, USA. Submitted to WHO by E.I. du Pont de Nemours & Co., Inc., DE, USA.

Keller, D.A. (1992b) Oncogenicity study with DPX-H6573-193 (flusilazole): 18-month feeding study in mice. Unpublished report No. HLR 35-92 from Haskell Laboratory for Toxicology and Industrial Medicine, DE, USA. Submitted to WHO by E.I. du Pont de Nemours & Co., Inc., DE, USA.

Keller, D.A. (1992c) Oncogenicity study with DPX-H6573-194 (flusilazole): 2-year feeding study in rats. Unpublished report No. HLR 527-92 from Haskell Laboratory for Toxicology and Industrial Medicine, DE, USA. Submitted to WHO by E.I. du Pont de Nemours & Co., Inc., DE, USA.

Lamontia, C.L., Staples, R.E. & Alvarez, L. (1984a) Embryo-fetal toxicity and teratogenicity study of INH-6573-39 by gavage in the rat. Unpublished report No. HLR 444-83 from Haskell Laboratory for Toxicology and Industrial Medicine, DE, USA. Submitted to WHO by E.I. du Pont de Nemours & Co., Inc., DE, USA.

Lamontia, C.L., Staples, R.E. & Alvarez, L. (1984b) Embryo-fetal toxicity and teratogenicity study of INH-6573-39 by gavage in the rat. Unpublished report No. HLR 142-84 from Haskell Laboratory for Toxicology and Industrial Medicine, DE, USA. Submitted to WHO by E.I. du Pont de Nemours & Co., Inc., DE, USA.

McCooey, K.T., Chromey, N.C., Sarrif, A.M. & Hemingway, R.E. (1983) CHO/HGPRT assay for gene mutation. Unpublished report No. HLR 449-83 from Haskell Laboratory for Toxicology and Industrial Medicine, DE, USA. Submitted to WHO by E.I. du Pont de Nemours & Co., Inc., DE, USA.

Mertens, J.J.W.M. (1999) DPX-H6573 technical: a 28-day mechanistic feeding study with 28 days of recovery in dogs. Unpublished report No. DuPont-1623 from WIL Research Laboratories Inc, OH, USA. Submitted to WHO by E.I. du Pont de Nemours & Co., Inc., DE, USA.

Mullin, L.S. (1990) Reproductive and fertility effects with flusilazole: multigeneration reproduction study in rats. Unpublished report No. HLR 424-90 from Haskell Laboratory for Toxicology and Industrial Medicine, DE, USA. Submitted to WHO by E.I. du Pont de Nemours & Co., Inc., DE, USA.

Munley, S.M. (2000) Flusilazole technical: developmental toxicity study in rats. Unpublished report No. Dupont-2287 from Haskell Laboratory for Toxicology and Industrial Medicine, DE, USA. Submitted to WHO by E.I. du Pont de Nemours & Co., Inc., DE, USA.

O'Neal, F.O., Rickard, R.W., Kaplan, A.M. & Gibson, J.R. (1985) One-year feeding study in dogs with INH-6573. Unpublished report No. HLR 461-85 from Haskell Laboratory for Toxicology and Industrial Medicine, DE, USA. Submitted to WHO by E.I. du Pont de Nemours & Co., Inc., DE, USA.

Pastoor, T.P., Wood, C.K., Krahn, D.F. & Gibson, J.R. (1983) Ninety-day feeding and one-generation reproduction study in rats with Silane [bis(4-fluorophenyl)](methyl) (1*H*-1,2,4-triazol-*1*-ylmethyl) (INH-6573). Unpublished report No. HLR 483-83 from Haskell Laboratory for Toxicology and Industrial Medicine, DE, USA. Submitted to WHO by E.I. du Pont de Nemours & Co., Inc., DE, USA.

Pastoor, T.P., Wood, C.K., Drahn, D.F. & Aftosmis, J.G. (1984) Four-week range finding and ninety-day feeding study in mice with Silane [bis(4-fluorophenyl)](methyl) (1*H*-1,2,4-triazol-*1*-ylmethyl) (INH-6573). Revised unpublished report No. HLR 341-83 1-83 from Haskell Laboratory for Toxicology and Industrial Medicine, DE, USA. Submitted to WHO by E.I. du Pont de Nemours & Co., Inc., DE, USA.

Pastoor, T.P., Rickard, R.W., Sykes, G.P., Kaplan, A.M. & Gibson, J.R. (1986) Long-term feeding (combined chronic toxicity/oncogenicity study) and two-generation, four-litter reproduction study in rats with INH-6573. Unpublished report No. HLR 32-86 from Haskell Laboratory for Toxicology and Industrial Medicine, DE, USA. Submitted to WHO bv E.I. du Pont de Nemours & Co., Inc., DE, USA.

Reynolds, V. (1991) Mutagenicity testing of DPX-H6573-194 in the *Salmonella typhimurium* plate incorporation assay. Unpublished report No. HLR 33-91 by Haskell Laboratory for Toxicology and Industrial Medicine, DE, USA. Submitted to WHO by E.I. du Pont de Nemours & Co., Inc., DE, USA.

Rickard, R.W., Wood, C.K., Krahn, D.F. & Aftosmis, J.G. (1983) Three-month feeding study in dogs with Silane [bis(4-fluorophenyl)](methyl) (1H-1,2,4-triazol-*1*-ylmethyl) (INH-6573). Unpublished report No. HLR 461-83 from Haskell Laboratory for Toxicology and Industrial Medicine, DE, USA. Submitted to WHO by E.I. du Pont de Nemours & Co., Inc., DE, USA.

Sarver, J.W., Vick, D.A., Valentine, R., Chromey, N.C. & Kaplan, A.M. (1986) Twenty-one dose dermal toxicity study with INH-6573-82 in rabbits. Unpublished report No. HLR 744-86 from Haskell Laboratory for Toxicology and Industrial Medicine, DE, USA. Submitted to WHO by E.I. du Pont de Nemours & Co., Inc., DE, USA.

Schardein, J. (1998). A dermal prenatal development toxicity study of flusilazole in rats. Unpublished report, No. HLO-1998-01504 Revision No. 1. from DuPont Haskell Laboratory, Newark, Delaware, USA and DuPont Experimental Station, Wilmington, Delaware, USA Submitted to WHO by E.I. du Pont de Nemours and Company, Wilmington, Delaware, U.S.

Solomon, H.M., Alvarez, L., Staples, R.E. & Hamill, J.C. (1984) Developmental toxicity study in rabbits given INH-6573 by gavage on days 7-19 of gestation. Unpublished report No. HLR 333-84 from Haskell Laboratory for Toxicology and Industrial Medicine, DE, USA. Submitted to WHO by E.I. du Pont de Nemours & Co., Inc., DE, USA.

Sorg, R.M., Naismith, R.W. & Mathews, R.J. (1984) Micronucleus test (MNT) - OECD H # 15,314. Unpublished report No. HLR 437-84 from Pharmakon Research International. Submitted to WHO by E.I. du Pont de Nemours & Co., Inc., DE, USA.

Turner, R.J., Kinney, L.A. & Chromey, N.C. (1985) Inhalation median lethal concentration (LC_{50}) of INH-6573 by EPA Guidelines. Unpublished report No. HLR 1-85 by Haskell Laboratory for Toxicology and Industrial Medicine, DE, USA. Submitted to WHO by E.I. du Pont de Nemours & Co., Inc., DE, USA.

Vlachos, D.A., Covell, D.L. & Sarrif, A.M. (1986) Evaluation of INH-6573-82 in the in vitro assay for chromosome aberrations in human lymphocytes. Unpublished report No. HLR 745-86 from Haskell Laboratory for Toxicology and Industrial Medicine, DE, USA. Submitted to WHO by E.I. du Pont de Nemours & Co., Inc., DE, USA.

Wylie, C.N., Henry, J.E., Ferenz, R.L., Burgess, B.A. & Kennedy, G.L. (1984a) Median lethal dose (LD_{50}) in rats - EPA proposed guidelines, Newark, DE, USA. Unpublished report No. HLR 433-83 by Haskell Laboratory for Toxicology and Industrial Medicine, DE, USA. Submitted to WHO by E.I. du Pont de Nemours & Co., Inc., DE, USA.

Wylie, C.N., Henry, J.E., Dashiell, O.L. & Kennedy, G.L. (1984b) Primary skin irritation and sensitization test on guinea pigs. Unpublished report No. HLR 626-82 from Haskell Laboratory for Toxicology and Industrial Medicine, DE, USA. Submitted to WHO by E.I. du Pont de Nemours & Co., Inc., DE, USA.

Wylie, C.N., Henry, J.E., Burgess, B.A. & Kennedy, G.L. (1984c) Ten-dose oral subacute test in rats. Unpublished report No. HLR 78-83 from Haskell Laboratory for Toxicology and Industrial Medicine, DE, USA. Submitted to WHO by E.I. du Pont de Nemours & Co., Inc., DE, USA.

Zellers, J.E., Staples, R.E., Alvarez, L. & Kaplan, A.M. (1985) Developmental toxicity study (supplemental) in rabbits dosed by gavage on davs 7-19 of gestation. Unpublished report No. HLR 669-85, from Haskell Laboratory for Toxicology and Industrial Medicine, DE, USA. Submitted to WHO by E.I. du Pont de Nemours & Co., Inc., DE, USA.

PROCYMIDONE

First draft prepared by
I. Dewhurst[1] & A. Boobis[2]

[1]Pesticides Safety Directorate, Department for Environment, Food and Rural Affairs, Kings Pool, York, England
[2]Experimental Medicine and Toxicology, Division of Medicine, Faculty of Medicine, Imperial College London, London, England

Explanation

Procymidone is the International Organization for Standardization (ISO) approved name for *N*-(3,5-dichlorophenyl)-1,2-dimethylcyclopropane-1,2-dicarboximide (International Union of Pure and Applied Chemistry, IUPAC), Chemical Abstracts Service, CAS No. 32809-16-8. It is a dicarboximide fungicide that is used on a range of vegetables, fruits, soya bean, sunflowers, tobacco and oil seed rape, as well as on ornamental plants and flower bulbs. The mechanism of pesticidal action involves the inhibition of triglyceride synthesis in fungi.

Procymidone was previously evaluated by JMPR in 1981, 1982 and 1989 (Annex 5, references *37*, *39*, *58*). No acceptable daily intakes (ADIs) were established when procymidone was evaluated by the JMPR in 1981 and 1982. In 1989, an ADI of 0–0.1 mg/kg bw was established based on the NOAEL of 12.5 mg/kg bw per day identified in studies of reproductive toxicity in rats. Procymidone was re-evaluated by the present Meeting as part of the periodic review programme of the Codex Committee on Pesticude Residues (CCPR). A range of new studies was submitted to the present Meeting; these studies addressed kinetics, developmental toxicity and hormonal effects in different species.

Many of the conventional studies of toxicity with procymidone were relatively old, were performed before the widespread use of good laboratory practice (GLP) and some contained relatively limited information. Overall, the Meeting considered that the database was adequate for the risk assessment.

Procymidone used in the main studies of toxicity was considered to be representative of current production material, which is typically of > 99% purity. The Food and Agriculture Organization (FAO) specification for procymidone specifies a minimum content of 98.5% w/w[1]. Procymidone has also been known under the development code S7131 or 'Sumilex'; 'Sumisclex' is a 50% formulation.

Evaluation for acceptable daily intake

1. Biochemical aspects

1.1 Absorption, distribution and excretion

(a) Oral route

Mice

Groups of five male ICR mice were given [phenyl-^{14}C]procymidone (radiochemical purity, > 99%; specific activity, 22.8 mCi/mmol (843.6 MBq/mmol) as a single oral dose at 100 mg/kg bw in corn oil. Urine and faeces were collected for 7 days from one group of mice. Other groups of mice were killed 2, 4, 6, 8, 12, 24 and 72 h after dosing and a range of tissues removed for radiochemical analysis. Rapid elimination was observed, with 92% of the administered dose excreted in the first 24 h, mainly in the urine (73.5%). Over a 7-day period, faeces and urine accounted for 22% and 82%, respectively, of the administered dose. There was no evidence of tissue accumulation in this study; the highest concentration of radioactivity (429 μg equivalent/g) was found in fat 2 h after dosing, but the concentration had declined to low concentrations (2 μg equivalent/g) by the end of the experimental period. Radioactivity in adrenals, blood, brain, kidneys, liver, prostate, epididymis and testes reached a maximum 2–8 h after dosing (133, 17, 27, 58, 67, 88, 48 and 17 μg equivalent/g respectively) and subsequently decreased with a half-life of 4–14 h. This study did not claim to be compliant with GLP (Kimura et al., 1988).

[1] http://www.fao.org/ag/AGP/AGPP/Pesticid/Specs/docs/Pdf/new/procymid.pdf

Rats

Groups of five male and female Wistar rats were given single oral doses of [phenyl-^{3}H]procymidone (radiochemical purity, > 99%; specific activity, 15.8 mCi/mmol (584.6 MBq/mmol) or [carbonyl-^{14}C]procymidone (radiochemical purity, > 99%; specific activity, 4.67 mCi/mmol (172.79 MBq/mmol) at 25 mg/kg bw, formulated in 10% Tween 80. A similar group received seven doses of [carbonyl-^{14}C]procymidone at 25 mg/kg bw per day. Samples of urine, faeces and carbon dioxide were taken for 7 days after the last dose. A limited number of tissue samples were taken from rats killed at 3, 6, 12, 24, 48 and 168 h after dosing and analysed for ^{14}C. Radioactivity was rapidly absorbed and eliminated. Almost identical patterns of excretion were found in both sexes and with both labelled forms, primarily in the urine (85–90%). Peak mean concentrations of radioactivity in tissues were reached at 6–12 h after dosing, the highest concentrations being in the kidneys (28 µg equivalent/g) and liver (19 µg equivalent/g) followed by muscle, stomach, lungs, and heart (15, 15, 12 and 12 µg equivalent/g respectively). Fat samples were not taken before 48 h. The tissue concentrations of radioactivity declined rapidly and except for fat (0.3 µg equivalent/g) were < 0.1 µg equivalent/g at 168 h after the last dose. Repeated dosing did not alter the pattern of excretion, but tissue residues were 3–10-fold those seen with a single dose at 168 h after the last dose. This study did not claim to be compliant with GLP (Mikami & Yamamoto, 1976).

Excretion and tissue distribution were investigated in groups of five male Sprague-Dawley rats given a single oral dose of [phenyl-^{14}C]procymidone (radiochemical purity, > 99%; specific activity, 22.8 mCi/mmol (843.6 MBq/mmol) at a dose of 100 mg/kg bw in corn oil. Urine and faeces were collected from five rats for 7 days after dosing. Tissues were removed from other groups at intervals up to 72 h after dosing. Radioactivity was rapidly and almost completely eliminated in the urine. Fifty-nine per cent of the dose was eliminated on the first day and 96% overall (urine, 84%; and faeces, 13%). The highest concentration of radioactivity was found in fat at 8 h (555 µg equivalent/g), declining to 5 µg equivalent/g at 72 h after dosing. Radioactivity in adrenals, blood, brain, kidneys, liver, prostate, epididymis and testes reached a maximum 8–12 h after dosing (77, 15, 26, 49, 67, 65, 28 and 11 µg equivalent/g, respectively) and subsequently decreased with an overall half-life of 7–12 h. This study did not claim to be compliant with GLP (Kimura et al., 1988).

Three groups of five male and five female Crl:CD®(SD)BR rats were given a single oral dose of [phenyl-^{14}C]procymidone (radiochemical purity, > 98%; specific activity, 242 µCi/mg (8.95 MBq/mg) at a dose of 1 or 250 mg/kg bw in corn oil. An additional group of rats received 14 consecutive daily doses of unlabelled procymidone at 1 mg/kg bw before being given a single dose of [phenyl-^{14}C]procymidone at 1 mg/kg bw. Urine and faeces were collected for 7 days after dosing; the rats were then killed and tissues removed for analysis. After oral administration of procymidone at a dose of 1 mg/kg bw, absorption was rapid and extensive, with approximately 80% of the administered dose being excreted in the urine in 24 h (Table 4). At the higher oral dose of 250 mg/kg bw, the proportion of radioactivity in the urine declined to 63–67% of the administered dose, with a concomitant increase in faecal radioactivity to 24–33% (Table 4). Radioactivity exhaled in the expired air accounted for < 0.03% of the administered dose. There was an indication of more rapid urinary excretion, but no substantial differences in the pattern of excretion for single or multiple lower doses. At either dose, radioactivity retained in the carcass and in the tissues 168 h after dosing accounted for < 0.3% of the dose and ≤ 0.01% of the administered dose respectively, demonstrating almost complete excretion. At 168 h, concentrations of radioactivity in all tissues of rats at 1 mg/kg bw were close to or less than 0.001 µg equivalent/g, except for fat (0.002–0.006 µg equivalent/g) and kidneys (0.001–0.002 µg equivalent/g). Fat also contained the highest concentration of radioactivity (5.0 µg equivalent/g) in rats at 250 mg/kg bw. This study claimed compliance with GLP and US EPA guidelines for studies of metabolism with pesticides (Struble, 1992a).

Comparative kinetics between species

To further investigate the species differences in findings in studies of developmental and reproductive toxicity, the kinetics and metabolism of procymidone were investigated in female rats, rabbits and monkeys, after single and repeated doses given to non-pregnant animals and to pregnant animals.

Single doses

Groups of four female Sprague-Dawley Crj:CD(SD)IGS rats were given a single oral dose of [phenyl-^{14}C]procymidone (specific activity, 13.7 MBq/mg; purity, > 97.8%) at 37.5, 62.5, 125, 250 or 500 mg/kg bw. Groups of three, female New Zealand White rabbits and female cynomolgus monkeys were given [phenyl-^{14}C]procymidone at a dose of 62.5, 125, 250 or 500 mg/kg bw. Corn oil was the vehicle in the studies in rats, while in studesi in rabbits and monkeys the vehicle was 0.5% methylcellulose. It is uncertain what effect the use of different vehicles would have had on the results. Each dose was given to two groups, blood was collected from one group at 1, 2, 4, 6, 8, 10, 12, 24, 48 and 72 h after dosing and excreta were collected from the other group at 6 (urine only), 24, 48, 72 and 120 h after dosing. The concentration of radioactivity was determined in excreta and plasma. Urine, faeces and plasma were analysed by thin-layer chromatography (TLC) either directly or after solvent extraction and metabolites were identified by co-chromatography with reference compounds.

The results (Table 1) showed marked species differences. Absorption in the rabbit was much more rapid than in rats and monkeys. Monkeys had much lower maximum concentration (C_{max}) values were much lower in monkeys than in rabbits and rats, but area under the curve of concentration–time (AUC) values were relatively low in rabbits given doses of up to 125 mg/kg bw owing to rapid elimination. In monkeys, the predominant compound in plasma was procymidone; in rabbits, acid metabolites and glucuronides of alcohol metabolites were the major components in plasma; in rats, free alcohol metabolites predominated. These results led to a hypothesis that the species differences in developmental effects observed were due to the high concentrations of free alcohol metabolites in rats, which lead to hypospadias (Sugimoto, 2005a, 2005b; Mogi, 2005a).

Repeated dosing

Groups of four female Sprague-Dawley Crj:CD(SD)IGS rats or groups of three female cynomolgus monkeys received 14 daily doses of [phenyl-^{14}C]procymidone (specific activity, 13.7 MBq/mg; purity, > 98.1%) at 37.5 (rats only), 62.5, 125, 250 or 500 (monkeys only) mg/kg bw per day. Corn oil was the vehicle in the studies in rats, for monkeys the vehicle was 0.5% methylcellulose. Blood was collected 2, 4, 8 and 24 h after the 1st, 3rd, 7th, 10th and 14th dose and additionally 48 and 72 h after the 14th dose. Urine and faeces were collected over a 24-h period after the 1st and 14th dose, at 2-day intervals at other times during the dosing period and at 24–72 h and 72–120 h after the final dose. Plasma, urine and faeces were analysed for radioactivity and for metabolites. The results are summarized in Table 2.

These results suggest that a steady state was reached after administration of repeated doses for approximately 3 days in rats, but only after 14 days or longer in monkeys. Although C_{max} and AUC were significantly higher in rats after a single dose, after 14 doses values were similar for rats and monkeys. These findings show that the doses used in the studies of developmental toxicity in cynomolgus monkeys would have produced plasma concentrations of total radioactivity that were similar to those produced in rats at doses resulting in hypospadias, but with different metabolite profiles. In monkeys, most of the radiolabel was unextractable and was not identified. On the data available, the metabolite patterns were similar to those seen with single doses, with unchanged procymidone beng the major component in plasma in monkeys, and alcohol metabolites predominating in rats (Sugimoto, 2005c; Mogi, 2005b).

Table 1. Pharmacokinetic parameters of radioactivity in plasma of female rats, rabbits and monkeys given a single oral dose of [phenyl-^{14}C]procymidone

Parameter	Dose (mg/kg bw per day)				
	37.5	62.5	125	250	500
Rats					
C_{max} (µg equivalent/ml)	11.48	16.42	30.06	35.67	50.33
T_{max} (h)	8	12	24	24	24
$T_{1/2}$: from C_{max} to 120 h (h)	18.3	15.6	16.3	15.9	14.6
$AUC_{0-120\ h}$ (µg equivalent/h per ml)	450	574	1108	1389	2317
$AUC_{0-\infty}$ (µg equivalent/h per ml)	454	578	1117	1401	2331
Radioactivity in urine (%)	83	84	81	67	47
Radioactivity in faeces (%)	12	11	15	29	49
Peak PCM (µg equivalent/ml)	3.3	6.1	11.4	10.6	19.4
Peak PCM-NH-COOH[a](µg equivalent/ml)	0.8	0.9	1.6	1.9	1.7
Peak PCM-CH_2OH (µg equivalent/ml)	5.4	7.1	14.4	8.2	19.4
Peak PA-CH_2OH (µg equivalent/ml)	2.9	7.3	8.3	16.7	8.4
Peak PA-COOH (µg equivalent/ml)	0.7	0.7	1.0	1.7	—
Peak glucuronides (µg equivalent/ml)[b]	0.5	0.3	0.4	0.7	—
Rabbits					
C_{max} (µg equivalent/ml)	—	19.36	29.96	37.22	50.38
T_{max} (h)	—	1	1	1	4
$T_{1/2}$: from C_{max} to 48 h (h)	—	7.0	6.9	7.6	17.6
$T_{1/2}$: from 48 to 120 h (h)	—	64.1	47.9	53.4	—
$AUC_{0-120\ h}$ (µg equivalent/h per ml)	—	180	251	457	1244
$AUC_{0-\infty}$ (µg equivalent/h per ml)	—	189	258	474	1260
Radioactivity in urine (%)	—	93	81	75	64
Radioactivity in faeces (%)	—	4	16	22	35
Peak PCM (µg equivalent/ml)	—	0	0.2	0.7	1.0
Peak PCM-COOH (µg equivalent/ml)	—	5.1	4.7	7.3	8.0
Peak PCM-CH_2OH (µg equivalent/ml)	—	0	0	0	0
Peak PA-CH_2OH (µg equivalent/ml)	—	0.3	0.8	1.8	2.5
Peak PA-COOH (µg equivalent/ml)	—	4.2	8.8	12.7	16.5
Peak glucuronides (µg equivalent/ml) [b]	-	7.0	10.7	11.9	17.8
Monkeys	—				
C_{max} (µg equivalent/ml)	—	5.92	4.43	8.76	8.69
T_{max} (h)	—	10	6	10	4
$T_{1/2}$: from C_{max} to 120 h (h)	—	81.7	78.9	58.5	84.5
$AUC_{0-120\ h}$ (µg equivalent/h per ml)	—	295	379	563	522
$AUC_{0-\infty}$ (µg equivalent/h per ml)	—	427	600	754	836

Radioactivity in urine (%)	—	57	36	29	12
Radioactivity in faeces (%)	—	42	66	72	87
Peak PCM (µg equivalent/ml)	—	2.8	2.3	3.4	3.6
Peak PCM-NH-COOH (µg equivalent/ml)	—	0.08	0.14	0.24	0.25
Peak PCM-CH_2OH[a] (µg equivalent/ml)	—	1.1	0.65	1.2	0.91
Peak PA-CH_2OH (µg equivalent/ml)	—	0.24	0.13	0.32	0.22
Peak PA-COOH (µg equivalent/ml)	—	0.26	0.16	0.40	0.71
Peak glucuronides (µg equivalent/ml)[b]	—	0.06	0.08	0.15	0.96

From Sugimoto (2005a, 2005b) and Mogi (2005a)
PA, procymidone acid; PA-CH_2OH, hydroxyprocymidone acid; PA-COOH, carboxyprocymidone acid; PCM, procymidone; PCM-CH_2OH, hydroxyprocymidone; PCM-NH-COOH, carboxyprocymidone.
[a] Included PCM-COOH.
[b] Glucuronides were of PA-CH_2OH and PCM-CH_2OH.

Table 2. Pharmacokinetic parameters of radioactivity in plasma of female rats and monkeys given 14 oral doses of [phenyl-^{14}C]procymidone

No. of doses	Parameter	Dose (mg/kg bw per day)				
		37.5	62.5	125	250	500
Rats						
Single dose	C_{max} (µg equivalent/ml)	11.7	15.7	29.6	42.1	—
	T_{max} (h)	8	8	24	24	—
	$AUC_{2–24\,h}$ (µg equivalent/h per ml)	189	325	520	599	—
Seven doses	C_{max} (µg equivalent/ml)	19.0	32.9	47.2	72.9	—
	T_{max} (h)	8	8	4	8	—
	$AUC_{2–24\,h}$ (µg equivalent/h per ml)	301	485	693	1187	—
Fourteen doses	C_{max} (µg equivalent/ml)	17.7	24.8	47.0	68.3	—
	T_{max} (h)	8	8	8	8	—
	$AUC_{2–24\,h}$ (µg equivalent/h per ml)	269	369	689	1038	—
	$AUC_{0–\infty}$ (µg equivalent/h per ml)	498	599	1111	1588	—
	Radioactivity in urine (%)	81	77	79	61	—
	Radioactivity in faeces (%)	15	18	16	35	—
	Peak PCM (µg equivalent/ml)	2.6	2.6	6.4	5.1	—
	Peak PCM-NH-COOH[a] (µg equivalent/ml)	—	—	—	—	—
	Peak PCM-CH_2OH (µg equivalent/ml)	9.9	9.8	32	51	—
	Peak PA-CH_2OH (µg equivalent/ml)	3.3	7.8	4.0	1.9	—
	Peak PA-COOH(µg equivalent/ml)	0.3	0.5	0.6	1.4	—
	Peak glucuronides (µg equivalent/ml)	0.6	0.9	0.7	2.9	—
Monkeys						
Single dose	C_{max} (µg equivalent/ml)	—	2.5	5.2	5.2	9.8
	T_{max} (h)	—	4	4	8	24
	$AUC_{2–24\,h}$ (µg equivalent/h per ml)	—	53.3	115	115	194

Seven doses	C_{max} (µg equivalent/ml)	—	10.1	17.3	23.8	31.6
	T_{max} (h)	—	2	8	4	4
	$AUC_{2–24\,h}$(µg equivalent/h per ml)	—	195	380	465	667
Fourteen doses	C_{max} (µg equivalent/ml)	—	15.1	26.9	30.0	46.9
	T_{max} (h)	—	2	2	2	2
	$AUC_{2–24\,h}$(µg equivalent/h per ml)	—	300	531	624	945
	$AUC_{0–\infty}$ (µg equivalent/h per ml)	—	1287	2195	2714	4167
	Radioactivity in urine (%)	—	39	25	19	18
	Radioactivity in faeces (%)	—	55	70	72	78
	Peak PCM (µg equivalent/ml)	—	3.6	6.6	6.8	11.5
	Peak PCM-NH-COOH (µg equivalent/ml)	—	0.1	0.2	0.1	0.2
	Peak PCM-CH_2OH[a] (µg equivalent/ml)	—	1.3	3.1	2.1	3.6
	Peak PA-CH_2OH (µg equivalent/ml)	—	0.3	0.8	0.5	0.5
	Peak PA-COOH(µg equivalent/ml)	—	0.1	0.3	0.1	1.1
	Peak glucuronides (µg equivalent/ml)	—	0.2	1.2	0.6	3.3

From Sugimoto (2005c) and Mogi (2005b)
AUC, area under the curve of concentration–time; PA, procymidone acid; PA-CH_2OH, hydroxyprocymidone acid; PA-COOH, carboxyprocymidone acid; PCM, procymidone; PCM-CH_2OH, hydroxyprocymidone; PCM-NH-COOH, carboxyprocymidone.
[a] Included PCM-COOH.

(b) Dermal route

The dermal absorption of procymidone from a formulated product has been measured in rats in vivo and in rat and human skin membranes in vitro.

Groups of four male Sprague-Dawley rats were given [phenyl-^{14}C]procymidone (radiochemical purity, 99.4%; specific activity, 449 µCi/mg (16.61 MBq/mg) formulated as Sumisclex SC at a dose of 0.002, 0.02 or 0.2 mg/cm^2 given as an application to the shaved back. Rats were exposed for 0.5, 1, 2, 4, 10 and 24 h, the application site was then washed and the rats were killed and radioactivity measured by liquid scintillation counting (LSC). An additional group of rats at each dose was washed 10 h after application and excreta were collected until 168 h after dosing. Recoveries of radioactivity were > 90%. Radioactivity was readily removed from the skin by washing and most of the excreted radioactivity was eliminated in the urine. In the groups exposed for 10 h and then killed 168 h after dosing, 1.2–2.5% of the administered dose remained in skin; this was considered to be bound radioactivity that was unavailable for absorption, as excretion was essentially complete at 120 h. Dermal absorption after a 10-h exposure was calculated to be 13%, 9% and 4% for dermal exposures of 0.002 mg/cm^2, 0.02 mg/cm^2 and 0.2 mg/cm^2 respectively. This study claimed compliance with GLP and complied with OECD test guideline 427 (Savides, 2002).

In a study of dermal penetration in vitro, [phenyl-^{14}C]procymidone (radiochemical purity, 99.4%; specific activity 449 µCi/mg), formulated as Sumisclex SC, was applied at a dose of 0.295, 1.48 or 500 g/l, corresponding to application concentrations of 0.003, 0.015 and 5 mg/cm^2, to rat and human epidermal membranes. Membranes were checked for integrity on the basis of electrical resistance. Samples of receptor fluid (ethanol/water; 1 : 1) were taken at intervals up to 24 h after application of the formulation and the membranes were then washed. Human, but not rat membranes were stripped with tape to remove the stratum corneum. Absorption was calculated as percentage of the applied dose in receptor fluid plus unstripped epidermal membrane. Twenty-four hours after the application of procymidone at 0.295, 1.48 or 500 g/l, absorption through rat epidermis was 79%, 45% and 0.8% of the applied radioactivity respectively. Absorption through human membranes was lower than rat membranes: 5.3%, 4.7% and 0.022% of applied doses of 0.295, 1.48 and 500 g/l. The study claimed GLP compliance and complied with OECD test guideline 428 (Owen, 2002).

1.2 Biotransformation

See also section on special studies.

Mice

Groups of five male ICR mice were given a single oral dose of [phenyl-^{14}C]procymidone (radiochemical purity, > 99%; specific activity 22.8 mCi/mmol (843.6 MBq/mmol) at a dose of 100 mg/kg bw in corn oil. Urine and faeces were collected over 2 days and analysed for metabolites by TLC and comparison with standards. A range of tissues was extracted and analysed for metabolites by TLC. The predominant metabolites in the urine were the acid derivatives of procymidone (Table 3). Procymidone was the main component in faeces. In tissues and blood, the major component was procymidone, with alcohol derivatives being present at higher concentrations than the acids. (Kimura et al., 1988)

Rats

Male Sprague-Dawley rats received a single oral dose of [phenyl-^{14}C]procymidone (purity, > 99%; specific activity, 22.8 mCi/mmol (843.6 MBq/mmol) at 100 mg/kg bw in corn oil. Urine and faeces were collected for 2 days after dosing. Tissues were removed from other groups of rats at intervals up to 72 h after dosing. Samples of excreta, blood and tissues were prepared and analysed for metabolites by TLC and comparison with standards. The predominant metabolites in the urine were the acid derivatives of procymidone (Table 3). Procymidone was the main component in the faeces. In tissues and blood, the major component was procymidone, with alcohol derivatives being present at higher concentrations than the acids. The metabolic profile in rats was similar to that in mice (Table 3) (Kimura et al., 1988). A similar pattern of acid metabolites in excreta and alcohol derivatives in tissues was reported after the administration of procymidone at a dose of 25 mg/kg bw (Mikami & Yamamoto, 1976).

Table 3. Metabolites in excreta from rats and mice given [phenyl-^{14}C]procymidone a single dose at 100 mg/kg bw

Metabolite[a]	Recovery (% of administered dose)			
	Mice		Rats	
	Urine	Faeces	Urine	Faeces
PCM	2.6	7.0	0.2	5.2
PCM-CH_2OH	7.1	2.1	3.2	1.1
PA-CH_2OH	1.4	0.5	1.5	0.3
PCM-COOH	20	1.4	22	0.4
PA-COOH	39	3	47	0.6
DCA	< 1	< 1	< 1	< 1

From Kimura et al. (1988)

DCA, 3,5-dichloroaniline; PA, procymidone acid; PA-CH_2OH, hydroxyprocymidone acid; PA-COOH, carboxyprocymidone acid; PCM, procymidone; PCM-CH_2OH, hydroxyprocymidone; PCM-NH-COOH, carboxyprocymidone.

[a] See Figure 1 for key.

Three groups of five male and five female Crl:CD®(SD)BR rats were given a single oral dose of [phenyl-^{14}C]procymidone (radiochemical purity, > 98%; specific activity, 242 µCi/mg (8.95 MBq/mg) at 1 or 250 mg/kg bw in corn oil. An additional group of rats received 14 consecutive daily doses of unlabelled procymidone at 1 mg/kg bw before receiving [phenyl-^{14}C]procymidone at a dose of 1 mg/kg bw. Urine and faeces were collected for 7 days after dosing, after which the rats were killed and tissues removed for analysis. Samples were processed, including incubation with glucuronidase. Metabolites in excreta collected for up to 48 h (for rats at 1 mg/kg bw) and to 72 h (for rats at 250 mg/kg bw) after dosing were isolated by TLC and identified by co-chromatography with standards by mass spectral analysis. Overall, up to 42 urinary and nine faecal metabolites were found. Unchanged procymidone in the faeces accounted for < 5% of the administered dose in the group at 1 mg/kg bw, but was a major faecal component (18–27% of the administered dose) at 250 mg/kg bw. There were no significant differences in metabolism between the sexes, or after repeated dosing (Table 4). The findings were consistent with the results of earlier studies, with acid metabolites predominating in the urine (Struble, 1992b).

The proposed metabolic pathway is shown in Figure 1. The major metabolic reactions for procymidone were oxidation of the methyl groups to hydroxymethyl or carboxylic acid derivatives; cleavage of the imide; and glucuronide conjugation of the resultant metabolites.

Human-derived tissues

In an initial study, [phenyl-^{14}C]procymidone (purity, > 98%; specific activity, 15.6 MBq/mg) at a concentration of approximately 1 µmol/l was incubated with human hepatocytes at 37 °C for 46 h . The reaction was terminated by the addition of acetonitrile. Metabolites were identified by

Table 4. Metabolites in excreta from rats given [phenyl-^{14}C]procymidone at a dose at 1 or 250 mg/kg bw

Metabolite[a]	Recovery (% of administered dose)					
	Dose					
	1 mg/kg bw				250 mg/kg bw	
	Single dose		Fourteen doses			
	Male	Female	Male	Female	Male	Female
Urine	71	70	67	75	58	61
PCM	0	0	0	0	0	0
PCM-CH_2OH	< 1	< 1	< 1	< 1	< 1	1
PCM-CH_2OH-glucuronide	2	1	2	3	—	—
PA-CH_2OH	< 1	1.8	< 1	< 1	< 1	< 1
PCM-COOH	2	0	3	1	2	1
PA-COOH	50	48	45	56	44	44
DCA	0.3	0.5	0.1	1	0.8	0.5
Unknown/unextracted	10	11	11	19	11	11
Faeces	15	10	11	6	32	23
PCM	3	1	< 1	< 1	26	18
PCM-CH_2OH	2	1	1	< 1	2	1
Unknown/unextracted	10	7	8	6	7	6

From Struble (1992b)

DCA, 3,5-dichloroaniline; PA, procymidone acid; PA-CH_2OH, hydroxyprocymidone acid; PA-COOH, carboxyprocymidone acid; PCM, procymidone; PCM-CH_2OH, hydroxyprocymidone; PCM-NH-COOH, carboxyprocymidone.

qualitative autoradiography of TLC plates run with two solvent systems. The polar components were reported to be glucuronides of hydroxyprocymidone (PCM-CH_2OH) and hydroxyprocymidone acid (PA-CH_2OH). No radioactive compounds other than procymidone were detected in samples from the controls. The hepatocytes came from four different donors and some variations were evident in the autoradiograph patterns from different preparations (Tarui, 2005b).

Groups of four male, chimeric mice with humanized livers (i.e. liver repopulated with human hepatocytes) (from PhoenixBio, Hiroshima, Japan) and four male mice (uPA$^{-/-}$SCID) serving as controls were given a single oral dose of [phenyl-^{14}C]procymidone (purity, > 95%; specific activity, 15.8 MBq/mg) at 37.5 mg/kg bw in corn oil. Urine and faeces were collected at 24-h intervals for 72 h after dosing and analysed for radioactivity and for metabolites using high-performance liquid chromatography (HPLC)/LSC and TLC/autoradiography. A glucuronidase/sulfatase preparation (± glucuronidase inhibitor) was used to release conjugates, but only qualitative data were presented. Radioactivity was excreted at similar rates in the urine and faeces of control and chimeric mice (Table 5). The main difference in metabolism was the higher concentration of glucuronides (of PCM-CH_2OH and PA-CH_2OH) and lower concentration of acid metabolites excreted in the chimeric mice (Table 5) (Ohzone, 2005a).

Comparison of metabolism between species

Procymidone [phenyl-^{14}C]procymidone (specific activity, 15.8 MBq/mg) and PCM-CH_2OH (specific activity, 3.3 MBq/mg) were incubated for 60 min at 37 °C with S9 fractions prepared from pooled livers of women (two Caucasians and one African-American), rats (Sprague-Dawley), rabbits (New Zealand White) and monkeys (cynomolgus). A single concentration of 50μmol/l was used. Metabolites were identified by co-chromatography with reference standards. Only limited information was given in the study report. The rates of hydroxylation of procymidone (to alcohol or acid derivatives) or oxidation of PCM-CH_2OH (to acid derivatives) were relatively constant over the assay period. S9 from rabbit liver had the highest activity for hydroxylation of procymidone, with activity in rat liver being lower than that in other species (Table 6). S9 from monkey liver had the highest activity for the oxidation of PCM-CH_2OH, with the activity of S9 from rats being much lower than that from other species tested (Table 6) (Matsui, 2005b).

Table 5. Metabolite excretion profiles in chimeric (humanized) mice (n = 4) and control mice (n = 4) given [phenyl-^{14}C]procymidone as a single dose at 37.5 mg/kg bw

Metabolite	Recovery (% of administered dose)			
	Chimeric mice[a]		Control mice	
	Urine	Faeces	Urine	Faeces
Excretion in 24 h	59	22	65	17
Excretion in 72 h	73	24	76	20
PCM (72 h)	< 1	1.8	1.4	< 1
PCM-COOH (72 h)	21	4.6	34	6
PA-CH_2OH (72 h)	2	< 1	2.5	1
PA-COOH (72 h)	12	2.6	26	3
Glucuronides (72 h)	35	7.5	7	2
Others/unextracted (72 h)	3	7	5	6

From Ohzone (2005a)

[a] Chimeric mice possessed humanized livers (i.e. liver repopulated with human hepatocytes).

PA-CH_2OH, hydroxyprocymidone acid; PA-COOH, carboxyprocymidone acid; PCM, procymidone; PCM-CH_2OH, hydroxyprocymidone; PCM-NH-COOH, carboxyprocymidone.

The excretion of radioactivity in bile, urine and faeces, and the metabolites present, were investigated in groups of bile-duct cannulated females given a single oral dose of [carbonyl-^{14}C]procymidone (purity, > 98.5%; specific activity, 8 MBq/mg). Groups of four Sprague-Dawley rats received [carbonyl-^{14}C]procymidone at a dose of 3.5 or 62.5 mg/kg bw in corn oil; a rabbit (New Zealand White) and a cynomolgus monkey received [carbonyl-^{14}C]procymidone at a dose of 125 mg/kg bw in 0.5% methylcellulose. Bile, urine and faeces were collected for 48 h after dosing; samples were analysed for radioactivity and for metabolites. At termination, the gastrointestinal tract and contents were removed and analysed for radioactivity. Metabolites were separated by TLC and identified by co-chromatography with reference standards either before or after enzyme hydrolysis. The results are summarized in Table 7.

Interpretation of the data from the rabbit and monkey were complicated by the high residues in the gastrointestinal contents and carcass. Excretion of glucuronides in the urine was lower in rats than the other two species, but with higher concentrations of acid derivatives. The indications were that the glucuronides in rat bile are subsequently deconjugated and reabsorbed (Sugimoto, 2005d; Mogi, 2005c, 2005d).

Table 6. Specific enzyme activities related to the metabolism of procymidone in S9 from liver of four species

Metabolic conversion	Enzyme activity (pmol/min per mg protein)			
	Human	Rat	Monkey	Rabbit
Procymidone hydroxylation	36	10	30	61
PCM-CH_2OH oxidation	13	1	33	26

From Matsui (2005b)
PCM-CH_2OH, hydroxyprocymidone.
S9, 9000 × *g* supernatant.

Table 7. Metabolite excretion profiles in bile-duct canulated females given a single dose of [carbonyl-^{14}C]procymidone

Excretion parameter	Recovery (% of administered dose)					
	Rat (62.5 mg/kg bw)		Rabbit (125 mg/kg bw)		Monkey (125 mg/kg bw)	
Faeces 0–48 h	7		3.5		<1	
Carcass	7		22		25	
Gastrointestinal tract/contents	4		50		53	
	Urine	Bile	Urine	Bile	Urine	Bile
Excretion in 48 h	59	19	24	1	15	6
PCM (48 h)	0	0	< 0.1	< 0.1	< 0.1	< 0.1
PCM-COOH/CH_2OH (72 h)	28	0	2	< 0.1	1	< 0.1
PA-CH_2OH (72 h)	2	0	< 0.1	< 0.1	< 0.1	< 0.1
PA-COOH (72 h)	21	0.5	1.4	< 0.1	0.4	< 0.1
Glucuronides (72 h)	4	19	20	1	14	6
Others/unextracted (72 h)	2	0	< 0.1	< 0.1	0	< 0.1

From Sugimoto (2005d) and Mogi (2005c, 2005d)
PA, procymidone acid; PA-CH_2OH, hydroxyprocymidone acid; PA-COOH, carboxyprocymidone acid; PCM, procymidone; PCM-CH_2OH, hydroxyprocymidone; PCM-NH-COOH, carboxyprocymidone.

Binding to plasma proteins

The plasma-protein binding of [phenyl-^{14}C]procymidone (specific activity, 15.8 MBq/mg) and its metabolite [phenyl-^{14}C]PCM-CH_2OH (specific activity, 3.3 MBq/mg) was determined in humans, Sprague-Dawley rats, cynomolgus monkeys and New Zealand White rabbits; plasma from females was used fo each species. Plasma-protein binding at pH 7.0 was investigated in vitro at concentrations of 1, 3, 10 and 30 µg/ml using an ultrafiltration method. Both compounds exhibited a high level of protein binding in plasma from all species (Table 8). Plasma-protein binding of procymidone was similar in all species tested and ranged from 92% to 98% over the range of concentrations tested. The plasma-protein binding of PCM-CH_2OH was slightly lower than that of procymidone; the value for human plasma (90–91%) was slightly greater than that for rats (82–90%), monkeys (77–85%) and rabbits (83–86%). Since rats make much more PCM-CH_2OH (tenfold), these results indicated that there are likely to be much higher levels of free PCM-CH_2OH in rats than in other species (Matsui, 2005a).

2. Toxicological studies

2.1 *Acute toxicity*

(a) Lethal doses

The results of studies of acute toxicity with procymidone administered by the oral, dermal, subcutaneous, intraperitoneal (i.p) and inhalation routes, are presented in Table 9. Procymidone was of low acute toxicity by all routes. All these studies were performed before GLP was instituted and are limited in detail; however, there are no reasons to doubt the findings.

Clinical signs of toxicity were typically increased respiration and reduced motor activity. These were evident at doses of ≥ 250 mg/kg bw orally and subcutaneously and ≥ 100 mg/kg bw i.p., but showed no clear trend with dose administered. There were no adverse findings at postmortem examination.

Table 8. Plasma-protein binding in vitro and total and calculated plasma concentrations of free procymidone and PCM-CH_2OH (µg equivalent/ml) in female mice, rats, monkeys and humans

Concentration (µg/ml) / dose (mg/kg bw per day)	Human		Rat		Monkey		Rabbit	
	PPB (%)	PPB (%)	Total (µg equivalent/ ml)	Free[a] (µg equivalent/ ml)	PPB (%)	Total (µg equivalent/ ml)	Free[a] (µg equivalent/ ml)	PPB (%)
Procymidone								
1 / 37	95.5	97.9	2.6	0.06	95.4	ND	ND	97.6
3 / 67	95.5	97.5	2.6	0.06	94.6	3.6	0.19	97.3
10 / 125	95.5	96.9	6.4	0.19	93.9	6.6	0.4	97.4
30 / 250	95.4	96.3	5.1	0.15	92.4	6.8	0.4	97.1
PCM-CH_2OH								
1 / 37	90.9	90.4	9.9	1.3	85.3	ND	ND	85.6
3 / 67	90.9	98.9	9.8	1.3	83.9	1.3	0.2	85.6
10 / 125	90.5	87.1	32	5.7	80.2	3.1	0.5	84.3
30 / 250	90.4	82.1	51	9.2	76.9	2.1	0.3	82.7

From Matsui (2005a)

ND, no data; PCM-CH_2OH, hydroxyprocymidone; PPB, plasma-protein binding.

[a] Concentration of free substance was calculated from binding percentage and total concentration.

Figure 1. Proposed metabolic pathways of procymidone

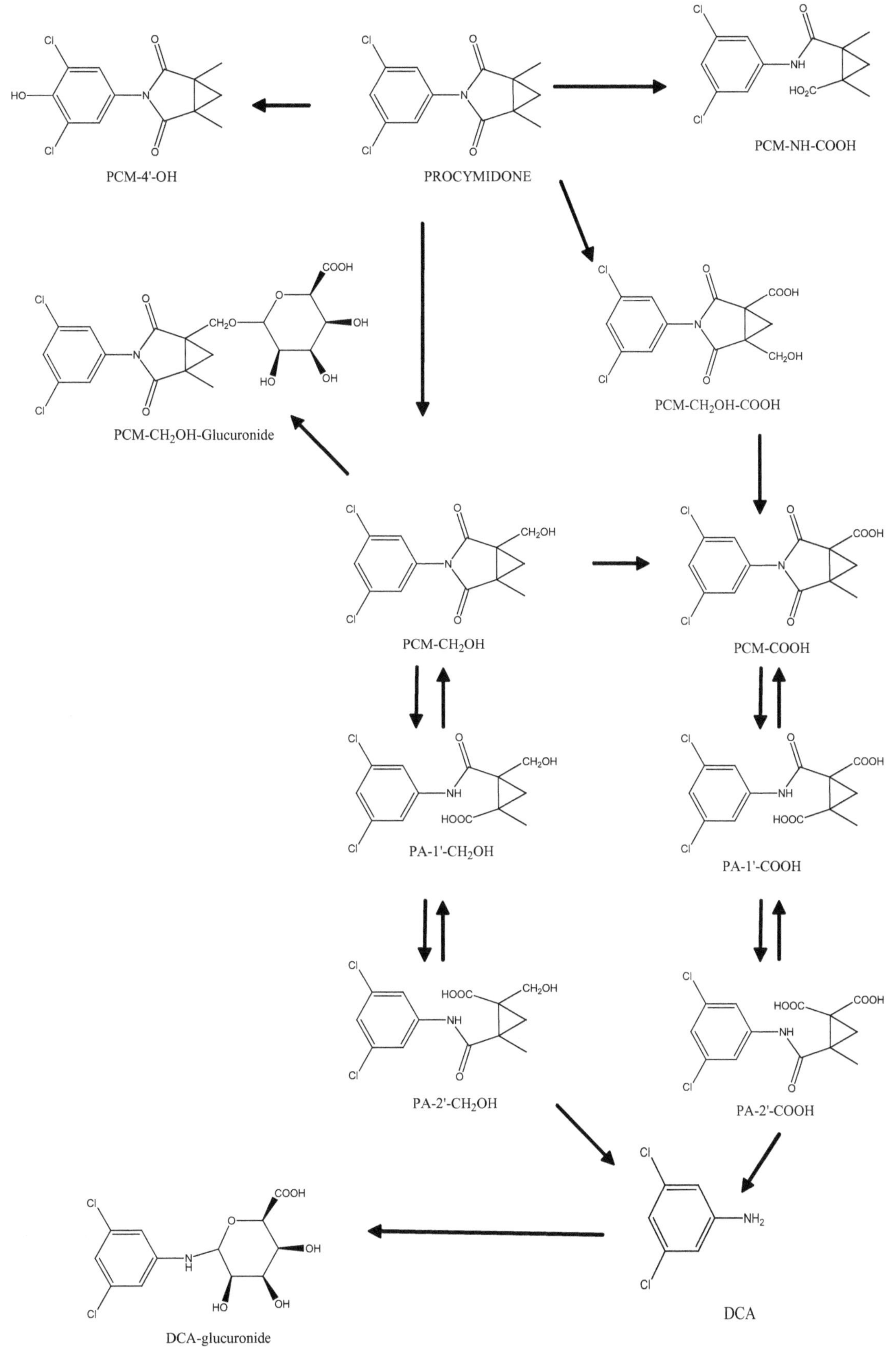

PCM-4'-OH

PROCYMIDONE

PCM-NH-COOH

PCM-CH_2OH-Glucuronide

PCM-CH_2OH-COOH

PCM-CH_2OH

PCM-COOH

PA-1'-CH_2OH

PA-1'-COOH

PA-2'-CH_2OH

PA-2'-COOH

DCA

DCA-glucuronide

DCA, 3,5-dichloroaniline; PA, procymidone acid; PA-CH_2OH, hydroxyprocymidone acid; PA-COOH, carboxyprocymidone acid; PCM, procymidone; PCM-CH_2OH, hydroxyprocymidone; PCM-NH-COOH, carboxyprocymidone.

Table 9. Results of studies of acute toxicity with procymidone

Species	Strain	Sex	Route	Vehicle	LD_{50} (mg/kg bw)	LC_{50} (mg/l air)	Purity (%)	Reference
Rat	SD	M & F	Oral	Corn oil	> 5 000	—	92.3	Kadota et al. (1976a)
							96.5	Segawa (1977)
Mouse	dd	M & F	Oral	Corn oil	> 5 000	—	92.3	Kadota et al. (1976a)
							96.5	Segawa (1977)
Rat	SD	M & F	Dermal	Corn oil	> 2 500	—	92.3	Kadota et al. (1976a)
							96.5	Segawa (1977)
Mouse	dd	M & F	Dermal	Corn oil	> 2 500	—	92.3	Kadota et al. (1976a)
							96.5	Segawa (1977)
Rat	SD	M & F	SC	Corn oil	> 10 000	—	92.3	Kadota et al. (1976a)
							96.5	Segawa (1977)
Mouse	dd	M & F	SC	Corn oil	> 10 000	—	92.3	Kadota et al. (1976a)
							96.5	Segawa (1977)
Rat	SD	M & F	IP	Corn oil	850 (M) 730 (F)	—	92.3	Kadota et al. (1976a)
Rat	SD	M & F	IP	Corn oil	1 440 (M) 1 450 (F)	—	96.5	Segawa (1977)
Mouse	dd	M & F	IP	Corn oil	1 560 (M) 1 900 (F)	—	92.3	Kadota et al. (1976a)
Mouse	dd	M & F	IP.	Corn oil	2 030 (M) 2 050 (F)	—	96.5	Kadota et al. (1976a); Segawa (1977)
Rat	SD	M & F	Inhalation	Fine dust	—	> 1.5 mg/l MMAD < 2.4 μm	99.5	Suzuki & Kato (1988); Tsuji (1994)

IP, intraperitoneal; F, female; M, male; MMAD, mass median aerodynamic diameter; SC; subcutaneous; SD, Sprague-Dawley.

(b) Dermal and ocular irritation and dermal sensitization

Procymidone was a slight transient irritant to rabbit skin (all signs had disappeared by 24 h; Kadota & Miyamoto, 1976), but produced no irritation to the rabbit eye (Kadota & Miyamoto, 1976). Procymidone was not a skin sensitizer in guinea-pigs in a repeated injection (10 + 1 doses) protocol using 1% or 5% w/v preparations in corn oil (Okuno, 1975), nor in a GLP-compliant Magnusson & Kligman maximization study using concentrations of 1% w/v for intradermal induction, 25% w/v for topical induction and 25% w/v for challenge application (Nakanishi et al., 1991).

2.2 Short-term studies of toxicity

(a) Oral route

Mice

Groups of 15 male and 15 female ICR mice received diets containing procymidone (purity, 96.9%) at a concentration of 0, 50, 150 or 500 ppm for 13 weeks. The mice were observed twice per day for clinical signs of toxicity and mortality and body weights were determined weekly. Food and water consumption were determined on three consecutive days each week. At the end of the treatment period, an ophthalmological examination was conducted and blood was taken for measurement of biochemical and haematological parameters. At necropsy, organs were weighed and tissues were examined microscopically.

Achieved intakes are given in Table 10. Behaviour, appearance and survival were not affected and no treatment-related effects were apparent in body weight, food and water intake, clinical pathology or macroscopic findings. An increase in serum glucose concentrations in females at the highest dose was not considered adverse as the absolute value was similar to typical mean values for the controls (114 vs 113 mg/dl) Absolute liver weights were increased for all males receiving procymidone and for females at 500 ppm, but relative liver weights were unaffected at ≤ 150 ppm (Table 10). Histopathology results showed a treatment-related centrilobular hypertrophy of hepatocytes in males at 500 ppm and hepatocyte granuloma in females (Table 10). Relative kidney weights were decreased in females at 150 and 500 ppm, but this was not considered to be treatment-related because there was no effect on absolute kidney weights or on histopathology.

The no-observed-adverse-effect level (NOAEL) was 150 ppm, equal to 22 mg/kg bw per day, on the basis of histopathological findings in the liver of males at 500 ppm. This study did not claim GLP compliance (Arai et al., 1980a).

Groups of 12 male and 12 female $B6C3F_1$ mice received diets containing procymidone (purity, 99.8%) at a concentration of 0, 100, 500, 2500 or 10 000 ppm for 13 weeks. All mice were observed daily for mortality and overt signs of toxicity, detailed clinical examinations were conducted weekly and body weight and food consumption were recorded at the same intervals. Blood and urine were collected for haematology, clinical chemistry and urine analysis at the start and end of the treatment period, respectively. Necropsies were performed at the end of the treatment period and organ weights were recorded. Tissues from the control group and from the group at the highest dose, and livers from the other groups, were processed for histopathology.

Actual intakes of procymidone were not calculated. A slightly depressed overall body weight (< 10%) was recorded for males at 10 000 ppm, although food intakes were not influenced by treatment. Behaviour, appearance and survival of the treated mice were not affected. Clinical pathology revealed higher activities of serum alanine aminotransferase (ALT) in males at 10 000 ppm and higher cholesterol concentrations in females at the highest dose. Blood urea nitrogen (BUN) was reduced in all groups of males, but the values were within the range for historical controls. Creatinine concentrations were also statistically lower than those of the controls in all groups (Table 11). At termination, organ-weight analysis revealed significantly higher liver-to-body-weight ratios and/or liver weights

Table 10. Liver weights and pathology findings in mice (n = 15) fed diets containing procymidone for 13 weeks

Finding	Dietary concentration (ppm)							
	0		50		150		500	
	Male	Female	Male	Female	Male	Female	Male	Female
Intake (mg/kg bw per day)	0	0	7	11	22	26	71	84
Liver weight (g)	1.4	1.3	1.5*	1.2	1.6*	1.2	1.6*	1.4*
Relative liver weight (%)	3.6	4.1	3.8	3.9	3.8	3.8	4.0	4.4
Hepatocyte hypertrophy	0	0	0	0	1	0	3	1
Hepatocytic granuloma	0	2	0	2	1	2	0	4

From Arai et al. (1980a)
* $p < 0.05$.

for males and females at 2500 or 10 000 ppm. Histopathology revealed changes in the liver of mice at 10 000 ppm and 2500 ppm. The changes consisted of multifocal coagulative necrosis of hepatic parenchyma, together with centrilobular cytoplasmic swelling, nuclear enlargement, coarsely dispersed chromatin and multinucleate hepatocytes (Table 11). Decreases in kidney and spleen weights were also noted at dietary concentrations of 2500 ppm and greater (Table 11).

The NOAEL was 100 ppm (reportedly equal to 19.6 mg/kg bw per day), on the basis of liver pathology (coagulative necrosis) in males at dietary concentrations of 500 ppm and greater. This study did not claim GLP compliance (Weir et al., 1984).

Groups of 20 male and 20 female ICR mice weer given diets containing procymidone (purity, 96.9%) at a concentration of 0, 50, 150 or 500 ppm for 26 weeks. The mice were observed twice per day for clinical signs of toxicity and mortality and body weights were determined weekly. Food and water consumption were determined on three consecutive days each week. At the end of the treatment period, an ophthalmological examination was conducted and blood was taken for measurement of biochemical and haematological parameters. All surviving mice were necropsied immediately after blood was taken, organs were weighed and tissues were examined microscopically.

Achieved intakes are presented in Table 12. There were 11 deaths during the study; these occurred in all groups and there was no indication of a relationship to treatment. No treatment-related effects were detected on behaviour, appearance, survival, body weights, food and water intakes, ophthalmological examination, or macroscopic pathology. There was a statistically significant decrease in the leukocyte count in males at 150 and 500 ppm (Table 12), the deficit being primarily in lymphocytes. Since no comparable change was found in other studies in mice, these effects were considered to be unrelated to treatment. There was also a statistically significant increase in creatinine concentration and cholinesterase activity in male mice at 500 ppm and a reduction in cholinesterase activity in female mice of that group; these are considered to be sporadic findings unrelated to treatment. Organ weights were similar across all groups; an increase in pituitary weight in all groups of males was without a dose–response relationship, not reproduced in females and was considered to be a chance

Table 11. Findings in mice (n = 12) fed diets containing procymidone for 13 weeks

Finding[a]	Dietary concentration (ppm)									
	0		100		500		2500		10000	
	M	F	M	F	M	F	M	F	M	F
Creatinine (mg/dl)	1.2	0.9	1.0	0.7	0.7*	0.6	0.6*	0.6	0.5*	0.8
Cholesterol (mg/dl)	139	116	139	139	158	141	158	128	154	164*
BUN (mg/dl)	41	29	31	25	24*	26	29*	27	23*	22
ALT (U/l)	268	229	353	260	256	125	531	216	795*	327
Liver weight (g)	1.7	1.5	1.7	1.5	1.7	1.7	2.0*	2.0*	2.4*	2.3*
Relative liver weight (%)	6.6	6.4	6.3	6.6	6.8	7.2*	8.2*	8.7*	10.1*	10.5*
Relative kidney weight (%)	2.4	2.1	2.3	2.0	2.3	2.0	2.2	1.9	1.9*	1.7*
Relative spleen weight (%)	0.33	0.45	0.35	0.43	0.32	0.47	0.35	0.40	0.31	0.39*
Centrilobular hepatocyte swelling	0	0	1	0	6	0	11	7	12	12
Liver, coagulative necrosis	0	0	0	0	1	1	5	0	10	7

From Weir et al. (1984)
ALT, alanine aminotransferase; BUN, blood urea nitrogen; F, females; M, males.
[a] Values are means for each group of 12 mice.
* $p < 0.05$.

finding. Histopathology revealed that the incidence of a testicular atrophy was higher than that among the controls, achieving statistical significance at 500 ppm (Table 12). Liver weights and histological findings were unaffected by treatment.

The NOAEL was 150 ppm, equal to 20 mg/kg bw per day, on the basis of the statistically significant increase in testicular atrophy at 500 ppm. This study did not claim GLP compliance (Arai et al., 1980b).

An additional 6-month study was conducted to further investigate the testicular atrophy reported by Arai et al. (1980b). Groups of 20 male Alpk/AP albino mice received diets containing procymidone (purity, 99.8%) at a concentration of 0, 10, 30, 100 or 300 ppm. Mice were checked daily and a detailed clinical examination was given weekly. Body weight was monitored weekly and food consumption was recorded weekly for the first 12 weeks and then for 7 days every 4 weeks for the remainder of the study. At termination, mice were given a full postmortem, the testes and epididymides from mice in all groups were weighed and examined by histopathology. Mean achieved intakes were 0, 1.1, 3.6, 11 and 37 mg/kg bw per day.

One mouse at the lowest dose died. No treatment-related clinical signs or deaths were reported. Body-weight gain, food consumption and food use were not influenced by treatment. After 26 weeks, organ-weight analysis and histopathology of testes and epididymides revealed no evidence of a treatment-related effect (Table 13). The NOAEL was 300 ppm, equal to 37 mg/kg bw per day, on the basis of the absence of dose-related effects. The study was checked by a quality assurance unit (Kinsey et al., 1985).

Rats

Groups of 12 male and 12 female Sprague-Dawley rats were fed diets containing procymidone (purity, 98.7%) at a concentration of 0, 150, 500 or 1500 ppm for 6 months. Additional groups of 15 males and 15 females were fed diets containing procymidone at 0 or 1500 ppm for 9 months, or 0 or 1500 ppm for 6 months and then placed on a control diet for another 3 months. Rats were observed

Table 12. Leukocyte and testicular findings in mice (n = 20) fed diets containing procymidone for 6 months

Finding	Dietary concentration (ppm)							
	0		50		150		500	
	Males	Females	Males	Females	Males	Females	Males	Females
Intake (mg/kg bw per day)	0	0	6.5	7.3	20	24	72	83
Leukocytes (10^2/mm^3)	48	41	47	46	39*	41	34*	37
Testicular atrophy (all grades;	2	—	5	—	6	—	10*	—
			($p = 0.2$)		($p = 0.1$)		($p = 0.01$)	

From Arai et al. (1980b)
* $p < 0.05$.

Table 13. Testes and epididymides weights in mice fed diets containing procymidone for 6 months

Organ weight	Dietary concentration (ppm)				
	0	10	30	100	300
Testes (mg)	213	240	225	257*	233
Epididymis (mg)	104	107	105	115	103

From Kinsey et al. (1985) * p<0.05

daily for signs of toxicity; body weight was measured two or three times during the first 2 weeks of treatment and then weekly; food consumption was also measured weekly. Urine analysis was conducted after 3, 6 and 9 months of treatment; haematological and blood chemistry examinations were made at 6 and 9 months. A complete necropsy was conducted on each rat at termination, organ weights were recorded and tissues were examined by histopathology.

Achieved intakes of procymidone were not calculated. No compound-related effects were observed on mortality, clinical signs, food and water consumption, urine analysis and clinical chemistry. Depression of body-weight gain was observed in males and females at 1500 ppm. There were slight decreases (< 10%) in packed cell volume (PCV; haematocrit) and haemoglobin values in males and females at 1500 ppm for 6 months, but not in those fed the same dose for 9 months. Serum ALT activity was increased at ≥ 500 ppm after 6 months but not after 9 months. Liver-weight to body-weight ratio was increased after 6 months at 1500 ppm in males and at 500 (< 10%) and 1500 ppm in females (Table 14), but this was reversible and had returned to normal in males after the 3-month recovery period. The spleen to body-weight ratio was also increased in male rats at 500 and 1500 ppm at 6 months; however, this increase was not apparent after treatment for 9 months and was therefore considered not related to treatment (Table 14). Relative adrenal weight was significantly increased at 1500 ppm in males and 500 and 1500 ppm in females, and relative pituitary weight at all doses in females after 6 months, but with no dose–response relationship; there were no histopathological changes in these organs and therefore alterations in weight were considered to be of no toxicological significance. After 9 months, absolute and relative testes weights were increased. The only histopathological lesion that could be attributed to treatment with procymidone was swelling and vacuolar degeneration of the liver cells in males fed 1500 ppm for 6 and 9 months. All the findings showed clear evidence of returning to normal after the recovery period.

Table 14. Findings in rats fed diets containing procymidone for 13 weeks

Finding	Dietary concentration (ppm)							
	0		150		500		1500	
	M	F	M	F	M	F	M	F
Body weight (g) at 6 months)	669	349	687	337	672	327	632	306*
ALT activity (U/l):								
6 months	15	18	17	19	21*	23	21	23
9 months	33	42	—	—	—	—	26	23
Relative liver weight (%):								
6 months	2.8	3.0	2.9	2.8	2.9	3.2*	3.2*	3.3*
9 months	2.9	2.8	—	—	—	—	3.3*	3.4*
Recovery	2.9	2.8	—	—	—	—	2.8	3.0*
Relative testes weight (%):								
6 months	0.55	—	0.56	—	0.54	—	0.59	—
9 months	0.42	—	—	—	—	—	0.58*	—
Recovery	0.42	—	—	—	—	—	0.49	—
Relative spleen weight (%)								
6 months	0.12	0.14	0.13	0.14	0.13	0.16*	0.13	0.18*
9 months	0.12	0.16	—	—	—	—	0.13	0.16

From Fujita et al. (1976) and Hosokawa (1984)
ALT, alanine aminotransferase; F, females; M, males.
* $p < 0.05$.

The NOAEL was 500 ppm, equivalent to 25 mg/kg bw per day, on the basis of reduced body-weight gain, a range of altered organ weights and other effects at 1500 ppm. The study did not claim GLP compliance (Fujita et al., 1976; Hosokawa, 1984).

(b) Dermal route

Rats

Groups of 10 male and 10 female Crj:CD(SD) rats were given procymidone (purity, 99.6%) at a dose of 0, 180, 450 or 1000 mg/kg bw per day as ground powder that was moistened with water and applied daily to the shaved backs for 28 days. Clinical observations were carried out daily during the treatment period and any skin reactions were noted. Body weight and food consumption were recorded weekly and an ophthalmological examination was conducted during the last week of treatment. Haematological and clinical parameters were measured on blood taken at termination and urine collected during week 4 was also analysed. A gross necropsy was performed at termination, organs were weighed and tissues were examined microscopically.

There was no evidence of local effects at the application site. Increases in adrenal and heart weights, and decreases in prostate and spleen weights were without a dose–response relationship and/or within typical values for controls. There were no effects on the liver or testes. Histopathology and clinical chemistry findings showed no treatment-related effects. The NOAEL was 1000 mg/kg bw per day, the highest dose tested. The study contained GLP compliance statements and complied with OECD test guideline 410 (Ogata, 2002).

Dogs

Groups of six male and six female beagle dogs were given gelatin capsules containing procymidone (purity, > 95%) at an oral dose of 0 (empty capsule), 20, 100 or 500 mg/kg bw per day for 26 weeks. Clinical signs and food consumption were recorded daily and body weight was measured weekly. Water consumption was measured over a 3-day period after 7, 16 and 25 weeks of treatment and urine analysis was performed at the same intervals. Ophthalmology, haematology, and clinical chemistry examinations were made after 4, 8, 12, 16, 21 and 25 weeks. At the end of the treatment period, the dogs were dissected, organ weights were recorded and microscopic investigations were made on stained tissues and organs.

All dogs survived the study. Vomiting and diarrhoea were seen in all groups, including controls, but was more prevalent in dogs at 500 mg/kg bw per day, the highest dose. The incidences of diarrhoea in males at the highest dose and emesis in both sexes was statistically significantly increased relative to controls ($p < 0.02$, Mann-Whitney U test subsequent to a Kruskal-Wallis ANOVA). Body weights, food and water intakes, ophthalmoscopy and urine analysis parameters were not affected by treatment. There were statistically significant changes in clinical chemistry. Serum alkaline phosphatase (ALP) activity was increased in male and female dogs at 500 mg/kg bw per day. At the same dose, BUN, glucose and calcium concentrations were increased in males (Table 15). BUN and calcium concentrations were also increased in dogs at 100 and 20 mg/kg bw per day (Table 15); the changes at 20 and 100 mg/kg bw per day were rather sporadic, possibly linked to pre-test values and were considered to be not related to administration of procymidone. Results of urine analysis and kidney pathology were unremarkable. Absolute and relative heart weights were significantly decreased in females at 20 or 500 mg/kg bw per day, but without any dose–response relationship or histopathological correlate.

The NOAEL was 100 mg/kg bw per day on the basis of the compound-related emesis, diarrhoea and serum chemistry changes at 500 mg/kg bw per day. The study was inspected by a quality assurance unit (Nakashima et al., 1984).

Groups of four male and four female beagle dogs received capsules containing procymidone technical (purity, 98.5%) once daily at a dose of 0 (empty capsule), 20, 100 or 500 mg/kg bw per day for 1 year. All dogs were observed twice per day for mortality, moribundity and clinical signs of toxicity and were given a detailed physical examination weekly. Body weight and food consumption were recorded weekly. Ophthalmoscopic assessments were made during weeks 26 and 52. Clinical pathology and haematology parameters were measured in blood sampled before dosing began and during weeks 13, 26, 39 and 52. All dogs were given a gross postmortem examination, selected organs were weighed and tissues were examined microscopically.

One male in the group at 100 mg/kg bw per day was killed following an irreparable intussusception of the large intestine in week 19. Emesis and soft faeces were observed in dogs in all groups, including the controls, but appeared to be more prevalent at 500 mg/kg bw per day; the incidence of diarrhoea/soft faeces in males at this, the highest dose, was statistically significantly increased relative to values for the controls (p = 0.02, Mann-Whitney U test subsequent to a Kruskal-Wallis ANOVA). Dogs in the groups receiving procymidone generally consumed more food and put on more weight than did the controls. There was an increase in serum globulin concentration in weeks 39 and 52 amongst male dogs at 500 mg/kg bw per day, resulting in a reduced albumin : globulin ratio; a less marked effect was present at 100 mg/kg bw per day. Serum ALP activity was increased in males at the highest dose (Table 16). Serum concentrations of creatinine, glucose, BUN and calcium were similar between groups. No treatment-related changes were detected on organ weights or by gross or microscopic pathology examination.

Table 15. Findings in dogs (n = 6) fed capsules containing procymidone for 26 weeks

Finding	Dose (mg/kg bw per day)							
	0		20		100		500	
	M	F	M	F	M	F	M	F
ALP activity (U/l)	149	143	164	133	160	142	196	242
BUN (mg/dl)	18	21	20*	21	20	20	21*	23
Glucose (mg/dl)	92	94	93	95	94	94	102*	89
Calcium (mg/dl)	11.1	11.3	11.3	11.5	11.4*	11.4	11.5*	11.3
(before dosing)	(11.1)	(11.2)	(11.0)	(11.2)	(11.2)	(11.3)	(11.2)	(11.4)
Heart weight (g)	103	93	99	83	100	97	102	85
Relative heart weight (%)	0.83	0.88	0.79	0.74*	0.78	0.86	0.83	0.76*

From Nakashima et al. (1984)
ALP, alkaline phosphatase; BUN, blood urea nitrogen; F, females; M, males.
* $p < 0.05$.

Table 16. Findings in dogs (n = 4) fed capsules containing procymidone for 52 weeks

Finding	Dose (mg/kg bw per day)							
	0		20		100		500	
	Males	Females	Males	Females	Males	Females	Males	Females
Alkaline phosphastase activity (U/l)	27	40	31	46	29	34	53*	44
Total protein (g/dl)	5.8	5.8	5.8	5.6	6.6*	6.1	6.6*	6.1
Globulin (g/dl)	2.4	2.3	2.3	2.1	2.9	2.5	3.2*	2.3
Albumin : globulin ratio	1.5	1.6	1.5	1.6	1.3	1.5	1.1	1.6

From Dalgard & Machotka (1992)
* $p < 0.05$.

The NOAEL was 100 mg/kg bw per day on the basis of increased emesis and diarrhoea/soft faeces, and clinical chemistry changes at 500 mg/kg bw per day. The study claimed GLP compliance and complied with OECD test guideline 409 (Dalgard & Machotka, 1992).

2.3 *Long-term studies of toxicity and carcinogenicity*

Mice

Groups of 50 male and 50 female B6C3F$_1$ mice received diets containing procymidone (purity, 99.8%) at a concentration of 0, 30, 100, 300 or 1000 ppm for 104 weeks. In addition, satellite groups of 10 males and 10 females were used for interim investigations at 26, 52 and 78 weeks. All mice were observed twice per day for mortality and signs of toxicity and a detailed examination, that included palpation, was conducted weekly. Body weights and food consumption were recorded weekly for the first 14 weeks and every 2 weeks thereafter. An ophthalmological examination was performed before the start of treatment and at weeks 26, 52, 78 and 105, and blood and urine samples were collected at the same intervals. A complete gross necropsy was performed on all mice, organs were weighed and samples of tissues and bone-marrow smears were taken for microscopic examination.

Mean intakes of procymidone over the study were presented in an addendum (Moore, 1993; Table 17). Treatment had no influence on survival. Survival for male mice at 0, 30, 100, 300 or 1000 ppm was 82%, 88%, 80%, 82% and 80%, respectively, and survival for females was 82%, 74%, 64%, 78% and 72%, respectively. There were no treatment-related effects on behaviour or appearance, body weights, food and water intakes or clinical pathology investigations. Organ-weight analysis revealed higher liver weights and liver-to-body-weight ratios for both sexes at 300 and 1000 ppm. Absolute and relative liver weights were statistically significantly increased at weeks 26 in mice at the highest dose, progressing with duration of dosing, such that absolute liver weights at week 105 were significantly higher than those of mice in the control group for males and females at 300 ppm and in males at 100 ppm. Histopathology revealed treatment-related changes only in the liver. After 52 weeks, there was an increase in a very mild hepatocellular change recorded as centrilobular cytologic alteration. After 105 weeks, high incidences of centrilobular cytomegaly were detected in males at 1000 ppm and to a lesser extent in males at 300 ppm and females at 1000 ppm. Focal or multifocal hepatocellular hyperplasia, eosinophilic foci and fatty change were also recorded at 1000 ppm. Neoplastic changes were an increased incidence of hepatocellular adenomas and carcinomas combined in males at 30, 100 and 1000 ppm, but without a dose–response relationship, and females at 300 and 1000 ppm. The incidence of hepatoblastomas was increased in males at 1000 ppm. Although the incidences were reported to be within the published range for control mice of this strain, clear increases are evident in the number of females at the highest dose with hepatocellular adenomas and the number of males at the highest dose with hepatoblastomas when compared with concurrent controls and groups at lower doses (Table 17). A subsequent statistical analysis (Thakur, 1988) concluded that there was a statistically significant trend in females for adenomas and adenomas plus carcinomas combined and in males for hepatoblastomas.

The NOAEL for toxicity was 100 ppm, equal to 15 mg/kg bw per day, on the basis of the overall range of liver pathologies at 300 ppm. The increased liver weight in males at 100 ppm was not considered to be adverse as there were no associated clinical chemistry or histopathological findings. Procymidone was carcinogenic to B6C3F$_1$ mice, inducing hepatocellular adenomas in females and hepatoblastomas in males at 1000 ppm; the NOAEL for neoplastic toxicity was 300 ppm, equal to 46 mg/kg bw per day. The study claimed GLP compliance and met the essential elements of OECD test guideline 453 (Filler & Parker, 1988).

Table 17. Hepatic findings in a long-term study of carcinogenicity in mice fed diets containing procymidone

Finding	Incidence (No. of mice per group of 50)									
	Dietary concentration (ppm)									
	Male					Female				
	0	30	100	300	1000	0	30	100	300	1000
Mean compound intake (mg/kg bw per day)	0	4.6	15	46	153	0	6.4	23	65	206
Liver weight (g)	1.6	1.8	2.1*	1.9*	2.4*	1.4	1.5	1.6	1.7*	2.2*
Centrilobular cytomegaly	0	0	1	6*	43*	0	0	0	0	5*
Multifocal hyperplasia	2	0	1	5	8	0	0	0	2	2
Eosinic foci	1	0	1	5	3	0	1	0	2	10*
Multifocal fatty change	2	1	3	5	11*	0	0	0	4	3
Hepatocellular adenomas	7	11	12	9	10	1	1	0	3	7*
Hepatocellular carcinomas	5	6	9	5	10	1	1	2	4	2
Hepatoblastomas	1	0	0	2	5	0	0	0	0	0
Malignant hepatocellular neoplasms	5	6	9	7	11	1	1	2	4	2

From Filler & Parker (1988) and Moore (1993)
* $p < 0.05$, Fisher exact test (one-sided).

Rats

Groups of 50 male and 50 female Osborne-Mendel rats were given diets containing procymidone (purity, 99.8%) at a concentration of 0, 100, 300, 1000 or 2000 ppm for 104 weeks. Satellite groups of 50 males and 50 females received the same treatment and were used for blood analyses and interim termination at 26, 52 or 78 weeks. Groups of 50 males and 50 females, and 60 males and 60 females, respectively, served as controls for the main and satellite groups. All rats were observed daily for mortality and overt signs of toxicity and were given a detailed clinical examination weekly. Body weights, food consumption and water consumption were recorded weekly for the first 13 weeks and at 2-week intervals thereafter. An ophthalmology assessment was performed on 10 males and 10 females per group at weeks 26, 52, 78 and 104, blood and urine were collected for clinical pathology from 10 males and 10 females per group at the same intervals. Testosterone analyses were performed on blood samples from males in the control group (n = 2–8) and males at the highest dose (n = 3–5) at 18 and 24 months. At termination, a necropsy was performed, organ weights were recorded and tissues were examined microscopically.

Survival at termination in the groups investigated for carcinogenicity was 22–34% for males and 50–70% for females; in males, 50% survival was maintained until week 90. There were no biologically significant or consistent compound-related effects on mortality, clinical signs, food and water consumption, ophthalmology, urine analysis, and haematology or blood chemistry. Reduced body weight (> 10%) was noted in males and females at 1000 and 2000 ppm (Table 18); body weight was also lower at 300 ppm during the early part of the study and, although occasionally statistically significant, varied by less than 5%. The incidence of palapable masses was higher in males at the highest dose than in other groups from week 26 onwards. Increased relative weights of liver in both sexes and testes and ovaries were recorded at 1000 and 2000 ppm. In females, the kidney to body-weight and the brain to body-weight ratios were also increased at 2000 ppm, but there was no pathological evidence of a treatment-related abnormality in either organ. Seminal vesicle weights were decreased at 2000 ppm. Histopathology revealed an increased incidence of hepatic centrilobular cytomegaly in males and females at 1000 and 2000 ppm and a statistically significant increased incidence of ovarian stromal hyperplasia in females at 2000 ppm. There was no increase in the total number of rats with benign or malignant tumours. An increased incidence of testicular interstitial-cell tumours and interstitial-cell hyperplasia was seen in males at 1000 and 2000 ppm (Table 18). Testosterone concentrations were increased in males at 2000 ppm (the only dose sampled) at 18 and 24 months.

The NOAEL for toxicity was 300 ppm, equal to 14.0 mg/kg bw per day, on the basis of reduced body weight and liver pathology at 1000 ppm and greater. The NOAEL for tumour incidence was 300 ppm, equal to 14 mg/kg bw per day, on the basis of the increase in interstitial-cell tumours of the testes at 1000 ppm and greater. The study claimed GLP compliance and met the requirements of OECD test guideline 453 (Keller & Cardy, 1986).

2.4 *Genotoxicity*

Procymidone was not genotoxic in a range of studies in vitro and in vivo (Table 19). Many of the studies were relatively old and contained moderate levels of detail, but the information they contained did not indicate that procymidone has any genotoxic potential. Three more modern studies in vitro, which complied with GLP and with the essential elements of contemporary test guidelines, and provided negative results.

2.5 *Reproductive toxicity*

(a) *Multigeneration studies*

Rats

Groups of 30 male and 30 female Alpk:APfSD (Wistar-derived) rats (F_0 generation) were given diets containing procymidone (purity, > 99%) at a concentration of 0, 50, 250 and 750 ppm. The doses

Table 18. Findings in a long-term study of carcinogenicity in rats fed diets containing procymidone

Finding	Dietary concentration (ppm)									
	Males					Females				
	0	100	300	1000	2000	0	100	300	1000	2000
Mean intake (mg/kg bw per day)	0	4.6	14	48	97	0	6	18	60	125
Body weight (g), week 81	582	572	563	535*	513*	313	297	289	278*	278*
Relative testes/ovary weight (%):										
12 months	0.84	0.86	0.84	0.87	0.92	0.046	0.048	0.050	0.053	0.058
18 months	0.78	0.81	0.85	0.90	0.90	0.041	0.056	0.054	0.046	0.049
24 months	0.71	0.80	0.84	1.06*	1.04*	0.042	0.043	0.042	0.048*	0.055*
Relative liver weight (%) at 24 months	4.1	3.7	4.4	4.6	4.9*	3.2	3.4	3.6	4.1*	4.4*
Liver centrilobular cytomegaly	0	0	0	11*	17*	0	0	2	25*	36*
Testes, interstitial-cell tumour	1	1	0	10*	20*	—	—	—	—	—
Interstitial-cell hyperplasia	2	0	1	7	12*	—	—	—	—	—
Ovaries, stromal hyperplasia	—	—	—	—	—	0	0	0	2	8*

From Keller & Cardy (1986)
* $p < 0.05$.

Table 19. Results of studies of genotoxicity with procymidone

End-point	Test object	Concentration	Purity (% w/w)	Result	Reference
In vitro					
Reverse mutation[a]	*S. typhimurium* strains TA98, TA100, TA1535, TA1537 and *E. coli* (WP2 *hcr)*	10–1 000 µg/plate	96.9	Negative +S9[d] Negative – S9	Moriya & Kato (1977)
Reverse mutation[b]	*S. typhimurium* strains TA98, TA100, TA1535, TA1537	10–10 000 µg/plate	96.3	Negative +S9[d] Negative –S9	Suzuki & Miyamoto (1976)
Reverse mutation[b,c,f]	*S. typhimurium* strains TA98, TA100, TA1535, TA1537, TA1538 and *E. coli* (WP2 *uvrA)*	5–5 000 µg/plate	99.4	Negative +S9[d] Negative –S9	Kogiso (1991)
Gene mutation (*Hprt* locus)	Chinese hamster lung cells (V79)	0.7–6.0 mmol/l	Sumisclex 50% granule	Negative +S9[d] Negative –S9	Principe et al. (1980) Nunziata (1982)
Chromosomal aberration[g]	Chinese hamster ovary cells (CHO-K1)	75–300 µg/ml	99.4	Negative +S9[d] Negative –S9	Hara (1991b)
DNA damage	*Bacillus subtilis* (H17 and M45*rec*⁻)	20–2 000 µg/disk	96.9	Negative	Moriya & Kato (1977)
DNA damage[b]	*Bacillus subtilis* (H17 and M45*rec*⁻)	10–10 000 µg/disk	96.3	Negative	Suzuki & Miyamoto (1976)
Sister chromatid exchange[h]	Primary cultures from ICR mouse embryos	1–100 µmol/l	96.9	Negative +S9[d] Negative –S9	Suzuki (1980)
Unscheduled DNA synthesis[i]	Heteroploid human epithelial cells	6 µmol/l–6.0 mmol/l	Sumisclex 50% granule	Negative +S9[d] Negative –S9	Principe et al. (1980) Nunziata (1982) (supplement)
Unscheduled DNA synthesis[c, f, j]	Primary rat hepatocytes	3–300 µg/ml	99.4	Negative	Hara (1991a); Hara (1992)
In vivo					
Chromosomal aberration[k]	Male ddY mice	400, 800 or 1 600 mg/kg bw intraperitoneally in corn oil	96.9	Negative	Hara & Suzuki (1980)
Reverse mutation, host-mediated[b]	Male ICR mice, *S. typhimurium* (G46)	1 000 or 2 000 mg/kg bw by gavage in DMSO	96.3	Negative	Suzuki & Miyamoto (1976)
Reverse mutation host-mediated[b, l]	Male ICR mice, *S. typhimurium* (G46)	2 × 200 and 2 × 500 mg/kg bw by gavage in 5% gum arabic	96.9	Negative	Moriya & Kato (1977)

Positive controls were included in all tests for genotoxicity and were performed adequately.
DMSO, dimethyl sulfoxide; S9, 9000 × *g* supernatant from rat livers.
[a] Each test was conducted in duplicate.

[b] Each test was conducted in triplicate.
[c] Two separate experiments were conducted.
[d] S9 was prepared from livers of Sprague Dawley rats pre-treated with polychlorinatedbiphenyl mixture..
[e] S9 was prepared from livers of Sprague Dawley rats pre-treated with phenobarbital and -naphthoflavone.
[f] The study was performed in compliance with good laboratory practice and according to United States Environmental Protection Agency (EPA) study guideline 84-2.
[g] One hundred cells from duplicate cultures were analysed at 10 and 18 h after treatment.
[h] Fifty metaphases were analysed at each concentration.
[i] Three replicates of 50 cells were analysed in each experiment.
[j] Hepatocytes were prepared from Sprague-Dawley rats; at each concentration, six replicates of 50 cells were analysed.
[k] Six mice were treated for 6, 24 and 48 h after intraperitoneal administration; 50 metaphases in bone-marrow cells were analysed for each mouse.
[l] Six mice were treated at each concentration.

were based on a preliminary study that showed decreased litter size and feminization of male pups at ≥ 1000 ppm. After 12 weeks, the rats were mated and allowed to rear the F_{1a} litters to weaning and then paired again to produce the F_{1b} litters. F_1 parents selected from the F_{1a} litters were mated to produce two further litters (F_{2a} and F_{2b}). Body weight, food intake and clinical condition were recorded weekly. Reproductive parameters (fertility, length of gestation, precoital interval, litter size, viability and number of live and dead pups) were recorded for each generation and litter. The parental rats were given a postmortem, the liver and reproductive organs were weighed and tissues were examined histopathologically. Clinical observations, body weights, organ weights and pathology findings were also recorded for the offspring. Additional pathological examinations were performed subsequently and were reported in addenda.

Achieved intakes of procymidone throughout the study were not tabulated; data for premating periods only were presented in Wickramaratne & Milburn (1990), but actual concentrations, stability and homogeneity in diet were confirmed as acceptable. All parental rats at the highest dose showed reductions in body-weight gain from the start of compound administration; statistically significant for F_0 rats of both sexes and F_1 and F_2 females. There were no consistent effects on body weight or body-weight gain at 250 ppm. Precoital interval was generally reduced in pairings at 750 ppm. In F_1 males at the highest dose (but not F_0), there was a significant reduction in fertility (Table 20), associated with hypospadias (Table 20). Relative liver and relative testes weights were increased at 250 and 750 ppm (Table 20). The duration of gestation was similar in all groups. Litter size was reduced in the F_{1b} mating at 750 ppm, but not in any other matings. Penile abnormalities were noted at both gross and microscopic examinations in rats at 750 ppm. No effects were observed on sperm parameters. Sexing pups at birth was made difficult by reductions in anogenital distance (qualitative assessment) in male offspring at 750 ppm. Liver weight and testes-weight findings in parents were repeated in offspring; the increase in testes weights at 50 ppm in the F_{1b} pups was < 10%, and not statistically significant in the F_{2b} pups and was not considered to be adverse. Reductions in weights of the prostate and epididymis were seen at 250 ppm (F_{1b}) and 750 ppm (Table 20). There were no adverse effects on the sex organs of female pups.

The NOAEL for pup development was 50 ppm, equivalent to 3 mg/kg bw per day, on the basis of alterations in testes, prostate and epididymis weights at 250 ppm. The NOAEL for reproductive effects was 250 ppm, equivalent to 17 mg/kg bw per day, on the basis of reduced male fertility at 750 ppm. The NOAEL for parental toxicity was 250 ppm on the basis of reduced body weights at 750 ppm. The study was inspected by a quality assurance unit and complied with OECD test guideline 416 (1983) (Wickramaratne et al., 1988a; Milburn & Moreland, 1991).

A modified study of reproduction in rats was performed to provide information on the development of reproductive organs of F_1 males aged 5 weeks and at sexual maturity (aged 10 weeks). Groups of 26 male and 26 female Alpk:APfSD (Wistar-derived) rats received diets containing procymidone (purity, 98.7–99.1%) at a constant dose equal to 0, 2.5, 12.5 or 37 mg/kg bw per day throughout the

Table 20. Findings in offspring and parental rats fed diets containing procymidone for two generations

Finding	Dietary concentration (ppm)							
	0		50		250		750	
	Males	Females	Males	Females	Males	Females	Males	Females
Fertile rats (%):								
F_{1b}	93	100	85	97	85	93	97	97
F_{1b}	76	97	88	86	81	90	89	100
F_{2a}	79	97	96	93	93	100	42*	86
F_{2b}	92	93	78	93	81	96	46*	97
Precoital time (days):								
F_{1a}		2.5		3.8		2.0		1.9
F_{1b}		2.0		2.0		2.1		1.9
F_{2a}		2.4		2.6		2.1		2.9
F_{2b}		3.2		2.0		2.0		2.0
Adjusted liver weight (g):								
F_o	20.6	12.2	20.3	12.1	21.7*	12.4	22.9*	12.9*
F_1	21.5	14.8	21.6	14.7	22.4*	15.5	24.9*	16.7*
F_2	19.8	10.2	19.6	10.3	20.5	10.9*	21.5*	11.2*
Adjusted testes weight (g):								
F_o	3.8	—	3.8	—	3.9*	—	4.1*	—
F_1	3.6	—	3.7	—	3.9*	—	4.0	—
F_2	3.7	—	3.6	—	3.8	—	3.9*	—
Penile abnormalities:								
F_0[a]	0/30	—	0/30	—	0/30	—	0/30	—
F_1[a]	0/30	—	0/30	—	0/30	—	12/30	—
F_2[a]	0/30	—	0/30	—	0/30	—	8/30	—
F_{2b}[b]	—	—	—	—	—	—	21/40	—
Adjusted prostate weight (g):								
F_{1b}	0.22	—	0.22	—	0.19*	—	0.17*	—
F_{2b}	0.22	—	0.20	—	0.19	—	0.17*	—
Adjusted testes weight (g):								
F_{1b}	1.22	—	1.29*	—	1.31*	—	1.39*	—
F_{2b}	1.22	—	1.29	—	1.34*	—	1.36*	—
Adjusted epididymis weight (g)								
F_{1b}	0.16	—	0.16	—	0.14*	—	0.13*	—
F_{2b}	0.17	—	0.15	—	0.15	—	0.12*	—

From Wickramaratne et al. (1988a) and Milburn & Moreland (1991)

[a] Parents.

[b] Pups, week 7.

* $p < 0.05$.

experimental period. Rats were observed daily for changes in clinical condition and behaviour and were given a detailed examination weekly. Body weight and food consumption were also recorded weekly. Rats were mated 12 weeks after the start of treatment and allowed to litter and wean the pups. Reproductive parameters were assessed. The clinical condition of pups was recorded daily and body weights were recorded at intervals after birth. F_1 males were separated from their mothers on postnatal day 22 and killed at either 5 or 10 weeks after birth. F_1 females were killed at postnatal day 22. F_1 males were given a postmortem examination, organs were weighed and reproductive tissues were examined by histopathology.

Treatment at 37.5 mg/kg bw per day slightly impaired the body-weight gain of F_0 males during the period before mating. Among females, reduced body-weight gain and food consumption was recorded during the same period at 37 and 12.5 mg/kg bw per day (Table 21); however, the decreases at 12.5 mg/kg bw per day were less than 5% and were not considered to be adverse. There were no significant effects on body weight during pregnancy and lactation. Doses of up to 37 mg/kg bw per day had no effect on fertility or mating success; however, at 37 mg/kg bw per day, litter size was lower than in the control group and the resulting superior pup weight at birth was maintained through to weaning. Total litter weights were unaffected. Body-weight gain in F_1 males was reduced in rats at 37 mg/kg bw per day. At age 5 and 10 weeks, F_1 males at 37 mg/kg bw per day showed an increase in absolute and relative testes weight and a decrease in the weight of the prostate with seminal vesicles at 10 weeks. At 37 mg/kg bw per day, hypospadias was observed in 1 out of 46 and 2 out of 47 F_1 males at age 5 and 10 weeks, respectively (typical incidence in the control group was zero). Histopathology did not reveal any significant differences between groups.

The NOAEL for parental toxicity was 12.5 mg/kg bw per day on the basis of the reductions in body weight and food consumption in females, before mating, at 37 mg/kg bw per day. The NOAEL for effects on reproduction was 12.5 mg/kg bw per day on the basis of the reduced litter size at 37 mg/kg bw per day. The NOAEL for offspring toxicity was 12.5 mg/kg bw per day on the basis of the incidence of hypospadias, increased testes weight and decreased weights of the accessory sex organ at 37 mg/kg bw per day. The study claimed GLP compliance (Hodge et al., 1991).

Table 21. Findings in rats fed diets containing procymidone in a study of reproduction/ developmental toxicity

Finding	Dose (mg/kg bw per day)							
	0		2.5		12.5		37	
	Males	Females	Males	Females	Males	Females	Males	Females
Food consumption before mating (g/day), weeks 1–12	28.8	20.9	28.6	20.4	28.8	19.9*	28.3	19.6*
Body weight (g), week 12	383	186	381	186	381	175*	370	169*
Litter size, day 1	11.3		11.8		10.9		9.6*	
F_1 body-weight adjusted testes weight (g):								
Week 5	1.03	—	1.06	—	1.03	—	1.12*	—
Week 10	3.11	—	3.12	—	3.08	—	3.30*	—
F_1 body-weight adjusted prostate weight (g):								
Week 5	0.182	—	0.192	—	0.179	—	0.175	—
Week 10	1.47	—	1.50	—	1.44	—	1.31*	—

From Hodge et al. (1991)
* $p < 0.05$.

(b) Developmental toxicity

Rats

Groups of 25 (presumed) pregnant female Sprague-Dawley rats received procymidone technical (purity, 99.6%) at a dose of 0, 30, 100 or 300 mg/kg bw per day in corn oil by oral gavage on days 6–15 of gestation. Doses were based on the results of a range-finding study (Pence et al., 1980a). Rats were observed twice per day and weighed at intervals during the study. Dams were terminated on day 20 of gestation, fetuses were delivered by caesarean section and a uterine examination was performed. The number, sex and weights of fetuses were recorded; visceral examinations (Wilson technique) were performed on 33% of the fetuses and skeletal examinations (alizarin staining) conducted on the remainder.

All dams survived the study, pregnancy rates were acceptable and there were no treatment-related clinical signs. Body-weight gains were reduced (60%) during days 6–11 of gestation at 300 mg/kg bw per day, although parity with the controls was regained by the end of the dosing period; food consumption was not measured. The proportion of males was reduced relative to controls (57.5%) at the intermediate (50%) and highest (46%) doses, but the values were within the normal ranges for studies of developmental toxicity in rats. There was no effect of treatment on litter values or the incidence of fetal abnormalities at any dose. The NOAEL for maternal toxicity was 100 mg/kg bw per day on the basis of reduced body-weight gain on days 6–11 of gestation. The NOAEL for developmental toxicity was 300 mg/kg bw per day, the highest dose tested, on the basis of the absence of a treatment-related effect on embryo-fetal development. The study claimed compliance with GLP and complies with the essential elements of OECD test guideline 414 of 1981 (Pence et al., 1980b).

To more fully evaluate the incidence and significance of the genital abnormalities observed in the multigeneration study of reproduction (Wickramaratne et al., 1988a), an additional study of developmental toxicity was conducted. For dose-range finding, groups of eight (presumed) pregnant female Sprague-Dawley rats received procymidone (purity, 99.4%) at a dose of 0, 12.5, 300, 500, 750 or 1000 mg/kg bw per day in corn oil by oral gavage on days 6–19 of gestation. The NOAEL for maternal toxicity was 12.5 mg/kg bw per day on the basis of clinical signs at 300 mg/kg bw per day, and the NOAEL for developmental toxicity was 12.5 mg/kg bw per day on the basis of hypospadias (71% vs 0% in controls), and reduced anogenital distance at 300 mg/kg bw per day (Hobermann, 1992a).

In the main study, groups of 45 (presumed) pregnant female Sprague-Dawley rats received procymidone (purity, 99.4%) at a dose of 0, 3.5, 12.5, 125 or 500 mg/kg bw per day in corn oil by oral gavage on days 6–19 of gestation. Approximately half the females in each group were terminated on day 20 of gestation. Uterine examinations were performed and litter parameters were evaluated for fetuses on day 20 of gestation. The fetuses were weighed and examined for visceral and skeletal abnormalities. The remaining females were allowed to deliver and rear their pups to weaning. Male offspring were retained until postnatal day 45 and females until postnatal days 51 to 53. Rats were observed daily for clinical signs of toxicity, body weights and food consumption were recorded daily in the periods during and after dosing. Pups delivered normally were weighed and anogenital distance measured on postnatal days 1, 21 and 45. At necropsy, these pups were given a detailed examination and reproductive organs were weighed and examined histopathologically.

There were no deaths during the study. Maternal toxicity was apparent at 125 and 500 mg/kg bw per day as clinical signs (ungroomed coat, urine-stained fur), initial body-weight loss and reduced food intake. Developmental toxicity was apparent at 125 and 500 mg/kg bw per day as lower fetal weights, reduced anogenital distance (Table 22), higher incidences of male offspring with undescended testes, hypospadias, testicular atrophy, distended preputial grand, inflammatory changes in the accessory sex organs (seminal vesicles, prostate and coagulating glands), and lower organ weights of testes and prostate. At 500 mg/kg bw per day, there was a significant increase in the incidence of bifid thoracic vertebral centra in fetuses and prostate lesions in male offspring. Male offspring at 12.5 and 3.5 mg/kg bw per day were not affected by treatment. No treatment-related influence on sexual differentiation and development of female offspring was observed at any dose.

The NOAEL for maternal toxicity was 12.5 mg/kg bw per day, on the basis of reduced weight gain and food intakes at 125 mg/kg bw per day and above. The NOAEL for developmental toxicity was 12.5 mg/kg bw per day, on the basis of the reduction of anogenital distance at 125 mg/kg bw per day and above in males. These studies contained GLP compliance statements (Hoberman, 1992a, 1992b, 1992c).

A study was performed to determine the minimum toxic dose of procymidone required to produce the external effects on male genitalia of rats. Groups of 20 pregnant Crj:Sprague-Dawley rats were given procymidone (purity, 99.6%) at a dose of 37.5 or 62.5 mg/kg bw per day during the critical period of development of fetal external genitalia (between days 6 and 19 of gestation). Dams were allowed to deliver and, together with female pups, were euthanized on postnatal day 21. Male pups were weaned and euthanized on postnatal day 56. Maternal rats and offspring were observed daily for clinical signs of toxicity and body weights were recorded at intervals. Maternal rats and offspring were given a gross necropsy at euthanasia. Observations on the offspring included macroscopic examination of the external genitalia for all males, organ weights were also measured. In maternal rats, samples of blood were taken for analysis of procymidone concentrations. Blood samples were collected from additional groups of rats on days 6 and 19 of gestation, at 2, 4, 8 and 24 h after dosing and analysed for procymidone.

There were no treatment-related effects on clinical or gross necropsy observations in the maternal rats. Maternal body-weight gains were suppressed in both groups. The blood concentration of procymidone was at a maximum (C_{max}) between 2 h and 4 h after dosing and higher C_{max} values were found at the end of the dosing period (day 19 of gestation) than at the beginning (day 6 of gestation), but the rate of elimination of procymidone from plasma appeared to be more rapid on day 19 of gestation than on day 6 of gestation. The kinetic data showed only slightly higher systemic exposures at the higher dose (Table 23). There were no effects on pregnancy outcome or pup viability. One pup in the group at 62.5 mg/kg bw per day died, but there were no effects on clinical signs or body weight in the surviving offspring. Gross necropsy observations on the offspring included a dose-related increase in misshapen penis (hypospadias, unseparated prepuce) and undescended and abnormal testes size. There were no statistically significant effects on organ weights in the male pups, although prostate and seminal vesicle weights were approximately 10% lower in the group at the highest dose. No abnormalities were detected in female offspring.

Table 22. Anogenital distances in male rats exposed to procymidone in utero

Time-point	Anogenital distance (mm)				
	Dose (mg/kg bw per day)				
	0	3.5	12.5	125	500
Postnatal day 1	2.6 ± 0.2	2.6 ± 0.2	2.6 ± 0.4	1.8 ± 0.2*	1.3 ± 0.1*
Postnatal day 21	12.3 ± 1.1	11.9 ± 0.7	11.8 ± 1.3	9.7 ± 1.4*	8.3 ± 1.1*
Termination (postnatal day 45)	33.0 ± 1.1	32.8 ± 1.6	32.3 ± 1.7	27.2 ± 2.2*	23.1 ± 2.2*

From Hoberman (1992a, 1992b, 1992c)
* $p < 0.05$.

The study did not identify a NOAEL as the lowest dose (37.5 mg/kg bw per day) produced effects on male external genitalia of rats (Inawaka, 2005).

Rabbits

Groups of 18 inseminated New Zealand White rabbits received procymidone (purity, 99.6%) at a dose of 0, 30, 150, 750 or 1000 mg/kg bw per day in corn oil by oral gavage on days 7–19 of gestation. Clinical observations, food consumption and body weight were monitored and dams were terminated on day 30 of gestation. Uteri were examined for live fetuses and intrauterine deaths. Fetuses were weighed and examined for external, visceral (dissection) and skeletal abnormalities (alizarin staining). Specific examinations of anogenital distance and external genitalia were not performed.

There was no compound-related maternal toxicity; pregnancy outcome was unaffected and no evidence of teratogenic potential of procymidone was recorded in any of the groups. The degree of ossification of the fifth and sixth sternebrae was reduced at 1000 mg/kg bw per day. The NOAEL for maternal toxicity and teratogenicity was 1000 mg/kg bw per day, the highest dose tested. The NOAEL for developmental toxicity was 750 mg/kg bw per day on the basis of reduced sternebrae ossification. The study and procedures were inspected by a quality assurance unit and complied with the essential elements of OECD test guideline 414 of 1981 (Wickramaratne, 1988; Wickramaratne et al., 1988b)

Groups of 26 pregnant New Zealand White rabbits were given procymidone (purity, 99.2%) at a dose of 0 or 125 mg/kg bw per day in a vehicle of 0.5% methylcellulose by gavage on days 6 to 28 of gestation. Clinical signs of toxicity, body weight and food consumption were recorded. The dams were terminated on day 29 of gestation and received an extensive uterine examination. Live fetuses were weighed, sexed and examined for abnormalities. A detailed examination of the external genitalia was performed, the anogenital distance and phallus boundary-genital distance were measured using a micrometer and the diameter and the ventral gap of the preputial lamella were measured using a microscope.

Signs of maternal toxicity (reduced food consumption, anorexia, abortion) were evident in the group receiving procymidone. There were no treatment-related effects on fetal parameters, although the litter size was lower in the procymidone group than controls (7.4 ± 2.7 vs 9.1 ± 2.8) this was within the normal range for rabbits. Anogenital distance, the phallus boundary-genital distance and the diameter of the preputial lamina in live fetuses were similar in both groups (Table 24). The ventral gap of the preputial lamina was unaffected in male fetuses, but was increased in female fetuses in the treated group. As a result, the ratio of the ventral gap to the diameter of preputial lamella was also increased. The relative ventral gap of the preputial lamina to body-weight ratio was not statistically significantly altered, indicating that this effect might be secondary to pup body weight, although pup weight was only slightly greater in the group receiving procymidone (Table 24). Only

Table 23. Blood plasma kinetic values in pregnant rats given procymidone by gavage

Parameter	Dose (mg/kg bw per day)			
	37.5	62.5	37.5	62.5
Gestation day	6	6	19	19
C_{max} (µg/ml)	3.11	3.91	4.01	4.43
T_{max} (h)	2	2	4	4
AUC (µg.h/ml)	33	46	38	36

From Inawaka (2005)
AUC, area under the curve of concentration–time.

a single dose, which produced overt maternal toxicity, was used in the study, and the fetal findings were considered to be of uncertain toxicological relevance. The NOAEL for developmental toxicity was 125 mg/kg bw per day, the highest dose tested. A NOAEL for maternal toxicity was not identified (Inawaka, 2003).

Primates

Groups of four pregnant cynomolgus monkeys (*Macaca fascicularis*) were given procymidone (purity, 99.0%) at a dose of 62.5 or 125 mg/kg bw per day by gavage in 0.5% methylcellulose during the critical period of differentiation of the fetal external genitalia (days 20–99 of gestation). Monkeys were observed twice per day during the dosing period and once per day thereafter. Body weight and food consumption were measured at intervals up to day 100 of gestation. Blood was taken for clinical chemistry and haematology on days 19 and 100 and 2, 4, 8 and 24 h after the first and last doses on days 20 and 99 for analysis of procymidone concentration. Fetal heartbeat and size were monitored by ultrasound at intervals during pregnancy. The fetuses and placentae were removed under anaesthesia by caesarean section on day 100 or 102 of gestation. The dams were then terminated and selected organs were removed, fetuses and placenta were weighed; the fetuses were then given an external examination followed by an internal examination of visceral organs and internal genitalia.

None of the monkeys died in either group and there were no signs of toxicity or treatment-related effects on body weight, food consumption, clinical chemistry or haematology. One dam in each group aborted. The major organs were unaffected. Blood concentrations of procymidone reached a maximum 2–4 h after dosing and C_{max} and AUC increased in a linear manner with dose on day 20 but were equal at day 99 (Table 25). There were two male and one female fetuses in the group at the lower dose and three males in the group at the higher dose. There were no effects on the weight of the fetus or placenta and no increase in adverse external or visceral findings including external genitalia of all five male fetuses treated with procymidone (Fukunishi, 2003a).

A more extensive examination of the fetuses was made in a second study in which groups of 16 pregnant cynomolgus monkeys (*Macaca fascicularis*) were given procymidone (purity, 99.0%) at a dose of 0 or 125 mg/kg bw per day by gavage in 0.5% methylcellulose during the critical period of differentiation of the fetal external genitalia (days 20–99 of gestation). Monkeys were observed twice per day for clinical signs of toxicity during the dosing period and once per day during the non-dosing period; body weight and food consumption were measured at intervals. Fetal heartbeat and size were monitored by ultrasound during the treatment period to confirm pregnancy status and fetal

Table 24. Findings in fetuses of female rabbits given procymidone by gavage

Finding[a]	Dose (mg/kg bw per day)			
	0		125	
	Males	Females	Males	Females
Pup body weight (g)	41.5 ± 6.5	39.8 ± 6.1	40.5 ± 8.2	40.8 ± 8.3
Anogenital distance (mm)	1.48 ± 0.18	1.34 ± 0.23	1.41 ± 0.16	1.35 ± 0.10
Phallus boundary-genital distance (µm)	434 ± 94	286 ± 91	368 ± 113	259 ± 76
Preputial lamella diameter (µm)	940 ± 58	836 ± 59	934 ± 60	833 ± 70
Ventral gap of preputial lamella (µm)	46 ± 40	421 ± 114	38 ± 19	498* ± 71
Ventral gap : preputial lamella ratio	0.05	0.50	0.04	0.60*
Relative ventral gap : body weight (µm/g)	1.1 ± 0.8	11 ± 3.4	1.0 ± 0.6	13 ± 3.5

From Inawaka (2003)
[a] Mean ± standard deviation.
* $p < 0.05$ relative to females in the control group.

Table 25. Blood plasma kinetic values in pregnant monkeys given procymidoneby gavage

Parameter	Dose (mg/kg bw per day)			
	62.5	125	62.5	125
Gestation day	20	20	99	99
C_{max} (µg/ml)	0.45	0.96	1.2	1.2
T_{max} (h)	3.3	2.7	3.3	3.3
AUC (µg.h/ml)	7.1	12.8	19.9	20.8

From Fukunishi (2003a)
AUC, area under the curve of concentration–time.

presence. Fetuses were delivered by caesarean section on days 100–102 of gestation, placentae were also removed. Each fetus was examined for viability and sexed. The placenta and fetus were weighed and measurements or observations were made of the crown–rump length, body form, symmetry of the head, facial form, mandibular formation, eyes and eyelids, hair on the head, nipple formation, anus, fingers, toes, finger and toe nails, ears, tail, upper and lower extremities, external genitalia, vertebral column, umbilical cord and palate. After the external examination each fetus was necropsied, a macroscopic assessment was made of the visceral organs and sex was determined from the internal genitalia.

No dams died or aborted in either group and there was no effect of treatment on clinical signs of toxicity, body weight, food consumption and necropsy. There were six male and ten female fetuses in the control groups, and eight male and eight female fetuses in the groups receiving procymidone. No test article-related effects were noted in external genitalia or in any of the other measurements or observations. The study claimed GLP compliance (Fukunishi, 2003b).

2.6 Special studies

(a) Neurotoxicity

No specific investigations of neurotoxic potential with procymidone had been performed. In studies of acute toxicity, there were some signs that might have been linked to a neurotoxic action at all doses (> 250 mg/kg bw per day), e.g. reduced motor activity, but these could also be related to general malaise after the administration of a bolus of corn oil. There was no indication from repeat-dose studies that procymidone has any neuropathic potential.

(b) Mechanism of action

(i) Binding to androgen receptors

Binding to androgen receptors in × 105 000 *g* prostate cytosols from castrated rats (Sprague-Dawley, age 10 weeks) and mice (ICR, age 14 weeks) was examined in a competitive assay using [^{3}H]dihydroxytestosterone (DHT; 135 Ci/mmol (4995 GBq/mmol). The affinity constants (Kd) for DHT were 0.23 nmol/l and 0.1 nmol/l in rats and mice respectively. The relative binding affinity—the ratio of the DHT concentration required to inhibit activity by 50% (IC_{50}) to the IC_{50} of the other compound—for procymidone was relatively low at 0.065% in rats and 0.07% in mice (DHT = 100%). Procymidone-3-Cl, a plant metabolite, was similar to parent compound with a relative affinity of 0.05%, while 2-(3,5-dichlorophenyl carbamoyl)-1,2 dimethylcyclopropane-1-carboxylic acid (carboxyprocymidone, PCM-NH-COOH) and cyclopropane dicarboxylate metabolites produced no inhibitory effects (Murakami et al., 1988b).

The anti-androgenic activities and androgen-receptor (AR) binding activities of procymidone (purity, 99.6%) and its metabolites PCM-CH_2OH (purity, 99.9%), PCM-COOH (purity, 96.4%), PA-CH_2OH (purity, 98.8%), carboxyprocymidone acid, PA-COOH (purity, 96.9%), PCM-NH-COOH (purity, 98.6%) and PCM-CH_2OH glucuronide were evaluated using androgen-receptor-mediated assays in vitro. Anti-androgenic activity was determined in HeLa cells transfected with an ARE×3/ luciferase reporter gene and full length human and rat AR expression vectors (AR reporter assay). AR-binding activity was also determined by a competitive ligand-binding assay on the basis of the fluorescence polarization method using a rat AR kit. Concentrations used were up to 30 µmol/l for the AR reporter assay and up to 150 µmol/l for the assay for competitive binding. Acceptable performance of the three assay systems was demonstrated by conducting reporter gene assays with the known androgen antagonists, flutamide, hydroxyflutamide and the binding assay with DHT.

In all the assays, procymidone exhibited almost identical binding to flutamide, with metabolites of procymidone having varying but lower levels of activity (Table 26). It is possible that the non-enzymic interconversion of the metabolites might have influenced the results of this assay (Suzuki, 2005).

(ii) Serum hormone concentrations

Groups of 30 male Sprague-Dawley rats received diets containing procymidone (purity, 99.1%) at concentrations of 100 to 6000 ppm for durations ranging from 14 days to 6 months, with and without recovery periods. The schedule of doses and termination times are shown below.

Ten male rats per group were used and the anti-androgen cadmium chloride (at a subcutaneous dose of 5.5 mg/kg bw) was used to treat a positive-control group. Body weights were determined at the start of treatment and at termination. Serum testosterone, luteinizing hormone (LH) and, in the first experiment, estradiol concentrations were measured at each termination. Reproductive organs were weighed and examined by microscopic pathology. In the first experiment, body weight was depressed and testes weights were slightly increased in rats at 2000 or 6000 ppm. Epididymis weights were reduced at 2000 and 6000 ppm after 14 days and 1 month, but there was no statistically significant effect after 3 months. Testosterone concentrations were increased in a dose-related

Table 26. IC_{50} values (µmol/l) for binding of procymidone and its metabolites to androgen receptors

Compound	IC_{50} (µmol/l)		
	Androgen-receptor reporter assay		Competitive-binding assay
	Rat	Human	
Flutamide	0.34	0.23	11
Procymidone	0.31	0.29	12
PCM-CH_2OH	3.4	2.4	26
PA-CH_2OH	4.1	3.5	227
PCM-NH-COOH	1.4	1.2	100
PCM-COOH	No activity	No activity	No activity
PA-COOH	No activity	No activity	No activity
PCM-CH_2OH-glucuronide	No activity	No activity	No activity

From Suzuki (2005)

IC50, concentration required to inhibit activity by 50%; PA, procymidone acid; PA-CH_2OH, hydroxyprocymidone acid; PA-COOH, carboxyprocymidone acid; PCM, procymidone; PCM-CH_2OH, hydroxyprocymidone; PCM-NH-COOH, carboxyprocymidone.

Table 27. Dosing schedule and termination times in a study of serum hormone concentrations in rats given diets containing procymidone

Parameter	Experiment 1	Experiment 2	Experiment 3
Dietary concentrations (ppm)	700, 2000, 6000	100, 300, 700 and 2000	6000
Duration	Up to 3 months	Up to 6 months	One month
Termination times	Fourteen days, 1 month and 3 months after the start of treatment	One, 3 and 6 months after the start of treatment	One month after the start of treatment, then 2 weeks, 1, 3 and 6 months after treatment

From Murakami et al (1986)

and significant manner at 700 ppm and greater; LH concentrations were significantly increased at 6000 ppm. In the second experiment, serum testosterone concentrations were increased at 700, and 2000 ppm. Serum concentrations of testosterone and LH returned to normal after 1 month of the recovery period. Serum estradiol concentrations were close to the limit of detection in all groups. In the positive-control group, administration of cadmium chloride produced a reduction in testes, seminal vesicle, epididymes and prostate weights, an increase in LH concentration and a decrease in testosterone indicating a different mechanism of action from procymidone. The NOAEL for hormonal effects was 300 ppm, equivalent to 30 mg/kg bw per day. The report was checked by a quality assurance unit (Murakami et al, 1986).

Tests for testicular function

Rodents

Groups of 30 male ICR mice received diets containing procymidone (purity, > 99.3%) at a concentration of 0, 1000, 5000 or 10 000 ppm for up to 13 weeks. Groups of 12 male Sprague-Dawley rats received diets containing procymidone at a concentration of 0, 700, 2000 or 6000 ppm for 13 weeks. Routine examinations were performed during the study. Rodents were terminated after 2, 4 and 13 weeks of treatment. Body weights and food consumption were recorded weekly. At termination, reproductive organs were removed and weighed. Concentrations of testosterone were determined in serum and testes and LH was determined in serum and pituitary. The testicular response to stimulation by human chorionic gonadotrophin (hCG) and the binding of hCG to the LH/hCG receptor in testes were determined in vitro.

In mice, mean achieved intakes over 13 weeks were 133, 638 and 1338 mg/kg bw per day. There was no consistent effect on body weight, food consumption or organ weights. Testosterone concentrations were elevated in serum (165%) and testes (248%) after 2 weeks of treatment at 10 000 ppm and levels returned to normal at 4 and 13 weeks. LH concentrations in serum and pituitary were elevated for the first 4 weeks at 1000 ppm and greater; returning to normal thereafter. There was a slight increase in hCG-stimulated testosterone production at 10 000 ppm after 2 weeks but this returned to control levels after 3 months. The binding affinity of hCG decreased in mice at 5000 and/or 10 000 ppm during the 3-month treatment period.

In rats, mean achieved intakes over 13 weeks were 47, 132 and 388 mg/kg bw per day. There was no consistent effect on body weight or food consumption. Seminal vesicle weight was decreased and testis weight increased at 6000 ppm after 2 weeks. An increase in testis weight was also apparent at all doses (approximately 10%, but no dose–response relationship) after 4 weeks and at 6000 ppm after 13 weeks (10%). In rats at 6000 ppm, testosterone concentrations in serum were increased (155%) throughout the experimental period and increases were noted in testes after 4 weeks (220%) and 13 weeks (160%). LH concentrations were only elevated in serum for the first 4 weeks at 6000 ppm (300%). In the pituitary, LH concentrations were elevated at all doses after 4 weeks (138–180%)

but only at 6000 ppm after 13 weeks (131%). The assay in vitro showed that basal production of testosterone was increased at all termination times at 6000 ppm. The hCG-stimulated production of testosterone was elevated at all times in rats at 2000 and 6000 ppm and after 13 weeks in rats at 700 ppm. There were no treatment-related changes in the number of binding sites or the binding affinity of hCG to the LH/hCG receptor (Murakami et al., 1988a).

Primates

In a dose range-finding study, no effects on ejaculate mass, sperm counts, testosterone and LH concentrations or organ weights were seen in groups of two male cynomolgus monkeys (*Macaca fascicularis*) that received procymidone (purity, 98.7%) at doses up to 1000 mg/kg bw per day by gavage for seven consecutive days (Zühlke, 1991).

In the main study, groups of five male cynomolgus monkeys (*Macaca fascicularis*) received procymidone (purity, 98.7%) at a dose of 0, 30 100 or 300 mg/kg bw per day by gavage as a suspension in 1% sodium carboxymethyl cellulose for 13 weeks. Clinical signs and food consumption were recorded daily and body weights weekly. Sperm was taken for analysis and testicular dimensions recorded six times before dosing and in weeks 4, 8 and 13. Serum testosterone and LH concentrations were measured in samples taken before dosing and weekly during the treatment phase. The monkeys were necropsied at the end of the treatment period and the epididymides, prostate gland, seminal vesicles and testes were weighed and examined microscopically.

All monkeys survived the study with no signs of reaction to treatment. Body-weight variations were similar across the groups. There was no clear evidence of a treatment-related effect on food intakes, testicular size or LH concentrations. Histopathology of the reproductive organs did not detect any treatment-related change. There were indications of effects on ejaculate mass, sperm counts (Table 28), and serum testosterone concentrations (Figure 2) particularly during the first 4 weeks, and on testes weights (Table 28). There was no statistically significant change when compared with values for concurrent controls, other than relative testes weights, which did not exhibit any dose–response relationship. When values after dosing were compared with values before dosing for each individual, using a repeated measures ANOVA with post-hoc comparisons using the least significant differences (LSD) test, ejaculate mass was found to be statistically significantly reduced ($p < 0.05$) at 100 and 300 mg/kg bw per day. The small group size, absence of a clear dose–response relationship and large degree of inter- and intra-group variation before and after dosing made it difficult to reach firm conclusions on the findings of this study.

The NOAEL was 30 mg/kg bw per day on the basis of statistically significant effects on testes weights and ejaculate mass at 100 mg/kg bw per day. The study contained statements of compliance with GLP (Bee, 1992).

(iii) Fetal distribution

Groups of three pregnant rats (Crj:CD(SD)IGS), New Zealand White rabbits or cynomolgus monkeys were given [phenyl-^{14}C]procymidone (purity, > 97.5%; specific activity, 15.8 MBq/mg) at a dose of 125 mg/kg bw on day 17, 21 or 54 of gestation, respectively. Groups of rats also received doses on: (i) day 16; (ii) days 16 and 17; or (iii) days 16, 17 and 18 of gestation. Corn oil was the vehicle in the studies in rats, for rabbits and monkeys the vehicle was 0.5% methylcellulose. The rats and monkeys were killed 6 and 24 h after dosing, rabbits after 2 and 24 h. Samples of maternal blood, amniotic fluid and fetal blood were taken. The maternal liver and kidneys, the placenta, fetus, fetal brain, fetal heart, fetal lung, fetal liver and fetal kidneys were removed at necropsy. These samples were analysed for radioactivity and the amounts of procymidone and its metabolites were also determined.

Table 28. Ejaculate mass and testes weights in monkeys (n = 5) given procymidone by gavage for up to 13 weeks

Parameter[a]	Dose (mg/kg bw per day)			
	0	30	100	300
Ejaculate mass (g):				
Week −6	0.28 ± 0.27	0.39 ± 0.27	0.65 ± 0.60	0.59 ± 0.47
Week −1	0.69 ± 0.56	0.55 ± 0.22	0.46 ± 0.34	0.63 ± 0.23
Week +4	0.52 ± 0.43	0.41 ± 0.21	0.16 ± 0.19*	0.18 ± 0.18*
Week +8	0.70 ± 0.49	0.26 ± 0.21	0.27 ± 0.22	0.31 ± 0.17
Minimum before dosing	0.28 ± 0.27	0.24 ± 0.14	0.46 ± 0.34	0.41 ± 0.16
Sperm count (10^6):				
Week −6	167 ± 223	118 ± 139	372 ± 535	133 ± 58
Week −1	174 ± 207	202 ± 151	201 ± 251	349 ± 195
Week +4	133 ± 143	187 ± 83	80 ± 87	75 ± 81
Week +8	121 ± 118	158 ± 149	141 ± 119	180 ± 138
Minimum before dosing	66 ± 54	73 ± 54	76 ± 52	133 ± 58
Relative testes weights (%), week 13	0.60 ± 0.11	0.87 ± 0.11**	0.79 ± 0.21	0.85 ± 0.04**

From Bee (1992)

[a] Mean ± standard deviation.

* $p < 0.05$ vs pre-test (ANOVA).

** $p < 0.01$ vs values for controls.

Figure 2. Serum testosterone concentrations in monkeys given procymidone by gavage for up to 13 weeks

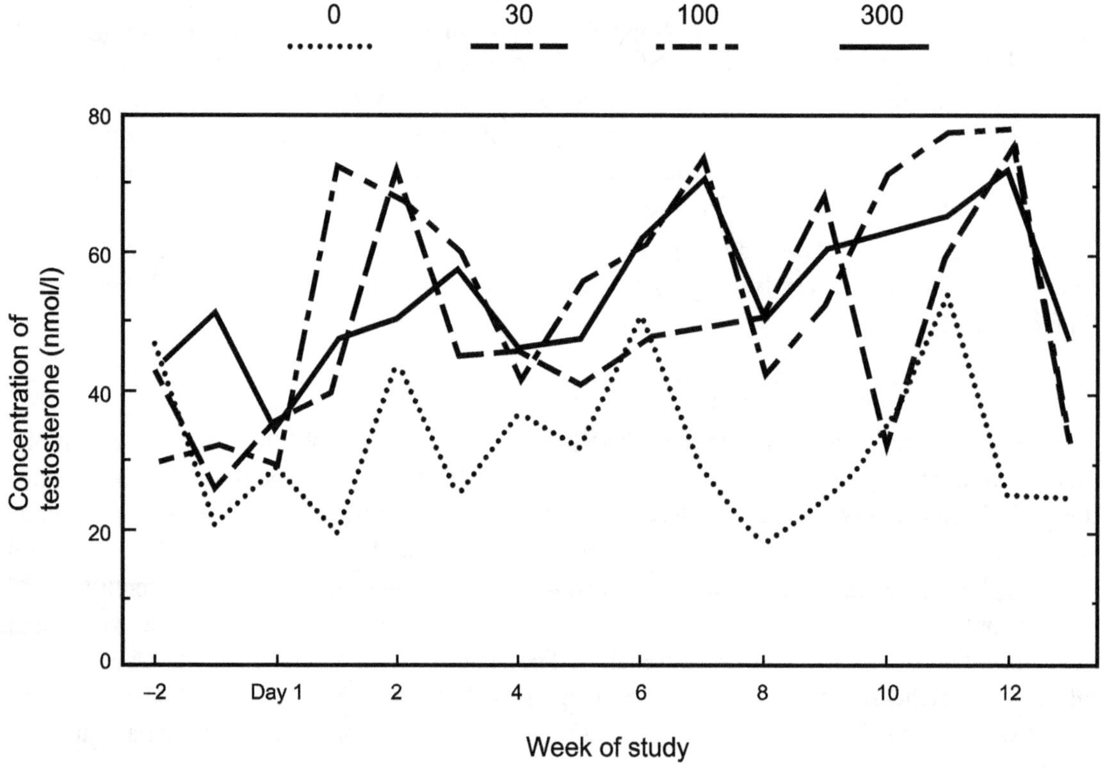

From Bee (1992)

The results are summarized in Table 29. The metabolic profile in fetal-rat tissues is the same as that in the plasma of non-pregnant and pregnant rats (Sugimoto, 2005e, 2005f). In rats, procymidone was the major component in maternal plasma 6 h after the first dose, followed by PCM-NH-COOH (and/or PCM-COOH) and PA-CH_2OH but at all other sampling times, the main plasma metabolites were PCM-NH-COOH (and/or PCM-COOH), PA-CH_2OH and PA-COOH. Unchanged procymidone was also the main component in the placenta and fetus 6 h after the first dose followed by PCM-CH_2OH. At other sampling times, however, PCM-CH_2OH was the major component. PA-CH_2OH was also the main metabolite in amniotic fluid and its concentration increased with time. The transfer ratios of metabolites showed that PCM-CH_2OH accumulated more readily in the fetus than did procymidone (Tomigahara, 2005d).

In rabbits, the main metabolites present in the fetus were procymidone and PCM-CH_2OH. PCM-COOH and PA-COOH, plus glucuronides of PCM-CH_2OH and PA-CH_2OH, were hardly transferred to the fetus, although PA-COOH was the major component of maternal plasma (Sugimoto, 2005g).

In monkeys, procymidone was also the major component in the placenta and fetus, PCM-CH_2OH and/or PCM-COOH and PCM-NH-COOH were lesser metabolites. Other metabolites were only detected at trace levels in fetuses (Mogi, 2005e).

These results show that the transfer of procymidone and metabolites to the fetus is significantly greater (approximately 10-fold) in rats than in rabbits or monkeys. In rats, there was also a transfer of PCM-CH_2OH, the postulated developmental toxin, from the dam to the fetus.

Table 29. Pharmacokinetic parameters in plasma and tissues of pregnant rats, rabbits and monkeys given oral doses of [phenyl-^{14}C]procymidone at 125 mg/kg bw

Parameter	Rats				Rabbits		Monkeys	
	Single dose		Three doses		Single dose		Single dose	
	6 h	24 h	6 h	24 h	2 h	24 h	6 h	24 h
Dams								
Concentration of ^{14}C (µg equivalent/g):								
Plasma	18	28	35	30	40	2.3	5.5	3.7
Liver	53	56	—	—	43	3.8	33	28
Kidney	32	59	—	—	143	8.7	30	21
Whole fetus								
Concentration of ^{14}C (µg equivalent/g):								
Single dose	10	16	7.4	10	2.5	0.36	1.2	1.2
Three doses	—	—	21	15	—	—	—	—
Ratio of concentrations of ^{14}C in whole fetus and in maternal plasma								
PCM	2.7	1.1	0.9	0.6	NM	NM	0.5	0.8
PCM-CH_2OH	16	10	48	22	2.1	NM	0.2	0.2
PCM-NH-COOH	< 0.1	< 0.1	< 0.1	< 0.1	< 0.1	0.4	0.2	0.2
PA-CH_2OH	0.8	0.8	0.2	0.5	< 0.1	NF	< 0.1	<0.1
PA-COOH	—	—	0.2	0.1	NF	< 0.1	< 0.1	<0.1
Glucuronides	—	—	0.4	0.5	< 0.1	< 0.1	<0.1	<0.1

From Tomigahara (2005d), Sugimoto (2005g) and Mogi (2005e)

NF, not in fetus; NM, not in maternal plasma but present in fetus; PA, procymidone acid; PA-CH_2OH, hydroxyprocymidone acid; PA-COOH, carboxyprocymidone acid; PCM, procymidone; PCM-CH_2OH, hydroxyprocymidone; PCM-NH-COOH, carboxyprocymidone.

3. Studies on metabolites

(a) 3,5-Dichloroaniline (DCA)

3,5-Dichloroaniline (DCA) in corn oil vehicle was of moderate acute toxicity to *dd* mice, having an oral median lethal dose (LD_{50}) of approximately 850 mg/kg bw and a subcutaneous LD_{50} of approximately 1270 mg/kg bw (Kohda et al., 1980).

(b) Carboxyprocymidone (PCM-NH-COOH)

PCM-NH-COOH (2-(3,5-dichlorophenyl carbamoyl)-1,2 dimethylcyclopropane-1-carboxylic acid) in corn oil, was of moderate acute toxicity to *dd* mice, having an oral LD_{50} of approximately 1450 mg/kg bw (Kohda et al., 1980).

(c) 1,2-Dimethylcyclopropane-dicarboxylic acid (DMCPA)

1,2-Dimethylcyclopropane-dicarboxylic acid (DMCPA), in Tween 80, was of low acute toxicity to *dd* mice, having an oral LD_{50} of approximately 4400 mg/kg bw and a subcutaneous LD_{50} of approximately 2400 mg/kg bw (Kohda et al., 1980). DMCPA (purity, 98.7%) gave negative results in an assay for gene mutation in *Salmonella typhimurium* (TA1535, TA1537, TA1538, TA98, TA100) and *E. coli* (WP2*uvr*A), in the presence and absence of metabolic activation, at concentrations of up to 5000 μg/plate (Kogiso et al., 1992). DMCPA (purity, 98.7%) gave negative results in an assay for micronucleus formation in ICR mice given doses of up to 320 mg/kg bw intraperitoneally. The study included three harvest times and used five male and five female mice per group (Murli,1992a, 1992b).

The absorption, distribution and metabolism of DMCPA was investigated in groups of five male and five female Sprague-Dawley rats. [Carboxy-^{14}C]DMCPA was administered orally at a dose of 0.56 mg/kg bw in corn oil to groups of five male and five female rats. Rats were killed 7 days after dosing. Excretion was rapid, with 87% (males) and 84% (females) of the administered dose being excreted within the first 24 h, and equally distributed between the urine and faeces. Retention of radioactivity in tissues was low, with concentrations of less than 2 ng equivalent/g at 7 days; the only tissue with quantifiable levels was bone. The only component identified in excreta corresponded to non-metabolized DMCPA accounting for 77% (male) and 74% (female) of the administered dose at 24 h (Saito, 1992).

(d) Hydroxyprocymidone (PCM-CH_2OH)

Two groups of 12 female Sprague-Dawley rats were given PCM-CH_2OH (purity, 100%) orally at dose of 62.5 or 125 mg/kg bw per day in corn oil by gavage on days 6–19 of gestation. The rats were allowed to deliver and wean the pups. Clinical signs were observed twice per day for dams during the treatment period and once daily for the remainder of the study. Offspring were observed once daily. Body weights of dams and offspring were recorded regularly. Live newborns were weighed, sexed and examined for external abnormalities and the anogenital distance was measured. At the time of weaning, on postnatal day 21, all dams and all female offspring were necropsied. All male offspring were necropsied on day 56 after birth.

Decreases in locomotor activity, and increases in bradypnoea and/or prone position were observed in the dams in both groups receiving PCM-CH_2OH from the first day of dosing. Body weights of dams and offspring were similar in all groups. No abnormal findings were seen in dams or female offspring. In the male offspring, a dose-related shortening of the anogenital distance was seen (3.35, 2.67 and 2.35 mm). Increases in the incidence of male offspring with abnormities of the external genitalia were observed in both groups receiving PCM-CH_2OH. Hypospadias were observed in 19 out of 86 males (8 out of 12 litters) in the group at 62.5 mg/kg bw per day and 75 out of 84 males (all litters) in the group at 125 mg/kg bw per day. In addition, small testes and epididymis and undescended testes and epididymis were observed occasionally in both groups. These results were similar to those seen with procymidone (Inawaka, 2005). A benchmark-dose (BMD) assessment of the incidence of hypospadia (Table 30) was performed by the Meeting, using a nested logistic model on incidences in individual litters. The 95% lower limit for a 10% response ($BMDL_{10}$) for hypospadias was 43.8 mg/kg bw per day for PCM-CH_2OH and 23.7 mg/kg bw per day for procymidone (Izumi, 2005).

(e) Interconversion of metabolites

The interconversion of metabolites of procymidone was investigated in female rats in vivo and and in vitro. After the subcutaneous administration of [phenyl-^{14}C]PA-CH_2OH at a dose of 62.5 mg/kg bw to a group of three Sprague-Dawley rats, imide-bond cleaved and uncleaved metabolites were found in tissues and plasma (Table 31) (Tomigahara, 2005a), indicating reversal of the cleavage reaction that produced PA-CH_2OH from procymidone (see Figure 1). In an identical study but with [phenyl-^{14}C]PCM-CH_2OH, both cleaved and uncleaved metabolites were seen in plasma and tissues (Table 31) (Tomigahara, 2005b). A comparative study with oral and subcutaneous administration showed that a subcutaneous dose of 125 mg/kg bw was equivalent, in terms of metabolite profile, C_{max} and AUC, to an oral dose of 62.5 mg/kg bw (Tomigahara, 2005c). The variations in the relative levels of metabolites between tissues (pH approximately 6.8) and plasma (pH approximately 7.4) were investigated in vitro. [phenyl-^{14}C]Procymidone and [phenyl-^{14}C]PCM-CH_2OH were incubated at pH 2.0, 4.0, 6.8, 7.4, 8.0 or 11.0 at room temperature for 16 h. The reaction mixtures were analysed by TLC and the ratio of procymidone to PCM-NH-COOH and of PCM-CH_2OH to PA-CH_2OH were calculated. Under acid conditions ($pH < 6.8$), procymidone and PCM-CH_2OH were stable, but under alkaline conditions ($pH > 7.4$) the imide bond was cleaved to give PCM-NH-COOH and PA-CH_2OH (Tarui, 2005a). These studies indicated that procymidone metabolites have the potential to interconvert non-enzymatically when transferred from plasma to tissues and vice versa.

4. Observations in humans

Annual medical records for workers involved with the packaging of procymidone were examined for 1979–1983. A total of 20 workers was included in the survey. Of these, 16 had been engaged in packing procymidone for more than 9 h per year for an average of 4 years. None of the workers had any symptoms that could be associated with exposure to procymidone (Harada, 1983).

A further review was completed in 2001. The medical records and operational details of 15 workers who had been engaged in the manufacture and packing of technical-grade procymidone into drums were examined. Operational details included number of hours spent packing per year, number of years involved in packing, age and use of personal protective equipment. Annual medical

Table 30. Incidence of hypospadia in pups from studies of developmental toxicity with procymidone and PCM-CH_2OH

Dose (mg/kg bw per day)	Day 22 (observation)		Day 56 (gross examination)	
	Procymidone	PCM-CH_2OH	Procymidone	PCM-CH_2OH
0	0/119	0/67	0/119	0/67
37.5	2/122	Not dosed	19/122	Not dosed
62.5	13/135	3/86	59/134	16/86
125	Not dosed	26/86	Not dosed	71/84

From Izumi (2005) and Inawaka (2005)
PCM-CH_2OH, hydroxyprocymidone.

Table 31. Metabolite concentrations (µg equivalent/g) in plasma and tissues from female rats (n = 3) receiving procymidone metabolites at 62.5 mg/kg bw

Metabolite	Concentration of metabolite (µg equivalent/g)			
	Plasma	Liver	Kidney	Ovary
Six h after dosing with PCM-CH_2OH				
PCM-CH_2OH	2.9	33.8	20.9	16.8
PCM-COOH	7.3	3.0	15.1	1.8
PA-CH_2OH	12.9	6.0	6.2	3.1
PA-COOH	2.3	1.4	2.8	0.5
Others	3.1	1.4	2.6	0.6
Four h after dosing with PCM-CH_2OH				
PCM-CH_2OH	0.6	16.4	12.7	10.6
PCM-COOH	2.7	5.1	10.4	2.0
PA-CH_2OH	12.8	2.0	5.4	0.5
PA-COOH	3.0	2.9	3.8	0.7
Others	4.2	0.6	1.3	0.3

From Tarui (2005a)

PA, procymidone acid; PA-CH_2OH, hydroxyprocymidone acid; PA-COOH, carboxyprocymidone acid; PCM, procymidone; PCM-CH_2OH, hydroxyprocymidone; PCM-NH-COOH, carboxyprocymidone.

examination records were checked for 1991–2000 for adverse effects on body weight, vital capacity, hearing acuity, chest X-ray, blood pressure, urine analysis, haematology and clinical biochemistry. An annual questionnaire that was completed at the time of the medical examination was also examined. This recorded headaches, dizziness, irritability, restlessness, fatigue, loss of appetite, full stomach, diarrhoea, cough, dyspnoea, hearing problems, tinnitus, abnormal taste, gastroenteric problems or any other abnormalities.

The number of hours spent packing ranged from 10 to 83 per year. Four of the workers had been engaged in packing for 10 years or more, four for between 5 and 9 years and seven for less than 4 years. Five of the workers were aged at least 50 years, two were aged between 40 and 49, five were aged between 30 and 39 and three were aged less than 29 years. All workers wore protective equipment consisting of long-sleeved cotton clothes, plastic helmets, masks, glasses and gloves. No abnormalities were apparent from the medical records that could be related to the manufacture or handling of technical-grade procymidone (Kurihara, 2001). A similar study in 2007 confirmed that there were no adverse health effects in the monitored workforce (Kumagai, 2007).

The sponsor reported there were no documented cases of intoxication with procymidone and no information on effects of exposure of the general population to procymidone.

Comments

Biochemical aspects

Studies with ^{14}C-labelled procymidone showed that radiolabel was rapidly absorbed (C_{max}, 2–8 h) and rapidly eliminated (> 80% in 24 h) in mice and rats. Absorption was extensive, as shown by the high level of urinary excretion (> 80% of the administered dose), with similar results obtained in mice and rats receiving doses of up to 100 mg/kg bw. At higher doses, the proportion of the administered dose excreted in the faeces increased: at 250 mg/kg bw, 24–33% was eliminated by that route after 168 h; the increase in faecal radioactivity was mainly attributable to an increase in unabsorbed procymidone (18–27% of the administered dose). Only low concentrations (< 0.3%) of radioactivity were retained 168 h after oral administration; the highest concentrations were found in fat. The major metabolic pathway for procymidone in mice and rats involved the oxidation of the methyl groups to hydroxymethyl or carboxylic acid derivatives, cleavage of the imide, and glucuronide formation of the resultant metabolites. In cynomolgus monkeys, rabbits and chimeric mice with humanized livers (i.e. liver repopulated with human hepatocytes), there was more extensive urinary excretion of glucuronide conjugates than in rats or normal mice. In rats, the glucuronides were formed, but were present in the bile and appear to be deconjugated and the aglycone was reabsorbed giving a prolonged elimination phase and relatively high AUC. Studies in vitro with liver preparations from humans, rabbits, cynomolgus monkeys and rats showed that the rate of metabolism of procymidone and PCM-CH_2OH was significantly lower in rats than in the other species studied.

In female cynomolgus monkeys, there was an increase in C_{max} and AUC values after repeated doses. In other species, the routes and rates of tissue distribution, biotransformation and excretion of procymidone were similar in males and females and after single or repeated administration.

Procymidone has a high binding affinity in vitro (92–98% bound) for plasma proteins when incubated with plasma from female rats, cynomolgus monkeys, rabbits and humans. The alcohol metabolite PCM-CH_2OH had a slightly lower binding affinity (77–91% in all species) than procymidone, with the highest affinity being for human plasma proteins.

Kinetic studies in rats, rabbits and cynomolgus monkeys (dams and fetuses of each species) have been performed as part of investigations of the effects on male rat reproductive tissues. After a single dose, cynomolgus monkeys had much lower C_{max} values for total radioactivity than did rabbits and rats, but at doses of up to 125 mg/kg bw rabbits had relatively low AUC values owing to rapid elimination. In cynomolgus monkeys, the predominant compound in plasma was procymidone; in rabbits, acid metabolites and glucuronides of alcohol metabolites were the major components in plasma; in rats, free alcohol metabolites predominated. After 14 doses at 37.5–500 mg/kg bw per day, C_{max} and AUC values for total radioactivity were similar for rats and cynomolgus monkeys. These findings show that the doses used in the studies of developmental toxicity in cynomolgus monkeys would have produced similar plasma concentrations of total radioactivity, but with different metabolite profiles, to those in rats at doses resulting in hypospadias. The investigations of the transfer of procymidone and metabolites to fetuses again showed species differences. After dosing at 125 mg/kg bw per day, the concentrations of procymidone and metabolites in the fetus were significantly greater (10-fold or greater) in rats than in rabbits or cynomolgus monkeys. In rats, the relative concentrations of PCM-CH_2OH in the fetus versus the dam were much higher than in rabbits or cynomolgus monkeys.

Toxicological data

Procymidone was of low acute toxicity by the oral (LD_{50} > 5000 mg/kg bw) and dermal (LD_{50} > 2500 mg/kg bw) routes, and after a 4-h exposure by inhalation (LC_{50} > 1.5 mg/l). Procymidone is not an eye irritant, but is a slight, transient skin irritant. Procymidone did not produce delayed contact hypersensitivity in guinea-pigs in either the maximization or repeat-injection tests.

The primary target organs in rats and mice exposed to procymidone in repeat-dose studies were the liver and testes. The major effect of short-term dietary administration of procymidone in ICR mice was on the liver. Centrilobular hepatocyte hypertrophy was noted in male mice and hepatocyte granuloma in females that received procymidone at 500 ppm, equal to 71 mg/kg bw per day, for 13 weeks; the NOAEL was 150 ppm, equal to 22 mg/kg bw per day. In a subsequent study with B6C3F_1 mice, multifocal coagulative necrosis of hepatic parenchyma, centrilobular cytoplasmic swelling, nuclear enlargement, coarsely-dispersed chromatin and multinucleate hepatocytes were noted in mice receiving procymidone at 10 000 ppm, equivalent to 1430 mg/kg bw per day. This was accompanied by increased liver weight and serum ALT activity in males and higher cholesterol concentrations in females. The histopathological effects were apparent to a lesser extent in animals treated with procymidone at 500 or 2500 ppm, equivalent to 71 and 355 mg/kg bw per day respectively. The NOAEL was 100 ppm, equal to 19.6 mg/kg bw per day. An increase in the incidence of testicular atrophy was noted in mice receiving procymidone at 500 ppm for 26 weeks, the NOAEL was 150 ppm, equal to 20 mg/kg bw per day. In an additional 6-month study, the NOAEL for effects on the testes was 300 ppm, equal to 37 mg/kg bw per day. The Meeting considered that the overall NOAEL in three short-term studies in mice was 300 ppm, equal to 37 mg/kg bw per day.

The only short-term study in rats included a 6-month exposure period with or without a 3-month recovery phase, and a 9-month exposure period. In Sprague-Dawley rats, there was a significant reduction in body-weight gain among females receiving procymidone at 1500 ppm, equivalent to 75 mg/kg bw per day, for 6 months (although this was not significant after an exposure of 9 months) and an increase in liver-to-body-weight ratio in both sexes. The liver-to-body-weight ratio was also increased in female rats at 500 ppm, equivalent to 25 mg/kg bw per day, and there was an increase in spleen-to-body-weight ratio at 500 and 1500 ppm. Absolute and relative weights of the testes were increased after 9 months at 1500 ppm. The only treatment-related histopathological effect was swelling of the liver cells in male rats at 1500 ppm. The findings showed clear evidence of reversal over the 3-month recovery phase. The NOAEL was 500 ppm, equivalent to 25 mg/kg bw per day.

In a 28-day study of dermal exposure, there were no treatment-related local or systemic changes when doses of up to 1000 mg/kg bw per day were applied to the shaved backs of rats.

Emesis and diarrhoea were the principal signs of toxicity in dogs given procymidone at 500 mg/kg bw per day for 26 weeks. At that dose, there was also an increase in serum ALP activity, BUN and calcium concentrations. The NOAEL was 100 mg/kg bw per day. In a 52-week study in dogs, there was an increase in emesis and soft faeces and increases in serum globulin and ALP activity at 500 mg/kg bw per day. The NOAEL was 100 mg/kg bw per day.

Procymidone gave negative results in an adequate range of assays for genotoxicity in vitro and in vivo. The Meeting concluded that procymidone was unlikely to be genotoxic.

The toxicity and carcinogenicity of procymidone had been investigated in long-term studies in mice and rats. In the study in mice, treatment-related effects were limited to the liver. Higher liver weights and liver-to-body-weight ratios were apparent in males at 100 ppm, equal to 15 mg/kg bw per day, and in both sexes at 300 and 1000 ppm, equal to 46 and 153 mg/kg bw per day respectively, and there were histopathological changes in the liver, comprising increased incidences of centrilobular cytomegaly in males at 1000 and 300 ppm and in females at 1000 ppm. In addition, focal or multifocal hepatocellular hyperplasia, and eosinophilic foci were noted in females at 1000 ppm. There was an increased incidence of hepatocellular adenomas in females at the highest dose. There was also an increase in the incidence of hepatoblastomas in males receiving procymidone at 1000 ppm. The NOAEL for toxicity was 100 ppm, equal to 15 mg/kg bw per day, on the basis of a range of liver effects at 300 ppm. The NOAEL for carcinogenicity was 300 ppm, equal to 46 mg/kg bw per day, on the basis of increases in liver tumours in males and females at 1000 ppm.

In the long-term study in rats, the liver was a target organ and reproductive organs were also affected. There was a reduction in body-weight gain among rats given procymidone at 1000 and 2000 ppm , equal to 48 and 97 mg/kg bw per day respectively. Effects on organ weights were noted at both these doses and consisted of increased relative and absolute weight of the liver in both sexes and increased weights of the testes and ovaries at dietary concentrations of 1000 ppm and greater. Histopathology revealed an increased incidence of centrilobular cytomegaly in the liver of both sexes at dietary concentrations of 1000 ppm and greater. At 1000 and 2000 ppm, there were increases in the incidence of interstitial cell tumours and interstitial cell hyperplasia in the testes; the incidence of ovarian stromal hyperplasia was statistically significantly increased at 2000 ppm. The NOAEL for general toxicity and for carcinogenicity was 300 ppm, equal to 14 mg/kg bw per day.

Procymidone induced liver tumours in mice and testicular tumours in rats. No specific mechanistic studies have been performed to investigate the liver tumours; however, hepatocellular hypertrophy was present in shorter-term studies, and mice are sensitive to the production of liver tumours in response to such effects produced by high concentrations of xenobiotics. Procymidone gave negative results in assays for genotoxicity in vitro and in vivo. A clear threshold for induction of liver tumours was identified in the study of carcinogenicity in mice. Investigative work on the endocrine effects of procymidone indicated that it binds to the androgen receptor with a similar potency to that of the prostate-cancer drug flutamide, and that in rats the mechanism of hormonal action appeared to be via binding to the androgen receptor, disrupting the feedback controls on LH and testosterone concentrations. Overall, the data suggest that procymidone may induce testicular interstitial cell tumours via an endocrine-mediated mechanism. Although this mode of action is relevant to humans, there is good evidence to suggest that humans are less sensitive to chemically-induced interstitial cell tumours of the testis than rats, owing to differences in sensitivity to LH on the basis of Leydig-cell receptor number and control of LH-receptor expression (e.g. via prolactin in rodents but not in humans).

The Meeting concluded that procymidone was unlikely to present a carcinogenic risk to humans at typical levels of dietary exposure.

The reproductive toxicity of procymidone had been investigated in two studies in rats. The developmental toxicity of procymidone had been studied in conventional studies in rats and rabbits and in special investigative studies in rats, rabbits and cynomolgus monkeys.

In the first generation of the two-generation study in rats, there was no effect on mating and reproduction, but at 750 ppm, equivalent to 50 mg/kg bw per day, there was an increase in the number of male pups with hypospadias and reduced anogenital distance. In the second generation, male fertility was reduced and the incidence of hypospadias was increased at 750 ppm. In parents and pups, there were increases in testes weights and decreases in body-weight gain, prostate and epididymis weights at 750 ppm, with the organ-weight changes also present in pups at 250 ppm. There were no adverse effects on female reproductive performance or on female offspring. The NOAELs for effects on parents and on reproduction were 250 ppm, equivalent to 17 mg/kg bw per day. The NOAEL for pup development was 50 ppm, equivalent to 3 mg/kg bw per day, on the basis of the alterations in weights of the testes, prostate and epididymis at 250 ppm. In a subsequent one-generation study, designed to investigate the effects on male pups in the previous study, there was a reduction in body-weight gain and reduced litter size at 37 mg/kg bw per day. Increases in the incidence of hypospadias and in testes weights, and decreases in weights of the prostate and seminal vesicles were seen in pups at 37 mg/kg bw per day. The NOAELs for parental toxicity, pup development and reproduction were 12.5 mg/kg bw per day.

Studies of developmental toxicity with procymidone have been performed in rats, rabbits and cynomolgus monkeys. In a conventional study of developmental toxicity in rats, there were significant (60%) reductions in maternal body-weight gain during the first 6 days of dosing at 300 mg/kg bw per day, but no adverse findings in pups. This study did not include specific investigations of anogenital distance or the external genitalia. The NOAEL for fetotoxicity and teratogenicity was 300 mg/kg bw per day, the highest dose tested. The NOAEL for maternal effects was 100 mg/kg bw per day. In a modified study of developmental toxicity, pregnant rats were dosed with procymidone on days 6–19 of gestation; on day 20, half the dams had caesarian sections and the other half were allowed to deliver normally. Examinations focused on the male reproductive tract. Maternal toxicity (body-weight loss and poor appearance) was evident at doses of 125 mg/kg bw per day and greater. A range of effects was seen on male fetuses and offspring from dams receiving doses of 125 mg/kg bw per day and greater: reduced anogenital distance, undescended testes, hypospadias, testicular atrophy, distended preputial gland, inflammatory changes in the accessory sex organs (seminal vesicles, prostate and coagulating glands), and lower organ weights of testes and prostate. At 500 mg/kg bw per day, there was a significant increase in the incidence of bifid thoracic vertebral centra in fetuses and prostate lesions in male offspring. The NOAEL for maternal and developmental toxicity was 12.5 mg/kg bw per day. A further investigative study of developmental toxicity, in which dams were allowed to deliver normally, confirmed the production of hypospadias and undescended testes at 37 mg/kg bw per day, the lowest dose tested.

In rabbits, an initial study of developmental toxicity found no evidence of maternal toxicity or teratogenicity, but there was a reduction in sternebrae ossification at 1000 mg/kg bw per day, the highest dose tested. The NOAEL for developmental toxicity was 750 mg/kg bw per day. In a subsequent study that concentrated on the external genitalia of fetuses, maternal toxicity (anorexia and abortions) was seen at 125 mg/kg bw per day, the only dose used. There were equivocal effects on measurements of the external genitalia in female offspring. Only a single dose, which produced overt maternal toxicity, was used in the study, and the findings are considered to be of uncertain toxicological relevance.

In an initial study focusing on effects on male fetuses, groups of four pregnant cynomolgus monkeys received doses of 62.5 or 125 mg/kg bw per day. There were no indications of maternal toxicity or fetotoxicity. In a more extensive investigation in groups of 16 pregnant cynomolgus monkeys, there was no maternal toxicity or effects on the external genitalia of offspring (six and eight male offspring in controls and in the test group, respectively) at 125 mg/kg bw per day, the only dose tested.

The characteristics of procymidone were similar in assays for binding to androgen receptors in rats and humans. The concentration required to inhibit activity by 50% (IC_{50}) values for procymidone (approximately 0.3 μmol/l) were similar to those of the anti-androgen prostate-cancer drug flutamide. The IC_{50} values for the procymidone metabolites PCM-CH_2OH, PA-CH_2OH and PCM-NH-COOH were approximately 5–10-fold that of procymidone; PCM-CH_2OH-glucuronide, PCM-COOH and PA-COOH showed no activity. However, these results might have been influenced by the non-enzymatic interconversion of procymidone metabolites. Under acid conditions ($pH < 6.4$), procymidone and PCM-CH_2OH are stable, but under alkaline conditions ($pH \geq 7.8$) the imide bond was shown to be cleaved to give PCM-NH-COOH and PA-CH_2OH.

The mechanism of the effect of procymidone on the reproductive organs has been investigated by measuring hormone concentrations in mice, rats and cynomolgus monkeys. Procymidone had no adverse effects on testes, sperm and serum or tissue levels of LH in cynomolgus monkeys receiving procymidone at doses of up to 1000 mg/kg bw per day for 7 days or up to 100 mg/kg bw per day for 91 days. There was evidence of a reduction in ejaculate mass in cynomolgus monkeys at a dose of 100 mg/kg bw per day or greater, but there was no statistically significant effect on ejaculate sperm counts. The small group size and extent of variation between animals and in values recorded before dosing makes it difficult to reach firm conclusions about the toxicological significance of these results. The NOAEL in this study was considered to be 30 mg/kg bw per day. Two studies in rats provided qualitatively similar results, but although the protocols were similar and the same strain was used, there were differences in the doses that produced effects. There were increases in hCG-stimulated production of testosterone and circulating testosterone concentrations in rats receiving diets containing procymidone at 700 ppm, equal to 47 mg/kg bw per day, for 13 weeks. LH concentrations were increased at 6000 ppm after 2 weeks. Epididymis weights were reduced after exposure to procymidone at a dietary concentration of 2000 ppm or greater for 2 weeks, but not after 3 months. Testes weights were increased at dietary concentrations of 700 ppm and greater. All effects showed signs of reversal during a recovery phase. Findings in mice were similar to those in rats, but the stimulation of testosterone production by hCG had returned to control levels after 3 months owing to a decrease in the binding affinity of hCG to the LH/hCG receptor.

The Meeting concluded that the existing database on procymidone was adequate to characterize the potential hazards to fetuses, infants and children

Procymidone has not been studied specifically for neurotoxicity, but there were no indications from the results of conventional studies that it has significant neurotoxic potential.

A number of studies have been performed with procymidone metabolites. DCA was of moderate acute toxicity in *dd* mice, having an oral LD_{50} of approximately 850 mg/kg bw. PCM-NH-COOH was also of moderate acute toxicity in *dd* mice, with an oral LD_{50} of approximately 1450 mg/kg bw. DMCPA was of low acute toxicity in *dd* mice, having an oral LD_{50} of approximately 4400 mg/kg bw and a subcutaneous LD_{50} of approximately 2400 mg/kg bw. DMCPA gave negative results in an Ames test and an assay for micronucleus formation in vivo. DMCPA was rapidly absorbed and excreted with minimal metabolism. Retention of radioactivity in tissues was low.

PCM-CH_2OH was investigated in a modified study of developmental toxicity in which rats were allowed to deliver normally. No abnormal findings were seen in dams or female offspring. In the male offspring, a dose-related shortening of the anogenital distance was seen and there were increases in the incidence of male offspring and litter with abnormities of the external genitalia. The incidence of hypospadias was increased in pups from dams receiving procymidone at 62.5 mg/kg bw per day, the lowest dose tested. In addition, small testis and epididymis and undescended testis and epididymis were observed occasionally in groups treated with PCM-CH_2OH. These results are similar to those seen with procymidone at similar doses. The $BMDL_{10}$ for hypospadias was 43.8 mg/kg bw per day for PCM-CH_2OH and 23.7 mg/kg bw per day for procymidone. The Meeting considered the data on the relative androgen receptor and plasma-protein binding affinities of hyroxy-procymidone and procymidone, and the marked species differences in formation and elimination of procymidone and its metabolites. The Meeting concluded that while PCM-CH_2OH might be a significant contributor to the effects seen in rats, due to the much higher systemic and fetal exposure, the contribution of procymidone could not be discounted in other species, particularly due to limitations in the studies in cynomolgus monkeys.

Surveys of the medical records of production-plant workers had not identified any symptoms or diseases related to the manufacture of procymidone. There were no documented cases of procymidone intoxication nor any significant effects associated with its use.

Toxicological evaluation

The Meeting established an ADI of 0–0.1 mg/kg bw based on a NOAEL of 12.5 mg/kg bw per day in a two-generation study of reproductive toxicity and a study of developmental toxicity in rats, on the basis of hypospadias and alterations in testes, prostate and epididymis weights, and a safety factor of 100. The ADI was supported by NOAELs of 14 mg/kg bw per day in the 2-year study in rats and 15 mg/kg bw per day in the 2-year study in mice.

The Meeting established an ARfD of 0.1 mg/kg bw based on a NOAEL of 12.5 mg/kg bw on the basis of hypospadias, which might have been a consequence of a single exposure, in a study of developmental toxicity in rats, and using a safety factor of 100. The Meeting concluded that the effects on organ weights observed in the multigeneration study were largely a consequence of postnatal exposure over a period of time and therefore not appropriate for the establishment of the ARfD. The Meeting considered that, on the basis of the observed differences between species in terms of kinetics, metabolism and toxicological sensitivity to procymidone, this ARfD might be conservative. However, uncertainties regarding potential responses in species other than rats were such that it was not possible to modify the default safety factor.

Levels relevant to risk assessment

Species	Study	Effect	NOAEL	LOAEL
Mouse	Two-year studies of toxicity and carcinogenicity[a]	Toxicity	100 ppm, equal to 15 mg/kg bw per day	300 ppm, equal to 46 mg/kg bw per day
		Carcinogenicity	300 ppm, equal to 46 mg/kg bw per day	1000 ppm, equal to 153 mg/kg bw per day
Rat	Two-year studies of toxicity and carcinogenicity [a]	Toxicity	300 ppm, equal to 14 mg/kg bw per day	1000 ppm, equal to 48 mg/kg bw per day
		Carcinogenicity	300 ppm, equal to 14 mg/kg bw per day	1000 ppm, equal to 48 mg/kg bw per day
	Multigeneration study of reproductive toxicity[ae]	Parental toxicity	250 ppm, equivalent to 17 mg/kg bw per day	750 ppm, equivalent to 50 mg/kg bw per day
		Offspring toxicity	12.5 mg/kg bw per day	250 ppm, equivalent to 17 mg/kg bw per day
	Developmental toxicity[b,e]	Maternal toxicity	100 mg/kg bw per day	125 mg/kg bw per day
		Embryo/fetotoxicity	12.5 mg/kg bw per day	37.5 mg/kg bw per day
Rabbit	Developmental toxicity[b,e]	Maternal toxicity	—	125 mg/kg bw per day[f]
		Embryo/fetotoxicity	750 mg/kg bw per day	1000 mg/kg bw per day
Monkey	Developmental toxicity[b,e]	Maternal toxicity	125 mg/kg bw per day[c]	—
		Embryo/fetotoxicity	125 mg/kg bw per day[c]	—
Dog	Six-month and 1-year studies of toxicity[d]	Emesis, soft faeces, increased alkaline phosphatase activity	100 mg/kg bw per day	500 mg/kg bw per day

[a] Dietary administration.
[b] Gavage administration.
[c] Highest dose tested.
[d] Capsule administration.
[e] More than one study combined.
[f] Lowest dose tested.

Estimate of acceptable daily intake for humans

0–0.1 mg/kg bw

Estimate of acute reference dose

0.1 mg/kg bw

Information that would be useful for the continued evaluation of the compound

Results from epidemiological, occupational health and other such observational studies of human exposure

Critical end-points for setting guidance values for exposure to procymidone

Absorption, distribution, excretion and metabolism in mammals	
Rate and extent of oral absorption	Rapid, C_{max} 2–8 h; extensive, > 80% excreted in the urine
Dermal absorption	4% concentrate, 13% for dilution in rats in vivo
Distribution	Extensive, highest concentrations in fat
Potential for accumulation	Low
Rate and extent of excretion	Rapid, > 80% in 24 h; enterohepatic recirculation in rats
Metabolism in animals	Hydroxylation, oxidation and conjugation. Some species differences.

Toxicologically significant compounds in animals, plants and the environment	Procymidone, hydroxyprocymidone
Acute toxicity	
Rat, LD_{50}, oral	> 5000 mg/kg bw
Rat, LD_{50}, dermal	> 2500 mg/kg bw
Rat, LC_{50}, inhalation	> 1.5 mg/l
Rabbit, skin irritation	Slight transient irritant
Rabbit, eye irritation	Not irritating
Guinea-pig, skin sensitization (test method used)	Negative (Magnusson & Kligman maximization test)
Short-term studies of toxicity	
Target/critical effect	Decreased body-weight gain, liver effects, testes
Lowest relevant oral NOAEL	25 mg/kg bw per day
Lowest relevant dermal NOAEL	1000 mg/kg bw per day
Lowest relevant inhalation NOAEC	No data
Genotoxicity	
	Not genotoxic
Long-term studies of toxicity and carcinogenicity	
Target/critical effect	Body weight, liver effects
Lowest relevant NOAEL	14 mg/kg bw per day (rat)
Carcinogenicity	Liver tumours in mice; testes tumours in rats.
Reproductive toxicity	
Reproduction target/critical effect	Hypospadias; altered testes and epididymis weight
Lowest relevant reproductive NOAEL	12.5 mg/kg bw per day
Developmental target/critical effect	Hypospadias (rat)
Lowest relevant developmental NOAEL	12.5 mg/kg bw per day
Neurotoxicity/delayed neurotoxicity	
	No evidence in conventional studies
Special studies	
	Kinetic and metabolism studies in rats, monkeys, rabbits, mice and with human tissues showed differences between species. Rats having a relatively high area-under-the curve and a low capability to metabolize procymidone and hydroxyprocymidone.
	Receptor-binding assays showed that procymidone and hydroxyprocymidone could bind to rat and human androgen receptors. Both procymidone and hydroxyprocymidone were extensively bound (≥ 77%) to plasma proteins.
	Procymidone produced increases in luteinizing hormone and human chorionic gonadotrophin-stimulated production of testosterone in rats at 6000 ppm and ≥ 700 ppm respectively.
	Ejaculate mass was reduced in monkeys exposed to procymidone for 13 weeks at ≥ 100 mg/kg bw per day. NOAEL was 30 mg/kg bw per day
	Hydroxyprocymidone induced hypospadias in rats at ≥ 62.5 mg/kg bw per day; $BMDL_{10}$ was 43.8 mg/kg bw per day.
Medical data	
	No adverse findings reported

Summary

	Value	Study	Safety factor
ADI	0–0.1 mg/kg bw	Rat, studies of developmental and reproductive toxicity	100
ARfD	0.1 mg/kg bw	Rat, study of developmental toxicity	100

References

Arai, M., Hasegawa, R. & Ito, N. (1980a) Three-month subacute toxicity study of S-7131 (Sumilex, Sumisclex) in mice. Unpublished report No. BT-00-0058 from Nagoya City University Medical School, Nagoya, Japan. Submitted to WHO by Sumitomo Chemical Co., Ltd, Osaka, Japan

Arai, M., Hasegawa, R. & Ito, N. (1980b) Six-month subacute toxicity study of S-7131 (Sumilex, Sumisclex) in mice. Unpublished report No. BT-00-0057 from Nagoya City University Medical School, Nagoya, Japan. Submitted to WHO by Sumitomo Chemical Co., Ltd, Osaka, Japan.

Bee, W. (1992) S-7131 90-day oral (gavage) subchronic study in the cynomolgus monkey. Unpublished report No BT-21-0147 from Hazleton Deutschland GmbH, Kesselfeld, Germany. Submitted to WHO by Sumitomo Chemical Co., Ltd, Osaka, Japan.

Dalgard, D. & Machotka, S. (1992) Chronic toxicity study in dogs with procymidone. Unpublished report No. BT 21-0162 from Hazleton Washington, Vienna, Virginia, USA. Submitted to WHO by Sumitomo Chemical Co., Ltd, Osaka, Japan.

Filler, R. & Parker, G. A. (1988) Oral chronic toxicity study and oncogenicity study in mice. Unpublished report No. BT-81-0126 from Hazleton Laboratories America Inc., Rockville, Maryland, USA. Submitted to WHO by Sumitomo Chemical Co., Ltd, Osaka, Japan.

Fujita, T., Fukuda, T., Kato, T. & Miyamoto, J. (1976) Toxicity of S-7131 technical material to rats in prolonged dietary administration over 9 months. Unpublished report No. BT-60-0003 from Sumitomo Chemical Co., Ltd, Osaka, Japan. Submitted to WHO by Sumitomo Chemical Co., Ltd, Osaka, Japan

Fukunishi, K. (2003a) A preliminary study of the effects of procymidone on external genitalia development in male fetuses by oral administration to cynomolgus monkeys. Unpublished report No. BT-0207 from Shin Nippon Biomedical Laboratories Ltd., Kagoshima, Japan. Submitted to WHO by Sumitomo Chemical Co., Ltd, Osaka, Japan.

Fukunishi, K. (2003b) A study of the effects of procymidone on external genitalia development in male fetuses by oral administration to cynomolgus monkeys. Unpublished report No. BT-0208 from Shin Nippon Biomedical Laboratories Ltd., Kagoshima, Japan. Submitted to WHO by Sumitomo Chemical Co., Ltd, Osaka, Japan.

Hara, M. (1991a) In vitro unscheduled DNA synthesis (UDS) assay of Sumilex in rat hepatocytes. Unpublished report No. BT-10-0142 from Sumitomo Chemical Co., Ltd, Osaka, Japan. Submitted to WHO by Sumitomo Chemical Co., Ltd.

Hara, M. (1991b) In vitro chromosomal aberration test of Sumilex in Chinese hamster ovary cells (CHO-K1). Unpublished report No.BT-10-0141 from Sumitomo Chemical Co., Ltd, Osaka, Japan. Submitted to WHO by Sumitomo Chemical Co., Ltd.

Hara, M. (1992) Comments on EPA's 'data evaluation report' of the report entitled in vitro unscheduled DNA synthesis (UDS) assay of Sumilex in rat hepatocytes. Unpublished report No. BT-20-0149 from Sumitomo Chemical Co., Ltd, Osaka, Japan. Submitted to WHO by Sumitomo Chemical Co., Ltd.

Hara, M. & Suzuki, H. (1980) Cytogenetic test of procymidone in mouse bone marrow. Unpublished report No. BT-00-0043 from Sumitomo Chemical Co., Ltd, Osaka, Japan. Submitted to WHO by Sumitomo Chemical Co., Ltd.

Harada, T. (1983) A review on medical examinations of factory workers possibly exposed to procymidone technical materials. Unpublished report No BG-31-0019 from Sumitomo Chemical Co., Ltd, Osaka, Japan. Submitted to WHO by Sumitomo Chemical Co. Ltd.

Hoberman, A. (1992a) Dose-range developmental toxicity (embryo-fetal toxicity and teratogenic potential) study of procymidone administered orally via gavage to Crl:CD®BR VAF/Plus® presumed pregnant rats. Unpublished report No BT-21-0160 from Argus Research Laboratories, Inc., Horsham, Pennsylvania, USA. Submitted to WHO by Sumitomo Chemical Co., Ltd, Osaka, Japan.

Hoberman, A. (1992b) Developmental toxicity (embryo-fetal toxicity and teratogenic potential including a postnatal evaluation) study of procymidone administered orally via gavage to Crl:CD®BR VAF/Plus® presumed pregnant rats. Unpublished report No BT-21-0161 from Argus Research Laboratories, Inc., Horsham, Pennsylvania, USA. Submitted to WHO by Sumitomo Chemical C., Ltd, Osaka, Japan.

Hoberman, A. (1992c) Dose-range developmental toxicity (embryo-fetal toxicity and teratogenic potential including a postnatal evaluation) study of procymidone administered orally via gavage to Crl:CD®BR VAF/Plus® presumed pregnant rats (amendment). Unpublished report No BT-21-0164 from Argus Research Laboratories, Inc., Horsham, Pennsylvania, USA. Submitted to WHO by Sumitomo Chemical Co., Ltd, Osaka, Japan.

Hodge, M., Williams, M., Hollis, K., Moreland, S., Greenwood, M., MacLean Head, L. & Foster, P. (1991) Procymidone: one-generation study in the rat. Unpublished report No. BT-11-0144 from Imperial Chemical Industries, Macclesfield, Cheshire, England. Submitted to WHO by Sumitomo Chemical Co., Ltd, Osaka, Japan.

Hosokawa, S. (1984) Toxicity of S-7131 technical material to rats in prolonged dietary administration over 9 months (BT-60-0003). addendum to final report. Unpublished report No. BT-40-0095 from Takarazuka Research Centre, Takarazuka, Japan. Submitted to WHO by Sumitomo Chemical Co. Ltd., Osaka, Japan.

Inawaka, K. (2003) Study for effects of procymidone on development of fetal external genitalia in rabbits. Unpublished report No. BT-0211 from Environmental Health Science Laboratory, Sumitomo Chemical Co., Ltd, Osaka, Japan. Submitted to WHO by Sumitomo Chemical Co., Ltd

Inawaka, K. (2005) Study for determination of minimum toxic dose of the effects on male external genitalia in rats with procymidone. Unpublished report No. BT-0217 from Environmental Health Science Laboratory, Sumitomo Chemical Co., Ltd, Osaka, Japan. Submitted to WHO by Sumitomo Chemical Co., Ltd.

Izumi, H. (2005) Teratogenicity study of hydroxylated procymidone in rats by oral administration (observations in male offspring). Unpublished report No. BT-0222 from Panapharm Laboratories Co., Ltd, Kumamoto, Japan. Submitted to WHO by Sumitomo Chemical Co., Ltd, Osaka, Japan.

Kadota K, Kohda H &Miyamoto J (1976a) Acute toxicity studies with S-7131 technical material in mice and rats; oral, subcutaneous, intraperitoneal, dermal. Unpublished report No. BT-60-0002 from Sumitomo Chemical Co., Ltd, Osaka, Japan. Submitted to WHO by Sumitomo Chemical Co., Ltd.

Kadota, K. & Miyamoto, J. (1976) Eye and skin irritation study with S-7131 technical material in albino rats. Unpublished report No. BT-60-0004 from the Institute for Biological Sciences, Hyogo, Japan. Submitted to WHO by Sumitomo Chemical Co., Ltd, Osaka, Japan.

Keller, J. & Cardy, R. H. (1986) Oral chronic toxicity and oncogenicity study in rats TB0100 Sumisclex. Unpublished report No. BT-61-0112 from Litton Bionetics Inc., Rockville, Maryland, USA. Submitted to WHO by Sumitomo Chemical Co., Ltd, Osaka, Japan.

Kimura, K., Shiba, K. & Iba, K. (1988) Comparative metabolism of procymidone in rats and mice. Unpublished report No. BM-80-0019 from Takarazuka Research Center, Sumitomo Chemical Co. Ltd., Takarazuka, Japan. Submitted to WHO by Sumitomo Chemical Co. Ltd.

Kinsey, D, Banham, P., Godley, M., Ishmael, J. & Blackburn, D. (1985) Procymidone: six-month feeding study in mice. Unpublished report No BT-51-0102 from Imperial Chemical Industries plc., Macclesfield, England. Submitted to WHO by Sumitomo Chemical Co., Ltd, Osaka, Japan.

Kogiso, S. (1991) Reverse mutation test of Sumilex in bacterial system. Unpublished report No. BT-10-0138 from Sumitomo Chemical Co., Ltd, Osaka, Japan. Submitted to WHO by Sumitomo Chemical Co., Ltd.

Kogiso, S., Kasaoka, Y., Kato, H. & Nakatsuka, I. (1992) Reverse mutation test of CCA in bacterial systems. Unpublished report No. BT-20-0148 from Sumitomo Chemical Co., Ltd, Osaka, Japan. Submitted to WHO by Sumitomo Chemical Co., Ltd.

Kohda, H., Nakamura, M. & Kadota, T. (1980) Acute oral and subcutaneous toxicity of three metabolites of procymidone. Unpublished report No. BT-00-0037 from Sumitomo Chemical Co., Ltd, Osaka, Japan. Submitted to WHO by Sumitomo Chemical Co., Ltd.

Kumagai, H. (2007). Statement on medical examinations at Sumitomo Osaka Works, dated 29 August, 2007. Submitted to WHO by Sumitomo Chemical Co., Ltd, Osaka, Japan.

Kurihara, T. (2001) A review on medical examination of factory workers possibly exposed to procymidone technical materials. Unpublished report No. BT-0197 from Sumitomo Chemical Co., Ltd, Osaka, Japan. Submitted to WHO by Sumitomo Chemical Co., Ltd.

Matsui, M, (2005a) In vitro plasma protein binding of procymidone and PCM-CH_2OH in human, rat, rabbit and monkey. Unpublished report No. BM-0109 from Environmental Health Science Laboratory, Sumitomo Chemical Co., Ltd, Osaka, Japan. Submitted to WHO by Sumitomo Chemical Co., Ltd.

Matsui, M. (2005b) In vitro metabolism of procymidone and PCM-CH_2OH in S9 fraction of human, rat, rabbit and monkey female livers. Unpublished report No. BM-0118 from Environmental Health Science Laboratory, Sumitomo Chemical Co., Ltd, Osaka, Japan. Submitted to WHO by Sumitomo Chemical Co., Ltd.

Mikami, N. & Yamamoto, H. (1976) Metabolism of *N*-(3,5-dichlorophenyl)-1,2-dimethylcyclopropane-1,2-dicarboximide in rats. Unpublished report No BM-60-0002 from Sumitomo Chemical Co., Ltd, Osaka, Japan. Submitted to WHO by Sumitomo Chemical Co., Ltd.

Milburn, G. & Moreland, S. (1991) Procymidone: multigeneration reproduction study in the rat (first supplement – additional histopathology). Unpublished report No. BT-11-0140 from Imperial Chemical Industries, Macclesfield, England. Submitted to WHO by Sumitomo Chemical Co., Ltd., Osaka, Japan.

Mogi, M. (2005a) Pharmacokinetics and excretion of [phenyl-^{14}C]procymidone after single oral administration to cynomolgus monkeys. Unpublished report No. BM-0103 from Shin Nippon Biomedical Laboratories, Wakayama, Japan. Submitted to WHO by Sumitomo Chemical Co. Ltd.

Mogi, M. (2005b) Pharmacokinetics and excretion of [phenyl-^{14}C]procymidone during and after repeated oral administration to cynomolgus monkeys. Unpublished report No. BM-0104 from Shin Nippon Biomedical Laboratories, Wakayama, Japan. Submitted to WHO by Sumitomo Chemical Co. Ltd.

Mogi, M. (2005c) Biliary excretion after single oral administration of [carbonyl-^{14}C]procymidone to monkey. Unpublished report No. BM-0106 from Shin Nippon Biomedical Laboratories, Wakayama, Japan. Submitted to WHO by Sumitomo Chemical Co. Ltd.

Mogi, M. (2005d) Biliary excretion after single oral administration of [carbonyl-^{14}C] procymidone to rabbit. Unpublished report No. BM-0107 from Shin Nippon Biomedical Laboratories, Wakayama, Japan. Submitted to WHO by Sumitomo Chemical Co, Ltd.

Mogi, M. (2005e) Tissue distribution of pregnant monkey and fetus after single oral administration of [phenyl-^{14}C]procymidone to pregnant monkey. Unpublished report No. BM-0125 from Shin Nippon Biomedical Laboratories, Wakayama, Japan. Submitted to WHO by Sumitomo Chemical Co. Ltd.

Moore, M.R. (1993) Two-year oral toxicity and carcinogenicity study in mice: supplement A to the final report. Unpublished report No. BT-31-0172 from Hazleton Laboratories America Inc., Rockville, Maryland, USA. Submitted to WHO by Sumitomo Chemical Co. Ltd.

Moriya, M, & Katom K, (1977) Mutagenicity of S-7131 in bacterial test systems. Unpublished report No. BT-71-0021 from Sumitomo Chemical Co., Ltd, Osaka, Japan. Submitted to WHO by Sumitomo Chemical Co., Ltd.

Murakami, M., Yoshitake, A. & Hosokawa, S. (1986) Study on serum hormone levels in rats treated with S-7131. Unpublished report No. BT-60-0113 from Sumitomo Chemical Co., Ltd, Osaka, Japan. Submitted to WHO by Sumitomo Chemical Co., Ltd.

Murakami, M., Ineyama, M., Yamada, T., Koyama, Y., Ito, M., Kimura, J., Hosokawa, S., Yoshitake, A. & Yamada, H. (1988a) Effect of testicular function of male rats and mice by subacute administration of procymidone. Unpublished report No. BT-80-0130 from Sumitomo Chemical Co., Ltd, Osaka, Japan. Submitted to WHO by Sumitomo Chemical Co. Ltd.

Murakami, M., Ineyama, M., Hosokawa, S., Yoshitake, A. & Yamada, H. (1988b) The affinity of procymidone to androgen receptor in rats and mice. Unpublished report No. BT-80-0131 from Sumitomo Chemical Co. Ltd., Osaka, Japan. Submitted to WHO by Sumitomo Chemical Co., Ltd.

Murli, H. (1992a) Dose range finding study for in vivo murine micronucleus assay on CCA. Unpublished report No. BT-21-0158 from Hazelton Washington Inc., Rockville, Maryland, USA. Submitted to WHO by Sumitomo Chemical Co. Ltd.

Murli, H. (1992b) Mutagenicity test on CCA: in vivo mammalian micronucleus test. Unpublished report No. BT-21-0159 from Hazelton Washington Inc., Rockville, Maryland, USA. Submitted to WHO by Sumitomo Chemical Co. Ltd.

Nakanishi, T., Nakatsuka, I. & Kogiso, S. (1991) Skin sensitization test of procymidone in guinea-pigs. Unpublished report No. BT-10-0146 from Environmental Health Science Laboratory, Sumitomo Chemical Co., Ltd, Osaka, Japan. Submitted by Sumitomo Chemical Co., Ltd.

Nakashima, N., Ebino, K., Tsuda, S., Kitazawa, T. & Harada, T. (1984) A 6-month subchronic study of Sumilex in beagles. Unpublished report No. BT-41-0091 from Institute of Environmental Toxicology, Tokyo, Japan. Submitted to WHO by Sumitomo Chemical Co., Ltd.

Nunziata, A. (1982) Supplementary information to the technical report (BT-00-0050). Unpublished report No. BT-21-0075 from Centro Ricerca Farmaceutica S.p.A., Rome, Italy. Submitted to WHO by Sumitomo Chemical Co., Ltd.

Ogata, H. (2002) 28-Day repeated-dose dermal toxicity study of procymidone TG in rats. Unpublished report No. BT-0200 from Panapharm Laboratories, Kumamoto, Japan. Submitted to WHO by Sumitomo Chemical Co., Ltd.

Ohzone, Y. (2005a) Excretions of procymidone in chimera mice. Unpublished report No. BM-0108 from Daiichi Pure Chemicals Co., Ltd, Ibaraki, Japan. Submitted to WHO by Sumitomo Chemical Co. Ltd.

Okuno, Y. (1975) Skin sensitization with S-7131 in guinea-pigs. Unpublished report No. BT-60-0005 from Sumitomo Chemical Co., Ltd, Osaka, Japan. Submitted to WHO from Sumitomo Chemical Co., Ltd.

Owen, H.M. (2002) Procymidone: in vitro absorption through human and rat epidermis. Unpublished report No. BM-0076 from Central Toxicology Laboratory, Macclesfield, Cheshire, England. Submitted to WHO by Sumitomo Chemical Co. Ltd.

Pence, D.H., Hoberman, A.M. & Spicer K.M. (1980a) Pilot teratology study in rats – S-7131 (technical). Unpublished report No. BT-01-0041 from Hazleton Laboratories America Inc., Rockville, Maryland, USA. Submitted to WHO by Sumitomo Chemical Co. Ltd.

Pence, D.H., Hoberman, A.M., Durloo, R.S.; Andrews, J.P. & Mossburg, P.A. (1980b) Teratology study in rats – S-7131 (technical). Unpublished report No. BT-01-0048 from Hazleton Laboratories America Inc., Rockville, Maryland, USA. Submitted to WHO by Sumitomo Chemical Co. Ltd.

Principe, P., Monaco, M. & Nunziata, A. (1980) Report on mutagenicity experiment on the substance Sumisclex (procymidone). Unpublished report No BT-01-0050 from Centro Ricerca Farmaceutica S.p.A., Rome, Italy. Submitted to WHO by Sumitomo Chemical Co. Ltd.

Saito, K. (1992) Metabolism of CCA in rats. Unpublished report No. BM-20-0032 from Sumitomo Chemical Co., Ltd, Osaka, Japan. Submitted to WHO by Sumitomo Chemical Co., Ltd.

Savides, M.C. (2002) The in vivo dermal absorption of [^{14}C]procymidone SC in the rat. Unpublished report No. BM-0075 from Ricerca, LLC, Concord, Ohio, USA. Submitted to WHO by Sumitomo Chemical Co. Ltd.

Segawa, T. (1977) Acute oral, subcutaneous, intraperitoneal and dermal toxicities of S-7131 in rats and mice. Unpublished report No. BT-71-0053 from Hiroshima University School of Medicine, Hiroshima, Japan. Submitted to WHO by Sumitomo Chemical Co. Ltd.

Struble, C. (1992a) Metabolism of [^{14}C]procymidone in rats (preliminary and definitive phases): volume 1 of 2: absorption distribution elimination and tissue residues. Unpublished report No BM21-0033 from Hazleton Wisconsin, Madison, Wisconsin, USA. Submitted by Sumitomo Chemical Co. Ltd.

Struble, C. (1992b) Metabolism of [^{14}C]procymidone in rats (preliminary and definitive phases): volume 2 of 2: metabolite identification. Unpublished report No BM21-0034 from Hazleton Wisconsin, Madison, Wisconsin, USA. Submitted by Sumitomo Chemical Co. Ltd.

Sugimoto, K. (2005a) Pharmacokinetics and excretions of [phenyl-^{14}C]procymidone in female rats (single oral administration). Unpublished report No. BM-0096 from Panapharm Laboratories Co., Ltd, Kumamoto, Japan. Submitted to WHO by Sumitomo Chemical Co. Ltd.

Sugimoto, K. (2005b) Pharmacokinetics and excretions of [phenyl-^{14}C]procymidone in female rabbits (single oral administration). Unpublished report No. BM-0102 from Panapharm Laboratories Co., Ltd, Kumamoto, Japan. Submitted to WHO by Sumitomo Chemical Co. Ltd.

Sugimoto, K. (2005c) Pharmacokinetics and excretions of [phenyl-^{14}C]procymidone in female rats (repeated oral administration for 14 days). Unpublished report No. BM-0098 from Panapharm Laboratories Co., Ltd, Kumamoto, Japan. Submitted to WHO by Sumitomo Chemical Co. Ltd.

Sugimoto, K. (2005d) Excretion of [carbonyl-^{14}C]procymidone in bile of female rats (single oral administration). Unpublished report No. BM-0099 from Panapharm Laboratories Co., Ltd, Kumamoto, Japan. Submitted to WHO by Sumitomo Chemical Co. Ltd.

Sugimoto, K. (2005e) Fetoplacental transfer of [phenyl-^{14}C]procymidone in rats (single oral administration). Unpublished report No. BM-0100 from Panapharm Laboratories Co., Ltd, Kumamoto, Japan. Submitted to WHO by Sumitomo Chemical Co. Ltd.

Sugimoto, K. (2005f) Fetoplacental transfer of [phenyl-^{14}C]procymidone in rats (single oral administration). Unpublished report No. BM-0122 from Panapharm Laboratories Co., Ltd, Kumamoto, Japan. Submitted to WHO by Sumitomo Chemical Co. Ltd.

Sugimoto, K. (2005g) Fetoplacental transfer of [phenyl-^{14}C]procymidone in rabbits (single oral administration). Unpublished report No. BM-0101 from Panapharm Laboratories Co., Ltd., Kumamoto, Japan. Submitted to WHO by Sumitomo Chemical Co. Ltd., Osaka, Japan.

Suzuki, H. (1980) Effect of procymidone on sister chromatid exchanges (SCE) in cultured mouse embryo cells. Unpublished report No. BT-00-0042 from Sumitomo Chemical Co., Ltd, Osaka, Japan. Submitted to WHO by Sumitomo Chemical Co, Ltd.

Suzuki, T. & Kato, T. (1988) Acute inhalation toxicity of Sumilex in rats. Unpublished report No. BT-60-0116 from Takarazuka Research Centre, Takarazuka, Japan. Submitted to WHO by Sumitomo Chemical Co., Ltd.

Suzuki, H. & Miyamoto, J. (1976) Studies on mutagenicity of S-7131 with bacterial systems. Unpublished report No.BT-60-0011 from Sumitomo Chemical Co., Ltd, Osaka, Japan. Submitted to WHO by Sumitomo Chemical Co., Ltd.

Suzuki, N. (2005) In vitro assays for procymidone and its metabolites with rat and human androgen receptors. Unpublished report No. BM-0120 from Environmental Health Science Laboratory, Sumitomo Chemical Co., Ltd, Osaka, Japan. Submitted to WHO by Sumitomo Chemical Co., Ltd.

Tarui, H. (2005a) Transformation of procymidone and its metabolites in various pH conditions. Unpublished report No. BM-0116 from Environmental Health Science Laboratory, Sumitomo Chemical Co., Ltd, Osaka, Japan. Submitted to WHO by Sumitomo Chemical Co., Ltd.

Tarui, H. (2005b) In vitro metabolism of procymidone in human hepatocyte. Unpublished report No. BM-0113 from Environmental Health Science Laboratory, Sumitomo Chemical Co., Ltd, Osaka, Japan. Submitted to WHO by Sumitomo Chemical Co., Ltd.

Thakur, A. K. (1988) Oral chronic toxicity and oncogenicity study in mice. Unpublished report amendment No. BT-81-0127 from Hazleton Laboratories America, Inc., Rockville, Maryland, USA. Submitted to WHO by Sumitomo Chemical Co. Ltd.

Tomigahara, Y. (2005a) Metabolism of [phenyl-^{14}C]PA-CH_2OH in female rats. Unpublished report No. BM-0115 from Environmental Health Science Laboratory, Sumitomo Chemical Co., Ltd, Osaka, Japan. Submitted to WHO by Sumitomo Chemical Co., Ltd.

Tomigahara, Y. (2005b) Metabolism of [phenyl-^{14}C]PCM-CH_2OH in female rats. Unpublished report No. BM-0119 from Environmental Health Science Laboratory, Sumitomo Chemical Co., Ltd, Osaka, Japan. Submitted to WHO by Sumitomo Chemical Co., Ltd.

Tomigahara, Y. (2005c) Pharmacokinetics and excretions of [phenyl-^{14}C]PCM-CH_2OH in female rats (single oral and subcutaneous administration. Unpublished report No. BM-0114 from Environmental Health Science Laboratory, Sumitomo Chemical Co., Ltd, Osaka, Japan. Submitted to WHO by Sumitomo Chemical Co., Ltd.

Tomigahara, Y. (2005d) Placental transfer of [phenyl-^{14}C]procymidone in rats (repeated oral administration). Unpublished report No. BM-0117 from Environmental Health Science Laboratory, Sumitomo Chemical Co., Ltd, Osaka, Japan. Submitted to WHO by Sumitomo Chemical Co., Ltd.

Tsuji, R. (1994) Acute inhalation toxicity of Sumilex in rats (addendum to Suzuki & Kato, 1988). Unpublished report No. BT-40-0178 from Sumitomo Chemical Co., Ltd, Osaka, Japan. Submitted to WHO by Sumitomo Chemical Co., Ltd.

Weir, R., Keller, J., Moe, J., Cardy, R., Kim, G., Paulin, H. & Fitzgerald, J. (1984) 13-week pilot study in mice: Sumilex. Unpublished report No. BT-41-0092 from Litton Bionetics, Kensington, Maryland, USA. Submitted to WHO by Sumitomo Chemical Co. Ltd.

Wickramaratne, G. (1988) Procymidone: teratogenicity study in rabbits – individual animal data supplement. Unpublished report No. BT-71-0177 from Imperial Chemical Industries Ltd., Macclesfield, Cheshire, England. Submitted to WHO by Sumitomo Chemical Co. Ltd.

Wickramaratne, G., Milburn, G., Banham, P., Greenwood, M. & Moreland, S. (1988a) Procymidone: multigeneration reproduction study in the rat. Unpublished report No. BT-81-0132 from Imperial Chemical Industries, Macclesfield, England. Submitted to WHO by Sumitomo Chemical Co. Ltd.

Wickramaratne, G., Kinsey, D., Banham, P. & Greenwood, M. (1988b) Procymidone: teratogenicity study in rabbits. Unpublished report No. BT-81-0128 from Imperial Chemical Industries Ltd., Macclesfield, Cheshire, England. Submitted to WHO by Sumitomo Chemical Co. Ltd.

Wickramaratne, G. & Milburn, G. (1990) Procymidone: multigeneration reproduction study in the rat (supplementary submission – substance intake data) Unpublished report No. BT-81-0169 from Imperial Chemical Industries, Macclesfield, England. Submitted to WHO by Sumitomo Chemical Co. Ltd.

Zühlke U. (1991) S-7131 Seven-day oral (gavage) range-finding study in the cynomolgus monkey. Unpublished report No BT-11-0143 from Hazleton Deutschland GmbH, Kesselfeld, Germany. Submitted to WHO by Sumitomo Chemical Co. Ltd.

PROFENOFOS

First draft prepared by
P.V. Shah[1] *and David Ray*[2]

[1]*United States Environmental Protection Agency,*
Office of Pesticide Programs, Washington, DC, USA; and
[2]*Medical Research Council, Applied Neuroscience Group,*
Biomedical Sciences,
University of Nottingham,
Queens Medical Centre, Nottingham, England

Explanation

Profenofos is the International Organization for Standardization (ISO) approved name for (*RS*)-*O*-4-bromo-2-chlorophenyl *O*-ethyl S-propyl phosphorothioate (International Union of Pure and Applied Chemistry, IUPAC). The Chemical Abstracts Service (CAS) chemical name for profenofos is *O*-(4-bromo-2-chlorophenyl) *O*-ethyl S-propyl phosphorothioate (CAS No. 41198-08-7). It is a broad-spectrum organophosphorus insecticide that is used to control insect pests in cotton, maize, sugar beet, soya bean, potato, vegetables and other crops. Its mode of action is by inhibition of acetylcholinesterase activity. Profenofos is a racemic mixture of the two optical isomers at the chiral phosphorus atom. The S (−) isomer is a markedly more potent inhibitor of acetylcholinesterase in vitro than the R (+) isomer.

Profenofos was previously evaluated by JMPR in 1990 (Annex 5, reference *171*) and an acceptable daily intake (ADI) of 0–0.01 mg/kg bw per day was established. The ADI was based on the no-observed-adverse-effect level (NOAEL) of 20 ppm, equal to 1.0 mg/kg bw per day, the highest dose tested, in a three-generation study of reproduction in rats.

Profenofos was re-evaluated by the present meeting within the periodic review programme of the Codex Committee on Pesticide Residues (CCPR). All pivotal studies with profenofos were certified as complying with good laboratory practice (GLP).

Evaluation for acceptable daily intake

1. Biochemical aspects

1.1 Absorption, distribution and excretion

RAI rats (four males and three females) received a single oral dose (gavage) of ring-labelled [^{14}C]profenofos (specific activity, 9.79 μCi/mg (362.23 kBq/mg) at approximately 4.8 mg/kg bw in ethanol : polyethylene glycol 200 : water (25 : 25 : 50). Urine and faeces were collected for 6 days after dosing and selected tissues were removed at necropsy. Essentially all the administered radioactivity was eliminated in the urine (males, 81.8%; and females, 96.4%) and faeces (males, 15.7%; and females, 2.5%) within 6 days. Most of the urinary and faecal excretion occurred within the first 24 h after dosing. The elimination half-life for excretion was less than 8 h for both sexes. Only minor amounts of $^{14}CO_2$ were eliminated in expired air (males, 0.08%; and females, 0.07%). When the rats were killed 6 days after dosing, detectable amounts of radioactivity were present only in the liver (males, 0.013 μg equivalent/g; and females, 0.023 μg equivalent/g) and kidneys (males, 0.007 μg equivalent/g; and females, 0.008 μg equivalent/g). The concentration of radioactivity in blood, fat, muscle, testis, ovary and brain was below the limit of quantification (for blood) or detection (Ifflaender & Mücke, 1974).

Figure 1. Chemical structure of profenofos

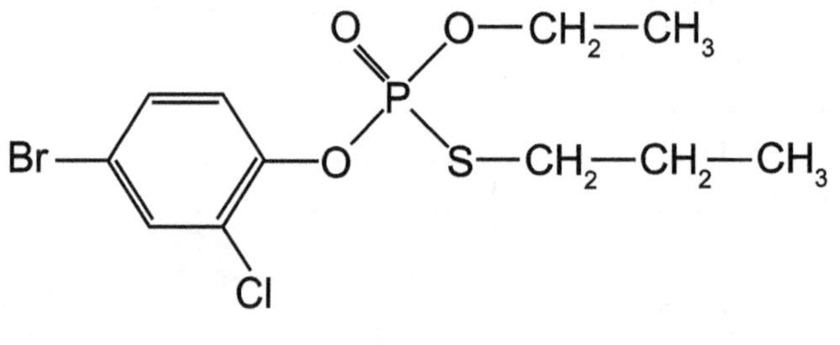

Similar results were obtained in a more recent study in which three groups of five male and five female Crl:CD®BR rats were given [^{14}C]profenofos (purity, 97.1–98.1%) orally by gavage as a single dose at 1 mg/kg bw or 100 mg/kg bw and another group was pre-treated with 14 consecutive daily oral doses of non-radiolabelled compound at 1 mg/kg bw followed by a single radiolabelled dose at 1 mg/kg bw on day 15. Treated rats were killed 7 days after administration of the radiolabelled dose.

No significant sex-related differences were observed in the elimination and/or distribution of radioactivity. Total radioactivity eliminated via the urine and faeces exceeded 99% of the administered dose for all groups. Most of the administered radioactivity (> 97%) was eliminated via the urine. Elimination was rapid with an average of 95% and 93% of the total radioactivity being excreted in the urine within the first 24 h for the groups at the lower dose and repeated dose, respectively. For the group at the higher dose, 97% of the dose was eliminated in the urine within 48 h. Less than 4% of the radioactivity was excreted in the faeces for all groups. Less than 0.2% of the administered dose was detected in the volatile and CO_2 traps combined. Less than 0.1% of the administered dose was recovered in tissues. At the lower dose, the residues in tissues were less than 0.002 µg equivalent/g, with the exception of the liver, which contained up to 0.02 µg equivalent/g. At the higher dose, residues ranged from not detectable in most tissues to 0.18 µg equivalent/g in the liver (Kennedy & Swain, 1992).

Male and female Harlan SD albino rats received single dermal applications of ring-labelled [^{14}C]profenofos at a dose of 0.5 mg/kg bw (specific activity, 9.34 µCi/mg (345.58 kBq/mg) and 10 mg/kg bw (specific activity, 2.6 µCi/mg (96.2 kBq/mg) in a 72-h balance study. Over the 72-h absorption period, the total recoveries of radiolabelled carbon averaged 92–95% of each applied dose in each sex (urine, 80–86%; faeces,, 2.2–3.9%; tissues, 0.09–1.8%; blood, 0.06% or less; treated skin, 3% or less; and cage-washings, 5% or less). Excretion in expired carbon dioxide (CO_2) was negligible (less than 0.02%, as determined from a preliminary study using the highest dermal dose). The dermal absorption was approximately 85% of the applied dose in 72 h, which indicated that [^{14}C]profenofos was absorbed at nearly the same rate for males and females regardless of the dose; t1/2 absorption values were 17.9 h and 15.0 h after treatment with the lower dose in males and females, respectively, and 16.7 h and 14.1 h after treatment with the higher dose in males and females, respectively. The calculated 50% excretion rates (urine was the major route of excretion) occurred 18.1 h and 17.4 h after treatment with the lower dose in males and females, respectively, and 23.2 h and 18.7 h after treatment with the higher dose in males and females, respectively. Fifty percent of [^{14}C]profenofos was excreted shortly after 50% had been absorbed, indicating that profenofos and its metabolites were rapidly excreted, i.e. there was no lag-time between absorption and excretion. Levels of radioactivity in selected tissues, organs (liver, kidney) and blood reached a maximum after 2 h, had reached a plateau by 8 h, and declined rapidly by 72 h after application. The total recovery after 72 h averaged 92–95% of the applied dose in males and females, 80–86% in urine and 2–4% in faeces (Williams et al., 1984).

1.2 Biotransformation

Urine collected in the 0–24 h after giving RAI rats a single oral dose of profenofos at approximately 4.8 mg/kg bw was analysed by thin-layer chromatography (TLC). Four metabolites were detected and no unchanged profenofos was present, indicating complete degradation of profenofos. The only metabolite identified by TLC was the phosphorous ester cleavage product, 4-bromo-2-chlorophenol. This metabolite did not appear in freshly-obtained urine, indicating that other labile metabolites are cleaved to this phenol (Ifflaender & Mücke, 1974).

The metabolism of ring-labelled [^{14}C]profenofos (specific activity, 26.2 µCi/mg (969 kBq/mg) was investigated using the urine and faeces collected over a 2-day period after giving eleven male RAI rats a single oral dose at 28.5 mg/kg bw; 90.4% and 3.6% of the administered dose was excreted in urine and faeces within 24 h. The major metabolites detected were CGA 65867 (7%), CGA 47196 (23%), CGA 55163 (26%) and metabolite C (34%). CGA 47197 was present as a minor metabolite (< 0.5%). Neither unchanged profenofos nor its corresponding phenol CGA 55960 was detected in freshly-obtained urine. The faeces contained small amounts of parent profenofos (2%), and CGA 55960 (1%). Additional metabolites were present in minute amounts (0.2% or less) but were not identified (Ifflaender & Mücke, 1976).

A small amount (approximately 7%) of CGA 55960 was found in the urine of male rats (Tif:RAI-f strain) dosed orally with single doses of ring-labelled [^{14}C]profenofos at 0.19 or 1.80 mg/kg bw. The main urinary metabolites identified in this study were similar to those reported previously by Mücke and colleagues (Mücke, 1986).

The pattern of metabolites was investigated in urine and faeces of groups of five male and five female Crl:CD®BR rats given [^{14}C]profenofos as a single dose at 1 or 100 mg/kg bw, or [^{14}C] profenofos as a single dose at 1 mg/kg bw after 14 consecutive daily oral doses of non-radiolabelled profenofos at 1 mg/kg bw. The metabolites identified in the urine were metabolite C – the sulfated phenol (37–48%), CGA 55163 (22–27%), CGA 47196 (14–25%), CGA 65867 (5–16%), and CGA 55960 (0.1–1.7%). Major faecal components consisted of small amounts of profenofos (< 0.1–1.3%) and CGA 55960 (0.7–1.9%) for all groups. The minor amounts of faecal radioactivity (combined, 0.2–1.3%) contained partially or all of the urinary metabolites (Kennedy & Swain, 1992).

An overview of the proposed metabolic pathways of profenofos in rats is given in Figure 2.

Figure 2. Proposed metabolic pathways of profenofos in the rat

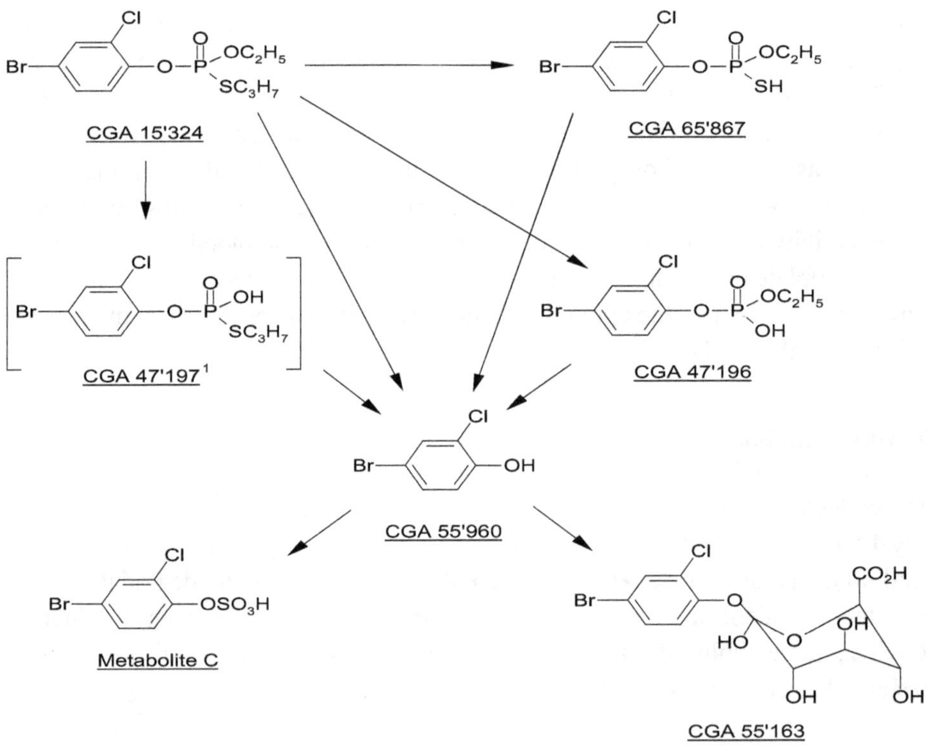

1.3 Effects on enzymes and other biochemical parameters

Profenofos is stereospecifically converted to a more potent inhibitor of acetylcholinesterase by oxidation of the sulfur on the phosphorus by a liver microsomal mixed-function oxidase system in the mouse. The net effect of this was that the chiral S (−) isomer was a 34 times more potent inhibitor of acetylcholinesterase in vitro, while the less toxic R (+) isomer was metabolically deactivated by a factor of 2, via esteratic cleavage. Prior treatment with inhibitors of mixed function oxidase markedly decreased the activation and also protected against inhibition of brain acetylcholinesterase activity and cholinergic symptoms resulting from administration of S-profenofos in chicks (Wing et al., 1983). Acetylcholinesterase activity inhibited by the S-isomer did not reactivate, probably as a result of rapid ageing (Glickman et al., 1984).

Table 1. Results of studies of acute toxicity with profenofos

Species	Strain	Sex	Route	LD_{50} (mg/kg bw)	LC_{50} (mg/l air)	Reference
Rat	Tif: RAI (SPF)	Males and females	Oral	358 (range, 318–403)	—	Bathe (1974a)
Rat	Tif: RAIf (SPF)	Males and females	Oral	502 (range, 391–727)	—	Kobel (1983)
Rat	HSD:(SD)	Males and females	Oral	Males: 492 (range, 363–666) Females: 809 (range, 600–1090) Males and females: 630	—	Kuhn (1990a)
Rat	Sprague-Dawley	Females	Oral	1178 (range, 630–2000	—	Merkel (2004a)
Rat	Sprague-Dawley	Females	Oral	621 (range, 350–1100)	—	Durando (2005a)
Mouse	Tif: MAG (SPF)	Males and females	Oral	298 (range, 268–332)	—	Bathe (1974b)
Rabbit	Russian	Males and females	Oral	700	—	Sachsse (1974a)
Rat	Tif: RAI (SPF)	Males and females	Dermal	3300	—	Bathe (1974c)
Rat	Sprague-Dawley	Males and females	Dermal	> 2000	—	Merkel (2004b)
Rat	Sprague-Dawley	Males and females	Dermal	> 2000	—	Durando (2005b)
Rabbit	Mixed breed	Males and females	Dermal	472 (range, 304–733)	—	Sachsse (1974b)
Rabbit	New Zealand White	Males and females	Dermal	Males: 111 (range, 84.5–145) Females: 155 (range, 124–193) Males and females: 131 (range, 110–156)	—	Cannelongo (1982a)
Rabbit	New Zealand White	Males and females	Dermal	2560 (range, 2190–2990)	—	Kuhn (1989)
Rat	Sprague-Dawley Crl:CD (SD) BR	Males and females	Inhalation (4-h, whole-body)	—	3.36	Horath (1982)
Rat	Wistar-derived Alpk:APfSD	Males and females	Inhalation (4-h, nose-only)	—	> 2.2	Rattray (2004)
Rat	Sprague-Dawley	Males and females	Inhalation (4-h, nose-only)	—	> 2.03	Durando (2005c)

2. Toxicological studies

2.1 *Acute toxicity*

Profenofos has a moderate toxic potential in laboratory animals via the oral route of exposure with median lethal dose (LD_{50}) values ranging from 298 to 1178 mg/kg bw. It is practically non-toxic when applied to the abraded or intact skin in rats and rabbits (LD_{50}, > 2000 mg/kg bw and 3000 mg/kg bw, respectively). However if the compound is massaged into the shaved skin and kept under occlusive dressing it can be moderately toxic to rabbits (LD_{50}, ≥ 131 mg/kg bw). Whole body or nose-only inhalation exposure for 4 h is not lethal to rats (LC_{50}, > 2.0 mg/l). In laboratory animals, single oral or dermal exposure and exposure by inhalation for 4 h causes clinical signs of toxicity that are characteristic of organophosphorus compounds. These include piloerection, salivation, lacrimation, hunched posture, sedation, prostration, tremors and ataxia. The results of studies of acute oral toxicity suggested that the clinical signs characteristic of exposure to organophosphorus compounds were observed at doses greater than 100 mg/kg bw.

(a) Oral administration

Rats

Five studies were available for review. In the first study, groups of five male and five female fasted adult Tif:RAI rats were given profenofos (purity unknown) as a single dose at 215, 278, 359 or 600 mg/kg bw by gavage in 2% aqueous carboxymethylcellulose. Treated rats were examined by gross necropsy at the end of a 14-day observation period. The study was not conducted in accordance with GLP regulations.

Within 48 h, 100% mortality was observed at 600 mg/kg bw, while only one male and two females died at 359 mg/kg bw. Clinical signs, seen in all dosed groups within 2 h of treatment, but more accentuated with increased doses, were sedation, dyspnoea, exophthalmos, ruffled fur, trismus tonic-clonic muscle spasms and hunched body posture. No other details of clinical signs (incidence or duration of observation) were provided in the study report. Surviving rats recovered within 6–7 days. No effects on body weights were observed within 14 days. Necropsy of rats that were killed or that died during the study did not show any substance-related gross changes to organs. The calculated oral LD_{50} (with 95% confidence interval) for males and females combined was 358 (95% CI, 318–403) mg/kg bw (Bathe, 1974a).

In a second study, groups of five male and five female fasted Tif:RAIf rats were given profenofos (purity, 88.5%) as a single dose at 100, 200, 450 or 800 mg/kg bw by gavage in 0.5% aqueous carboxymethylcellulose. Rats were observed for a 14 days. This study was not conducted in accordance with GLP regulations.

All rats of the group at 800 mg/kg bw and one male and one female from the group at 450 mg/kg bw died within 3 days, while surviving rats recovered from all clinical signs within 13 days. Clinical signs seen within 1 h of dosing in all groups were similar to those observed in the previous study. At 100 and 200 mg/kg bw, the clinical signs were non-specific. Clinical signs in rats in the control group were not reported. Rats in the groups at 450 and 800 mg/kg bw exhibited tremors, sedation and loss of postural tone. The surviving rats gained weight during the study period. The oral LD_{50} in male and female rats was calculated to be 491 (95% CI, 325–1016) and for males and females combined it was 502 (95% CI, 391–727) mg/kg bw (Kobel, 1983).

In the third study, groups of five male and five female fasted Harlan Sprague-Dawley rats were given undiluted profenofos technical (purity unknown) as a single dose at 200, 400 or 1000 mg/kg bw by gavage. An additional five females received profenofos at a dose of 1500 mg/kg bw. After a 14-day observation period, gross necropsy was performed. This study was conducted in accordance with GLP regulations.

Most rats in the groups at 1500 (five out of five) and 1000 mg/kg bw (nine out of ten) died within the first 3 days after treatment. At 200 and 400 mg/kg bw, piloerection was seen in all rats within 0.5 h after dosing. Diarrhoea was seen in one male at 400 mg/kg bw, 6 h after dosing. No other clinical signs were observed during a 14-day observation period in the groups at 200 and 400 mg/kg bw. After 0.5 h from dosing, prominent clinical signs seen in the groups that received 1000 and 1500 mg/kg bw included piloerection, decreased activity, aggression, ataxia, tremors, diarrhoea, exophthalmus, gasping, lacrimation, nasal discharge, polyuria, prolapsed penis and salivation. These symptoms disappeared within 6 days in surviving rats. Body weights were unaffected by the treatment. At gross necropsy, discoloration of the gastrointestinal contents and gastrointestinal tract distended with gas were further findings. The acute oral LD50 values in rats were 492 (95% CI, 363–666) for males and 809 (95% CI, 600–1090) mg/kg bw for females. For males and females combined, the oral LD_{50} was calculated to be 630 mg/kg bw (95% CI undefined) (Kuhn, 1990a).

Two studies had been conducted in fasted rats using the "up-and-down" procedure.

In the first study, there were no signs of toxicity in one female Sprague-Dawley rat that received profenofos (purity, 92.1%) as a single dose at 199 mg/kg bw. Three female rats were subsequently given profenofos as single dose at 630 or 2000 mg/kg bw and observed for 14 days. The study was conducted in accordance with GLP regulations.

All rats at 2000 mg/kg bw died with 1 day of dosing. Rats at 2000 and 630 mg/kg bw exhibited clinical signs of toxicity that consisted of ocular discharge, facial staining, hypoactivity, hunched posture, anogenital staining and reduced faecal volume. There were no mortalities and no clinical signs of toxicity 10 days after treatment. All rats gained weight over the 14-day experimental period but at 630 mg/kg bw one rats lost weight between days 0 and 7. Gross necropsy of rats that died in the group at 2000 mg/kg bw revealed discoloration of the intestines. The LD_{50} for males and females combined was 1178 (95% CI, 630–2000) mg/kg bw (Merkel, 2004a).

In the second study using the "up-and-down" procedure, three female Sprague-Dawley rats, fasted overnight, were given profenofos (purity, 89.0%) as a single dose at 350 or 1100 mg/kg bw . Treated rats were examined by gross necropsy at the end of a 14-day observation period. This study was conducted in accordance with GLP regulations. All rats that received 1100 mg/kg bw died within 2 days of dosing. Ocular discharge, hypoactivity, abnormal posture, piloerection, anogenital staining and reduced faecal volume were noted before death and necropsy revealed discoloration of the intestines. There were no signs of toxicity amongst the rats at 350 mg/kg bw; all rats gained weight The LD_{50} for female rats was calculated to be 621 (95% CI, 350–1100) mg/kg bw (Durando, 2005a).

Mice

Groups of five male and five female fasted Tif:MAG mice were given profenofos (purity unknown) as a single dose at 215, 278, 317 or 464 mg/kg bw by gavage formulated in 2% aqueous carboxymethyl cellulose. Mice were observed for 14-days. The study was not conducted in accordance with GLP regulations. At 215, 278, 317 and 464 mg/kg bw, 1 out of 10, 3 out of 10, 6 out of 10 and 10 out of 10 mice died within 48 h. Surviving rats showed sedation, dyspnoea, and exophthalmus, curved position and ruffled fur within 2 h of dosing. At 464 mg/kg bw, tonic-clonic muscle spasms were observed. No other details of clinical signs (incidence and duration of observation) were provided in the study report. Surviving rats recovered within 4–12 days. At necropsy, no gross changes in organs were seen. The oral LD50 for male and female mice was calculated to be 298 (95% CI, 268–332) mg/kg bw (Bathe, 1974b).

Rabbits

Groups of two male and two female Russian-breed rabbits were given profenofos (purity unknown) as a single dose at 100, 600, 1000 or 2150 mg/kg bw in 2% aqueous carboxymethyl cellulose by oral intubation. Rabbits were observed for only 7 days. The study was not conducted in accordance with GLP regulations. Within 24 h, all rabbits at 2150 mg/kg bw and one male and two females at 600 mg/kg bw had died; only one male at 600 mg/kg bw died at 48 h after dosing. None of the rabbits in the group at the lowest dose (100 mg/kg bw) died or showed any signs of toxicity. In other groups, within 2–3 h after dosing, rabbits showed salivation, ataxia, exophthalmus, lateral position and sedation, these being more accentuated in groups at higher doses. No other details of clinical signs (incidence and observation duration) were provided in the study report. The surviving rabbits recovered within 4–12 days. Necropsy of these animals revealed no gross changes to organs; congested organs were noted in rabbits that died during the study. The oral LD50 for male and female rabbits was calculated to be approximately 700 mg/kg bw (95% confidence interval undefined) (Sachsse, 1974a).

(b) Dermal administration

Rats

Groups of three male and three female Tif:RAI rats received a single application of profenofos (purity unknown) at a dose of 2150, 2780 or 3170 mg/kg bw. Profenofos was applied as a concentrate under occlusive conditions (with a plaster and aluminium foil) to the shaved back of the rat. After 24 h, the plaster and aluminium foil were peeled off and the skin was washed with warm water. Rats were observed for clinical signs and mortality for 14 days. This study was not conducted in accordance with GLP regulations. Two male rats at the highest dose died within 7 days after application of the substance. There were no deaths in other groups. Within 24 h after treatment, rats at all doses showed sedation, dyspnoea, loss of postural tone, and ruffled fur. In the group at the highest dose, slight erythema was observed. Surviving rats recovered within 5–8 days. At necropsy, no substance-related gross changes to organs were seen. The dermal LD_{50} was calculated to be approximately 3300 mg/kg bw in males and females (Bathe, 1974c).

In two more recent studies, groups of five male and five female Sprague-Dawley rats were given profenofos (two batches; purity, 92.1%; and purity, 89.0%) at a dose of 2000 mg/kg bw applied once to the skin for 24 h. These studies were conducted in accordance with GLP regulations. In the first study, there were no signs of toxicity or dermal irritation and no treatment-related gross abnormalities were noted in any rat at necropsy 14 days after treatment (Merkel, 2004b). All rats gained weight during the observation period. Similar results were obtained in the second study except that dermal irritation (erythema and oedema) was noted in four of the female rats 1 and 2 days after treatment. Neither the severity of irritation nor the scoring method was given in the study report. In the second study, the lot number and purity of profenofos was different to that used in the first study. The dermal LD_{50} was > 2000 mg/kg bw in male and female rats (Durando, 2005b).

Rabbits

Dermal toxicity was investigated in mixed-breed rabbits. Groups of three male and three female rabbits were given profenofos (purity unknown) as a single dose at 215, 464 and 1000 mg/kg bw. The test substance was applied as a concentrate under occlusive conditions (plaster and aluminium foil) to the shaved back of the rats for 24 h. After 24 h, the plaster and aluminium foil were peeled off and the skin was washed with warm water. Rabbits were observed for 14 days. The study was not conducted in accordance with GLP regulations. Three males and two females at the highest dose (1000 mg/kg bw) and two males and two females at the intermediate dose (464 mg/kg bw) died within 7 days after application of the substance. Twenty-four hours after treatment, rabbits at the intermediate and highest dose showed ataxia, tremor, salivation, loss of postural tone, and sedation. Surviving rabbits recovered within 12 days. No other details of clinical signs (incidence and observation duration) were provided in the study report. At necropsy, congested organs and bleeding along the gut were seen. There were no adverse effects in rabbits that received 215 mg/kg bw. The dermal LD_{50} was calculated to be 472 (range, 304–733) mg/kg bw (Sachsse, 1974b).

In another study, dermal toxicity was determined in New Zealand White rabbits with intact and with abraded skin. Doses ranged between 72.6 and 2000 mg/kg bw. Profenofos (unknown purity) was applied once as a concentrate and gently massaged into the shaved back of the trunk of test animals and then kept under occlusive conditions for 24 h. In all, 45 females and 50 males were used for this study (five males and five females per group, or five males or five females per group). Gross necropsy was performed after a 14-day observation period. This study was not conducted in accordance with GLP regulations. In addition to some of the symptoms noted in the previous study, signs of activity decrease, constricted pupils, decreased or no defecation, decreased or no urination, diarrhoea, emaciation, lacrimation, ptosis, and small faeces were noted throughout the observation period beginning with day 2 after treatment. Findings on gross necropsy were discoloured organs and again haemorrhagic areas along the gastrointestinal tract. There was no difference between groups with abraded vs intact skin. The acute dermal LD50 for male and female rabbits were 111 (95% CI, 84.8–145) mg/kg bw for rabbits with intact skin and 155 (95% CI, 124–193) mg/kg bw for rabbits with abraded skin. The overall dermal LD50 was 131 (95% CI, 110–156) mg/kg bw (Cannelongo, 1982a).

In a further study, groups of five male and five female albino rabbits were given a single application of profenofos (purity unknown) at a dose of 250, 2010, 2300 and 2600 mg/kg bw applied as a concentrate under a semi-occlusive dressing to the dorsal surface of the trunk for 24 h. Rabbits were observed for 14 days. This study was conducted in accordance with GLP regulations. Beginning on day 1 through 12 days after treatment, one out of five, two out of five, four out of five, five out of five rabbits died at doses of 250, 2010, 2300 and 2600 mg/kg bw, respectively. Clinical signs found were similar to those observed previously and gross examination at necropsy revealed discoloured and gas-distended gastrointestinal tract as well as empty intestinal tract. The dermal LD50 was calculated to be 2560 (range, 2190–2990) mg/kg bw (Kuhn, 1989).

These apparently divergent results in studies of acute dermal toxicity in rabbits were likely to be due to the differences in dressings (occlusive vs semi-occlusive) and application methods used (abraded skin vs massaging the substance into the skin). In rats, there was no indication of increased toxicity via dermal administration.

(c) Exposure by inhalation

Four groups of five male and five female Sprague-Dawley (Crl:CD(SD)BR) rats were exposed (whole body) to aerosol atmospheres of profenofos (purity, 90.5%) at a nominal concentration of 2.23, 2.77, 4.57 or 6.30 mg/l (analytical concentrations of 2.31, 3.10, 4.51, and 6.30 mg/l, respectively) for 4 h. The particle size of the test aerosol was 2.04 μm (mass median aerodynamic diameter, MMAD) with geometric standard deviation of 1.84. Rats in the control group were exposed to air only. At the end of the 14-day observation period, rats were examined by gross necropsy. The study was conducted in accordance with GLP regulations. The mortalities observed at 2.31, 3.10, 4.51, and 6.30 mg/l were three out of ten, four out of ten, five out of ten and ten out of ten, respectively. During exposure, shortly after exposure and after exposure, damp fur, irregular breathing, unkempt fur, yellow/brown stained fur, alopecia, crusty muzzle, salivation, crusty nose, lacrimation, prostration, ataxia, exophthalmos, gasping and tremors were observed among the rats. At necropsy, abnormalities of the lung, eyes, stomach, skin, kidneys, spleen, intestine and external surface were seen. The median lethal concentration (LC50) (4-h) for males and females combined was calculated to be 3.36 mg/l air (Horath, 1982).

Groups of five male and five female Wistar-derived Alpk:AP_fSD rats were exposed to profenofos (purity, 92.1%) at a nominal concentration of 2 mg/l (analytical concentration, 2.2 mg/) by nose-only inhalation for 4 h. The particle size of the test aerosol ranged from 3.01 to 3.51 µm MMAD, with geometric standard deviation in the range of 1.59–1.65. This study was conducted in accordance with GLP regulations. There were no deaths and only transient signs of respiratory irritation (wet fur, salivation, piloerection, haunched posture) were observed in all rats during and immediately after exposure. All males and all except three females had gained weight by the end of the 14-day experimental period and no treatment-related effects were found at necropsy. The LC_{50} (4-h) of profenofos for males and females combined was > 2.2 mg/l air (Rattray, 2004).

In a similar design of study to the above, groups of five male and five female Sprague-Dawley rats were exposed (nose only) to profenofos (purity, 89.0%) at an analytical concentration of 2.03 mg/l for 4 h. The particle size of the test material was 2.5 µm (MMAD). This study was conducted in accordance with GLP regulations. During the 14-day observation period, there were no signs of gross toxicity, abnormal behaviour, mortality and no treatment-related effects on body weight or on gross findings at necropsy. The LC_{50} (4-h) for male and female rats was > 2.03 mg/l air (Durando, 2005c).

(d) Dermal irritation

In a study of primary skin irritation, 0.5 ml of undiluted technical profenofos (purity, 90.4%) was applied under semi-occlusive dressing to the intact skin of New Zealand White rabbits for 4 h. This study was conducted in accordance with GLP regulations. Erythema was present at each observation time until day 21; and oedema was present until day 17 after application. The mean score for erythema formation at 24, 48 and 72 h was 1.94; that for oedema was 1.61. The primary irritation index was 3.3 (Kuhn, 1990b).

Three New Zealand Albino rabbits received 0.5 ml of undiluted profenofos (purity, 92.1%) applied to the skin for 4 h. This study was conducted in accordance with GLP regulations. Very slight to well-defined erythema and very slight oedema were found at the application site in all rabbits 1 h after the end of treatment, but the effects had disappeared within 3 days. The primary dermal irritation index was 1.2 (Merkel, 2004c). Profenofos was not irritating to the rabbit skin.

In a third study, a single application of 0.5 ml of profenofos (purity, 89.0%) was made to the skin of three New Zealand Albino rabbits for 4 h. This study was conducted in accordance with GLP regulations. Well-defined erythema and slight oedema were again noted 1 h after the treatment period. There were no signs of irritation on day 7, but desquamation was observed at all application sites. The primary dermal irritation index was 3.1 and profenofos was classified as moderately irritating to the skin (Durando, 2005d).

(e) Ocular irritation

In a study of primary eye irritation, 0.1 ml of undiluted profenofos (purity, 90.5%) was instilled into the conjunctival sac of the eyes of New Zealand White rabbits. A statement of quality assurance was provided in the study report. However, no statements were made in the report regarding GLP compliance. At 1 h after instillation, the maximum average irritation score was 4.3 and 4.0 for the unwashed and washed eyes. The reactions were limited to a slight transient conjunctivitis (Cannelongo, 1982b).

Similar results were obtained in two more recent studies.

Conjunctivitis was observed in all three New Zealand Albino rabbits 1 h after instillation of 0.1 ml of profenofos (purity, 92.1%). In this study, the conjunctivitis had disappeared within 24 h of treatment. Corneal opacity was noted in one rabbit, but only at 24 h after treatment and there was no evidence of iritis. Profenofos was classified as non-irritating (Merkel, 2004d).

In the other study, neither corneal opacity nor iritis was noted at any time after treatment with profenofos (purity, 89.0%) Profenofos was classified as mildly irritating to the eye (Durando, 2005e). Both studies were conducted in accordance with GLP regulations.

(f) Dermal sensitization

Guinea-pigs

A study of dermal sensitization was performed on Pirbright White guinea-pigs (Tif:DHP) according to the maximization test protocol of Magnusson & Kligman, which uses addition of Freund adjuvant. This study was conducted in accordance with GLP regulations. The induction phase was performed with an intradermal injection of a 3% solution of profenofos (purity, 91.2%) and an epidermal application of a 30% solution of profenofos, both in peanut oil. The guinea-pigs were challenged with a 5% solution of profenofos in Vaseline. Under these conditions, profenofos showed a sensitization rate of 45% (Winkler, 1996). The Meeting therefore concluded that profenofos may cause sensitization by skin contact.

Mice

The skin sensitization potential of profenofos has also been investigated in the local lymph-node assay. This study was conducted in accordance with GLP regulations. Profenofos (purity, 92.1%) was applied as a 1%, 2.5% or 5% preparation in acetone : olive oil (4 : 1) to the ears of groups of four female CBA/Ca mice. The application was repeated on three consecutive days. Application of the 1% preparation in acetone caused an increase of greater than threefold in the incorporation of [3H-methyl]thymidine into lymph-node cells and profenofos was therefore a sensitizer under the conditions of the study. The groups of mice given the 2.5% and 5% preparations were killed early due to toxicity (Betts, 2005).

In a similar study, the ears of three groups of five female CBA/J mice were treated with 0.5%, 1% or 2% profenofos (purity, 97.3%) in acetone : olive oil (4 : 1) for 3 days. This study was conducted in accordance with GLP regulations. A stimulation index of > 3 was found in mice treated with 2% profenofos, and profenofos was therefore considered to be a sensitizer under the conditions of the study (Kuhn, 2005).

2.2 *Short-term studies of toxicity*

Erythrocyte acetylcholinesterase activity was found to be significantly more sensitive to profenofos than was brain acetylcholinesterase activity in rats, mice, rabbits, and dogs. However, in no species were any signs of toxicity seen at doses that did not also produce significant inhibition of brain acetylcholinesterase activity. Hence it was concluded that brain acetylcholinesterase activity was the more appropriate indicator of the toxic potential of profenofos.

Rats

Thirteen groups of 25 male and 25 female COBS®CD®F/Cr1BR F344 rats were fed diets containing profenofos (purity, 90.6%) at a concentration of 0 (two groups), 0.01, 0.03, 0.1, 0.3, 1.0, 3.0, 10, 30, 100, 300 or 1000 ppm, equal to intakes of 0, 0.001, 0.0025, 0.009, 0.025, 0.090, 0.24, 0.87, 2.4, 8.4, 22 and 85 mg/kg bw per day, respectively. Five males and five females in each test group and in both control groups were killed after 2 and 4 weeks. The remaining rats in one control group and in the groups at 0.01, 0.03, 0.1, 1, 10, 100 and 1000 ppm were killed after 8 weeks. Rats in the other control group and in the test groups at 0.3, 3.0, 30 and 300 ppm were killed after 13 weeks. Stability and dietary concentrations were confirmed analytically. Rats were observed daily for mortality and signs of moribundity. Clinical signs, body weights and food consumption were monitored weekly. Blood and urine was collected from rats in all groups at weeks 2 and 4 for haematology and clinical chemistry investigation. Cholinesterase activity was determined in plasma, erythrocytes and brain (2, 4 and 13 weeks), and differences were analysed statistically only for those rats treated for 13 weeks. Histopathological examination was not conducted in this study.

The test compound was stable in the diet over the duration of the feeding period and the concentrations were within 20% of the target. However, at < 1 ppm the concentrations generally exceeded the target.

There were no treatment-related mortalities or clinical signs in any of the test groups. Reduced food consumption and a dose-related growth depression occurred at doses of 100–1000 ppm.. At week 4, the treatment-related depression on growth was slight at 100 ppm (males, 10%; and females, 14%), but marked (males, 45%; and females, 38%) at 1000 ppm. At week 13, reduced body-weight gains were noted at 300 ppm (males, 15%; and females, 17%). Results of haematology, clinical chemistry (without cholinesterase measurements) and urine analysis were unremarkable and did not indicate any relationship to treatment. The activities of plasma and erythrocyte acetylcholinesterase showed a dose-related inhibition in males and females at 10, 30, 300 and 1000 ppm, being more pronounced in females, and generally reached a plateau after 4 weeks. Plasma cholinesterase activity was inhibited by 6–70% in males and 26–90% in females at 10–1000 ppm. The corresponding figures for inhibition of erythrocyte acetylcholinesterase activity were 33–84% in males and 30–84% in males. Slightly lower brain acetylcholinesterase activity was present in the males at 300 ppm (14%) and in the females (11%) at week 13, but this was not statistically significant. At 1000 ppm, brain acetylcholinesterase activity was inhibited by 30% in males and 29% in females. All inhibition data were been calculated based on values for concurrent controls. Necropsies performed during or at the end of the study did not reveal any treatment-related changes that were observable on gross examination.

The NOAEL for inhibition of brain acetylcholinesterase activity was 300 ppm (equal to 22 mg/kg bw per day) on the basis of inhibition of brain acetylcholinesterase activity (males, 30%; and females, 29%) at 1000 ppm (equal to 85 mg/kg bw per day). The NOAEL for other signs of toxicity (excluding cholinesterase activity) was 100 ppm (equal to 8.4 mg/kg bw per day) on the basis of reduced food intake and reduced growth rate at 300 ppm (equal to 22.0 mg/kg bw per day). The study author established a no-observed-effect level (NOEL) of 3 ppm (equal to 0.24 mg/kg bw per day) on the basis of depression of plasma and erythrocyte acetylcholinesterase activity (Piccirillo, 1978). The study was not conducted in accordance with GLP regulations and did not conform to any particular regulatory guideline.

In a 21-day study of toxicity after exposure by inhalation, groups of nine male and nine female RAI albino rats were exposed by nose-only inhalation to profenofos (purity, 89.9%) for 6 h per day, 5 days a week for 3 weeks. The nominal exposure concentrations were 0, 68, 219 and 449 mg/m3. Four males and four females in the control group and the group at 219 mg/m^3 were kept for a 21-day recovery period without further treatment. The particle size of the test aerosol (45–70%) ranged from < 1 to 3 µm (MMAD). Mortalities, appearance, behaviour, signs of toxicity and body weight were monitored daily. Food consumption was recorded weekly. Haematology and blood chemistry measurements were carried out at the end of the treatment and recovery periods. The blood chemistry determinations included cholinesterase activity in plasma and erythrocytes, which was calculated relative to the values for the control group. Rats were killed at the end of the treatment or recovery periods and subjected to a macroscopic examination that included weighing selected organs; tissues were taken for microscopic examination and acetylcholinesterase activity was determined in brain.

All rats of the group at the highest dose and one female of the group at the intermediate dose died during the first week. Clinical signs such as exophthalmos, dyspnoea, tremor, ruffled fur, lateral position and irritation/secretion of the mucous membranes of the eyes and nose were seen after 2 days in the rats at the highest dose (449 mg/m^3). In the group at the intermediate dose and, to a lesser extent also, in the group at the lowest dose, food intake and body-weight gain was impaired, but returned to normal during the recovery period. The results of the haematological and blood chemistry analysis were generally unremarkable for treated and control rats. In rats exposed to 68 and 219 mg/m^3 plasma cholinesterase activity was inhibited by > 35% and > 47% in males and females, respectively. At these same doses, erythrocyte acetylcholinesterase activity was inhibited by > 64% and > 53% in males and females, respectively, and brain acetylcholinesterase activity was similarly inhibited by > 61% and > 50% in males and females respectively. A minor but statistically significant depression in plasma protein concentration was also observed. The effect on erythrocyte acetylcholinesterase activity was reversible, although plasma and brain acetylcholinesterase activity in rats exposed at 219 mg/m3 remained at about 25% below control levels 21 days after the end of treatment. Compound-related pathological findings in the group at the highest dose included acute conjunctivitis, congestion of the nasal mucous membrane, and, in the majority of rats, severe interstitial or purulent keratitis. All surviving rats of the group at 219 mg/m^3 were slightly emaciated. Rats in the group at the lowest dose showed only incidental macroscopic and microscopic findings not related to the inhalation of profenofos (Sachsse, 1977).

On the basis of a significant reduction of brain acetylcholinesterase activity in all treated groups a no-observable-adverse concentration (NOAEC) could not be identified for this study. This study was not conducted in accordance with GLP standards, but the study design was broadly similar to the OECD guideline 412 for a "repeated dose inhalation toxicity: 21 or 14-day study".

Rabbits

Three short-term studies of dermal toxicity were available for review.

In the first study, groups of three male and three female Himalayan rabbits were given a mixture containing profenofos (purity, 89.8%) diluted with a polyethylene glycol and saline (70 : 30, w/w) applied daily to the shaved skin. Doses of 0, 5, 20 and 100 mg/kg bw per day were applied under occlusive dressing during 24 h on 5 days per week for 3 weeks. One male and one female per dose was reserved for a 3-week recovery period. Rabbits were observed daily for clinical signs of toxicity and skin irritation; food consumption and body weight were determined weekly. Acetylcholinesterase activity was measured in plasma and erythrocytes at 4, 10 and 21 days after the start of treatment and, in the rabbit in the recovery group at 5, 11 and 21 days after cessation of treatment. Haematology and other blood chemistry parameters were determined before testing and at the end of the treatment and recovery periods. The rabbits were subjected to a gross necropsy, and selected organs were weighed and processed for microscopic examination. Brain acetylcholinesterase activity was also determined. Cholinesterase activity was calculated as percent of values for the concurrent control group and no statistical analysis was conducted.

All of the rabbits treated with profenofos at 100 mg/kg bw per day died within 6 days after the start of dosing. These rabbits displayed moderate erythema and oedema of the skin, reduced food intake and body-weight gain and various clinical signs of toxicity (dyspnoea, salivation, tremors, ataxia, sedation, and curved position) within 3 days after the start of treatment. Inhibition of plasma cholinesterase activity was increased by 58% and 91% at 5 and 20 mg/kg bw per day, respectively. Inhibition of erythrocyte acetylcholinesterase activity was 20–36% in males and 24–28% in females at 5 mg/kg bw per day and 38–61% in males and 20–49% in females at 20 mg/kg bw per day. At the end of the treatment period, inhibition of brain acetylcholinesterase activity in the rabbits at 5 mg/kg bw per day was 20% and 21% in males and females, respectively. Inhibition was 42% and 30% in males and females, respectively, in rabbits at 20 mg/kg bw per day. Cholinesterase activity recovered rapidly and was close to control levels 21 days after the end of treatment. At necropsy, congestion of the internal organs, particularly liver, was noted in the rabbits at the highest dose. Histopathological examination of the rabbits at the highest dose showed focal hypertrophy, haemorrhages and fatty changes in the hepatocytes accompanied by necrosis of the liver parenchyma, and slight to moderate atrophy of lymphoid and thymic tissue. Microscopic examination of the skin at the site of dermal application showed oedema, minute haemorrhages, small intradermal pustules, focal acanthosis and parakeratosis. In the rabbits at the lowest and intermediate doses (5 and 20 mg/kg bw per day) the observed changes consisted of slight erythema and oedema at the application site. These effects disappeared during the recovery period. No other unusual findings were observed in these two groups.

Although the number of rabbits used was small, the lowest-observed-adverse-effect level (LOAEL) for inhibition of brain acetylcholinesterase activity was 5 mg/kg bw per day. A NOAEL was not identified in this study (Sachsse, 1976). This study was not conducted in accordance with GLP regulations or any regulatory guideline.

In the second study, profenofos (purity, 92%) suspended in purified water containing 0.5% Tween 80, was applied to the shaved intact skin of groups of five male and five female HAR:PF/CF New Zealand White rabbits. Doses of 0, 0.05, 1 or 10 mg/kg bw per day were applied once under semi-occlusive dressing made of a porous gauze, during 6 h for 5 days per week during a 3-week period. The rabbits were examined daily for clinical signs of toxicity and skin reactions. Body weights and food consumption were recorded weekly. Haematology and clinical chemistry parameters were measured before dosing and at study termination. Inhibition of cholinesterase activity was calculated as percentage of values for the concurrent control group. Gross postmortem examination was conducted at the end of the study; selected tissues were weighed and/or taken for histopathology.

Treatment-related observations among the rabbits at the highest dose included hyperactivity, diarrhoea and soft faeces. Slight local reactions (erythema) were noted in all groups including controls and a well-defined erythema was found in the groups at the intermediate and highest dose during the third week of treatment. Treatment with profenofos at 10 mg/kg bw per day resulted in a 27% and 41% inhibition of plasma cholinesterase activity in males and females, respectively. Erythrocyte acetylcholinesterase activity was statistically significantly inhibited in both sexes, although the inhibition in males (12%) was much less than that in females (28%). Inhibition of brain acetylcholinesterase activity was statistically significant in males of the group at the highest dose (30%) but not in females (15%). Brain acetylcholinesterase activity was also inhibited in rabbits at 1 mg/kg bw per day (15%; not statistically significant). Minor alterations in serum total bilirubin and sodium concentrations as well as gamma-glutamyl transferase activities were also noted in this group.

There were no toxicologically significant effects in the rabbits at the intermediate and lowest dose. The NOAEL was 1 mg/kg bw per day on the basis of inhibition of brain acetylcholinesterase activity in males (30%) and hyperactivity at 10 mg/kg bw per day (Johnson et al., 1984). This study was conducted in accordance with GLP regulations and was also in compliance with the then current version of the OECD guidelines for testing of chemicals.

In a third study of dermal toxicity, groups of 10 male and 10 female New Zealand White rabbits were given repeated dermal applications of profenofos (purity, 91.2%) at a dose of 0, 2.5, 5 or 10 mg/kg bw per day diluted with distilled water containing 0.5% Tween 80, applied under semi-occlusive conditions for 22 days for 5 days per week during weeks 1 and 2 and daily on days 15–22. The exposure period was 6 h per day. Clinical signs, body weight, food consumption and mortality were monitored throughout the study. Ophthalmological examinations were performed before the start and on the last day of treatment. Haematology and clinical chemistry analyses were also carried out on the last day of treatment. These included determination of acetylcholinesterase activity in erythrocytes and brain, and butyrylcholinesterase activity in plasma; inhibition was calculated as percentge of values for the concurrent control group. At termination, rabbits were killed and examined macroscopically; selected organs were weighed and/or examined microscopically.

The treatment produced no significant clinical signs, no effects on mean body weights or on food consumption. No signs of irritation occurred at the application site. No ocular changes were seen and no effects were recorded on haematological parameters. The evaluation of organ weights did not reveal any treatment-related effects. Macroscopic postmortem examination did not indicate any treatment-related changes. Upon microscopic examination, all treated groups were observed to have an increased incidence of acanthosis at the skin application site.

A treatment-related inhibition of cholinesterase activities in plasma, erythrocytes and brain was seen at all doses. At the end of the study there was a statistically significant decrease in plasma cholinesterase activity of 81–86% and in erythrocyte acetylcholinesterase activity of 47–58% at the lowest dose. There was a 33% inhibition of brain acetylcholinesterase in the group at 5 mg/kg bw per day and a 47% inhibition in the group at 10 mg/kg bw per day. The effect was statistically significant at both doses. Brain acetylcholinesterase activity was inhibited by only 14–15% at 2.5 mg/kg bw per day. All other blood chemistry parameters were unaffected by treatment.

The NOAEL for brain acetylcholinesterase activity was 2.5 mg/kg bw per day on the basis of statistically significant inhibition of brain acetylcholinesterase activity in males and females at 10 mg/kg bw per day. The study author concluded that NOAEL for erythrocyte acetylcholinesterase activity could not be determined in this study owing to biologically significant inhibition of erythrocyte acetylcholinesterase activity at the lowest dose (Cantoreggi, 1998). The study was conducted in accordance with GLP regulations and US EPA Health Effects Test Guidelines OPPTS 870.3200 (1996).

Dogs

In a 90-day feeding study, groups of four male and four female beagle dogs were given diets containing profenofos (purity, 94.8%) at a concentration of 0, 2, 20 or 200 ppm (equivalent to 0, 0.05, 0.5 and 5.0 mg/kg bw per day). In addition, one male and one female were added to the untreated control group and to the test group at the highest dose to study recovery during a 28-day period. Stability and test concentrations were confirmed analytically. Body weight and food consumption were recorded weekly, haematology, clinical chemistry and urine analysis was conducted before testing and after 1, 2 and 3 months treatment. At the conclusion of treatment, the dogs were given a complete necropsy; selected organs were weighed and taken for microscopic pathology. Owing to the small sample size, no statistical analysis was conducted. Percentage inhibition of cholinesterase activity was not included in the report, but was been calculated by the Meeting from the tabulated data.

The concentration analyses of the test diets indicated mean concentrations of 96%, 94% and 84% at 2, 20 and 200 ppm, respectively. The stability of the test compound in the diet was demonstrated in a 6-month study in dogs.

There were no treatment-related significant changes observed in food consumption, body weight, haematology, clinical chemistry, urine analysis, ophthalmology, organ weights, organ ratios or macro- and micro-pathology. Plasma cholinesterase activity was inhibited by 32–76% at all dietary concentrations, with no evidence of a dose–response relationship, but returned to normal values during the 28-day recovery period. Inhibition of erythrocyte acetylcholinesterase activity was < 22% in the dogs at 2 ppm. In the group at 20 ppm, inhibition of erythrocyte acetylcholinesterase activity was 13–20% and 30–50% in males and females, respectively. At 200 ppm, there was a 68–80% inhibition of erythrocyte acetylcholinesterase activity in males and females. In dogs in the recovery group, erythrocyte cholinesterase activity remained depressed at about 50% of values before testing (although some recovery was seen with respect to values at 90 days on test). Brain acetylcholinesterase activity in males at 200 ppm was depressed by 21%, but returned to normal at the end of the 28-day recovery period; inhibition of brain acetylcholinesterase activity was < 20% in females at 200 ppm, as well as in males and females at lower dietary concentrations of profenofos. No other treatment-related effects were observed at any dose.

The NOAEL for brain acetylcholinesterase was 20 ppm, equivalent to 0.5 mg/kg bw per day, on the basis of reduced brain acetylcholinesterase activity at 200 ppm, equivalent to 5.0 mg/kg bw per day; the highest dose tested. The NOAEL for other systemic toxicity was 200 ppm, equivalent to 5.0 mg/kg bw per day; the highest dose tested. The study was conducted before development of Federal Insecticide, Fungicide, and Rodenticide Act (FIFRA) test guidelines and GLP regulations. The data had not been independently validated (Burtner et al., 1975).

Since a clear NOAEL was not established in the 90-day study, a follow-on 6-month study was conducted. Groups of seven male and seven female beagle dogs were fed diets containing profenofos (purity, 89.3%) at a concentration of 0, 0.2, 2, 100 or 500 ppm, equivalent to 0, 0.0072, 0.05, 2.9 and 14.4 mg/kg bw per day, for 6 months. One male and one female in each dose group was kept for a 4-week recovery period on control diet. The diet analyses were conducted before initiation and throughout the duration of the study. Dogs were observed daily for clinical signs of toxicity. Food consumption was measured daily, while body weight was monitored weekly. Haematology, clinical chemistry (including plasma and erythrocyte cholinesterase activity and plasma and liver carboxylesterase activities) and urine analysis parameters were determined at 4, 9, 13, 18, 22 and 26 weeks, and also in dogs in the recovery group at 31 weeks. Ophthalmoscopy was performed after 26 and 31 weeks. The dogs in the group at 500 ppm were additionally given a battery of tests to examine central and peripheral neural responses (e.g. muscle strength and tone, reflexes). At necropsy, organs and tissues were examined for gross changes, and subsequently by histopathology. Acetylcholinesterase activity was determined in the brain. Data on inhibition of cholinesterase activity were calculated relative to values for dogs in the concurrent control group.

At study initiation, the concentrations of the test material in the diet at 0.02, 2. 100 and 500 ppm were 115%, 74%, 98% and 98%, respectively. The test compound was stable in the diet at –20°C for > 4 weeks.

Besides a transiently-decreased food intake in the males of the group at 500 ppm, no effects were observed with respect to mortality, clinical signs, including special neurological examination, body-weight development, ophthalmoscopy, hearing tests, urine analysis, organ weights, gross examination and histopathology. Slightly decreased erythrocyte parameters (erythrocyte count, haemoglobin concentration and erythrocyte volume fraction) in the group at 500 ppm were the only haematological findings. These haematological findings were not considered to be toxicologically significant since the magnitude of changes were minor, and were within the range for historical controls. Clinical chemistry results indicated that plasma cholinesterase activity was inhibited by 45–64% in both sexes at 2 ppm, 62–79% at 100 ppm and 80–90% at 500 ppm. Erythrocyte acetylcholinesterase activity was inhibited at 100 and 500 ppm by 52–80% and 68–95%, respectively. The inhibition of plasma or erythrocyte acetylcholinesterase activity was essentially unchanged from week 4 until the end of the treatment period. Plasma and liver carboxylesterase activities were decreased in males of the groups at 100 and 500 ppm. The effects on plasma and erythrocyte acetylcholinesterase activity showed a tendency to reverse during the 4-week recovery period. Brain acetylcholinesterase activity was inhibited by only 5% at 2.0 ppm in males; in females, inhibition was 8%, 10%, 11%, and 5% at dietary concentrations of 0.2, 2.0, 100 or 500 ppm, respectively. No data for the recovery group were available for inhibition of brain acetylcholinesterase activity.

The potential oculotoxic hazard of profenofos was assessed further through re-evaluation of relevant data from this 6-month study in dogs. No ophthalmological findings related to treatment were observed. Histological examination of the eyes was unremarkable and some of the findings were confined to dogs in the control group. Re-evaluation of haematology and blood chemistry data confirmed the results outlined above, that profenofos is not toxic to the ocular or haematological system in dogs.

The NOAEL for inhibition of brain acetylcholinesterase activity was 500 ppm, equivalent to 14.4 mg/kg bw per day; the highest dose tested. The study was not conducted in accordance with GLP regulations and there was no claim of compliance with a regulatory guideline. However, a protocol was available and the study design was consistent with OECD guideline 409 (Gfeller, 1981).

In another study, groups of four male and four female beagle dogs were given gelatin capsules containing profenofos (purity, 91.2%) at a dose of 0, 0.015, 0.05, 1 or 12.5 mg/kg bw per day mixed with lactose for 1 year. Six mixtures of each concentration were prepared for the 1-year period and were stored at 0–5°C. Clinical signs and food consumption was monitored daily and the dogs were weighed weekly. Eye examinations were conducted and tonometry measurements made before testing, and at weeks 7 (tonometry only), 13, and 26, and towards the end of the treatment period. Neurological, clinical and haematological investigations were made at weeks 13, 26 and 52. This included determination of acetylcholinesterase activity in erythrocytes and butyrylcholinesterase activity in plasma. The dogs were given a gross examination at necropsy and selected organs were weighed. A microscopic examination was also conducted and acetylcholinesterase activity was determined in brain. Cholinesterase activity was analysed statistically and inhibition calculated relative to the values measured before testing.

The analyses of dietary concentration were performed at higher concentrations. The results showed that the mean concentrations at 1000 and 50 000 ppm (29 and 1450 mg/kg bw per day, respectively) were 90% and 97.6% of the nominal concentrations, respectively. The test compound was homogeneously distributed and was stable in lactose powder (bulk or in capsules) for 10 weeks under the conditions of the study.

There were no treatment-related deaths in this study. No treatment-related changes were observed in behaviour, food consumption, eye examination, intraocular pressure, neurological examination, urine analysis, organ weights, and macroscopic examinations at necropsy. Mean body-weight gain was slightly depressed at 12.5 mg/kg bw per day, mainly due to results for one female.

Haematology parameters, namely erythrocyte count, haemoglobin concentration and erythrocyte volume fraction, were reduced in males and females at 12.5 mg/kg bw per day and in males at 1 mg/kg bw per day. Mean erythrocyte volume was elevated in males at 12.5 mg/kg bw per day. At weeks 26 and 52, a slightly higher platelet count was recorded in males at 1 and 12.5 mg/kg bw per day. The changes in the haematological parameters were not considered to be toxicologically significant since there was no clear dose–response relationship and the magnitude of changes observed were within the range for historical controls. Small changes in blood chemistry were noted in the groups at 1 or 12.5 mg/kg bw per day. They included reduced concentrations of plasma protein and albumin and, as a result, a reduced albumin : globulin ratio. Reductions in calcium concentration paralleled those of plasma albumin owing to its calcium-binding properties. Plasma glucose concentrations were decreased, and alkaline phosphatase activities increased. Changes in the clinical chemistry parameters were not considered to be toxicologically relevant.

There was a 54–68% inhibition of erythrocyte acetylcholinesterase activity in dogs at 1 mg/kg bw per day and an 82–86% inhibition at 12.5 mg/kg bw per day. At both doses the inhibition was statistically significant. There was no sex difference in cholinesterase inhibition and no significant change in inhibition after the first determination in week 13. There was also a dose-dependent inhibition of plasma cholinesterase activity in all groups; at the lowest dose; the inhibition was 26–55% in males and 13–42% in females, although the group mean differences did not achieve statistical significance. Brain acetylcholinesterase activity in the group at the highest dose was slightly (not statistically significantly) lower than that of dogs in the control group (males, 11%; and females, 6%). Neurotoxic esterase activity was not affected by the treatment.

Microscopic examination revealed an increased number of binucleated perilobular hepatocytes in males and females at 1 and 12.5 mg/kg bw per day, accompanied by hyperplasia of bile-duct epithelium in the males. Concentrations of bile pigments were increased in the convoluted renal tubules in dogs at 12.5 mg/kg bw per day. These pathological findings (Table 2) were minimal in severity and were not observed in the 90-day or 6-month studies of toxicity in dogs given similar doses. In addition, they were not associated with any biological correlates. Therefore, a NOAEL was not identified on the basis of these findings.

Table 2. Histopathological findings in dogs fed capsules containing profenofos for 1 year

Finding	Severity	No. of dogs affected (n = 4)[a]									
		Dose (mg/kg bw per day)									
		Males					Females				
		0	0.015	0.05	1.0	12.5	0	0.015	0.05	1.0	12.5
Kidney											
Deposition of bile pigment	Grade 1	1	3	1	1	1	2	0	2	2	2
	Grade 2	1	0	1	1	1	0	2	0	0	1
	Grade 3	0	0	0	0	3	0	0	0	0	0
	Total[b]	2	3	2	2	4	2	2	2	2	3
	Average[c]	—	1.5	1.0	1.5	2.0	1.0	2.0	1.0	1.0	1.3
Liver											
Increase in binucleated hepatocytes	Grade 1	0	0	0	3	0	0	0	0	4	0
	Grade 2	0	0	0	0	1	0	0	0	0	2
	Grade 3	0	0	0	0	3	0	0	0	0	2
	Total[b]	0	0	0	3	4	0	0	0	4	4
	Average[c]	—	—	—	1.0	2.8	—	—	—	1.0	2.5
Hyperplasia of bile duct	Grade 1	0	0	0	1	1	0	0	0	0	0
	Grade2	0	0	0	0	1	0	0	0	0	0
	Total[b]	0	0	0	1	2	0	0	0	0	0
	Average[c]	—	—	—	1.0	1.5	—	—	—	—	—

From Altmann (1999)
[a] Average values for four dogs.
[b]Total number of dogs with tissues affected
[c]Average grade.

The NOAEL was 12.5 mg/kg bw per day (Altmann, 1999). The study was conducted in accordance with GLP regulations and followed the US EPA FIFRA pesticide assessment guidelines, subdivision F (1982), section 83-1 and OECD guideline No. 452 (1981).

The results of three studies of toxicity in dogs indicated that brain acetylcholinesterase activity was significantly inhibited in males at 5 mg/kg bw per day in the 90-day study, but not in males or females at 2.9 or 14.4 mg/kg bw per day in the 6-month study, or at 1 or 12.5 mg/kg bw per day (the highest dose tested) in the 1-year study. Hence, the overall NOAEL for inhibition of brain acetylcholinesterase activity, in these three studies in dogs was 2.9 mg/kg bw per day.

2.3 Long-term studies of toxicity and carcinogenicity

Mice

In a 2-year study of carcinogenicity, groups of 65 male and 65 female HaM/ICR Swiss, CR CD-1 albino mice were given diets containing profenofos (purity, 90.6%) at a concentration of 0, 1, 30 or 100 ppm (equal to ingested doses of 0, 0.14, 4.5 and 14.2 mg/kg bw per day in males and 0, 0.19, 5.77 and 19.18 mg/kg bw per day in females, respectively) for 85 weeks (males) or 96 weeks (females). The test diets were prepared weekly and shipped to the sponsor for analysis. However, the results of analyses were not provided by the sponsor. Mortality was checked daily, while clinical signs, individual body weights and food consumption were measured once every 4 weeks. The presence of nodules or masses was checked by palpation monthly during the first 41 weeks of the study and weekly thereafter. Five males and five females per group were killed at week 53 and subjected to gross necropsy. All surviving mice were examined by gross necropsy at the termination of feeding. Selected organs from all necropsied mice were weighed and examined histopathologically. Determinations of plasma, erythrocyte, and brain acetylcholinesterase activity were carried out on mice killed at week 53 and on five or six males and ten females per group at the end of the study. Inhibition of cholinesterase activity was calculated using the values for concurrent controls.

Treatment did not adversely affect survival. Survival of male mice treated with profenofos at 0, 1, 30 and 100 ppm was 63%, 55%, 65%, 63%, respectively. The corresponding figures for female mice were 65%, 70%, 75% and 78%, respectively. There were no indications of a treatment-related effect with respect to clinical signs of toxicity, body weights, food consumption, incidence of tumours, gross pathology, or histopathology. In treated mice, plasma cholinesterase activity was inhibited by 46–76% at 100 ppm and by 38–68% at 30 ppm. Erythrocyte acetylcholinesterase activity was also statistically significantly inhibited by 66–74% and 49–68% in mice at 100 and 30 ppm respectively. There was no sex difference in inhibition of either plasma or erythrocyte acetylcholinesterase activity. In females, there was a statistically significant inhibition of brain acetylcholinesterase activity (25%) at termination of the group at 100 ppm. Brain acetylcholinesterase activity was also decreased in male mice at 100 ppm at week 53 (15% inhibition); but there was an increase in acetylcholinesterase activity in this group at termination. Hence, the effect in males at week 53 was considered not to be biologically relevant.

The results of this study suggested that profenofos is not carcinogenic in mice. The NOAEL was 30 ppm, equal to 4.5 mg/kg bw per day, on the basis of inhibition of brain acetylcholinesterase activity at 100 ppm, equal to 14.2 mg/kg bw per day, in female mice (Burdock, 1981b).

Rats

In a long-term study of toxicity, groups of 60 male and 60 female Fischer 344 albino rats were fed diets containing profenofos technical (purity, 90.6%) at a concentration of 0, 0.3, 10 or 100 ppm (equal to 0, 0.017, 0.56, and 5.7 mg/kg bw per day in males and 0, 0.02, 0.69, and 6.9 mg/kg bw per day in females) for 2 years. An additional 10 males and 10 females were also included in the control group and the group at the highest dose, respectively. Of the latter, five males and five females per group were killed at 52 weeks (interim kill), and five males and five females per group were placed on control feed after 52 weeks so that recovery could be investigated; these rats were then killed during week 63 (recovery group). The study report stated that the test diets were shipped weekly for analyses. However, the results were not available for review. The rats were observed daily for morbidity and moribundity. Body weights, food consumption and clinical signs were recorded weekly for the first 14 weeks and monthly thereafter. The rats were examined and palpated weekly for tissue masses. Blood and urine were collected for clinical chemistry and haematological investigations from 10 males and 10 females at weeks 13, 26, 52, 78 and 105 and from five males and five females in the recovery group after 63 weeks. Plasma and erythrocyte acetylcholinesterase activity was determined at 13, 26, 52, 78, and 105 weeks (main study groups) and in the recovery group at week 57. Brain acetylcholinesterase activity was determined at interim and terminal kill at weeks 53 and 105. Inhibition of cholinesterase activity was calculated in comparison with concurrent control values. Gross necropsy was performed for all rats. Selected organs from all rats necropsied were weighed and examined histopathologically.

There was no treatment-related increase in mortality. Survival in groups of male rats at 0, 0.3, 10 and 100 ppm was 85%, 90%, 80% and 83%, respectively. The corresponding survival in female rats at 0, 0.3, 10 and 100 ppm was 78%, 72%, 77% and 68%, respectively. No clinical signs of toxicity were observed, and body weights were not affected at any dose. However, at the highest dose, there was an increase in food consumption in females. At the highest dose, there was an increase in relative thyroid weight in males (in rats at the interim kill and in the recovery group but not at terminal kill), an increase in thyroid gland perifollicular-cell hyperplasia in males (4 out of 70 in the control group vs 10 out of 70), and an increase in liver neoplastic nodules in females (control group,

1 out of 70; lowest dose, 3 out of 60; intermediate dose, 2 out of 60; and highest dose, 6 out of 70). These histopathological findings were not considered to be treatment-related, or to be suggestive of an oncogenic effect. No increase in the incidence of liver carcinomas occurred.

Inhibition of plasma cholinesterase activity in rats in the group at 100 ppm was statistically significant at all sampling times and ranged from 30% to 62% in males and 50% to 62% in females. There was a statistically significant inhibition of plasma cholinesterase activity at some sampling times in males at 0.3 ppm and in males and females at 10 ppm. However, inhibition never exceeded 20% and 28% in rats at 0.3 ppm and 10 ppm, respectively. There was a statistically significant inhibition of erythrocyte acetylcholinesterase activity that ranged from 58% to 71% in rats at 100 ppm and from 12% to 31% in rats at 10 ppm. Inhibition of erythrocyte acetylcholinesterase activity was noted at some time

Table 3. Results of studies of genotoxicity with profenofos

End-point	Test object	Concentration	Purity (%)	Results	Reference
In vitro					
Reverse mutation	*Salmonella typhimurium* TA98, TA100, TA1535, TA1537, *Escherichia coli* WP2 *uvr*A	312.5–5000 µg/plate in DMSO	90.7	Negative ± S9	Ogorek (1991)
Gene mutation	Mouse lymphoma L5178Y/ *Tk*+/-cells	0.078–0.625 µg/ml in DMSO	91.8	Negative ± S9	Strasser (1982)
Chromosomal aberration[a]	Chinese hamster ovary cells (CCL 61)	4.69–75 µg/ml in DMSO	90.7	Negative ± S9	Strasser (1990)
Chromosomal aberration[b]	Chinese hamsters	13–52 mg/kg bw in CMC	88.1	Negative	Hool (1981)
Micronucleus formation[c]	Mouse (Tif:MAGf) bone marrow	50–200 mg/kg bw	90.7	Negative	Hertner (1990)
		0–216 mg/kg bw	72	Equivocal	El Nahas et al. (1988)
Dominant lethal mutation	Male mice (NMRI)	35 and 100 mg/kg bw	NR	Negative	Fritz (1974a)
Mitotic gene conversion, mitotic crossing over, reverse mutation	*Saccharomyces cerevisiae* D7	12.5–500 µg/ml without activation, 640–1000 µg/ml with activation	90	Negative	Arni (1982)
Mitotic disjunctione	*Saccharomyces cerevisiae* D61.M	39.06–10000 µg/ml with and without activation	91.8	Negative	Hool (1986)
Unscheduled DNA synthesis	Rat hepatocytes	0.02–2.91 µg/ml in DMSO	91.8	Negative	Puri (1982a)
Unscheduled DNA synthesis	Human fibroblasts	0.46–58.2 µg/ml in DMSO	91.8	Negative	Puri (1982b)

CMC, caboxymethyl cellulose; DMSO, dimethyl sulfoxide, NR, not reported; S9, 9000 × g supernatant prepared from Arochlor-induced rat liver.

[a] Cells were evaluated after: (i) 3-h incubation without metabolic activation at 18.75–75 µg/ml; (ii) 3-h incubation with metabolic activation at 4.69–18.75 µg/ml; and (iii) 24-h incubation without metabolic activation at 9.38–37.5 µg/ml.

[b] Bone-marrow cells were examined for anomalies such as single Jolly bodies, polyploid cells, fragments of nuclei in erythrocytes, micronuclei in erythroblasts and leukopoietic cells.

[c] Micronuclei were counted in bone-marrow erythrocytes.

[d] A formulation was tested that contained unknown formulation components. Under the experimental conditions (small number of animals, only one sex, low reproducibility, unusually high background values, no overall clear dose–response relationship, only one sampling time in some tests), some mutagenic activity of the formulation could not be excluded but the result was not confirmed in guideline-compliant studies.

[e] Monosomic colonies resulting from chromosomal loss were recognized as leucine-requiring white colonies that were resistant to cycloheximide.

intervals at 0.3 ppm, but never exceeded 20%. The effects on plasma and erythrocyte acetylcholinesterase activity were fully reversible after 4 weeks of recovery following a 52-week dosing period. Brain acetylcholinesterase activities were statistically significantly inhibited in females at 100 and 10 ppm at week 105, but the inhibition was only 12 and 9% respectively. Statistical re-evaluation of erythrocyte acetylcholinesterase activities at 0.3 ppm with values for the study controls after combination with values for historical controls revealed that these differences were not statistically significant.

Although the highest dose did not produce overt toxicity, the dosing was considered adequate for testing for carcinogenicity because there was significant inhibition of plasma and erythrocyte cholinesterase activity, minimal toxicity in the thyroid and liver and minimal inhibition of brain acetylcholinesterase activity in females at 10 and 100 ppm at week 105. In addition, the steepness of the dose–response relationship characterizing the effects of this compound on cholinesterase inhibition posed a limitation on the ability to test for the effects of profenofos at higher doses.

The results of this study lead to the conclusion that profenofos is not carcinogenic to rats. The NOAEL was 100 ppm, 5.7 mg/kg bw per day, the highest dose tested. The report author concluded that the NOEL was 0.3 ppm, 0.017 mg/kg bw per day, on the basis of inhibition of plasma and erythrocyte acetylcholinesterase activities at 10 ppm, 0.56 mg/kg bw per day (Burdock, 1981a).

2.4 *Genotoxicity*

In previous evaluations, the Meeting had concluded that after reviewing the results of short-term tests in vitro and in vivo, there was no evidence of genotoxicity. The results of additional new studies, in which the genotoxic potential of profenofos was investigated in eukaryotic and prokaryotic systems in vivo and in vitro, did not alter the previous conclusions of the Meeting. The results are listed in Table 3.

2.5 *Reproductive toxicity*

(a) Multigeneration studies

In a study of reproductive toxicity, Crl:CD®(SD)BRVAF/Plus™ rats received diets containing profenofos (purity, 92%) at a concentration of 0, 5, 100 or 400 ppm (equivalent to 0, 0.4, 7 and 35 mg/kg bw per day) for two generations. Administration started when rats of the P generation were age 43 days and was continued until termination. Rats were observed daily for clinical signs. Body weights and food consumption were determined weekly. After treatment for 10 weeks, rats were randomly paired within each group for up to 21 days to yield the F_1 generation. Stability, homogeneity and dietary concentrations were confirmed analytically. Body weights and food consumption were determined on days 0, 6, 13 and 20 of gestation and on days 0, 4, 7, 14 and 21 of lactation. Litter size, number of live and dead pups, individual sexes, weights, and external observations were recorded for pups on the same days of lactation. On day 4 of lactation, litters were culled to four males and four females wherever possible. Culled pups underwent a soft tissue examination with the focus on brain and heart. F_1 pups were weaned at age 21 days. Within each group, rats were randomly selected to continue on treatment as parent animals of the F_2 generation. They underwent the same study phases as P animals. All non-parental F_1 weanlings and all F_2 weanlings were examined as indicated for the culled pups. Adult rats were necropsied and reproductive tissues were examined histologically.

The mean concentrations of the test compound in the diet were 5.02, 99.3 and 398 ppm for the diets at 5, 100, and 400 ppm , respectively. Profenofos was homogeneously distributed in the diet. It was stable in the diet (bulk storage) at 4°C or at room temperature for 35 days and in feed jars at room temperature it was stable for 16 days. At 400 ppm, reduced body weights (4–11% decrease), body-weight gains (6–16% decrease) and food consumption (7–15% decrease) were noted in male and female parental animals. At 400 ppm, reduced pup body weights (2–9% decrease) and body-weight gains (3–10% decrease) on days 14 and 21 of lactation. There were no treatment-related clinical signs, necropsy findings or histopathological observations at any dose. Reproduction was not affected at any dose. There were no effects on mating behaviour, duration of gestation, the number of litters with live-born pups, the total number of pups per litter, pre-weaning losses, survival indices and other reproductive indices. No macroscopic findings were noted in pups.

The NOAEL for reproductive toxicity was 400 ppm, equivalent to 35 mg/kg bw per day. On the basis of reduced body-weight gains and food consumption at 400 ppm, the NOAEL for parental and pup systemic toxicity was 100 ppm (7 mg/kg bw per day) (Minor & Richter, 1994).

In an earlier three-generation study of reproduction in CD strain Charles River albino rats, dietary treatment with profenofos (purity, 95.5%) at 0, 0.2, 1.0 and 20 ppm (equivalent to 0, 0.01, 0.05, and 1.0 mg/kg bw per day) did not affect brain acetylcholinesterase activity, reproductive performance, or the development and survival of the offspring through three generations. The NOAEL for reproductive effects was at least 20 ppm, equal to 1 mg/kg bw per day; the highest dose tested. This study was conducted before development of FIFRA guidelines. The data had not been independently validated (Phillips, 1978).

(b) Developmental toxicity

Rats

In a study of developmental toxicity, groups of 25 pregnant Sim:(SD)BR Sprague-Dawley rats were given profenofos (purity, 88.0%) at a dose of 0, 10, 30, 60, 90, or 120 mg/kg bw per day by oral gavage on days 6–15 of gestation. The study was not conducted in accordance with GLP regulations and there was no claim for compliance with any particular guidelines. However, the study design conforms to the OECD guideline No 414. The rats were examined daily for changes in clinical signs and were weighed on days 0, 6 to 15 and 20 of gestation. Food consumption was measured on days 6, 13 and 20 of gestation. On day 21 of gestation, all dams were killed and fetuses were delivered by caesarean section. Uterine contents were examined and fetuses were weighed, sexed and examined for abnormalities. Intracranial structures were examined in approximately half of the fetuses from each litter.

Maternal toxicity was evident at a dose of 120 mg/kg bw per day. Four rats died or were terminated and food consumption was reduced on days 6–13. Various clinical signs of toxicity (e.g. hypoactivity or tremors, ocular porphyrin discharge, dyspnoea, diuresis, and hypothermia) were noted. Two of the four dams that died displayed these clinical signs, while the other two did not. In addition, two of the dams that died also showed scattered haemorrhages in the stomach upon gross necropsy. One rat in the group at 60 mg/kg bw per day also died, but there were no clinical signs of toxicity at doses of up to 90 mg/kg bw per day. Measures of prenatal toxicity such as litter size, percentage of live fetuses, number of resorbed fetuses, number of dead fetuses and the mean sex ratio were not significantly different between the control and treated groups. Malformations and developmental variants observed in fetuses in the control group and those in groups receiving profenofos were within the normal range.

Profenofos was not embryotoxic or teratogenic in rats given doses of up to 120 mg/kg bw per day. The NOAEL for maternal toxicity was 90 mg/kg bw per day on the basis of mortality and clinical signs of toxicity seen at 120 mg/kg bw per day. The NOAEL for developmental toxicity was 120 mg/kg bw per day, the highest dose tested (Harris, 1982).

Another study of developmental toxicity in rats given profenofos at lower doses arrived at similar conclusions. In this study, 23–25 pregnant JCL-SD rats were given profenofos (purity, 95.8%) at a dose of 0, 18, 35, or 70 mg/kg bw per day via intubation on days 7–17 of gestation. The doses selected for testing were based upon the results of a preliminary range-finding study in which a dose of 140 mg/kg bw per day caused death in five out of six treated rats during days 8–15 of gestation.

The study was not conducted in accordance with GLP regulations and no compliance with guidelines was claimed.

Treatment with profenofos was associated with increases in body weight in the dams at doses of 35 and 70 mg/kg bw per day on days 20 and/or 21, increased water consumption on days 17 and 20–21, respectively, and an increase in food consumption at 70 mg/kg bw per day on days 14–21. Although increases in the weights of several organs (heart, spleen, liver, and right kidney) were seen at 70 mg/kg bw per day, these were small in magnitude. No treatment-related changes in mortality or behaviour were observed, and no abnormal findings were observed at gross necropsy. There were no adverse effects on the offspring with respect to resorptions, sex ratios, placental weights, body weights and lengths, or distribution of fetuses within the uterine horns. External and visceral examinations of fetuses were unremarkable. Skeletal examination of fetuses showed increased incidences of progeny with holes in the xiphoid at the intermediate and highest doses (controls, 0%; lowest dose, 0%; intermediate dose, 18.8%; and highest dose, 15.6%) and delayed ossification of vertebral arches at the highest dose (controls, 8.8%; lowest dose, 6.7%; intermediate dose, 0.5%; and highest dose, 26.7%). No data for historical controls were provided and it could not be determined whether the findings were all from one litter or from multiple litters.

The NOAEL for maternal and developmental toxicity was 70 mg/kg bw per day, the highest dose tested (Sugiya et al., 1982).

In an earlier study conducted before establishment of FIFRA test guidelines, profenofos (technical grade; purity unspecified) was administered to groups of 20–27 pregnant rats (strain unspecified) at a dose of 0, 10, 30, or 60 mg/kg bw per day by oral gavage on days 6–15 of gestation. On day 21 of gestation, all dams were killed and fetuses were delivered by caesarean section. Maternal toxicity was indicated by a slight decrease in food consumption at 30 mg/kg bw per day and a marked decrease in food consumption at 60 mg/kg bw per day during the period of treatment. No other adverse effects occurred in the dams. Similarly, profenofos did not appear to affect embryonic or fetal development and no teratogenic effects were observed at any dose tested.

The NOAEL for maternal toxicity was 30 mg/kg bw per day on the basis of markedly decreased food consumption seen at 60 mg/kg bw per day. The NOAEL for fetotoxicity/teratogenicity was 60 mg/kg bw per day; the highest dose tested (Fritz, 1974b).

Rabbits

In a study of developmental toxicity, groups of 16 pregnant New Zealand White rabbits were given profenofos (technical grade; purity, 90.8%) at a dose of 0, 30, 60, 90, and 175 mg/kg bw per day by gavage on days 6–18 of gestation. Rabbits were observed daily for clinical signs; body weights were measured on days 0, 6, 9, 12, 15, 18, 25 of gestation and before caesarean section. On day 30 of gestation, all does were euthanized and fetuses delivered by caesarean section. Uterine contents were examined and the fetuses were examined for external abnormalities. Visceral dissections and skeletal evaluations were performed on all fetuses. The doses of profenofos selected for testing were based upon the results of a preliminary range-finding study in which doses up to 150 mg/kg bw per day did not produce any signs of toxicity. The study was not conducted in accordance with GLP regulations and there was no claim for compliance with any particular guidelines, although the study design conformed to the OECD guideline No. 414.

Treatment with profenofos was associated with anorexia in all groups, but particularly in the group at 175 mg/kg bw per day. Clinical signs of toxicity (including diarrhoea, soft stools, oral/perianal discharges) were noted at 175 mg/kg bw per day and nine of the does at 175 mg/kg bw per day died. Many of the does that died exhibited the above clinical signs of toxicity as well as signs of pin-point stomach haemorrhages and yellow-discoloured areas in the mesentery in the gastric region upon gross necropsy. No statistically significant differences were observed between the control and treated groups for body-weight gains during gestation or mean number of corpora lutea. No significant differences were observed for the following measures of prenatal toxicity: mean number of implantations, litter size, fetal body weight or embryolethality. No significant differences were detected between the control group and groups receiving profenofos for malformations or variations.

The NOAEL for maternal toxicity was 30 mg/kg bw per day on the basis of reduced maternal-weight gain and food consumption seen at doses of 60 mg/kg bw per day and greater. The NOAEL for developmental toxicity was 175 mg/kg bw per day; the highest dose tested (Holson, 1983).

Previously, the 1990 JMPR had reviewed a study of developmental toxicity in Chinchilla rabbits given profenofos at a dose of 0, 5, 15 or 30 mg/kg bw per day. In this study, the NOAEL for parental and developmental toxicity was greater than 30 mg/kg bw per day, the highest dose tested (Fritz, et al., 1979).

2.6 *Special studies*

(a) *Acute neurotoxicity*

This study was conducted in two phases to determine the NOAEL for clinical signs and inhibition of cholinesterase activity after single doses of profenofos (purity, 92%) by oral gavage. In each phase, test groups consisted of five males and five females per group (Crl:CD®BR/VAF/Plus®). The aim of phase 1 was to identify a NOAEL for clinical signs and body-weight effects, while the aim of phase 2 was to identify a NOEL/NOAEL for effects on plasma, erythrocyte and brain acetylcholinesterase activity.

In phase 1, undiluted profenofos (purity, 92%) was administered at a dose of 100, 200, 300 or 400 mg/kg bw in female rats and 100, 200, 400, 600 or 800 mg/kg bw in male rats. There was no control group. Clinical signs were recorded 1 2, 4 and 8 h after treatment and daily thereafter. Mortalities were checked twice per day, body weights were determined on days 0, 7 and 14 or at death when survival exceeded 1 day. All surviving rats were necropsied on day 14 and tissues with macroscopic lesions were retained.

Three males at 800 mg/kg bw, one male at 600 mg/kg bw and one female at 400 mg/kg bw died within the first 5 days. Body-weight gains appeared to be unaffected by treatment with profenofos. Clinical signs of toxicity such as hypoactivity, red-stained face, dark-stained urogenital area, soft stool or few faeces were seen at doses of 200 mg/kg bw and greater. Rats surviving to study termination exhibited no macroscopic lesions. Test material-related findings observed in the males at 800 mg/kg bw were red ocular discharge, erosion/ulceration of the stomach, nasal discharge and diffuse reddening of the stomach. One male at 600 mg/kg bw showed dark fluid in the intestinal tract and one female at 400 mg/kg bw showed yellow perineal staining and dark-red eroded areas of the stomach and duodenum.

The NOAEL for single oral application of profenofos was 100 mg/kg bw on the basis of clinical signs at 200 mg/kg bw and greater.

In phase 2, groups of five male and five female rats were given a single dose of profenofos (purity, 92%) at 0, 0.1, 0.5, 25, 100, or 400 mg/kg bw in corn oil. Clinical signs were recorded 1, 2 and 4 h after treatment. Rats were weighed approximately 4 h after dosing, blood samples were taken for determination of plasma and erythrocyte cholinesterase activity and all rats were necropsied. Inhibition of cholinesterase activity was calculated relative to values for the respective control group. Tissues with macroscopic lesions were examined histologically. The brain of each rat was flash-frozen for evaluation of brain acetylcholinesterase activity. The study was conducted in accordance with GLP regulations and followed the US EPA FIFRA subdivision F, addendum 10-neurotoxicity series 81-1) guidelines.

The only clinical sign was soft stool, observed in one or a few rats in each group. Findings at necropsy were regarded as incidental. Plasma cholinesterase activity was significantly reduced (71–98%) in males and in females at 25, 100 or 400 mg/kg bw. Erythrocyte acetylcholinesterase activity was inhibited by 21% and 30% in males at 100 or 400 mg/kg bw, although only at 400 mg/kg bw was the inhibition statistically significant. There was a statistically significant and dose-dependent inhibition of cholinesterase activity in erythrocytes of females at 25, 100 or 400 mg/kg bw, which ranged from 31% to 46%. Brain acetylcholinesterase activity was significantly inhibited by 37% and 43%, respectively, in males and females at 400 mg/kg bw.

The NOAEL was 100 mg/kg bw on the basis of the inhibition of brain acetylcholinesterase activity at 400 mg/kg bw. The report author considered the NOAEL for erythrocyte acetylcholinesterase in males to be 25 mg/kg bw because there was a 21% inhibition in rats at 100 mg/kg bw (Glaza, 1994).

In another study, groups of 15 male and 15 female Hsd:Sprague Dawley SD™ rats were given profenofos (purity, 92%) at a dose of 0, 95, 190 and 380 mg/kg bw once by gavage. Rats int eh positive-control group were given a single dose of either propoxur (6 mg/kg bw) or triadimefon (males, 100 mg/kg bw; and females, 150 mg/kg bw). Rats were observed twice per day for clinical signs and mortality and given a physical examination each week. Body weight and food consumption were also monitored weekly. Ten males and ten females per group were monitored in a functional observation battery (FOB) and observed for motor activity in a figure-of-eight maze. These tests were performed 1 week before exposure to profenofos, approximately 4–6 h after dosing (estimated time of peak effect) and on days 7 and 14 after dosing. Blood samples from the remaining five males and females per group were taken at the time of peak effect (approximately 4 h after dosing) and on day 14 for determination of plasma and erythrocyte acetylcholinesterase activity. In addition, brain acetylcholinesterase activity was determined in these rats 14 days after exposure. At necropsy, 10 males and 10 females per group were examined in a special histopathological investigation; tissues were fixed by whole-body perfusion and the brain, spinal cord, peripheral nerves, skeletal muscles and the eyes with the optic nerve were examined in detail. Further organs were examined if gross lesions occurred. This study was conducted in accordance with GLP regulations and the US EPA FIFRA subdivision F guideline, addendum 10-neurotoxicity series 81, 82 and 83.

One male in the group at 380 mg/kg bw died on study day 2. A transient decrease in body weight, cumulative body-weight gain, food consumption and food conversion efficiency was recorded in the first week after exposure in males in the groups at 190 and 380 mg/kg bw, but was compensated for by higher values in the second week. Food consumption was reduced in females at 380 mg/kg bw in week 1, but increased in week 2. As a result, food conversion efficiency was increased in females in week 2. Effects on the autonomic nervous system (diarrhoea, lacrimation, slight impairment of respiration, and miosis), bizarre behavioural effects, neuromuscular effects (ataxia, abnormal gait, impaired reflexes etc.) and effects on the central nervous sytsem (tremors, altered posture, and ease of handling) were only observed in rats of the group at 380 mg/kg bw at the time of peak effect. Motor activity of these animals was reduced at this time-point, but was not apparent at 7 and 14 days after treatment. At the time of peak effect in the group at the highest dose, no effects on the FOB and motor activity tests were observed in the two groups at the lower doses. There was an 84–97% inhibition of plasma cholinesterase activity and 68–96% inhibition of erythrocyte acetylcholinesterase activity across all three groups. The inhibition of erythrocyte cholinesterase activity was statistically significant at all doses. At the 14-day time-point, erythrocyte acetylcholinesterase activity showed some recovery while plasma cholinesterase activity had returned to control levels. Brain acetylcholinesterase activity was not inhibited at the 14-day time-point. All necropsy observations as well as all microscopic findings were interpreted as incidental. There was no evidence of neuropathological alterations attributable to treatment with profenofos in any of the tissues examined.

The dose of 380 mg/kg bw was toxic with rats showing autonomic and functional changes (on FOB) consistent with inhibition of cholinesterase activity. The dose of 190 mg/kg bw was identified as the NOAEL for central nervous system effects. Pathological examination revealed no evidence of compound-related neurotoxicity at any dose (Pettersen & Morrissey, 1993, 1994a).

(b) *Short-term studies of neurotoxicity*

In a 90-day study of neurobehavioural toxicity, groups of 15 male and 15 female Hsd:Sprague-Dawley rats were given diets containing profenofos (purity, 88.4%) at a concentration of 0, 30, 135 or 600 ppm (equal to 0, 1.7, 7.7 and 36 mg/kg bw in males and 0, 1.84, 8.4 and 37.9 mg/kg bw in females) for 13 weeks. Ras in the positive-control group were treated with acrylamide and trimethyltin chloride. Concentration, homogeneity and stability of the test compound in the diet were confirmed. Mortality and clinical signs were monitored twice per day and the rats were given a detailed physical examination weekly. Body weight and food consumption were also determined weekly. The first ten rats from each group were given a battery of tests to assess neurological functions (including FOB and motor activity) approximately 1 week before exposure to profenofos and during weeks 3, 7, and 12. For necropsy, these rats were killed by whole-body perfusion and examined by a special histopathological investigation focusing on the brain, spinal and peripheral nerves, skeletal muscles and the eyes with the optic nerve. Further organs were examined if gross lesions occurred. Blood samples were taken from the other five rats of each group during weeks 3, 7, and 12 for determination of serum and erythrocyte acetylcholinesterase activity. At week 12, the whole brain was removed from these rats for determination of brain acetylcholinesterase activity. Inhibition of cholinesterase activity was calculated as a reduction relative to values for the appropriate control group. This study was conducted in accordance with GLP requirements and the US EPA FIFRA subdivision F guideline, addendum 10-neurotoxicity series 81, 82 and 83.

Dietary analysis indicated that the desired concentrations were achieved (range, 98–100% of nominal) and that the homogeneity was acceptable. The test compound was stable in the diet for up to 35 days when stored in closed containers at room temperature or in the refrigerator.

Overall body-weight gain was reduced at week 13 in the males and females at 600 ppm by about 7% and 11%, respectively. Reduced food consumption and reduced feed efficiencies were also observed at 600 ppm. At all dietary concentrations and at all time-points, a statistically significant inhibition of plasma and erythrocyte acetylcholinesterase activity was observed. There was a statistically significant inhibition of approximately 30% and 60% in plasma cholinesterase activity in male and female rats, respectively, in the group at 30 ppm and the inhibition increased to approximately 80% (males) and 93% (females) at the highest dose of 600 ppm. Erythrocyte acetylcholinesterase activity was also significantly inhibited by approximately 60% in male and female rats at 30 ppm and by approximately 78% at 135 ppm. Examination of the time-course of cholinesterase activity for rats at 30 and 135 ppm indicated that maximal inhibition was achieved after treatment for 3 weeks and that this level of inhibition was generally maintained throughout the study. At 600 ppm, there was a statistically significant inhibition of brain acetylcholinesterase activity of 12% in males and 20% in females at week 13. No clinical observations, ophthalmoscopic findings, FOB findings or motor activity effects related to profenofos were noted at any dietary concentration. Pathological examination revealed no evidence of compound-related neurotoxicity at any dose.

The NOAEL was 135 ppm, equal to 7.7 mg/kg bw per day, on the basis of statistically significant inhibition of brain acetylcholinesterase activity and reduced body weights, body-weight gains, reduced food consumption and reduced feed efficiency at 600 ppm, equal to 36.0 mg/kg bw per day. The report authors considered that depression of erythrocyte acetylcholinesterase activity at lower dietary concentrations was not an adverse effect but rather a sign of exposure, but the Meeting considered that this may simply have been a reflection of the relatively low sensitivity of their FOB end-point. Rats treated with the substances serving as positive controls gave the expected results (Pettersen & Morrissey, 1994b).

(c) *Developmental neurotoxicity*

In a preliminary study, groups of 15 time-mated Alpk:AP_fSD Wistar rats were given diets containing profenofos (purity 91.8%) at a concentration of 0, 4, 200, 400 or 600 ppm from day 7 of gestation until postnatal day 22. The achieved doses were 0, 0.3, 15.5, 30.2 and 46.1 mg/kg bw per day during gestation and 0, 0.7, 33.9, 66.0 and 97.6 mg/kg bw per day during lactation and weaning. Day 1 of gestation was the day of confirmation of mating and postnatal day 1 was the day of littering. Rats were examined daily, clinical signs of toxicity and body weight were recorded on days 7, 8, 15 and 22 of gestation and on postnatal days 1, 5, 8, 12, 15 and 22. Food consumption was recorded at 3–4-day intervals during gestation and lactation. Concentration, homogeneity and stability of the test compound in the diet were confirmed. The sex, weight and clinical condition of each pup were recorded on postnatal days 1, 5, 8, 12, 15 and 22 . Plasma, erythrocyte and brain acetylcholinesterase activity was determined in dams on day 22 of gestation and postnatal day 22. Cholinesterase activity was also determined in pups on day 22 of gestation and postnatal days 5, 12 and 22. Staements of compliance with GLP and QA statement were provided, but as this was a preliminary study compliance with a regulatory guideline was not applicable.

Profenofos was stable in the diet at room temperature for 13 days and in the freezer for 28 days. The test compound was homogenously distributed in the diet. The mean concentrations were within 8% of nominal concentrations.

During gestation, body weight was slightly reduced (6%) in rats at 600 ppm and food consumption was reduced in rats at 400 or 600 ppm. On postnatal day 22, decreases in male (13%) and female pup weights (9%) and total litter weights (13%) were evident in the group at 600 ppm compared with values for controls. There were no other effects on litter or pup parameters.

In the parental rats, there were statistically significant reductions in plasma and erythrocyte acetylcholinesterase activity in rats at 200, 400 and 600 ppm at day 22 of gestation and postnatal day 22 . The inhibition was not dose-dependent and ranged from 76–86% for plasma cholinesterase and 42–53% for erythrocyte acetylcholinesterase activity. Brain acetylcholinesterase activity was significantly inhibited in the groups at 200, 400 and 600 ppm on postnatal day 22. The inhibition appeared to be dose-dependent and ranged from 21% at 200 ppm to 52% at 600 ppm. On day 22 of gestation, there was an 18% and 17% inhibition of brain acetylcholinesterase activity in rats at 400 and 600 ppm, respectively, but the reduction was not statistically significant. In pups there were no statistically significant effects on plasma, erythrocyte or brain cholinesterase activity on day 22 of gestation. A statistically significant inhibition of erythrocyte and brain acetylcholinesterase activity was only found on postnatal day 22, but there was no clear dose–response relationship. Brain acetylcholinesterase activity was inhibited by 25% in females in the group at 400 ppm and by 16% in males in the group at 600 ppm; erythrocyte acetylcholinesterase activity was inhibited by 33–46% in pups in the groups at 400 and 600 ppm. Inhibition of plasma cholinesterase activity increased during weaning and there was a 36–65% inhibition in pups exposed to profenofos at 200–400 ppm by postnatal day 22.

The NOAEL for parental toxicity was 4 ppm (0.3 mg/kg bw per day) on the basis of significant inhibition of brain acetylcholinesterase activity at doses of 200 ppm (15.5 mg/kg bw per day) and greater, and reduced body-weight gain and food consumption at 400 and 600 ppm. The NOAEL for offspring toxicity was 200 ppm (33.9 mg/kg bw per day) on the basis of reductions in brain acetylcholinesterase activity at doses of 400 ppm (66.0 mg/kg bw per day) and greater (Milburn, 2002).

In the main study of developmental neurotoxicity, groups of 30 mated Alpk:APfSD Wistar-derived rats were fed diets containing profenofos (purity, 91.8%) at a concentration of 0, 3, 60 or 600 ppm from day 7 of gestation to postnatal day 29. These dietary concentrations corresponded to daily doses of 0, 0.3, 5.1 and 50.6 mg/kg bw per day during gestation and 0, 0.5, 10.7 and 103.4 mg/kg bw per day during lactation and weaning. Concentration, stability and homogeneity of the test compound in the diet were determined. Rats were examined daily; clinical signs of toxicity, body weight and food consumption were recorded at intervals during gestation and up to termination on postnatal day 29. The parental animals were assessed in a FOB on days 10 and 17 of gestation and on postnatal days 2 and 9. The sex, weight and clinical condition of each pup were recorded on postnatal days 1 and 5. Where possible, litters were culled on postnatal day 5 to eight pups per litter with sexes represented as equally as possible. Pups were selected on postnatal day 5 for the F1 generation; they were separated on day 29 and allowed to grow to adulthood. Rats in the F1 generation were examined daily, clinical examinations, food consumption and body weights were measured at intervals, and evaluations of motor activity, auditory startle response and assessments of learning and memory were also carried out. Rats were killed on postnatal day 63 and tissues were removed for neuropathological investigations. Additional satellite groups (five dams per group, five male and five female fetuses per group and five male and five female pups per group) were used to determine plasma, erythrocyte and brain acetylcholinesterase activity in dams on day 22 of gestation and postnatal day 22, in fetuses on day 22 of gestation and in pups on postnatal days 5, 12 and 22. The study was conducted in accordance with GLP regulations and the protocol complied with US EPA OPPTS guideline No 870.6300.

The results of dietary analyses indicated that the test compound was stable in the diet for 27 days when stored at room temperature or at nominally − 20 °C. The mean concentration and homogeneity of the test compound in the diet were within limits of acceptability (± 9% and 3% of nominal, respectively).

Body weights were slightly reduced in the parent females during late gestation (15%) and early in the postnatal period (5%) in rats at 600 ppm. Food consumption was also reduced during late gestation and during late lactation in rats of this group. There were no treatment-related effects on clinical observations, FOB measurements, reproductive parameters, litter losses or pup survival. On day 22 of gestation and on postnatal day 22 there was a statistically significant reduction in plasma cholinesterase activity of approximately 60% and 80% in the parental rats receiving profenofos at 60 and 600 ppm, respectively. There was a similar reduction, of approximately 55% of values for controls, in erythrocyte acetylcholinesterase activity in these groups at both time-points. At 600 ppm, brain acetylcholinesterase activity was decreased by 44% on day 22 of gestation, and by 26% (not statistically significant) on postnatal day 22.

Pup and total litter weights were reduced on postnatal day 5 in the group at 600 ppm and group mean body weights for F_1 rats in this group were lower than those of the controls from day 5 to day 29 in males and until day 36 in females. In the F_1 rats, there were no treatment-related effects on motor activity, auditory startle response, learning and memory or neurohistopathology. No statistically significant effects were noted on cholinesterase activity in pups in the group at 3 ppm. In the group at 60 ppm, only plasma cholinesterase activity was significantly inhibited (23%) at postnatal day 22. In males and females in the group at 600 ppm, plasma cholinesterase activity was significantly inhibited by up to approximately 50% at postnatal day 22 and erythrocyte cholinesterase activity was inhibited by up to 40% at the same sampling time. There was a statistically significant difference in brain acetylcholinesterase activity in female pups at day 5 (11% lower) but not at later sacrifice times. On postnatal day 12, absolute brain weight was decreased (by 4%) in the males at the highest dose only. The differences were no longer evident when the weights were adjusted to lower body weights of these animals. No treatment-related gross or microscopic pathological findings were noted in any treated group. Significant differences in various morphometric measurements were seen in males and females at the highest dose on postnatal days 12 and 63.

The NOAEL for maternal toxicity was 60 ppm, equal to 5.1 mg/kg bw per day, on the basis of inhibition of brain acetylcholinesterase activity on day 22 of gestation and day 22 of lactation, and reductions in body weight and food consumption at 600 ppm, equal to 50.6 mg/kg bw per day. The NOAEL for developmental neurotoxicity was 600 ppm, equal to 50.6 mg/kg bw per day. The NOAEL for offspring toxicity was 60 ppm, equal to 5.1 mg/kg bw per day, on the basis of inhibition of brain acetylcholinesterase activity, reduced body weight, decreased brain weights in males on postnatal day 12 and changes in the brain morphometric parameters at 600 ppm, equal to 50.6 mg/kg bw per day (Milburn, 2003).

(d) Delayed neurotoxicity

A first study was conducted with three groups of two male and two female White Leghorn chickens given profenofos (purity undefined) at a dose of 21.7, 46.4 or 60 mg/kg bw in polyethylene glycol as vehicle. Surviving chickens (only in the groups at 21.7 and 46.4 mg/kg bw) were given a second dose 21 days later. For protection against the acute toxic effects of profenofos, atropine was administered intramuscularly 60 min before the second dose. Tri-orthocresyl phosphate was given to chickens in a positive-control group. Clinical signs of toxicity and mortality were recorded up to 42 days after the second dose. At the end of the observation period, the remaining birds were killed and samples of muscle and spinal cord were taken for histopathology. The study was not conducted in accordance with GLP requirements, but the procedure used was as recommended by the United Kingdom Ministry of Agriculture, Fisheries and Food (MAFF), Pesticide Safety Precaution Scheme, working document No 2, appendix B, 1967.

All the chickens at 60 mg/kg bw died after the first dose and only those in the group at the lowest dose survived the two treatments. Clinical signs of toxicity were salivation, asynchronisms of the extremities, curved position, apathy and ruffled feathers. Neither symptoms of delayed neurotoxicity nor histological changes in spinal cord or peripheral nerve could be detected. The birds in the concurrent positive-control groups treated with tri-orthocresyl phosphate showed the expected reactions (ataxia, deterioration of the reflexes, swelling, fragmentation and disruption of myelin sheaths) by day 18 after treatment and the histopathology findings were also as expected (Krinke et al., 1974).

In a second study, chickens of the White Leghorn strain (sex not specified) were given profenofos (purity, 89.5%) at a dose of 30 or 45.7 mg/kg bw. The doses were based on an initial assessment of the acute toxicity of profenofos, which showed the LD_{50} to be 45.7 mg/kg bw. Surviving birds were given a second dose 21 days after the initial dose, but owing to the unexpectedly high mortality after the first dose, the second was reduced to 17.1 mg/kg bw, which was the revised LD_{50} on the basis of mortality after the first dose. Birds in a positive-control group were given tri-orthocresyl phosphate. The birds were observed daily for mortality and possible neurotoxic signs for up to 21 days after the second dose. Body weight was measured on days 0, 21 and 42 and food consumption was measured weekly. All birds that died during the study were given a gross necropsy. At the end of the study, the sciatic nerves, brain and spinal cord were examined in surviving birds. The results of this study had not been independently validated.

Forty-one of the 50 birds at 45.7 mg/kg bw and 28 of the 40 birds at 30 mg/kg bw died within 21 days of dosing, mostly within 24 h. There was only one further death when surviving birds of both groups were given the second dose 21 days after the first, even without pre-treatment with atropine. Clinical signs noted after the first dose were lethargy and salivation. Slight lethargy was also apparent for 24–48 h after the second dose. There were no behavioural signs of delayed neurotoxicity. In particular, no signs of locomotor disturbances were observed. Reduced body-weight gains and food consumption were apparent in treated birds, particularly after the first dose. Gross and histopathological studies of neural tissues from these birds did not reveal any treatment-related findings. The concurrent positive-control group showed the expected established signs of neurotoxicity by day 8 of study (Reinart, 1978).

These studies demonstrate that profenofos caused no delayed neurotoxic effects in adult chickens even at very high (sublethal) doses.

(e) Antagonistic agents

A protective effect of atropine given early after the oral administration of profenofos in rats or intraperitoneal administration in chicks and mice was demonstrated by a reduction in mortality and signs of toxicity (e.g. salivation, tremors, sedation, and convulsions) typical of exposure to anti-cholinesterase. The effect of oximes was limited, probably due to rapid ageing (Sachsse & Bathe, 1976; Gfeller & Kobel, 1984; Glickman et al., 1984). These investigations were not conducted to GLP or to any regulatory guideline.

(f) Potentiation studies

Potentiation studies were divided into two phases. First, the oral LD_{50} for each compound was experimentally determined in rats. In the second phase, the LD_{50} values for the equitoxic mixtures of the insecticides were evaluated and compared with the theoretical LD_{50} values derived from an assumption of strictly additive toxicity. The procedure for the determination of the acute oral LD_{50} for each compound and the equitoxic mixtures was essentially the same. The compounds were suspended in aqueous carboxymethyl cellulose and given to groups of five male and five female Tif-RAIf (SPF) rats that were then observed for 14 days. None of the studies were conducted to GLP or to any regulatory guideline.

No potentiation effects were found when mixtures of profenofos with methidathion (GS 13'005), methacrifos (CGA 20'168) or diazinon (G 24'480) were given to rats in equitoxic doses (Sachsse & Bathe, 1977, Sachsse & Bathe, 1978). In contrast to this, a potentiation experiment with profenofos Q (active ingredient of higher purity; from the toxicological point of view regarded as equivalent to CGA 15'324) and malathion showed a strong potentiation of the acute oral toxicity.

The experimental LD_{50} values of the single compounds were 377 (range, 282–545) mg/kg bw for profenofos (purity, 96.3%) and 4658 (3320–8504) mg/kg bw for malathion (purity not stated). The equitoxic mixture led to an experimental LD_{50} of 76 (range, 45–125) mg/kg bw, while the theoretical (calculated) LD_{50} was 2517 mg/kg bw (Sarasin, 1981).

Profenofos (purity, > 95%) at doses of 0.5 to 5.0 mg/kg bw given intraperitoneally to mice strongly inhibited the liver microsomal esterase(s) responsible for the hydrolysis of *trans*-permethrin. At an intraperitoneal dose of 25 mg/kg bw, profenofos increased the toxicity of fenvalerate by more than 25-fold, and that of malathion by more than 100-fold. This potentiation did not occur with *trans*-permethrin (Gaughan et al., 1980).

3. Studies on metabolites

No studies have been conducted with metabolites of profenofos. However, studies conducted in 1981 with analogous alkyl phosphate metabolites in vitro demonstrated a complete lack of anticholinesterase activity (Chukwudebe et al., 1984).

4. Observations in humans

Workers engaged in the manufacture of profenofos were given a medical examination annually between 1980 and 1998. No health effects related to exposure to profenofos were identified (Novartis Crop Protection AG Assessment, 1998).

In a biological monitoring study, cotton was sprayed with different formulations of profenofos, using manual equipment. Six workers in Multan, Central Pakistan, were monitored daily during a 4-day spraying campaign. Cholinesterase activities were determined in whole blood and were found to be slightly below the values determined before exposure. Individual cholinesterase activities declined from 100% (pre-test) to an average of 81%, with a lowest value of 73%. None exceeded the threshold of 30% inhibition that was considered to be biologically significant. The handling of the active ingredient and of its formulations is subject to the usual precautionary measures recommended for the handling of insecticidal organophosphates (Loosli, 1989).

There are seven reports of effects after exposure to profenofos during application of the product. In one, considered of moderate severity, the operator used only marginally protective equipment. He was given atropine and fully recovered. The other six reports described adverse incidents that were of minor severity and also involved people using the product without protective equipment or with marginally protective equipment. The symptoms were transient and resolved spontaneously. The exposed persons fully recovered.

Finally, one accidental exposure had been notified. This involved a child (bystander) who inhaled spray mist. Again the symptoms were transient, the effects disappeared without the need for treatment and the child fully recovered.

Comments

Biochemical aspects

[Phenyl-^{14}C]profenofos was rapidly absorbed and eliminated after oral administration to rats. Total radioactivity eliminated via the urine and faeces exceeded 99% of the administered dose for a single dose of 1 or 100 mg/kg bw by gavage and repeated doses of 1 mg/kg bw by gavage. Elimination was rapid, with about 95% of the total radiolabel being excreted in the urine within the first 24 h in all treated groups. For all doses, less than 4% of the radiolabel was excreted in the faeces. The concentration of radiolabel in tissues and organs reached a maximum after 2 h and remained at similar levels until 8 h after dosing. By 72 h, the tissue concentration of radiolabel was minimal. The absorption, distribution and excretion of 14C-labelled profenofos was not sex- or dose-dependent in the range of 1 to 100 mg/kg bw and was unaffected by pre-treatment with unlabelled profenofos for 14 days. Unchanged profenofos was detected in the faeces, but the amount was very small (approximately 1–2% of the administered dose), and this was probably the proportion of the dose that was not absorbed. Four major metabolites were present in urine and no unchanged profenofos was detected. The major metabolites were the sulfate and glucuronide conjugates of 4-bromo-2-chlorophenol that were formed by hydrolysis of the aryloxy–phosphorus bond followed by conjugation with sulfate or glucuronic acid. The other two metabolites were formed by cleavage of the phosphorus–sulfur bond either by loss of the propyl group or hydrolysis. The 4-bromo-2-chloro-phenol was detected in some urine samples, but probably arose as a result of hydrolysis of the conjugates after excretion.

Toxicological data

The oral LD_{50} for profenofos ranged from 358 to 1178 mg/kg bw in rats. The oral LD_{50} for profenofos was 298 mg/kg bw in mice and 700 mg/kg bw in rabbits. The clinical signs detected in all the studies of acute toxicity were typical of cholinergic poisoning, which appeared at doses greater than 100 mg/kg bw. Profenofos was of low toxicity when administered by the dermal route to rats (LD_{50}s, > 2000 and 3300 mg/kg bw). More varied results were obtained after dermal application to rabbits, with LD_{50}s ranging from 131 to 2560 mg/kg bw depending on method of application (semi-occlusive, abraded skin or massaging). Profenofos was of low toxicity on exposure by inhalation, the LC_{50} being > 3.36 mg/l air. Profenofos was moderately irritating to skin and mildly irritating to the eye and was shown to be a sensitizer under the conditions of the Magnusson & Kligman test and in the local lymph-node assay.

The primary effect of profenofos in studies of acute toxicity and short- and long-term studies of toxicity was inhibition of acetylcholinesterase activity and this was associated with signs of neurotoxicity at high levels of inhibition. Profenofos is a racemic mixture of the two optical isomers at the chiral phosphorus atom. The S (−) isomer is a markedly more potent inhibitor of acetylcholinesterase in vitro than the R (+) isomer. The inhibited acetylcholinesterase ages rapidly, an effect that prevents spontaneous reactivation. Rapid ageing would lead to a cumulative inhibitory effect after repeated exposures to profenofos, and would also render reactivation therapy with oximes ineffective.

In a short-term repeat-dose study, no clinical signs of toxicity were observed in rats given diet containing profenofos at a concentration of 1000 ppm, equal to 85 mg/kg bw per day, for 8 weeks. Reduced food intake and body-weight gain were apparent at this dose and also at a dose of 100 ppm, equal to 8.4 mg/kg bw per day, which was given for 13 weeks. Inhibition of cholinesterase activity was the only other effect noted. Erythrocyte cholinesterase activity was inhibited by more than 20% at doses of 30 ppm, equal to 2.4 mg/kg bw per day, and greater. Brain acetylcholinesterase activity was inhibited at 1000 ppm, equal to 85 mg/kg bw per day. The NOAEL for inhibition of brain acetylcholinesterase activity was 300 ppm, equal to 22.0 mg/kg bw per day.

Inhibition of brain acetylcholinesterase activity and clinical signs consistent with neurotoxicity were observed in rats exposed to profenofos at a concentration of 0.07 mg/l per day by inhalation for 21 days.

In three studies of dermal toxicity in rabbits, the overall NOAEL for inhibition of brain acetylcholinesterase was 2.5 mg/kg bw per day on the basis of significantly reduced activity at 5 mg/kg bw per day.

Three studies were carried out in dogs given profenofos orally for 90 days, 6 months, or 1 year. Profenofos was given in the diet in the 90-day and 6-month studies, and daily in gelatin capsules in the 1-year study. No clinical signs of toxicity were recorded in these studies, the 6-month and 1-year studies including neurological examinations (NOAEL for clinical signs, 12.5 mg/kg bw per day). Brain acetylcholinesterase activity was significantly inhibited in males at 5 mg/kg bw per day in the 90-day study, but not in either sex at 2.9 or 14.4 mg/kg bw per day in the 6-month study, or at 1 or 12.5 mg/kg bw per day (the highest dose tested) in the 1-year study. Hence, for brain acetylcholinesterase inhibition, the overall NOAEL in these three studies in dogs was 2.9 mg/kg bw per day. Haematology parameters (erythrocyte count, haemoglobin concentration and erythrocyte volume fraction) were reduced; however, they were not considered to be toxicologically significant since there was no clear dose–response relationship, and the small changes observed were within the range for historical controls. Treatment of dogs with profenofos at 12.5 mg/kg bw per day for 1 year was also associated with an increase in binucleated perilobular hepatocytes, bile-duct hyperplasia and an increase in bile pigments in kidney tubules. These pathological findings were minimal in severity and were not observed in the 90-day or 6-month studies of toxicity.

Profenofos was not mutagenic in an adequate battery of studies of genotoxicity.

The Meeting concluded that profenofos is unlikely to be genotoxic.

In long-term studies, treatment of mice and rats with profenofos did not adversely affect survival; there were no clinical signs of toxicity, no increase in the incidence of tumour formation and no treatment-related changes in either gross pathology or histopathology. Plasma and erythrocyte cholinesterase activity were significantly reduced in mice given diet containing profenofos at 30 ppm, equal to 4.5 mg/kg bw per day, and in rats at 100 ppm, equal to 5.7 mg/kg bw per day. In female mice, there was a statistically significant inhibition of brain acetylcholinesterase activity (25%) at termination of the group at 100 ppm, equal to 14.2 mg/kg bw per day, resulting in a NOAEL of 30 ppm, equal to 4.5 mg/kg bw per day. The NOAEL in the 2-year study of carcinogenicity in rats was 100 ppm, equal to 5.7 mg/kg bw per day, the highest dose tested. Profenofos was not carcinogenic in mice and rats up to the highest dose tested. Although overt toxicity was not observed in the study in rats, the Meeting considered that the available database was sufficient to evaluate the carcinogenic potential of profenofos.

In view of the lack of genotoxicity, the absence of carcinogenicity in mice and rats, and any other indication of carcinogenic potential, the Meeting concluded that profenofos is unlikely to pose a carcinogenic risk to humans.

Multigeneration studies have shown that profenofos has no effect on reproduction at doses of up to 400 ppm, equivalent to 35 mg/kg bw per day. The NOAEL for parental and pup toxicity was 100 ppm, equivalent to 7.0 mg/kg bw per day, on the basis of reduced body-weight gains and food consumption at 400 ppm, equivalent to 35 mg/kg bw per day, and the NOAEL for reproductive toxicity was 400 ppm, the highest dose tested.

Profenofos did not cause developmental effects in rats or rabbits. Clinical signs typical of cholinesterase inhibition were noted in rabbits given profenofos at 175 mg/kg bw per day and approximately 50% of the animals died. There were no treatment-related effects on the mean number of implantations, litter size, fetal body weight or embryolethality and there were no significant increases in variations or malformations in the fetuses. The NOAEL for maternal toxicity was 30 mg/kg bw per day and the NOAEL for developmental toxicity was 175 mg/kg bw per day, the highest dose tested.

Studies of developmental toxicity in rats, maternal toxicity, which included clinical signs typical of cholinesterase inhibition, and deaths were observed at the highest dose of 120 mg/kg bw per day. There was no evidence for prenatal toxicity at either of these doses and the type and incidence of fetal malformations and variations was unaffected by treatment. The NOAEL for maternal toxicity was 90 mg/kg bw per day and the NOAEL for developmental toxicity was 120 mg/kg bw per day, the highest dose tested.

The Meeting concluded that profenofos is not teratogenic.

The potential for profenofos to cause developmental neurotoxicity had also been investigated in rats. In a preliminary range-finding study, rats were given diets containing profenofos at a concentration of 0, 4, 200, 400 or 600 ppm, equal to 0, 0.7, 33.9, 66.0 or 97.6 mg/kg bw per day. In this study, dose-dependent inhibition of the brain acetylcholinesterase activity was observed in dams at ≥ 200 ppm on postnatal day 22. The NOAEL for inhibition of brain acetylcholinesterase activity in dams was 4 ppm, equal to 0.7 mg/kg bw per day. A statistically significant inhibition of brain acetylcholinesterase activity of > 20% and 16% was found in female pups at ≥400 ppm and male pups at 600 ppm, respectively. In the main study of developmental neurotoxicity, rats were given diets containing profenofos at a concentration of 0, 3, 60 or 600 ppm (equal to 0, 0.3, 5.1 or 50.6 mg/kg bw per day). At 600 ppm in dams, brain acetylcholinesterase activity was decreased by 44% on day 22 of gestation, and by 26% (not statistically significant) on day 22 of lactation, and body weights and food consumption were reduced. A statistically significant inhibition of brain acetylcholinesterase activity was observed in female pups at 600 ppm compared with controls on day 5 (11% lower) but not at later times. At 600 ppm, there was a statistically significant reduction in pup body weights (11–12%). No effects on functional parameters or neurohistopathology were observed. The NOAEL for maternal toxicity was 60 ppm, equal to 5.1 mg/kg bw per day, on the basis of inhibition of brain acetylcholinesterase activity on day 22 of gestation and day 22 of lactation and reductions in body weight and food consumption at 600 ppm, equal to 50.6 mg/kg bw per day. The overall NOAEL for inhibition of brain acetylcholinesterase in pups was 60 ppm, equal to 5.1 mg/kg bw per day. The NOAEL for developmental neurotoxicity was 600 ppm, equal to 50.6 mg/kg bw per day, the highest dose tested.

In two studies of acute neurotoxicity in rats, there were reversible signs typical of poisoning with acetylcholinesterase inhibitors (diarrhoea, miosis, lacrimation, tremor), peaking 4 h after administration of profenfos at 380 mg/kg bw by gavage. Lesser effects were seen at 200 mg/kg bw (hypoactivity, soft faeces), and there were no effects in the FOB at 190 mg/kg bw (the NOAEL for clinical signs). There was significant inhibition of brain acetylcholinesterase activity (by 37% in males and 43% in females) at 4 h after dosing at 400 mg/kg bw, with a NOAEL of 100 mg/kg bw. Inhibition was absent after a recovery period of 14 days.

There were also no clinical signs of toxicity, and no adverse findings in a FOB or effects on motor activity in a 90-day study of neurotoxicity in rats. Pathological investigation revealed no evidence of treatment-related toxicity. At the highest dose of 600 ppm, equal to 36 mg/kg bw per day, there was a reduction of approximately 10% in body-weight gain. At 600 ppm, there was a statistically significant inhibition of brain acetylcholinesterase activity of 12% in males and 20% in females at week 13. The NOAEL for brain acetylcholinesterase inhibition was 135 ppm, equal to 7.7 mg/kg bw per day.

Profenofos did not induce delayed neuropathy in chickens given two doses at 45.7 mg/kg bw (maximum tolerated dose) and then at 17.1 mg/kg bw, separated by an interval of 21 days (atropine protection being given as soon as clinical signs appeared).

No cases of adverse effects have been reported among workers involved in the manufacture of profenofos. In a biological monitoring study, whole-blood cholinesterase activity was inhibited by less than 30% in six workers who were monitored daily for 4 days during spraying of profenofos.

The Meeting concluded that the existing database on profenofos was adequate to characterize the potential hazards to fetuses, infants and children.

Toxicological evaluation

Erythrocyte acetylcholinesterase activity was found to be significantly more sensitive to profenofos than was brain acetylcholinesterase activity in rats, mice, rabbits, and dogs. However, in no species were any signs of toxicity seen at doses that did not also produce significant inhibition of brain acetylcholinesterase. The Meeting thus concluded that inhibition of brain acetylcholinesterase activity was the more appropriate end-point for risk assessment of profenofos.

The Meeting established an ADI of 0–0.03 mg/kg bw per day based on an overall NOAEL of 2.9 mg/kg bw per day identified on the basis of inhibition of brain acetylcholinesterase activity in three short-term studies in dogs and using a safety factor of 100. This ADI was supported by the NOAEL of 5.1 mg/kg bw per day identified on inhibition of maternal and pup brain acetylcholinesterase activity in a study of developmental neurotoxicity in rats and a NOAEL of 4.5 mg/kg bw per day identified on the basis of inhibition of brain acetylcholinesterase activity in a 2-year study in mice.

The Meeting established an ARfD of 1 mg/kg bw based on a NOAEL of 100 mg/kg bw in studies of acute neurotoxicity in rats, identified on the basis of clinical signs of neurotoxicity seen at ≥ 200 mg/kg bw and inhibition of brain acetylcholinesterase activity at 400 mg/kg bw and using a safety factor of 100. The appropriate study for establishing the ARfD was the study of acute neurotoxicity since there was no evidence of developmental effects. This ARfD was considered to be protective against any clinical signs of acetylcholinesterase inhibition seen in studies of acute oral toxicity.

Levels relevant to risk assessment

Species	Study	Effect	NOAEL	LOAEL
Mouse	Two-year studies of toxicity and carcinogenicity[a]	Toxicity	4.5 mg/kg bw per day	14.2 mg/kg bw per day
		Carcinogenicity	14.2 mg/kg bw per day[c]	—
Rat	Two-year studies of toxicity and carcinogenicity[a]	Toxicity	5.7 mg/kg bw per day[c]	—
		Carcinogenicity	5.7 mg/kg bw per day[c]	—
	Multigeneration study of reproductive toxicity[a]	Parental	7.0 mg/kg bw per day	35.0 mg/kg bw per day
		Reproductive toxicity	35.0 mg/kg bw per day[c]	—
		Offspring toxicity	7.0 mg/kg bw per day	35.0 mg/kg bw per day
	Developmental toxicity[b]	Maternal toxicity	90.0 mg/kg bw per day	120.0 mg/kg bw per day
		Embryo/fetotoxicity	120.0 mg/kg bw per day[c]	—
	Developmental neurotoxicity[a]	Parental toxicity	5.1 mg/kg bw per day	50.6 mg/kg bw per day
		Offspring toxicity	5.1 mg/kg bw per day	50.6 mg/kg bw per day
	Acute neurotoxicity[b,d]	Toxicity	100.0 mg/kg bw	400.0 mg/kg bw per day
Rabbit	Developmental toxicity[b]	Maternal toxicity	30.0 mg/kg bw per day	60.0 mg/kg bw per day
		Embryo/fetotoxicity	175.0 mg/kg bw per day[C]	—
Dog	Studies of toxicity[d]	Toxicity	2.9 mg/kg bw per day	12.5 mg/kg bw per day

[a] Dietary administration.

[b] Gavage administration.

[c] Highest dose tested.

[d] The results of two or more studies were combined.

Estimate of acceptable daily intake for humans

0–0.03 mg/kg bw

Estimate of acute reference dose

1 mg/kg bw

Information that would be useful for the continued evaluation of the compound

Results from epidemiological, occupational health and other such observational studies of human exposures

Critical end-points for setting guidance values for exposure to profenofos

Absorption, distribution, excretion and metabolism in mammals	
Rate and extent of oral absorption	About 94% within 24 h
Dermal absorption	Approximately 90%
Distribution	Widely distributed
Potential for accumulation	Low, no evidence of accumulation
Rate and extent of excretion	94% in urine within 24 h
Metabolism in animals	> 95% by conversion of the phosphorothiolate group to a variety of hydrolysis products
Toxicologically significant compounds in animals, plants and the environment	Parent
Acute toxicity	
Rat, LD_{50}, oral	358–1178 mg/kg bw
Rat, LD_{50}, dermal	3300 mg/kg bw
Rat, LC_{50}, inhalation	3.36 mg/l
Skin irritation	Moderately irritating
Eye irritation	Mildly irritating
Guinea-pig, skin sensitization (test method used)	Sensitizer (Magnusson & Kligman and local lymph-node assay)
Short-term studies of toxicity	
Target/critical effect	Inhibition of brain acetylcholinesterase activity
Lowest relevant oral NOAEL	2.9 mg/kg bw per day (dogs)
Lowest relevant dermal NOAEL	2.5 mg/kg bw per day
Lowest relevant inhalation NOAEC	< 0.07 mg/l air
Genotoxicity	
	No genotoxic potential
Long-term studies of toxicity and carcinogenicity	
Target/critical effect	Inhibition of brain acetylcholinesterase activity
Lowest relevant NOAEL	4.5 mg/kg bw per day (2-year study in mice)
Carcinogenicity	Not carcinogenic
Reproductive toxicity	
Reproduction target/critical effect	No reproductive effects
Lowest relevant reproductive NOAEL	400 ppm (35 mg/kg bw per day) (rats)
Developmental target/critical effect	No developmental effects
Lowest relevant developmental NOAEL	120 mg/kg bw per day (rats)

Neurotoxicity/delayed neurotoxicity	
Acute neurotoxicity	Inhibition of brain acetylcholinesterase activity, NOAEL was 100 mg/kg bw per day (rats)
Developmental neurotoxicity	Inhibition of brain acetylcholinesterase activity, NOAEL was 5.1 mg/kg bw per day (rats)
Delayed neuropathy	No delayed neurotoxicity, NOAEL was 45.7 mg/kg bw (chickens)
Medical data	
	No detrimental effects on agricultural workers

Summary

	Value	Study	Safety factor
ADI	0–0.03 mg/kg bw	Dog, studies of oral toxicity	100
ARfD	1 mg/kg bw	Rat, study of acute neurotoxicity	100

References

Altmann, B. (1999) CGA 15'324 tech. - 12-month chronic toxicity study in beagle dogs. Unpublished report No. 962002 from Novartis Crop Protection AG, Stein, Switzerland. Submitted to WHO by Syngenta Crop Protection Greensboro, North Carolina, USA.

Arni, P. (1982) *Saccharomyces cerevisiae* D7/mammalian microsome mutagenicity test in vitro with CGA 15324 (test for mutagenic properties in yeast cells). Unpublished report No. 811557 from Ciba-Geigy Ltd, Basel, Switzerland. Submitted to WHO by Syngenta Crop Protection, Greensboro, North Carolina, USA.

Bathe, R. (1974a) Acute oral LD_{50} of technical CGA 15324 in the rat. Unpublished report No. Siss 3647 from Ciba-Geigy Ltd, Basel, Switzerland. Submitted to WHO by Syngenta Crop Protection, Greensboro, North Carolina, USA.

Bathe, R. (1974b) Acute oral LD50 in the mouse of technical CGA 15324. Unpublished report No. Siss 3647 from Ciba-Geigy Ltd, Basel, Switzerland. Submitted to WHO by Syngenta Crop Protection, Greensboro, North Carolina, USA.

Bathe, R. (1974c) Acute dermal LD50 of technical CGA 15324 in the rat. Unpublished report No. Siss 3647 from Ciba-Geigy Ltd, Basel, Switzerland. Submitted to WHO by Syngenta Crop Protection, Greensboro, North Carolina, USA.

Betts, C.J. (2005) Profenofos technical Goa origin: local lymph node assay. Unpublished report No. GM7966-REG from Syngenta Central Toxicology Laboratory, Macclesfield, England. Submitted to WHO by Syngenta Crop Protection, Greensboro, North Carolina, USA.

Burdock, G.A. (1981a) Two-year chronic oral toxicity study in albino rats with technical CGA 15324. Unpublished report No. 483-134 from Hazleton, Inc., Vienna, Virginia, USA. Submitted to WHO by Syngenta Crop Protection, Greensboro, North Carolina, USA.

Burdock, G.A. (1981b) Twenty-four month carcinogenicity study in mice with technical CGA 15324. Unpublished report No. 483-133 from Hazleton Inc., Vienna, Virginia, USA. Submitted to WHO by Syngenta Crop Protection, Greensboro, North Carolina, USA.

Burtner, B.R., Kennedy, G.L., Keplinger, M.L. & Nelson, R. (1975) 90-day subacute oral toxicity study with CGA 15324 technical in beagle dogs. Unpublished report No. IBT 611-05122-B from Industrial Bio-Test Laboratories, Inc., Northbrook, Illinois, USA. Submitted to WHO by Syngenta Crop Protection, Greensboro, North Carolina, USA.

Cannelongo, B.F. (1982a) CGA 15324 technical: rabbit acute dermal toxicity. Unpublished report No. 2460-81 from Stillmeadow Inc., Houston, Texas, USA. Submitted to WHO by Syngenta Crop Protection, Greensboro, North Carolina, USA.

Cannelongo, B.F. (1982b) CGA 15’324 technical: rabbit eye irritation. Unpublished report No. 2461-81 from Stillmeadow Inc., Houston, Texas, USA. Submitted to WHO by Syngenta Crop Protection, Greensboro, North Carolina, USA.

Cantoreggi, S. (1998) CGA 15324 tech - 21-Day repeated-dose dermal toxicity study in rabbits. Unpublished report No. 972418 from Novartis Crop Protection AG, Stein, Switzerland. Submitted to WHO by Syngenta Crop Protection, Greensboro, North Carolina, USA.

Chukwudebe, A.C., Hussain, M.A. & Oloffs, P.C. (1984) Hydrolytic and metabolic products of acephate in water and mouse liver. *J. Environ. Sci. Health B.*, **19**, 501–522.

Durando, J. (2005a) Profenofos technical: acute oral toxicity up and down procedure in rats. Unpublished report No. T018339-04 from Product Safety Laboratories, Dayton, New Jersey, USA. Submitted to WHO by Syngenta Crop Protection, Greensboro, North Carolina, USA.

Durando, J. (2005b) Profenofos technical: acute dermal toxicity study in rats. Unpublished report No. T018345-04 from Product Safety Laboratories, Dayton, New Jersey, USA. Submitted to WHO by Syngenta Crop Protection, Greensboro, North Carolina, USA.

Durando, J. (2005c) Profenofos technical: acute inhalation toxicity study in rats. Unpublished report No. T018352-04 from Product Safety Laboratories, Dayton, New Jersey USA. Submitted to WHO by Syngenta Crop Protection, Greensboro, North Carolina, USA.

Durando, J. (2005d) Profenofos technical: primary skin irritation study in rabbits. Unpublished report No. T018318-04 from Product Safety Laboratories, Dayton, New Jersey, USA. Submitted to WHO by Syngenta Crop Protection, Greensboro, North Carolina, USA.

Durando, J. (2005e) Profenofos technical: primary eye irritation study in rabbits. Unpublished report No. T018166-04 from Product Safety Laboratories, Dayton, New Jersey, USA. Submitted to WHO by Syngenta Crop Protection, Greensboro, North Carolina, USA.

El Nahas, S.M., de Hondt, H.A. & Ramadan,. A.I. (1988) In vivo evaluation of the genotoxic potential of curacron in somatic cells of mice. *Environ. Mol. Mutagen.*, **11**, 515–522.

Fritz, H. (1974a) Dominant lethal study in the mouse of technical CGA 15’324 (test for cytotoxic or mutagenic effects on male germinal cells). Unpublished report No. 327438 from Ciba-Geigy Ltd, Basel, Switzerland. Submitted to WHO by Syngenta Crop Protection, Greensboro, North Carolina, USA.

Fritz, H. (1974b) Reproductive study - technical CGA 15324, rat, segment II (test for teratogenic or embryotoxic effects). Unpublished report No. 22741900 from Ciba-Geigy Ltd, Basle, Switzerland. Submitted to WHO by Ciba-Geigy Ltd, Basle, Switzerland.

Fritz, H., Becker, H. & Hess, R. (1979) Segment. II reproductive study in rabbits, CGA 15324 technical. Unpublished report No. 785565 from Ciba-Geigy Ltd, Basle, Switzerland. Submitted to WHO by Ciba-Geigy Ltd, Basle, Switzerland.

Gaughan, L.C., Engel, J.L., & Casida, J.E. (1980) Pesticide interactions: effects of organophosphorus pesticides on the metabolism, toxicity and persistence of selected pyrethroid insecticides. *Pestic, Biochem. Physiol.*, **14**, 81–85.

Gfeller, W. (1981) CGA 15324 technical: 6-month toxicity study with dogs. Unpublished report No. 790804, (reevaluated 2 February 1993) from Ciba-Geigy Ltd,, Basel, Switzerland. Submitted to WHO by Syngenta Crop Protection, Greensboro, North Carolina, USA.

Gfeller, W. & Kobel, W.(1984) Tentative antilethal study in the rat of technical CGA 15324. Unpublished report No. 820034 from Ciba-Geigy Ltd, Basel, Switzerland. Submitted to WHO by Syngenta Crop Protection, Greensboro, North Carolina, USA.

Glaza, S.M. (1994) Acute oral toxicity study of profenofos technical in rats. Unpublished report No. HWI 6117-236 from Hazleton Wisconsin Laboratories Inc., Madison, Wisconsin, USA. Submitted to WHO by Syngenta Crop Protection, Greensboro, North Carolina, USA.

Glickman, H., Wing, K.D. & Casida, J.E. (1984) Profenofos insecticide bioactivation in relation to antidote action and the stereospecificity of acetylcholinesterase inhibition, reactivation, and ageing. *Toxicol. Appl. Pharmacol.*, **73**, 16–22.

Harris, S.B. (1982) A teratology study of technical CGA 15324 in albino rats. Unpublished report No. 282009 from Science Applications Inc., San Diego, California, USA. Submitted to WHO by Syngenta Crop Protection, Greensboro, North Carolina, USA.

Hertner, T. (1990) CGA 15324 tech.: micronucleus test, mouse. Unpublished report No. 891212 from Ciba-Geigy Ltd, Basel, Switzerland. Submitted to WHO by Syngenta Crop Protection, Greensboro, North Carolina, USA.

Holson, J.F. (1983) Teratology study (seg II) in albino rabbits with technical CGA 15324. Unpublished report No. 283003 from Science Applications, Inc., San Diego, California, USA. Submitted to WHO by Syngenta Crop Protection, Greensboro, North Carolina, USA.

Hool, G. (1981) Nucleus anomaly test in somatic interphase nuclei in Chinese hamster of technical CGA 15'324 (test for mutagenic effects on bone marrow cells). Unpublished report No. 791557 from Ciba-Geigy Ltd, Basel, Switzerland. Submitted to WHO by Syngenta Crop Protection, Greensboro, North Carolina, USA.

Hool, G. (1986) Test for non-disjunction on *Saccharomyces cerevisiae* D 61.M in vitro with CGA 15'324. Unpublished report No. 850811 from Ciba-Geigy Ltd, Basel, Switzerland. Submitted to WHO by Syngenta Crop Protection, Greensboro, North Carolina, USA.

Horath, L.L. (1982) Acute aerosol inhalation toxicity study in rats of CGA 15324 technical. Unpublished report No. 420-0921 from ToxiGenics, Inc., Decatur, Illinois, USA. Submitted to WHO by Syngenta Crop Protection, Greensboro, North Carolina, USA.

Ifflaender, U. & Mücke, W. (1974) Distribution, degradation and excretion of CGA 15'324 in the rat. Unpublished report No. 16/74 from Ciba-Geigy Ltd, Basel, Switzerland. Submitted to WHO by Syngenta Crop Protection, Greensboro, North Carolina, USA.

Ifflaender, U. & Mücke, W. (1976) The metabolism of CGA 15'324 in the rat. Unpublished report No. 27/76 from Ciba-Geigy Ltd, Basel. Switzerland. Submitted to WHO by Syngenta Crop Protection, Greensboro, North Carolina, USA.

Johnson, S., Tai, C.M. & Katz, R. (1984) Profenofos technical: 21-day dermal toxicity study in rabbits. Unpublished report No. 842008 from Ciba-Geigy Pharmaceuticals Division, Summit, New Jersey, USA. Submitted to WHO by Syngenta Crop Protection, Greensboro, North Carolina, USA.

Kennedy, E. & Swain, W.E., Jr (1992) CGA 15324: metabolism of ^{14}C-profenofos in rats. Unpublished report No. ABR-92026 from Ciba-Geigy Corporation, Greensboro, North Carolina, USA. Submitted to WHO by Syngenta Crop Protection, Greensboro, North Carolina, USA.

Kobel, W. (1983) CGA 15324 tech: acute oral LD_{50} in the rat. Unpublished report No. 830188 from Ciba-Geigy Ltd, Basel, Switzerland. Submitted to WHO by Syngenta Crop Protection, Greensboro, North Carolina, USA.

Krinke, G., Ullmann, L. & Sachsse, K. (1974) Acute oral LD_{50} and neurotoxicity study in the domestic fowl (*Gallus domesticus*) of technical CGA 15324. Unpublished report No. 2850 from Ciba-Geigy Ltd, Basel, Switzerland. Submitted to WHO by Syngenta Crop Protection, Greensboro, North Carolina, USA.

Kuhn, J.O. (1989) Acute dermal toxicity study in rabbits. Unpublished report No. 5522-88 from Stillmeadow, Inc., Houston, Texas, USA. Submitted to WHO by Syngenta Crop Protection, Greensboro, North Carolina, USA.

Kuhn, J.O. (1990a) CGA 15324 tech.: acute oral toxicity study in rats. Unpublished report No. 7312-90 from Stillmeadow Inc., Houston, Texas, USA. Submitted to WHO by Syngenta Crop Protection, Greensboro, North Carolina, USA.

Kuhn, J.O. (1990b) CGA 15324: primary dermal irritation study in rabbits. Unpublished report No. 7313-90 from Stillmeadow Inc., Houston, Texas, USA. Submitted to WHO by Syngenta Crop Protection, Greensboro, North Carolina, USA.

Kuhn, J.O. (2005) Profenofos technical (batch P03): skin sensitization: local lymph-node assay in mice. Unpublished report No. T003191-05 from Stillmeadow Inc, Sugar Land, Texas, USA. Submitted to WHO by Syngenta Crop Protection, Greensboro, North Carolina, USA.

Loosli, R. (1989) Cholinesterase activity in field workers during cotton spraying with Polytrin C. Unpublished report No. 850910 from Ciba-Geigy Ltd, Basel, Switzerland. Submitted to WHO by Syngenta Crop Protection, Greensboro, North Carolina, USA.

Merkel, D.J. (2004a) Acute oral toxicity up and down procedure in rats with profenofos technical (CGA15324). Unpublished report No. T013184-04 from Product Safety Laboratories, Dayton, New Jersey, USA. Submitted to WHO by Syngenta Crop Protection, Greensboro, North Carolina, USA.

Merkel, D.J. (2004b) Acute dermal toxicity study in rats – limit test with profenofos technical (CGA15324). Unpublished report No. T013183-04 from Product Safety Laboratories, Dayton, New Jersey, USA. Submitted to WHO by Syngenta Crop Protection, Greensboro, North Carolina, USA.

Merkel, D.J. (2004c) Primary skin irritation study in rabbits with profenofos technical (CGA15324). Unpublished report No. T013181-04 from Product Safety Laboratories, Dayton, New Jersey, USA. Submitted to WHO by Syngenta Crop Protection, Greensboro, North Carolina, USA.

Merkel, D.J. (2004d) Primary eye irritation study in rabbits with profenofos technical (CGA15324). Unpublished report No. T013182-04 from Product Safety Laboratories, Dayton, New Jersey, USA. Submitted to WHO by Syngenta Crop Protection, Greensboro, North Carolina, USA.

Milburn, G.M. (2002) Profenofos (CGA 15324): preliminary developmental neurotoxicity study in rats. Unpublished report No. RR0927 from Syngenta Central Toxicology Laboratories, Macclesfield, England. Submitted to WHO by Syngenta Crop Protection, Greensboro, North Carolina, USA.

Milburn, G.M. (2003) Profenofos: developmental neurotoxicity study in rats. Unpublished report No. RR0928 from Syngenta Central Toxicology Laboratories, Macclesfield, England. Submitted to WHO by Syngenta Crop Protection, Greensboro, North Carolina, USA.

Minor, J.L. & Richter, A.G. (1994) A two-generation reproduction study in rats with CGA 15324 technical. Unpublished report No. F-00102 from the Environmental Health Center, Farmington, Connecticut, USA. Submitted to WHO by Syngenta Crop Protection, Greensboro, North Carolina, USA.

Mücke, W. (1986) The renal excretion of [U-^{14}C]phenyl CGA 15324 by male rats after oral administration (exposure monitoring). Unpublished report No 13/86 from Ciba-Geigy Ltd, Basel, Switzerland. Submitted to WHO by Syngenta Crop Protection, Greensboro, North Carolina, USA.

Novartis Crop Protection AG (1998) Assessment and medical data on CGA 15'324 (profenofos). Unpublished report from Novartis Crop Protection Submitted to WHO by Syngenta Crop Protection, Greensboro, North Carolina, USA.

Ogorek, B. (1991) CGA 15324: *Salmonella* and *Escherichia*/liver-microsome test. Unpublished report No. 901526 from Ciba-Geigy Ltd, Basel, Switzerland. Submitted to WHO by Syngenta Crop Protection, Greensboro, North Carolina, USA.

Pettersen, J.C. & Morrissey, R.L. (1993) Acute neurobehavioral toxicity study with CGA 15'324 in rats. Unpublished report No. F-00166 from the Environmental Health Center, Farmington, Connecticut, USA. Submitted to WHO by Syngenta Crop Protection, Greensboro, North Carolina, USA.

Pettersen, J.C. & Morrissey, R.L. (1994a) Acute neurobehavioral toxicity study with CGA 15'324 in rats. Unpublished report No F-00166 (amendment No. 1) from the Environmental Health Center, Farmington, Connecticut, USA. Submitted to WHO by Syngenta Crop Protection, Greensboro, North Carolina, USA.

Pettersen, J.C. & Morrissey, R.L. (1994b) 90-Day subchronic neurobehavioral toxicity study with CGA 15324 technical in rats. Unpublished report No. F-00167 from the Environmental Health Center, Farmington, Connecticut, USA. Submitted to WHO by Syngenta Crop Protection, Greensboro, North Carolina, USA.

Phillips, B.M. (1978) Three-generation reproduction study in albino rats of technical CGA15324. Unpublished report No. IBT 623-07924 from Industrial BIO-TEST Laboratories, Northbrook, Illinois, USA. Submitted to WHO by Syngenta Crop Protection, Greensboro, North Carolina, USA.

Piccirillo, V.J. (1978) 90-Day subacute oral toxicity study in rats with CGA 15324 technical. Unpublished report No. 483-135 from Hazleton, Inc., Vienna, Virginia, USA. Submitted to WHO by Syngenta Crop Protection, Greensboro, North Carolina, USA.

Puri, E. (1982a) Autoradiographic DNA repair test on rat hepatocytes with CGA 15324. Unpublished report No. 811490, supplemented in April 1991, from Ciba-Geigy Ltd, Basel, Switzerland. Submitted to WHO by Syngenta Crop Protection, Greensboro, North Carolina, USA.

Puri, E. (1982b) Autoradiographic DNA repair test on human fibroblasts with CGA 15324. Unpublished report No. 811658 from Ciba-Geigy Ltd, Basel, Switzerland. Submitted to WHO by Syngenta Crop Protection, Greensboro, North Carolina, USA.

Rattray, N.J. (2004) Profenofos echnical (CGA15324): 4-hour acute inhalation toxicity study in rats. Unpublished report No. CTL/HR2503/REG/REPT from Syngenta Central Toxicology Laboratory, Macclesfield, England. Submitted to WHO by Syngenta Crop Protection, Greensboro, North Carolina, USA.

Reinart, D. (1978) 42-Day neurotoxicity study in adult chickens of technical CGA 15324. Unpublished report No. IBT 8580-11187 from Industrial BIO-TEST Laboratories, Inc., Northbrook, Illinois, USA. Submitted to WHO by Syngenta Crop Protection, Greensboro, North Carolina, USA.

Sachsse, K. (1974a) Acute oral LD_{50} of technical CGA 15’324 in the rabbit. Unpublished report No. Siss 3647 from Ciba-Geigy Ltd, Basel, Switzerland. Submitted to WHO by Syngenta Crop Protection, Greensboro, North Carolina, USA.

Sachsse, K. (1974b) Acute dermal LD_{50} of technical CGA 15’324 in the rabbit. Unpublished report No. 3647 from Ciba-Geigy Ltd, Basel, Switzerland. Submitted to WHO by Syngenta Crop Protection, Greensboro, North Carolina, USA.

Sachsse, K. (1976) CGA 15324 technical: 21-day dermal toxicity study in rabbits. Unpublished report No Siss 5119 from Ciba-Geigy Ltd, Basel, Switzerland. Submitted to WHO by Syngenta Crop Protection, Greensboro, North Carolina, USA.

Sachsse, K. (1977) CGA 15324 technical: 21-day inhalation study on the rat. Unpublished report No. Siss 5119 from Ciba-Geigy Ltd, Basel, Switzerland. Submitted to WHO by Syngenta Crop Protection, Greensboro, North Carolina, USA.

Sachsse, K .& Bathe, R. (1976) The therapeutic activity of pralidoxime (PAM) and Toxogonin® with regard to CGA 15324 in the rat. Unpublished report from Ciba-Geigy Ltd, Basel, Switzerland. Submitted to WHO by Syngenta Crop Protection, Greensboro, North Carolina, USA.

Sachsse, K. & Bathe, R. (1977) Potentiation study: CGA 15324 versus 2 insecticides, GS 13005 (methidathion) and G 24480 (diazinon) in the rat. Unpublished report from Ciba-Geigy Ltd, Basel, Switzerland. Submitted to WHO by Syngenta Crop Protection, Greensboro, North Carolina, USA.

Sachsse, K. & Bathe, R. (1978) Potentiation study: CGA 20168 versus 6 insecticides, C 177 (DDVP), C 570 (phosphamidon), GS 13005 (methidathion), G 24480 (diazinon), CGA 15324 and malathion. Unpublished report No. Siss 6526 from Ciba-Geigy Ltd, Basel, Switzerland. Submitted to WHO by Syngenta Crop Protection, Greensboro, North Carolina, USA.

Sarasin, G. (1981) Potentiation study CGA 15324 Q versus malathion. Unpublished report No. 810249 from Ciba-Geigy Ltd, Basel, Switzerland. Submitted to WHO by Syngenta Crop Protection, Greensboro, North Carolina, USA.

Strasser, F. (1982) L5178Y/TK+/- mouse lymphoma mutagenicity test with CGA 15324 (in vitro test for mutagenic properties of chemical substances in mammalian cells). Unpublished report No. 811491 from Ciba-Geigy Ltd, Basel, Switzerland. Submitted to WHO by Syngenta Crop Protection, Greensboro, North Carolina, USA.

Strasser, F. (1990) CGA 15324 tech.: chromosome studies on Chinese hamster ovary cell line CCL 61 in vitro. Unpublished report No. 891251 from Ciba-Geigy Ltd, Basel, Switzerland. Submitted to WHO by Syngenta Crop Protection, Greensboro, North Carolina, USA.

Sugiya, Y., Yoshida, K., Tamaki, Y., Yokota, M., Abo, Y. & Kawakami, S. (1982) Teratogenicity in rats administered CGA 15’324 (profenofos) prenatally during the major organogenic period (translation). Unpublished report from Medical Research Laboratories Co. Ltd, Saitama, Japan. Submitted to WHO by Syngenta Crop Protection, Greensboro, North Carolina, USA.

Williams, S.C., Marco, G.J., Simoneaux, B.J. & Ballantine, L. (1984) Percutaneous absorption of ^{14}C-profenofos in rats. Unpublished report No. ABR-84023 from Ciba-Geigy Corporation, Greensboro, North Carolina, USA. Submitted to WHO by Syngenta Crop Protection, Greensboro, North Carolina, USA.

Wing, K.D., Glickman, A.H. & Casida, J.E. (1983) Oxidative bioactivation of S-alkyl phosphorothiolate pesticides: stereospecificity of profenofos insecticide activation. *Science*, **219**, 63–65.

Winkler, G. (1996) CGA 15'324 tech. - skin sensitization test in the guinea-pig - maximization test. Unpublished report No. 962054 from Ciba-Geigy Ltd, Basel, Switzerland. Submitted to WHO by Syngenta Crop Protection, Greensboro, North Carolina, USA.

PYRIMETHANIL

First draft prepared by
P.V. Shah and Vicki Dellarco

United States Environmental Protection Agency,
Office of Pesticide Programs,
Washington, DC, USA

Explanation

Pyrimethanil is the approved International Organization for Standards (ISO) name for *N*-(4,6-dimethylpyrimidin-2-yl)aniline (International Union of Pure and Applied Chemistry, IUPAC), also known as 4,6-dimethyl-*N*-phenyl-2-pyrimidinamine (Chemical Abstracts Service, CAS No. 53112-28-0). Pyrimethanil is an anilinopyrimidine fungicide that inhibits the secretion of fungal enzymes. It is a fungicide that is intended for the control of *Botrytis cinerea* on grapes and strawberries.

Pyrimethanil has not been evaluated previously by JMPR and was evaluated by the present Meeting at the request of the Thirty-ninth Session of the Codex Committee on Pesticide Residues (CCPR). All pivotal studies with pyrimethanil were certified as complying with good laboratory practice (GLP).

Evaluation for acceptable daily intake

Unless otherwise stated, studies evaluated in this monograph were performed by GLP-certified laboratories and complied with the relevant Organisation for Economic Co-operation and Development (OECD) and/or United States Environment Protection Agency (EPA) test guideline(s).

1. Biochemical aspects: absorption, distribution, and excretion

Rats

The absorption, distribution, and elimination of pyrimethanil was studied after oral dosing with pyrimethanil radiolabelled with ^{14}C as shown in Figure 1.

In a toxicokinetic study, groups of five male and five female Crl:CD(SD)BR rats were given [U-phenyl ring ^{14}C]pyrimethanil (purity, 98.1–98.4%) as a single oral dose at 11.8 or 800 mg/kg bw by gavage in an aqueous solution of 1% (w/v) gum tragacanth. Treated rats were housed in Radley all-glass metabolism cages for 96 h. Urine and faeces were collected at 24, 48, 72 and 96 h after dosing. Cages were washed at 24 h and 96 h after dosing. At 96 h, various organs and tissues were removed and blood was sampled and all were analysed for radioactivity. Tissues were either combusted or solubilized. All samples were analysed for radioactivity by liquid scintillation counting (LSC).

Pyrimethanil was rapidly excreted in both groups, with more than 95% of the lower dose and 63–67% of the higher dose being excreted in the first 24 h (Table 1). The major route of excretion was in the urine that, together with the cage wash, accounted for about 80% of the administered dose in both groups, indicating at least 80% of the orally administered dose was absorbed in 96 h. There were no sex-related differences in the route or rate of elimination in the two dosed groups.

At the lower dose, most of the tissues examined had radioactive residues that were below the limit of detection at 96 h, except the carcass and the liver (0.082–0.223 mg pyrimethanil equivalent/kg). At the higher dose, the highest residues were found in the liver (10.33–11.27 mg pyrimethanil equivalent/kg) and kidneys (8.845–8.566 mg pyrimethanil equivalent/kg). The concentrations of residue in the bone, brain, eyes, muscle, pituitary, plasma, renal fat and gonads were less than 1 mg pyrimethanil equivalent/kg. There were no significant differences between the residue concentrations seen in the tissues of male and female rats. There were no significant sex-related differences in the rate, route or tissue distribution were observed in male and female rats (Needham & Hemmings, 1991).

In a kinetic study in plasma, two groups of 24 male Crl:CD(SD)BR rats were dosed orally with an aqueous suspension of [U-phenyl ring ^{14}C]pyrimethanil (purity, 100%)at a nominal dose of 10 or 1000 mg/kg bw by gavage. The actual achieved doses were 9.99 and 748 mg/kg bw for the lower and higher dose, respectively. Groups of three rats were killed at 20 min, 1 h, 1.5 h, 3 h, 5 h, 7 h, 12 h and

Figure 1. Position of the pyrimethanil radiolabel used in pharmacokinetic studies in rats

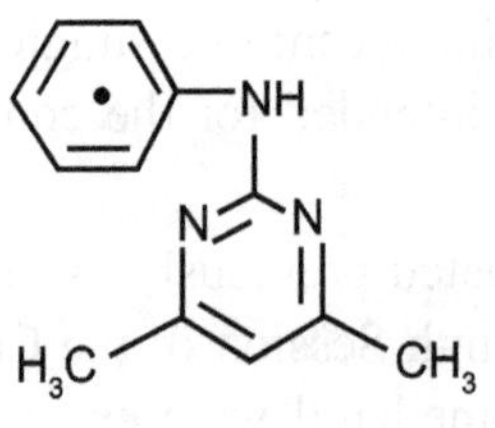

NH
N N
H_3C CH_3

* Position of radiolabel.

Table 1. Excretion profile in rats given radiolabelled pyrimthanil as a single dose by gavage

Time-point	Recovery of administered dose (%)[a]			
	Lower dose (11.8 mg/kg bw)		Higher dose (800 mg/kg bw)	
	Male	Female	Male	Female
Urine:				
24 h	73.60	70.97	42.45	43.30
48 h	1.25	1.64	22.47	18.82
72 h	0.43	0.59	1.61	2.04
96 h	0.43	0.48	0.33	0.56
Total	75.70	73.69	66.86	64.71
Faeces:				
24 h	19.38	20.32	8.86	9.91
48 h	1.04	1.98	5.71	6.08
72 h	0.28	0.27	0.66	1.84
96 h	0.19	0.20	0.22	0.37
Total	20.88	22.78	15.46	18.20
Cage wash[b]	5.69	4.91	12.32	14.60
Total recovery	102.27	101.37	94.63	97.51

From Needham & Hemmings (1991) and Jardinet (2006)
[a] Each group contained an average of five rats.
[b] Combined results for 24 h and 96 h.

24 h after dosing, blood was removed for analysis and a sample was centrifuged to produce plasma. The concentration of radioactivity was measured in the blood by combustion and in the plasma by direct addition to scintillation fluid.

After a single oral dose of radiolabelled pyrimethanil at 9.99 mg/kg bw, the concentrations of radioactivity in plasma reached a measured maximum of 3.89 mg/kg at 1 h after dosing, declining to 0.04 mg/kg by 24 h with a calculated half-life of 4.8 h (Table 2). The blood : plasma ratio showed that the radioactivity was mainly associated with the plasma at the early time-points when the phenol metabolite, SN 614 276 and unchanged pyrimethanil were the major components present. After dosing with radiolabelled pyrimethanil at 748 mg/kg bw, concentrations of radioactivity showed an initial peak (49.477 mg equivalent/kg) at 1 h after dosing. After an initial decline, there was a second peak of radioactivity (52.580 mg equivalent/kg) at 5 h after dosing and the residues then declined with a half-life of 11.8 h. The calculated maximum plasma concentration (C_{max}) was 49.58 mg/kg at a T_{max} of 3.94 h after dosing and the area under the curve of concentration–time (AUC) was 96 times greater than that of the rats receiving the lower dose.

The blood : plasma ratio of the rats at the higher dose remained constant throughout the study (0.72 ± 0.05) indicating that the radioactivity remained mainly associated with the plasma, and high-performance liquid chromatography (HPLC) analysis showed that unchanged pyrimethanil and SN 614 276 were the major components present. At both doses, clearance of radiolabelled residues was rapid. This combined with a rapid plasma half-life indicates that pyrimethanil residues would not accumulate in tissues after repeated doses (Challis, 1995).

In a distribution study, groups of three male and three female Crl:CD(SD)BR rats were given a single oral dose of [U-phenyl ring ^{14}C]pyrimethanil (purity, 98.1–98.4%) at 10 or 800 mg/kg bw by gavage in an aqueous solution of 1% (w/v) gum tragacanth. The dose volume was 5 ml/kg bw for the lower dose and 10 ml/kg for the higher dose. Treated rats were housed in steel cages (three males and

Table 2. Calculated pharmacokinetic parameters in male rats receiving a single dose of radiolabelled pyrimethanil

Radioactivity	Parameter				
	C_{max} (mg/kg)	T_{max} (h)	Half-life (t1/2 h)	AUC (mg/kg per h)	Clearance (kg/h)
Lower dose (9.99 mg/kg bw)[a]					
Total radioactivity	4.62	0.735	4.80	11.3	0.881
Pyrimethanil	0.625	0.607		1.24	8.08
SN 614 276	2.39	0.957		5.67	1.76
Higher dose (748 mg/kg bw)[a]					
Total radioactivity	56.5	3.94	11.8	1080	0.707
Pyrimethanil	19.0	1.33		76.6	9.95
SN 614 276	15.3	1.29		71.5	10.7

From Challis (1995) and Jardinet (2006)
[a] An average of three rats were used for each time-point.
AUC, area under the plasma concentration–time curve extrapolated to infinity; C_{max}, calculated maximum plasma concentration; T_{max}, calculated time of maximum plasma concentration.

three females per cage). For the group at the lower dose, treated rats were killed (three males and three females per group) at 1, 2, 4, 8, 12 and 24 h after dosing, while for the group at the higher dose, rats were killed (three males and three females per group) at 2, 6, 10, 18, 24 and 48 h after dosing. Various organs and tissues were removed and analysed for radioactivity. Blood was analysed for radioactivity. Tissues were either combusted or solubilized. All samples were analysed for radioactivity by LSC.

Radiochemical purity was verified by thin-layer chromatography (TLC) and HPLC. The absorption of pyrimethanil was rapid after the administration of a single low dose at 10 mg/kg bw by gavage and pyrimethanil was widely distributed into the tissues. At the lower dose, peak concentrations of radioactive residues were achieved by 1 h after dosing. With the exception of the gastrointestinal tract, the highest concentrations were associated with the renal fat, thyroid, adrenals, liver, kidney and ovaries, which also had tissue : plasma ratios of > 1 throughout the study. With the exception of the gonads, there were no consistent significant differences in tissue concentrations of pyrimethanil between male and female rats. At the higher dose (800 mg/kg bw), the peak concentrations of radioactivity were achieved in tissues between 2 and 10 h after dosing and then declined slowly thereafter. With the exception of the 2-h time-point, there were no consistent significant differences between the concentration of pyrimethanil residues in the tissues of male and female rats at this dose. The highest concentrations of residues were found in the renal fat, thyroid, adrenals, liver, kidney and ovaries. These tissues had tissue : plasma ratios of > 1. The concentration of radioactive residues was one to three orders of magnitude greater than those found in the rats at the lower dose, although the tissue distribution ratios were 1.5–8 times lower.

The half-lives of radiolabelled pyrimethanil and its metabolites in selected tissues (Table 3) were independent of sex but were greater at the higher dose. The elimination half-lives in the plasma were 5.41 and 5.23 and 8.22 and 11.69 in males and females at the lower and higher dose, respectively (Whitby, 1995a, 1995b).

In a study of whole-body radioautography, groups of male and female Crl:CD(SD)BR rats were given a single oral dose of [U-phenyl ring ^{14}C]pyrimethanil (purity, 98.8%) at 10 or 800 mg/kg bw by gavage in an aqueous solution of 1% (w/v) gum tragacanth. One male and one female from each group was killed at 45 min, 90 min, 3 h, 6 h, 12 h, 24 h or 48 h after dosing. The rats were frozen and prepared for whole-body radioautography. Sagittal sections (about 25 μm thickness) were then taken at six levels in order to visualize the major tissues and organs of the rat. After 21 or 42

Table 3. Half-life of pyrimethanil and its metabolites in selected tissues of rats given radiolabelled pyrimethanil as a single dose by gavage

Tissue	Half-life (h)			
	Higher dose (10 mg/kg bw)		Lower dose (800 mg/kg bw)	
	Males	Females	Males	Females
Blood	10.18	7.51	18.37	22.49
Kidney	5.57	4.79	10.33	14.02
Liver	7.76	8.64	19.60	21.26
Plasma	5.41	5.23	8.22	11.69
Renal fat	3.89	3.70	10.42	20.94

From Whitby (1995a, 1995b) and Jardinet (2006)

days of exposure, the autoradiograms were developed and analysed using the Seescan Solitaire Plus Quantitative Whole-body Autography System (QWBAS).

With the exception of the gonads, the distribution of radioactivity was similar in male and female rats given a single oral dose of pyrimethanil at 10 mg/kg bw. Absorption was rapid, with maximum concentrations of radioactivity in most tissues being found at the first time-point (45 min after dosing) when concentrations of > 10 mg equivalent/kg were present in the lachrymal glands, Harderian gland (females), kidney, liver and white fat. Residue concentrations were > 1 mg equivalent/kg tissue in all other tissues except for the bone and eye in males. The residues were rapidly cleared from the tissues and, by 6 h after dosing, concentrations of > 1 mg equivalent/kg were detected only in the kidney and liver (the organs of excretion), and the white fat of females. After administration of a single oral dose of radiolabelled pyrimethanil at 800 mg/kg bw, the peak concentrations of radioactivity in the tissues occurred at 3–6 h in males and 6–12 h in females. Maximum concentrations of > 100 mg equivalent/kg were detected in the adrenal gland, brown fat, Harderian gland, lachrymal glands, kidney, liver, spleen and white fat of males and females, in the seminal vesicles of males, and in the blood, lung, lymph nodes, myocardium, ovary, pancreas and salivary gland of females. As with the lower dose, the concentrations fell rapidly by 24 h after dosing. The concentrations of the residues at this time were higher in females (in which the adrenal, kidney, liver, ovary and pancreas contained concentrations of greater than 50 mg equivalent/kg) than in the males (only the liver, kidney and thyroid contained residues at a concentration of >10 mg equivalent/kg). Absorption was more protracted at the higher dose, and the concentration of residues in the blood and highly perfused tissues did not increase in a dose-related manner. It is therefore possible that the absorption of pyrimethanil was limited by the dissolution rate of the compound. Irrespective of the dose, the radioactive residues were rapidly cleared from all rats such that by 48 h after dosing there was virtually no quantifiable radioactivity in any of the tissues (Whitby, 1993).

Groups of three male Crl:CD(SD)BR rats were given [U-phenyl ring ^{14}C]pyrimethanil (purity, > 98%) at a dose of 10 mg/kg bw orally once per day for 28 days, with periodic sacrifices at days 1, 3, 5, 8, 11, 17, 23, and 28. An additional group of rats was placed in all-glass metabolism cages after the final dose on day 28 for collection of urine and faeces for 4 days. At necropsy, various tissues and organs were removed and analysed for radioactivity. Plasma was prepared from blood and analysed. Tissues were either combusted or solubilized. All samples were analysed for radioactivity by LSC.

Recovery of radioactivity was not reported. Detectable concentrations of radiolabel were found in the adrenals, blood, kidney, liver, spleen, and thyroid after exposure. At 24 h, after a single dose, only the blood (0.16 mg/kg bw) and liver (0.40 mg/kg bw) displayed detectable concentrations of radiolabel that were higher than the limit of detection. Four days after the last dose, detectable concentrations of radiolabel were found only in the liver, kidney, and thyroids. It appeared that the concentrations in the blood, kidney, and thyroid would continue to increase with increased exposure time, while the concentration in the adrenal appeared to have reached a plateau, and that in the liver appeared to be declining. Although urine and faeces were collected from one group after 28 days, no data were provided in the study report (Hemmings, 1991a).

In an another study with repeated doses, groups of five male and five female Crl:CD(SD)BR rats were given 14 daily doses of unlabelled pyrimethanil at 10 mg/kg bw per day by gavage. They were then given a single oral dose of [U-phenyl ring ^{14}C]pyrimethanil (purity, 99.00% or greater) at 10 mg/kg bw by gavage in an aqueous solution of 1% (w/v) gum tragacanth. Treated rats were placed in all-glass metabolism cages. Urine, faeces and cage wash were collected at 6 h and 24 h after dosing in cooled containers to prevent bacterial degradation of the metabolites present. At necropsy (24 h after dosing with the radiolabel) the following tissues/organs were removed for analysis: adrenals, blood, bone, brain, residual carcass, eyes, gastrointestinal tract, heart, kidneys, liver, lungs, muscle, ovaries, renal fat, spleen and testes. Plasma was prepared from the blood. Tissues were either combusted or solubilized. All samples were analysed for radioactivity by LSC.

Pyrimethanil was rapidly excreted, with 91% of the administered dose being excreted within 24 h. The major route of elimination was via the urine that, together with the cage wash, accounted for 72% of the administered dose. There were no sex differences in the rate or route of elimination. Tissue concentrations of residue at necropsy were low, with concentrations in most tissues being below the limits of quantification. Quantifiable residues were only found in the liver (0.30 and 0.44 mg equivalent/kg for male and females, respectively), kidney (0.13 and 0.17 mg equivalent/kg for males and females, respectively) and blood (0.044 and 0.058 mg equivalent/kg for males and females, respectively). The excretion profile and magnitude of the tissue residues were similar to those seen 24 h after a single oral dose of pyrimethanil at 11.8 mg/kg bw (Hemmings, 1993).

In a study of metabolite identification, the excreta from rats at the lower dose (11.8 mg/kg bw) and higher dose (800 mg/kg bw) were collected at 24, 48, 72 and 96 h after dosing and subjected to metabolic identification. The excreta from rats receiving the repeated dose at 10 mg/kg bw for 14 days was also subjected to metabolic identification. The proposed metabolic scheme of pyrimethanil in rats is shown in Figure 2. The excreta contained more than 95% and 63–67% of the administered dose after a single dose at 11.8 mg/kg bw or 800 mg/kg bw, respectively, in the first 24 h. Approximately 89–90% of the administered dose was recovered in the excreta in the first 24 h after repeated dosing at 10 mg/kg bw for 14 days. Many metabolites of pyrimethanil were excreted as glucuronide or sulfate conjugates that could be released after incubation with *Helix pomatia* juice, a mixture of beta D-glucuronidase and aryl sulfatase obtained from snails. Only low concentrations of unchanged pyrimethanil were detected in faeces. The main pathways of metabolism involved aromatic oxidation to form phenols in either or both rings, which were then excreted as glucuronide and/or sulfate conjugates. After deconjugation to release the aglycones, the major metabolite peak was identified as 4-(4,6-dimethylpyrimidin-2-ylamino)phenol (SN 614 276, which was also detected as a sulfate conjugate). Aromatic oxidation on the pyrimidine ring produced 2-anilino-4, 6-dimethylpyrimidin-5-ol (SN 614 277), and oxidation on both rings gave 2-(4-hydroxyanilino)-4, 6-dimethylpyrimidin-5-ol (SN 615 244). A minor metabolic pathway resulted in oxidation of the methyl group on the pyrimidine ring to produce an alcohol with the proposed structure of 2-anilino-6-methylpyrimidine-4-methanol (SX 614 278). This was present in trace amounts in the excreta, but was further oxidized to the corresponding phenol, 2-(4-hyroxyanilino)-6-methylpyrimidine-4-methanol (SN 614 800).

The mass spectra of the polar metabolites did not contain any recognizable fragments. The polar fraction was therefore collected and treated with diazomethane to methylate any acidic protons. The resulting mixture then separated on HPLC into four peaks, which were collected and examined by mass spectrometry. One of the peaks contained the diphenol SN 615 224, two other peaks contained a monomethylated derivative of SN 615 224, and the final peak contained the pyrimidine phenol

Table 4. Identification of metabolites in excreta and cage wash of rats given radiolabelled pyrimethanil by gavage

Sample	Recovery in excreta (% of adminis-tered dose)	Major metabolites present in excreta (mean % of administered dose)									
		Polar metabolite	SN 614 800	SN 615 224	SN 614 276 sulfate	SN 614 276	SX 614 278	SN 614 277	SN 100 309	Sum of metabolites	
										Characterized	Identified
Single dose at 10 mg/kg bw:											
Males											
Urine 0–24 h	73.6	26.1	5.7	ND	8.9	30.3	0.8	ND	ND	26.1	45.7
Faeces 0–24 h	19.4	5.3	0.3	0.7	1.1	3.8	0.3	2.2	2.1	5.3	10.6
Cage wash 0–24 h	5.1	1.8	0.4	ND	0.6	2.1	0.1	ND	ND	1.8	3.2
Sum	98.1	—	—	—	—	—	—	—	—	33.2[a]	59.4
Total[b]	—	—	—	—	—	—	—	—	—	—	92.6
Females											
Urine 0–24 h	71.0	29.7	3.1	ND	12.3	24.8	1.2	ND	ND	29.7	41.4
Faeces 0–24 h	20.3	6.4	0.8	1.1	1.4	5.2	0.3	1.8	0.3	6.4	11.0
Cage wash 0–24 h	4.3	1.8	0.2	ND	0.8	1.5	0.1	ND	ND	1.8	2.5
Sum	95.6	—	—	—	—	—	—	—	—	37.9 [a]	54.9 [a]
Total[b]	—	—	—	—	—	—	—	—	—	—	92.8
Single dose at 800 mg/kg bw:											
Males											
Urine 0–48 h	64.9	19.9	2.5	1.0	4.5	19.3	2.3	7.9	ND	19.9	37.5
Faeces 0–48 h	14.6	4.5	0.7	ND	1.3	2.9	0.3	0.8	2.7	4.5	8.6
Cage wash 0–24 h	11.6	3.5	0.4	0.2	0.8	3.4	0.4	1.4	ND	3.5	6.7
Sum	91.1	—	—	—	—	—	—	—	—	28.0 [a]	52.8 [a]
Total[b]	—	—	—	—	—	—	—	—	—	—	80.8
Females											
Urine 0–48 h	62.1	19.4	4.0	5.0	6.0	14.9	ND	6.8	ND	19.4	36.7
Faeces 0–48 h	16.0	6.8	0.8	ND	1.2	4.4	0.2	0.4	0.6	6.8	7.5
Cage wash 0–24 h	13.4	4.2	0.9	1.1	1.3	3.2	ND	1.5	ND	4.2	7.9

Sum	91.5	—	—	—	—	—	—	—	—	30.5 [a]	52.1 [a]
Total[b]	—	—	—	—	—	—	—	—	—	—	82.5
Repeated dosing at10 mg/kg bw per day:											
Males											
Urine 0–24 h	61.4	30.6	4.7	ND	5.7	7.8	0.9	1.9	ND	30.6	21.0
Faeces 0–24 h	17.9	9.4	ND	1.4	1.5	0.8	1.5	1.7	0.5	9.4	7.4
Cage wash 0–24 h	10.2	5.1	0.8	ND	0.9	1.3	0.2	0.3	ND	5.1	3.5
Sum	89.6	—	—	—	—	—	—	—	—	45.2 [a]	31.9 [a]
Total[b]	—	—	—	—	—	—	—	—	—	—	77.1
Females											
Urine 0–24 h	58.8	31.7	3.6	ND	8.0	4.9	0.9	ND	ND	31.7	17.5
Faeces 0–24 h	16.8	9.6	ND	1.2	1.5	1.5	ND	1.5	0.7	9.6	6.4
Cage wash 0–24 h	13.6	7.3	0.8	ND	1.8	1.1	0.2	ND	ND	7.3	4.0
Sum	89.1	—	—	—	—	—	—	—	—	48.6[a]	28.0[a]
Total[b]	—	—	—	—	—	—	—	—	—	—	76.6

Data from Grosshans (2003)

[a] Sum (urine, faeces and cage wash).

[b] Characterized (classified as either polar or non polar) and identified (identity confirmed).

SN 614 277. The presence of these compounds suggested that the original polar metabolites were possibly polymers or conjugates of these metabolites, which were now breaking down in the mass spectrometer to yield recognizable fragments (Needham & Hemmings, 1993; Needham, 1996).

In response to an enquiry from the country preparing the EC draft assessment (Austria), the quantity of metabolites identified was recalculated as a percentage of the administered dose. It was also stated in the study report that the radioactivity found in the cage wash most probably consisted of spoiled urine. Therefore the calculation of individual metabolites in the cage wash as a percentage of administered dose was based on the concentration of metabolites in the urine (as a percentage of total activity) stated in the dossier table and the percentage of activity found in the cage wash. The amount of metabolites identified in various fractions after single low, single high and repeated dosing is shown in Table 4 (Grosshans, 2003).

Figure 2. Metabolic pathway of pyrimethanil in rats

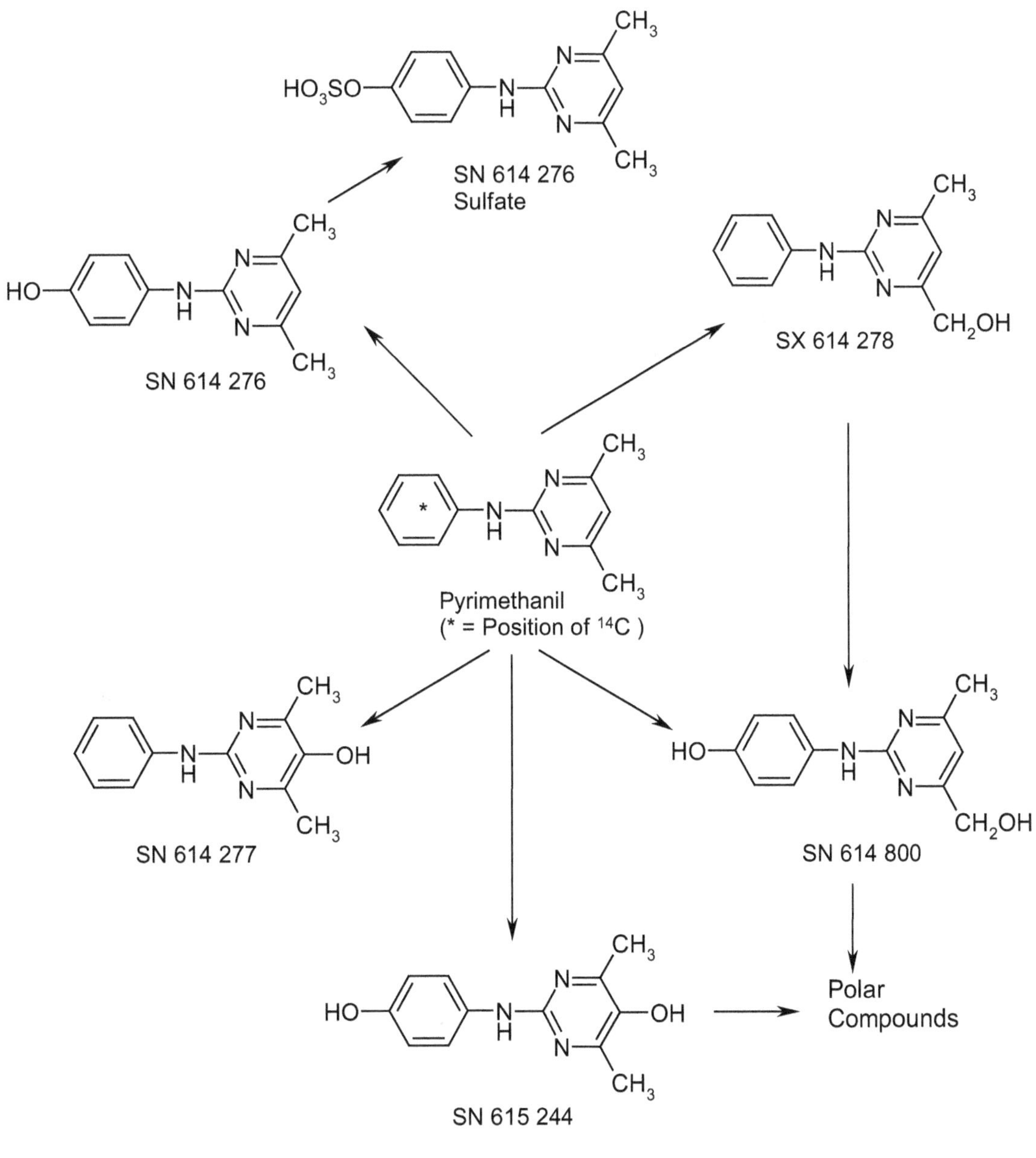

Mice

In a study of absorption, distribution and excretion, groups of five male and five female fasted CD1 mice were treated with [U-phenyl ring ^{14}C]pyrimethanil (purity not reported) at a dose of 10 mg/kg bw by gavage in water containing 1% gum tragacanth. Mice were housed individually in glass metabolism cages with food and water freely available. Urine and faeces were collected at 24, 48, 72 and 96 h after dosing. Cage washes were collected at 24 and 96 h after dosing. At the end of the collection period, mice were killed and blood collected. Several tissues and organs were removed and analysed for radioactivity. Carcass were weighed, homogenized and analysed. All samples were analysed for radioactivity by LSC.

One male mouse died in first 24 h after dosing, probably due to gavage error. Total recovery of radioactivity ranged from 91.96% to 123.03%, with an average recovery of 108.8%. Pyrimethanil was absorbed quickly and excreted rapidly from the body, with excretion being nearly complete within 24 h. Urinary excretion in male mice was 60.5%, 2.83%, 0.96% and 0.15% of the administered dose at 24, 48, 72 and 96 h after dosing, respectively. Urinary excretion in female mice was 66.32%, 2.44%, 0.80% and 0.21% of the administered dose at 24, 48, 72 and 96 h after dosing, respectively. Faecal excretion in male mice was 21.00%, 1.71%, 0.67% and 0.39% of the administered dose at 24, 48, 72 and 96 h after dosing, respectively. Faecal excretion in female mice was 13.41%, 2.12%, 0.58% and 0.47% of the administered dose at 24, 48, 72 and 96 h after dosing, respectively. Radioactivity in cage wash at 96 h was 21.08% and 22.12% in male and female mice, respectively. Urine (urine + cage wash combined) was the major route of excretion, containing 85.5 ± 10.4% of the administered dose in males and 91.9 ± 21.0% in females. The remainder of the administered dose was excreted via the faeces. At 96 h after dosing, only the liver, kidney, blood and carcass contained measurable residues (0.003–0.040 mg pyrimethanil equivalent/kg) that were above the limit of detection (Hemmings, 1991b).

Dogs

In a study of absorption, distribution and excretion, groups of three male and three female beagle dogs were given a gelatin capsule containing [U-phenyl ring ^{14}C]pyrimethanil (purity not reported) at a dose of 10 mg/kg bw. Dogs were housed individually in stainless steel metabolism cages with water freely available and were fed pelleted diet. Urine was collected at 6, 24, 48, and 72 h after dosing and cages were washed and collected for quantification. Faeces were collected at 24, 48 and 72 h after dosing. Blood samples were taken from the jugular vein at 0.5, 1, 2, 3, 4, 6, 8, 10, 12, 24, 30, 48 and 72 h after dosing and centrifuged to provide plasma. At the end of the collection period, the dogs were killed. Several tissues and organs were removed and analysed for radioactivity. The carcass was not examined for radioactivity. All samples were analysed for radioactivity by LSC.

Total recovery of the orally administered dose of pyrimethanil in excreta, cage wash and tissues was 94.6% and 94.5% in male and female dogs, respectively. More than 70% of the orally administered dose was excreted in 24 h and excretion was nearly complete in 48 h after dosing. The main route of excretion was via the faeces, which accounted for 59.4% and 53.2% of the administered dose in males and females, respectively, in 72 h. Urinary excretion in 72 h was 29.2 and 37.7% of the administered in males and females, respectively. Study authors suggested that there was a strong evidence to support biliary excretion of orally administered pyrimethanil since higher concentrations of radioactivity were detected in the bile than in the liver or kidney at necropsy. With the exception of the bile and gastrointestinal tract, the concentration of radioactive residues in the tissues at necropsy, 72 h after dosing, was low with only the adrenals, liver, thyroid and male spleen containing concentrations of radioactivity of > 0.1 mg equivalent/kg. The concentration of pyrimethanil-derived residues in the blood and plasma of the dogs reached a maximum at 0.5–3 h after dosing. The C_{max} was calculated as 3.68 ± 2.86 and 5.82 ± 0.13 mg equivalent/kg, and the elimination half-life ($t_{½}$) was 11.6 ± 1.9 h and 13.5 ± 2.7 h for male and female dogs, respectively. The results suggested that pharmacokinetic parameters were comparable in male and female dogs (Reynolds & Swalwell, 1992).

2. Toxicological studies

2.1 *Acute toxicity*

Results of studies of acute toxicity with pyrimethanil are summarized in Table 5. Young adult CD-1 mice were given pyrimethanil as a single dose at 0, 1250, 2500 or 5000 mg/kg bw. Two females and three males at 5000 mg/kg bw were killed in extremis within five h of dosing. One male died on day 6 after dosing. The death of male on day 6 was attributed to fighting. Rats dosed at 1250 mg/kg bw and greater generally showed reduced activity, reduced muscle tone, prostration and coolness to the touch, which were generally observed within about 2 h after dosing. Recovery was complete within 5 h and within 6 days in the group at the highest dose. Groups of young adult Sprague-Dawley rats were given pyrimethanil as a single dose at 0, 800, 1600 or 6400 mg/kg bw. Five males and three females at 6400 mg/kg bw were sacrificed in extremis 6 h after dosing. Some clinical signs such as reduced activity, reduced muscle tone, body soiling, urogenital soiling, ataxia and hunched posture were seen within 1 h after dosing and recovery was complete within 4 h at 800 mg/kg bw, and within 14 days at higher doses. No abnormal findings were observed on necropsy. No clinical signs of toxicity were observed in the study of toxicity with pyrimethanil delivered by inhalation and also in the studies of eye and skin irritation.

(a) Oral administration

Mice

Groups of five male and five female young adult CD-1 mice were given pyrimethanil (purity, 98.4%) as a single dose at 0, 1250, 2500 or 5000 mg/kg bw by gavage in 1% w/v methyl cellulose in distilled water. Treated mice were subjected to gross necropsy at the end of a 14-day observation period. Two females and three males at 5000 mg/kg bw were killed in extremis within 5 h after dosing. One male died on day 6 after dosing. The death of one male on day 6 was attributed to fighting. Mice at 5000 mg/kg generally showed reduced activity, reduced muscle tone, prostration and coolness to the touch. Clinical observations were generally observed 2 h after dosing in males and 2–5 h after dosing in females. The surviving male was recovered by 6 days after dosing. Surviving females recovered within 5 h after dosing. At 2500 mg/kg bw, reduced activity and reduced muscle tone were observed within 2 h after dosing and recovery was complete within 5 h. At 1250 mg/kg bw, reduced activity and reduced muscle tone was observed in four out of five females within 2 h after dosing and they recovered within 5 h. No effects on body weights were observed within 14 days. No abnormal

Table 5. Results of studies of acute toxicity with pyrimethanil

Species	Strain	Route	Purity (%)	LD_{50} (mg/kg bw)	LC_{50} (mg/l air)	Reference
Mice	CD1, Crl:CD-1-(ICR)BR	Oral	98.4	Males: 4665 (range, 3322–6552) Females: 5359 (range, 3816–7526)	—	Malarkey (1990)
Rat	Sprague-Dawley, COBS CD	Oral	98.4	Males: 4149 (range, 3341–5153) Females: 5971 (range, 4252–8386)	—	Markham (1989a)
Rat	Sprague-Dawley, COBS CD	Dermal	NS	> 5000 (males and females)	—	Markham (1989b)
Rat	Sprague-Dawley, Crl:CD(SD)BR	Inhalation (4-h, nose only)	NS	—	> 1.98 (males and females)	Jackson & Hardy (1992)

NS, not stated.

findings were observed on necropsy at termination. The oral median lethal dose (LD50) values in mice (with 95% confidence interval, CI) were 4665 (95% CI, 3322–6552) and 5359 (3816–7526) mg/kg bw for males and females, respectively (Malarkey, 1990).

Rats

Groups of five male and five female young adult Sprague Dawley (COBS CD) rats were given pyrimethanil (purity, 98.4%) as a single dose at 0, 800, 1600 or 6400 mg/kg bw by gavage in 1% w/v methyl cellulose in distilled water. Treated rats were subjected to gross necropsy at the end of a 14-day observation period. Five males and three females at 6400 mg/kg bw were killed in extremis 6 h after dosing. Two surviving females at 6400 mg/kg bw showed slight signs of reduced muscle tone, body soiling, urogenital soiling and hunched posture until termination. All rats at 3200 mg/kg bw showed slight signs of reduced activity and muscle tone, ataxia and hunched posture, 1 h after dosing and recovery was complete after 14 days. At 1600 mg/kg bw, four out of five males and two out of five females showed slightly reduced muscle tone or reduced activity. Females showed complete recovery within 24 h, and males within 3 days. At 800 mg/kg bw, one male showed slightly reduced activity immediately after dosing, but appeared normal within 4 h. No effects on body weights were observed within 14 days. No abnormal findings were observed on necropsy at termination. The oral LD_{50} values in rats were 4149 (95% CI, 3341–5153) and 5971 (95% CI, 4252–8386) mg/kg bw for males and females, respectively (Markham, 1989a).

(b) Dermal irritation

In a study of primary dermal irritation, young adult female New Zealand White rabbits were dermally exposed to 0.5 g of pyrimethanil (purity, 98.4%), moistened with distilled water, for 4 h. the rabbits were then observed for 3 days. Irritation was scored by the Draize method at 30–60 min, 24, 48, and 72 h after removal of dressing. No erythema or oedema was noted on any rabbits during the study. Pyrimethanil was considered to be a non-irritant to the intact rabbit skin (Markham, 1989c).

(c) Ocular irritation

In a study of primary ocular irritation, 0.1 ml of pyrimethanil (purity, 98.4%), was instilled into the conjunctival sac of one eye of three male and three female young adult New Zealand White rabbits. Irritation was scored by the Draize method at 1 h, and 1, 2, 3 and 4 days after exposure. One rabbit showed very slight redness of the conjunctiva, 1 h after instillation, and recovered within 24 h. Pyrimethanil was considered to be minimally irritating to the eyes of rabbits under the conditions of this study (Markham, 1989d).

(d) Dermal sensitization

In a study of dermal sensitization with pyrimethanil (purity, 99.7%), 10 young female Dunkin Hartley guinea-pigs were tested using the Buehler method. The main study was conducted using 10 guinea-pigs each in the control and treated group. On the basis of results of the preliminary range-finding study, a maximum non-irritant concentration of 60% (w/v) of pyrimethanil in Alembicol D was selected for the induction and challenge phases. In the treatment group, each guinea-pig received an induction application of 60% w/v of the test material on the dorsal skin under an occlusive dressing for 6 h. Another control group of five males and five females received Alembicol D (vehicle) only. Two further induction applications were made at weekly intervals. Two weeks after the final induction application, test guinea-pigs were challenged with pyrimethanil at a concentration of 60% w/v under an occlusive dressing for 6 h. Assessments of the dermal reactions were made 24, 48, and 72 h after the challenge application. Guinea-pigs in the control group were challenged with the vehicle. One guinea-pig in the control group was killed in extremis on study day 11. Necropsy findings included intussusception of the jejunum, with the stomach and duodenum being distended with fluid and gas. Body weights and body-weight gains were not affected by the treatment. In the induction phase, three treated rabbits showed slight erythema (score 1), which was not persistent. No irritation was observed in the treated group during the challenge phase. No irritation was observed in the control group either in the induction or challenge phase. Positive controls were treated using 1-chloro-2,4-dinitrobenzene (DNCB), which indicated that the test system was capable of detecting a sensitizer (Davies, 1990a, 1990b). Thus pyrimethanil was not a skin sensitizer in guinea-pigs as determined by the Buehler method (Davies, 1990a, 1990b).

In a study of dermal sensitization with pyrimethanil (purity, 96.5%), young male and female Dunkin-Hartley guinea-pigs were tested using the maximization method of Magnusson & Kligman. For the main study, five males and five females were assigned to the control group, and ten males and ten females to the treatment group. The test substance was mixed with paraffin oil. In this study, the test concentrations chosen were 20% for intradermal induction, 50% for topical induction, and 50% for the challenge. Skin reactions at the challenge sites were observed at 24 h and 48 h after removal of the patch. The positive-control group was treated with DNCB. No mortalities or clinical signs of toxicity were observed. There were no signs of dermal reactions in any guinea-pigs in the induction or challenge phase. Thus pyrimethanil was not a skin sensitizer in guinea-pigs as determined by the maximization method (Clouzeau, 1994; Healing, 1996a, 1996b).

2.2 *Short-term studies of toxicity*

Mice

In a 28-day study of oral toxicity, groups of five male and five female Charles River CD-1 mice were given diets containing pyrimethanil (purity, 95.3–95.6%) at a concentration of 0, 1000, 3000, 10 000, or 30 000 ppm (equivalent to 0/0, 167/236, 567/667, 1960/2357 mg/kg bw per day for males/females) for 28 days. Mice were observed twice per day for mortality and moribundity and once on weekends and holidays. Body weight and food consumption were measured each week. During the study, diets were analysed for homogeneity, stability and concentration. Blood was collected from the retro-orbital sinus of each mouse, under light ether anaesthesia, on day 28 for haematology and days 29 (males) or 30 (females) for clinical chemistry. An ophthalmoscopy examination and urine analysis were not performed. All mice that died or were sacrificed in extremis and those sacrificed at study termination were given a gross pathological examination. Selected organs were weighed. The selected tissues were collected from all mice, preserved in 10% neutral buffered formalin, and stained with haematoxylin and eosin. Tissues from the liver, kidneys, thyroid gland, and urinary bladder were examined in all groups (except mice at 30 000 ppm). A bone-marrow smear was taken from all mice surviving until scheduled sacrifice.

Data on homogeneity, stability and concentrations were not provided in the study report (a separate report that was not submitted. At 30 000 ppm, all mice died within the first week of treatment (days 5–7). Before death, these mice experienced severe emaciation. Clinical signs such as altered activity, prostration, ataxia, pallor and hypothermia associated with emaciation were observed. At necropsy, the males exhibited the various macroscopic non-specific abnormalities. No tissues were examined microscopically from these mice. At 10 000 ppm, emaciation was observed in two females for an average of 5 days. Body weights were decreased by 7–19% in males and females throughout treatment. Overall body-weight gains were decreased (46–55%) in males and females in 29 days. Also in the first 2 weeks of the study, decreases of 17–28% were observed in food consumption in the females, with correlating decreases in food-conversion ratio. Minor alterations in haematological and clinical chemistry parameters were observed, but were considered to be not toxicologically relevant. There were no dose-related macroscopic findings in the mice surviving to scheduled sacrifice. Relative to body weight, liver weight was increased (+19%; $p \leq 0.01$) in females, probably owing to decreases in body weight. Increased incidence and/or severity of the following microscopic findings was observed (relative to controls): (a) minimal to slight pigmentation of the follicular cells in the thyroid in males (five out of five) and females (two out of five); (ii) minimal to moderate urothelial hyperplasia (three out of five) and uroliths present in the lumen of the urinary bladder (one out of five) in males; and (iii) minimal to slight basophilic tubules (two out of five) and slight tubular degeneration (two out of five) in the kidneys in females. At 3000 ppm, relative adrenal weights in males were significantly increased (+40–43%; $p \leq 0.05$). But there were no significant effects at 10 000 ppm; thus the significance of this effect seems questionable. No other treatment-related effects were observed at 3000 ppm. No treatment-related effects were observed in the group at 1000 ppm.

The lowest-observed-adverse-effect level (LOAEL) was 10 000 ppm (equivalent to 1960/2357 mg/kg bw per day for males/females) on the basis of decreased body weights, body-weight gains, food consumption, and food-conversion ratio and on increased incidences of clinical signs (emaciation, general pallor, and piloerection) and microscopic findings in the thyroid, kidneys, and urinary bladder. The NOAEL was 3000 ppm, equivalent to 567/667 mg/kg bw per day for males/females. The study author concluded that the no-observed-effect level (NOEL) was 1000 ppm (Harvey, 1991b).

In a 90-day feeding study, groups of 20 male and 20 female Crl:CD-1(ICR)BR mice were fed diets containing pyrimethanil (purity, 97.7–97.9%) at a concentration of 0, 80, 900, or 10 000 ppm for 13 weeks. Those doses were equivalent to 0, 12, 139 or 1864 mg/kg bw per day for males and 0, 18, 203 or 2545 mg/kg bw per day for females, respectively. Diets were prepared twice each week. Stability, homogeneity and dietary concentrations were confirmed analytically. Mice were inspected twice per day for signs of toxicity and mortality, with detailed cage-side observations were made once per day. Body weight and food consumption were measured each week. At termination, blood was taken for haematological and clinical chemistry analyses. Urine analysis and ophthalmoscopic examinations were not performed. All mice that died and those sacrificed on schedule (10 males and 10 females per group) were subjected to gross pathological examination and selected organs were weighed. Selected tissues were collected for histological examination.

The results on test article homogeneity were reported in a separate report (not submitted to the Meeting). The test article was stable in the diet for 4 days at room temperature. The concentration analysis of the test substance indicated that the measured concentration ranged from 87.1% to 100.8% of the target concentrations. There were no treatment-related effects on mortality, clinical signs or water intake. The body weights measured each week did not show statistically significant difference between groups. At 10 000 ppm, total body-weight gains were 12.4% lower in males and 7.2% lower in females than in the control group. There was an overall increase in food consumption (14.1% in males, 9.8% in females) compared with controls, resulting in reduced food-conversion ratios in treated mice. Clinical chemistry parameters showed significant increases in serum cholesterol and total bilirubin concentration (in females only). No treatment-related effects on haematological parameters were observed. Treatment-related necropsy findings included dark thyroid glands in eight out of ten males and a bladder stone in one out of ten females. Significantly increased liver weight in females and increased relative-liver-weight-to-body-weight ratios in males and females were observed at the highest dose. Histopathological changes were detected in the kidneys, liver, thyroid glands, and urinary bladder. Tubular dilation was seen in the kidneys of three out of ten males. Uroliths were detected in urinary bladders of one out of ten males and four out of ten females (one bladder-stone was detected at necropsy), with three out of ten females showing hyperplasia of the bladder epithelium at the highest dose. There was marked depletion of glycogen in the liver of males and females as indicated by decreased margination of cytoplasm and reduced intensity of periodic acid Schiff (PAS) staining (performed for males only). In the thyroid gland, exfoliative necrosis of follicular cells was seen in eight out of ten males and one out of ten females. Pigmentation of follicular cells was seen in ten out of ten males and nine out of ten females at 10 000 ppm. Special staining techniques indicated that the pigment was lipofuscin. At 900 ppm, some glycogen depletion was still evident in the liver; however, no significant histopathological finding was reported at this dose. This condition is associated with the nutritional and/or physiological status of the mice and has no toxicological significance. At 80 ppm, no treatment-related effects were detected.

The NOAEL was 900 ppm, equal to 139 and 203 mg/kg bw per day for males and females, respectively, on the basis of decreased body-weight gains, slightly increases in cholesterol and total bilirubin concentration, an increase in liver weight, and histopathological findings in the thyroid, kidney and kidney stones seen at the LOAEL of 10 000 ppm, equal to 1864 and 2545 mg/kg bw per day for males and females, respectively (Harvey & Rees, 1991).

Rats

In a 10-day dose range-finding study, groups of three male and three female CR1:COBS CD(SD)BR rats were given pyrimethanil (purity, 99.7%) at a dose of 0, 10, 100, or 1000 mg/kg bw per day by gavage in 1% aqueous methyl cellulose for 10 days. All dose suspensions were prepared fresh each day. Stability and homogeneity were confirmed analytically. Rats were inspected twice per day for signs of toxicity and mortality, with detailed cage-side observations performed once each day. Body weight and food consumption were measured 3 days before start, at start and at 3-day intervals during the study. At termination, blood was taken for haematological and clinical chemistry analyses. Rats were killed and examined macroscopically. Selected tissues were collected for histopathological examinations.

There were no treatment-related effects on mortality, clinical signs, food consumption or on haematological parameters. There was a slight decrease in body-weight gain (10%) in males observed initially at the highest dose only. At termination, the body weights of treated groups were comparable to those of the control group. At 1000 mg/kg bw per day, the plasma cholesterol concentration was slightly increased in males (2.86 vs 2.12 in controls) and moderately increased in females (2.92 vs 1.87 in controls). Necropsy findings were comparable in the control group and treated groups. The highest dose of 1000 mg/kg bw per day for 10 days was tolerated by rats (Harvey, 1992b, 1992c). At 1000 mg/kg bw per day, colloid depletion was seen in the thyroid gland (two out of three males and one out of three females) and some follicular-cell hypertrophy was evident. At 100 mg/kg bw per day, for one male only, some moderate colloid depletion was observed. Moderate follicular-cell hypertrophy was evident in one out of three males. No histopathological changes were observed at 10 mg/kg bw per day (Harvey, 1992b, 1992c).

In a 28-day dose range-finding study, groups of five male and five female Crl:CD(SD)BR rats were fed diet containing pyrimethanil (purity, 99.2–99.7%) at a dose of 0, 10 000, 15 000, or 30 000 ppm (equivalent to 0, 844, 1161 or 2701 mg/kg bw per day for 28 consecutive days. An additional group of five males and five females received pyrimethanil at a dose of 1500 mg/kg bw per day in 1% aqueous methyl cellulose by oral gavage for 11 days, after which the dose was reduced to 1000 mg/kg bw per day owing to adverse clinical signs. Diets were prepared each week. Stability, concentrations and homogeneity were confirmed analytically. Rats were inspected twice per day for signs of toxicity and mortality, with detailed cage-side observations performed once per day. Body weight was measured twice per week. Food consumption was measured each week. Ophthalmological examinations were performed before the start of the study and at termination. At termination, blood was taken for haematological and clinical chemistry analyses. All surviving rats were killed and examined macroscopically. Selected organs were weighed. Selected tissues were collected for histopathological examinations.

The diets were reported to be homogenous and analytical concentrations were within the range -15% to +10% of the nominal. The study report containing the results of analytical measurement was not available to the JMPR. There was no treatment-related effect on mortality in the group treated by gavage nor in the groups given diets containing pyrimethanil. Two females (day 7 and day 9) and one male and one female (day 15) in the group treated by gavage were killed in extremis.

The deaths were considered to be related to gavage errors and one female was cannibalized. Clinical signs such as general nasal soiling, reduced activity, piloerection, hunched posture and severe emaciation were seen in the majority of rats in the group at 30 000 ppm and urogenital soiling was seen in most males. Emaciation and general soiling was also observed in the group at 15 000 ppm. Most rats treated by gavage showed general soiling, facial soiling, emaciation, piloerection and reduced activity. Most males also showed urogenital soiling and reduced muscle tone, and most females showed hunched posture. These signs were first seen from day 7 to 17 and had disappeared by day 26. There was a dose-related decrease in body weight in all dietary groups and in males and females for the first 4 days and this persisted throughout the study in rats at the highest dose only. A reduction in mean body-weight gains were observed in all dietary groups and also in the group treated by gavage compared with controls. Decreased food consumption was observed in all dietary groups and the group treated by gavage with marked effects during the first week of treatment. There was decrease in water consumption at 30 000 ppm in males and females and a slight reduction in males at 10 000 ppm. No effect on water consumption was observed in the group treated by gavage or the other dietary groups. Ophthalmoscopic examination revealed no treatment-related effects. Statistically significant increases in erythrocyte count and haemoglobin concentration and reductions in mean corpuscular volume (MCV) and mean cell haemoglobin (MCH) were observed at the highest dietary concentration of 30 000 ppm and in the group treated by gavage. Statistically significant changes were also observed in some haematological parameters in other dietary groups, but changes were small and of doubtful toxicological significance. At 30 000 ppm, statistically significant increases in urea, total bilirubin and gamma-glutamyl transferase (GGT) were recorded in males and females. Also at this dose, a statistically significant decrease in total protein, albumin, total globulin, glucose and an increase in alanine aminotransferase and carbon dioxide were recorded in males, and reductions in alkaline phosphatase and glucose were recorded in females. At 15000 ppm, there were statistically significant increases in GGT (males and females), cholesterol (males), alkaline phosphatase (males), phosphate (females), total bilirubin (females), and decreased glucose and creatinine phosphokinase (CPK) (females). At 10 000 ppm, there was a statistically significant increase in cholesterol concentration (males and females).

Statistically significant increases in potassium and cholesterol (males and females), a significant increase in carbon dioxide and alanine aminotransferase (males) and a decrease in glucose (females) were observed in the group treated by gavage. At 30 000 ppm, there were marked decreases in absolute and relative (to brain weight) liver, kidney, heart, spleen, adrenal and gonad weights compared with controls. However, kidney and liver weights relative to body weights of males and females were increased, probably due to the emaciated condition of the rats. Liver and kidney weights were affected in all treatment groups. At necropsy, an accentuated lobular pattern of the liver was seen in all treated groups. At 30 000 ppm, multiple yellow foci in the kidneys were seen in one out of five male and four out of five females. Histopathology revealed a treatment-related effect in the kidneys (increase in luminal diameter, degenerative change to basophilic tubules, tubular necrosis, increased mitotic division, karyomegaly in tubular epithelium, inflammatory infiltration), liver (accentuated lobular pattern), oesophagus (focal acute inflammation of the muscularis in rats treated by gavage), stomach (diffuse epithelial hyperplasia in rats treated by gavage, mucosal ulceration) and thyroid (colloid depletion, agglomeration of colloid, follicular-cell hypertrophy, follicular epithelial hyperplasia in rats treated by gavage). In rats given diets containing pyrimethanil, effects were predominately recorded at 30 000 ppm, although some effects in the kidney and liver were also recorded at 10 000 and 15000 ppm.

Dietary concentrations of 10 000 ppm and greater and the dose of 1500/1000 mg/kg bw per day administered by gavage produced treatment-related effects on clinical signs, body weight and body-weight gains, food consumption, haematology, clinical chemistry, organ weights, gross pathology and histopathology (Harvey, 1991a).

In a 90-day feeding study, groups of 10 male and 10 female Crl:CD(SD)BR Sprague-Dawley rats were fed diets containing pyrimethanil (purity not stated) at a concentration of 0, 80, 800, or 8000 ppm for 13 weeks (equal to 0/0, 5.4/6.8, 54.5/66.7 or 529.1/625.9 mg/kg bw per day, for males/females, respectively). Additional groups of 10 males and 10 females were given diets containing pyrimethanil at a concentration of 0 or 8000 ppm, the highest dose, for 13 weeks, and then maintained on diets without pyrimethanil for 4 weeks to investigate the reversibility of any findings. Diets were prepared each week. Stability, homogeneity and dietary concentrations were confirmed analytically. The rats were inspected twice per day for signs of toxicity and mortality, with detailed cage-side observations performed once per day. Body weight and food consumption were measured before treatment, at start of treatment, each week during treatment and at necropsy. Ophthalmoscopic examination was performed on all rats at start of treatment and on the rats in the control group and at the highest dose at termination (91 days). Urine analysis was performed on all rats after 90 days and also during the first week for rats in the reversibility group. Blood was taken for haematology and clinical chemistry measurements at week 4 and 13, and during the final week before termination for groups in the reversibility study. At termination, all rats were necropsied and examined histopathologically. Selected organs were weighed. Selected tissues were collected for histological examination.

Test diets were stable for 4 days at room temperature, so diets were deep-frozen and thawed before use. The analytical concentrations were in the range of −15% to +10% of nominal concentrations. Results on homogeneity were reported elsewhere and were not submitted to the JMPR. There were no treatment-related effects on mortality, clinical signs, water intake, ophthalmology, haematology, clinical chemistry or macroscopic findings. Mean body weight and body-weight gains of males and females at 8000 ppm were consistently lower than those of the controls. The overall decreases in body-weight gains were 28% in males and 33% in females. Body-weight gains of rats at 80 or 800 ppm was similar to those of controls. During the reversibility period, increases in body-weight gains were apparent in males and female previously given diets containing pyrimethanil at 8000 ppm. However, the total body weights were still lower than those of the controls (15% in males and 13% in females). During the first week of treatment, marked reductions (33%) in food intake of males and females at 8000 ppm were noted when compared with those of rats in the control group. From week 2 of the study onwards, the food intake of males and females at 8000 ppm was slightly reduced (overall, by 16% in males and 17% in females) when compared with those of controls. Food intake of rats at 80 or 800 ppm was comparable to that of controls throughout the study. During the reversibility period, rats previously at 8000 ppm ate similar amounts to rats in the control group. Food conversion efficiencies were similar between treated and control groups, except for the first week when they were decreased for rats at 8000 ppm. Urine analysis showed dark brown coloration in males and females at 8000 ppm and 800 ppm. At 8000 ppm, the group showed increases in urinary protein compared with the controls. At the end of the reversibility period, the urine samples of males and females previously at 8000 ppm were comparable to those of controls. Significantly lower absolute weights of the heart and adrenals in males and females, and the spleen, kidney, and thymus weight in females were observed at 8000 ppm. Significant increases in relative organ weight to body weight ratios were seen in the brain, liver, gonads, and kidneys for males, and the brain, liver, and kidney for females. These changes could be considered to reflect the growth retardation observed in rats at 8000 ppm. No changes in organ weights were apparent in rats previously at 8000 ppm and killed at the end of reversibility period. It was stated in the study report that organ-weight changes in the groups at 800 and 80 ppm were within the range of values for control groups (although ranges for organ weights of rats in the control groups were not presented in the report). At 8000 ppm, microscopic pathology showed an increased incidence of liver hypertrophy in centrilobular hepatocytes (males, nine out of ten; females, three out of ten), increases in the incidence and severity of follicular epithelial hypertrophy (males, nine out of ten; females, six out of ten), and epithelial brown pigment (males, eight out of ten; females, seven out of ten) were seen. At 800 ppm, an increased incidence (two out of ten) of centrilobular hepatocyte hypertrophy was seen. After the 28-day recovery period, centrilobular hepatocyte hypertrophy was not apparent in rats previously at 8000 ppm.

The LOAEL was 8000 ppm (529.1 and 625.9 mg/kg bw per day for males and females, respectively) on the basis of decreased body weights, body-weight gains and food consumption; brown urine and increased urinary protein in males; decreased absolute heart, adrenal, spleen, thymus weights and increased relative liver, kidney, gonad weights and hypertrophy in the liver and thyroid. The NOAEL was 800 ppm (54.5/66.7 mg/kg bw per day in males/females). The study author concluded that the NOEL was 80 ppm (5.4/6.8 mg/kg bw per day in males/females) (Higham, 1990; Harvey, 1994; Reader, 2002).

In an amendment to the study report, the study author concluded that the treatment-related changes in the incidence and severity of follicular epithelial hypertrophy and incidence of follicular epithelial brown pigment observed at 8000 ppm that stained positively for lipofucin by the Schmorl method. No treatment-related findings were apparent in either males or females at 800 or 80 ppm. After a 28-day recovery period, treatment-related findings were no longer apparent, indicating reversibility of the effects (Husband, 1992).

In a position paper (Reader, 2003a) prepared as a part of the ongoing review of the Toxicology and Metabolism Tier II dossier for pyrimethanil (under EC directive 91/414/EEC), the Rapporteur Member State (Austria) requested a statement regarding the thyroid changes seen in the 90-day and 104-week studies in rats and their possible relevance to man. The study author suggested that changes in the thyroid pathology (e.g. minimal to slight colloid depletion, hypertrophy of the follicular epithelium and deposition of intra-cytoplasmic brown pigment in the follicular epithelium) were noted at high doses of pyrimethanil. Mechanistic data suggested that an imbalance in thyroid hormones due to increased thyroid-hormone clearance by the induction of liver enzymes resulted in increased thyroid-stimulating hormone (TSH) activity and persistent thyroid stimulation. Such effects may lead to changes in thyroid functionality and morphology. Since rodents are known to be particularly sensitive to this effect, it is considered that this mechanism has little relevance to humans (Reader, 2003a).

Dogs

In a 13-day dose range-finding study of oral toxicity, groups of one male and one female beagle dog were given pyrimethanil (purity, 98.5–100%) at a dose of 10 mg/kg bw per day by gavage in 1% methyl cellulose in distilled water, the dose being increased each day up to 2000 mg/kg bw per day in this the first phase of the study. No treatment was administered on days 6, 7 and 13–15. In the second phase of the study, the male, after a 4-day period without treatment, was treated with pyrimethanil at a dose of 1200 mg/kg bw per day for 3 days (days 24–26) then at 800 mg/kg bw per day for 10 days (days 27–36). The female, after a 4-day period without treatment, was treated with 1200 mg/kg bw per day for 14 days (days 24–36). A previously untreated male (number 1602) was treated with pyrimethanil at a dose of 1200 mg/kg bw per day for 3 days (days 29–31) then at 1000 mg/kg bw per day for 5 days (days 32–36) and a previously untreated female (No. 1604) was left untreated. Dogs were inspected twice per day for morbidity or mortality, with clinical signs being checked daily. A detailed physical examination was performed before commencement of treatment and at the end of treatment. Body weight and food consumption were measured daily during the treatment. Electrocardiograms were made on all dogs pre-test and at the end of the study. Blood was collected from all dogs at pre-test and at termination for measurement of haematological and clinical parameters. At the end of the study, a complete gross postmortem was performed. The adrenals, brain, kidneys, liver, thyroid and parathyroid, heart, and gonads were weighed. The organs specified were examined microscopically.

All dosing solutions were prepared daily. The dosing solutions were analysed for concentrations and homogeneity on days 24, 29 and 32. However, the results of the suspension concentrations and homogeneity were not provided in the study report. No deaths occurred during the study. Salivation was observed from dosing for 6 h in the male dog at 800 mg/kg bw and greater during the first phase of the study, and liquid faeces were recorded. The male dog vomited within 1–3 h after dosing after treatment at 640 mg/kg and greater. Vomiting also occurred in males treated in the second phase of the study at 1200 mg/kg bw per day, although this ceased when the dose was reduced to 1000 mg/kg bw per day or 800 mg/kg bw per day. The female dog also vomited within 1–3 h after dosing and liquefied faeces were recorded between 1 h and 6 h after dosing during the first phase of the study. Vomiting also occurred on one occasion in a female treated at 1200 mg/kg bw per day during the second phase of the study; liquefied and white soft faeces were also recorded. Slight decreases in body weight and food consumption were noted at 1200 mg/kg bw per day and greater. There were no treatment-related effects on haematological and clinical chemistry parameters, electrocardiograms, organ weights and necropsy findings. The study author concluded that the maximum tolerated dose was between 640 and 1000 mg/kg bw per day (Harvey, 1992a).

In a 28-day study of oral toxicity, groups of two male and two female beagle dogs were given pyrimethanil (purity, 99.0%) at a dose of 0, 100, 500, or 1000 mg/kg bw per day by gavage in 1% methyl cellulose in distilled water. The highest dose of 1000 mg/kg bw per day was reduced to 800 mg/kg bw per day on day 7 owing to persistent vomiting. The dogs were inspected twice per day for morbidity or mortality, with clinical signs being checked daily. A detailed physical examination was performed before commencement of treatment and at week 4 before the end of treatment. Body weight and food consumption were measured each week. An ophthalmoscopic examinations and electrocardiography were performed pre-test and at the end of the study. Blood was collected from all animals at pre-test, and after 15 and 28 days of treatment for measurement of haematological and clinical parameters. Urine analysis was performed on all animals at termination. At the end of the study, a complete gross postmortem was done. The adrenals, brain, kidneys, liver, thyroid and parathyroid, heart, and gonads were weighed. The organs specified were examined microscopically.

All dosing solutions were prepared daily. The dosing solutions were analysed for concentrations and homogeneity on the first day of each week. However, the results of analysis of the suspension concentrations and homogeneity were not provided in the study report.

No deaths occurred during the study. There were no treatment-related findings on urine analysis, electrocardiography, ophthalmoscopic examinations and haematological parameters. Frequent vomiting, 2 to 4 h after dosing, was recorded in all animals treated with 1000 mg/kg bw per day. This occurred less frequently when the dose was reduced to 800 mg/kg bw. Vomiting also occurred in one male and one female at 500 mg/kg bw per day. No other treatment-related symptoms were recorded. No treatment-related changes were noted on detailed clinical examination conducted at the end of treatment. Slight decreases in body-weight gain (24% decrease compared with controls) were observed in females at 1000 mg/kg bw per day. After reduction of the highest dose, overall body weights were slightly reduced (4%) in females. There was no effect on body weights in the groups at 100 and 500 mg/kg bw per day. No treatment-related effect was observed on food consumption in males at any dose. Slight decreases in food consumption was noted in females at 500 and 1000/800 mg/kg bw per day in the first 3 weeks and were comparable to controls during the last week of the treatment. There was a trend towards slight increases in plasma cholesterol concentration in the treatment groups compared with controls; however, the increase in cholesterol concentration was not considered to be treatment-related in the absence of any other histopathological findings. No treatment-related effects were noted on organ weights, necropsy findings or histopathological examinations.

The NOAEL was 500 mg/kg bw per day and the LOAEL was 1000 mg/kg bw per day on the basis of vomiting, decreased body-weight gain and slight decreases in food consumption in females (Harvey & Davies, 1990).

In a 90-day study of oral toxicity, groups of four male and four female beagle dogs were given pyrimethanil (purity, 97.9%) at a dose of 0, 6, 80 or 1000/800 mg/kg bw per day by gavage in 0.5% methyl cellulose in distilled water for 13 weeks. The highest dose was reduced from 1000 to 800 mg/kg bw per day on day 7 owing to persistent vomiting seen in all dogs. The dogs were inspected twice per day for morbidity or mortality, with clinical signs being checked daily. A detailed physical examination was performed before commencement of treatment and at termination. Body weight and food consumption were measured weekly. Water consumption was measured over a 4-day period during weeks 3–4, 7–8 and 11–12. An ophthalmoscopic examination and electrocardiography were performed at pre-test and at the end of the study. Blood was collected from all dogs at pre-test, and after 4 and 13 weeks of treatment for measurement of haematological and clinical parameters. Urine analysis was performed on all dogs at termination. At the end of the study, a complete gross postmortem was done. The adrenals, brain, kidneys, liver, pituitary, thyroid and parathyroid, lungs, heart, spleen and testes/ovaries were weighed. The organs specified were examined microscopically. The dosing solutions were prepared daily for the first 7 days and then the day before use for the remaining treatment period. Aliquots of the dosing solutions were analysed on mixing, before, during and after dosing, and mean concentrations were reported to be 82.5–121.7% of nominal doses (study report was not provided).

There were no treatment-related effects on mortality, organ weights, necropsy findings, histopathological, ophthalmological, electrocardiography, urine analysis or haematological parameters. At 1000 mg/kg bw per day, there was an increased incidence of vomiting and diarrhoea in all dogs during the first 6 days of treatment. The vomiting was observed between 0.75 h and 3.75 h after dosing. Vomiting was also observed in one male in the control group. Vomiting was not considered to be a toxicologically relevant effect since it may have been induced by local irritation of the gastrointestinal tract. When the dose was reduced to 800 mg/kg bw per day, occasional to frequent vomiting was observed in the majority of animals (about 9% of all doses). Other clinical signs observed included salivation, cream coloration of faeces and hypoactivity. Very slight weight loss (about 4%) occurred at the highest dose, probably due to vomiting. Food consumption was slightly decreased in the first week of treatment. There was a marked reduction in water consumption in males (30%) and females (19%). Clinical chemistry analysis indicated that there were statistically significant reductions in serum phosphate concentration in males and serum total protein concentration in females after 4 weeks of treatment. The toxicological significance of these findings is uncertain. At 80 mg/kg bw per day, infrequent vomiting was observed in all females and three out of four males, the overall incidence of vomiting was less than 2% of all doses was not considered to be a toxicologically relevant effect. Water consumption was decreased (males, 9%; and females, 17%). Clinical chemistry analysis indicated that there was a small reduction in serum phosphate concentration in males at week 4 and in females at week 1. No treatment-related effects were reported at 6 mg/kg bw per day.

The NOAEL was 80 mg/kg bw per day and the LOAEL was 1000/800 mg/kg bw per day on the basis of diarrhoea, salivation, hypoactivity and decreased water consumption (Harvey, 1991c).

In a 1-year study of oral toxicity, groups of four male and four female beagle dogs were given pyrimethanil (purity, 96.3–96.9%) at a dose of 0, 2, 30 or 400/250 mg/kg bw per day by gavage in 0.5% methyl cellulose in distilled water for 12 months. The highest dose was reduced from 400 to 250 mg/kg bw per day on day 8 of the study owing to persistent vomiting seen in all dogs. The dogs were inspected twice per day for morbidity or mortality, with clinical signs being checked daily. Body weights were recorded weekly and food consumption was measured daily. Water consumption was measured over a 4-day period during weeks of 13, 26 and 52 of treatment. An ophthalmoscopic examination and electrocardiography were performed at pre-test and on dogs in the control group and in the group at the highest dose at termination. Clinical examinations, including heart rate and rectal temperature, were conducted for all dogs pre-test and at termination. Blood was collected from all dogs at pre-test, and during 13, 26 and 52 weeks of treatment for measurement of haematological and clinical parameters. Urine analysis was performed on all dogs during weeks 13, 26 and at termination. At the end of the study, a complete gross postmortem was performed. The adrenals, brain, kidneys, liver, pituitary, thyroid and parathyroid, lungs, heart, spleen and testes/ovaries were weighed. The organs specified were examined microscopically.

The dosing solutions were prepared the day before use. Aliquots of the dosing solutions on 1 day from week 1, 2, 3, 4, 5, 8 and every 4 weeks thereafter, were analysed. The mean results of all the analysed suspensions used were in the range 76.6–116.1% of nominal doses (the study report was not provided). There were no treatment-related effects on mortality, organ weights, necropsy findings, histopathological, ophthalmological, electrocardiography, urine analysis or clinical chemistry parameters. At 400/250 mg/kg bw per day, there was an increased incidence of vomiting that occurred within 30 min to 6 h after dosing (35.7% of doses in all males and 75% of doses in females) during week 1 of the study. When the dose was reduced to 250 mg/kg bw per day, vomiting was decreased to about 1% in all dogs. Incidents of coloured faeces or diarrhoea occurred in most of the dogs receiving pyrimethanil at intervals throughout the study. At 30 mg/kg bw per day, the overall incidence of vomiting was 0.4% of the doses in males only, which was not considered to be a toxicologically relevant effect since it may have been caused by local irritation of the gastrointestinal tract. No treatment-related clinical sign including vomiting was recorded at 2 mg/kg bw per day. After treatment at 400 mg/kg bw per day, a slight mean body-weight loss occurred in males (0.8%) and females (2.5%) in the first week of treatment. There was also an overall very slight reduction in body-weight gain (males, 6%; and females, 17%) throughout the study at the highest dose. Overall mean body-weight gain at the highest dose (400/250 mg/kg bw) was reduced by 50% in males and 73% in females when compared with those of the controls. Overall food-conversion efficiency was reduced in males and females at the highest dose by about 50% and 98%, respectively. Overall water consumption at 400/250 mg/kg bw was significantly decreased by 35% and 26% in males and females, respectively, at the highest dose. There was a slight but statistically significant increase in leukocytes (mainly neutrophils) in males after 3, 6 and 12 months after treatment at 400/250 mg/kg bw per day. After 12 months at this dose, a slight reduction in clotting time was recorded in males and females. There were no significant treatment-related effects at 2 or 30 mg/kg bw per day.

The NOAEL was 30 mg/kg bw per day and the LOAEL was 400/250 mg/kg bw per day on the basis of decreased in body-weight gains, food consumption and feed efficiency, water consumption, reduced clotting times and increases in neutrophils. The study author concluded that the NOEL was 30 mg/kg bw per day and the NOAEL was 250 mg/kg bw per day (Rees, 1992). The difference between the NOAEL identified by the Meeting and that identified by the study author was attributed to the fact that the study author did not consider decreases in body-weight gains, decreases in feed efficiency and reduced water consumption as adverse effects.

2.3 *Long-term studies of toxicity and carcinogenicity*

Mice

In a study of carcinogenicity in mice, groups of 51 male and 51 female Crl:CD-1(ICR)BR mice were given diets containing pyrimethanil (purity, 96–97.3%) at a concentration of 0, 16, 160, or 1600 ppm (equal to 0, 2.0, 20.0 or 210.9 and 0, 2.5, 24.9 or 253.8 for males and females, respectively) for up to 80 weeks. Diets were prepared twice per week. Stability, homogeneity and dietary concentrations were confirmed analytically. The mice were inspected twice per day for mortality and morbidity. Changes in clinical condition or behaviour were recorded daily. Detailed clinical observations were recorded weekly. Body weight and food consumption were measured weekly for the first 16 weeks, then every 4 weeks until termination. Water consumption was not measured. An ophthalmoscopic examination was not performed. Blood was collected from 10 mice per group during weeks 26, 52 and 80. All mice that died and those that were killed on schedule were subjected to gross pathological examination and selected organs were weighed. Tissues were collected for histological examination from mice in the control group and in the group at the highest dose, mice that died prematurely, and mice that were killed in extremis.

Pyrimethanil was homogenously distributed in the diet and was stable for a storage period of 3 days at room temperature. The measured concentrations were within the range of 85–100% of the target concentrations except on a few occasions. At the end of the study, the survival of males was 67%,

43%, 71% and 59% (control, lowest, intermediate and highest dose, respectively). The difference in survival between males in the control group and males at the lowest dose was statistically significant. The survival of treated and control females was similar (76%, 80%, 76% and 78% for control group to group receiving the highest dose, respectively). There was no dose-related adverse effect on survival of males or females and adequate numbers of male and female mice were available at study termination. No adverse effects were observed on body weight or body-weight gain, food or water consumption, clinical or haematological parameters, or organ weights. No treatment-related differences in palpable masses were recorded. The incidence of macroscopic lesions was comparable among the groups (males and females), with the exception of an increase in the incidence of urinary bladder distension in the treated males. Males receiving pyrimethanil displayed a higher incidence of urinary bladder distension at necropsy, and the incidence of lesions of the urogenital tract were increased at the highest dose compared with the control values. In the first 52 weeks of treatment, an increase in the incidence of lesions of the urogenital tract was seen in the treated males, 31% occurring in the groups at the lowest and intermediate doses, and 57% in the group at the highest dose, while the incidence in the control group was 18%. There was no evidence for an increase in the incidence of any tumour type in either sex.

The NOAEL for systemic effects was 160 ppm, equal to 20.0 mg/kg bw per day, on the basis of increased incidences of lesions of the urinary tract, including bladder distension and thickening seen at 1600 ppm (210.9 mg/kg bw per day), in males. The NOAEL in female mice was 1600 ppm (equal to 253.8 mg/kg bw per day, the highest dose tested). The study authors (Clay & Healing, 1993; Clay, 1994; Healing, 1995c; Clay, 1995; Clay, 1996; Patton, 1995) concluded that the NOEL for males and females was 160 ppm, equivalent to 20.0 and 24.9 mg/kg bw per day for males and females, respectively.

A report on survival data for historical controls and, in relation to this, some necropsy/histopathology findings (primarily concerning the urinary bladder) were provided to the country preparing the EC draft assessment (Austria). The overall survival at 80 weeks was approximately 58% for males and 78% for females in the study of carcinogenicity in mice. The data for historical controls from the performing laboratory for 1993–2000 indicate a mean survival of 66.5% for males and 73.3% for females at week 80. The study author (Reader, 2003b) concluded that the overall pattern of mortality observed in the study of carcinogenicity in mice would not suggest an effect of treatment with pyrimethanil compared with data for historical controls. The slightly lower survival of male mice in the study of carcinogenicity compared with females may have been primarily attribuatble to male aggression exacerbated by co-housing. The study authors suggested that lesions in the urogenital region of males led to a large proportion of the cases of morbidity and mortality. The incidence of bladder distension in the historical controls for the male decedents was 40% at week 52 and 21% for the terminal kill compared with 43% observed in the study of carcinogenicity. The study authors suggested that the pattern seen in the study of carcinogenicity was caused by an underlying inflammatory response in the male mice, which was confirmed at terminal necropsy that showed more than 40% of the males in the control group had inflammatory cell foci. The study authors further concluded that at the highest dose of pyrimethanil, 1600 ppm, the incidence of bladder distention slightly exceeded the value for historical controls and may suggest an effect at this dose. The Meeting considered that, on the basis of study results, it was not clear that the effects in males were treatment-related since values were within the range for historical controls and the dose–response relationship could not be evaluated since tissues from the urogentital tract were not examined in mice at the lowest and intermediate doses (Reader, 2003b).

Rats

In a combined long-term study of toxicity/carcinogenicity, groups of 50 male and 50 female Sprague-Dawley Crl:CD(SD)BR rats were given diets containing pyrimethanil (purity, 95.5.2–97.6%) at a nominal concentration of 0, 32, 400, or 5000 ppm (equal to 0/0, 1.3/1.8, 17.0/22.0, or 221/291 mg/kg bw per day in males/females, respectively) for up to 104 weeks. Additional groups of 20 males and 20 females were sacrificed at 52 weeks (interim kill). Diets were prepared twice per week throughout the study. Stability, homogeneity and dietary concentrations were confirmed analytically. Rats were inspected twice per day for mortality and morbidity. Changes in clinical condition and behaviour were recorded daily. Detailed clinical observations were recorded weekly. Body weight and food consumption were measured weekly for the first 13 weeks, then monthly for the rest of the study. Water consumption for each cage was measured over 4-day periods ending on weeks 9, 16, 32 and 48 of treatment. An ophthalmoscopic examination was performed before treatment, during week 50 and during week 104. Blood and urine were collected during weeks 13, 26, 52, 78 and week 102. An additional bleed was conducted during week 72 for measurement of thyroid hormones, thyrotrophin, thyroxine (T4) and triiodothyronine (T3). Haematological parameters examined were erythrocyte count, leukocyte count, erythrocyte volume fraction, haemoglobin, platelet count, differential leukocyte count and cell morphology. Standard clinical-chemistry and urine-analysis parameters were examined. At week 52, 20 males and 20 females from each group were necropsied and examined histopathologically. At termination, all rats were necropsied and examined histopathologically. Liver, kidney, spleen, testes, ovaries, heart, brain and adrenals were removed and weighed. Selected tissues were examined histopathologically.

Pyrimethanil was uniformly distributed in the diet and was stable at room temperature for 4 days (less than 11% of nominal dose lost over 8 days storage at room temperature). The measured test concentrations ranged from –15% to +10% of the nominal concentrations, except for some results at 32 ppm and 400 ppm (Bright, 1993).

There was no treatment-related mortality or clinical signs of toxicity. A decreased incidence of palpable masses (24%) in females at 5000 ppm was observed, and the mean time of onset was approximately 9 weeks later than that of controls. However, there was no indication that the delayed onset was caused by treatment with pyrimethanil. Throughout the study, males and females at 5000 ppm consistently displayed significantly lower mean body weights than did the controls. The body-weight gains of rats at 5000 ppm for 0–13 weeks were lower than those of the controls (males, 10%; and females, 16%). Overall body-weight gains of rats at 5000 ppm were lower (males, 5%; and females, 42%) than those of the controls. There was no significant difference in the mean body weights of rats at 400 ppm or 32 ppm. Food consumption was reduced by approximately 5% in males and 11% females in the group at 5000 ppm compared with rats in the control group. However, the food-conversion ratios were not significantly different between rats receiving pyrimethanil and rats in the control group. There was no significant effect on water consumption. Ophthalmoscopic examination did not revealed any treatment-related lesions at any dose. At 5000 ppm, statistically significant differences from control values were observed in several parameters at various intervals.

Leukocyte counts were decreased at weeks 78 and 102 for males only and erythrocyte counts were decreased at weeks 13 and 26 for females only. Decreased MCH concentrations were observed at week 52 and 78 for males and females, and also at weeks 13 and 26 for females. Increased platelet counts were observed at weeks 13 and 26 for males and females, and at weeks 52 and 78 for females only. At 400 ppm, a statistically significant increase in platelet counts was observed at week 26 in females only, while a decreased platelet count was observed at week 102 in males. Other group mean values for haematological parameters in rats at 32 ppm and 400 ppm were essentially similar to those of the controls. All haematological data were within the range for historical controls, there was a lack of dose–response relationship and effects tended to be transient and inconsistent over time. Thus, these haematological findings were considered to be of no toxicological significance. At 5000 ppm, slightly elevated gamma glutamyl transpeptidase (GGT) activity were observed after 78 and 102 weeks in males. There was slight to moderate increased in cholesterol concentrations in males at 13 and 26 weeks and in females at all sampling points. Decreases in serum phosphate concentrations were observed between weeks 13 and 78 in males and females (at all doses in males and at the lowest and highest doses in females). Phosphate concentrations were within the range for historical controls and the noted decreases were transient and not always dose-related (Reader, 2003a). Concentrations of thyroid hormones were normal at 72 weeks. There were no significant treatment-related effects on urine-analysis parameters. However, a brownish-black urine coloration, which darkened on standing, was reported in certain males and females at 5000 ppm throughout the treatment period.

At interim kill, no significant treatment-related effect was reported for rats at 400 or 32 ppm. At interim kill, the absolute liver weights were 13% higher in males and 5% higher in females in the groups at 5000 ppm compared with controls. At interim kill, at 5000 ppm, minimal to moderate hypertrophy of centrilobular hepatocytes were observed in all males and 1 out of 20 females, but was not seen in the control group of either sex. A significantly higher kidney to body-weight ratio was observed in males only at 5000 ppm. At terminal kill, the absolute liver weight of males (22%) at 5000 ppm was higher than in the controls. The relative liver to body-weight ratios were significantly increased in males (26%) and females (21%) at 5000 ppm. At 5000 ppm, dark thyroids were observed in 1 out of 20 males and 4 out of 20 females at the interim kill. At this dose, dark thyroids were also observed in 6 out of 50 males and 17 out of 50 females at terminal kill. In the thyroid gland, there were minimal to moderate intra-epithelial depositions of brown pigment (which stained positively for lipofuscin by the Schmorl method) in 18 out of 20 males and 19 out of 20 females compared with 1 out of 19 in the controls for males and females. There were higher incidences of colloid depletion and hypertrophy of the follicular epithelium (males, 18 out of 20; and females, 13 out of 20) than controls (males, 9 out of 19; and females, 6 out of 19). At terminal kill, minimal to slight hypertrophy of centrilobular hepatocyte were observed in rats at 5000 ppm only (males, 32 out of 50; and female 6 out of 50). Also, there were higher incidences of minimal to severe eosinophilic foci (males, 19 out of 50; and females, 12 out of 50) than in controls (males, 2 out of 51; and females, 7 out of 51). In the thyroid gland, minimal to severe colloid depletion and hypertrophy of follicular epithelium were observed at 5000 ppm in males and females at higher incidences (males, 36 out of 50; and females, 38 out of 50) than in the controls (males, 25 out of 51; and females, 15 out of 51). Focal hyperplasia of the follicular epithelium was seen at 5000 ppm in 9 out of 50 males and 7 out of 50 females, but in the control group in 2 out of 51 males and 1 out of 51 females). Minimal to moderate depositions of intra-cytoplasmic brown pigment (lipofuscin) in the thyroid follicular epithelium were detected at 5000 ppm in males and females (males, 38 out of 50; and females, 47 out of 50), but was not present in the control group for either sex. No treatment-related effects on microscopic findings were noted in rats given pyrimethanil at 400 or 32 ppm after 52 or 104 weeks.

Overall, the only tissue showing a higher incidence of tumours than did the controls was the thyroid gland; benign follicular-cell adenomas were observed in 9 out of 70 males and 7 out of 70 females at 5000 ppm, compared with 3 out of 70 males and 0 out of 70 females in the control group. A pair-wise Fischer test indicated that the incidence of tumours in males at the highest dose (9 out of 70) was not statistically significant ($p = 0.06$) when compared with that of males in the control group (3 out of 70). However, the incidence in males at the highest dose (12.9%) was higher than the upper limit of the range for historical controls (1.5–5.9%). In females, the incidence of tumours at the highest dose (7 out of 70, or 10%) was higher than that of females in the control group (0 out of 70) and exceeded the uppr limit of the range for historical controls (0–3.0%). A positive trend in the incidence of these tumours in males and females was also noted. In addition, thyroid follicular-cell adenocarcinomas were seen in male rats at 32 ppm (males, 1 out of 70) and at 5000 ppm (males, 1 out of 70); however, the incidence (1.4%) was within the range for historical controls.

The NOAEL was 400 ppm, equal to 17 and 22 mg/kg bw per day for males and females, respectively. The LOAEL was 5000 ppm, equal to 221 and 291 mg/kg bw per day for males and females, respectively, on the basis of decreased body-weight gains, increased serum cholesterol concentration and GGT activity, necropsy, and histopathological findings (Rees, 1993; Simpson, 2003; Healing, 1994). In a position paper (Reader, 2003a) prepared as a part of the ongoing review of the Toxicology and Metabolism Tier II dossier for pyrimethanil (under EC directive 91/414/EEC), the Rapporteur Member State requested additional information on the 104-week combined long-term study of carcinogenicity and toxicity in rats regarding historical data on survival and some clinical chemistry parameters (phosphate and urea). The author concluded that the data on survival, blood phosphate and urea concentrations were essentially similar to those for historical controls.

2.4 Genotoxicity

Pyrimethanil was evaluated in tests for mutagenicity in bacteria and mammalian cells in vitro, for chromosome damage (clastogenicity) in vitro and in vivo, and for unscheduled DNA synthesis (UDS). The results of these studies indicated that pyrimethanil is not genotoxic. In vivo, pyrimethanil was assessed for the formation of micronuclei in mice. The result of this study showed that pyrimethanil does not exhibit a chromosome-damaging potential. The Meeting concluded that pyrimethanil has no mutagenic or genotoxic properties either in vitro or in vivo.

Table 6. Results of studies of genotoxicity with pyrimethanil

End-point	Test object	Concentration	Purity (%)	Result	Reference
In vitro					
Reverse mutation[a] (Ames test)	*E. coli* strains CM 881 and CM 891	15–1500 µg/plate ± S9, in DMSO	96.2	Negative	Jones & Gant (1991)
Reverse mutation[b] (Ames test)	*S. typhimurium* strains TA98, TA100, TA1535, TA1538 and TA1537	15–1500 µg/plate ± S9, in DMSO	98.7	Negative	Jones & Gant (1990)
Reverse mutation (Ames test)	*S. typhimurium* strains TA98, TA100, TA1535, TA1538 and TA1537 and *E. coli* strain CM 891 (AE F132593, metabolite of pyrimethanil)	5–5000 µg/plate ± S9, in DMSO	98.7	Negative	Kitching (1998)
Gene mutation	Chinese hamster ovary cells (*Hprt* locus)	Initial assay: 10–400 µg/ml ± S9 in DMSO Independent repeat assay: 10–240 µg/ml −S9 10–280 µg/ml + S9, in DMSO	97.2	Negative	Adams et al. (1992); Jackson & Everett (1994)
Mammalian cytogenetic test [d]	Human lymphocytes	Initial assay: 7.8–62.5 µg/ml −S9 31.3–250 µg/ml + S9, in DMSO Independent repeat assay: 125 µg/ml −S9 250 µg/ml + S9, in DMSO	99.4	Negative	Brooker et al. (1990)
Unscheduled DNA synthesis	Rat hepatocytes	100, 300, 1000 mg/kg	97.6	Negative	Proudlock & Howard (1991)
DNA damage (Rec assay)	*B. subtilis* strains H17 and M45	50–5000 µg/disk ±S9, in DMSO	96.2	Equivocal	Gant & Jones (1994)
In vivo					
Micronucleus formation[c]	Mouse (CD-1 Swiss)	225, 450, 900 mg/kg bw	97.6	Negative	Proudlock (1991)

[a] In a previous test, precipitation and some cytotoxicity were observed at 5000 µg per plate.

[b] In the preliminary assay, precipitation and cytotoxicity were observed at 5000 µg per plate. Cytotoxicity was observed in all strains ± metabolic activation at ≥ 1500 µg/ml. In the second test, cytotoxicity was observed at ≥ 500 µg/ml in strains TA98 and TA100.

[c] Fifteen males and fifteen females per group, with twenty males and twenty females in the group at the highest dose. Pyrimethanil was given by gavage in 1% methylcellulose in distilled water. Bone marrow was collected at 24, 48 and 72 h. Five males and five females were given mitomycin C at a dose of 12 mg/kg bw, serving as positive controls. Statements of good laboratory practices and quality assurance were included. Mortality was observed in 1 out of 20 females at 900 mg/kg bw. Clinical signs after dosing included hunched posture, lethargy, piloerection at the intermediate and highest doses, and coma and decreased respiratory rate at the highest dose.

[d] The assay gave negative results in the absence of metabolic activation. Assays conducted with 2-aminofluorene with metabolic activation demonstrated preferential killing of the rec-strain at three of the six concentrations used in the first test and at none of the concentrations used in the second test.

2.5 *Reproductive toxicity*

(a) *Multigeneration studies*

Rats

In a two-generation study of reproductive toxicity, groups of 30 male and 30 female (parental generation, P) or 25 male and 25 female (F_1) Sprague-Dawley Crl:CD(SD)BR rats were given diets containing pyrimethanil (purity, 96.2–97.3%) at a concentration of 0, 32, 400, or 5000 ppm (equal to 0/0, 1.9/2.2, 23.1/27.4, and 293.3/342.8 mg/kg bw per day in the P rats and 0/0, 2.3/2.7, 29.1/34.0, and 389.0/449.6 mg/kg bw per day in the F_1 for males/females, respectively). Exposure of the P generation began at age approximately 8 weeks and continued for 14 weeks, and throughout the pairing (mating) period, gestation and lactation to necropsy. The groups of 25 male and 25 female F_1 pups selected to produce the F_2 generation were exposed to the same doses as their parents, beginning on postnatal day 21. F_1 rats were dosed with pyrimethanil for 14 weeks before mating produce the F_2 litters, and throughout pairing (mating) period, gestation and lactation to necropsy. Mating to produce a second F_2b generation was not performed.

Diets were prepared weekly and stored at −20 °C. Stability, homogeneity and dietary concentrations were confirmed analytically. Analysis of homogeneity, stability and achieved concentrations were not reported in the study report. The sponsor reported that "results were within limits considered acceptable for administration to the study animal".

Rats were inspected twice per day for mortality and morbidity and daily for clinical observations. Body weights of males were recorded weekly. Females were weighed weekly during pre-mating and on days 0, 6, 12, 15 and 20 of gestation, and days 1, 4, 7, 14, and 21 of lactation. Food consumption was measured weekly for males and females during the pre-mating period and daily during gestation and lactation for the females, but was reported at body-weight intervals; food consumption for males was recorded on a weekly basis after mating and until scheduled necropsy. Estrous cycles were monitored, with vaginal smears taken during the mating and until mating was confirmed. Gestation duration was calculated. Females were allowed to deliver normally and rear young to weaning on day 21. Litters were examined after delivery and pups were sexed, examined for gross abnormalities and the number of stillborn and live pups recorded. Litters were then examined twice per day for survival. Number, sex and weight of pups were recorded on postnatal days 1, 4, 7, 14 and 21. All surviving pups were assessed for the age at which pinnae unfolding, incisor eruption and eye opening took place, and on day 1 for surface-righting reflex, day 17 for air-righting reflex, and day 21 for grip reflex, pupilliary reflex and auditory startle response. All P and F_1 rats and those found dead and killed in extremis were necropsied and examined macroscopically. Selected tissues (ovaries, testes, uterus, epididymides, cervix, seminal vesicles, vagina, prostate pituitary gland, and coagulating glands lesions) were examined histopathologically in untreated rats and those treated at 5000 ppm. Organ weights were not recorded in this study.

There were no treatment-related clinical signs or mortalities observed in the P or F_1. Rats killed in a moribund condition included two parental females (one at the intermediate dose on day 15 of gestation and one at the highest dose on day 1 of lactation), as well as two F_1 females (one at the intermediate dose on day 1 of lactation and one at the highest dose on day 24 of gestation). One F^1 male from the group at the lowest dose was found dead on week 11 of the pre-mating period. One F_1 female from the group at the lowest dose died as a result of accidents. None of the deaths in either generation were attributed to treatment. Clinical signs, consisting of increased incidence of fur staining, were limited to females at the highest dose (F_0, and F_1) during gestation and lactation. A high incidence of food scattering was also observed in females at the highest dose (F_0, and F_1). Treatment-related decreases in mean body weights were limited to parental rats at the highest dose and their offspring. During weeks 0–7 of the pre-mating period, F_0 males and females had significantly lower mean body-weight gains. After mating, mean body weights of F_0 females at the highest dose were significantly lower during gestation (6.4%) and lactation (3.7%). The intergroup differences in mean body weights of the F_1 pups at birth were minimal. However, throughout lactation, F_1 pups at the highest dose had significantly decreased (approximately 20%) body-weight gains, which resulted in a significant decrease (16.8%) in the mean pup body weight at weaning. For the F_1 generation, statistically and biologically significant decreases in mean body weights were observed during the pre-mating period (males, 14–17%; females, 11–16%), gestation (14%), and lactation (9–16%) periods. At the highest dose, F_2 pups had significantly lower mean body weights at birth (both sexes combined). The mean body-weight gains for these rats were significantly lower (16.9%) than those of the controls, resulting in a significant decrease in mean body weight at weaning. Mean food consumption was decreased in rats at the highest dose (F_0 males, F_1 males and females) during the pre-mating period; food consumption was significantly increased during gestation (F_1 females) and lactation (F_0 and F_1 females). Food-conversion efficiency in the controls and treated groups were comparable. Statistically significant decreases in the F_0 female fertility and fecundity indices were observed at the highest dose; the indices were, however, within the range for historical controls. No other significant treatment-related effects were noted in any other reproductive parameters. At the highest dose, there was a statistically significant reduction in mean pup body weights (F_1 offspring) from day 1 of lactation to weaning compared with controls. The combined mean body weights at birth of F_2 pups at the highest dose were significantly lower than those of the controls. From day 1 of lactation until day 21, the differences in the mean body weights between F_1 and F_2 pups at the highest dose and in the control group progressively decreased. Compared with controls, the mean body weights at the highest dose were 18% lower for F_1 pups and 16% lower for F_2 pups on day 21 of lactation. Mean body-weight gains were also significantly reduced compared with those of the controls. There was no treatment-related effect on total number of litters and implantation sites, pup numbers, sex ratio, survival and clinical conditions of offspring. Developmental milestones of the F_1 generation pups did not show any treatment-related effects. Intergroup differences in the attainment time for eye opening, incisor eruption and pinna opening were minimal and not related to treatment; surface-righting, air-righting, auditory response, pupillary reflex and grip strength were similarly unaffected by treatment. No treatment-related abnormalities were observed at necropsy.

The NOAEL for parental systemic toxicity was 400 ppm, equal to 23.1 and 27.4 mg/kg bw per day in males and females, respectively. The LOAEL parental systemic toxicity was 5000 ppm, equal to 293.3 and 342.8 mg/kg bw per day in males and females, respectively, on the basis of decreased body weights (11–13%) and body-weight gains (11–17%). The NOAEL for reproductive toxicity was 5000 ppm, equal to 293.3 and 342.8 mg/kg bw per day in males and females, respectively. The NOAEL for offspring toxicity was 400 ppm, equal to 23.1 and 27.4 mg/kg bw per day in males and females, respectively. The LOAEL for offspring toxicity was 5000 ppm, equal to 293.3 and 342.8 mg/kg bw per day in males and females, respectively, on the basis of decreased pup body weights (17%) on day 21 of lactation (Clark, 1993a, 1993b, 1995).

(b) Developmental toxicity

Rats

In a study of developmental toxicity, groups of 30 pregnant Sprague-Dawley Crl:COBS CD (SD) rats were given pyrimethanil (purity, 96.3–97.0%) at a dose of 0, 7, 85, or 1000 mg/kg bw per day by gavage in 0.1% methyl cellulose in distilled water on days 6–15 of gestation, inclusive. Stability, homogeneity and dose concentrations were confirmed analytically. All rats were observed twice per day for mortality and moribundity and daily for clinical signs of toxicity. Maternal body weights were recorded on days 1 and 3 of gestation, daily from day 6 to day 16 of gestation, and on day 18 and 20 of gestation. Food consumption was measured between days 3–5, 6–8, 9–11, 12–15, 16–17 and 18–19 of gestation. On day 20 of gestation, all surviving dams were sacrificed and subjected to gross necropsy. The uterus and ovaries were exposed and the number of corpora lutea on each ovary was recorded. Gravid uteri were weighed, opened, and the location and number of viable and nonviable fetuses, early and late resorptions, and the total number of implantations were recorded. Uteri from females that appeared nongravid were opened and placed in 10% ammonium sulfide to detect any early implantation loss. All fetuses were weighed, sexed, and examined for external malformations/variations. Each fetus was examined viscerally by fresh dissection and the sex verified. Visceral examination was performed on fetuses preserved in Bouin fixative using a modified Wilson technique. For skeletal examination, fetuses were preserved in methanol and stained with Alizarin Red.

Analysis of preparations for the lowest, intermediate and highest dose were within 114.0–138.6%, 94.7–96.9% and 96.2–106.8% of the target doses, respectively. All rats survived to terminal sacrifice except one in the control group and one at the lowest dose that died on days 13 and 18 of gestation, respectively. The cause of death of the female at the lowest dose was gavage error. At the highest dose, treatment-related clinical signs of toxicity included hair loss, slight to moderate emaciation and hunched posture. No clinical signs of toxicity were observed at 7 or 85 mg/kg bw per day. Treatment-related, statistically significant decreases (9.8% on day 16 of gestation) in body weights and body-weight gains (42.3%) were observed in rats at the highest dose. These decreases were accompanied by 16.7–18.2% decreases in food consumption from day 6 to 15 of gestation. No effects on body weights, body-weight gains and food consumption were observed at 7 or 85 mg/kg bw per day. At necropsy, no treatment-related gross pathological findings were observed except hair loss in rats at the highest dose.

No treatment-related effects were seen in the mean number of corpora lutea, implantation sites, pre- and post-implantation loss, or early and late resorptions. There were no statistically significant differences in litter size, number of fetuses, number of implantations or fetal sex ratio. Statistically significant decreases in mean litter weight (14%) and mean fetal weight (7%) was observed in rats at the highest dose. Data provided for historical controls indicated that the mean litter and fetal weights were essentially similar. No treatment-related external, visceral, or skeletal malformations/variations were observed in any fetuses.

The NOAEL for maternal toxicity was 85 mg/kg bw per day. The LOAEL for maternal toxicity was 1000 mg/kg bw per day on the basis of abnormal clinical signs, reduced body weights and body-weight gains and reduced food consumption. The NOAEL for developmental toxicity was 1000 mg/kg bw per day. The LOAEL for developmental toxicity was not identified (Jackson & Bennett, 1991; Reader, 2003c).

Rabbit

In a study of developmental toxicity, groups of 18 or 19 pregnant New Zealand White rabbits were given pyrimethanil (purity, 97.1%) at a dose of 0, 7, 45, or 300 mg/kg bw per day by gavage in 1% methylcellulose in sterile water on days 7–19 of gestation, inclusive. Stability, homogeneity and dose concentrations were confirmed analytically. All rabbits were observed twice per day for mortality and moribundity and once daily for clinical signs of toxicity. Rats were also observed for signs of toxicity at approximately 1 h after dosing. Maternal body weights were recorded on days 0, 3 and daily from days 7–20 of gestation, and on days 22, 25 and 28 of gestation. Food consumption was measured every 2 days from day 3 to day 27 of gestation and daily from day 27 and day 28 of gestation. A gross postmortem examination was done on all females aborting during the study. On day 28 of gestation, all surviving does were killed and subjected to gross necropsy. The uterus and ovaries were excised and the number of corpora lutea on each ovary was recorded. Gravid uteri were weighed, opened, and the location and number of viable and nonviable fetuses, early and late resorptions, and the total number of implantations were recorded. All fetuses were weighed and examined for external malformations/variations. Crown–rump measurements were recorded for late resorptions and the tissues were discarded. Each fetus was examined viscerally by fresh dissection and the sex was determined. The brain from each fetus was examined by mid-coronal slice. All carcasses were eviscerated and processed for skeletal examination.

Analysis of dosing solutions indicated that the formulations were homogenous and the measured concentrations were within 15% of the target concentrations. Maternal toxicity at 300 mg/kg bw per day included significant decreases in body weight (6–8%), body-weight gain (29%) and decreases in food consumption (35–42%). Decreased food consumption resulted in increased incidence of rabbits with reduced production and size of faecal pellets. In addition, three rabits became emaciated, due to the decreased consumption, and were subsequently sacrificed on days 24, 15 and 22 of gestation. At 45 mg/kg bw per day, a slight increase in the number of dams with reduced production and size of faecal pellets (less frequent and of shorter duration) was seen, but was not considered to be an adverse treatment-related effect. At 7 mg/kg bw per day, no treatment-related maternal effects were noted. Treatment-related developmental findings at 300 mg/kg bw per day, when compared with those for concurrent controls, included decreases in fetal weight (10%), increases in fetal runts, increases in the incidence of retarded fetal bone ossification, and increases in the incidence of extra thoracic vertebrae and ribs. The increased number of fetal runts was outside the range for historical controls. The Meeting requested a clarification on the criteria used by the study authors in describing fetal runts to the submitter during the meeting. Fetal runts were defined based on reduced weights only and not by fetal size. Since the decreased in fetal body weights were observed in the presence of severe maternal toxicity (moribund condition, decreased body weights), the Meeting considered that these effects were secondary to maternal toxicity and were not treatment-related. No treatment-related developmental effects were observed at 45 or 7 mg/kg bw per day.

The NOAEL for maternal toxicity was 45 mg/kg bw per day on the basis of deaths in three rabbits (kiled), decrease in body-weight gain (29%), a decrease in food consumption (35–42% and reduced production and size of faecal pellets seen at the LOAEL of 300 mg/kg bw per day. The NOAEL for developmental toxicity was 300 mg/kg bw per day, the highest dose tested (Irvine, 1991).

2.6 *Special studies*

(a) Acute neurotoxicity

In a study of acute oral neurotoxicity, groups of 12 male and 12 female non-fasted Sprague-Dawley CD (Crl:CD [SD] IGS BR) rats were given a single dose of pyrimethanil (purity, 99.8%) at a dose of 0, 30, 100, or 1000 mg/kg bw by gavage in 0.5% (w/v) methylcellulose (5 ml/kg). In a separate study, the time of peak effects for pyrimethanil was estimated to be 1.5–4 h after dosing. Functional observational battery (FOB) and motor activity were evaluated on days −6, 1 (approximately 1.5–2 h after dosing), 8, and 15. On day 16, five males and five females per group were perfused in situ for neurohistological examination, and tissues from rats in the control group and group at the highest dose were examined microscopically. Data for positive controls were provided and were judged to be adequate, but were several years old.

No unscheduled deaths occurred. Clinical signs, body weights, body-weight gains, food consumption, gross pathology, brain weights and measurements, and neuropathology were unaffected by treatment. At 1000 mg/kg bw, the following transient FOB effects were noted on day 1: (i) slight to moderate overall gait incapacity, indicated by slight to moderate ataxia; (ii) moderately dilated pupils in females; (iii) decreased hindlimb-grip strength (24%; $p < 0.01$) in the males; and (iv) decreased body temperature (0.56–1.49 °C; $p < 0.01$). Total motor activity was also decreased by 52–110% ($p < 0.001$) at 1000 mg/kg bw on day 1 in males and females compared with controls. Habituation of motor activity was unaffected by treatment. All rats appeared to be normal on days 8 and 15. These effects were considered to be non-specific, transient and occurred at high doses.

The NOAEL was 100 mg/kg bw on the basis of decreased motor activity, ataxia, and decreased body temperature in males and females, decreased hindlimb-grip strength in males, and dilated pupils in females on day 1 seen at 1000 mg/kg bw (Beyrouty, 2001a, 2001b).

(b) Short-term studies of neurotoxicity

In a short-term study of neurotoxicity, groups of 12 male and 12 female Sprague-Dawley CD (Crl:CD[SD]IGS BR) rats were given diets containing pyrimethanil (purity, 99.8%) at a concentration of 0, 60, 600, 6000 ppm (equal to 0/0, 4.0/4.6, 38.7/44.3, or 391.9/429.9 mg/kg bw per day for males/females) for 13 weeks. FOB and motor activity were evaluated during weeks −1 (before dosing), 4, 8, and 13. At termination, five males and five females from each group were perfused in situ, and tissues from rats in the control group and in the group at 6000 ppm were examined microscopically. Data for the positive controls were provided and were judged to be adequate, but were several years old.

No treatment-related mortalities were observed. Clinical signs, FOB, motor activity, brain measurements (weight, length, and width), gross necropsy, and neurohistology were unaffected by treatment. In females at 6000 ppm, there were decreases in body weight of 8% at the end of the study, an overall decrease of body-weight gain of 21%, and corresponding decreases in food consumption (9–15%) in many weeks throughout the study. In males, significant decreases in body-weight gain of 21% and food consumption of 12% were seen only in the first week of the study. No treatment-related effects were observed at 600 ppm or less.

The NOAEL for males was 6000 ppm (equal to 391.9 mg/kg bw per day, the highest dose tested). The NOAEL for females was 600 ppm (equivalent to 44.3 mg/kg bw per day) on the basis of decreased body weight (8%), body-weight gain (21%), and food consumption (9–15%) seen at 6000 ppm (equal to 429.9 mg/kg/day, the highest dose tested) (Beyrouty, 2001c).

Mice

A study in mice was designed to investigate the effect of pyrimethanil on the activity of hepatic enzymes and the estrus cycle of female mice after 4 days dietary exposure. In a 4-day feeding study, groups of 15 female Crl:CD-1(ICR)BR mice were given diets containing pyrimethanil (purity, 96.2%) at a concentration of 0 or 900 ppm (equal to 162 mg/kg bw per day) for 4 days. All mice were weighed and killed on day 5. Livers were removed, weighed and five sets of three livers were pooled and used to prepare microsomes. The protein content and a measurement of the concentration of cytochrome P450 were carried out. The microsomes were then used to determine the level of the activity of 7-pentoxyresorufin-*O*-depentylase (PROD). Vaginal smears were prepared on day 1 and on day 4.

Treatment with pyrimethanil for 4 days did not affect the terminal body weights, or the liver weights. The concentration of microsomal protein/g liver rose by 8% in the treated mice and the level of cytochrome P450/mg protein rose by 5%. Each of these results was statistically significantly as was the increase in the activity of PROD (38% when expressed per mg protein). There were no overt differences between the cell types or the stages of the estrus cycle between the treated and control groups over the treatment period. The study author concluded that treatment with pyrimethanil at a dietary concentration of 900 ppm for 4 days resulted in a low potency induction of cytochrome P450, including the Cyp2b subfamily, with a profile of induction that was similar to that expected with phenobarbitone (Barker, 1998).

Rats

A study in rats was conducted to investigate whether the thyroid effects (colloid depletion, follicular-cell hypertrophy and follicular epithelial hyperplasia) seen after dietary administration of pyrimethanil at repeated high doses occurs via a direct effect on the thyroid or an indirect extra-thyroidal mechanism (mediated by the liver). These two mechanisms can be differentiated by implementing the perchlorate discharge test, which is based on: (a) the uptake of radiolabelled iodine; and (b) the ability of perchlorate to discharge any non-organified thyroidal iodine into the circulation. Normal uptake of iodine will be inhibited by a direct thyroid-blocking agent. Propylthiouracil (known to inhibit thyroid function directly) and phenobarbital (acts indirectly) were used as positive controls for each type of mechanism, respectively.

In a 7-day feeding study, groups of six male Sprague-Dawley Crl:CD(SD)BR rats were fed diets containing pyrimethanil (purity, 96.2%) at a concentration of 0 or 5000 ppm (equal to 1007 mg/kg bw per day), or propylthiouracil at a concentration of 2000 ppm (equal to 353 mg/kg bw per day), or phenobarbital at a concentration of 1000 ppm (equal to 217 mg/kg bw per day) for 7 days. On day 8, the rats were injected intraperitoneally with radiolabelled ^{125}I; 6 h later, all rats were killed exactly 2.5 min after intraperitoneal injection of either saline or potassium perchlorate.

The rats were observed daily for abnormal behaviour and physical condition. Individual body weights were recorded before treatment, at the start of treatment, on days 4 and 7 and at necropsy. The weekly food consumption was recorded. At termination, all rats were necropsied and examined macroscopically. The weight of the thyroid was recorded and radioactivity in the thyroid was measured. The volume of blood was measured and radioactivity in the blood measured.

There were no mortalities during the study. Rats treated with pyrimethanil had no adverse clinical effects. Rats treated with propylthiouracil showed reduced activity and piloerection while treatment with phenobarbital resulted in reduced activity, unsteady gait, reduced muscle tone, piloerection, and wasted body condition. Group mean body weights of rats treated with either pyrimethanil or propylthiouracil were lower than those for the controls at days 4, 7, and 8 and total body-weight gains (19.6% and 12.2%, respectively) were reduced compared with those of the controls (35.4%). Food consumption was decreased by 36% at day 3 in rats treated with pyrimethanil compared with those of the controls. Reduced food consumption in rats treated with propylthiouracil was seen at days 3 and 7 (−46% and −26%, respectively). Uptake of ^{125}I was significantly increased in the thyroid of rats treated with either pyrimethanil or phenobarbital (150% or 221% of the corresponding value for concurrent controls receiving saline, respectively). There was no significant discharge of ^{125}I after administration of perchlorate in either the phenobarbital or pyrimethanil group. The group treated with propylthiouracil showed a significant reduction in uptake of ^{125}I uptake (65% of the value for controls) and a significant discharge of ^{125}I (61%) after administration of perchlorate. At necropsy, increased thyroid weights (+76%) and relative thyroid : body weight ratio (+113%) were seen in rats treated with propylthiouracil when compared with the controls. No treatment-related effects on thyroid weights were observed after exposure to pyrimethanil or phenobarbital. The results indicated that findings in rats treated with pyrimethanil were similar to findings in rats treated with phenobarbital and were different from those reported after treatment with propylthiouracil. The study results suggested that pyrimethanil is not a direct-acting thyroid blocker (Healing, 1992a).

A 14-day feeding study in rats was conducted to characterize the thyroid effects seen in short- and long-term studies with pyrimethanil at high doses. Groups of 10 male Sprague-Dawley Crl:CD(SD)BR rats were given diets containing pyrimethanil (purity, 96.2%) at a concentration of 0 or 5000 ppm (equivalent to 378.5 mg/kg bw per day) for 14 days. Five rats from each group were killed on day 15. The remaining five males per group received untreated diet for a further 14 days before necropsy. Rats were observed daily for abnormal behaviour and physical condition. Individual body weights were recorded before treatment, at the start of treatment, twice per week thereafter and at necropsy. The weekly food consumption was recorded. Blood samples for clinical chemistry were collected on study days 1, 2, 4, 8, 15 and 29. At termination, all rats were necropsied and examined macroscopically. The weights of the liver, pituitary and thyroid were recorded and the organs examined histopathologically.

There were no mortalities or treatment-related clinical signs reported. A reduction in body-weight gain (23%) was seen in treated rats at the first week only. Rats treated with pyrimethanil at 5000 ppm for 14 days showed a markedly increased activity of liver enzyme uridine diphosphoglucuronyl transferase (UDPGT) (446% of the value for controls). Also, significantly increased concentrations of TSH were seen at days 1, 4, and 15 of the study (162%, 155%, and 215% of values for respective controls) while decreased concentrations of T3 and T4 (82% and 76% of respective controls) were noted at day 4. Liver weights of the treated rats were significantly higher (23%) than those of the controls at interim kill, while thyroid weights were significantly lower (56%) than those of the controls. After a 14-day recovery period, liver weight and plasma concentrations of T3, T4, TSH, and reverse triiodothyronine were back to normal, while the UDPGT value remained elevated (163% of that of the controls). There were no treatment-related gross pathological effects. Histopathologically, 14 days of treatment with pyrimethanil resulted in centrilobular hepatocyte enlargement in the liver and colloid depletion, follicular-cell hypertrophy and follicular epithelial hyperplasia in the thyroid gland. After a 14-day recovery period, these effects returned to normal.

It was suggested by the study author that the elevated level of liver enzyme (UDPGT) led to increased clearance of thyroid hormones via enhanced hepatic metabolism. This in turn resulted in an extrathyroidal, indirect effect on thyroid-hormone homeostasis, which via normal pituitary feedback led to increased concentrations of TSH, and stimulation of the thyroid gland. Hence, the elevation of TSH shown to result from treatment with pyrimethanil may be responsible for the histopathological changes in the thyroid glands of rats (Healing, 1992b; Reader, 2003d).

Groups of rats were given pyrimethanil (technical) orally to investigate the effect on hepatic mixed function oxidase system after 4 days dietary exposure. Groups of six male Sprague-Dawley CD-1 rats were dosed orally with pyrimethanil (purity, 99.4%) at a dose of 0, 100 or 200 mg/kg bw twice per day in 0.5% gum tragacanth for 4 days. Additional groups were given 0.1% phenobarbitone in drinking-water for a minimum of 14 days, an intraperitoneal dose of phenobarbitone at 80 mg/kg bw per day in water for 4 days, an intraperitoneal dose of β-naphthaflavone at 80 mg/kg bw per day in corn oil for 4 days and an intraperitoneal dose of clofibrate at 400 mg/kg bw per day in corn oil for 4 days. Rats were necropsied 17 h after the last dose and microsomal suspensions prepared from their livers. Activities of liver enzymes were determined.

Pyrimethanil had only a minor effect on the liver enzymes, showing very low levels of ethoxyresorufin-*O*-deethylase and pentoxyresorufin-*O*-dealkylase. Pyrimethanil also led to a statistically significant increase in liver weight (at 100 mg/kg bw per day only) and concentration of cytochrome b5 (at 200 mg/kg bw per day only). The type of induction pattern was similar to that for phenobarbitone. The increase in ethoxyresorufin-*O*-deethylase was not considered to be caused by an increase in cytochrome P448. The magnitude of the increases with pyrimethanil was less than for phenobarbitone and much less than for β-naphthaflavone. However, the increases in pentoxyresorufin-O-dealkylase were less than for phenobarbitone but more than for β-naphthaflavone. Pyrimethanil did not increase the activity of lauric acid hydroxylase and so was not considered to be a clofibrate-type inducer of cytochrome P_{452}. The study results suggest that treatment with pyrimethanil at 100 or 200 mg/kg bw per day by gavage for 4 days resulted in a marginal induction of the hepatic mixed function oxidase system, predominantly of the phenobarbitone type (Needham, 1991).

3. Studies with metabolites

Groups of five male and five female young adult Sprague-Dawley Crl:CD(IGS)BR rats were given AE F 132593, 2-amino-4, 6 dimethylpyrimidine, a pyrimethanil soil-photolysis metabolite (purity, 98.7%) as a single dose of 0, 100, 200, 400, 800, or 1600 mg/kg bw by gavage in 1% w/v methyl cellulose in distilled water. Treated rats were subjected to gross necropsy at the end of a 14-day observation period. At 1600 mg/kg bw, all rats were culled owing to their moribund condition between 30 min and 4 h after treatment. At 800 mg/kg bw, three males and two females were culled between 1 and 6 h after dosing. At 400 mg/kg bw, one female was culled in about 5 h after dosing. No mortality was observed in the groups at 100 or 200 mg/kg bw. Clinical signs such as salivation, prostration, laboured respiration, unsteady gait, reduced activity, reduced muscle tone, occasional findings in the eye (water, red discharge) and hair loss were observed in rats at 1600, 800, 400 and in females only at 200 mg/kg bw. These clinical signs were observed immediately after dosing, with recovery within 7 days except for hair loss. No treatment-related effects on body weights were observed. Necropsy of surviving rats did not reveal any macroscopic findings. The oral LD_{50} value was 735 mg/kg bw (95% confidence interval, 575–939 mg/kg bw) in male and female rats (Weir & Sindle, 1998). Observations in humans

A toxicological monograph on pyrimethanil prepared by the sponsor for the present Meeting stated that there had been no reported cases of clinical signs or poisoning incidents with pyrimethanil. It also stated that no epidemiological studies with pyrimethanil had been conducted (Jardinet, 2006).

Comments

Biochemical aspects

In rats given radiolabelled pyrimethanil orally, about 80% of the administered dose was absorbed (for the lower dose, 11.8 mg/kg bw, and for the higher dose, 800 mg/kg bw) on the basis of urinary excretion (cage-wash included) in 96 h. About 72% of the dose was absorbed after pretreatment with pyrimethanil at a dose of 10 mg/kg bw per day for 14-days, on the basis of urinary excretion (cage-wash included). Pyrimethanil was rapidly excreted at both doses, with more than 95% of the lower dose and 63–67% of the higher dose being excreted within the first 24 h. At the lower dose, plasma concentrations of radioactivity peaked at 1 h after dosing. At the higher dose, plasma concentrations of radioactivity initially peaked at 1 h after dosing. After an initial decline, a second peak of plasma radioactivity was observed at 5 h after dosing. The elimination half-life was about 4.8 h and 11.8 h at the lower and higher dose, respectively. Most of a radiolabelled dose was eliminated in the urine (79–81%) with the remainder in faeces (15–23%) at the lower and higher doses. No bioaccumulation of pyrimethanil was observed. A similar excretion pattern was observed in mice and dogs.

Systemically absorbed pyrimethanil was extensively metabolized. The major metabolites of pyrimethanil in the urine and faeces resulted from aromatic oxidation to form phenols in either or both rings and conjugation with glucuronic acid and sulfate. A minor pathway included oxidation of the methyl group on the pyrimidine ring to produce alcohol. The same six metabolites were identified in the urine and faeces. Unchanged pyrimethanil was isolated only in the faeces of males and females (0.3% and 2.1% of the faecal radioactivity at 10 and 1000 mg/kg bw, respectively). Distribution, metabolite profiles and excretion were essentially independent of pre-treatment with unlabelled compound and of sex.

Toxicological data

Pyrimethanil has low acute toxicity when administered by oral, dermal or inhalation routes. The LD_{50} in rats treated orally was 4149 mg/kg bw in males and 5971 mg/kg bw in females. The LD_{50} in rats treated dermally was > 5000 mg/kg bw. The LC_{50} in rats treated by inhalation (nose only) was > 1.98 mg/l (dust). Pyrimethanil was minimally irritating to the eyes of rabbits and not irritating to the skin of rabbits. Pyrimethanil was not a skin sensitizer as determined by Buehler and Magnusson & Kligman (maximization) tests in guinea-pigs. Clinical signs after oral administration consisted of reduced activity, reduced muscle tone, urogenital soiling, coolness to touch, which generally resolved within 1 day. There were no pathological findings.

In short-term and long-term studies in mice, rats and dogs, the major toxicological findings included decreased body weight and body-weight gains, often accompanied by decreased food consumption. The major target organs in mice and rats were liver and thyroid organs as evidenced by organ-weight changes, histopathological alterations, and clinical chemistry parameters (including increased cholesterol, and GGT activity).

In a 90-day dietary study of toxicity in mice, decreased body-weight gains, slightly increased concentrations of cholesterol and total bilirubin, an increase in liver weights and histopathological findings in thyroid, kidney and kidney stones were seen at 10 000 ppm, equal to1864 mg/kg bw per day. Increases in thyroid weights were associated with exfoliative necrosis and pigmentation of follicular cells. The NOAEL was 900 ppm, equal to 139 mg/kg bw per day).

In a 90-day dietary study of toxicity in rats, decreased body weights, body-weight gains (28–33%) and decreased food consumptions, brown urine and increased urinary proteins, decreased organ weights (heart, adrenal, spleen, thymus), increased liver, kidney, gonad weights, and hypertrophy in liver and thyroid were seen at 8000 ppm, equal to 529.1 mg/kg bw per day, in males and females. Thyroid effects in rats were manifested as increased incidence and severity of follicular epithelial hypertrophy and follicular brown pigment. The NOAEL was 800 ppm, equal to 54.5 mg/kg bw per day.

Gavage administration of pyrimethanil at > 600 mg/kg bw per day induced vomiting in dogs within 4 h after dosing, suggesting local irritation of the gastrointestinal tract. This was not considered to be a toxicologically relevant effect for establishing an acute reference dose (ARfD). In a 90-day study of toxicity in dogs, diarrhoea, salivation hypoactivity (within 3 h after dosing) and slightly decreased water consumption was observed at 800 mg/kg bw per day. The NOAEL was 80 mg/kg bw per day. In a 52-week study of toxicity in dogs, decreases in body-weight gains (6% and 17% in males and females, respectively), food consumption and feed-conversion efficiency, water consumption, reduced clotting time and increased count of neutrophils were observed at 250 mg/kg bw per day. The NOAEL was 30 mg/kg bw per day. The overall NOAEL was 80 mg/kg bw per day when results of 90-day and 1-year studies of toxicity in dogs were combined.

Pyrimethanil was not mutagenic in an adequate battery of studies of genotoxicity in vitro and in vivo.

The Meeting concluded that pyrimethanil is unlikely to be genotoxic.

The carcinogenicity potential of pyrimethanil was studied in mice and rats. In a study of carcinogenicity in mice, an increased incidence of urinary tract lesions including bladder distension and thickening were observed in male mice during the first weeks at 1600 ppm, equal to 210.9 mg/kg bw per day. The NOAEL was 160 ppm, equal to 20.0 mg/kg bw per day. There were no treatment-related neoplastic findings in the bioassay in mice.

In the study of carcinogenicity in rats, decreased body-weight gains, increased serum cholesterol and GGT levels, necropsy (dark thyroids), and histopathological findings (increases in centrilobular hepatocyte hypertrophy, and increased incidence of colloid depletion and hypertrophy of the follicular epithelium in thyroids) were observed at 5000 ppm, equal to 221 mg/kg bw per day). The NOAEL was 400 ppm, equal to 17 mg/kg bw per day. In rats given pyrimethanil, the thyroid was the only tissue to show a higher incidence of tumours than the controls. The number of benign follicular-cell adenomas in males and females at the highest dose was higher than in concurrent controls and historical controls.

Special studies were conducted to evaluate the toxicity seen in the liver and thyroid. Mechanistic data suggest that thyroid hormone imbalance caused by increased thyroid hormone clearance by the induction of liver enzymes resulted in increased TSH activity and persistent stimulation of the thyroid. Such effects may lead to changes in thyroid homeostasis and alterations in morphology. Rodent thyroid tumours induced by this mode of action are not relevant to humans because rats are much more sensitive to thyroid-hormone imbalance and elevations in TSH concentrations. Thus, the results of bioassays in rats do not raise a cancer concern for humans.

In view of the lack of genotoxicity and the absence of relevant carcinogenicity in rats and mice, the Meeting concluded that pyrimethanil is unlikely to pose a carcinogenic risk to humans.

In a two-generation study of reproduction in rats, reproductive parameters were not affected at the highest dose tested (5000 ppm, equal to 293.4 mg/kg bw per day). The NOAEL for parental systemic toxicity was 400 ppm (equal to 23.1 mg/kg bw per day) on the basis of decreases in body-weight (11–13%) and body-weight gains (11–17%). Offpring toxicity was manifested as a decrease in pup body weights (17%) on postnatal day 21 at 5000 ppm, equal to 293.3 mg/kg bw per day. The NOAEL for offspring toxicity was 400 ppm, equal to 23.1 mg/kg bw per day. Pyrimethanil was not embryotoxic, fetotoxic or teratogenic at doses of up to 1000 mg/kg bw per day in rats. Pyrimethanil was not teratogenic in rabbits. Decreases in fetal body weights were observed at 300 mg/kg bw per day. These decreases in fetal weights (described as "runts" in the study report) were observed in the presence of severe maternal toxicity manifested as a significant decrease in body-weight gain and food consumption, reduced production and size of faecal pellets and death of three rabbits (moribund condition) at 300 mg/kg bw per day. The NOAEL for maternal toxicity in rabbits was 45 mg/kg bw per day and the NOAEL for developmental toxicity was 300 mg/kg bw per day, the highest dose tested.

The Meeting concluded that pyrimethanil is not teratogenic.

In a study of acute neurotoxicity in rats, transient functional observational battery (FOB) effects (gait, ataxia, decreased hindlimb-grip strength in males, decreased body temperature) were observed at 1000 mg/kg bw on day 1. Total motor activity was also decreased by ≥ 52% at 1000 mg/kg bw on day 1 in males and females compared with controls. All rats appeared normal on days 8 and 15. As these transient and non-specific effects occurred at a high dose administered by gavage, the Meeting concluded that they were not an appropriate basis for establishing an ARfD. The NOAEL was 100 mg/kg bw. In a short-term study of neurotoxicity in rats, no treatment-related changes in mortality, clinical signs, FOB, motor activity, brain measurements (weight, length, and width), gross necropsy, or neurohistopathology were observed at doses of up to 6000 ppm, equal to 391.9 mg/kg bw per day. In females, an overall decrease in body-weight gain of 21% was observed at 6000 ppm, equal to 429.9 mg/kg bw per day. The NOAEL in females was 600 ppm, equal to 38.7 mg/kg bw per day, and 6000 ppm, equal to 319.9 mg/kg bw per day, in males.

The Meeting considered that pyrimethanil is not neurotoxic on the basis of the available data.

No significant adverse effects were reported in personnel working in production plants.

The Meeting concluded that the existing database on pyrimethanil was adequate to characterize the potential hazards to fetuses, infants and children.

Toxicological evaluation

The Meeting established an acceptable daily intake (ADI) of 0–0.2 mg/kg bw based on a NOAEL of 400 ppm (equal to 17.0 mg/kg bw per day) on the basis of increased cholesterol and GGT levels, and histopathological changes in the liver and thyroid at 5000 ppm (equal to 221 mg/kg bw per day) in a 2-year study in rats, and using a safety factor of 100. This ADI is supported a by two-generation study of reproduction in rats in which the NOAEL for parental systemic toxicity was 400 ppm, equal to 23.1 mg/kg bw per day, on the basis of decreased body weights and body-weight gains at 5000 ppm, equal to 293.3 mg/kg bw per day. This ADI is also supported by the NOAEL of 160 ppm, equal to 20.0 mg/kg bw per day, in males in a 2-year study of toxicity in mice; this NOAEL was identified on the basis of increased incidences of urinary tract lesions including bladder distension and thickening seen at 1600 ppm, equal to 210.9 mg/kg bw per day.

The Meeting concluded that it was not necessary to establish an ARfD for pyrimethanil because no toxicity could be attributable to a single exposure in the available database, including a study of developmental toxicity in rats and rabbits. Observations in the study of acute toxicity in rats and clinical signs of toxicity in the pyrimethanil database appeared at doses of 640 mg/kg bw per day and greater were not considered to be relevant for establishing an ARfD since they were transient, non-specific and occurred at high doses. The Meeting also considered clinical signs (vomiting) in several studies of toxicity in dogs; these were considered to be local effects and therefore not relevant in establishing an ARfD.

Levels relevant to risk assessment

Species	Study	Effect	NOAEL	LOAEL
Mouse	Eighty-week study of toxicity and carcinogenicity[a]	Toxicity	160 ppm, equal to 20.0 mg/kg bw per day	1600 ppm, equal to 210.9 mg/kg bw per day
		Carcinogenicity	1600 ppm, equal to 210.9 mg/kg bw per day[c]	—
Rat	Two-year study of toxicity and carcinogenicity[a]	Toxicity	400 ppm, equal to 17 mg/kg bw per day	5000 ppm, equal to 221 mg/kg bw per day
		Carcinogenicity	5000 ppm, equal to 221 mg/kg bw per day[c]	—
	Multigeneration study of reproductive toxicity[a]	Parental toxicity	400 ppm, equal to 23.1 mg/kg bw per day	5000 ppm, equal to 293.3 mg/kg bw per day
		Offspring toxicity	400 ppm equal to 23.1 mg/kg bw per day	5000 ppm, equal to 293.3 mg/kg bw per day
	Developmental toxicity[b]	Maternal toxicity	85 mg/kg bw per day	1000 mg/kg bw per day
		Embryo/fetotoxicity	1000 mg/kg bw per day[c]	—
Rabbit	Developmental toxicity[b]	Maternal toxicity	45 mg/kg bw per day	300 mg/kg bw per day
		Embryo/fetotoxicity	45 mg/kg bw per day	300 mg/kg bw per day
Dog	Ninety-day and 1-year study of toxicity[b]	Toxicity	80 mg/kg bw per day	400/250 mg/kg bw per day

a Dietary administration.
[b] Gavage administration.
[c] Highest dose tested.

Estimate of acceptable daily intake for humans

0–0.2 mg/kg bw per day

Estimate of acute reference dose

Unnecessary

Information that would be useful for continued evaluation of the compound

Results from epidemiological, occupational health and other such observational studies of human exposure

Critical end-points for setting guidance values for exposure to pyrimethanil

Absorption, distribution, excretion, and metabolism in mammals	
Rate and extent of oral absorption	Rapid and nearly complete absorption; maximum plasma concentration reached by 1 h
Distribution	Widely distributed in tissues
Potential for accumulation	Low, no evidence of accumulation
Rate and extent of excretion	Approximately 97% (77% in urine and 20% in faeces) within 24 h at 11.8 mg/kg bw per day
Metabolism in animals	Extensive; metabolic pathways include aromatic oxidation to form phenols and conjugation with glucuronic acid and sulfate, minor pathway included oxidation of methyl group to produce alcohol
Toxicologically significant compounds in animals, plants and the environment	Pyrimethanil

Acute toxicity	
Rat, LD_{50}, oral	4149 mg/kg bw for males
Rat, LD_{50}, dermal	> 5000 mg/kg bw
Rat, LC_{50}, inhalation	> 1.98 mg/l dust (4-h exposure, nose only)
Rabbit, skin irritation	Not an irritant
Rabbit, eye irritation	Minimal irritation
Guinea-pig, skin sensitization	Not a sensitizer (Magnussen & Kligman and Buehler test)
Short-term studies of toxicity	
Target/critical effect	Liver and thyroid hypertrophy
Lowest relevant oral NOAEL	54.5 mg/kg bw per day (90-day study in rats)
Lowest relevant dermal NOAEL	No data
Lowest relevant inhalation NOAEC	No data
Genotoxicity	
	No genotoxic potential
Long-term studies of toxicity and carcinogenicity	
Target/critical effect	Liver and thyroid
Lowest relevant NOAEL	17 mg/kg bw per day (2-year study of carcinogenicity in rats)
Carcinogenicity	No relevant carcinogenicity in mice and rats
Reproductive toxicity	
Reproduction target/critical effect	No toxicologically relevant effects
Lowest relevant reproductive NOAEL	239.9 mg/kg bw per day (rats; highest dose tested)
Developmental target/critical effect	No developmental toxicity in rats and rabbits
Lowest relevant developmental NOAEL	300 mg/kg bw per day (rabbits; highest dose tested)
Neurotoxicity/delayed neurotoxicity	
Acute neurotoxicity	No sign of specific neurotoxicity
Mechanistic data	
	Studies on hepatic clearance and thyroid-hormone perturbations
Medical data	
	No significant adverse health effects reported

Summary

	Value	Study	Safety factor
ADI	0–0.2 mg/kg bw per day	Rats, 2-year study of toxicity	100
ARfD	Unnecessary	—	—

References

Adams, K., Hensly, S.M., & Godfrey, A. (1992) Technical SN 100 309: in vitro Chinese hamster ovary/HPRT locus gene mutation assay. Unpublished report No. A81827 from Huntingdon Research Centre, Huntingdon, UK. Submitted to WHO by BASF, France.

Barker, M (1998) Pyrimethanil - Investigation of liver enzyme induction following dietary administration to female CD-1 mice for 4 days. Unpublished report No. C001378 from Huntingdon Life Sciences Ltd, Huntingdon Cambridgeshire, UK Submitted to WHO by BASF, France.

Beyrouty, P. (2001a) An acute oral neurotoxicity study of pyrimethanil technical in rats. Unpublished report dated September 24 from ClinTrials BioResearch Ltd, Senneville, Quebec, Canada. Laboratory Project No. 97567. With permission from Bayer CropScience.

Beyrouty, P. (2001b) A time of peak behavioral effects study of pyrimethanil technical in the rat. Unpublished report dated 22 August from ClinTrials BioResearch Ltd, Senneville, Quebec, Canada. Laboratory Project No. 97572. With permission from Bayer CropScience.

Beyrouty, P. (2001c) A 13-week dietary neurotoxicity study of pyrimethanil technical in rats. Unpublished report dated 19 October from ClinTrials BioResearch Ltd, Senneville, Quebec, Canada. Laboratory Project No. 97568. With permission from Bayer CropScience.

Bright, J.H. (1993) Technical SN 100309: 104-week rat combined chronic toxicity and oncogenicity study. Unpublished report No. A54965 from Schering Agrochemicals Ltd, Saffron Walden, Essex, UK. Submitted to WHO by BASF, France.

Brooker, P.C., Akhurst, L.C., King, J.D., & Howell, A. (1990) Technical SN 100309: metaphase chromosome analysis of human lymphocytes cultured in vitro. Unpublished report No. A81789 from Huntingdon Research Centre, Huntingdon, UK. Submitted to WHO by BASF, France.

Challis, I.R. (1995) Pyrimethanil - code: SN 100 309 - rat pharmacokinetics. Unpublished report No. A88956 from AgrEvo UK Ltd, Chesterford Park, Saffron Walden, Essex, UK. Submitted to WHO by BASF, France.

Clark, R. (1993a) Technical SN 100309: two generation oral (dietary administration) reproduction toxicity study in the rat. Unpublished report No. A81822 from Hazleton UK, Harrogate, North Yorkshire, UK. Submitted to WHO by BASF, France.

Clark, R. (1993b) 1st amendment to report No. TOX/91/223-49: technical SN 100309: rat two generation dietary reproduction toxicity study. Unpublished report No. A89218 from Hazleton UK, Harrogate, North Yorkshire, UK. Submitted to WHO by BASF, France.

Clark, R. (1995) 2nd amendment to report No. TOX/91/223-49: technical SN 100309: rat two generation dietary reproduction toxicity study. Unpublished report No. A89219 from Hazleton UK, Harrogate, North Yorkshire, UK. Submitted to WHO by BASF, France.

Clay, H. (1994) 1st amendment to report No. TOX/92/223-68 - technical SN 100309: 80 week oral (dietary administration) carcinogenicity study in the mouse. Unpublished report No. A81814 from Hazleton UK, Harrogate, North Yorkshire, UK. Submitted to WHO by BASF, France.

Clay, H. (1995) 3rd amendment to report No. TOX/92/223-68: technical SN 100309: 80 week oral (dietary administration) carcinogenicity study in the mouse. Unpublished report No. A89480 from Hazleton UK, Harrogate, North Yorkshire, UK. Submitted to WHO by BASF, France.

Clay, H. (1996) 4th amendment to report No. TOX/92/223-68: technical SN 100309: 80 week oral (dietary administration) carcinogenicity study in the mouse. Unpublished report No. A89481 from Corning Hazleton Europe, Harrogate, UK. Submitted to WHO by BASF, France.

Clay, H. & Healing, G. (1993) Technical SN 100 309: 80 week oral (dietary administration) carcinogenicity study in the mouse. Unpublished report No. A81811 from Hazleton UK, Harrogate, North Yorkshire, UK. Submitted to WHO by BASF, France.

Clouzeau, J. (1994) Pyrimethanil (code: SN 100309) - guinea-pig skin sensitization study. Unpublished report No. A81848 from Centre International de Toxicologie, Evreux, France. Submitted to WHO by BASF, France.

Codex Alimentarius Commission (2007) *Report of the Thirty-ninth Session of the Codex Committee on Pesticide Residues, 7–12 May 2007, Beijing, China* (ALINORM07/30/24).

Davies, M. (1990a) SN 100 309: skin sensitization in guinea pig (Buehler test). Unpublished report No. A81778 from Schering Agrochemicals Ltd, Saffron Walden, Essex, UK. Submitted to WHO by BASF, France.

Davies, M. (1990b) T10, addendum #1: SN 100309: skin sensitization in guinea pig - positive control results Unpublished report No. A55313 from Schering Agrochemicals Ltd, Saffron Walden, Essex, UK Submitted to WHO by BASF, France.

Gant, R.A. & Jones, E. (1994) Pyrimethanil: bacterial (REC) assay for DNA damage. Unpublished report No. A81859 from Huntingdon Research Centre, Huntingdon, UK. Submitted to WHO by BASF, France.

Grosshans, F. (2003) Pyrimethanil: review of the dossier - answer to questions from the Rapporteur Member State. Unpublished report No. 2003/1023037 from BASF AG, Agrarzentrum Limburgerhof, Limburgerhof, Germany. Submitted to WHO by BASF, France.

Harvey, P.W. (1991a) Technical SN 100 309: 28 day dietary and gavage repeated dose study in rats Unpublished report No. A81779 from Schering Agrochemicals Ltd, Saffron Walden, Essex, UK. Submitted to WHO by BASF, France.

Harvey, P.W. (1991b) Technical SN 100 309: Mouse 28-day dietary repeat dose study. Unpublished report No. A81781 from Schering Agrochemicals Ltd, Saffron Walden, Essex, UK. Submitted to WHO by BASF, France.

Harvey, P.W. (1991c) Technical SN 100309: dog 90-day oral (gavage) repeat dose study. Unpublished report No. A81790 from Schering Agrochemicals Ltd, Saffron Walden, Essex UK. Submitted to WHO by BASF, France.

Harvey, P.W. (1992a) Technical SN 100309: dog oral maximum tolerated dose - 13 day range-finding study. Unpublished report No. A81763 from Schering Agrochemicals Ltd, Saffron Walden, Essex , UK. Submitted to WHO by BASF, France.

Harvey, P.W. (1992b) Technical SN 100309: rat 10-day oral range finding study. Unpublished report No. A81776 from Schering Agrochemicals Ltd, Saffron Walden, Essex, UK. Submitted to WHO by BASF, France.

Harvey, P.W. (1992c) Addendum No. 1: technical SN 100 309: rat 10-day oral range finding study. Unpublished report No. A81777 from Schering Agrochemicals Ltd, Saffron Walden, Essex, UK. Submitted to WHO by BASF, France.

Harvey, P.W. (1994) 2nd amendment to report No. TOX90/223-25 - SN 100309: 13 weeks oral (dietary) toxicity study in the rat followed by a 4 week reversibility period. Unpublished report No. A81785 from Toxicol Laboratories Ltd, Ledbury Herefordshire, UK. Submitted to WHO by BASF, France.

Harvey, P.W. & Davies, M. (1990) Technical SN 100 309: 28-day repeat dose study in dogs. Unpublished report No. A81764 from Schering Agrochemicals Ltd, Saffron Walden, Essex, UK. Submitted to WHO by BASF, France.

Harvey, P.W. & Rees, S.J. (1991) SN 100309 (CR 19325/3): mouse 90-day dietary repeat dose study. Unpublished report No. A81792 from Schering Agrochemicals Ltd, Saffron Walden, Essex, UK. Submitted to WHO by BASF, France.

Healing, G. (1992a) Technical SN 100 309: rat 7-day dietary thyroid function test using perchlorate discharge as a diagnostic test. Unpublished report No. A81829 from Schering Agrochemicals Ltd, Saffron Walden, Essex, UK. Submitted to WHO by BASF, France.

Healing, G. (1992b) Technical SN 100 309: rat 14-day dietary study to investigate the mechanism of thyroid response. Unpublished report No. A81828 from Schering Agrochemicals Ltd, Saffron Walden, Essex, UK. Submitted to WHO by BASF, France.

Healing, G. (1994) 1st addendum to report No. TOX/92/223-62: technical SN 100309: 104 week rat combined chronic toxicity and oncogenicity study - report of individual pathology findings. Unpublished report No. A81808 from AgrEvo UK Ltd, Chesterford Park, Saffron Walden, Essex, UK. Submitted to WHO by BASF, France.

Healing, G. (1995c) 2nd amendment to report No. TOX/92/223-68: technical SN 100309: 80 week oral (dietary administration) carcinogenicity study in the mouse. Unpublished report No. A89479 from Hazleton UK, Harrogate, North Yorkshire, UK. Submitted to WHO by BASF, France.

Healing, G. (1996a) 1st amendment to report TOX/94/223-82: technical SN 100309 - guinea-pig skin sensitization study (Magnusson & Kligman maximisation assay). Unpublished report No. A89294 from Centre International de Toxicologie, Evreux, France. Submitted to WHO by BASF, France.

Healing, G. (1996b) 2nd amendment to report TOX/94/223-82: technical SN 100309 - guinea-pig skin sensitization study (Magnusson & Kligman maximisation assay). Unpublished report No. A89340 from Centre International de Toxicologie, Evreux, France. Submitted to WHO by BASF, France.

Hemmings, P.A. (1991a) Residue levels in rat tissues following repeated daily oral dosing with (^{14}C) SN 100 309 at 10 mg/kg bodyweight. Unpublished report No. A81622 from Schering Agrochemicals Ltd, Saffron Walden, Essex, UK. Submitted to WHO by BASF, France.

Hemmings, P.A. (1991b) The distribution and excretion of radiolabelled residues in the mouse following oral dosing with [^{14}C] SN 100309 at 10 mg/kg bodyweight. Unpublished report No. A81624 from Schering Agrochemicals Ltd, Saffron Walden, Essex , UK. Submitted to WHO by BASF, France.

Hemmings, P.A. (1993) SN 100 309: excretion and tissue residues of a radiolabelled oral dose in rats following pre-dosing for 14 days with unlabelled SN 100 309. Unpublished report No. A81631 from Schering Agrochemicals Ltd, Saffron Walden, Essex, UK. Submitted to WHO by BASF, France.

Higham, A.T. (1990) SN 100 309: 13-week oral (dietary) toxicity study in the rat followed by a 4-week regression period. Unpublished report No. A81783 from Toxicol Laboratories Ltd, Ledbury, Herefordshire, UK. Submitted to WHO by BASF, France.

Husband, R.F. (1992) Amendment 1 - conforming amendment number one Schering Agrochemicals Limited - SN 100309 - 13 week oral (dietary) toxicity study in the rat followed by a 4 week reversibility period. Unpublished report No. A81784 from Toxicol Laboratories Ltd, Ledbury, Herefordshire, UK. Submitted to WHO by BASF, France.

Irvine, L.F. (1991) Technical SN 100 309: oral (gavage) development toxicity (teratogenicity) study in the New Zealand White rabbit. Unpublished report No. A81798 from Toxicol Laboratories Ltd, Ledbury, Herefordshire, UK. Submitted to WHO by BASF, France.

Jackson, C.M. & Bennett, L.K. (1991) Technical SN 100309: rat oral developmental toxicity (teratogenicity) study. Unpublished report No. A81800 from Schering Agrochemicals Ltd, Saffron Walden, Essex, UK. Submitted to WHO by BASF, France.

Jackson, C.M., Everett, D.J. (1994) 1st amendment to report No. TOX/92/223-57 - technical SN 100309: in vitro Chinese hamster ovary/HPRT locus gene mutation assay. Unpublished report No. A89515 from Huntingdon Research Centre, Huntingdon, UK. Submitted to WHO by BASF, France.

Jackson, G.C., Hardy, C.J. (1992) Technical SN 100309: rat acute (4-hour exposure) inhalation toxicity study. Unpublished report No. A81820 from Huntingdon Research Centre, Huntingdon, UK. Submitted to WHO by BASF, France.

Jardinet, F. (2006) Pyrimethanil (BAS 605 F) JMPR evaluation. Toxicology monograph. Unpublished report No. 2006/1039449 from BASF Agro SAS, France. Submitted to WHO by BASF, France.

Jones, E. & Gant, R.A. (1990) Technical SN 100309: bacterial mutation assay. Unpublished report No. A81788 from Huntingdon Research Centre, Huntingdon, UK. Submitted to WHO by BASF, France.

Jones, E. & Gant, R.A. (1991) Technical SN 100309: bacterial mutation assay with *Escherichia coli*. Unpublished report No. A81805 from Huntingdon Research Centre, Huntingdon, UK. Submitted to WHO by BASF, France.

Kitching, J. (1998) AE F132593 (soil metabolite of pyrimethanil) - code: AE F132593 00 ID99 0001 - bacterial reverse mutation assay. Unpublished report No. C000864 from Huntingdon Life Sciences Ltd, Huntingdon, Cambridgeshire, UK. Submitted to WHO by BASF, France.

Malarkey, P. (1990) Mouse acute oral study. Unpublished report No. A81775 from Schering Agrochemicals Ltd, Saffron Walden, Essex, UK. Submitted to WHO by BASF, France.

Markham, L.P. (1989a) Technical SN 100 309: rat acute oral toxicity study. Unpublished report No. A81766 from Schering Agrochemicals Ltd, Saffron Walden, Essex, UK. Submitted to WHO by BASF, France.

Markham, L.P. (1989b) Technical SN 100 309: rat acute dermal toxicity study. Unpublished report No. A81769 from Schering Agrochemicals Ltd, Saffron Walden, Essex, UK. Submitted to WHO by BASF, France.

Markham, L.P. (1989c) Technical SN 100 309: rabbit skin irritancy study. Unpublished report No. A81771 from Schering Agrochemicals Ltd, Saffron Walden, Essex, UK. Submitted to WHO by BASF, France.

Markham, L.P. (1989d) Technical SN 100 309: rabbit eye irritancy study. Unpublished report No. A81773 from Schering Agrochemicals Ltd, Saffron Walden, Essex, UK. Submitted to WHO by BASF, France.

Needham, D. (1991) The effect of SN 100 309 on the hepatic mixed function oxidase system of male rats following oral administration at 100 or 200 mg/kg bodyweight. Unpublished report No. A81625 from Schering Agrochemicals Ltd, Saffron Walden, Essex, UK. Submitted to WHO by BASF, France.

Needham, D. (1996) 1st amendment to report No. TOX/93/223-70: SN 100309: metabolism in the rat. Unpublished report No. A89419 from Schering Agrochemicals Ltd, Saffron Walden, Essex, UK. Submitted to WHO by BASF, France.

Needham, D. & Hemmings, P.A. (1991) The distribution and excretion of radiolabelled residues in the rat following oral dosing with SN 100 309 at 11.8 or 800 mg/kg bodyweight. Unpublished report No. A81623 from Schering Agrochemicals Ltd, Saffron Walden, Essex, UK. Submitted to WHO by BASF, France.

Needham, D. & Hemmings, P.A. (1993) Pyrimethanil: metabolism in the rat. Unpublished report No. A81626 from Schering Agrochemicals Ltd, Saffron Walden, Essex, UK. Submitted to WHO by BASF, France.

Patton, D.S. (1995) 1st addendum to report SN 100309/T27/2 - pyrimethanil - 80 week oral (dietary administration) carcinogenicity study in the mouse. Unpublished report No. A81813. Hazleton UK, Harrogate, North Yorkshire, UK. Submitted to WHO by BASF, France.

Proudlock, R.J. (1991) Technical SN 100309: mouse micronucleus test Unpublished report No. A81802 from Huntingdon Research Centre, Huntingdon, UK. Submitted to WHO by BASF, France.

Proudlock, R.J. & Howard, W.R. (1991) Technical SN 100 309: unscheduled DNA synthesis assay in rat hepatocytes treated in vivo. Unpublished report No. A81803 from Huntingdon Research Centre, Huntingdon, UK. Submitted to WHO by BASF, France.

Reader, S. (2002) Position paper: pyrimethanil code: AE B100309: background control data for liver weights generated from sub-chronic toxicity studies in the rat - prepared at the request of the European Rapporteur reviewing the Tier 2 summary of toxicological and metabolism studies for pyrimethanil code: AE B100309. Unpublished report No. C023548 from Bayer CropScience France; Sophia Antipolis, France. Submitted to WHO by BASF, France.

Reader, S. (2003a) Pyrimethanil (AE B100309) - position paper: pyrimethanil: statements on the 104-week combined chronic toxicity and oncogenicity study in the rat. Unpublished report No. 2003/1023033 from Bayer CropScience France, Sophia Antipolis, France. Submitted to WHO by BASF, France.

Reader, S. (2003b) Pyrimethanil (AE B100309) - position paper: pyrimethanil: statements on the 80-week carcinogenicity study in the mouse. Unpublished report No. 2003/1023034 from Bayer CropScience France, Sophia Antipolis, France. Submitted to WHO by BASF, France.

Reader, S. (2003c) Pyrimethanil (AE B100309) - position paper: pyrimethanil: statements on the rat developmental toxicity study. Unpublished report No. 2003/1023035 from Bayer CropScience France, Sophia Antipolis, France. Submitted to WHO by BASF, France.

Reader, S. (2003d) Pyrimethanil (AE B100309) - position paper: pyrimethanil: statement: The non-relevance to man of thyroid changes seen in rat toxicology studies. Unpublished report No. 2003/1023036 from Bayer CropScience France, Sophia Antipolis, France. Submitted to WHO by BASF, France.

Rees, S.J. (1992) Technical SN 100309: dog 12 month oral (gavage) repeat dose study. Unpublished report No. A81809 from Schering Agrochemicals Ltd, Saffron Walden, Essex, UK. Submitted to WHO by BASF, France.

Rees, S.J. (1993) Technical SN 100309: 104 week rat combined chronic toxicity and oncogenicity study. Unpublished report No. A81806 from Schering Agrochemicals Ltd, Saffron Walden, Essex, UK. Submitted to WHO by BASF, France.

Reynolds, C.M. & Swalwell, L.M. (1992) The distribution and excretion of radiolabelled residues in the dog following oral dosing with [^{14}C]-SN 100309 at 10 mg/kg bodyweight. Unpublished report No. A81630 from Schering Agrochemicals Ltd, Saffron Walden, Essex, UK. Submitted to WHO by BASF, France.

Simpson, E. (2003) Historical control data. Unpublished report No. 2003/1023032 from Covance Laboratories Ltd, North Yorkshire, UK. Submitted to WHO by BASF, France.

Weir, L.R. & Sindle, T. (1998) AE F132593 (soil photolysis metabolite of pyrimethanil) - code: AE F132593 00 1D99 0001 - rat acute oral toxicity. Unpublished report No. C001117 from AgrEvo UK Ltd, Chesterford Park, Saffron Walden, Essex, UK. Submitted to WHO by BASF, France.

Whitby, B. (1993) (^{14}C)-SN 100 309: quantitative whole-body autoradiography following oral administration to the rat. Unpublished report No. A81632 from Hazleton UK, Harrogate, North Yorkshire, UK. Submitted to WHO by BASF, France.

Whitby, B. (1995a) Pyrimethanil - (^{14}C)-pyrimethanil: clearance of a radiolabelled dose from the tissues of rats following a single oral dose of 10 or 800 mg/kg body weight. Unpublished report No. A84647 from Corning Hazleton Europe, Harrogate, UK. Submitted to WHO by BASF, France.

Whitby, B. (1995b) 1st amendment to report No. TOX/95/223-91 - (^{14}C)-pyrimethanil: clearance of a radiolabelled dose from the tissues of rats following a single oral dose of 10 or 800 mg/kg body weight. Unpublished report No. A89286 from Corning Hazleton Europe, Harrogate, UK. Submitted to WHO by BASF, France.

ZOXAMIDE

First draft prepared by
I. Dewhurst,[1] E. Efa[1] & A. Moretto[2]

[1]Pesticides Safety Directorate, Department for Environment, Food and Rural Affairs, Mallard House, Kings Pool, York, England;
[2] Department of Environmental and Occupational Health, University of Milan, Milan, Italy

Explanation

Zoxamide is the International Organization for Standards (ISO) approved name for (*RS*)-3,5-dichloro-*N*-(3-chloro-1-ethyl-1-methyl-2-oxopropyl)-4-methylbenzamide (Chemical Abstracts Service; CAS No. 156052-68-5). Zoxamide is a chlorinated benzamide fungicide that acts against late blight (*Phytophthera infestans)* and powdery mildew (*Plasmopara viticola)*. The mechanism of fungicidal action involves disruption of microtubule formation by binding to β-tubulin.

Zoxamide has not been evaluated previously by the JMPR and was reviewed at the present Meeting at the request of the Codex Committee on Pesticide Residues (CCPR).

All the pivotal studies met the basic requirements of the relevant Organisation for Economic Co-operation and Development (OECD) or national test guideline and contained certificates of compliance with good laboratory practice (GLP).

Evaluation for acceptable daily intake

1. Biochemical aspects

1.1 Absorption, distribution and excretion

(a) Oral route

Rats

In a GLP-compliant study, the balance and excretion patterns of ^{14}C- and ^{13}C-phenyl ring-labelled zoxamide (Figure 1), and the amount of residual radioactivity in blood, organs and tissues were determined in male and female Sprague-Dawley Crl:CDBR rats (body weight, 220–426 g) treated orally.

Groups of three to six males and three to six females were given [^{14}C]zoxamide (radiochemical purity, 97.6–99.5%; 45.8–90.2 mCi/g or 1.69–3.34 GBq/g) at nominal dose levels of 10 (lower dose) or 1000 (higher dose) mg/kg bw by gavage—details of the main groups are outlined in Table 1. The test materials were prepared as suspensions in corn oil and administered in a volume of 5 ml/kg bw. For all groups, [^{14}C]zoxamide was combined with appropriate amounts of non-radiolabelled zoxamide (purity, 92.9–94.2%). In addition, ^{13}C-labelled zoxamide (purity, 96.8%) was added to the dosing solution for groups A, B, Q and R to assist in metabolite determinations.

The following experiments were conducted:

- Determination of the excretion, distribution and mass balance of radioactivity (up to 120 h after dosing) in groups A and B (lower dose) and C and D (higher dose);
- Evaluation of the pharmacokinetics of radioactivity in blood (to determine the time to C_{max} and ½ C_{max}) (groups 1, 2 (also investigated for expired air) (higher dose) and 3 (lower dose);
- Determination of the tissue distribution of radioactivity at C_{max} and ½ C_{max} (groups I to P; both dose levels),
- Investigation of effects after repeated doses (groups E, F, G and H). Rats in groups E and F received diets containing non-radiolabelled zoxamide technical at a concentration of 200 ppm for 2 weeks before receiving [^{14}C]zoxamide as a single oral dose at 10 mg/kg bw. Rats in groups G and H received [^{14}C]zoxamide as five consecutive daily doses at 10 mg/kg bw/day by gavage.

Figure 1. Structure of zoxamide

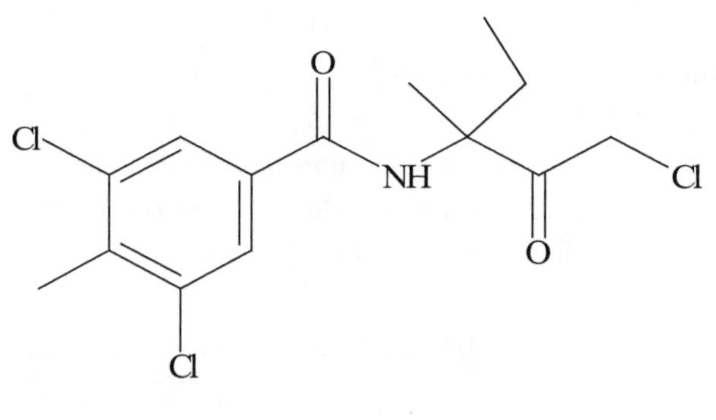

The excretion of radioactivity in rats given a single oral dose of [^{14}C]zoxamide followed similar patterns, with a higher (approximately threefold) proportion in the urine at the lower dose (Table 1). More than 85% of the administered radioactivity was excreted during the first 24–48 h after dosing. The major route for the elimination of radiolabel was in the faeces, which contained 74–92% of the administered dose. The remaining radioactivity (4–27%) was excreted in the urine, with females tending to have a higher level of urinary excretion than males. A significant proportion of the absorbed dose (approximately 45%) was excreted in the bile. Biliary excretion of radioactivity in rats given [^{14}C]zoxamide as a single oral dose at 10 mg/kg bw was rapid. Most of the dose was recovered within 12 h after dosing. Based on the recovery of radioactivity from the bile, blood, urine, tissues and carcasses, 59–63% of the administered oral dose of [^{14}C]zoxamide was considered to be systemically absorbed. Very little radioactivity remained in tissues (0.04–0.17% of the administered dose) or carcass (0.34–1.9% of the administered dose) at 5 days after dosing, indicating that zoxamide has a low potential for accumulation. No radioactivity was recovered as either $^{14}CO_2$ or volatile organic compounds. Pre-treatment of animals with diets containing non-radiolabelled zoxamide for 2 weeks or with five daily doses of radiolabelled zoxamide did not significantly alter the absorption

Table 1. Recovery of radiolabel in excreta, blood, tissues and residual carcass of rats treated with radiolabelled zoxamide

Group	Dose (mg/kg bw)	Route[a]	Sex	Time of sacrifice	Recovery (% of administered dose)						
					Urine[d]	Bile	Faeces	Blood	Tissues	Carcass	Total[e]
2	1000	Oral	Male	7 days	6.4	—	94.1	0.01	0.02	0.30	100.9[f]
			Female	7 days	11.2	—	84.4	0.01	0.02	0.33	96.0[f]
C	1000	Oral	Male	5 days	3.5	—	92.4	0.00	0.04	0.34	96.2
D	1000	Oral	Female	5 days	8.1	—	88.8	0.01	0.05	0.55	97.5
A	10	Oral	Male	5 days	10.3	—	87.8	0.01	0.16	1.86	100
B	10	Oral	Female	5 days	26.8	—	73.5	0.02	0.17	1.87	102.4
Q	10	Oral (bile); bile-duct cannulated	Male	72 h	9.5	45.8	32.2	0.01	0.18	2.98	90.8[g]
R	10	Oral (bile) bile-duct cannulated	Female	72 h	12.0	47.8	33.9	0.01	0.14	2.79	96.7[g]
E	10	Oral, pulse[b]	Male	5 days	16.3	—	78.6	0.02	0.58	3.41	98.9
F	10	Oral, pulse[b]	Female	5 days	28.7	—	71.0	0.02	0.19	1.55	101.5
G	10	Oral, repeat[c]	Male	C_{max} after fifth dose	7.9	—	60.3	0.03	2.69	17.77	96.3[g]
H	10	Oral, repeat[c]	Female	C_{max} after fifth dose	19.9	—	51.6	0.04	3.27	12.65	93.0[g]

From Swenson et al. (1998)

[a] All rats except those in groups G and H were administered a single dose by gavage in a constant volume of 5 ml/kg bw.

[b] These rats received diets containing non-radiolabelled zoxamide at a concentration of 200 ppm for 2 weeks before receiving radiolabelled zoxamide as a single oral dose at 10 mg/kg bw.

[c] These animals received five consecutive daily oral doses by gavage and were sacrificed at the C_{max} time-point (8 h after dosing) on day 5 of dosing.

[d] Includes urine, urine funnel wash and urine cage wash.

[e] Mean total percentage of administered dose, reflects the mean and standard deviation for the individual animals.

[f] No radioactivity was recovered as either $^{14}CO_2$ or volatile organic compounds.

[g] Stomach contents, stomach wash, and intestinal tract contents and wash accounted for 7.55% and 5.42% of the administered dose for males and females, respectively.

or distribution of [^{14}C]zoxamide when compared with rats that were not pre-treated and that received a single dose of [^{14}C]zoxamide (Table 1).

In pharmacokinetic studies, [^{14}C]zoxamide was observed to be rapidly absorbed by rats. The maximum concentrations of radioactivity in plasma were observed at 8 h after dosing (C_{max} in plasma, 8 h; ½ C_{max}, 22 h). Peak blood and tissue concentrations were noted to be low. Elimination of radiolabel from plasma followed a biphasic pattern. The overall elimination half-life of the ^{14}C radiolabel in plasma was essentially similar (12–14 h) in male and female rats at the lower and at the higher dose (Table 2).

The concentrations of radioactivity in the tissues were highest in the organs associated with oral absorption—liver, stomach, intestines, and carcass (which included the caecum). The results for tissue distribution were consistent with the pharmacokinetic data and indicated that radioactivity was rapidly cleared from the tissues (Table 3). A comparison of repeated-doses (five) and single-dose C_{max} values indicated slightly higher (approximately twofold) values after repeated administration; these results were within typical variability for a study of this type and were not seen as a clear indication of bioaccumulation (Swenson et al., 1998).

In an investigation of the distribution of zoxamide in bone marrow, [^{14}C]zoxamide as a single dose at 2000 mg/kg bw was administered orally by gavage in corn oil to groups of four male and four female CD-1 mice. Mice were killed at 4, 8, 24 and 48 h after administration of the test material, bone marrow tissue samples were collected and analysed for radiolabel. The study was certified to be compliant with GLP and was conducted to support a study of micronucleus formation in mice that was conducted in accordance with OECD 474 guidelines. At all time-points (i.e., 4, 8, 24 and 48 h), radiolabel derived from [^{14}C]zoxamide was present in the bone marrow of male and female mice. Peak concentrations of ^{14}C, 55 and 39 µg equivalents/g in males and females respectively were seen at 4 h; declining to approximately 9 µg equivalents/g by 24 h (Swenson & Frederick, 1998).

(b) Dermal route

Rats

In a study of dermal absorption, male rats received single dermal applications of two formulations of [^{14}C]zoxamide (each of two different batches):

- [^{14}C]zoxamide 80WP (lot No. TEM-2627, 80% active ingredient (a.i.); radiochemical purity, 96.3%; specific activity, 0.67 mCi/g (24.79 MBq/g); or lot No. TEM-2629; 0.15% a.i., radiochemical purity, 96.3%; specific activity, 0.05 mCi/ml (1.85 MBq/ml);

Table 2 Pharmacokinetic half-lives ($t_{1/2}$) of ^{14}C-radiolabel and elimination and peak plasma concentration for rats given oral doses of radiolabelled zzoxamide[a]

Group	Dose (mg/kg bw)	Sex	Elimination half-life[b] (h)	Alpha-Phase half-life[c] (h)	Beta-Phase half-life[d] (h)	Peak concentration[e] (ppm)	AUC[f] (ppm.h)
1	1000	Male	11.7	5.5	100	32	1360
1	1000	Female	13.8	6.3	107	43	1882
3	10	Male	14.0	5.6	70	0.62	26
3	10	Female	13.1	6.6	164	0.98	45

From Swenson et al. (1998)

AUC, area under the curve of concentration–time.

[a] Elimination rates calculated for a two-compartment pharmacokinetic model (PK Analyst® Model 13) with first-order input and first-order output (elimination).

Table 3. Mean concentration of radiolabel in blood, carcass, plasma and tissues of rats at 8 h or 22 h after oral doses of radiolabelled zzoxamide

Tissue	Dose (mg/kg bw)									
	1000 mg/kg		1000 mg/kg		10 mg/kg		10 mg/kg		5 × 10 mg/kg	
	C_{max}		1/2 C_{max}		C_{max}		1/2 C_{max}		C_{max} at 8 h	
	8 h		22 h		8 h		22 h		Fifth dose[b]	
	M	F	M	F	M	F	M	F	M	F
Adrenals	59	185	16	33	0.51	0.66	0.87	0.93	1.2	1.8
Bone marrow	14	22	6.5	7.5	0.19	0.20	0.13	0.17	0.22	0.39
Brain	2.5	4.2	0.48	1.2	0.04	0.03	0.01	0.02	0.04	0.07
Carcass (residual)	805	727	88	159	5.8	3.9	2.5	4.0	11	8.9
Fat	7.5	14	3.7	9.5	0.11	0.23	0.09	0.14	0.17	0.52
Heart	19	31	4.4	8.5	0.20	0.25	0.10	0.16	0.30	0.53
Intestinal tract	1779	1826	257	375	33	33	6.4	6.1	29	20
Kidneys	120	177	18	31	1.7	2.1	0.41	0.59	1.9	4.2
Liver	879	1131	71	175	15	25	2.5	4.1	15	32
Lungs	29	41	6.4	11	0.29	0.39	0.13	0.19	0.41	0.69
Muscle (thigh)	6.7	9.6	2.0	3.5	0.08	0.07	0.04	0.06	0.12	0.18
Ovaries	NA	34	NA	12	NA	0.65	NA	0.21	NA	0.56
Plasma	50	64	11	17	0.53	0.73	0.20	0.30	0.76	1.0
Spleen	21	28	5.3	8.2	0.19	0.21	0.09	0.16	0.28	0.53
Stomach	4846	1750	23	93	14	20	1.0	0.44	8.5	9.1
Testes	7.7	NA	2.1	NA	0.09	NA	0.06	NA	0.16	NA
Thyroid	30	39	13	23	0.34	0.43	0.24	0.29	0.57	0.81
Whole blood	38	49	12	18	0.49	0.55	0.22	0.31	0.73	1.0

From Swenson et al. (1998)

F, female; M, male; NA, not applicable.

[a] These rats received diets containing non-radiolabelled zoxamide at a concentration of 200 ppm for 2 weeks before receiving a single oral dose (pulse) of radiolabelled zoxamide.

[b] These rats received fiveconsecutive daily oral doses of ^{14}C-labelled zoxamide by gavage and were sacrificed at the C_{max} time-point (8 h after dosing) on day 5 of dosing.

- [^{14}C]zoxamide 2F (lot No. TEM-2616; 24% a.i.; radiochemical purity, 96.3%; specific activity, 0.068 mCi/ml (2.52 MBq/ml); and lot No. TEM-2618; 0.015% a.i.; radiochemical purity, 96.3%; specific activity, 0.005 mCi/ml (0.19 MBq/ml).

The study was certified to be compliant with GLP and conducted in accordance with OECD guideline 417.

The [^{14}C]zoxamide 80WP formulation was given either as a 10 mg aliquot of the undiluted wettable powder (80% a.i.) or as 100 µl aliquots diluted in water. This gave concentrations of 0.15% active substance (a.s.) and 0.015% a.s. respectively. Similarly, the [^{14}C]zoxamide 2F formulation was given at concentrations of 24% and 0.015% a.s. These concentrations were chosen to represent exposure to the concentrated product and a typical in-use dilution.

After administration of undiluted [^{14}C]zoxamide 80WP or [^{14}C]zoxamide 2F formulations to male rats, approximately 1% of the administered dose was absorbed within 24 h. After administration of the 0.015% a.i. dilutions of either formulation, 5–6% was absorbed after 24 h. These studies indicated that zoxamide is very poorly absorbed after dermal exposure, irrespective of formulation. The low rate of dermal absorption was considered to be attributable to the very low solubility of zoxamide in water (log Kow 3.8; water solubility, < 1 mg/l) (Frederick & Swenson, 1998).

(c) Biotransformation

Metabolites in samples of urine, faeces and bile from the studies described above were identified and quantified. Samples from rats given [^{14}C]zoxamide were pooled by sex and dose to create composite samples.

Faeces were studied by extracting homogenates with methanol, then analysing the methanol fractions by reverse-phase high-performance liquid chromatography (RP-HPLC) and normal-phase thin-layer chromatography (TLC). Samples of urine were filtered and examined directly by HPLC and TLC. Bile was analysed directly by RP-HPLC and liquid chromatography-mass spectroscopy (LC-MS). The amounts of metabolites were quantified using liquid scintillation counting (LSC) of collected HPLC fractions, except for two benzoic acid metabolites RH-141,455 and RH-141,452, which were quantified from samples from rats at the highest dose by gas chromatography-electron capture detection (GC-ECD) using a method not requiring radiolabel. Isolated metabolites were characterized and identified by TLC, GC-MS analyses (either as derivatized or underivatized metabolites), and/or LC-MS with electrospray (ESI) or atmospheric chemical ionization (APCI), and/or by comparison to authentic reference standards.

Zoxamide was found to be extensively metabolized. Including parent compound, a total of 36 metabolites were found in the faeces and urine; 24 of these were identified. In bile, 17 products were detected and 13 were identified. Altogether, 32 structures were determined. No single metabolite other than parent zoxamide accounted for more than 10% of the administered dose. Zoxamide was observed in faeces at 12% and 23% of the administered dose at 10 mg/kg bw and 72% and 74% at 1000 mg/kg bw for females and male, respectively, probably from unabsorbed material. The levels of major metabolites are given in Table 4 and a proposed metabolic pathway is given in Figure 2. The main reactions occurring were hydrolysis and dehalogenation with subsequent oxidation and conjugation.

The primary metabolites in bile were M14A/B (10–12%), a glutathione conjugate; M25 (8%), the glucuronide of M4; and M26 (6%), also a glutathione conjugate. The pattern of biliary metabolites was very similar in males and females.

Overall metabolism was similar irrespective of dose or sex. The metabolic profile in samples of faeces from rats at the highest dose was qualitatively similar to that in rats at the lowest dose, except for the large amount of parent compound that was considered to be suggestive of incomplete absorption for the higher dose. The amount of parent compound found in rats given repeated doses in a dietary 14-day study (5.56–5.84%) were reduced compared with rats at the lower dose (12.10–22.96%) suggesting increased metabolism after repeat dosing (Swenson et al., 1998).

In a supplementary study, bile samples from males and urine samples from females that had been retained from the above study of metabolism were analysed in order to determine whether the benzamide metabolite RH-139432 was present. Quantitative analysis was by two-dimensional TLC with reverse-phase and normal-phase coatings. Radioanalysis was performed before chromatography to allow for material balance calculations. For bile, RH-139432 was found to account for 0.21–0.25% of the radioactivity applied to the TLC plate, depending on the TLC phase used. For urine, RH-139432 accounted for 0.02–0.03% of the applied radioactivity, depending on the TLC phase used. After purification of the material obtained from the bile sample, the identity of RH-139432 was confirmed by GC-MS and HPLC-MS. This study showed there was no significant production of RH-139432 in rats treated with zoxamide (Reibach & Detweiler, 2001).

Table 4. Distribution of metabolites in rats given radiolabelled zoxamide as a low dose at 10 mg/kg bw

Metabolite	Recovery (% of applied dose)											
	Single dose						Repeated doses					
	Male			Female			Male			Female		
	Faeces	Urine	Subtotal	Faeces	Urine	Subtotal	Faeces	Urine	Subtotal	Faeces	Urine	Subtotal
M-1 (zoxamide)	22.96	—	22.96	12.10	—	12.10	5.56	—	5.56	5.84	—	5.84
M-2 (RH-127,450)	2.75	—	2.75	2.43	—	2.43	2.64	—	2.64	2.28	—	2.28
M-3 (RH-141,643)	7.44	—	7.44	4.93	—	4.93	5.30	—	5.30	5.37	—	5.37
M-4 (RH-141,288)	2.84	—	2.84	3.15	—	3.15	4.02	—	4.02	6.43	—	6.43
M-5	2.46	—	2.46	1.54	—	1.54	2.09	—	2.09	1.76	—	1.76
M-7 (RH-141,454)	5.74	—	5.74	3.73	—	3.73	6.19	—	6.19	2.79	—	2.79
M-8A, M-8B, and M-15	3.36	0.70	4.06	2.94	5.05	7.99	2.22	1.38	3.60	3.84	3.08	6.92
M-9	4.14	—	4.14	3.58	—	3.58	5.12		5.12	3.78		3.78
M-10A, M-10B, M-16, M-17 (RH-141,452), and M-18	7.47	4.30	11.77	9.14	4.93	14.07	9.42	5.90	15.32	8.12	3.64	11.76
M-12	—	0.11	0.11	—	1.09	1.09	—	0.75	0.75	—	1.58	1.58
M-13	—	—	—	—	5.06	5.06	—	0.32	0.32	—	9.64	9.64
M-14A and M-14B	—	0.55	0.55	—	1.27	1.27	—	1.67	1.67	—	1.71	1.71
M-19 and M-20	—	1.78	1.78	—	2.65	2.65	—	2.46	2.46	—	2.26	2.26
M-21A and M-21B	—	0.94	0.94	—	1.02	1.02	—	1.34	1.34	—	1.48	1.48
Sum of identified metabolites	59.16	8.38	67.54	43.54	21.07	64.61	42.56	13.82	56.38	40.21	23.39	63.60
% of administered dose submitted for analysis	87.78	8.84	96.62	73.50	23.17	96.67	78.57	15.30	93.87	71.05	26.45	97.50
Total % of administered dose	87.78	10.29[a]	98.07	73.50	26.85[c]	100.35	78.57	16.27[a]	94.84	71.05	28.72[a]	99.77

[a] This includes urine, urine funnel wash and urine cage wash. Urine funnel wash and urine cage wash were not analysed for metabolites.

Metabolite RH141,452 (3,5-dichloro-4-hydroxymethyl benzoic acid)

The absorption, distribution, metabolism and elimination of [^{14}C]RH-141,452 (lot No. 955.0005, specific activity, 75.38 mCi/g (2.79 GBq/g) were studied in male rats. Four male rats were each given a single oral dose of [^{14}C]RH-141,452 in pH-adjusted water at a nominal dose of 1000 mg/kg bw. The rats were sacrificed 78 h after dosing and the total recovery of radiolabelled residues was determined.

Most of the radioactivity was eliminated in the urine (approximately 98% in urine and cage rinse), with only a small amount being excreted from faeces (< 2%). Approximately 0.01% of the administered dose of radioactivity was found in the expired air. The excretion of the ^{14}C was rapid, with more than 97% being excreted within 24 h. Studies of metabolism showed that most of the RH-141,452 was eliminated unchanged in the urine, accounting for > 94% of the administered dose. Three minor conjugates, M-2, M-3 (glucuronide conjugates), and M-4 (glycine conjugate)

Figure 2. Summary of the metabolic pathway of zoxamide in the rat

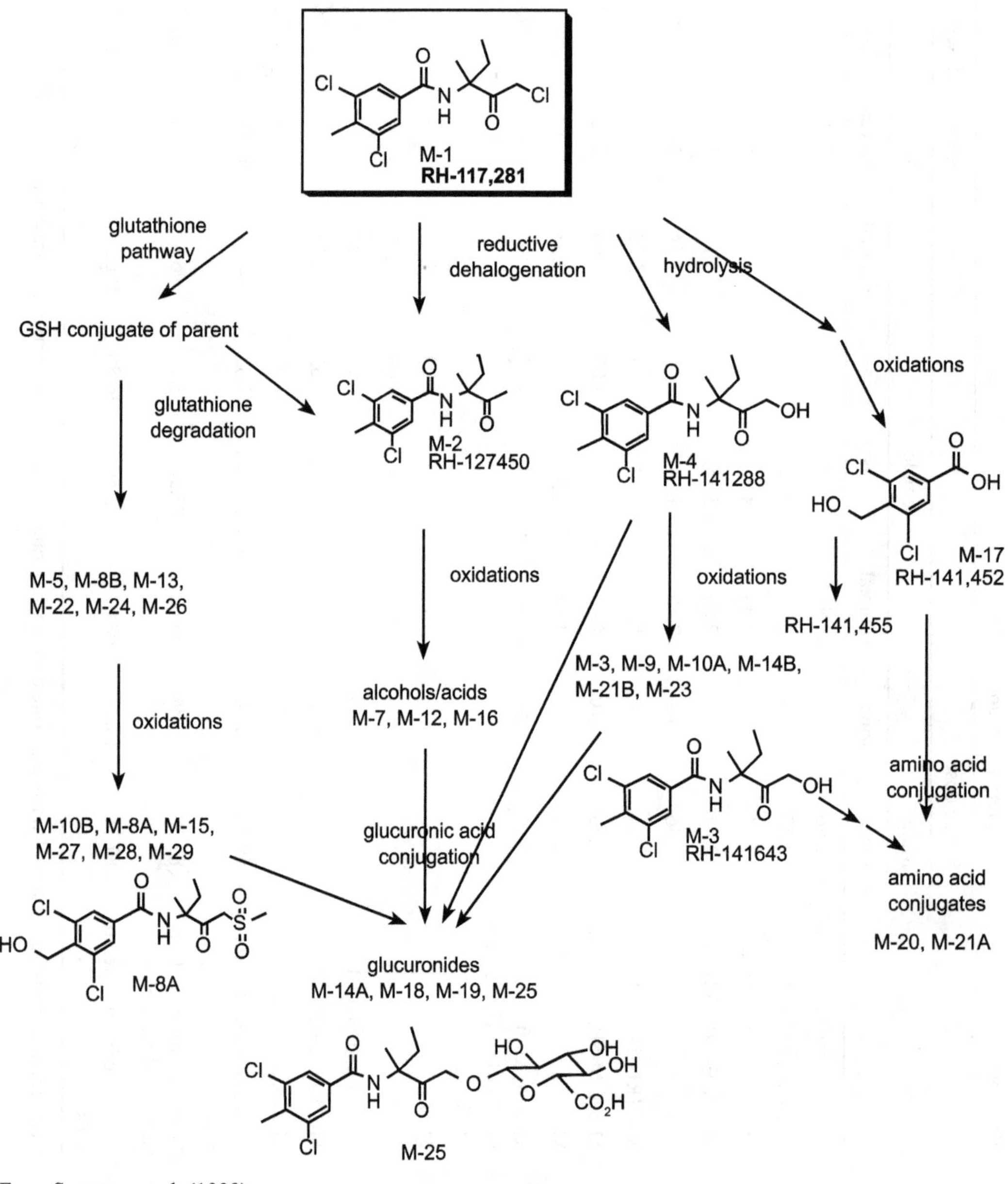

From Swenson et al. (1998)

were also found in the urine, accounting for approximately 3% of the administered dose. An additional 1.6% of the administered radioactivity was excreted in the faeces as the parent chemical (Wu & Gu, 1998a).

Metabolite RH141,455 (3,5-dichloro-terphthalic acid)

The absorption, distribution, metabolism and elimination of [^{14}C]RH-141,455 (lot No. 958.0005; specific activity, 74.46 mCi/g) in male rats were studied. Four male rats were given a single oral dose of [^{14}C]RH-141,455 in pH-adjusted water at a nominal dose of 1000 mg/kg bw. The rats were sacrificed 168 h after dosing and the total recovery of radiolabelled residues was determined. More than 96% of radioactivity excreted from faeces (72%) and urine plus cage wash (20%) was identified as unchanged RH-141,455. Some minor metabolites were also observed in urine samples, but were not identified owing to their low concentrations (Wu & Gu, 1998b).

2. Toxicological studies

2.1 *Acute toxicity*

(a) Lethal doses

The results of studies of acute toxicity with zoxamide administered by the oral, dermal and inhalation routes, and two of its metabolites, are presented in Table 5. Zoxamide was of low acute toxicity by all routes. The metabolites RH 141-452 & 141,455 were of low acute oral toxicity.

(b) Dermal and ocular irritation and dermal sensitization

Zoxamide was not irritating to the skin of rabbits (Gingrich & Parno, 1996c) but exhibited slight, transient irritation to the rabbit eye (Gingrich & Parno, 1996d). Zoxamide produced delayed contact hypersensitivity in the guinea-pig in the Magnusson & Kligman maximization study (Glaza, 1997) and in Buehler skin sensitization tests (Robison et al., 1998a). Low concentrations (< 0.25% w/w) were shown to be not sensitizing to the skin of guinea-pigs (Robison et al., 1998b).

Table 5. Acute toxicity with zoxamide and its metabolites

Test substance	Species	Strain	Sex	Route	LD_{50} (mg/kg bw)	LC_{50} (mg/l air)	Purity (%)	Vehicle	Reference
Zoxamide	Rat	Crl:CDBR	Males & females	Oral	> 5000	—	92.3	Corn oil	Gingrich & Parno (1996a)
Zoxamide	Mouse	Crl:CD-1-(ICR)BR	Males & females	Oral	> 5000	—	94.4	Corn oil	Ferguson & Lutz (1998)
Zoxamide	Rat	Crl:CDBR	Males & females	Dermal	> 2000	—	92.3	Corn oil	Gingrich & Parno (1996b)
Zoxamide	Rat	Crl:CDBR	Males & females	Inhalation (4-h, nose-only)	—	> 5.3 (MMAD 4.3 μm)	92.3	None (dust aerosol)	Bernacki,& Ferguson (1996)
RH-141,452	Mouse	Crl:CD-1-(ICR)BR	Males & females	Oral	> 5000	—	97.7	Corn oil	Ferguson et al. (1998c)
RH-141,452	Mouse	Crl:CD-1-(ICR)BR	Males & females	Oral	> 5000	—	98.7	Corn oil	Ferguson et al. (1998a)

MMAD, mass median aerodynamic diameter.

2.2 *Short-term studies of toxicity*

Mice

Groups of 10 male and 10 female Crl:CD-1(ICR)BR mice were given diets containing zoxamide (purity, 94.2%) at a concentration of 0, 70, 700, 2500 or 7000 ppm (equal to 0, 12, 123, 436 and 1212 mg/kg bw per day respectively in males, and 0, 17, 174, 574, 1666 mg/kg bw per day in females) for 90 days. All mice were observed daily for signs of ill health or reaction to treatment. Physical examinations were performed each week. Body weight and feed consumption were determined each week. At the end of the study period, all surviving mice were bled for haematology and clinical chemistry investigations, killed and necropsied. Selected organ weights were recorded and tissues were collected for histopathological evaluation. The study was certified to be compliant with GLP and satisfied the essential criteria of OECD guideline 408.

No treatment-related deaths or clinical signs of toxicity were observed. There were no treatment-related effects on body weight or body-weight change in males at any dose. After 4 weeks of treatment, there was a persistent and apparently treatment-related but statistically non-significant decrease in body weight and cumulative body-weight gain in females at 7000 ppm compared with concurrent controls (Table 6). In males, a number of statistically significant decreases in mean body weight and cumulative body-weight gain were observed in the groups at 700 ppm and 2500 ppm throughout the treatment period, but these decreases were not considered to be treatment-related

Table 6. Body weights, body-weight gain and liver weight in mice fed diets containing zoxamide for 90 days

Parameter	Dietary concentration (ppm)				
	0	70	700	2500	7000
Males					
Body weight (g):					
Week 0	30.8	30.4	30.2	29.2	29.8
Week 4	34.6	33.7	33.2*	32.0	32.8
Week 8	37.9	36.3	35.8*	34.3*	35.4
Week 13 (mean ± SD)	41.1 ± 4.9	38.5 ± 2.2	38.0 ± 1.9*	36.7 ± 3.5	38.3 ± 3.9
Cumulative body-weight gain, week 13 (g, mean ± SD)	12.3 ± 4.2	9.5 ± 1.9	9.1 ± 1.1*	8.9 ± 2.7	10.0 ± 2.7
Absolute liver weight (g, mean ± SD)	2.1 ± 0.24	2.0 ± 0.20	1.9 ± 0.13	2.1 ± 0.25	2.2 ± 0.21
Relative liver weight (%, mean ± SD)	5.2 ± 0.41	5.2 ± 0.39	5.2 ± 0.25	5.8 ± 0.49*	5.8 ± 0.55*
Females					
Body weight (g):					
Week 0	23.3	23.3	22.2	22.8	20.9
Week 4	26.8	26.8	25.6	25.1	23.6
Week 8	28.1	28.9	27.3	26.8	25.0
Week 13 (mean ± SD)	30.6 ± 4.8	31.1 ± 3.2	29.3 ± 3.8	28.7 ± 3.3	25.8 ± 2.1
Cumulative body-weight gain, week 13 (g, mean ± SD)	7.4 ± 3.1	7.7 ± 2.4	7.0 ± 2.6	5.9 ± 2.0	4.9 ± 1.7*
Absolute liver weight (g, mean ± SD)	1.5 ± 0.16	1.6 ± 0.16	1.7 ± 0.21	1.6 ± 0.24	1.5 ± 0.17
Relative liver weight (%, mean ± SD)	5.2 ± 0.51	5.3 ± 0.26	5.7 ± 0.52	5.7 ± 0.57	6.0 ± 0.51*

From Shuey et al. (1996)
SD, standard deviation.
* Significant difference from controls ($p < 0.05$).

owing to the absence of a dose–response relationship. There was no statistically significant treatment-related effect on feed consumption in either sex at any dose; but mice at the highest dose regularly consumed less food than did other groups. However, in females there was a statistically non-significant but dose-related reduction in body-weight gain (34%) and in body weight in females at 7000 ppm compared with controls. The body-weight effects in females at the highest dose did not appear to be directly related to the lower body weight at the start of the study. Haematology and clinical chemistry parameters did not reveal any significant treatment-related differences. There were no treatment-related effects on absolute or relative organ weights in either sex at any dose, other than a statistically significant increase in relative liver weights in males receiving at_2500 ppm or greater and in females at 7000 ppm; there was evidence of a dose response in the females (Table 6). No treatment-related gross pathological changes or histopathological findings were observed in any tissues. In the absence of clinical chemistry and histopathological correlates, the increase in liver weight per se is not considered to be an adverse effect.

The no-observed-adverse-effect level (NOAEL) was 2500 ppm, equal to 574 mg/kg bw per day, on the basis of reduction in body-weight gain and in overall body weight in female mice at 7000 ppm (1666 mg/kg bw per day) (Shuey et al., 1996)

Rats

Groups of 15 male and 15 female Crl:CD®BR rats were given diets containing zoxamide (purity, 92.9%) at a concentration of 0, 1000, 5000, or 20,000 ppm (equal to 0, 74, 372, and 1509 mg/kg bw per day in males, and 0, 80, 401, and 1622 mg/kg bw per day in females) for 90 days. All rats were observed daily for signs of ill health or reaction to treatment. Physical examinations were performed each week. Body weight and feed consumption were monitored each week. Ten males and ten females per group were randomly selected for in-life testing for neurotoxicity via a standard battery of behavioural observations and motor activity before treatment and at weeks 4, 8, and 13; five of these were randomly selected for whole-body perfusion and special neuropathology evaluation. Haematology, urine analysis and clinical chemistry tests were conducted on samples taken from the remaining rats during week 13. The rats were then killed and necropsied. Selected organ weights were recorded and tissues were collected for routine histopathological evaluation. All animals were given ophthalmological examinations before testing and during week 13. The study was certified to be compliant with GLP and satisfied the essential criteria of OECD guideline 408.

No treatment-related mortalities or clinical signs of toxicity were observed during the study period. Body-weight gain and feed consumption showed no treatment-related inter-group differences. There were no treatment-related effects on haematology, clinical chemistry, or urine-analysis parameters. The functional observation battery (FOB) and motor activity assessments did not show any indications of neurotoxicity. No treatment-related effects on organ weights or ophthalmological changes were observed. There were no observations of treatment-related macroscopic or histopathological changes after routine necropsy or perfusion in males or females selected for neuropathology examinations.

The NOAEL was 20000 ppm, equal to 1509 mg/kg bw per day, the highest dose tested (Morrison & Gillette, 1996)

Groups of 10 male and 10 female Crl:CD®BR rats received zoxamide (purity, 93.83%) as 22 dermal doses at 0, 150, 400 or 1000 mg/kg bw per day. The material was moistened with tap water (1 : 2 w/v) and applied topically under an occlusive dressing to the shaved intact skin (an area approximately of 10% of the total body surface area) for at least 6 h per day, 5 days per week. The study was certified to be compliant with GLP and was conducted in accordance with OECD guideline 410.

The only findings of note were a dose-related and time-related increase in the incidence of reddening and scabbing at the application sites, a dose-related increase in leukocytes and a decrease in albumin : globulin ratio (A : G) in females,; and an altered differential count in males (Table 7). The study investigators suggested that the alteration in A : G were secondary to the skin irritation, but it is of note that reductions in globulin levels were also seen in the dietary studies in dogs (see Table 11) and do not explain the absence of an equivalent effect in males. The albumin and globulin values in treated animals were within ranges for historical controls and would not be considered to be adverse in isolation.

Zoxamide produced significant local effects at doses of 107 mg/kg bw per day and greater. Findings of systemic toxicity were most likely secondary to the local effects and the NOAEL for systemic effects was 714 mg/kg bw per day (corrected for administration on 5 days/week) (Robison et al., 1998d).

Dogs

Groups of two male and two female beagle dogs were given diets containing zoxamide (purity, 92.9%) at a concentration of 0, 500, 5000, 15 000 or 30 000 ppm (equal to 0, 20, 175, 542 and 1045 mg/kg bw per day for males and 0, 20, 191, 579 and 1085 mg/kg bw per day for females) for 28 days. All dogs were observed daily for signs of ill health or reaction to treatment. During pre-test and throughout treatment, feed consumption was monitored daily, physical examinations were performed and body weights were determined weekly. During the first week of pre-test and after 2 and 4 weeks of treatment, blood samples were collected from all dogs (fasted) for haematology and clinical chemistry analysis. After 4 weeks, all dogs were killed and organs and tissues were grossly examined. Selected organ weights were recorded at necropsy. Microscopic examination was conducted on all gross lesions and on all tissues and organs collected from all dogs in the group receiving the highest dose and in the control group.

There were no deaths during the study. The only treatment-related clinical signs were soft faeces present throughout the dosing period in one out of four dogs at 5000 ppm and in all dogs at 15 000 ppm and greater. There were no overt treatment-related effects on any other measured parameter; but the dogs varied markedly in size at the start of treatment.

Table 7. Findings in rats treated dermally with zoxamide

Finding	Dose (nominal mg/kg bw per day)			
	0	150	400	1000
Males				
Application site reaction	0/10	4/10	9/10	10/10
Leukocytes	6.3	5.7	6.7	6.0
Lymphocytes (%)	88	90	83	80*
Segmented neutrophils (%)	11	9	15	18*
Females				
Application site reaction	4/10	9/10	10/10	10/10
Leukocytes	3.6	4.5	5.1	5.8*
Albumin (g/dl)	4.2	3.9	3.7*	3.7*
Globulin (g/dl)	1.7	1.9	2.0*	2.1*
Albumin : globulin ratio	2.5	2.1*	1.9*	1.8*

From Robison et al. (1998d)
* Significant difference from control ($p < 0.05$).

The NOAEL was 30 000 ppm, equivalent to 1045 or 1085 mg/kg bw per day. The only finding was the occurrence of a minimal incidence of soft stools at 5000 ppm (175 or 191 mg/kg bw per day) and an increased incidence at doses of 15 000 ppm and greater; however, soft faeces was not a consistent finding in other studies in dogs with a longer duration and using the same formulated diet (Vandenberghe et al., 1996).

Groups of four male and four female beagle dogs were given diets containing zoxamide (lot No. DSR-9510, purity 92.3%) at a concentration of 0, 1500, 7500 or 30 000 ppm (equal to 0, 55, 281, and 1139 mg/kg bw per day in males, and 0, 62, 322 and 1055 mg/kg bw per day respectively in females) for at least 90 days. One female in the group at 1500 ppm was moved to a concurrent 1-year study and replaced with another dog during the third week of dosing, hence the treatment period was extended to 16 weeks. All dogs were observed daily for signs of ill health or toxicity. Feed consumption was determined daily for all dogs beginning 2 weeks before treatment and throughout the treatment period. Physical examinations were performed and body weights were determined each week. Blood samples were collected from all dogs for haematology and clinical chemistry analyses during pre-test, and after 8 and 16 weeks of treatment. Urine analysis and ophthalmology examinations were performed on all dogs during pre-test and after the treatment period. Selected organs were weighed and tissues were collected for histopathological evaluation. The study complied with GLP and with OECD guideline 409.

The study investigators considered that there were no treatment-related deaths or clinical signs indicative of systemic toxicity during the treatment period. Two dogs were killed humanely during the study, a female at 30 000 ppm and a male at 7500 ppm. The female was diagnosed with

Table 8. Body weight and feed consumption in dogs fed diets containing zoxamide for 90 days

Parameter	Dietary concentration (ppm)			
	0	1500	7500	30 000
Males				
Body weight (g):				
Week 0	7911	7821	8017	7537
Week 1	8339	8091	8118	7515
Week 8	9861	9714	9584	7803*
Body-weight gain (g), week 1–16 (mean ± SD)	3296 ± 477	3290 ± 437	2718 ± 894	1296 ± 813*
Food consumption (g), week 1	290	300	285	226*
Females				
Body weight (g):				
Week 0	6562	6846	6659	6665
Week 1	6885	7116	6753	6529
Week 8	8043	7749	7728	6669
Body-weight gain (g), week 1–16 (mean ± SD)	2314 ± 891	1658 ± 928	1899 ± 398	824 ± 528[a]
Food consumption (g/day), week 1	270	278	280	187*

From Ferguson et al. (1997)
SD, standard deviation.
* Significant difference from controls ($p < 0.05$).
[a] $p = 0.052$ (two-sided t-test).

canine juvenile polyarteritis syndrome (CJPS[1]); the male with pneumonia. Two additional males (at 7500 ppm and 30 000 ppm) exhibited signs consistent with CJPS. Sporadic instances of soft faeces were recorded pre-test and during treatment. There was a treatment-related decrease in mean body weight, cumulative body-weight gain, and feed consumption in males and females at 30 000 ppm (Table 8). These findings were evident in the first weeks of dosing and did not appear to be related to the sick animals in the groups at the intermediate and highest dose.

Haematology investigations did not reveal any treatment-related changes in either sex at doses of 7500 ppm or less (Table 9). Statistically significant changes, including decreased erythrocyte count, increased mean cell haemoglobin, and increased mean cell haemoglobin concentration were seen in females at 30 000 ppm. A similar but statistically non-significant trend was seen in males. A dose-related decrease in lymphocytes and increased segmented neutrophils were also observed in males from week 8 to 16; the total leukocyte count was very variable but did not show any treatment-related differences. The toxicological significance of these haematological changes was unclear, as they were not reproduced in the 1-year study in dogs.

Many clinical chemistry parameters exhibited considerable variation between dogs, but did not show any changes in males or females at doses of 7500 ppm or less. At 30 000 ppm, apparent treatment-related decreases in albumin and in the albumin to globulin ratio were seen in both sexes after 8 and 16 weeks of treatment and an increase in serum gamma glutamyl transferase activity was seen in males (less than twofold). Urine analysis parameters did not show any treatment-related differences. Ophthalmology at 16 weeks showed no treatment related effects. Organ weights showed a treatment-related increase in the absolute and relative liver weights in females at 7500 ppm and greater and in males at 30 000 ppm. The increase in absolute (≥ 23%) and relative liver weight (29%) of the females at 7500 ppm was considered to be treatment-related but not an adverse effect, since no corresponding clinical pathology or histopathological findings were observed at this dose (Table 10).

There were no treatment-related findings on gross examination. Treatment-related microscopic changes were observed in the group at 30 000 ppm and consisted of a diffuse hepatocellular hypertrophy affecting all males and females, and hypertrophy of the thyroid follicular epithelium in one male and one female. Neither of these findings were reported in the data for historical controls; however, hypertrophy of the thyroid was not reported in the 1-year study in dogs.

The NOAEL was 7500 ppm, equal to 281 mg/kg bw per day, on the basis of treatment-related changes including reduction in body weight and in body-weight gain, thyroid hypertrophy, changes in erythrocyte and leukocyte parameters and reduced albumin concentrations. Increases in absolute and relative liver weights were noted at 7500 ppm, but the changes in organ weights were not accompanied by any clinical chemistry histopathological changes and were not considered to be adverse effects (Ferguson et al., 1997).

Groups of four male and four female beagle dogs were given diets containing zoxamide (purity, 92.3%) at a concentration of 0, 1500, 7500 or 30 000 ppm (0, 50, 255, and 1016 mg/kg bw per day respectively in males, and 0, 48, 278 and 994 mg/kg bw per day respectively in females) for 1 year. All dogs were observed daily for signs of ill health or reaction to treatment. Feed consumption was determined daily for all animals beginning 2 weeks before treatment (i.e. pre-test) and continued until the end of week 13 of treatment. Thereafter, feed consumption was measured for 1 week every 4 weeks until the end of the study. Body weights were determined each week for all animals, beginning 2 weeks before treatment (i.e. pre-test) and continued until the end of week 13 of treatment. Thereafter, body weights were measured once every 4 weeks until the end of the study. Physical examinations

[1] CJPS is reported to be a spontaneous disease of uncertain etiology (Ruben et al., 1989; Snyder et al., 1995; Son, 2004).

Table 9. Haematological findings in dogs fed diets containing zoxamide for 90 days (mean)

Finding	Dietary concentration (ppm)			
	0	1500	7500	30 000
Males				
Erythrocytes (10^6/mm^3):				
Week 8	5.92	6.05	6.11	5.72
Week 16	6.61	6.75	6.45	5.88
MCH (pg):				
Week 8	21.4	22.0	21.3	22.1
Week 16	21.5	21.9	21.6	22.6
MCHC (%):				
Week 8	34.0	34.2	34.3	34.7
Week 16	34.2	34.1	34.4	34.9
Segmented neutrophils (%):				
Week 8	64	62	64	69
Week 16	56	71	77	83*
Lymphocytes (%):				
Week 8	31	35	30	30
Week 16	38	26	20	14*
Females				
Erythrocytes (10^6/mm^3):				
Week 8	6.33	6.07	5.89	5.71*
Week 16	7.01	6.57	6.63	5.98
MCH (pg) :				
Week 8	21.7	22.1	22.0	23.5*
Week 16	21.7	22.3	22.2	24.3*
MCHC (%):				
Week 8	34.0	34.3	34.2	35.1
Week 16	33.7	34.6	34.5	35.8*

From Ferguson et al. (1997)
MCH, mean cell haemoglobin; MCHC, mean cell haemoglobin concentration.
* Significant difference from control ($p < 0.05$).

were performed weekly beginning 2 weeks before treatment. Clinical chemistry, haematology, and urinary parameters were evaluated for all dogs during pre-test, after 3, 6, and 9 months of treatment, and for all dogs surviving to necropsy. Ophthalmological examinations were performed on all dogs during pre-test and just before the end of treatment. Selected organs were weighed and tissues were collected for histopathological evaluation. The study was certified to be compliant with GLP and was conducted in accordance with OECD guideline 452.

There were no deaths or clinical signs indicative of systemic toxicity during the study. A female from the group at 1500 ppm was found dead (presumed cause, bilateral haemorrhagic pneumonia) and replaced with an equivalent animal from the 90-day study (see above). A female at the highest dose was humanely killed on week 38; the clinical signs and post-mortem findings were reported to be consistent with CJPS. A male at the lowest dose showed evidence of CJPS at post mortem examination. Soft faeces were seen in all groups and in some animals before testing, but appeared to be more persistent in animals at 30 000 ppm. There was a treatment-related reduction in mean body

Table 10. Organ weight and histopathological findings in dogs receiving zoxamide

Finding	Dietary concentration (ppm)			
	0	1500	7500	30 000
Mean organ weights (mean ± SD)				
Males				
Absolute liver weight (g)	304 ± 14	319 ± 33	310 ± 116	371 ± 45*
Relative liver weight (%)	2.66 ± 0.24	2.80 ± 0.22	2.80 ± 0.87	4.14 ± 0.06*
Females				
Absolute liver weight (g)	244 ± 49	269 ± 39	300 ± 48	329 ± 45*
Relative liver weight (%)	2.68 ± 0.20	3.11 ± 0.40	3.44 ± 0.30*	4.37 ± 0.31*
Histopathology (No. affected/No. examined)				
Males				
Liver hypertrophy	0/4	0/4	0/4	4/4
Thyroid hypertrophy	0/4	0/4	0/4	1/4
Females				
Liver hypertrophy	0/4	0/4	0/4	4/4
Thyroid hypertrophy	0/4	0/4	0/4	1/4

From Ferguson et al. (1997)
* Significant difference from controls ($p < 0.05$).

weight, cumulative body-weight gain, and feed consumption in both sexes at 30 000 ppm; it was not until week 7 that all the females returned to their pre-test body weight. Females at 7500 ppm had reduced body-weight gain in the early stages of the study and this early deficit persisted (Table 11). Haematology and urine analysis parameters did not reveal any treatment-related differences. Clinical chemistry parameters showed treatment-related decreases in albumin and increases in alkaline phosphatase activity in both sexes at 30 000 ppm (Table 11). Ophthalmology at termination did not reveal any treatment-related ocular changes.

Organ weights showed a treatment-related increase in absolute and relative liver weights in both sexes at 30 000 ppm and in females at 7500 ppm (Table 12). A dose-related increase in absolute and relative thyroid weights was also evident (Table 12). No treatment-related gross findings were made at necropsy. Treatment-related microscopic changes were observed in the livers of some animals from the group at 30 000 ppm and consisted of diffuse hepatocellular hypertrophy in two males and one female; another female had multifocal haemorrhage and necrosis; a third female had congestion and mononuclear cell infiltration, a finding also seen in a male without hypertrophy (Table 12). None of the animals in this study had thyroid cell hyperplasia.

The NOAEL was 1500 ppm, equal to 48 mg/kg bw per day, on the basis of reduction in body-weight gain in females at 7500 ppm, equal to 255 mg/kg bw per day (Ferguson et al., 1998b).

2.3 *Long-term studies of toxicity and carcinogenicity*

Mice

In a long-term study of toxicity/carcinogenicity, groups of 60 male and 60 female Crl:CD-1 (ICR) BR (VAF/+) mice were given diets containing zoxamide (purity, 92.3%) at a concentration of 0, 350, 1750, or 7000 ppm (equal to 0, 51.1, 251 or 1021 mg/kg bw per day and 0, 60.4, 326 or 1289 mg/kg bw per day in males and females respectively) for 18 months. All mice were observed

Table 11. Body weight, food consumption and clinical chemistry findings in dogs fed diets containing zoxamide for 1 year

Finding (means)	Dietary concentration (ppm)			
	0	1 500	7 500	30 000
Males				
Body weight (g):				
Week 0	7 846	8 036	7 576	7 882
Week 4	8 939	9 094	8 374	8 136
Week 13	10 640	10 773	10 349	9 466
Week 25	11 623	12 063	11 933	10 718
Week 52	11 734	12 318	12 247	11 041
Body-weight gain (g):				
Week 1	169	180	9	36*
Week 4	1 093	1 058	799	254*
Week 13	2 794	2 737	2 773	1 584*
Week 25	3 777	4 027	4 357	2 836
Week 52	3 888	4 283	4 672	3 159
Food consumption (g):				
Week 1	290.1	285.7	241.2	217.1
Week 25	324.4	343.1	330.8	359.8
Albumin (g/dl):				
Pre-test	3.7	3.3	3.4	3.6
Day 93	3.7	3.7	3.5	3.4*
Day 183	3.7	3.5	3.6	3.3*
Day 274	3.7	3.4	3.4	3.2*
Day 365	3.8	3.7	3.7	3.5
Alkaline phosphatase (U/l):				
Pre-test	260	402	498	325
Day 93	190	241	388	367
Day 183	122	168	275	434*
Day 274	117	174	236	412*
Day 365	97	159	200	414*
				Females
Body weight (g):				
Week 0	7 004	6 781	6 755	6 980
Week 4	7 898	7 658	7 137	6 839*
Week 13	9 242	9 008	8 322	8 035
Week 25	10 176	9 511	9 179	8 503*
Week 52	10 171	9 576	9 293	8 912
Body-weight gain (g):				
Week 1	251	415	139	−175*
Week 4	895	877	382*	−141*
Week 13	2 238	2 227	1 567	1 056*
Week 25	3 172	2 730	2 425	1 523

Week 52	3 167	2 795	2 538	2 052
Food consumption (g/day):				
Week 1	259.9	254.4	250.5	196.4*
Week 25	281.6	252.6	313.5	274.2
Albumin (g/dl):				
Pre-test	3.4	3.6	3.5	3.3
Day 93	3.6	3.8	3.6	3.2*
Day 183	3.6	3.8	3.5	3.1
Day 274	3.6	3.8	3.5	3.3
Day 365	3.7	3.6	3.6	3.3*
Alkaline phosphatase (U/l):				
Pre-test	283	314	311	290
Day 93	208	177	258	311*
Day 183	170	116	242	297*
Day 274	176	106	219	304*
Day 365	152	108	183	330*

From Ferguson et al. (1998b)
* Significant difference from controls ($p < 0.05$); ANOVA, Dunnett t-test.
a Mainly due to one dog that lost 397 g in body weight, but consumed typical amounts of food.

Table 12. Organ weights and histopathological findings in dogs fed diets containing zoxamide for 1 year (means)

Finding	Dietary concentration (ppm)			
	0	1500	7500	30 000
Mean organ weights:				
Males				
Absolute liver weight (g)	293	330	353	412*
Relative liver weight (%)	2.5	2.7	2.9	3.7*
Absolute thyroid weight (g)	0.81	0.92	1.07	1.04
Relative thyroid weight (%)	0.007	0.007	0.009	0.010
Females				
Absolute liver weight (g)	278	276	309	341
Relative liver weight (%)	2.7	2.9	3.4*	3.8*
Absolute thyroid weight (g)	0.73	0.75	0.81	0.98*
Relative thyroid weight (%)	0.007	0.008	0.009	0.011*
Histopathology (No. affected/No. examined):				
Males				
Liver hypertrophy	0/4	0/4	0/4	2/4
Hepatocellular necrosis	0/4	0/4	0/4	0/4
Females				
Liver hypertrophy	0/4	0/4	0/4	1/4
Hepatocellular necrosis	0/4	0/4	0/4	1/4

From Ferguson et al. (1998b)
* Significant difference from controls ($p < 0.05$); ANOVA, Dunnett t-test.

daily for signs of moribundity, mortality, ill health or reaction to treatment. Physical examinations were performed each week. Body weights and feed consumption were monitored weekly beginning 1 week before initiation of treatment until week 13 and then every fourth week thereafter for the duration of the study. After 12 and 18 months, all mice were bled for leukocyte differential counts. After 18 months, all mice were killed, and organs and tissues grossly examined at necropsy. Selected organ weights were recorded and histopathological examinations were conducted on tissues. The study was certified to be compliant with GLP and was conducted in accordance with OECD guideline 451.

There were no effects on survival in mice at any dose; absolute survival was more than 50% in all groups. No treatment-related deaths or clinical signs indicative of systemic toxicity were observed in any of the treatment groups. Males at 350 and 7000 ppm had lower body weights than did controls (< 10% reduction) for much of the study, but there was no clear dose–response relationship. There were no other notable effects on body weight, cumulative body-weight gain, or feed consumption. Leukocyte differential counts from all mice at the highest dose were similar to those of controls at 12 or 18 months of treatment. Organ-weight measurements at necropsy did not show any treatment-related intergroup differences. Relative liver weights in all treated groups of males were approximately 10% higher than in controls, but there was no dose–response relationship and no associated histopathological findings, therefore this was not considered to be an adverse effect of treatment. Gross examination at necropsy and microscopic examination of organs and tissues did not reveal any treatment-related changes. There were no treatment-related effects on the type or incidence of any of the neoplasms observed in this study.

The Meeting concluded that zoxamide is not carcinogenic. The NOAEL was 7000 ppm, equal to 1021 mg/kg bw per day, the highest dose tested (Robison et al., 1998c; Gillette & Brown, 1998).

Rats

In a combined long-term study of toxicity and carcinogenicity, groups of 70 male and 70 female Sprague-Dawley Crl:CD®BR rats were given diets containing zoxamide (purity, 92.0%) at a concentration of 0 (control), 1000, 5000, or 20 000 ppm (equal to 0, 51, 260, and 1058 mg/kg bw per day in males and 0, 65, 328, and 1331mg/kg bw per day in females) for 2 years. Ten males and 10 females per group were randomly selected before treatment for interim sacrifice after 52 weeks. Routine observations were performed for survival, clinical signs, body weights and food consumption. During weeks 13, 26, 52, 78, and 104, blood samples were collected for haematology and clinical chemistry tests from 20 males and 10 females per group, and urine samples were collected from 10 males and 10 females per group. The same animals were bled at each interval when possible. After either 52 (10 males and 10 females per group) or 104 (all survivors) weeks of treatment, the animals were weighed, anaesthetized, killed, and necropsied. At necropsy, macroscopic observations were recorded, selected organs were weighed, and selected tissues were collected and preserved. Animals that died during test or were sacrificed at an unscheduled interval were also necropsied; however, organs were not weighed. Microscopic examinations were performed on tissues from each rat in the control group and in the group at the highest dose and from each rat that died or was sacrificed at an unscheduled interval. The lung, liver, kidney, and any gross lesions were also examined microscopically from each rat at the lowest and intermediate dose. The study was certified to be compliant with GLP and satisfied the essential requirements of OECD guideline 453.

There were no test material-related effects on survival or in clinical signs; overall survival was > 44% in all groups. Clinical chemistry and haematology did not reveal any adverse, treatment-related differences. Observed changes in the haematology and serum chemistry parameters were considered incidental to the administration of the test material owing to the lack of dose–response relationship, the low magnitude of the change, the lack of histopathological correlation, or the inconsistent occurrence of the differences at the different sampling times. There were no treatment-related effects in any urine analysis parameter or in ophthalmical findings in either sex at any dose. Liver weights were increased in a dose-related manner in females at the interim kill (Table 13), but not at termination, nor in males. There were no indications of treatment-related histopathological changes in the liver. Gross and histopathological examinations of organs and tissues did not reveal any treatment-related abnormalities. An apparent increase in thyroid C-cell lesions in males at the highest dose was not statistically significant ($p > 0.05$), did not exhibit a dose–response relationship, was not reproduced in females and was within the range for historical controls (Giknis & Clifford, 2001).

Zoxamide was not carcinogenic and did not produce any evidence of significant systemic toxicity at doses of up to 20 000 ppm (1058 mg/kg bw per day) in this 2-year dietary study in rats. The NOAEL was 20 000 ppm (1058 mg/kg bw per day) on the basis of the absence of toxicity at the highest dose tested (Ivett, 1998a; 1998b)

2.4 Genotoxicity

The genotoxic potential of zoxamide and two of its metabolites has been investigated in a range of studies (Table 14). Zoxamide was not genotoxic in an assay for reverse mutation in *Salmonella typhimurium*; an assay for gene mutation in mammalian cells nor in an assay for micronucleus formation in mice in vivo. In a study of chromosome aberration in Chinese hamster ovary cells in vitro, there was no increase in structural aberrations, but mitotic accumulation was observed at concentrations that inhibited cell growth in tests with and without metabolic activation. The increases observed in all cultures were noted to be predominantly attributable to increases in the frequency of cells with polyploidy. Such a finding is consistent with the mode of pesticidal action of zoxamide, which involves binding to β-tubulin (Young, 1998). To investigate this finding, a study of micronucleus formation in bone marrow with kinetochore staining was performed in rats in vivo; this produced negative results for both micronucleus formation and kinetochore staining.

Table 13. Liver weights and thyroid pathology findings in rats given diets containing zoxamide for 2 years

Finding	Dietary concentration (ppm)			
	0	1000	5000	20,000
Females				
Week 53 (n = 10):				
Body weight (g)	398	380	355	383
Absolute liver weight (g)	9.5	10.0	0.1	11.1
Relative liver weight (%)	2.38	2.64	2.86*	2.91*
Week 105 (n = 26–30):				
Body weight (g)	406	389	383	388
Absolute liver weight (g)	11.5	11.7	11.9	11.8
Relative liver weight (%)	2.90	3.02	3.15	3.06
Thyroid C-cell hyperplasia	8/58	2/31	4/33	5/60
Thyroid C-cell carcinoma	4/58	0/31	2/33	2/60
Thyroid C-cell adenoma	4/58	2/31	3/33	5/60
Males				
Thyroid C-cell hyperplasia	6/58	2/33	0/23	8/58
Thyroid C-cell carcinoma	2/58	2/33	0/23	3/58
Thyroid C-cell adenoma	2/58	2/33	2/23	6/58

From Giknis & Clifford (2001)

* Significant difference from controls ($p < 0.05$).

Table 14 Results of studies of genotoxicity with zoxamide and its metabolites

Test substance	End-point	Test object	Concentration	Purity (%)	Result	Reference
In vitro						
Zoxamide	Reverse mutation	*S. typhimurium* strains TA98, TA100, TA102, TA1535, TA1537	16–5000 µg/plate (precipitation at $\geq$ 160 µg/plate)	92.3	Negative +S9 Negative –S9	Sames & Ciaccio (1996)
Zoxamide	Chromosome aberrations	Chinese hamster ovary cells	1–100 µg/ml (1–4 µg/ml –S9 and 1–16 µg/ml +S9 scored) Cytotoxicity at 4–µg/ml – S9; and 11–µg/ml +S9	92.3	Negative +S9 Negative –S9 Positive for polyploidy ± S9	Riley (1998)
Zoxamide	Gene mutation (*HPRT*)	Chinese hamster ovary cells (CHO-K1, BH4)	–65 µg/ml (precipitation at 50 µg/ml)	94.2	Negative +S9 Negative –S9	Pant (1994)
In vivo						
Zoxamide	Micronucleus formation	Bone marrow from CD-1 mice (five males and five females per group per time-point)	200, 1000 or 2000 mg/kg bw by gavage in corn oil (sacrifice at 24 or 48 h)	92.3	Negative	Sames & Vandenberghe (1996)
Zoxamide	Micronucleus formation with kinetochore staining	Bone marrow from Crl:CD-BR rats (five males per group)	Two administrations at 500, 1000 or 2000 mg/kg bw in corn oil. All doses scored for micronucleus formation; highest dose only for kinetochore staining.	97.35	Negative	Gudi & Krsmanovich (2003)
Metabolites						
RH141,452	Reverse mutation	*S. typhimurium* strains TA98, TA100, TA102, TA1535, TA1537	50–5000 µg/plate	97.7	Negative +S9 Negative –S9	Sames & Ciaccio (1998a)
RH141,455	Reverse mutation	*S. typhimurium* strains TA98, TA100, TA102, TA1535, TA1537	50–5000 µg/plate	98.7	Negative +S9 Negative –S9	Sames & Ciaccio (1998b)

S9, 9000 × *g* supernatant from livers of rats.

The zoxamide metabolites, RH 141,452 and RH 141,455 gave negative results in an assay for reverse mutation in *S. typhimurium*.

2.5 *Reproductive toxicity*

(a) Multigeneration studies

Rats

Groups of 30 male and 30 female rats were given diets containing zoxamide (purity, 92.3%) at a concentration of 0, 1000, 5000 or 20 000 ppm (active ingredient) over two generations. Parental rats of the first generation (P_1) were exposed from age approximately 6 weeks and of the second generation (P_2) from weaning (21 days). P_1 and P_2 rats were mated initially (to produce F_{1a} and F_{2a} litters) after at least 10 weeks of exposure to treated diets. Treatment continued throughout gestation, lactation, and until terminal necropsy. P_1 and P_2 rats were each mated a second time to produce a second set of litters (designated F_{1b} and F_{2b}, respectively). Rats were re-mated approximately 1 week after weaning of all litters from the first mating was complete. A second mating of the P_1 rats was conducted to verify apparent findings among the F_{1a} offspring. The P_2 rats were re-mated to ascertain if the apparent findings observed in the F_{1a}, F_{1b} and F2a litters were related to decreased feed intake, direct toxicity of the test substance, and/or from maternally mediated effects. To accomplish this, litters resulting from the second mating of P_2 rats (F_{2b} pups) were randomly divided into two subgroups within each dose group on postnatal day 0. From day 0 to day 14, the treatment regimen and evaluation of the rats remained unchanged. On days 14 to 21, all dams and their litters were removed from their respective diets and placed on untreated feed. All dams were given zoxamide (suspended in corn oil and dosed at a constant volume of approximately 5 ml/kg) at a dose of 0, 50, 250 or 1000 mg/kg bw per day (corresponding to 0, 1000, 5000, or 20 000 ppm, respectively) daily by oral gavage. Pups in litters assigned to one subgroup (at each dose) remained untreated during this period. Pups in litters assigned to the second subgroup (at each dose) were dosed daily by oral gavage at the same dose as their dams. All dams were placed back on treated feed on day 21 and maintained on treated feed for 3–5 weeks until terminal necropsy.

Body weight, feed consumption, and clinical signs were monitored in parental animals throughout treatment. Estrous cycling was evaluated in P_1 and P_2 females for 3 weeks before initial mating. Male parental rats were killed and necropsied after the second mating. Female parental rats were killed and necropsied after weaning of their second litters. Sperm evaluation was performed for all P_1 and P_2 males at the time of necropsy. Selected tissues (including reproductive tissues) were weighed from all P_1 and P_2 rats at necropsy. Histopathological evaluation was performed for all selected tissues for rats in the control group and at the highest dose, and rats found dead or sacrificed during the course of the study. Reproductive tissues were also examined in all rats suspected of reduced fertility.

Survival and growth of offspring were monitored throughout lactation. Litters were culled to eight pups (four males and four females per litter where possible) on day 4 of lactation. Stillborn pups, pups that died during lactation, pups culled at day 4, or sacrificed at weaning were examined grossly. Selected tissues were collected, weighed and preserved from F_{1a}, F_{1b}, F_{2a}, and F_{2b} litters at weaning (one male and one female per litter) for possible histopathology. Microscopic evaluation was performed on spleens of the F_{1a}, F_{2a} and F_{2b} weanlings (both subgroups) and on stomachs of the F_{2a} and F_{2b} weanlings. Sexual maturation (age at preputial separation in males, vaginal patency in females) was evaluated in F_{1a} offspring selected as P_2 parental animals.

The achieved intakes of zoxamide for the P_1 and P_2 parents during the pre-mating period were 71–100, 360–489 and 1474–2091 mg/kg bw per day, in males and 82–108 , 409–534 , and 1624–2239 mg/kg bw per day, in females at 1000, 5000 or 20 000 ppm, respectively. Similar intakes were achieved during gestation; with intakes during lactation being higher than during the pre-mating phase. The study was certified to be compliant with GLP and satisfied the essential criteria of OECD guideline 416.

There were no treatment-related deaths or clinical signs of systemic toxicity. At 20 000 ppm, a treatment-related decrease in cumulative body-weight gain was noted in P_1 rats during the first week of treatment in both sexes and the decrease was evident throughout the pre-mating period in P_1 females. Female (P_1) feed consumption was decreased at 20 000 ppm during the first week of treatment, but there were no consistent effects on food consumption over the whole study. There were no treatment-related effects on estrous cycling nor on sperm motility, morphology, epididymal sperm count or concentration, or testicular sperm count and concentration of P_1 or P_2 males. Reproductive performance, gestation and birth parameters were similar in all groups. The viability of offspring at birth and the ratio of male to female pups in either generation did not show any treatment-related differences. There was a slight delay in sexual maturation in F_{1a} animals at the highest dose (approximately 1 day), but this was considered to be associated with lower body weights (Table 15).

Gross examination and histopathology of P_1 and P_2 rats did not reveal any treatment-related abnormalities. Statistically significant increases in relative liver weight were seen in males and females at 5000 and 20 000 ppm, but these findings were not considered to be adverse as there were no associated gross or histopathological changes.

There were no treatment-related gross findings in F_{1a}, F_{1b}, F_{2a} or F_{2b} pups. There was an apparent decrease in body-weight gain in the F_{1a}, F_{1b}, and F_{2a} pups at all doses, particularly during the latter stages of lactation (Table 14), which was accompanied by a decrease in spleen weights and histopathological changes in the spleen (decreased extramedullary hematopoiesis) (Table 16). These effects were not evident in the F_{2b} litters that were placed on untreated diets from day 14 to 21 of lactation and dosed by gavage during that time with the equivalent dose of zoxamide. In these litters, no effects on body weight, changes in spleen weight or histopathology of the spleen were seen (Tables 15 and 16). The decrease in body weight seen in the F_{1a}, F_{1b} and F_{2a} litters was judged to be a secondary effect related to the palatability of the treated diets and not a systemic toxic effect of zoxamide. The effects noted in the spleen (F_{1a}, F_{1b}, F_{2a}) were considered to be potentially secondary to the decreased body-weight gain since these effects were not seen in the F_{2b} litters where body-weight gain was not affected although an equivalent dose was given by gavage.

There were no adverse reproductive effects in rats given diets containing zoxamide. Some reductions in pup body-weight gain, spleen weight and spleen pathology were seen, but the evidence indicated that these were potentially related to palatability as they were not present in the F_{2b} litters that received equivalent doses of zoxamide via gavage rather than in the diet. A slight delay in sexual maturation was seen in pups, but this was considered to be secondary to the reduced body weight. An increase in relative liver weight was noted at doses of 5000 ppm and greater in both sexes and in absolute liver weight in males only at 20 000 ppm. These effects were not associated with any histopathological changes and appear to be consistent with the metabolic burden associated with high doses of a xenobiotic rather than direct toxicity of zoxamide.

The NOAEL for reproductive toxicity, parental toxicity and toxicity in pups was 20 000 ppm (equal to 1474–2091 mg/kg bw per day in males and 1624–2239 mg/kg bw per day in females), the highest dose tested. The findings seen in the study were not considered to be adverse effects of treatment with zoxamide (O'Hara et al., 1998).

(b) Developmental toxicity

Rats

Groups of 25 mated female rats were given zoxamide (purity, 94.2%) at a dose of 0 (control), 100, 300, or 1000 mg/kg bw per day in corn oil by gavage on days 6–15 of presumed gestation. All rats were examined daily for signs of ill health or reaction to treatment. Body weight and feed

Table 15 Pup body weights during lactation in a two-generation study of reproductive toxicity in rats exposed to zoxamide

Mean weights (g)	Dietary concentration (ppm)			
	0	1000	5000	20 000
P_1/F_{1a}				
Pup weight:				
Day 0	6.6	6.7	6.7	6.6
Day 7	18.3	17.8	18.5	17.4
Day 14	37.8	35.9	37.3	35.2*
Day 21	60.2	54.5*	54.8*	51.0*
P_1/F_{1b}				
Pup weight:				
Day 0	6.4	6.4	6.5	6.3
Day 7	17.6	17.2	16.5	16.6
Day 14	36.0	34.6	34.6	33.8
Day 21	58.3	54.4	52.8*	51.2*
P_2/F_{2a}				
Pup weight [litter weight]:				
Day 0	6.4 [86]	6.3 [92]	6.2 [91]	6.2 [91]
Day 7	17.3 [137]	16.2 [126]	15.8 [117]	15.6* [123]
Day 14	35.4 [280]	34.1 [252]	33.7 [236]	32.4* [243]
Day 21	55.9 [442]	52.1* [386]	49.0* [343]	47.8* [358]
P_2/F_{2b} a				
Pup weight [litter weight]:				
Day 0	6.6 [80]	6.3 [92]	6.5 [92]	6.2 [95]
Day 4 – before culling	11.2 [131]	10.0* [140]	9.6* [126]	9.3* [140]
Day 4 – after culling	11.2 [77]	10.0* [76]	9.6* [74]	9.3* [74]
Day 7	18.1 [125]	16.1* [121]	15.7* [113]	15.3* [119]
Day 14	36.1 [242]	35.0 [220]	34.7 [230]	33.4* [247]
Day 14 – pups treated by gavage	35.0	34.1	34.3	33.1
Day 14 – pups on control diet	37.2	36.0	35.1	33.7
Day 21 – pups treated by gavage	57.3	55.3	56.2	54.1
Day 21 – pups on control diet	60.0	56.9	57.6	55.3

From O'Hara et al. (1998)

* Significant difference from control ($p<0.05$); ANOVA, Dunnett's t-test

[a] Rats were exposed to diets containing zoxamide until day 14 of lactation (postpartum). From day 14 to day 20 of lactation, doses of 0, 50, 250 or 1000 mg/kg bw per day (corresponding to dietary concentrations of 0, 1000, 5000, or 20 000 ppm, respectively) were administered by gavage. All dams received the appropriate dose by gavage. Pups in half of the litters received the dose by gavage; pups in the remaining litters were not treated during this period.

Table 16. Spleen weights and histology for pups in a two-generation study of reproductive toxicity in rats exposed to zoxamide

Generation	Finding	Dietary concentration (ppm)			
		0	1000	5000	20 000
F_{1a}	*Males*				
	Absolute spleen weight (g)	0.283	0.223*	0.236*	0.205*
	Relative spleen weight (%)	0.473	0.415*	0.420*	0.394*
	Reduced extramedullary haematopoiesis (No. affected/No. examined)	0/24	6/26*	7/24*	9/26*
	Females				
	Absolute spleen weight (g)	0.286	0.248*	0.226*	0.208*
	Relative spleen weight (%)	0.489	0.461	0.424*	0.408*
	Reduced extramedullary haematopoiesis (No. affected/No. examined)	1/24	5/25	12/25*	15/27*
F_{1b}	*Males*				
	Absolute spleen weight (g)	0.263	0.242	0.220*	0.203*
	Relative spleen weight (%)	0.442	0.431	0.404	0.385*
	Females				
	Absolute spleen weight (g)	0.267	0.235	0.216*	0.199*
	Relative spleen weight (%)	0.457	0.442	0.419	0.395*
F_{2a}	*Males*				
	Absolute spleen weight (g)	0.253	0.229	0.217*	0.172*
	Relative spleen weight (%)	0.439	0.425	0.425	0.354*
	Reduced extramedullary haematopoiesis	1/23	5/24	6/24	13/22*
	Females				
	Absolute spleen weight (g)	0.267	0.234	0.211*	0.184*
	Relative spleen weight (%)	0.479	0.460	0.429*	0.380*
	Reduced extramedullary haematopoiesis	0/23	5/24	12/23*	12/22*
F_{2b}	*Males (dams and pups dosed by gavage)*				
	Absolute spleen weight (g)	0.269	0.280	0.270	0.252
	Relative spleen weight (%)	0.457	0.534	0.481	0.448
	Reduced extramedullary haematopoiesis	0/10	—	—	0/11
	Females (dams and pups dosed by gavage)				
	Absolute spleen weight (g)	0.251	0.265	0.273	0.258
	Relative spleen weight (%)	0.453	0.455	0.483	0.481
	Reduced extramedullary haematopoiesis (n=)	0/9	—	—	0/10
	Males (dams only dosed by gavage)				
	Absolute spleen weight (g)	0.290	0.295	0.282	0.285
	Relative spleen weight (%)	0.474	0.484	0.468	0.478
	Reduced extramedullary haematopoiesis (n=)	0/10	—	—	1/11
	Females (dams only dosed by gavage)				
	Absolute spleen weight (g)	0.286	0.266	0.283	0.267
	Relative spleen weight (%)	0.483	0.485	0.510	0.476
	Reduced extramedullary haematopoiesis (n=)	0/9	—	—	0/11

From O'Hara et al. (1998)
* Significant difference from controls ($p < 0.05$);

consumption were monitored throughout gestation. Dams were killed on day 20 of presumed gestation, and subjected to a gross necropsy and uterine examination. Each live fetus was removed, weighed individually, and examined for external abnormalities. One half of each litter was examined for soft-tissue and head alterations; skeletal examinations were performed on all fetuses. The study was certified to be compliant with GLP and satisfied the essential criteria of OECD guideline 414.

There were no treatment-related mortalities, or clinical signs of toxicity at any dose. Maternal body weight, body-weight gain, feed consumption or gravid uterine weights showed no treatment-related differences. Pregnancy rates were > 90% in all groups. There were no dams aborting or treatment-related effects on the numbers of early or late resorptions, live fetuses per litter, fetal body weight or sex ratio. There were no notable increases in external, soft-tissue, head or skeletal malformations, variations, or developmental retardations observed at any dose.

The NOAEL was 1000 mg/kg bw per day for maternal and fetotoxicity; on the basis of the absence of effects at 1000 mg/kg bw per day, the highest dose tested. Under the conditions of this study of developmental toxicity in rats, zoxamide was not teratogenic (Kane & Shuey, 1995).

Rabbits

Groups of 16 artificially-inseminated New Zealand White rabbits were given zoxamide (purity, 92.3%) at a dose of 0, 100, 300 or 1000 mg/kg bw per day suspended in 0.5% aqueous methylcellulose solution by gavage on days 7–19 of presumed gestation. All doses were administered in a constant volume of 20 ml/kg. Clinical signs were recorded routinely. The does were weighed on days 0, 7, 9, 11, 14, 17, 20 and 29; feed consumption was recorded daily. On day 29, the does were killed and the thoracic and abdominal cavities were examined for gross changes and a full uterine examination was performed. All live fetuses were sexed, weighed and examined for external and visceral alterations (Staples' technique). After the visceral examinations, fetuses were stained with Alizarin Red S and examined for skeletal alterations. The study was certified to be compliant with GLP and satisfied the essential criteria of OECD guideline 414.

There were no treatment-related deaths or clinical signs of toxicity in does. No treatment-related effects were noted for maternal body weight, body-weight change, feed consumption or gravid uterine weights at any dose. No treatment-related gross lesions were observed in does during postmortem examinations at any dose. Pregnancy rates were 100% in all groups; no dams aborted. The number of viable litters, mean numbers of resorptions, live or dead fetuses, and sex ratio per litter did not reveal any treatment-related intergroup differences. No treatment-related differences were observed in fetal body weights. External, visceral and skeletal examinations of fetuses did not reveal any treatment-related abnormalities; all findings were typical of the background observations in studies of developmental toxicity in rabbits.

The NOAEL for fetotoxicity and maternal toxicity with zoxamide in rabbits was 1000 mg/kg bw per day, the highest dose tested, on the basis of the absence of treatment-related toxicity in dams or fetuses. Zoxamide was not teratogenic in rabbits (Shuey, 1997).

2.6 *Special studies*

(a) *Neurotoxicity*

In a study of acute neurotoxicity, groups of 10 male and 10 female Crl:CD®BR rats were given zoxamide (purity, 92.9% active ingredient) at a dose of 0 (control), 125, 500 or 2000 mg/kg bw suspended in corn oil by gavage in a volume of 10 ml/kg). All rats were observed daily for signs of ill health or reaction to treatment. Each rat received a pre-test functional observational battery (FOB) and motor activity assessment 7 days before dosing. FOB and motor activity testing was repeated approximately 5 h after dosing (day 0) and on days 7 and 14 after dosing. On the day after the final

FOB/motor activity assessment, rats were anaesthetized, perfused with neutral buffered formalin and given a limited gross necropsy. Twelve randomly selected control rats (six males and six females) and twelve randomly selected rats at the highest dose (six males and six females) received a special neuropathology evaluation that included microscopic examination of the brain, spinal cord, peripheral nerves of the hindlimb and selected ganglia. It was claimed that the study was compliant with GLP and met the basic requirements of EPA guideline 81-8.

Dosing solutions were confirmed to be within 3% of nominal values. There were no mortalities, treatment-related clinical signs of toxicity or body-weight effects observed during the study period. There were no treatment-related effects on motor activity or any of the FOB parameters. Motor activity was lower in males at the highest dose on the day of dosing, but as this did not achieve statistical significance and was not evident in females this was not considered to be an adverse effect of treatment. No treatment-related morphological alterations occurred in any of the examined areas of the central or peripheral nervous systems.

The NOAEL for general toxicity and neurotoxicity was 2000 mg/kg bw, the highest dose tested (Danberry & Gillette, 1997).

A study of neurotoxicity/general toxicity with repeated doses of zoxamide in rats is reported in section 2.3.

(b) Mechanism of action

A range of studies have been performed (Young, 1998) in order to characterize the antifungal mechanism of action of zoxamide and the potential relevance to mammals. These have included investigations in fungi, plant cells, mammalian cells and on isolated tubulin. In some studies, a compound encoded RH-54032 was also investigated: RH-54032 is a dichlorobenzamide compound like zoxamide, but without the 4-methyl group on the aromatic ring.

Investigations on the effects of zoxamide and RH-54032 on morphology and nuclear division in *Phytophthora capsici* showed that the encystment of *P. capsici* zoospores and germination were unaffected by treatment with zoxamide at high concentrations (25 ppm, 74.3 μmol/l) and RH-54032 (25 ppm, 77.5 μmol/l). However, germ-tube elongation was completely arrested shortly after germination even by treatment at very low concentrations (zoxamide at 0.5 ppm, 1.49 μmol/l), and this was accompanied by swelling of the germ tube. Mithramycin staining of treated cells revealed that the first cycle of nuclear division, which occurred in untreated cells at between 1 and 2.5 h, failed to occur in the presence of zoxamide or RH-54032. Treated cells contained a single nucleus located in the cyst whereas untreated cells contained multiple nuclei, which migrated into the growing germ tubes. When the effects of zoxamide and RH-54032 at various concentrations on the first cycle of nuclear division were compared with their inhibition of radial growth in poison-agar assays, it was shown that concentrations that inhibited growth strongly inhibited nuclear division. Inhibition of nuclear division was found to occur very rapidly when zoxamide, RH-54032 or the anti-microtubule agent colchicine were added to germlings at the onset of the first cycle of nuclear division.

In an investigation of the effect of RH-54032 on mitosis in mouse lymphoma cells, RH-54032 inhibited the growth of mouse lymphoma cells, and at concentrations that inhibited growth produced an accumulation of cells in arrested metaphase as reflected by an increase in the mitotic index. Normal mitotic figures were absent in treated cells, and the cells contained scattered chromosome pairs that failed to align at the equatorial metaphase plate. RH-54032 was considerably less potent than colchicine: RH-54032 produced a mitotic index of 27.4 at 2 μmol/l (0.642 ppm), while colchicine produced a mitotic index of 45.6 at 0.125 μmol/l (0.05 ppm).

In investigations on the effects of RH-54032 and zoxamide on the assembly of microtubules in vitro, RH-54032 and zoxamide inhibited the assembly of bovine tubulin into microtubules in vitro in a dose-dependent manner. In these experiments, purified bovine tubulin was incubated with the anti-tubulin compounds at 37 °C for an appropriate time, then cooled on ice. Microtubule assembly was initiated by addition of guanosine triphosphate (GTP) and incubation at 37 °C. The ability of anti-microtubule agents to inhibit microtubule assembly was characterized by the need for a relatively long pre-incubation period with tubulin before initiation of assembly, indicating a low potency.

A comparison of RH-54032, zoxamide, and other anti-microtubule agents with respect to their ability to inhibit microtubule assembly in vitro and the growth of mouse lymphoma cells is shown in Table 17. EC_{50} values for inhibition of growth of mouse lymphoma cells by RH-54032, zoxamide and carbendazim fell in the range of 1.1 to .9.0 µmol/l), while colchicine and taxol were more than two orders of magnitude more potent, and vinblastine was more than three orders of magnitude more potent. RH-54032, zoxamide and carbendazim were also less potent inhibitors of microtubule assembly in vitro than were colchicine and vinblastine.

Binding of RH-54032 to isolated bovine tubulin was shown to involve the β-subunit of tubulin, in agreement with results from labelling experiments using whole cells. Further information about the anti-tubilin benzamide binding site on tubulin was obtained by testing the effect of other anti-microtubule agents on binding of [^{3}H]RH-54032 to bovine tubulin. Binding was strongly inhibited by colchicine, podophyllotoxin and the benzimidazole nocodazole, while vinblastine had little effect. Since it is known that colchicine, podophyllotoxin and benzimidazoles bind to a common site on β-tubulin, while vinblastine binds to a different region, these results are consistent with binding of antitubilin benzamides to the β-tubulin subunit (Young, 1998).

3. Observations in humans

No specific surveys of personnel of manufacturing plants were available. There were two adverse reports linked to exposure to a diluted formulation of mancozeb/zoxamide. One relates to dermal irritation, the other to non-specific "flu-like" symptoms. The Meeting considered it to be unlikely that these findings were directly related to exposure to zoxamide.

Comments

Biochemical aspects

In rats given zoxamide, approximately 60% of a dose of 10 mg/kg bw was absorbed, with peak plasma concentrations of radioactivity occurring at 8 h after dosing. Zoxamide was extensively distributed among organs and tissues with highest concentrations reported in the liver.

Table 17. Inhibition of microtubule assembly and growth of mouse lymphoma cells by RH-54032, zoxamide and other antimicrotubule agents

Compound	Growth of mouse lymphoma cells		Microtubule assembly	
	EC_{50} (µmol/l)	EC_{50} (ppm)	IC_{50} (µmol/l)	IC_{50} (ppm)
Zoxamide	3.7	1.24	23.5	7.90
RH-54032	1.1	0.35	9.0	2.89
Colchicine	0.01	0.004	1.0	0.40
Taxol	0.007	0.006	NT	NT
Vinblastine	0.0008	0.0007	1.8	1.64
Carbendazim	9.0	1.72	29.2	5.58

From Young (1998)

EC_{50}, concentration having an effect on half the sample; IC_{50}, concentration inhibiting half the sample; NT, not tested.

Excretion was primarily in the faeces, via the bile. The overall elimination half-life was 13–14 h. At 1000 mg/kg bw, there was some evidence of saturation of absorption, with C_{max} and area under the curve of concentration–time (AUC) values being approximately 40–50 times those at 10 mg/kg bw, but with a similar elimination half-life. Females excreted approximately twice as much radiolabel in the urine as did males. Very little radioactivity remained in tissues (< 0.2% of the administered dose) or carcass (< 2% of the administered dose) at 5 days after dosing. Pre-treatment of animals with diets containing zoxamide for 2 weeks or with five daily gavage doses of radiolabelled zoxamide did not significantly alter the absorption or distribution of radiolabel compared with that in untreated animals.

The metabolism of zoxamide was extensive, involving a variety of pathways including hydrolysis, glutathione-mediated reactions, and reductive dehalogenation, secondary oxidation on both the aromatic methyl and the aliphatic side-chain, limited deamidation; and terminal glucuronic acid and amino-acid conjugation. Thirty-two separate metabolites were identified; no single metabolite accounted for more than 10% of the administered dose. After repeated doses, there was an indication of an increase in glutathione-mediated metabolism.

Toxicological data

Zoxamide was of low acute toxicity when administered orally (median lethal dose, LD_{50}, > 5000 mg/kg bw), dermally (LD_{50}, > 2000 mg/kg bw) or after a 4-h exposure by inhalation (LC_{50}, > 5.3 mg/l). Zoxamide is not a skin irritant, but is a slight, transient eye irritant. Zoxamide produced delayed contact hypersensitivity in guinea-pigs in the maximization and Buehler tests.

In repeat-dose studies, the main effects of zoxamide were reduced body-weight gain and liver hypertrophy. The reductions in body-weight gain were not consistent across studies. Investigative work performed as part of the study of reproductive toxicity indicated there might be palatability problems with diet containing zoxamide. However, food consumption was not reduced consistently in studies in which reduced body-weight gain was reported. Liver hypertrophy was not associated with any histopathological or clinical chemistry changes that indicated damage to liver cells. Therefore, in line with the guidance developed by the 2006 JMPR, increased liver weight and hepatocyte hypertrophy were considered to be adaptive rather than adverse effects of exposure to zoxamide.

In a 90-day study of toxicity in mice, the NOAEL was 2500 ppm, equal to 574 mg/kg bw per day, on the basis of reduced body-weight gains in females at 7000 ppm, equal to 1606 mg/kg bw per day. Increases in relative liver weights (by approximately 10%) were not associated with any pathological or clinical chemistry changes and are not considered to be adverse. In a 90-day study of toxicity and neurotoxicity in rats, the NOAEL was 20 000 ppm, equal to 1509 mg/kg bw per day, the highest dose tested.

In a 28-day study of toxicity in dogs, the NOAEL was 30 000 ppm, equal to 1045 mg/kg bw per day, the highest dose tested. Soft stools were present at an increased incidence at doses of 5000 ppm, equal to 175 mg/kg bw per day, and above, but as this finding was not seen consistently in other studies in dogs given similar doses and the same formulated diet, this finding was not considered to be an adverse effect of treatment. In the 90-day study of toxicity in dogs, the NOAEL was 7500 ppm, equal to 281 mg/kg bw per day, on the basis of reductions in body-weight gain, serum albumin concentrations and erythrocyte counts in both sexes at 30 000 ppm, equal to 1055 mg/kg bw per day. Increases in liver weights (by approximately 25%) in females at 7500 ppm were not associated with any histopathological or clinical chemistry changes and were not considered to be adverse. In the 1-year study of toxicity in dogs, reduced body-weight gain (45%) was present from the beginning of the study in females at 7500 ppm, equal to 255 mg/kg bw per day, and a deficit in body-weight gain (20%) was still present at the end of the study. Males receiving zoxamide at 7500 ppm also had reduced body-weight gain during the early stages of the study, but these animals had terminal body weights that were higher than those of the controls. Although food consumption was reduced transiently, there was no clear link between body weights of individual animals and food consumption. At the highest dose of 30 000 ppm, there were marked effects on body weight and food consumption, with females taking up to 7 weeks to regain their pre-test body weight. Reduced concentrations of serum albumin, and increases in liver and thyroid weights and serum alkaline phosphatase activities were also seen in both sexes at 30 000 ppm. The NOAEL in the 1-year study was 1500 ppm, equal to 48 mg/kg bw per day.

In the 90-day and 1-year studies in dogs, cases of CJPS were seen in the groups receiving zoxamide, but not in the controls. CJPS is reported to be specific to beagle dogs, occurring spontaneously but with unknown etiology. A genetic link has been postulated, which might explain the occurrence in the 90-day and 1-year studies, which were started at the same time and used animals from the same supplier. Therefore, CJPS was not considered to be related to exposure to zoxamide.

In a 28-day study of dermal toxicity in rats, zoxamide produced significant local effects at doses of 107 mg/kg bw per day and greater. Findings of systemic toxicity were most likely to be secondary to the local effects and the NOAEL for systemic effects was 714 mg/kg bw per day

Negative results were obtained in assays for gene mutation in vitro and in assays for micronucleus formation in bone marrow of rats and mice in vivo. Zoxamide was found to induce polyploidy in an assay for chromosomal aberration in Chinese hamster ovary cells in vitro. These findings are consistent with the mechanism of fungicidal action of zoxamide, involving binding to the β-subunit of tubulin. Zoxamide also inhibits microtubule assembly in mouse lymphoma cells (IC_{50}, 23.5 μmol/l). The induction of polploidy after inhibition of tubulin polymerization and disruption of microtubule formation has been investigated for other compounds and is considered to be a threshold-mediated effect. The assay for micronucleus formation in rats included kinetochore staining and produced negative results for micronuclei and chromosomal damage. A supplementary kinetic study in mice demonstrated that there was exposure of the bone marrow after administration of zoxamide.

The Meeting concluded that zoxamide was unlikely to pose a genotoxic risk to humans at levels typical of dietary exposures.

In long-term studies of toxicity in mice and rats, zoxamide exhibited no general toxicity and was not carcinogenic in either species. Increased liver weights (approximately 20%) in female rats killed after a 1-year exposure to zoxamide at a dietary concentration of 5000 ppm and greater were not considered to be adverse as there were no associated histopathological or clinical chemistry findings at any time during the study. An apparent increase in thyroid C-cell lesions in male rats at the highest dose was not statistically significant, did not exhibit a dose–response relationship, was not reproduced in females and was within the range for historical controls. The NOAEL in mice was 7000 ppm, equal to1021 mg/kg bw per day, and the NOAEL in rats was 20 000 ppm, equal to 1058 mg/kg bw per day, both values being identified on the basis of the absence of treatment-related toxicity at the highest doses tested.

In view of the absence of carcinogenic potential in rodents and the lack of genotoxicity in vivo, the Meeting concluded that zoxamide was unlikely to pose a carcinogenic risk to humans.

The reproductive toxicity of zoxamide has been investigated in a two-generation study in rats and studies of developmental toxicity in rats and rabbits. In the study of reproductive toxicity in rats, the NOAEL for effects on fertility, parental toxicity and pup development was 20 000 ppm, equal to 1474 mg/kg bw per day. Reductions in pup body-weight gain and spleen weights and reduced extramedullary haematopoiesis in the spleen were seen in F_{1a} F_{1b} and F_{2a} offspring, but these effects appeared to be related to palatability as they were not evident in the F_{2b} generation, when pups and dams received equivalent exposures of zoxamide by gavage, rather than from the diet, from postnatal days 14 to 21. Increased relative liver weight was noted at doses of 5000 ppm and greater in males and females, and in absolute liver weight only in males at 20 000 ppm. The changes in liver weight were not associated with any histopathological or clinical chemistry change and were not considered to be adverse.

There was no evidence of toxicity in the studies of prenatal developmental toxicity in rats or rabbits. The NOAEL in both studies was 1000 mg/kg bw per day on the basis of absence of toxicity to dams or fetuses at the highest dose tested. Zoxamide was not teratogenic in rats or rabbits.

Zoxamide was not neurotoxic in a study of acute neurotoxicity at doses of up to 2000 mg/kg bw. No adverse effects were seen during neurological and behavioural examinations performed during routine repeat-dose studies with zoxamide.

Studies on two plant metabolites of zoxamide—RH-141,452 (3,5-dichloro-4-hydroxymethyl benzoic acid) RH-141,455 (3,5-dichloro-1,4-benzene-dicarboxylic acid)—formed to a limited extent in rats, showed them to be rapidly absorbed and rapidly excreted, essentially unchanged; to have low acute oral toxicities to mice (LD_{50}s, > 5000 mg/kg bw), and to give negative results in assays for gene mutation with strains of *Salmonella typhimurium*.

There were two reports of mild adverse effects after exposure to a diluted formulation containing zoxamide and mancozeb. In one case there was a report of skin irritation, in the other "flu-like" symptoms were reported. The Meeting considered it to be unlikely that these effects were related directly to exposure to zoxamide.

Toxicological evaluation

An acceptable daily intake (ADI) of 0–0.5 mg/kg bw was established for zoxamide based on the NOAEL of 48 mg/kg bw per day in the 1-year study in dogs, on the basis of reduced body-weight gain in females at 255 mg/kg bw per day.

An acute reference dose (ARfD) was considered to be unnecessary for zoxamide as zoxamide is of low acute toxicity, did not produce developmental effects and did not produce any other significant effects after acute exposures.

Levels relevant to risk assessment

Species	Study	Effect	NOAEL	LOAEL
Mouse	Two-year studies of toxicity and carcinogenicity[a]	Toxicity	7000 ppm, equal to 1021 mg/kg bw per day[c]	—
		Carcinogenicity	7000 ppm, equal to 1021 mg/kg bw per day[c]	—
Rat	Two-year studies of toxicity and carcinogenicity[a]	Toxicity	20 000 ppm, equal to 1058 mg/kg bw per day[c]	—
		Carcinogenicity	20 000 ppm, equal to 1058 mg/kg bw per day[c]	—
	Multigeneration study of reproductive toxicity[a]	Reproductive toxicity	30 000 ppm, equal to 1474 mg/kg bw per day[c]	—
		Parental toxicity	30 000 ppm, equal to 1474 mg/kg bw per day[c]	—
		Offspring toxicity	30 000 ppm, equal to 1474 mg/kg bw per day[c]	—
	Developmental toxicity[b]	Maternal toxicity	1000 mg/kg bw per day[c]	—
		Embryo/fetotoxicity	1000 mg/kg bw per day[c]	—

	Acute neurotoxicity[b]		2000 mg/kg bw per day[c]	—
Rabbit	Developmental toxicity[a]	Maternal toxicity	1000 mg/kg bw per day[c]	—
		Embryo/fetotoxicity	1000 mg/kg bw per day[c]	—
Dog	One-year study of toxicity[a]	Reduced body-weight gain	1500 ppm, equal to 48 mg/kg bw per day	7500 ppm, equal to 255 mg/kg bw per day

[a] Dietary administration.
[b] Gavage administration.
[c] Highest dose tested.

Estimate of acceptable daily intake for humans

0–0.5 mg/kg bw

Estimate of acute reference dose

Unnecessary

Studies that would be useful for the continued evaluation of the compound

Results from epidemiological, occupational health and other such observational studies of human exposure

Critical end-points for setting guidance values for exposure to zoxamide

Absorption, distribution, excretion and metabolism in mammals	
Rate and extent of oral absorption	Moderate (C_{max}, 8 h); approximately 60% absorbed at 10 mg/kg bw
Dermal absorption	Approximately 1% from concentrate; 6% from dilution
Distribution	Extensive. Highest levels in liver.
Potential for accumulation	Low
Rate and extent of excretion	> 85% in 48 h. Urine (approximately 10–20%); bile (approximately 45%); faeces (approximately 50–80%).
Metabolism in animals	Extensive. Primarily via hydrolysis, dehalogenation, oxidation and conjugation.
Toxicologically significant compounds in animals, plants and the environment	Zoxamide.
Acute toxicity	
Rat, LD_{50}, oral	> 5000 mg/kg bw
Rat, LD_{50}, dermal	> 2000 mg/kg bw
Rat, LC_{50}, inhalation	> 5.3 mg/l
Rabbit, skin irritation	Not irritating
Rabbit, eye irritation	Slight transient irritant
Guinea-pig, skin sensitization (test method used)	A skin sensitizer (Buehler; Magnusson & Kligman)
Short-term studies of toxicity	
Target/critical effect	Body-weight gain
Lowest relevant oral NOAEL	1500 ppm (48 mg/kg bw per day) in a 1-year study in dogs
Lowest relevant dermal NOAEL	< 107 mg/kg bw for local effects; 714 mg/kg bw per day for systemic effects.

Lowest relevant inhalation NOAEC	No data (not required)
Genotoxicity	
	Not genotoxic in vivo
Long-term studies of toxicity and carcinogenicity	
Target/critical effect	None
Lowest relevant NOAEL	7000 ppm (1021mg/kg bw per day) in mice (highest dose tested)
Carcinogenicity	Not carcinogenic
Reproductive toxicity	
Reproduction target/critical effect	None
Lowest relevant reproductive NOAEL	20 000 ppm (1047 mg/kg bw per day) in rats (highest dose tested)
Developmental target/critical effect	None
Lowest relevant developmental NOAEL	1000 mg/kg bw per day in rats and rabbits (highest dose tested)
Neurotoxicity/delayed neurotoxicity	
	No indications of neurotoxicity in studies of acute toxicity or repeat-doses
Acute neurotoxicity	NOAEL was 2000 mg/kg bw in rats (highest dose tested)
Other toxicological studies	
	RH-141,452 Rapid excretion; essentially unmetabolized. Oral LD_{50} in mice, > 5000 mg/kg bw Negative in an Ames test.
	RH-141,455 Rapid excretion; essentially unmetabolized. Oral LD_{50} in mice, > 5000 mg/kg bw Negative in an Ames test
Medical data	
	Two reports (one case of irritation & one of flu-like symptoms) after exposure to a diluted formulation of mancozeb/zoxamide. Unlikely to be directly related to zoxamide.

Summary

	Value	Study	Safety factor
ADI	0–0.5 mg/kg bw	Dog, 1-year study	100
ARfD	Unnecessary	—	—

References

Bernacki, H.J. & Ferguson, J.S. (1996) RH-117,281 technical: acute inhalation toxicity study in rats. Unpublished report No. 95R-266 (ER Ref.. No. 2.2, US Ref. No. 95R-266) from Rohm and Haas Co. Submitted to WHO by Dow AgroSciences, Indianapolis, USA.

Danberry, T.L. & Gillette, D.M. (1997) RH-117,281 technical: acute oral (gavage) neurotoxicity study in rats. Unpublished report No. 95R-182 (ER Ref. No. 10.1, US Ref. No. 95R-182) from Rohm and Haas Co. Submitted to WHO by Dow AgroSciences, Indianapolis, USA.

Ferguson, J.S. & Lutz, M.F. (1998) RH-117,281 technical: acute oral toxicity study in male and female mice. Unpublished report No. 98R-165 (ER Ref. No. 24.3, US Ref. No. 98R-165) from Rohm and Haas Co. Submitted to WHO by Dow AgroSciences, Indianapolis, USA.

Ferguson, J.S., Morrison, R.D. & Kemmerer, M.G. (1997) RH-117,281 technical: three-month dietary toxicity study in dogs. Unpublished report No. 96R-030 (ER Ref. No. 9.1, US Ref. No. 96R-030) from Rohm and Haas Co. Submitted to WHO by Dow AgroSciences, Indianapolis, USA.

Ferguson, J.S., Lutz, M.F. & Procopio, K.R. (1998a) RH-141,455: acute oral toxicity study in male and female mice. Unpublished report No. 98R-047 (ER Ref. No. 27.3, US Ref. No. 98R-047) from Rohm and Haas Co. Submitted to WHO by Dow AgroSciences, Indianapolis, USA.

Ferguson, J.S., Morrison, R.D. & Davidson, B.F. (1998b) RH-117,281 technical: one-year chronic dietary toxicity study in dogs. Unpublished report No. 95R-277 (ER Ref. No. 25.1, US Ref. No. 95R-277) from Rohm and Haas Co. Submitted to WHO by Dow AgroSciences, Indianapolis, USA.

Ferguson, J.S., Procopio, K.R. & Lutz, M.F. (1998c) RH-141,452: acute oral toxicity study in male and female mice. Unpublished report No. 98R-049 (ER Ref. No. 25.2, US Ref. No. 98R-049) from Rohm and Haas Co. Submitted to WHO by Dow AgroSciences, Indianapolis, USA.

Frederick, C.B. & Swenson, R.E. (1998) 14C-RH-117,281 80WP AND 14C-RH-117,281 2F formulations: dermal absorption study in male rats. Unpublished report No. 97R-076 from Rohm and Haas Co. Submitted to WHO by Dow AgroSciences, Indianapolis, USA.

Giknis, M.L.A & Clifford, C.B (2001) Compilation of spontaneous neoplastic lesions and survival in Crl:CD(SD) BR rats from control groups. Unpublished report from Charles River Laboratories. http://www.criver.com/flex_content_area/documents/rm_rm_r_lesions_survival_crlcd_sd_rats.pdf

Gillette, D.M. & Brown, W.R. (1998) RH-117,281 technical: eighteen-month dietary oncogenicity study in mice. Photomicrographs. Unpublished report No. 96R-094A (ER Ref. No. 20.1, US Ref. No. 96R-094A) from Rohm and Haas Co. Submitted to WHO by Dow AgroSciences, Indianapolis, USA.

Gingrich, S.L. & Parno, J.R. (1996a) RH-117,281 technical: acute oral toxicity study in male and female rats. Unpublished report No. 95R-268 (ER Ref. No. 1.3, US Ref. No. 95R-268) from Rohm and Haas Co. Submitted to WHO by Dow AgroSciences, Indianapolis, USA.

Gingrich, S.L. & Parno, J.R. (1996b) RH-117,281 technical: acute dermal toxicity study in male and female rats. Unpublished report No. 95R-269 (ER Ref. No. 1.4, US Ref. No. 95R-269) from Rohm and Haas Co. Submitted to WHO by Dow AgroSciences, Indianapolis, USA.

Gingrich, S.L. & Parno, J.R. (1996c) RH-117,281 technical: skin irritation study in rabbits. Unpublished report No. 95R-270 (ER Ref. No. 1.5, US Ref. No. 95R-270) from Rohm and Haas Co. Submitted to WHO by Dow AgroSciences, Indianapolis, USA.

Gingrich, S.L. & Parno, J.R. (1996d) RH-117,281 technical: eye irritation study in rabbits. Unpublished report No. 95R-271 (ER Ref. No. 1.6, US Ref. No. 95R-271) from Rohm and Haas Co. Submitted to WHO by Dow AgroSciences, Indianapolis, USA.

Glaza, S.M. (1997) Dermal sensitization study of RH-117,281 technical in guinea pigs - maximization test. Covance Laboratories Project No. 6228-112. Unpublished report No. 95RC-170 (ER Ref. No. 4.2, US Ref. No. 95RC-170) from Rohm and Haas Co. Submitted to WHO by Dow AgroSciences, Indianapolis, USA.

Gudi, R. & Krsmanovic, L. (2003) Amended report for zoxamide: mammalian erythrocyte micronucleus test with kinetochore analysis. Unpublished report No. 021122R, AA65WR.126.BTL from Bioreliance Laboratories, Rockville, Maryland. Submitted to WHO by Dow AgroSciences, Indianapolis, USA.

Ivett, J. (1998a) RH-117,281 technical: 24-month dietary chronic/oncogenicity study in rats. Covance project No. 417-505. Unpublished report No. 94RC-236 (ER Ref. No. 21.1, US Ref. No. 94RC-236) from Rohm and Haas Co. Submitted to WHO by Dow AgroSciences, Indianapolis, USA.

Ivett, J. (1998b) RH-117,281 technical: 24-month dietary chronic/oncogenicity study in rats. Photomicrographs on selective tissues. Covance project No. 417-505. Unpublished report No. 94RC-236A (ER Ref. No. 21.1, US Ref. No. 94RC-236A) from Rohm and Haas Co. Submitted to WHO by Dow AgroSciences, Indianapolis, USA.

Kane, W.W. & Shuey, D.L. (1995) RH-7281 technical: oral (gavage) developmental toxicity study in rats. Unpublished report No. 94R-079 (ER Ref. No. 6.1, US Ref. No. 94R-079) from Rohm and Haas Co. Submitted to WHO by Dow AgroSciences, Indianapolis, USA.

Morrison, R.D. & Gillette, D.M. (1996) RH-117,281: three-month dietary toxicity/ neurotoxicity study in rats. Unpublished report No. 94R-233 (ER Ref. No. 3.1, US Ref. No. 94R-233) from Rohm and Haas Co. Submitted to WHO by Dow AgroSciences, Indianapolis, USA.

O'Hara, G.P., Craig, L.P. & Romanello, A.S. (1998) RH-117,281 technical: two-generation reproductive toxicity study in rats. Unpublished report No. 95R-272 (ER Ref. No. 26.1, US Ref. No. 95R-272) from Rohm and Haas Co. Submitted to WHO by Dow AgroSciences, Indianapolis, USA.

Pant, K. (1994) RH-117,281: test for chemical induction of gene mutation at the HGPT locus in cultured Chinese hamster ovary cells with and without metabolic activation. Sitek Study No. 0282-2510. Unpublished report No 94RC-077 (ER Ref. No. 23.4, US Ref. No. 94RC-077) from Rohm and Haas Co. Submitted to WHO by Dow AgroSciences, Indianapolis, USA.

Reibach P. H. & Detweiler K. (2001) Identification of RH-139432 from zoxamide (RH-7281) pharmacokinetic study samples. Unpublished report No. TR-34-00-105 (ER Ref. No. 45.3) from Rohm and Haas Co. Submitted to WHO by Dow AgroSciences

Riley, S. (1998) RH-117,281: Test for chemical induction of chromosome aberrations in cultured Chinese hamster ovary (CHO) cells. Covance Laboratories, UK Project No. 616/20-D5140. Unpublished report No. 96RC-125 (ER Ref. No. 23.6, US Ref. No. 96RC-125). from Rohm and Haas Co. Submitted to WHO by Dow AgroSciences, Indianapolis, USA.

Robison, P., Anderson, D.M., & Ecke, B.F. (1998a) RH-117,281 technical: Delayed contact hypersensitivity study in guinea pigs. Unpublished report No. 97R-074 (ER Ref. No. 23.2, US Ref. No. 97R-074) from Rohm and Haas Co. Submitted to WHO by Dow AgroSciences, Indianapolis, USA.

Robison, P., Anderson, D.M. & Ecke, B.F. (1998b) RH-117,281 technical: delayed contact hypersensitivity (dilution) study in guinea pigs. Unpublished report No. 98R-154 (ER Ref. No. 24.4, US Ref. No. 98R-154) from Rohm and Haas Co. Submitted to WHO by Dow AgroSciences, Indianapolis, USA.

Robison, P., Anderson, D.M. & Ecke, B.F. (1998c) RH-117,281 technical: Eighteen-month dietary oncogenicity study in mice. Unpublished report No. 96R-094 (ER Ref. No. 20.1, US Ref. No. 96R-094) from Rohm and Haas Co. Submitted to WHO by Dow AgroSciences, Indianapolis, USA.

Robison, P., Morrison, R.D., Bannister, R.M. & Eberly, S.L. (1998d) RH-117,281 technical: twenty-eight day dermal toxicity study in rats. Unpublished report No. 97R-075 (ER Ref. No. 23.3, US Ref. No. 97R-075) from Rohm and Haas Co. Submitted to WHO by Dow AgroSciences, Indianapolis, USA.

Ruben, Z., Deslex, P., Nash, G. et al. (1989). Spontaneous disseminated panarteritis in laboratory beagle dogs in a toxicity study: a possible genetic predilection. *Toxcologic. Pathology*, **17**, 145–152.

Sames, J.S. & Ciaccio, P.C. (1996) RH-117,281 technical: *Salmonella typhimurium* gene mutation assay (Ames test). Unpublished report No. 95R-262 (ER Ref. No. 2.7, US Ref. No. 95R-262) from Rohm and Haas Co. Submitted to WHO by Dow AgroSciences, Indianapolis, USA.

Sames, J.L. & Ciaccio, P.J. (1998a) RH-141,452: *Salmonella typhimurium* gene mutation assay (Ames test). Unpublished report No. 98R-050 (ER Ref. No. 25.3, US Ref. No. 98R-050) from Rohm and Haas Co. Submitted to WHO by Dow AgroSciences, Indianapolis, USA.

Sames, J.L. & Ciaccio, P.J. (1998b) RH-141,455: *Salmonella typhimurium* gene mutation assay (Ames test). Unpublished report No. 98R-048 (ER Ref. No. 27.4, US Ref. No. 98R-048) from Rohm and Haas Co. Submitted to WHO by Dow AgroSciences, Indianapolis, USA.

Sames, J.S. & Vandenberghe, Y.L. (1996) RH-117,281 technical: micronucleus assay in CD-1 mouse bone marrow cells. Unpublished report No. 95R-264 (ER Ref. No. 1.9, US Ref. No. 95R-264) from Rohm and Haas Co. Submitted to WHO by Dow AgroSciences, Indianapolis, USA.

Shuey, D.L., Kaminski, E.J., Anderson, D.M. & Lomax, L.G. (1996) RH-117,281: three-month dietary toxicity study in mice. Unpublished report No. 94R-075 (ER Ref. No.5.3, US Ref. No. 94R-075) from Rohm and Haas Co. Submitted to WHO by Dow AgroSciences, Indianapolis, USA.

Shuey, D.L. (1997) RH-117,281 technical: oral (gavage) developmental study in rabbits. Unpublished report No. 95R-267 (ER Ref. No. 8.2, US Ref. No. 95R-267) from Rohm and Haas Co. Submitted to WHO by Dow AgroSciences, Indianapolis, USA.

Snyder, P.W., Kazacos, E.A., Scott-Moncreiff, J.C. et al. (1995) Pathologic features of naturally occurring juvenile polyarteritis in beagle dogs. *Vet. Pathol.,* **32**, 337–345.

Son, W-C. (2004). Idiopathic canine polyarteritis in control beagle dogs from toxicity studies. *J. Vet. Sci.*, **5**, 147–150

Swenson, R.E. & Frederick, C.B. (1998) Distribution of ^{14}C-RH-117,281 to the bone marrow of mice. Unpublished report No. 97R-173 ER Ref. No. 24.2, US Ref. No. 97R-173) from Rohm and Haas Co. Submitted to WHO by Dow AgroSciences

Swenson, R.E., Frederick, C.B. & Graves, D.D. (1998) ^{14}C-RH-117,281: pharmacokinetic and metabolism study in rats. Unpublished report No. 94R-235 ER Ref. No. 24.1, US Ref. No. 94R-235) from Rohm and Haas Co. Submitted to WHO by Dow AgroSciences

Vandenberghe, Y.L., Kaminski, E.J. & Lomax, L.G. (1996) RH-117,281 technical: four-week range-finding toxicity study in dogs. Unpublished report No. 94R-234 (ER Ref. No. 2.3, US Ref. No. 94R-234) from Rohm and Haas Co. Submitted to WHO by Dow AgroSciences, Indianapolis, USA.

Wu, D. & Gu, Z. (1998a) ^{14}C-RH-141,452: rat metabolism study, tier I testing. Unpublished report No. RPT00410 (ER Ref. No. 27.1, US Ref. No. 97RC-154) by Xenobiotic Laboratories Inc, New Jersey for Rohm and Haas Co. Submitted to WHO by Dow AgroSciences, Indianapolis, USA.

Wu, D. & Gu, Z. (1998b) ^{14}C-RH-141,455: rat metabolism study, tier I testing. Unpublished report No. RPT00411 (ER Ref.. No. 27.2, US Ref. No. 98RC-017) by Xenobiotic Laboratories Inc, New Jersey for Rohm and Haas Co. Submitted to WHO by Dow AgroSciences, Indianapolis, USA.

Young, D.H. (1998) Mechanism of action of the oomycete fungicides RH-54032 and RH-117281 on *Phytophthora capsici*, tobacco, mouse lymphoma cells and isolated bovine tubulin. Unpublished report No. 98R-1098 ER Ref. No. 23.5, US Ref. No. 98R-1098) from Rohm and Haas Co. Submitted to WHO by Dow AgroSciences, Indianapolis, USA.

ANNEX 1

Reports and other documents resulting from previous Joint Meetings Of the FAO Panel of Experts on Pesticide Residues in Food and the Environment and WHO Expert Groups on Pesticide Residues

1. Principles governing consumer safety in relation to pesticide residues. Report of a meeting of a WHO Expert Committee on Pesticide Residues held jointly with the FAO Panel of Experts on the Use of Pesticides in Agriculture. FAO Plant Production and Protection Division Report, No. PL/1961/11; WHO Technical Report Series, No. 240, 1962.

2. Evaluation of the toxicity of pesticide residues in food. Report of a Joint Meeting of the FAO Committee on Pesticides in Agriculture and the WHO Expert Committee on Pesticide Residues. FAO Meeting Report, No. PL/1963/13; WHO/Food Add./23, 1964.

3. Evaluation of the toxicity of pesticide residues in food. Report of the Second Joint Meeting of the FAO Committee on Pesticides in Agriculture and the WHO Expert Committee on Pesticide Residues. FAO Meeting Report, No. PL/1965/10; WHO/Food Add./26.65, 1965.

4. Evaluation of the toxicity of pesticide residues in food. FAO Meeting Report, No. PL/1965/10/1; WHO/Food Add./27.65, 1965.

5. Evaluation of the hazards to consumers resulting from the use of fumigants in the protection of food. FAO Meeting Report, No. PL/1965/10/2; WHO/Food Add./28.65, 1965.

6. Pesticide residues in food. Joint report of the FAO Working Party on Pesticide Residues and the WHO Expert Committee on Pesticide Residues. FAO Agricultural Studies, No. 73; WHO Technical Report Series, No. 370, 1967.

7. Evaluation of some pesticide residues in food. FAO/PL:CP/15; WHO/Food Add./67.32, 1967.

8. Pesticide residues. Report of the 1967 Joint Meeting of the FAO Working Party and the WHO Expert Committee. FAO Meeting Report, No. PL:1967/M/11; WHO Technical Report Series, No. 391, 1968.

9. 1967 Evaluations of some pesticide residues in food. FAO/PL:1967/M/11/1; WHO/Food Add./68.30, 1968.

10. Pesticide residues in food. Report of the 1968 Joint Meeting of the FAO Working Party of Experts on Pesticide Residues and the WHO Expert Committee on Pesticide Residues. FAO Agricultural Studies, No. 78; WHO Technical Report Series, No. 417, 1968.

11. 1968 Evaluations of some pesticide residues in food. FAO/PL:1968/M/9/1; WHO/Food Add./69.35, 1969.

12. Pesticide residues in food. Report of the 1969 Joint Meeting of the FAO Working Party of Experts on Pesticide Residues and the WHO Expert Group on Pesticide Residues. FAO Agricultural Studies, No. 84; WHO Technical Report Series, No. 458, 1970.

13. 1969 Evaluations of some pesticide residues in food. FAO/PL:1969/M/17/1; WHO/Food Add./70.38, 1970.

14. Pesticide residues in food. Report of the 1970 Joint Meeting of the FAO Working Party of Experts on Pesticide Residues and the WHO Expert Committee on Pesticide Residues. FAO Agricultural Studies, No. 87; WHO Technical Report Series, No. 4574, 1971.

15. 1970 Evaluations of some pesticide residues in food. AGP:1970/M/12/1; WHO/Food Add./71.42, 1971.

16. Pesticide residues in food. Report of the 1971 Joint Meeting of the FAO Working Party of Experts on Pesticide Residues and the WHO Expert Committee on Pesticide Residues. FAO Agricultural Studies, No. 88; WHO Technical Report Series, No. 502, 1972.

17. 1971 Evaluations of some pesticide residues in food. AGP:1971/M/9/1; WHO Pesticide Residue Series, No. 1, 1972.

18. Pesticide residues in food. Report of the 1972 Joint Meeting of the FAO Working Party of Experts on Pesticide Residues and the WHO Expert Committee on Pesticide Residues. FAO Agricultural Studies, No. 90; WHO Technical Report Series, No. 525, 1973.

19. 1972 Evaluations of some pesticide residues in food. AGP:1972/M/9/1; WHO Pesticide Residue Series, No. 2, 1973.

20. Pesticide residues in food. Report of the 1973 Joint Meeting of the FAO Working Party of Experts on Pesticide Residues and the WHO Expert Committee on Pesticide Residues. FAO Agricultural Studies, No. 92; WHO Technical Report Series, No. 545, 1974.

21. 1973 Evaluations of some pesticide residues in food. FAO/AGP/1973/M/9/1; WHO Pesticide Residue Series, No. 3, 1974.

22. Pesticide residues in food. Report of the 1974 Joint Meeting of the FAO Working Party of Experts on Pesticide Residues and the WHO Expert Committee on Pesticide Residues. FAO Agricultural Studies, No. 97; WHO Technical Report Series, No. 574, 1975.

23. 1974 Evaluations of some pesticide residues in food. FAO/AGP/1974/M/11; WHO Pesticide Residue Series, No. 4, 1975.

24. Pesticide residues in food. Report of the 1975 Joint Meeting of the FAO Working Party of Experts on Pesticide Residues and the WHO Expert Committee on Pesticide Residues. FAO Plant Production and Protection Series, No. 1; WHO Technical Report Series, No. 592, 1976.

25. 1975 Evaluations of some pesticide residues in food. AGP:1975/M/13; WHO Pesticide Residue Series, No. 5, 1976.

26. Pesticide residues in food. Report of the 1976 Joint Meeting of the FAO Panel of Experts on Pesticide Residues and the Environment and the WHO Expert Group on Pesticide Residues. FAO Food and Nutrition Series, No. 9; FAO Plant Production and Protection Series, No. 8; WHO Technical Report Series, No. 612, 1977.

27. 1976 Evaluations of some pesticide residues in food. AGP:1976/M/14, 1977.

28. Pesticide residues in food—1977. Report of the Joint Meeting of the FAO Panel of Experts on Pesticide Residues and Environment and the WHO Expert Group on Pesticide Residues. FAO Plant Production and Protection Paper 10 Rev, 1978.

29. Pesticide residues in food: 1977 evaluations. FAO Plant Production and Protection Paper 10 Suppl., 1978.

30. Pesticide residues in food—1978. Report of the Joint Meeting of the FAO Panel of Experts on Pesticide Residues and Environment and the WHO Expert Group on Pesticide Residues. FAO Plant Production and Protection Paper 15, 1979.

31. Pesticide residues in food: 1978 evaluations. FAO Plant Production and Protection Paper 15 Suppl., 1979.

32. Pesticide residues in food—1979. Report of the Joint Meeting of the FAO Panel of Experts on Pesticide Residues in Food and the Environment and the WHO Expert Group on Pesticide Residues. FAO Plant Production and Protection Paper 20, 1980.

33. Pesticide residues in food: 1979 evaluations. FAO Plant Production and Protection Paper 20 Suppl., 1980

34. Pesticide residues in food—1980. Report of the Joint Meeting of the FAO Panel of Experts on Pesticide Residues in Food and the Environment and the WHO Expert Group on Pesticide Residues. FAO Plant Production and Protection Paper 26, 1981.

35. Pesticide residues in food: 1980 evaluations. FAO Plant Production and Protection Paper 26 Suppl., 1981.

36. Pesticide residues in food—1981. Report of the Joint Meeting of the FAO Panel of Experts on Pesticide Residues in Food and the Environment and the WHO Expert Group on Pesticide Residues. FAO Plant Production and Protection Paper 37, 1982.

37. Pesticide residues in food: 1981 evaluations. FAO Plant Production and Protection Paper 42, 1982.

38. Pesticide residues in food—1982. Report of the Joint Meeting of the FAO Panel of Experts on Pesticide Residues in Food and the Environment and the WHO Expert Group on Pesticide Residues. FAO Plant Production and Protection Paper 46, 1982.

39. Pesticide residues in food: 1982 evaluations. FAO Plant Production and Protection Paper 49, 1983.

40. Pesticide residues in food—1983. Report of the Joint Meeting of the FAO Panel of Experts on Pesticide Residues in Food and the Environment and the WHO Expert Group on Pesticide Residues. FAO Plant Production and Protection Paper 56, 1985.

41. Pesticide residues in food: 1983 evaluations. FAO Plant Production and Protection Paper 61, 1985.

42. Pesticide residues in food—1984. Report of the Joint Meeting on Pesticide Residues. FAO Plant Production and Protection Paper 62, 1985.

43. Pesticide residues in food—1984 evaluations. FAO Plant Production and Protection Paper 67, 1985.

44. Pesticide residues in food—1985. Report of the Joint Meeting of the FAO Panel of Experts on Pesticide Residues in Food and the Environment and a WHO Expert Group on Pesticide Residues. FAO Plant Production and Protection Paper 68, 1986.

45. Pesticide residues in food—1985 evaluations. Part I. Residues. FAO Plant Production and Protection Paper 72/1, 1986.

46. Pesticide residues in food—1985 evaluations. Part II. Toxicology. FAO Plant Production and Protection Paper 72/2, 1986.

47. Pesticide residues in food—1986. Report of the Joint Meeting of the FAO Panel of Experts on Pesticide Residues in Food and the Environment and a WHO Expert Group on Pesticide Residues. FAO Plant Production and Protection Paper 77, 1986.

48. Pesticide residues in food—1986 evaluations. Part I. Residues. FAO Plant Production and Protection Paper 78, 1986.

49. Pesticide residues in food—1986 evaluations. Part II. Toxicology. FAO Plant Production and Protection Paper 78/2, 1987.

50. Pesticide residues in food—1987. Report of the Joint Meeting of the FAO Panel of Experts on Pesticide Residues in Food and the Environment and a WHO Expert Group on Pesticide Residues. FAO Plant Production and Protection Paper 84, 1987.

51. Pesticide residues in food—1987 evaluations. Part I. Residues. FAO Plant Production and Protection Paper 86/1, 1988.

52. Pesticide residues in food—1987 evaluations. Part II. Toxicology. FAO Plant Production and Protection Paper 86/2, 1988.

53. Pesticide residues in food—1988. Report of the Joint Meeting of the FAO Panel of Experts on Pesticide Residues in Food and the Environment and a WHO Expert Group on Pesticide Residues. FAO Plant Production and Protection Paper 92, 1988.

54. Pesticide residues in food—1988 evaluations. Part I. Residues. FAO Plant Production and Protection Paper 93/1, 1988.

55. Pesticide residues in food—1988 evaluations. Part II. Toxicology. FAO Plant Production and Protection Paper 93/2, 1989.

56. Pesticide residues in food—1989. Report of the Joint Meeting of the FAO Panel of Experts on Pesticide Residues in Food and the Environment and a WHO Expert Group on Pesticide Residues. FAO Plant Production and Protection Paper 99, 1989.

57. Pesticide residues in food—1989 evaluations. Part I. Residues. FAO Plant Production and Protection Paper 100, 1990.

58. Pesticide residues in food—1989 evaluations. Part II. Toxicology. FAO Plant Production and Protection Paper 100/2, 1990.

59. Pesticide residues in food—1990. Report of the Joint Meeting of the FAO Panel of Experts on Pesticide Residues in Food and the Environment and a WHO Expert Group on Pesticide Residues. FAO Plant Production and Protection Paper 102, Rome, 1990.

60. Pesticide residues in food—1990 evaluations. Part I. Residues. FAO Plant Production and Protection Paper 103/1, Rome, 1990.

61. Pesticide residues in food—1990 evaluations. Part II. Toxicology. World Health Organization, WHO/PCS/91.47, Geneva, 1991.

62. Pesticide residues in food—1991. Report of the Joint Meeting of the FAO Panel of Experts on Pesticide Residues in Food and the Environment and a WHO Expert Group on Pesticide Residues. FAO Plant Production and Protection Paper 111, Rome, 1991.

63. Pesticide residues in food—1991 evaluations. Part I. Residues. FAO Plant Production and Protection Paper 113/1, Rome, 1991.

64. Pesticide residues in food—1991 evaluations. Part II. Toxicology. World Health Organization, WHO/PCS/92.52, Geneva, 1992.

65. Pesticide residues in food—1992. Report of the Joint Meeting of the FAO Panel of Experts on Pesticide Residues in Food and the Environment and a WHO Expert Group on Pesticide Residues. FAO Plant Production and Protection Paper 116, Rome, 1993.

66. Pesticide residues in food—1992 evaluations. Part I. Residues. FAO Plant Production and Protection Paper 118, Rome, 1993.

67. Pesticide residues in food—1992 evaluations. Part II. Toxicology. World Health Organization, WHO/PCS/93.34, Geneva, 1993.

68. Pesticide residues in food—1993. Report of the Joint Meeting of the FAO Panel of Experts on Pesticide Residues in Food and the Environment and a WHO Expert Group on Pesticide Residues. FAO Plant Production and Protection Paper 122, Rome, 1994.

69. Pesticide residues in food—1993 evaluations. Part I. Residues. FAO Plant Production and Protection Paper 124, Rome, 1994.

70. Pesticide residues in food—1993 evaluations. Part II. Toxicology. World Health Organization, WHO/PCS/94.4, Geneva, 1994.

71. Pesticide residues in food—1994. Report of the Joint Meeting of the FAO Panel of Experts on Pesticide Residues in Food and the Environment and a WHO Expert Group on Pesticide Residues. FAO Plant Production and Protection Paper 127, Rome, 1995.

72. Pesticide residues in food—1994 evaluations. Part I. Residues. FAO Plant Production and Protection Paper 131/1 and 131/2 (2 volumes), Rome, 1995.

73. Pesticide residues in food—1994 evaluations. Part II. Toxicology. World Health Organization, WHO/PCS/95.2, Geneva, 1995.

74. Pesticide residues in food—1995. Report of the Joint Meeting of the FAO Panel of Experts on Pesticide Residues in Food and the Environment and the Core Assessment Group. FAO Plant Production and Protection Paper 133, Rome, 1996.

75. Pesticide residues in food—1995 evaluations. Part I. Residues. FAO Plant Production and Protection Paper 137, 1996.

76. Pesticide residues in food—1995 evaluations. Part II. Toxicological and Environmental. World Health Organization, WHO/PCS/96.48, Geneva, 1996.

77. Pesticide residues in food—1996. Report of the Joint Meeting of the FAO Panel of Experts on Pesticide Residues in Food and the Environment and the WHO Core Assessment Group. FAO Plant Production and Protection Paper, 140, 1997.

78. Pesticide residues in food—1996 evaluations. Part I. Residues. FAO Plant Production and Protection Paper, 142, 1997.

79. Pesticide residues in food—1996 evaluations. Part II. Toxicological. World Health Organization, WHO/PCS/97.1, Geneva, 1997.

80. Pesticide residues in food—1997. Report of the Joint Meeting of the FAO Panel of Experts on Pesticide Residues in Food and the Environment and the WHO Core Assessment Group. FAO Plant Production and Protection Paper, 145, 1998.

81. Pesticide residues in food—1997 evaluations. Part I. Residues. FAO Plant Production and Protection Paper, 146, 1998.

82. Pesticide residues in food—1997 evaluations. Part II. Toxicological and Environmental. World Health Organization, WHO/PCS/98.6, Geneva, 1998.

83. Pesticide residues in food—1998. Report of the Joint Meeting of the FAO Panel of Experts on Pesticide Residues in Food and the Environment and the WHO Core Assessment Group. FAO Plant Production and Protection Paper, 148, 1999.

84. Pesticide residues in food—1998 evaluations. Part I. Residues. FAO Plant Production and Protection Paper, 152/1 and 152/2 (two volumes).

85. Pesticide residues in food—1998 evaluations. Part II. Toxicological and Environmental. World Health Organization, WHO/PCS/99.18, Geneva, 1999.

86. Pesticide residues in food—1999. Report of the Joint Meeting of the FAO Panel of Experts on Pesticide Residues in Food and the Environment and the WHO Core Assessment Group. FAO Plant Production and Protection Paper, 153, 1999.

87. Pesticide residues in food—1999 evaluations. Part I. Residues. FAO Plant Production and Protection Paper, 157, 2000.

88. Pesticide residues in food—1999 evaluations. Part II. Toxicological. World Health Organization, WHO/PCS/00.4, Geneva, 2000.

89. Pesticide residues in food—2000. Report of the Joint Meeting of the FAO Panel of Experts on Pesticide Residues in Food and the Environment and the WHO Core Assessment Group. FAO Plant Production and Protection Paper, 163, 2001.

90. Pesticide residues in food—2000 evaluations. Part I. Residues. FAO Plant Production and Protection Paper, 165, 2001.

91. Pesticide residues in food—2000 evaluations. Part II. Toxicological. World Health Organization, WHO/PCS/01.3, 2001.

92. Pesticide residues in food—2001. Report of the Joint Meeting of the FAO Panel of Experts on Pesticide Residues in Food and the Environment and the WHO Core Assessment Group. FAO Plant Production and Protection Paper, 167, 2001.

93. Pesticide residues in food—2001 evaluations. Part I. Residues. FAO Plant Production and Protection Paper, 171, 2002.

94. Pesticide residues in food—2001 evaluations. Part II. Toxicological. World Health Organization, WHO/PCS/02.1, 2002.

95. Pesticide residues in food—2002. Report of the Joint Meeting of the FAO Panel of Experts on Pesticide Residues in Food and the Environment and the WHO Core Assessment Group. FAO Plant Production and Protection Paper, 172, 2002.

96. Pesticide residues in food—2002 evaluations. Part I. Residues. FAO Plant Production and Protection Paper 175/1 and 175/2 (two volumes).

97. Pesticide residues in food—2002 evaluations. Part II. Toxicological. World Health Organization, WHO/PCS/03.1, 2003.

98. Pesticide residues in food—2003. Report of the Joint Meeting of the FAO Panel of Experts on Pesticide Residues in Food and the Environment and the WHO Core Assessment Group. FAO Plant Production and Protection Paper, 176, 2004.

99. Pesticide residues in food—2003 evaluations. Part I. Residues. FAO Plant Production and Protection Paper, 170, 2004.

100. Pesticide residues in food—2003 evaluations. Part II. Toxicological. World Health Organization, WHO/PCS/04.1, 2004.

101. Pesticide residues in food—2004. Report of the Joint Meeting of the FAO Panel of Experts on Pesticide Residues in Food and the Environment and the WHO Core Assessment Group. FAO Plant Production and Protection Paper, 178, 2004.

102. Pesticide residues in food—2004 evaluations. Part I. Residues. FAO Plant Production and Protection Paper, 182/1 and 182/2 (two volumes), 2005.

103. Pesticide residues in food—2005. Report of the Joint Meeting of the FAO Panel of Experts on Pesticide Residues in Food and the Environment and the WHO Core Assessment Group. FAO Plant Production and Protection Paper, 183, 2005.

104. Pesticide residues in food—2004 evaluations. Part II. Toxicological. World Health Organization, 2006

105. Pesticide residues in food—2005 evaluations. Part I. Residues. FAO Plant Production and Protection Paper, 184/1 and 184/2, 2006.

106. Pesticide residues in food - 2005 evaluations. Part II - Toxicological. World Health Organization, 2006.

107. Pesticide residues in food - 2006. Report of the Joint Meeting of the FAO Panel of Experts on Pesticide Residues in Food and the Environment and WHO the Core Assessment Group. FAO Plant Production and Protection Paper, 187, 2006.

108. Pesticide residues in food - 2006 evaluations. Part I - Residues. FAO Plant Production and Protection Paper, 189/1 and 189/2, 2007.

109. Pesticide residues in food - 2007. Report of the Joint Meeting of the FAO Panel of Experts on Pesticide Residues in Food and the Environment and WHO the Core Assessment Group. FAO Plant Production and Protection Paper, 191, 2007.

110. Pesticide residues in food - 2006 evaluations. Part II - Toxicological. World Health Organization, 2008.

111. Pesticide residues in food - 2008. Report of the Joint Meeting of the FAO Panel of Experts on Pesticide Residues in Food and the Environment and WHO the Core Assessment Group. FAO Plant Production and Protection Paper, 193, 2008.

www.ingramcontent.com/pod-product-compliance
Lightning Source LLC
Chambersburg PA
CBHW061103210326
41597CB00021B/3962

9789241665230